TREATISE ON INVERTEBRATE PALEONTOLOGY

Prepared under Sponsorship of
The Geological Society of America, Inc.

The Paleontological Society *The Society of Economic Paleontologists and Mineralogists*
The Palaeontographical Society *The Palaeontological Association*

Directed and Edited by
RAYMOND C. MOORE

Part U
ECHINODERMATA 3

By J. WYATT DURHAM, K. E. CASTER, HARRIET EXLINE, H. B. FELL, A. G. FISCHER, D. L. FRIZZELL, R. V. KESLING, P. M. KIER, R. V. MELVILLE, R. C. MOORE, D. L. PAWSON, GERHARD REGNÉLL, W. K. SPENCER, GEORGES UBAGHS, CAROL D. WAGNER, and C. W. WRIGHT

VOLUME 1

THE GEOLOGICAL SOCIETY OF AMERICA, INC.
and
THE UNIVERSITY OF KANSAS PRESS

1966

Second Printing 1986*

Library of Congress Catalogue Card Number 53-12913
ISBN 0-8137-3022-8

Published 1966

*Distributed by the Geological Society of America, Inc., P.O. Box 9140, Boulder, Colorado 80301, from which current price lists of Parts in print may be obtained and to which all orders and related correspondence should be directed. Editorial office for the *Treatise:* Paleontological Institute, 121 Lindley Hall, The University of Kansas, Lawrence, Kansas 66045.

The *Treatise on Invertebrate Paleontology* has been made possible by (1) grants of funds from The Geological Society of America through the bequest of Richard Alexander Fullerton Penrose, Jr., for preparation of illustrations and partial defrayment of organizational expense and the United States National Science Foundation, in December, 1959, for completion of the *Treatise* project; (2) contribution of the knowledge and labor of specialists throughout the world, working in co-operation under sponsorship of The Geological Society of America, The Paleontological Society, The Society of Economic Paleontologists and Mineralogists, The Palaeontographical Society, and The Palaeontological Association; and (3) acceptance by The University of Kansas Press of publication without cost to the Societies concerned and without any financial gain to the Press.

TREATISE ON INVERTEBRATE PALEONTOLOGY

Directed and Edited by
RAYMOND C. MOORE

Assistants: LAVON MCCORMICK, ROGER B. WILLIAMS

Advisers: R. C. BECKER, MARTIN RUSSELL (The Geological Society of America), BERNHARD KUMMEL, J. WYATT DURHAM (The Paleontological Society), N. D. NEWELL, R. M. JEFFORDS (The Society of Economic Paleontologists and Mineralogists), C. J. STUBBLEFIELD, H. DIGHTON THOMAS (The Palaeontographical Society), M. R. HOUSE, C. H. HOLLAND (The Palaeontological Association), CLYDE K. HYDER (The University of Kansas Press, Editor).

PARTS

Parts of the *Treatise* are distinguished by assigned letters with a view to indicating their systematic sequence while allowing publication of units in whatever order each may be made ready for the press. The volumes are cloth-bound with title in gold on the cover. Copies are available on orders sent to the Publication Department of The Geological Society of America at 231 East 46th Street, New York 17, N.Y. The prices quoted very incompletely cover costs of producing and distributing the several volumes, but on receipt of payment the Society will ship copies without additional charge to any address in the world. Special discounts are available to members of sponsoring societies under arrangements made by appropriate officers of these societies, to whom inquiries should be addressed.

VOLUMES ALREADY PUBLISHED

(Previous to 1965)

Part C. PROTISTA 2 (Sarcodina, chiefly "Thecamoebians" and Foraminiferida), xxxi+900 p., 5311 fig., 1964.

Part D. PROTISTA 3 (chiefly Radiolaria, Tintinnina), xii+195 p., 1050 fig., 1954.

Part E. ARCHAEOCYATHA, PORIFERA, xviii+122 p., 728 fig., 1955.

Part F. COELENTERATA, xvii+498 p., 2700 fig., 1956.

Part G. BRYOZOA, xii+253 p., 2000 fig., 1953.

Part H. BRACHIOPODA, xxxii +927 p., 5198 fig., 1965.

Part I. MOLLUSCA 1 (Mollusca General Features, Scaphopoda, Amphineura, Monoplacophora, Gastropoda General Features, Archaeogastropoda, mainly Paleozoic Caenogastropoda and Opisthobranchia), xxiii+351 p., 1732 fig., 1960.

Part K. MOLLUSCA 3 (Cephalopoda General Features, Endoceratoidea, Actinoceratoidea, Nautiloidea, Bactritoidea), xxviii+519 p., 2382 fig., 1964.

Part L. MOLLUSCA 4 (Ammonoidea), xxii+490 p., 3800 fig., 1959.

Part O. ARTHROPODA 1 (Arthropoda General Features, Protarthropoda, Euarthropoda General Features, Trilobitomorpha), xix+560 p., 2880 fig., 1959.

Part P. ARTHROPODA 2 (Chelicerata, Pycnogonida, Palaeoisopus), xvii+181 p., 565 fig., 1955.

Part Q. ARTHROPODA 3 (Crustacea, Ostracoda), xxiii+442 p., 3476 fig., 1961.

Part V. GRAPTOLITHINA, xvii+101 p., 358 fig., 1955.

Part W. MISCELLANEA (Conodonts, Conoidal Shells of Uncertain Affinities, Worms, Trace Fossils, Problematica), xxv+259 p., 1058 fig., 1962.

THIS VOLUME

Part U. ECHINODERMATA 3 (Asterozoans, Echinozoans), xxx+695 p., 3485 fig., 1966.

VOLUMES IN PREPARATION (1965)

Part A. INTRODUCTION.
Part B. PROTISTA 1 (Chrysomonadida, Coccolithophorida, Charophyta, Diatomacea, etc.).
Part J. MOLLUSCA 2 (Caenogastropoda, Opisthobranchia).
Part M. MOLLUSCA 5 (Coleoidea).
Part N. MOLLUSCA 6 (Bivalvia).
Part R. ARTHROPODA 4 (Crustacea, Branchiopoda, Cirripedia, Malacostraca; Myriapoda; Hexapoda).
Part S. ECHINODERMATA 1 (Echinodermata General Features, Cystoidea, Cyclocystoidea, Paracrinoidea, Edrioblastoidea, Blastoidea, Edrioasteroidea, Eocrinoidea, Carpoidea).
Part T. ECHINODERMATA 2 (Crinoidea).
Part X. ADDENDA, INDEX.

CONTRIBUTING AUTHORS

(Arranged by countries and institutions. Accompanying numbers are for cross reference from alphabetically arranged list, which follows.)

AUSTRALIA

1. University of **Adelaide:** M. F. Glaessner.
2. **National** University (Canberra): D. A. Brown.
3. University of **Queensland** (Brisbane): Dorothy Hill.
4. **South Australia** Geological Survey (Adelaide): N. H. Ludbrook.

BELGIUM

5. Université de **Liège:** Georges Ubaghs.
6. Université de **Louvain:** Marius Lecompte.

CANADA

7. University of **British Columbia** (Vancouver): V. J. Okulitch.
8. Geological Survey of **Canada** (Ottawa): J. A. Jeletzky, D. J. McLaren, G. W. Sinclair.
9. **National** Museum (Ottawa): A. H. Clarke, Jr.

DENMARK

10. Universitet **København**: Chr. Poulsen, H. W. Rasmussen.

FRANCE

11. **British Embassy** (Paris): R. V. Melville.
12. Université de **Paris** (Sorbonne): Colette Dechaseaux.
13. Unattached: André Chavan, Chantemerle, Seyssel (Ain).

GERMANY

14. Universität **Bonn**: H. K. Erben, K. J. Müller.
15. **Hamburg** Staatsinstitut: Walter Häntzschel.
16. **Senckenbergische** Museum (Frankfurt): Herta Schmidt, Wolfgang Struve.
17. Universität **Tübingen**: O. H. Schindewolf.
18. Universität **Würzberg**: Klaus Sdzuy.

ITALY

19. Universitá **Modena**: Eugenia Montanaro Gallitelli.

JAPAN

20. **Tohoku** University (Sendai): Kotora Hatai.
21. University of **Tokyo**: Tetsuro Hanai.

NETHERLANDS

22. **Rijksmuseum** van **Natuurlijke Historie** (Leiden): H. Boschma, L. B. Holthuis.

NEW ZEALAND

23. **Auckland** Museum: A. W. B. Powell.
24. **Dominion** Museum (Wellington): R. K. Dell.
25. **New Zealand** Geological Survey (Lower Hutt): C. A. Fleming, J. Marwick.

NORWAY

27. Universitet **Oslo**: Gunnar Henningsmoen, T. Soot-Ryen, Leif Størmer.

POLAND

28. **Państwowe Wydawnictwo Naukowe** (Warszawa): Gertruda Biernat.

SWEDEN

29. Universitet **Lund**: Gerhard Regnéll.
30. Universitet **Stockholm:** Ivar Hessland, R. A. Reyment.
31. Universitet **Uppsala**: Valdar Jaanusson.

SWITZERLAND

32. Universität **Basel**: Manfred Reichel.

UNITED KINGDOM

33. University of **Birmingham**: L. J. Wills.
34. **British Museum (Natural History)**: Leslie Bairstow, L. R. Cox, Isabella Gordon, S. M. Manton, N. J. Morris, H. M. Muir-Wood.
35. **British Petroleum Company** (Middlesex): F. E. Eames.
36. University of **Cambridge**: O. M. B. Bulman, M. J. S. Rudwick.
37. University of **Glasgow**: W. D. I. Rolfe, C. M. Yonge.
38. Geological Survey of **Great Britain** (London): Raymond Casey, C. J. Stubblefield.
39. **Iraq Petroleum Company** (London): G. F. Elliott.
40. University of **Leicester**: P. C. Sylvester-Bradley.
41. University of **London**: D. V. Ager.
42. University of **Nottingham**: A. J. Rowell.
43. **Queen's** University of **Belfast**: Margaret Jope, Alwyn Williams, A. D. Wright.
44. University of **Reading**: H. L. Hawkins.
45. University of **Swansea**: F. H. T. Rhodes.
46. Unattached: (England) Dennis Curry, Middlesex; R. P. Tripp, Seven Oaks, Kent; C. W. Wright, London; (Scotland) John Weir, Tayport, Fife.

UNITED STATES OF AMERICA

47. **American Museum** of **Natural History** (New York): R. L. Batten, W. K. Emerson, L. H. Hyman, N. D. Newell.
48. **California Academy** of **Sciences** (San Francisco): G. D. Hanna, L. G. Hertlein, A. G. Smith.
49. **California Institute** of **Technology** (Pasadena): A. J. Boucot, J. G. Johnson, H. A. Lowenstam.
50. **California Research Corporation** (La Habra): A. R. Loeblich, Jr.
51. University of **California** (Berkeley): J. W. Durham, C. D. Wagner.
52. University of **California** (Los Angeles): W. D. Popenoe, Helen Tappan.
53. University of **California** (San Diego, La Jolla): M. N. Bramlette, A. R. Loeblich, III.
54. **Chicago Natural History Museum:** Fritz Haas.
55. University of **Chicago**: J. M. Weller.
56. University of **Cincinnati** (Cincinnati, O.): K. E. Caster.
57. **Cornell** University (Ithaca, N.Y.): W. S. Cole, J. M. Wells.
58. **Florida** Geological Survey (Tallahassee): H. S. Puri.
59. University of **Florida** (Gainesville): H. K. Brooks.
60. **Florida** State University (Tallahassee): W. H. Heard.
61. **Harvard** University (Cambridge, Mass.): F. M. Carpenter, W. J. Clench, H. B. Fell, Bernhard Kummel, W. A. Newman, R. D. Staton, Ruth Turner, H. B. Whittington.
62. **Humble Oil & Refining Company** (Houston, Texas): H. H. Beaver, J. A. Eyer, R. M. Jeffords, S. A. Levinson, Joan Stough.
63. **Illinois** Geological Survey (Urbana): M. L. Thompson.
64. University of **Illinois** (Urbana): H. W. Scott.
65. **Indiana** Geological Survey (Bloomington): R. H. Shaver.
66. State University of **Iowa** (Iowa City): W. M. Furnish, B. F. Glenister.
67. **Jersey Production Company** (Tulsa, Okla.): J. S. Van Sant.
68. **John Hopkins** University (Baltimore, Md.): Franco Rasetti.
69. **Kansas** Geological Survey (Lawrence): D. E. Nodine Zeller.
70. University of **Kansas** (Lawrence): A. B. Leonard, R. C. Moore, Curt Teichert, R. H. Thompson.
71. **Louisiana** State University (Baton Rouge): W. A. van den Bold, H. V. Howe.
72. University of **Massachusetts** (Amherst): C. W. Pitrat.
73. University of **Michigan** (Ann Arbor): R. V. Kesling, E. C. Stumm.
74. University of **Minnesota** (Minneapolis): F. M. Swain.
75. University of **Missouri** (Columbia): R. E. Peck.

76. **Missouri** School of Mines (Rolla): Harriet Exline, D. L. Frizzell.
77. **New Mexico** Institute Mining & Geology (Socorro); Christina Lochman-Balk.
78. **New York** State Museum (Albany): D. W. Fisher.
79. **Ohio** State University (Columbus): Aurèle La Rocque, W. C. Sweet.
80. **Oklahoma** Geological Survey (Norman): T. W. Amsden, R. O. Fay.
81. University of **Oklahoma** (Norman): C. C. Branson.
82. College of the **Pacific** (Dillon Beach, Calif.): Joel Hedgpeth.
83. **Paleontological Research** Institute (Ithaca, N.Y.): K. V. W. Palmer.
84. **Philadelphia** Academy of Natural Sciences: A. A. Olsson, Robert Robertson.
85. **Princeton** University (Princeton, N.J.): A. G. Fischer, B. F. Howell.
86. **Radford** College (Blacksburg, Va.): R. L. Hoffman.
87. **St. Mary's** College (St. Mary's College, Calif.): A. S. Campbell.
88. **Shell Development** Company (Houston, Tex.): R. W. Barker, H. B. Stenzel, John Wainwright.
89. **Sinclair Oil & Gas** Company (Tulsa, Okla.): A. L. Bowsher.
90. **Smithsonian Institution** (Washington, D.C.): F. M. Bayer, R. H. Benson, R. S. Boardman, G. A. Cooper, T. G. Gibson, P. M. Kier, R. B. Manning, David Pawson, H. A. Rehder.
91. **Stanford** University (Stanford, Calif.): Eugene Coan, A. Myra Keen.
92. Unattached: H. J. Harrington, Houston, Tex.
93. **Tulane** University (New Orleans, La.): Emily Vokes, H. E. Vokes.
94. **United States** Geological Survey (Washington, D.C.): J. M. Berdan, R. C. Douglass, Mackenzie Gordon, Jr., R. E. Grant, K. E. Lohman, A. R. Palmer, I. G. Sohn, E. L. Yochelson.
95. **Western Reserve** University (Cleveland, O.): F. G. Stehli.
96. University of **Wichita** (Wichita, Kans.): Paul Tasch.
97. **Woods Hole Oceanographic** Institution (Woods Hole, Mass.): R. R. Hessler, V. A. Zullo.
98. **Yale** University (New Haven, Conn.): A. L. McAlester, Alexander Petrunkevitch.

DECEASED

99. W. J. Arkell, R. S. Bassler, L. R. Cox, L. M. Davies, Julia Gardner, W. H. Hass, J. B. Knight, M. W. de Laubenfels, A. K. Miller, Emma Richter, Rudolf Richter, W. K. Spencer, M. A. Stainbrook, L. W. Stephenson, O. W. Tiegs, Johannes Wanner, T. H. Withers, Arthur Wrigley.

Alphabetical List

(Numbers refer to preceding list arranged by countries and institutions.)

Ager, D. V. (41)
Amsden, T. W. (80)
Arkell, W. J. (99)
Bairstow, Leslie (34)
Barker, R. W. (88)
Bassler, R. S. (99)
Batten, R. L. (47)
Bayer, F. M. (90)
Beaver, H. H. (62)
Benson, R. H. (90)
Berdan, J. M. (94)
Biernat, Gertruda (28)
Boardman, R. S. (90)
Bold, W. A. van den (71)
Boschma, H. (22)
Boucot, A. J. (49)
Bowsher, A. L. (89)
Bramlette, M. N. (53)
Branson, C. C. (81)
Brooks, H. K. (59)
Brown, D. A. (2)
Bulman, O. M. B. (36)
Campbell, A. S. (87)
Carpenter, F. M. (61)
Casey, Raymond (38)
Caster, K. E. (56)
Chavan, André (13)
Clarke, A. H., Jr. (9)
Clench, W. J. (61)
Coan, Eugene (91)
Cole, W. S. (57)
Cooper, G. A. (90)
Cox, L. R. (99)
Curry, Dennis (46)
Davies, L. M. (99)
Dechaseaux, Colette (12)
Dell, W. K. (24)
Douglass, R. C. (94)
Durham, J. W. (51)
Eames, F. E. (35)
Elliott, G. F. (39)
Emerson, W. K. (47)
Erben, H. K. (14)
Exline, Harriet (76)
Eyer, J. A. (62)
Fay, R. O. (80)
Fell, H. B. (61)
Fischer, A. G. (85)
Fisher, D. W. (78)
Fleming, C. A. (25)
Frizzell, D. L. (76)
Furnish, W. M. (66)
Gardner, Julia (99)
Gibson, T. G. (90)
Glaessner, M. F. (1)
Glenister, B. F. (66)
Gordon, Isabella (34)
Gordon, Mackenzie, Jr. (94)
Grant, R. E. (94)
Haas, Fritz (54)
Hanai, Tetsuro (21)
Hanna, G. D. (48)
Häntzschel, Walter (15)
Harrington, H. J. (92)
Hass, W. H. (99)
Hatai, Kotora (20)
Hawkins, H. L. (44)
Heard, W. H. (60)
Hedgpeth, Joel (82)

Henningsmoen, Gunnar (27)
Hertlein, L. G. (48)
Hessland, Ivar (30)
Hessler, R. R. (97)
Hill, Dorothy (3)
Hoffman, R. L. (86)
Holthuis, L. B. (22)
Howe, H. V. (71)
Howell, B. F. (85)
Hyman, L. H. (47)
Jaanusson, Valdar (31)
Jeffords, R. M. (62)
Jeletzky, J. A. (8)
Johnson, J. G. (49)
Jope, Margaret (43)
Keen, A. Myra (91)
Kesling, R. V. (73)
Kier, P. M. (90)
Knight, J. B. (99)
Kummel, Bernhard (61)
La Rocque, Aurèle (79)
Laubenfels, M. W. de (99)
Lecompte, Marius (6)
Leonard, A. B. (70)
Levinson, S. A. (62)
Lochman-Balk, Christina (77)
Loeblich, A. R., Jr. (50)
Loeblich, A. R., III (53)
Lohman, K. E. (94)
Lowenstam, H. A. (49)
Ludbrook, N. H. (4)
McAlester, A. L. (98)
McLaren, D. J. (8)
Manning, R. B. (90)
Manton, S. M. (34)
Marwick, J. (25)
Melville, R. V. (11)
Miller, A. K. (99)
Montanaro Gallitelli, Eugenia (19)
Moore, R. C. (70)
Morris, N. J. (34)
Muir-Wood, H. M. (34)
Müller, K. J. (14)
Newell, N. D. (47)
Newman, W. A. (61)
Okulitch, V. J. (7)
Olsson, A. A. (84)
Palmer, A. R. (94)
Palmer, K. V. W. (83)
Pawson, David (90)
Peck, R, E. (75)
Petrunkevitch, Alexander (98)
Pitrat, C. W. (72)
Popenoe, W. D. (52)
Poulsen, Chr. (10)
Powell, A. W. B. (23)
Puri, H. S. (58)
Rasetti, Franco (68)
Rasmussen, H. W. (10)
Regnéll, Gerhard (29)
Rehder, H. A. (90)
Reichel, Manfred (32)
Reyment, R. A. (30)
Rhodes, F. H. T. (45)
Richter, Emma (99)
Richter, Rudolf (99)
Robertson, Robert (84)
Rolfe, W. D. I. (37)
Rowell, A. J. (42)
Rudwick, M. J. S. (36)
Schindewolf, O. H. (17)
Schmidt, Herta (16)
Scott, H. W. (64)
Sdzuy, Klaus (18)
Shaver, R. H. (27)
Sinclair, G. W. (8)
Smith, A. G. (48)
Sohn, I. G. (94)
Soot-Ryen, T. (27)
Spencer, W. K. (99)
Stainbrook, M. A. (99)
Staton, R. D. (61)
Stehli, F. G. (95)
Stenzel, H. B. (88)
Stephenson, L. W. (99)
Størmer, Leif (27)
Stough, Joan (62)
Struve, Wolfgang (16)
Stubblefield, C. J. (38)
Stumm, E. C. (73)
Swain, F. M. (74)
Sweet, W. C. (79)
Sylvester-Bradley, P. C. (40)
Tappan, Helen (52)
Tasch, Paul (96)
Teichert, Curt (70)
Thompson, M. L. (63)
Thompson, R. H. (70)
Tiegs, O. W. (99)
Tripp, R. P. (46)
Turner, Ruth (61)
Ubaghs, Georges (5)
Van Sant, J. F. (67)
Vokes, Emily (93)
Vokes, H. E. (93)
Wagner, C. D. (51)
Wainwright, John (88)
Wanner, Johannes (99)
Weir, John (46)
Weller, J. M. (55)
Wells, J. M. (57)
Whittington, H. B. (61)
Williams, Alwyn (43)
Wills, L. J. (33)
Withers, T. H. (99)
Wright, A. D. (43)
Wright, C. W. (46)
Wrigley, Arthur (99)
Yochelson, E. L. (94)
Yonge, C. M. (37)
Zeller, D. E. Nodine (69)
Zullo, V. A. (97)

EDITORIAL PREFACE

The aim of the *Treatise on Invertebrate Paleontology,* as originally conceived and consistently pursued, is to present the most comprehensive and authoritative, yet compact statement of knowledge concerning invertebrate fossil groups that can be formulated by collaboration of competent specialists in seeking to organize what has been learned of this subject up to the mid-point of the present century. Such work has value in providing a most useful summary of the collective results of multitudinous investigations and thus should constitute an indispensable text and reference book for all persons who wish to know about remains of invertebrate organisms preserved in rocks of the earth's crust. This applies to neozoologists as well as paleozoologists and to beginners in study of fossils as well as to thoroughly trained, long-experienced professional workers, including teachers, stratigraphical geologists, and individuals engaged in research on fossil invertebrates. The making of a reasonably complete inventory of present knowledge of invertebrate paleontology may be expected to yield needed foundation for future research and it is hoped that the *Treatise* will serve this end.

The *Treatise* is divided into parts which bear index letters, each except the initial and concluding ones being defined to include designated groups of invertebrates. The chief purpose of this arrangement is to provide for independence of the several parts as regards date of publication, because it is judged desirable to print and distribute each segment as soon as possible after it is ready for press. Pages in each part bear the assigned index letter joined with numbers beginning with 1 and running consecutively to the end of the part.

The outline of subjects to be treated in connection with each large group of invertebrates includes (1) description of morphological features, with special reference to hard parts, (2) ontogeny, (3) classification, (4) geological distribution, (5) evolutionary trends and phylogeny, and (6) systematic description of genera, subgenera, and higher taxonomic units. In general, paleoecological aspects of study are omitted or little emphasized because comprehensive treatment of this subject is given in the *Treatise on Marine Ecology and Paleoecology* (H. S. LADD, Editor, Geological Society of America, Memoir 67, 1957), prepared under auspices of a committee of the United States National Research Council. A selected list of references is furnished in each part of the *Treatise*.

Features of style in the taxonomic portions of this work have been fixed by the Editor with aid furnished by advice from representatives of the societies which have undertaken to sponsor the *Treatise*. It is the Editor's responsibility to consult with authors and co-ordinate their work, seeing that manuscript properly incorporates features of adopted style. Especially he has been called on to formulate policies in respect to many questions of nomenclature and procedure. The subject of family and subfamily names is reviewed briefly in a following section of this preface, and features of *Treatise* style in generic descriptions are explained.

A generous grant of $35,000 has been made by the Geological Society of America for the purpose of preparing *Treatise* illustrations. Administration of expenditures has been in charge of the Editor and most of the work by photographers and artists has been done under his direction at the University of Kansas, but sizable parts of this program have also been carried forward in Washington and London.

In December, 1959, the National Science Foundation of the United States, through its Division of Biological and Medical Sciences and the Program Director for Systematic Biology, made a grant in the amount of $210,000 for the purpose of aiding the completion of yet-unpublished volumes of the *Treatise*. Payment of this sum was provided to be made in installments distributed over a five-year period, with administration of disbursements handled by the University of Kansas. Expenditures planned are primarily for needed assistance to authors and may be arranged through approved institutions located anywhere. Important help for the Director-Editor of the *Treatise* has been made available from the grant, but no part of his stipend has come from it. Grateful acknowledgment to the Foundation is expressed on behalf of the societies sponsoring the *Treatise,* the University of Kansas, and innumerable individuals benefited by the *Treatise* project.

ZOOLOGICAL NAMES

Many questions arise in connection with zoological names, especially including those that relate to their acceptability and to alterations of some which may be allowed or demanded. Procedure in obtaining answers to these questions is guided and to a large extent governed by regulations published (1961) in the *International Code of Zoological Nomenclature* (hereinafter cited simply as the *Code*). The prime object of the *Code* is to promote stability and universality in the scientific names of animals, ensuring also that each name is distinct and unique while avoiding restrictions on freedom of taxonomic thought or action. Priority is a basic principle, but under specified conditions its application can be modified. This is all well and good, yet nomenclatural tasks confronting the zoological taxonomist are formidable. They warrant the complaint of some that zoology, including paleozoology, is the study of animals rather than of names applied to them.

Several ensuing pages are devoted to aspects of zoological nomenclature that are judged to have chief importance in relation to procedures adopted in the *Treatise*. Terminology is explained, and examples of style employed in the nomenclatural parts of systematic descriptions are given.

TAXA GROUPS

Each taxonomic unit (taxon, pl., taxa) of the animal and protistan kingdoms belongs to some one or another rank in the adopted hierarchy of classificatory divisions. In part, this hierarchy is defined by the *Code* to include a species-group of taxa, a genus-group, and a family-group. Units of lower rank than subspecies are excluded from zoological nomenclature and those higher than superfamily of the family-group are not regulated by the *Code*. It is natural and convenient to discuss nomenclatural matters in general terms first and then to consider each of the taxa groups separately. Especially important is provision that within each taxa group classificatory units are coordinate (equal in rank), whereas units of different taxa groups are not coordinate.

FORMS OF NAMES

All zoological names are divisible into groups based on their form (spelling). The first-published form (or forms) of a name is defined as original spelling (*Code,* Art. 32) and any later-published form (or forms) of the same name is designated as subsequent spelling (Art. 33). Obviously, original and subsequent spellings of a given name may or may not be identical and this affects consideration of their correctness. Further, examination of original spellings of names shows that by no means all can be distinguished as correct. Some are incorrect, and the same is true of subsequent spellings.

Original Spellings

If the first-published form of a name is consistent and unambiguous, being identical wherever it appears, the original spelling is defined as correct unless it contravenes some stipulation of the *Code* (Arts. 26-31), unless the original publication contains clear evidence of an inadvertent error, in the sense of the *Code,* or among names belonging to the family-group, unless correction of the termination or the stem of the type-genus is required. An unambiguous original spelling that fails to meet these requirements is defined as incorrect.

If a name is spelled in more than one way in the original publication, the form adopted by the first reviser is accepted as the correct original spelling, provided that it complies with mandatory stipulations of the *Code* (Arts. 26-31), including its provision for automatic emendations of minor sort.

Incorrect original spellings are any that fail to satisfy requirements of the *Code,* or that represent an inadvertent error, or that are one of multiple original spellings not adopted by a first reviser. These have no separate status in zoological nomenclature and therefore cannot enter into homonymy or be used as replacement names. They call for correction wherever found. For example, a name originally published with a diacritic mark, apostrophe, diaeresis, or hyphen requires correction by deleting such features and uniting parts of the name originally separated by them, except that deletion of an umlaut from a vowel is accompanied by inserting "e" after the vowel.

Subsequent Spellings

If a name classed as a subsequent spelling is identical with an original spelling, it is distinguishable as correct or incorrect on the same criteria that apply to the original spelling. This means that a subsequent spelling identical with a correct original spelling is also correct, and one identical with an incorrect original spelling is also incorrect. In the latter case, both original and subsequent spellings require correction wherever found (authorship and date of the original incorrect spelling being retained).

If a subsequent spelling differs from an original spelling in any way, even by the omission, addition, or alteration of a single letter, the subsequent spelling must be defined as a different name (except that such changes as altered terminations of adjectival specific names to obtain agreement in gender with associated generic names, of family-group names to denote assigned taxonomic rank, and corrections for originally used diacritic marks, hyphens, and the like are excluded from spelling changes conceived to produce a different name).

Altered subsequent spellings other than the exceptions noted may be either intentional or unintentional. If demonstrably intentional, the change is designated as an emendation. Emendations are divisible into those classed as justifiable and those comprising all others classed as unjustifiable. Justifiable emendations are corrections of incorrect original spellings, and these take the authorship and date of the original spellings. Unjustifiable emendations are names having their own status in nomenclature, with author and date of their publication; they are junior objective synonyms of the name in its original form.

Subsequent spellings that differ in any way from original spellings, other than previously noted exceptions, and that are not classifiable as emendations are defined as incorrect subsequent spellings. They have no status in nomenclature, do not enter into homonymy, and cannot be used as replacement names.

AVAILABLE AND UNAVAILABLE NAMES

Available Names

An available zoological name is any that conforms to all mandatory provisions of the *Code*. Such names are classifiable in groups which are usefully recognized in the *Treatise,* though not explicitly differentiated in the *Code*. They are as follows:

(1) So-called *"inviolate names"* include all available names that are not subject to any sort of alteration from their originally published form. They comprise correct original spellings and commonly include correct subsequent spellings, but include no names classed as emendations. Here belong most generic and subgeneric names, some of which differ in spelling from others by only a single letter.

(2) Names may be termed *"perfect names"* if, as originally published (with or without duplication by subsequent authors), they meet all mandatory requirements, needing no correction of any kind, but nevertheless are legally alterable in such ways as changing the termination (e.g., many species-group names, family-group names, suprafamilial names). This group does not include emended incorrect original spellings (e.g., *Oepikina,* replacement of *Öpikina*).

(3) *"Imperfect names"* are available names that as originally published (with or without duplication by subsequent authors) contain mandatorily emendable defects. Incorrect original spellings are imperfect names. Examples of emended imperfect names are: among species-group names, *guerini* (not *Guérini*), *obrienae* (not *O'Brienae*), *terranovae* (not *terra-novae*), *nunezi* (not *Nuñezi*), *Spironema rectum* (not *Spironema recta,* because generic name is neuter, not feminine); among genus-group names, *Broeggeria* (not *Bröggeria*), *Obrienia* (not *O'Brienia*), *Maccookites* (not *McCookites*; among family-group names, Oepikidae (not Öpikidae), Spironematidae (not Spironemidae, incorrect stem), Athyrididae (not Athyridae, incorrect stem). The use of "variety" for named divisions of fossil species, according to common practice of some paleontologists, gives rise to imperfect names, which generally are emendable (*Code,* Art. 45e) by omitting this term so as to indicate the status of this taxon as a subspecies.

(4) *"Vain names"* are available names consisting of unjustified intentional emendations of previously published names. The emendations are unjustified because they are not demonstrable as corrections of incorrect original spellings as defined by the *Code* (Art. 32,c). Vain names have status in nomenclature under their own authorship and date. They constitute junior objective synonyms of names in their original form. Examples are: among species-group names, *geneae* (published as replacement of original unexplained masculine, *geni,* which now is not alterable), *ohioae* (invalid change from original *ohioensis*); among genus-group names, *Graphiodactylus* (invalid change from original *Graphiadactyllis*); among family-group names, Graphiodactylidae (based on junior objective synonym having invalid vain name).

(5) An important group of available zoological names can be distinguished as *"transferred names."* These comprise authorized sorts of altered names in which the change depends on transfer from one taxonomic rank to another, or possibly on transfers in taxonomic assignment of subgenera, species, or subspecies. Most commonly the transfer calls for a change in termination of the name so as to comply with stipulations of the *Code* on endings of family-group taxa and agreement in gender of specific names with associated generic names. Transferred names may be derived from any of the preceding groups except the first. Examples are: among species-group names, *Spirifer ambiguus* (masc.) to *Composita ambigua* (fem.), *Neochonetes transversalis* to *N. granulifer transversalis* or vice versa; among genus-group names, *Schizoculina* to *Oculina (Schizoculina)* or vice versa; among family-group names, Orthidae to Orthinae or vice versa, or superfamily Orthacea derived from Orthidae or Orthinae; among supra-

familial taxa (not governed by the *Code*), order Orthida to suborder Orthina or vice versa. The authorship and date of transferred names are not affected by the transfers, but the author responsible for the transfer and the date of his action may appropriately be recorded in such works as the *Treatise.*

(6) Improved or *"corrected names"* include both mandatory and allowable emendations of imperfect names and of suprafamilial names, which are not subject to regulation as to name form. Examples of corrected imperfect names are given with the discussion of group 3. Change from the originally published ordinal name Endoceroidea (TEICHERT, 1933) to the presently recognized Endocerida illustrates a "corrected" suprafamilial name. Group 6 names differ from those in group 5 in not being dependent on transfers in taxonomic rank or assignment, but some names are classifiable in both groups.

(7) *"Substitute names"* are available names expressly proposed as replacements for invalid zoological names, such as junior homonyms. These may be classifiable also as belonging in groups 1, 2, or 3. The glossary appended to the *Code* refers to these as "new names" *(nomina nova)* but they are better designated as substitute names, since their newness is temporary and relative. The first-published substitute name that complies with the definition here given takes precedence over any other. An example is *Mareita* LOEBLICH & TAPPAN, 1964, as substitute for *Reichelina* MARIE, 1955 (*non* ERK, 1942).

(8) *"Conserved names"* include a relatively small number of species-group, genus-group, and family-group names which have come to be classed as available and valid by action of the International Commission on Zoological Nomenclature exercising its plenary powers to this end or ruling to conserve a junior synonym in place of a rejected "forgotten" name *(nomen oblitum)* (Art. 23,b). Currently, such names are entered on appropriate "Official Lists," which are published from time to time.

It is useful for convenience and brevity of distinction in recording these groups of available zoological names to employ Latin designations in the pattern of *nomen nudum* (abbr., *nom. nud.*) and others. Thus we may recognize the preceding numbered groups as follows: (1) *nomina inviolata* (sing., *nomen inviolatum,* abbr., *nom. inviol.*), (2) *nomina perfecta (nomen perfectum, nom. perf.),* (3) *nomina imperfecta (nomen imperfectum, nom. imperf.),* (4) *nomina vana (nomen vanum, nom. van.),* (5) *nomina translata (nomen translatum, nom. transl.),* (6) *nomina correcta (nomen correctum, nom. correct.),* (7) *nomina substituta (nomen substitutum, nom. subst.),* (8) *nomina conservata (nomen conservatum, nom. conserv.).*

Unavailable Names

All zoological names which fail to comply with mandatory provisions of the *Code* are unavailable names and have no status in zoological nomenclature. None can be used under authorship and date of their original publication as a replacement name *(nom. subst.)* and none preoccupies for purposes of the Law of Homonymy. Names identical in spelling with some, but not all, unavailable names can be classed as available if and when they are published in conformance to stipulations of the *Code* and they are then assigned authorship and take date of the accepted publication. Different groups of unavailable names can be discriminated, as follows.

(1) *"Naked names"* include all those that fail to satisfy provisions stipulated in Article 11 of the *Code,* which states general requirements of availability, and in addition, if published before 1931, that were unaccompanied by a description, definition, or indication (Arts. 12, 16), and if published after 1930, that lacked accompanying statement of characters purporting to serve for differentiation of the taxon, or definite bibliographic reference to such a statement, or that were not proposed expressly as replacement *(nom. subst.)* of a pre-existing available name (Art. 13,a). Examples of "naked names" are: among species-group taxa, *Valvulina mixta* PARKER & JONES, 1865 (=*Cribrobulimina mixta* CUSHMAN, 1927, available and valid); among genus-group taxa, *Orbitolinopsis* SILVESTRI, 1932 (=*Orbitolinopsis* HENSON, 1948, available but classed as invalid junior synonym of *Orbitolina* D'ORBIGNY, 1850); among family-group taxa, Aequilateralidae D'ORBIGNY,

1846 (lacking type-genus), Hélicostègues d'ORBIGNY, 1826 (vernacular not latinized by later authors, Art. 11,e,iii), Poteriocrinidae AUSTIN & AUSTIN, 1843 (=fam. Poteriocrinoidea AUSTIN & AUSTIN, 1842) (neither 1843 or 1842 names complying with Art. 11,e, which states that "a family-group name must, when first published, be based on the name then valid for a contained genus," such valid name in the case of this family being *Poteriocrinites* MILLER, 1821).

(2) *"Denied names"* include all those that are defined by the *Code* (Art. 32,c) as incorrect original spellings. Examples are: Specific names, *nova-zelandica, mülleri, 10-brachiatus;* generic name, *M'Coyia, Størmerella, Römerina, Westgårdia;* family name, Růžičkinidae. Uncorrected "imperfect names" are "denied names" and unavailable, whereas corrected "imperfect names" are available.

(3) *"Impermissible names"* include all those employed for alleged genus-group taxa other than genus and subgenus (Art. 42,a) (e.g., supraspecific divisions of subgenera), and all those published after 1930 that are unaccompanied by definite fixation of a type-species (Art. 13,b). Examples of impermissible names are: *Martellispirifer* GATINAUD, 1949, and *Mirtellispirifer* GATINAUD, 1949, indicated respectively as a section and subsection of the subgenus *Cyrtospirifer; Fusarchaias* REICHEL, 1949, without definitely fixed type-species (=*Fusarchaias* REICHEL, 1952, with *F. bermudezi* designated as type-species).

(4) *"Null names"* include all those that are defined by the *Code* (Art. 33,b) as incorrect subsequent spellings, which are any changes of original spelling not demonstrably intentional. Such names are found in all ranks of taxa.

(5) *"Forgotten names"* are defined (Art. 23,b) as senior synonyms that have remained unused in primary zoological literature for more than 50 years. Such names are not to be used unless so directed by ICZN.

Latin designations for the discussed groups of unavailable zoological names are as follows: (1) *nomina nuda* (sing., *nomen nudum,* abbr., *nom. nud.*), (2) *nomina negata (nomen negatum, nom. neg.),* (3) *nomina vetita (nomen vetitum, nom. vet.),* (4) *nomina nulla (nomen nullum, nom. null.),* (5) *nomina oblita (nomen oblitum, nom. oblit.).*

VALID AND INVALID NAMES

Important distinctions relate to valid and available names, on one hand, and to invalid and unavailable names, on the other. Whereas determination of availability is based entirely on objective considerations guided by Articles of the *Code,* conclusions as to validity of zoological names partly may be subjective. A valid name is the correct one for a given taxon, which may have two or more available names but only a single correct name, generally the oldest. Obviously, no valid name can also be an unavailable name, but invalid names may include both available and unavailable names. Any name for a given taxon other than the valid name is an invalid name.

A sort of nomenclatorial no-man's-land is encountered in considering the status of some zoological names, such as *"doubtful names," "names under inquiry,"* and *"forgotten names."* Latin designations of these are *nomina dubia, nomina inquirenda,* and *nomina oblita,* respectively. Each of these groups may include both available and unavailable names, but the latter can well be ignored. Names considered to possess availability conduce to uncertainty and instability, which ordinarily can be removed only by appealed action of ICZN. Because few zoologists care to bother in seeking such remedy, the "wastebasket" names persist.

SUMMARY OF NAME GROUPS

Partly because only in such publications as the *Treatise* is special attention to groups of zoological names called for and partly because new designations are now introduced as means of recording distinctions explicitly as well as compactly, a summary may be useful. In the following tabulation valid groups of names are indicated in boldface type, whereas invalid ones are printed in italics.

DEFINITIONS OF NAME GROUPS

nomen conservatum (nom. conserv.). Name unacceptable under regulations of the *Code* which is made valid, either with original or altered spelling, through procedures specified by the *Code* or by action of ICZN exercising its plenary powers.

nomen correctum (nom. correct.). Name with intentionally altered spelling of sort required or allowable by the *Code* but not dependent on transfer from one taxonomic rank to another ("improved name"). (*See Code*, Arts. 26-b, 27, 29, 30-a-3, 31, 32-c-i, 33-a; in addition change of endings for suprafamilial taxa not regulated by the *Code*.)

nomen imperfectum (nom. imperf.). Name that as originally published (with or without subsequent identical spelling) meets all mandatory requirements of the *Code* but contains defect needing correction ("imperfect name"). (*See Code*, Arts. 26-b, 27, 29, 32-c, 33-a.)

nomen inviolatum (nom. inviol.). Name that as originally published meets all mandatory requirements of the *Code* and also is not correctable or alterable in any way ("inviolate name").

nomen negatum (nom. neg.). Name that as originally published (with or without subsequent identical spelling) constitutes invalid original spelling, and although possibly meeting all other mandatory requirements of the *Code*, cannot be used and has no separate status in nomenclature ("denied name"). It is to be corrected wherever found.

nomen nudum (nom. nud.). Name that as originally published (with or without subsequent identical spelling) fails to meet mandatory requirements of the *Code* and having no status in nomenclature, is not correctable to establish original authorship and date ("naked name").

nomen nullum (nom. null.). Name consisting of an unintentional alteration in form (spelling) of a previously published name (either available name, as *nom. inviol., nom. perf., nom. imperf., nom. transl.*; or unavailable name, as *nom. neg., nom. nud., nom. van.,* or another *nom. null.*) ("null name").

nomen oblitum (nom. oblit.). Name of senior synonym unused in primary zoological literature in more than 50 years, not to be used unless so directed by ICZN ("forgotten name").

nomen perfectum (nom. perf.). Name that as originally published meets all mandatory requirements of the *Code* and needs no correction of any kind but which nevertheless is validly alterable by change of ending ("perfect name").

nomen substitutum (nom. subst.). Replacement name published as substitute for an invalid name, such as a junior homonym (equivalent to "new name").

nomen translatum (nom. transl.). Name that is derived by valid emendation of a previously published name as result of transfer from one taxonomic rank to another within the group to which it belongs ("transferred name").

nomen vanum (nom. van.). Name consisting of an invalid intentional change in form (spelling) from a previously published name, such invalid emendation having status in nomenclature as a junior objective synonym ("vain name").

nomen vetitum (nom. vet.). Name of genus-group taxon not authorized by the *Code* or, if first published after 1930, without definitely fixed type-species ("impermissible name").

Except as specified otherwise, zoological names accepted in the *Treatise* may be understood to be classifiable either as *nomina inviolata* or *nomina perfecta* (omitting from notice *nomina correcta* among specific names) and these are not discriminated. Names which are not accepted for one reason or another include junior homonyms, senior synonyms classifiable as *nomina negata* or *nomina nuda,* and numerous junior synonyms which include both objective *(nomina vana)* and subjective types; rejected names are classified as completely as possible.

NAME CHANGES IN RELATION TO TAXA GROUPS

SPECIES-GROUP NAMES

Detailed consideration of valid emendation of specific and subspecific names is unnecessary here because it is well understood and relatively inconsequential. When the form of adjectival specific names is changed to obtain agreement with the gender of a generic name in transferring a species from one genus to another, it is never needful to label the changed name as a *nom. transl.* Likewise, transliteration of a letter accompanied by a diacritical mark in manner now called for by the *Code* (as in changing originally published *bröggeri* to *broeggeri*) or elimination of a hyphen (as in changing originally published *cornu-oryx* to *cornuoryx*) does not require *"nom. correct."* with it.

GENUS-GROUP NAMES

So rare are conditions warranting change of the originally published valid form of generic and subgeneric names that lengthy discussion may be omitted. Only elimination of diacritical marks of some names in this category seems to furnish basis for valid emendation. It is true that many changes of generic and subgeneric names have been published, but virtually all of these are either *nomina vana* or *nomina nulla.* Various names which formerly were classed as homonyms are not now, for two names that differ only by a single letter (or in original publication by presence or absence of a diacritical mark) are construed to be entirely distinct.

Examples in use of classificatory designations for generic names as previously given are the following, which also illustrate designation of type-species, as explained later.

Kurnatiophyllum THOMPSON, 1875 [**K. concentricum;* SD GREGORY, 1917] [=*Kumatiophyllum*

THOMPSON, 1876 *(nom. null.)*; *Cymatophyllum* THOMPSON, 1901 *(nom. van.)*; *Cymatiophyllum* LANG, SMITH & THOMAS, 1940 *(nom. van.)*].

Stichophyma POMEL, 1872 [**Manon turbinatum* RÖMER, 1841; SD RAUFF, 1893] [=*Stychophyma* VOSMAER, 1885 *(nom. null.)*; *Sticophyma* MORET, 1924 *(nom. null.)*].

Stratophyllum SMYTH, 1933 [**S. tenue*] [=*Ethmoplax* SMYTH, 1939 *(nom. van. pro Stratophyllum)*; *Stratiphyllum* LANG, SMITH & THOMAS, 1940 (*nom. van. pro Stratophyllum* SMYTH) (non *Stratiphyllum* SCHEFFEN, 1933)].

Placotelia OPPLIGER, 1907 [**Porostoma marconi* FROMENTEL, 1859; SD DELAUBENFELS, herein] [=*Plakotelia* OPPLIGER, 1907 *(nom. neg.)*].

Walcottella DELAUBENFELS, 1955 [*nom. subst., pro Rhopalicus* SCHRAMM, 1936 (*non* FÖRSTER, 1856)].

Cyrtograptus CARRUTHERS, 1867 [*nom. correct.* LAPWORTH, 1873 (*pro Cyrtograpsus* CARRUTHERS, (1867), *nom. conserv.* proposed BULMAN, 1955 (ICZN pend.)].

FAMILY-GROUP NAMES; USE OF *"NOM. TRANSL."*

The *Code* specifies the endings only for subfamily (-inae) and family (-idae) but all family-group taxa are defined as coordinate, signifying that for purposes of priority a name published for a taxon in any category and based on a particular type-genus shall date from its original publication for a taxon in any category, retaining this priority (and authorship) when the taxon is treated as belonging to a lower or higher category. By exclusion of -inae and -idae, respectively reserved for subfamily and family, the endings of names used for tribes and superfamilies must be unspecified different letter combinations. These, if introduced subsequent to designation of a subfamily or family based on the same nominate genus, are *nomina translata,* as is also a subfamily that is elevated to family rank or a family reduced to subfamily rank. In the *Treatise* it is desirable to distinguish the valid alteration comprised in the changed ending of each transferred family-group name by the abbreviation *"nom. transl."* and record of the author and date belonging to this alteration. This is particularly important in the case of superfamilies, for it is the author who introduced this taxon that one wishes to know about rather than the author of the superfamily as defined by the *Code,* for the latter is merely the individual who first defined some lower-rank family-group taxon that contains the nominate genus of the superfamily. The publication of the author containing introduction of the superfamily *nomen translatum* is likely to furnish the information on taxonomic considerations that support definition of the unit.

Examples of the use of *"nom. transl."* are the following.

Subfamily STYLININAE d'Orbigny, 1851
[*nom. transl.* EDWARDS & HAIME, 1857 (*ex* Stylinidae D'ORBIGNY, 1851)]

Superfamily ARCHAEOCTONOIDEA Petrunkevitch, 1949
[*nom. transl.* PETRUNKEVITCH, 1955 (*ex* Archaeoctonidae PETRUNKEVITCH, 1949)]

Superfamily CRIOCERATITACEAE Hyatt, 1900
[*nom. transl.* WRIGHT, 1952 (*ex* Crioceratitidae HYATT, 1900)]

FAMILY-GROUP NAMES; USE OF *"NOM. CORRECT."*

Valid name changes classed as *nomina correcta* do not depend on transfer from one category of family-group units to another but most commonly involve correction of the stem of the nominate genus; in addition, they include somewhat arbitrarily chosen modification of ending for names of tribe or superfamily. Examples of the use of *"nom. correct."* are the following.

Family STREPTELASMATIDAE Nicholson, 1889
[*nom. correct.* WEDEKIND, 1927 (*pro* Streptelasmidae NICHOLSON, 1889, *nom. imperf.*)]

Family PALAEOSCORPIIDAE Lehmann, 1944
[*nom. correct.* PETRUNKEVITCH, 1955 (*pro* Palaeoscorpionidae LEHMANN, 1944, *nom. imperf.*)]

Family AGLASPIDIDAE Miller, 1877
[*nom. correct.* STØRMER, 1959 (*pro* Aglaspidae MILLER, 1877, *nom. imperf.*)]

Superfamily AGARICIICAE Gray, 1847
[*nom. correct.* WELLS, 1956 (*pro* Agaricioidae VAUGHAN & WELLS, 1943, *nom. transl. ex* Agariciidae GRAY, 1847)]

FAMILY-GROUP NAMES; USE OF *"NOM. CONSERV."*

It may happen that long-used family-group names are invalid under strict application of the *Code.* In order to retain the otherwise invalid name, appeal to ICZN is needful. Examples of use of *nom. conserv.* in this connection, as cited in the *Treatise,* are the following.

Family ARIETITIDAE Hyatt, 1874
[*nom. correct.* HAUG, 1885 (*pro* Arietidae HYATT, 1875) *nom. conserv.* proposed ARKELL, 1955 (ICZN pend.)]

Family STEPHANOCERATIDAE Neumayr, 1875
[*nom. correct.* FISCHER, 1882 (*pro* Stephanoceratinen NEUMAYR, 1875, invalid vernacular name), *nom. conserv.* proposed ARKELL, 1955 (ICZN pend.)]

FAMILY-GROUP NAMES; REPLACEMENTS

Family-group names are formed by adding letter combinations (prescribed for family and subfamily but not now for others) to the stem of the name belonging to genus (nominate genus) first chosen as type of the assemblage. The type-genus need not be the oldest in terms of receiving its name and definition, but it must be the first-published as name-giver to a family-group taxon among all those included. Once fixed, the family-group name remains tied to the nominate genus even if its name is changed by reason of status as a junior homonym or junior synonym, either objective or subjective. Seemingly, the *Code* (Art. 39) requires replacement of a family-group name only in the event that the nominate genus is found to be a junior homonym, and then a substitute family-group name is accepted if it is formed from the oldest available substitute name for the nominate genus. Authorship and date attributed to the replacement family-group name are determined by first publication of the changed family group-name, but for purposes of the Law of Priority, they take the date of the replaced name. Numerous long-used family-group names are incorrect in being *nomina nuda,* since they fail to satisfy criteria of availability (Art. 11,e). These also demand replacement by valid names.

The aim of family-group nomenclature is greatest possible stability and uniformity, just as in case of other zoological names. Experience indicates the wisdom of sustaining family-group names based on junior subjective synonyms if they have priority of publication, for opinions of different workers as to the synonymy of generic names founded on different type-species may not agree and opinions of the same worker may alter from time to time. The retention similarly of first-published family-group names which are found to be based on junior objective synonyms is less clearly desirable, especially if a replacement name derived from the senior objective synonym has been recognized very long and widely. To displace a much-used family-group name based on the senior objective synonym by disinterring a forgotten and virtually unused family-group name based on a junior objective synonym because the latter happens to have priority of publication is unsettling.

Replacement of a family-group name may be needed if the former nominate genus is transferred to another family-group. Then the first-published name-giver of a family-group assemblage in the remnant taxon is to be recognized in forming a replacement name.

FAMILY-GROUP NAMES; AUTHORSHIP AND DATE

All family-group taxa having names based on the same type-genus are attributed to the author who first published the name for any of these assemblages, whether tribe, subfamily, or family (superfamily being almost inevitably a later-conceived taxon). Accordingly, if a family is divided into subfamilies or a subfamily into tribes, the name of no such subfamily or tribe can antedate the family name. Also, every family containing differentiated subfamilies must have a nominate *(sensu stricto)* subfamily, which is based on the same type genus as that for the family, and the author and date set down for the nominate subfamily invariably are identical with those of the family, without reference to whether the author of the family or some subsequent author introduced subdivisions.

Changes in the form of family-group names of the sort constituting *nomina correcta,* as previously discussed, do not affect authorship and date of the taxon concerned, but in publications such as the *Treatise* it is desirable to record the authorship and date of the correction.

SUPRAFAMILIAL TAXA

International rules of zoological nomenclature as given in the *Code* (1961) are limited to stipulations affecting lower-rank categories (infrasubspecies to superfamily). Suprafamilial categories (suborder to phylum) are either unmentioned or explicitly placed outside of the application of zoological rules. The *Copenhagen Decisions on Zoological Nomenclature* (1953, Arts. 59-69) proposed to adopt rules for naming suborders and higher taxonomic divisions up to and including phylum, with provision for designating a type-genus for each, hopefully in such manner as not to interfere with the taxonomic freedom of workers. Procedures for applying the Law of Priority and Law of Homonymy to suprafamilial taxa were outlined and for dealing with the names for such units and their authorship, with assigned dates, when they should be transferred on taxonomic grounds from one rank to another. The adoption of terminations of names, different for each category but uniform within each, was recommended.

The Colloquium on zoological nomenclature which met in London during the week just before the XVth International Congress of Zoology convened in 1958 thoroughly discussed the proposals for regulating suprafamilial nomenclature, as well as many others advocated for inclusion in the new *Code* or recommended for exclusion from it. A decision which was supported by a wide majority of the participants in the Colloquium was against the establishment of rules for naming taxa above family-group rank, mainly because it was judged that such regulation would unwisely tie the hands of taxonomists. For example, if a class or order was defined by some author at a given date, using chosen morphologic characters (e.g., gills of pelecypods), this should not be allowed to freeze nomenclature, taking precedence over another later-proposed class or order distinguished by different characters (e.g., hinge-teeth of pelecypods). Even the fixing of type-genera for suprafamilial taxa might have small value, if any, hindering taxonomic work rather than aiding it. At all events, no legal basis for establishing such types and for naming these taxa has yet been provided.

The considerations just stated do not prevent the Editor of the *Treatise* from making "rules" for dealing with suprafamilial groups of animals described and illustrated in this publication. At least a degree of uniform policy is thought to be needed, especially for the guidance of *Treatise*-contributing authors. This policy should accord with recognized general practice among zoologists, but where general practice is indeterminate or nonexistent our own procedure in suprafamilial nomenclature needs to be specified as clearly as possible. This pertains especially to decisions about names themselves, about citation of authors and dates, and about treatment of suprafamilial taxa which on taxonomic grounds are changed from their originally assigned rank. Accordingly, a few "rules" expressing *Treatise* policy are given here, some with examples of their application.

(1) The name of any suprafamilial taxon must be a Latin or latinized uninominal noun of plural form, or treated as such, (a) with a capital initial letter, (b) without diacritical mark, apostrophe, diaeresis, or hyphen, and (c) if component consisting of a numeral, numerical adjective, or adverb is used, this must be written in full (e.g., Stethostomata, Trionychi, Septemchitonina, Scorpiones, Subselliflorae). No uniformity in choice of ending for taxa of a given rank is demanded (e.g., orders named Gorgon*acea,* Millepor*ina,* Rugos*a,* Scleractin*ia,* Stromatopor*oidea,* Phalang*ida*).

(2) Names of suprafamilial taxa may be constructed in almost any way, (a) intended to indicate morphological attributes (e.g., Lamellibranchiata, Cyclostomata, Toxoglossa), (b) based on the stem of an included genus (e.g., Bellerophontina, Nautilida, Fungiina), or (c) arbitrary combinations of letters, (e.g., Yuania), but none of these can be allowed to end in -idae or -inae, reserved for family-group taxa. A class or subclass (e.g., Nautiloidea), order (e.g., Nautilida), or suborder (e.g., Nautilina) named from the stem of an included genus may be presumed to have that genus (e.g., *Nautilus*) as its objective type. No suprafamilial name identical in form to that of a genus or to another published suprafamilial name should be employed (e.g., order Decapoda Latreille, 1803, crustaceans, and order Decapoda Leach, 1818, cephalopods; suborder Chonetoidea Muir-Wood, 1955, and genus *Chonetoidea* Jones, 1928). Worthy of notice is the classificatory and nomenclatural distinction between suprafamilial and family-group taxa which respectively are named from the same type-genus, since one is not considered to be transferable to the other (e.g., suborder Bellerophontina Ulrich & Scofield, 1897; superfamily Bellerophontacea M'Coy, 1851; family Bellerophontidae M'Coy, 1851). Family-group names and suprafamilial names are not coordinate.

(3) The Laws of Priority and Homonymy lack any force of international agreement as applied to suprafamilial names, yet in the interest of nomenclatural stability and the avoidance of confusion these laws are widely accepted by zoologists above the

family-group level wherever they do not infringe on taxonomic freedom and long-established usage.

(4) Authors who accept priority as a determinant in nomenclature of a suprafamilial taxon may change its assigned rank at will, with or without modifying the terminal letters of the name, but such change(s) cannot rationally be judged to alter the authorship and date of the taxon as published originally. (a) A name revised from its previously published rank is a "transferred name" *(nom. transl.)*, as illustrated in the following.

Order CORYNEXOCHIDA Kobayashi, 1935

[*nom. transl.* MOORE, 1955 (*ex* suborder Corynexochida KOBAYASHI, 1935)]

(b) A name revised from its previously published form merely by adoption of a different termination, without changing taxonomic rank, is an "altered name" *(nom. correct.)*. Examples follow.

Order DISPARIDA Moore & Laudon, 1943

[*nom. correct.* MOORE, 1952 (*pro* order Disparata MOORE & LAUDON, 1943)]

Suborder AGNOSTINA Salter, 1864

[*nom. correct.* HARRINGTON & LEANZA, 1957 (*pro* suborder Agnostini SALTER, 1864)]

(c) A suprafamilial name revised from its previously published rank with accompanying change of termination (which may or may not be intended to signalize the change of rank) is construed to be primarily a *nom. transl.* (compare change of ending for family-group taxa -idae to -inae, or vice versa, and to superfamily) but if desired it could be recorded as *nom. transl. et correct.*

Order ORTHIDA Schuchert & Cooper, 1931

[*nom. transl.* MOORE, 1952 (*ex* suborder Orthoidea SCHUCHERT & COOPER, 1931)]

(5) The authorship and date of nominate subordinate and superordinate taxa among suprafamilial taxa are considered in the *Treatise* to be identical since each actually or potentially has the same type. Examples are given below.

Subclass ENDOCERATOIDEA Teichert, 1933

[*nom. transl.* TEICHERT, 1964 (*ex* superorder Endoceratoidea SHIMANSKIY & ZHURAVLEVA, 1961, *nom. transl. ex* order Endoceroidea TEICHERT, 1933)]

Order ENDOCERIDA Teichert, 1933

[*nom. correct.* TEICHERT, 1964 (*pro* order Endoceroidea TEICHERT, 1933)]

Suborder ENDOCERINA Teichert, 1933

[*nom. correct.* TEICHERT, 1964 (*pro* suborder Endoceracea SCHINDEWOLF, 1935, *nom. transl. ex* order Endoceroidea TEICHERT, 1933)]

(6) A suprafamilial taxon may or may not contain a family-group taxon or taxa having the same type-genus, and if it does, the respective suprafamilial and family-group taxa may or may not be nominate (having names with the same stem). The zoological *Code* (Art. 61) affirms that "each taxon [of any rank] has, actually or potentially, its type." Taxa above the family-group level which may be designated as having the same type-genus (such designations not being stipulated or recognized by any articles of the zoological *Code*) are considered to have identical authorship and date if the stem of names employed is the same (illustrated in preceding paragraph), but otherwise their authorship and date are accepted as various. Examples showing both suprafamilial and familial taxa in a group of spiders follow.

Class ARACHNIDA Lamarck, 1801

[*nom. correct.* NEWPORT, 1830 (*pro* class—not family—Arachnidae LAMARCK, 1801) (type, *Araneus* CLERCK, 1757, validated ICZN, 1948)]

Subclass CAULOGASTRA Pocock, 1893

[type, *Araneus* CLERCK, 1757]

Superorder LABELLATA Petrunkevitch, 1949

[type, *Araneus* CLERCK, 1757]

Order ARANEIDA Clerck, 1757

[*nom. correct.* DALLAS, 1864 (*pro* Araneidea BLACKWALL, 1861, *pro* Araneides LATREILLE, 1801, *pro* Aranei CLERCK, 1757, validated ICZN, 1948) (type, *Araneus* CLERCK, 1757)]

Suborder DIPNEUMONINA Latreille, 1817

[*nom. correct.* PETRUNKEVITCH, 1955 (*pro* Dipneumones LATREILLE, 1817) (type, *Araneus* CLERCK, 1757)]

Division TRIONYCHI Petrunkevitch, 1933

[type, *Araneus* CLERCK, 1757]

Superfamily ARANEOIDEA Leach, 1815

[*nom. transl.* PETRUNKEVITCH, 1955 (*ex* Araneides LEACH, 1815) (type, *Araneus* CLERCK, 1757)]

Family ARANEIDAE Leach, 1815

[*nom. correct.* PETRUNKEVITCH, 1955 (*pro* Araneadae LEACH, 1819, *pro* Araneides LEACH, 1815) (type, *Araneus* CLERCK, 1757)]

Subfamily ARANEINAE Leach, 1815

[*nom. transl.* SIMON, 1892 (*ex* Araneidae LEACH, 1815) (type, *Araneus* CLERCK, 1757)]

TAXONOMIC EMENDATION

Emendation has two measurably distinct aspects as regards zoological nomenclature. These embrace (1) alteration of a name itself in various ways for various reasons, as has been reviewed, and (2) alteration of taxonomic scope or concept in application of a given zoological name, whatever its hierarchical rank. The latter type of emendation primarily concerns classification and inherently is not associated with change of name, whereas the other type introduces change of name without necessary expansion, restriction, or other modification in applying the name. Little attention generally has been paid to this distinction in spite of its significance.

Most zoologists, including paleozoologists, who have signified emendation of zoological names refer to what they consider a material change in application of the name such as may be expressed by an importantly altered diagnosis of the assemblage covered by the name. The abbreviation *"emend."* then may accompany the name, with statement of the author and date of the emendation. On the other hand, a multitude of workers concerned with systematic zoology think that publication of *"emend."* with a zoological name is valueless, because more or less alteration of taxonomic sort is introduced whenever a subspecies, species, genus, or other assemblage of animals is incorporated under or removed from the coverage of a given zoological name. Inevitably associated with such classificatory expansions and restrictions is some degree of emendation affecting diagnosis. Granting this, still it is true that now and then somewhat radical revisions are put forward, generally with published statement of reasons for changing the application of a name. To erect a signpost at such points of most significant change is worthwhile, both as aid to subsequent workers in taking account of the altered nomenclatural usage and as indication that not to-be-overlooked discussion may be found at a particular place in the literature. Authors of contributions to the *Treatise* are encouraged to include records of all specially noteworthy emendations of this nature, using the abbreviation *"emend."* with the name to which it refers and citing the author and date of the emendation.

In Part G (Bryozoa) and Part D (Protista 3) of the *Treatise,* the abbreviation *"emend."* is employed to record various sorts of name emendations, thus conflicting with usage of *"emend."* for change in taxonomic application of a name without alteration of the name itself. This is objectionable. In Part E (Archaeocyatha, Porifera) and later-issued divisions of the *Treatise,* use of *"emend."* is restricted to its customary sense, that is, significant alteration in taxonomic scope of a name such as calls for noteworthy modifications of a diagnosis. Other means of designating emendations that relate to form of a name are introduced.

STYLE IN GENERIC DESCRIPTIONS

CITATION OF TYPE-SPECIES

The name of the type-species of each genus and subgenus is given next following the generic name with its accompanying author and date, or after entries needed for definition of the name if it is involved in homonymy. The originally published combination of generic and trivial names for this species is cited, accompanied by an asterisk (*), with notation of the author and date of original publication. An exception in this procedure is made, however, if the species was first published in the same paper and by the same author as that containing definition of the genus which it serves as type; in such case, the initial letter of the generic name followed by the trivial name is given without repeating the name of the author and date, for this saves needed space. Examples of these two sorts of citations are as follows:

Diplotrypa NICHOLSON, 1879 [**Favosites petropolitanus* PANDER, 1830].

Chainodictyon FOERSTE, 1887 [**C. laxum*].

If the cited type-species is a junior synonym of some other species, the name of this latter also is given, as follows:

Acervularia SCHWEIGGER, 1819 [**A. baltica* (=**Madrepora ananas* LINNÉ, 1758)].

It is judged desirable to record the manner of establishing the type-species, whether by original designation or by subsequent designation.

Fixation of type-species originally. The type-species of a genus or subgenus, according to provisions of the *Code,* may be fixed in various ways originally (that is, in the publication containing first proposal of the generic name) or it may be fixed in specified ways subsequent to the original publication. Fixation of a type-species originally may be classified as automatic if the new genus was introduced for a single species (monotypy), or if the names of species referred to the genus are objectively synonymous. In addition, fixation of a type-species originally may be established in several ways by original designation, as by explicit statement given by an author, by use of *typus* or *typicus* as a new specific name, and by absolute tautonymy (e.g., *Mesolobus mesolobus*). According to convention adopted in the *Treatise,* the absence of indication as to the manner of fixing the type-species is to be understood as signifying fixation of the type-species in one way or another originally. Where an author wishes to specify the mode of original fixation, however, this may be done by such abbreviations as "M" (monotypy), "OS" (objective synonymy), and "OD" (original designation), the first- and last-mentioned being most common and the other very rare.

Fixation of type-species subsequently. The type-species of many genera are not determinable from the publication in which the generic name was introduced and therefore such genera can acquire a type-species only by some manner of subsequent designation. Most commonly this is established by publishing a statement naming as type-species one of the species originally included in the genus, and in the *Treatise* fixation of the type-species in this manner is indicated by the letters "SD" accompanied by the name of the subsequent author (who may be the same person as the original author) and the date of publishing the subsequent designation. Some genera, as first described and named, included no mentioned species and these necessarily lack a type-species until a date subsequent to that of the original publication when one or more species are assigned to such a genus. If only a single species is thus assigned, it automatically becomes the type-species and in the *Treatise* this subsequent monotypy is indicated by the letters "SM." Of course, the first publication containing assignment of species to the genus which originally lacked any included species is the one concerned in fixation of the type-species, and if this named 2 or more species as belonging to the genus but did not designate a type-species, then a later "SD" designation is necessary. Examples of the use of "SD" and "SM" as employed in the *Treatise* follow.

Hexagonaria GÜRICH, 1896 [**Cyathophyllum hexagonum* GOLDFUSS, 1826; SD LANG, SMITH & THOMAS, 1940].

Muriceides STUDER, 1887 [**M. fragilis* WRIGHT & STUDER, 1889; SM WRIGHT & STUDER, 1889].

Another mode of fixing the type-species of a genus that may be construed as a special sort of subsequent designation is action of the International Commission on Zoological Nomenclature using its plenary powers. Definition in this way may set aside application of the *Code* so as to arrive at a decision considered to be in the best interest of continuity and stability of zoological nomenclature. When made, it is binding and commonly is cited in the *Treatise* by the letters "ICZN," accompanied by the date of announced decision and (generally) reference to the appropriate numbered Opinion.

HOMONYMS

Most generic names are distinct from all others and are indicated without ambiguity by citing their originally published spelling accompanied by name of the author and date of first publication. If the same generic name has been applied to 2 or more distinct taxonomic units, however, it is necessary to differentiate such homonyms, and this calls for distinction between junior homonyms and senior homonyms. Because a junior homonym is invalid, it must be replaced by some other name. For example, *Callopora* HALL, 1851, introduced for Paleozoic trepostome bryozoans, is invalid because GRAY in 1848 published the same name for Cretaceous-to-Recent cheilostome bryozoans, and BASSLER in 1911 introduced the new name *Hallopora* to replace HALL's homonym. The *Treatise* style of entry is:

Hallopora BASSLER, 1911, *nom. subst.* [*pro Callopora* HALL, 1851 (*non* GRAY, 1848)].

In like manner, a needed replacement generic name may be introduced in the *Treatise* (even though first publication of generic names otherwise in this work is avoided). The requirement that an exact bibliographic reference must be given for the replaced name commonly can be met in the *Treatise* by citing a publication recorded in the list of references, using its assigned index number, as shown in the following example.

Mysterium DE LAUBENFELS, *nom. subst.* [*pro Mystrium* SCHRAMMEN, 1936 (ref. 40, p. 60) (*non* ROGER, 1862)] [**Mystrium porosum* SCHRAMMEN, 1936].

For some replaced homonyms, a footnote reference to the literature is necessary. A senior homonym is valid, and in so far as the *Treatise* is concerned, such names are handled according to whether the junior homonym belongs to the same major taxonomic division (class or phylum) as the senior homonym or to some other; in the former instance, the author and date of the junior homonym are cited as:

Diplophyllum HALL, 1851 [*non* SOSHKINA, 1939] [**D. caespitosum*].

Otherwise, no mention of the existence of a junior homonym generally is made.

Homonyms by misidentification. When an author uses a generic name for species not congeneric with the type-species, it is needful to record the misuse of the generic name, even though this is only determinable subjectively. In the *Treatise* homonyms by misidentification are cited in synonymies as illustrated in the following example.

Asmussia PACHT, 1849 [**A. membranacea*] [=*Posidonomya* PACHT, 1852 (*non* BRONN, 1834); *Estheria* JONES, 1856 (*non* ROBINEAU-DESVOIDY, 1830; *nec* RUEPPELL, 1837)].

Synonymic homonyms. An author sometimes publishes a generic name in two or more papers of different date, each of which indicates that the name is new. This is a bothersome source of errors for later workers who are unaware that a supposed first publication which they have in hand is not actually the original one. Although the names were separately published, they are identical and therefore definable as homonyms; at the same time they are absolute synonyms. For the guidance of all concerned, it seems desirable to record such names as synonymic homonyms and in the *Treatise* the junior one of these is indicated by the abbreviation "jr. syn. hom."

Identical family-group names not infrequently are published as new names by different authors, the author of the later-introduced name being ignorant of previous publication(s) by one or more other workers. In spite of differences in taxonomic concepts as indicated by diagnoses and grouping of genera and possibly in assigned rank, these family-group taxa are nomenclatural homonyms, based on the same type-genus, and they are also synonyms. Wherever encountered, such synonymic homonyms are distinguished in the *Treatise* as in dealing with generic names.

SYNONYMS

Citation of synonyms is given next following record of the type-species and if 2 or more synonyms of differing date are recognized, these are arranged in chronological order. Objective synonyms are indicated by accompanying designation "(obj.)," others being understood to constitute subjective synonyms. Examples showing *Treatise* style in listing synonyms follow.

Calapoecia BILLINGS, 1865 [**C. anticostiensis;* SD LINDSTRÖM, 1883] [=*Columnopora* NICHOLSON, 1874; *Houghtonia* ROMINGER, 1876].

Staurocyclia HAECKEL, 1882 [**S. cruciata* HAECKEL, 1887] [=*Coccostaurus* HAECKEL, 1882 (obj.); *Phacostaurus* HAECKEL, 1887 (obj.)].

A synonym which also constitutes a homonym is recorded as follows:

Lyopora NICHOLSON & ETHERIDGE, 1878 [**Palaeopora? favosa* M'COY, 1850] [=*Liopora* LANG, SMITH & THOMAS, 1940 (*non* GIRTY, 1915)].

Some junior synonyms of either objective or subjective sort may take precedence desirably over senior synonyms wherever uniformity and continuity of nomenclature are served by retaining a widely used but technically rejectable name for a generic assemblage. This requires action of ICZN using its plenary powers to set aside the unwanted name and validate the wanted one, with placement of the concerned names on appropriate official lists. In the *Treatise* citation of such a conserved generic name is given in the manner shown by the following example.

Tetragraptus SALTER, 1863 [*nom. correct.* HALL, 1865 (*pro Tetragrapsus* SALTER, 1863), *nom. conserv.* proposed BULMAN, 1955, ICZN pend.] [**Fucoides serra* BRONGNIART, 1828 (=*Graptolithus bryonoides* HALL, 1858)].

ABBREVIATIONS

Abbreviations used in this division of the *Treatise* are explained in the following alphabetically arranged list.

Abbreviations

Aalen., Aalenian
abbrev., abbreviation
Abhandl., Abhandlung(en)
adj., adjective
aff., *affinis* (related to)
Afr., Africa, -an
Ala., Alabama
Alb., Albian
Alg., Algeria
Am., America, -n
Ann., Anñaes, Annales, Annual
ant., anterior
approx., approximately
Apt., Aptian
Aquitan., Aquitanian
Arbeit., Arbeiten
Arch., Archipelago, Archives, Archivos
archibenth., archibenthal
Arenig., Arenigian
Arg., Argentina
Argov., Argovian
art., article
Ashgill., Ashgillian
Atl., Atlantic
auctt., *auctorum* (of authors)
Aus., Austria
Austin., Austinian
Auvers., Auversian

Bajoc., Bajocian
Barrem., Barremian
Batesford., Batesfordian
Bathon., Bathonian
B.C., British Columbia
Bd., Band
Beil., Beilage
Belg., Belgium, Belgique
Berrias., Berriasian
Blackriv., Blackriveran
Boh., Bohemia
Br.I., British Isles
Brit., Britain, British
Bulg., Bulgaria
Bull., Bulletin
Burdigal., Burdigalian

C., Central
ca., *circa*
Cab., Cabinet
Calif., California
Callov., Callovian
Cam., Cambrian
Campan., Campanian
Can., Canada
Carb., Carboniferous
Carib., Caribbean
Carn., Carnian
Cenoman., Cenomanian
cf., *confer* (compare)
Charmouth., Charmouthian
Chemung., Chemungian
Chester., Chesterian
cm., centimeter
Coll., Collection(s)
Coniac., Coniacian
Contrib., Contribution(s)
cosmop., cosmopolitan
Couvin., Couvinian
Cret., Cretaceous
Czech., Czechoslovakia

Dan., Danian
D.C., District of Columbia
dec., decade
Denkschr., Denkschrift(en)
Denm., Denmark
Dev., Devonian
Devon., Devonshire
diagram., diagrammatic
diam., diameter
Distr., District
Doc., Document
Domer., Domerian

E., East
ed., editor
edit., edition
e.g., *exempli gratia* (for example)
emend., *emendatus(-a)*
Eng., England
enl., enlarged
Eoc., Eocene
err., *errore* (by error)
Est., Estonia
et al., *et alii* (and other persons)
etc., *et cetera* (and others, objects)
Eu., Europe
Ex., Executive
ext., exterior

F., Formation
fam., family
Famenn., Famennian
fig., figure, -s
Fla., Florida
Förhandl., Förhandlingar
Forhandl.. Forhandlinger
Fr., France, French, Français, -e

Ga., Georgia
Gaj., Gajian
G.Brit., Great Britain
Geol., Geology, Geological, Geologische, etc.
Ger., Germany, German
Glos., Gloucestershire
Gotl., Gotland
Gr., Group, Great
Gt., Great

Handl., Handlingar
Hauteriv., Hauterivian
Helvet., Helvetian
Herefords., Herefordshire
Hettang., Hettangian
hom., homonym
horiz., horizontal
Hung., Hungary, Hungarica

I., Isle
ICZN, International Commission of Zoological Nomenclature
i.e., *id est* (that is)
Ill., Illinois
incl., including
Ind., Indiana
indet., indetermined
Ind.O., Indian Ocean
Indon., Indonesia
int., interior
interamb., interambulacral
Internatl., International
Ire., Ireland
Is., Island, -s

Jahrb., Jahrbuch
Jahrg., Jahrgang
Jour., Journal
jr., junior
Jur., Jurassic
juv., juvenile

Kimmeridg., Kimmeridgian
Kinderhook., Kinderhookian
Ky., Kentucky

L., **Low.**, Lower
lat., lateral
Lias., Liassic
litt., letters
Llandover., Llandoverian
long., longitudinal
Ls., Limestone
Ltd., Limited
Ludlov., Ludlovian
Lutet., Lutetian

m., meter
M., **Mid.**, Middle
M., monotypy
Maastricht., Maastrichtian
Madag., Madagascar
mag., magnification
Maurit., Mauritius
Maysvill., Maysvillian
Md., Maryland
Medd., Meddelanden, Meddelelser
Medit., Mediterranean
Mem., Memoir -s, Memoria, Memorie
Mém., Mémoire, -s
Mex., Mexico
Minn., Minnesota
Mio., Miocene
Miss., Mississippi, Mississippian
Mitt., Mitteilungen
mm., millimeter, -s
Mo., Missouri

Mon., Monograph, Monographia
Mont., Montana

n., new
N., North
N.Am., North America, -n
Nat., Natural
NE., Northeast
Necom., Neocomian
Neog., Neogene
Neth., Netherlands
Nev., Nevada
Newf., Newfoundland
N.J., New Jersey
no., number
nom. conserv., *nomen conservatum* (conserved name)
nom. correct., *nomen correctum* (corrected or intentionally altered name)
nom. dub., *nomen dubium* (doubtful name)
nom. imperf., *nomen imperfectum* (imperfect name)
nom. neg., *nomen negatum* (denied name)
nom. nov., *nomen novum* (new name)
nom. nud., *nomen nudum* (naked name)
nom. null., *nomen nullum* (null, void name)
nom. oblit., *nomen oblitum* (forgotten name)
nom. subst., *nomen substitutum* (substitute name)
nom. transl., *nomen translatum* (transferred name)
nom. van., *nomen vanum* (vain, void name)
nom. vet., *nomen vetitum* (impermissible name)
Nor., Norian
Notizbl., Notizblatt
Nouv., Nouvelle
N.Scotia, Nova Scotia
NW., Northwest
N.Y., New York
N.Z., New Zealand

O., Ocean
obj., objective
Occas., Occasional
OD, original designation
Okla., Oklahoma
Oligo., Oligocene
Ont., Ontario
op. cit., *opere citato* (in the work cited)
Opin., Opinion
opp., opposite
Ord., Ordovician
Ore., Oregon
Oxford., Oxfordian
Oxfords., Oxfordshire

p., page, -s
Pa., Pennsylvania
Pac., Pacific
Pak., Pakistan
Paleoc., Paleocene
Paleog., Paleogene
pend., pending
Penn., Pennsylvanian
Perm., Permian
Permocarb., Permocarboniferous
Philip., Philippines
pl., plate, -s, plural
Pleist., Pleistocene
Pliensbach., Pliensbachian
Plio., Pliocene
Pol., Poland
Port., Portugal
Portland., Portlandian
post., posterior
Proc., Proceedings
Prof., Professional
Prov., Province
pt., part, -s
publ., publication, published

Quart., Quarterly
Que., Quebec

Ranikot., Ranikotian
Raurac., Rauracian
Rec., Recent, Record(s)
reconstr., reconstructed, -ion
Rept., Report
Rhaet., Rhaetian
Richmond., Richmondian

S., South, Sea
S.Am., South America
Santon., Santonian
S.Car., South Carolina
Scot., Scotland
SD, subsequent designation
SE., Southeast
sec., section(-s)
Senon., Senonian
sep., separate
Sequan., Sequanian
ser., series, serial, etc
Sess., Session
Sh., Shale
Shrops., Shropshire
Sil., Silurian
sing., singular
Sitzungsber., Sitzungsberichte
s.l., *sensu lato* (in the wide sense, broadly defined)
SM, subsequent monotypy
Somal., Somaliland
sp., species (spp., plural)
spec., special, specimen
sr., senior
s.s., *sensu stricto* (in the strict sense, narrowly defined)
Str., Strait, -s
subfam., subfamily
subj., subjective
subtrop., subtropical
superfam., superfamily
suppl., supplement
SW., Southwest
Swed., Sweden
Switz., Switzerland
syn., synonym

tech., technical
Tenn., Tennessee
Tert., Tertiary
Tex., Texas
Tithon., Tithonian
Toarc., Toarcian
Torton., Tortonian
Trans., Transactions
transl., translated, translation
transv., transverse
Tremadoc., Tremadocian
Trenton., Trentonian
Trias., Triassic
trop., tropical
Turon., Turonian

U., Up., Upper
Univ., Universidad, Università, Universität, Université, Universitets, University
Urgon., Urgonian
U.S., United States
USA, United States (America)
USSR, Union of Soviet Socialist Republics

v., volume, -s
Valangin., Valanginian
var., variety
Venez., Venezuela
Verhandl., Verhandlungen
Vesul., Vesulian
Vict., Victoria

W., West
Wash., Washington
Wenlock., Wenlockian

Yorks., Yorkshire
Ypres., Ypresian

Z., Zone
Zeitschr., Zeitschrift
Zool., Zoological, Zoology

REFERENCES TO LITERATURE

Each part of the *Treatise* is accompanied by a selected list of references to paleontological literature consisting primarily of recent and comprehensive monographs available but also including some older works recognized as outstanding in importance. The purpose of giving these references is to aid users of the *Treatise* in finding detailed descriptions and illustrations of morphological features of fossil groups, discussions of classifications and distribution, and especially citations of more or less voluminous literature. Generally speaking, publications listed in the *Treatise* are not original sources of information concerning taxonomic units of various rank but they tell the student where he may find them; otherwise it is necessary to turn to such aids as the *Zoological Record* or NEAVE's *Nomenclator Zoologicus.* References given in the *Treatise* are arranged alphabetically by authors and accompanied by index numbers which serve the purpose of permitting citation most concisely in various parts of the text; these citations of listed papers are enclosed invariably in parentheses and except in Part C, are distinguishable from dates because the index numbers comprise no more than 3 digits. The systematic descriptions given in Part C are accompanied by a reference list containing more than 2,000 entries; the index numbers for them are marked by an asterisk.

The following is a statement of the full names of serial publications which are cited in abbreviated form in the *Treatise* lists of references. The information thus provided should be useful in library research work. The list is alphabetized according to the serial titles which were employed at the time of original publication. Those following it in brackets are those under which the publication may be found currently in the *Union List of Serials,* the United States Library of Congress listing, and most library card catalogues. The names of serials published in Cyrillic are transliterated; in the reference lists these titles, which may be abbreviated, are accompanied by transliterated authors' names and titles, with English translation of the title. The place of publication is added (if not included in the serial title).

The method of transliterating Cyrillic letters that is adopted as "official" in the *Treatise* is the so-called Anglo-American method given by the Geographical Society of London. It follows that names of some Russian authors in transliterated form derived in this way differ from other forms, possibly including one used by the author himself. In *Treatise* reference lists the alternative (unaccepted) form is given enclosed by square brackets (e.g., Chernyshev [Tschernyschew], T. N.; Gekker [Hecker], R. F.).

List of Serial Publications

Académie des Sciences de Paris, Comptes Rendus, Mémoires. Paris.

Académie des Sciences de l'URSS, Comptes Rendus; Institut Paléontologique, Travaux; Institut Paléozoologique, Travaux [Akademiya Nauk SSSR, Doklady]. Leningrad.

Academy of Natural Sciences of Philadelphia, Journals; Proceedings.

[R.] Accademia dei Lincei, Classe di Scienze Fisiche, Matematiche e Naturali, Atti; Memórie; Rendiconti. Roma.

[R.] Accademia delle Scienze dell' Istituto di Bologna, Memórie.

Acta Palaeontologia Polonica [Polska Akademiya Nauk, Komitet Geologiczny]. Warszawa.

[K.] Akademie der Wissenschaften zu Wien, mathmatische-naturwissenschaftliche Klasse, Denkschriften; Sitzungsberichte.

Akademiya Nauk SSR, Laboratoire Zoologique et Station Biologique de Sébastopol, Travaux. (St. Petersbourg) Leningrad.

Akademiya Nauk SSSR Leningrad, Izvestiya, Geological Series.

Allgemeinen Schweizerischen Gesellschaft für die gesamten Naturwissenschaften, Neue Denkschriften. Zurich.

American Journal of Science. New Haven, Conn.

Annales de Paléontologie. Paris.

Annales des Sciences Naturelles, Zoologie. Paris.

Annales de la Société Géologique de Belgique. Liège.

Annals and Magazine of Natural History. London.

Annals of Philosophy. London.

Annotationes Zoologicae Japonenses. Tokyo.

Archiv für Biontologie (Gesellschaft Naturforschender Freunde). Berlin.

Archives de Zoologie, Expérimentale et Générale. Paris.

Archivos do Museu Nacional. Rio de Janeiro.

Arquivos de Geologia, Gabinete de Micropaleontologia, Universidade do Recife (Brazil).

Bayerische Staatssammlung Paleontologie, Historische, Geologische, Mitteilungen. München.

Beiträge zur Naturkundlichen Forschung im Sudwestdeutschland. Karlsruhe.

Beiträge zur Paläontologie und Geologie Österreich-Ungarns und des Orients. Wien.

Biological Reviews (see Cambridge Philosophical Society).

Boston Society of Natural History, Memoirs; Proceedings.

Bulletin of American Paleontology. Ithaca, N.Y.

Cairo: Gouvernement Egyptien.

California, University of, Publications in Geological Sciences, Bulletins. Berkeley and Los Angeles.

California, University of, Publications in Mathematics and Physical Sciences. Los Angeles.

California, University of, Publications in Zoology. Berkeley.

Cambridge Philosophical Society, Biological Reviews and Biological Proceedings. Cambridge, Eng.
Canada, Geological Survey of, Department of Mines and Resources, Mines and Geology Branch, Bulletins; Memoirs; Museum Bulletins; Victoria Memorial Museum Bulletins. Ottawa.
Carnegie Institution of Washington, Papers; Publications. Washington, D.C.
Challenger. Report on the Scientific Results of the Exploring Voyage of HMS Challenger, Zoology. Edinburgh.
Cincinnati Society of Natural History, Journals.
Commission Géologiques de Portugal, Travaux. Lisbon.
Couronnés et autres memoires p. p. l'Académie R. des Sciences, des lettres et des beaux-arts de Belgique. Bruxelles.
Danish Biological Station, Reports of, to the Board of Agriculture. Köbenhavn.
Danmarks Geologiske Undersøgelse, Skrifter, Köbenhavn.
Dansk Geologisk Forening, Meddelelser. Köbenhavn.
[K.] Danske Videnskabernes Selskabs, Matematiska-Fysiske, Skrifter. Köbenhavn.
Denison University, Scientific Laboratories, Bulletins; Journals. Granville, Ohio.
Deutsche Geologische Gesellschaft, Zeitschrift. Berlin & Hannover.
"Discovery" Reports. Cambridge, Eng.
Eclogae Geologicae Helvetiae (*see* Schweizerische Geologische Gesellschaft). Basel.
Eesti NSV Teaduste Akadeemia, Geoloogia Instituudi, Uurimused [Akademiya Nauk Estonskoi SSR, Instituta Geologii, Trudy]. Tallinn.
Egypt, Geological Survey of, Paleontology. Cairo.
Erdoel-Zeitschrift. Wien, Hamburg.
Evolution. Lancaster, Pa.
Exploration scientifique de la Tunisie. Paris.
France, Bureau de Recherches Géologiques et Minières, Bulletins; Mémoires.
[K.] Fysiografiska Sällskapet i Lund, Förhandlingar; Handlingar.
Geological Magazine. London, Hertford.
Geological Society of America, Bulletins; Memoirs; Special Papers. New York.
Geological Society of Glasgow, Transactions.
Georogical Society of London; Memoirs; Proceedings; Quarterly Journals; Transactions.
Geological Survey of [see under name of country, state, or province].
Geologie (Zeitschrift für das Gesamtgebiet der Geologie und Mineralogie sowie der angewandten Geophysik). Berlin.
[K.K.] Geologische Bundesanstalt Wien, Abhandlungen; Jahrbuch; Verhandlungen.
[K.K.] Geologische Reichsanstalt Wien (*see* Geologische Bundesanstalt Wien).
Geologische Spezialkarte Preussen und der Thuringischen Staaten, Abhandlungen. Berlin.
Geologiska Föreningen, Stockholm, Förhandlingar.
Geologists' Association, Proceedings. London.
Gesellschaft Geologie- und Bergbaustudenten, Mitteilungen. Wien.
Great Britain, Geological Survey of, Palaeontology, Bulletins; Memoirs. London.
Grenoble, Université de, Faculté des Sciences, Travaux du Laboratoire de Géologie.
Harvard University, Museum of Comparative Zoology, Bulletins; Breviora; Memoirs. Cambridge, Mass.
Hessisches Landesamt für Bodenforschung; Abhandlungen; Notizblatt.
Hunterian Museum of Glasgow, Geological Department, Memoirs.
Illinois State Museum of Natural History, Bulletins. Springfield.
Institut Égyptien (d'Égypte), Bulletin; Mémoires. Cairo.
Institut Océanographique de Monaco, Annales; Bulletin. Monte Carlo, Paris.
Institut Royal des Sciences Naturelles de Belgique, Bulletin; Mémoires. Bruxelles. (*See* Musée Royal d'Histoire Naturelle de Belgique.)
International Geological Congress, 21st Session, Reports. Köbenhavn.
Journal of Geology. Chicago.
Journal Marine Research. New Haven, Conn.
Journal of Paleontology. Tulsa, Okla.
Kansas State Geological Survey, Bulletins. Lawrence, Kansas.
Kansas, The University of, Paleontological Contributions, Articles. Lawrence, Kans.
Leidsche Geologische Mededeelingen.
Linnean Society of London (Botany), Journals; Transactions.
Linnean Society of London (Zoology), Journals; Transactions.
Lunds Geologisk-Mineralogiska Institution, Meddelanden.
Lyon, Université de, Laboratoire de Géologie, Travaux.
Marine Biological Association of the United Kingdom, Journals. Cambridge, Eng.
Michigan, University of, Museum of Paleontology, Contributions. Ann Arbor.
Micropaleontology. American Museum of Natural History. New York.
Missouri, University of, School of Mines and Metallurgy, Bulletins. Rolla, Mo.
Musée Royal d'Histoire Naturelle de Belgique, Annales; Bulletin; Mémoires (continued as Institut Royal des Sciences Naturelles de Belgique). Bruxelles.
Museu de Ciencias Naturales, Treballs. Barcelona.
Muséum d'Histoire Naturelle, Nouvelles Archives. Paris.
Nassauischen Vereins für Naturkunde, Jahrbuch. Wiesbaden.
National Museum, Melbourne, Memoirs.
Nature. London.
Neues Jahrbuch für Geologie und Paläontologie (Before 1950, Neues Jahrbuch für Mineralogie, Geologie, und Paläontologie), Abhandlungen; Beilage-Bande; Monatshefte. Stuttgart.
Neues Jahrbuch für Mineralogie (Before 1950, Neues Jahrbuch für Mineralogie, Geologie, und Paläontologie), Abhandlungen; Beilage-Bande; Monatshefte. Stuttgart.
New South Wales, Geological Survey of, Ethnology, Memoirs; Geology, Memoirs; Paleontology, Memoirs. Sydney.
New York State Museum of Natural History, Annual Reports; Bulletins. Albany.
New Zealand Geological Survey, Paleontological Bulletins. Wellington.

Niedersachsische Geologische Vereins, Jahresbericht. Hannover.
Nigeria, Geological Survey of, Occasional Papers. London.
Norsk Geologisk Tidsskrift (Norsk Geologisk Forening). Oslo.
Ohio Journal of Science. Columbus.
Österreichische Akademie der Wissenschaften, Mathematisch-naturwissenschaftliche Klasse, Sitzungsberichte. Wien.
Palaeontographica Italica. Pisa.
Palaeontographica. Stuttgart & Kassel.
Palaeontographical Society, Memoirs; Monographs. London.
Palaeontologia Indica (Memoirs of the Geological Survey of India). Calcutta.
Paläontologie von Timor. Stuttgart.
Paläontologische Zeitschrift. Berlin & Stuttgart.
Palaeontology (Palaeontological Association). London.
[K.] Preussische Geologische Landesanstalt, Abhandlungen; Jahrbuch. Berlin.
Publications du Service de la Carte Géologique de l'Algérie, new series, Paleontologie, Mémoires. Alger.
Queensland Museum, Memoirs. Brisbane.
Records of Oceanographic Works in Japan, National Research Council. Tokyo.
Revue Suisse de Zoologie, Annales de la Société Zoologique Suisse et du Museum d'Histoire Naturelle de Genève.
Revue de Micropaléontologie (Laboratoire de Micropaléontologie). Paris.
Royal Physical Society of Edinburgh, Proceedings.
Royal Society of Edinburgh, Memoirs; Transactions.
Royal Society of London, Philosophical Transactions; Proceedings.
Royal Society of New South Wales, Journals and Proceedings. Sydney.
Royal Society of New Zealand, Proceedings; Transactions. Wellington.
Royal Society of Victoria, Proceedings. Melbourne.
Russisch-Kaiserliche Mineralogische Gesellschaft zu St. Petersburg, Verhandlungen [Vserossiyskoe Mineralogicheskoe Obshchestvo]. Leningrad.
Schweizerische Geologische Gesellschaft. Basel.
Schweizerische Paläontologische Gesellschaft, Abhandlungen. Zurich.
Science. New York.
Senckenbergiana Lethaea* (Senckenbergische Naturforschende Gesellschaft Wissenschaftliche Mitteilungen) [*"Lethaea" added to title, 1954]. Frankfurt.
Service des Mines Madagascar, Annales Géologiques. Tananarive.
Seto Marine Biological Laboratory, Publications. Sirahama.
Smithsonian Institution, Annual Reports. Washington, D.C.
Smithsonian Miscellaneous Collections. Washington, D.C.
Sociedad Geológica del Perú, Boletín. Lima.
Società Geologica Italiana, Bolletini; Memorie. Rome.
Société Académique d'agriculture, des sciences, arts et belles-lettres du départemente de l'Aube, Mémoires. Troyes.
Société des Amis Sciences Naturelles de Rouen, Année.
Société Belge de Géologie, de Paléontologie et d'Hydrologie, Bulletin. Bruxelles.
Société d'étude des Sciences Naturelles d'Elbeuf, Bulletins.
Société Géologique de France, Comptes Rendus des Séances; Bulletin; Mémoires. Paris.
Société Géologique et Minéralogique de Bretagne, Mémoires. Rennes.
Société Géologique de Normandie, Bulletin. Le Havre.
Société d'Histoire Naturelle et de Palethnographie de la Haute-Marne. Chaumont.
Société Linnéenne de Bordeaux, Actes.
Société Linnéenne de Normandie, Bulletin; Mémoires. Caen.
Société Neuchâteloise Sciences Naturelles, Mémoires.
Société Paléontologique de la Suisse, Mémoires (*see* Schweizerische Paläontologische Gesellschaft). Zurich.
Société Paléontologique de la Russie, Annuaire. Leningrad.
Société de Physique et d'Histoire Naturelle de Genève, Mémoires.
Société des Sciences Historiques et Naturelles de l'Yonne, Bulletin. Auxerre, France.
Société des Sciences Naturelles de la Charente-Inférieure, Annales. La Rochelle.
Société Scientifique de l'Isère, Bulletin. Grenoble.
Société Zoologique de France, Bulletin; Mémoires. Paris.
Society for Experimental Biology, Symposia. Cambridge, Eng.
South African Museum, Annals. Capetown.
Spolia Zoologica Musei Hauniensis. Köbenhavn.
[K.] Svenska Vetenskapsakademien, Arkiv för Mineralogi och Geologi; Arkiv för Zoologi; Handlingar. Stockholm.
Systematic Zoology. Lawrence, Kans.
Systematics Association, Publications. London.
Tohoku University, Science Reports. Sendai, Japan.
Tokyo, University of, Faculty of Science, Journals. (From 1887-1925, Tokyo Imperial University, College of Science.)
Tuatara. Victoria University College, Biological Society, Journals. Wellington, N.Z.
[R.] Ufficio Geologico d'Italia, Bolletini. Roma.
Undersökningar över Öresund. Lund.
United Kingdom, Geological Survey of, Memoirs. London.
United States Geological Survey, Bulletins; Monographs; Professional Papers. Washington, D.C.
United States National Museum, Bulletins; Proceedings. Washington, D.C.
Victoria University of Wellington, Zoological Publications.
Videnskabelige Meddelelser fra Dansk Naturhistorisk Forening. Köbenhavn.
Washington Academy of Sciences, Journals. Washington, D.C.
Wissenschaftliche Ergebnisse der Deutschen Tiefsee-Expedition auf dem Dampfer "Valdivia," 1898-1899. Jena.

Wissenschaftliche Ergebnisse der Schwedischen Süd-polarexpedition, 1901-1903. Stockholm.
Zeitschrift für Wissenschaftliche Zoologie. Leipzig.
Zentralblatt für Geologie und Paläontologie (Before 1950, Neues Jahrbuch für Mineralogie, Geologie, und Paläontologie). Stuttgart.
Zentralblatt für Mineralogie (Before 1950, Neues Jahrbuch für Mineralogie, Geologie und Paläontologie), Stuttgart.
Zoological Society of London, Proceedings.
Zoologische Jahrbücher, Abteilung 2—Anatomie und Ontogenie. Jena.
Zoologiska Bidrag från Uppsala. Uppsala & Stockholm.

SOURCES OF ILLUSTRATIONS

At the end of figure captions an index number is given to supply record of the author of illustrations used in the *Treatise,* reference being made either (1) to publications cited in reference lists or (2) to the names of authors with or without indication of individual publications concerned. Previously unpublished illustrations are marked by the letter "n" (signifying "new") with the name of the author.

Classification of rocks forming the geologic column as commonly cited in the *Treatise* in terms of units defined by concepts of time is reasonably uniform and firm throughout most of the world as regards major divisions (e.g., series, systems, and rocks representing eras) but it is variable and unfirm as regards smaller divisions (e.g., substages, stages, and subseries), which are provincial in application. Users of the *Treatise* have suggested the desirability of publishing reference lists showing the stratigraphic arrangement of at least the most commonly cited divisions. Accordingly, a tabulation of European and North American units, which broadly is applicable also to other continents, is given here.

Generally Recognized Divisions of Geologic Column

EUROPE	NORTH AMERICA
ROCKS OF CENOZOIC ERA	**ROCKS OF CENOZOIC ERA**
NEOGENE SYSTEM[1]	**NEOGENE SYSTEM**[1]
Pleistocene Series (including Recent)	Pleistocene Series (including Recent)
Pliocene Series	Pliocene Series
Miocene Series	Miocene Series
PALEOGENE SYSTEM	**PALEOGENE SYSTEM**
Oligocene Series	Oligocene Series
Eocene Series	Eocene Series
Paleocene Series	Paleocene Series
ROCKS OF MESOZOIC ERA	**ROCKS OF MESOZOIC ERA**
CRETACEOUS SYSTEM	**CRETACEOUS SYSTEM**
Upper Cretaceous Series	**Gulfian Series (Upper Cretaceous)**
Maastrichtian Stage[2]	Navarroan Stage
Campanian Stage[2]	Tayloran Stage
Santonian Stage[2]	Austinian Stage
Coniacian Stage[2]	
Turonian Stage	
Cenomanian Stage	Woodbinian (Tuscaloosan) Stage
	Comanchean Series (Lower Cretaceous)
	Washitan Stage
Lower Cretaceous Series	
Albian Stage	Fredericksburgian Stage
	Trinitian Stage
Aptian Stage	
	Coahuilan Series (Lower Cretaceous)
Barremian Stage[3]	Nuevoleonian Stage
Hauterivian Stage[3]	
Valanginian Stage[3]	Durangoan Stage
Berriasian Stage[3]	
JURASSIC SYSTEM	**JURASSIC SYSTEM**
Upper Jurassic Series	**Upper Jurassic Series**
Portlandian Stage[4]	Portlandian Stage
Kimmeridgian Stage	Kimmeridgian Stage
Oxfordian Stage	Oxfordian Stage
Middle Jurassic Series	**Middle Jurassic Series**
Callovian Stage (or Upper Jurassic)	Callovian Stage (or Upper Jurassic)
Bajocian Stage	Bathonian Stage
Bathonian Stage	Bajocian Stage

Lower Jurassic Series (Liassic)

Toarcian Stage
Pliensbachian Stage
Sinemurian Stage
Hettangian Stage

TRIASSIC SYSTEM

Upper Triassic Series

Rhaetian Stage[5]
Norian Stage
Carnian Stage

Middle Triassic Series

Ladinian Stage
Anisian Stage (Virglorian)

Lower Triassic Series

Scythian Series (Werfenian)

ROCKS OF PALEOZOIC ERA

PERMIAN SYSTEM

Upper Permian Series

Tartarian Stage[6]
Kazanian Stage [7]
Kungurian Stage

Lower Permian Series

Artinskian Stage[8]
Sakmarian Stage

CARBONIFEROUS SYSTEM

Upper Carboniferous Series

Stephanian Stage

Westphalian Stage

Namurian Stage

Lower Carboniferous Series

Visean Stage

Tournaisian Stage
Strunian Stage

DEVONIAN SYSTEM

Upper Devonian Series

Famennian Stage

Frasnian Stage

Lower Jurassic Series (Liassic)

Toarcian Stage
Pliensbachian Stage
Sinemurian Stage
Hettangian Stage

TRIASSIC SYSTEM

Upper Triassic Series

(Not recognized)
Norian Stage
Carnian Stage

Middle Triassic Series

Ladinian Stage
Anisian Stage

Lower Triassic Series

Scythian Stage

ROCKS OF PALEOZOIC ERA

PERMIAN SYSTEM

Upper Permian Series

Ochoan Stage
Guadalupian Stage

Lower Permian Series

Leonardian Stage
Wolfcampian Stage

PENNSYLVANIAN SYSTEM

Kawvian Series (Upper Pennsylvanian)

Virgilian Stage
Missourian Stage

Oklan Series (Middle Pennsylvanian)

Desmoinesian Stage
Bendian Stage

Ardian Series (Lower Pennsylvanian)

Morrowan Stage

MISSISSIPPIAN SYSTEM

Tennesseean Series (Upper Mississippian)

Chesteran Stage

Meramecian Stage

Waverlyan Series (Lower Mississippian)

Osagian Stage
Kinderhookian Stage

DEVONIAN SYSTEM

Chautauquan Series (Upper Devonian)

Conewangoan Stage
Cassadagan Stage

Senecan Series (Upper Devonian)

Chemungian Stage
Fingerlakesian Stage

Middle Devonian Series	**Erian Series (Middle Devonian)**
Givetian Stage	Taghanican Stage
	Tioughniogan Stage
Couvinian Stage	Cazenovian Stage
Lower Devonian Series	**Ulsterian Series (Lower Devonian)**
Coblenzian Stage	Onesquethawan Stage
	Deerparkian Stage
Gedinnian Stage	Helderbergian Stage
SILURIAN SYSTEM	**SILURIAN SYSTEM**
Ludlow Series	**Cayugan Series** Includes age equivalents of middle and upper Ludlow (in New York)
Wenlock Series	**Niagaran Series** Includes age equivalents of upper Llandovery, Wenlock, and lower Ludlow (in New York)
Landovery Series	**Medinan Series** Includes age equivalents of lower and middle Llandovery (in New York)
ORDOVICIAN SYSTEM	**ORDOVICIAN SYSTEM**
	Cincinnatian Series (Upper Ordovician)
	Richmondian Stage
Ashgill Series	Maysvillian Stage
	Edenian Stage
Caradoc Series	**Champlainian Series (Middle Ordovician)**
	Mohawkian Stage
	Trentonian Substage
	Blackriveran Substage
Llandeilo Series	Chazyan Stage
Llanvirn Series	
Arenig Series	**Canadian Series (Lower Ordovician)**
Tremadoc Series[9]	
CAMBRIAN SYSTEM	**CAMBRIAN SYSTEM**
Upper Cambrian Series	**Croixian Series (Upper Cambrian)**
	Trempealeauan Stage
	Franconian Stage
	Dresbachian Stage
Middle Cambrian Series	**Albertan Series (Middle Cambrian)**
Lower Cambrian Series	**Waucoban Series (Lower Cambrian)**
EOCAMBRIAN SYSTEM	**EOCAMBRIAN SYSTEM**
ROCKS OF PRECAMBRIAN AGE	**ROCKS OF PRECAMBRIAN AGE**

RAYMOND C. MOORE

---

[1] Considered by some to exclude post-Pliocene deposits.

[2] Classed as division of Senonian Subseries.

[3] Classed as division of Neocomian Subseries.

[4] Includes Purbeckian deposits.

[5] Interpreted as lowermost Jurassic in some areas.

[6] Includes some Lower Triassic and equivalent to upper Thuringian (Zechstein) deposits.

[7] Equivalent to lower Thuringian (Zechstein) deposits.

[8] Equivalent to upper Autunian and part of Rotliegend deposits.

[9] Classed as uppermost Cambrian by some geologists.

PART U
ECHINODERMATA 3
ASTEROZOA — ECHINOZOA

By J. Wyatt Durham, K. E. Caster, Harriet Exline, H. B. Fell, A. G. Fischer, D. L. Frizzell, R. V. Kesling, P. M. Kier, R. V. Melville, R. C. Moore, D. L. Pawson, Gerhard Regnéll, W. K. Spencer, Georges Ubaghs, Carol D. Wagner, and C. W. Wright

VOLUME 1

CONTENTS

[Volume 1, p. i-xxx, 1-366a; Volume 2, p. 367-695]

INTRODUCTION

By Raymond C. Moore

The publication of *Treatise* Echinodermata 3 in advance of two other volumes allotted to this phylum, respectively numbered 1 and 2, seems anomalous, especially when it is learned that a discussion broadly dealing with characteristics and relationships of echinoderms as a whole has been assigned place at the beginning of Echinodermata 1. Such a chapter is logical as introduction to all others concerned with individual groups however these may be defined and arranged. Accordingly, a comprehensive section entitled "General Features of Echinodermata" has been prepared by Georges Ubaghs, of the Université de Liège. Completed in 1961, it has subsequently been revised twice, mostly in minor ways but with some important additions and subtractions. In my opinion the contribution by UBAGHS is one of outstanding value, one which will be studied with profit by neozoologists as well as paleontologists when it becomes available. Unhappy delay in its appearance is due to the tardy pace in completing fossil groups—all Paleozoic echinoderms—which belong in the same volume.

A disadvantage of the *Treatise* project is the "packaged" nature of volumes planned to contain designated taxonomic assemblages. So long as required parts of a package are unavailable, others remain in files with cumulative needs for revision. The fact that authors are not compensated financially calls for other incentives in order to finish tasks that have been accepted and a burden is placed on the editor (also unpaid) to plead for accomplishment of assignments. Thus, some *Treatise* units move at snail's pace. On the other hand, an advantage of arrangements is the possibility of sending to the press any volume as soon as it is ready. This explains the appearance of Echinodermata 3 in advance of Echinodermata 1 and 2.

At the time many years ago (1948) when the echinoderms were divided into three groups for purposes of the *Treatise,* the view was accepted that the phylum was divisible into two subphyla based essentially on mode of life—forms prevailingly sessile on one hand and others prevailingly free-moving on the other. These have been named Pelmatozoa and Eleutherozoa, respectively. Pelmatozoans were planned for treatment in Echinodermata 1 and 2, Eleutherozoa in Echinodermata 3. Eleutherozoans were interpreted to include starfishes, ophiuroids, echinoids, and holothuroids, with ophiocistioids as an afterthought.

The assignment to prepare descriptions and plan illustrations of fossil asteroids and ophiuroids was accepted in 1950 by W. K. SPENCER, who had published extensive studies of them in monographs of the Palaeontographical Society. By 1951 he had brought together rough notes, including directions for securing desired figures, mostly new drawings to be made under my direction. The notes were organized by me into typescript suitable for use in the *Treatise* and with minor changes this was approved in conferences at Dr. SPENCER's residence at Beaulieu-sur-Mer in southern France. This was in 1951 and 1952. At this time C. W. WRIGHT, of London, was invited to contribute information and some written discussions of post-Paleozoic asterozoans and later SPENCER and WRIGHT worked together to reshape classification that would coordinate previously divergent arrangements of suprageneric taxa based on the separate worlds of fossil and extant forms. In this important effort they were successful, as recounted briefly in introductory paragraphs of the contribution by SPENCER and WRIGHT on asterozoans in this volume. Not until ten years later, however, was a final version made ready, with numerous changes and additions for which WRIGHT is to be credited and thanked.

Initially, a considerable part of the *Treatise* presentation of Echinoidea was planned to be secured from H. L. HAWKINS, of the University of Reading, in part with the collaboration of R. V. MELVILLE, of the Geological Survey of Great Britain, one of his former students. MELVILLE in 1951 produced an excellent account of the general morphology of echinoids, with notes for preparation of figures, but HAWKINS found it infeasible to contribute. Then J. WYATT DURHAM, of the University of California (Berkeley) was invited to help and he accepted willingly. In 1954-55 he was awarded a Guggenheim Fellowship for echinoid studies in Europe and used this opportunity as a primary means of advancing the *Treatise* project in the realm of echinoids. One result of this study was the development of a revised classification of Echinoidea, published by DURHAM and MELVILLE in a *Journal of Paleontology* paper (1957) and proposed as the framework for taxonomic arrangement in the *Treatise.*

In 1960 I asked Durham to assume leadership in organizing the various needed chapters on these echinoderms, some of general scope and others for coverage of major systematic groups. With help from him various assignments were made, with result that the team of workers was enlarged to include P. M. KIER, H. B. FELL, D. L. PAWSON, C. D. WAGNER, and A. G.

FISCHER. Subsequent chapters in this volume by these paleontologists and by DURHAM and MELVILLE speak for themselves, but very much unseen and unrecorded effort is represented by correspondence on many problems and by editorial coordination of typescript and illustrations. An example of behind-the-scenes labor is preparation of an exhaustive list of all nominal genera of echinoids, fossil and Recent, with authorships, dates, and literature references—work done by DURHAM and WAGNER. Obviously, such a list is indispensable for achievement in reasonable degree of the *Treatise* aims of comprehensiveness and authoritativeness.

In connection with KIER's chapters on noncidaroid Paleozoic echinoids and cassiduloids in this volume, it is appropriate to mention two grants from the National Science Foundation to him for museum and field studies in Europe and work pursued in Washington, because, as recorded by him (KIER, 1962, p. 2), the underlying purpose of his researches was to provide a firm foundation for his *Treatise* chapters. Also, acknowledgment to the National Science Foundation for help to other *Treatise* contributors, in part through funds allotted under my direction, should be made. In aggregate the aid has been considerable and thus extremely important.

Judgment that the division of Echinodermata into subphyla named Pelmatozoa and Eleutherozoa, long accepted in textbooks and various monographic works, is untenable has become firm in the minds of at least several *Treatise* contributors who are specialists in the study of various echinoderm groups. This conclusion is not new, but to date it has received little attention and is not yet generally accepted. In the *Treatise* four subphyla are recognized—Homalozoa, Crinozoa, Asterozoa, and Echinozoa. The last two of those named are assigned to the present volume. The asterozoans contain a single class, Stelleroidea, composed of three subclasses, Somasteroidea, Asteroidea, Ophiuroidea; the subphylum is well delimited. The echinozoans are less easily discriminated, though the classes Echinoidea and Holothuroidea unquestionably belong here. The content of Echinozoa is discussed in a chapter by FELL and MOORE on "General Features and Relationships" of the subphylum and the inclusion of chapters devoted to Helicoplacoidea (by DURHAM and K. E. CASTER), Edrioasteroidea (by GERHARD REGNÉLL), Ophiocistioidea (by UBAGHS), and Cyclocystoidea (by R. V. KESLING) indicates classification of these groups in Echinozoa.

The homologies of morphological features observed in the subphyla, classes, and subclasses of echinoderms are very interesting from the standpoints of efforts to adopt uniform orientation, and reasonable correspondence in designation of parts, and especially of inquiries as to "natural" classification, evolution, and phylogeny of all groups. Difficulties and uncertainties are many. Even so, a chapter on the "Homology of Echinozoan Rays" is given by MOORE and FELL.

Description of the morphological features of holothurian sclerites and typescript and illustrations for systematic treatment of dissociated fossil holothurian sclerites were completed by DON L. FRIZZELL and HARRIET EXLINE (Mrs. FRIZZELL) as long ago as 1955. This material lodged ?patiently and ?peacefully in the editor's files for nine years, until it was returned at request of the authors for updating. Little change was needed, but at last this contribution was sent to the press. Meanwhile, D. L. PAWSON, a specialist on holothurians, who is an associate and former student of H. B. FELL at Victoria University of Wellington, New Zealand, had joined the staff of the Smithsonian Institution in Washington, D. C. He has assisted the FRIZZELLS and has been assisted by them on some points and has contributed a chapter of his own on "Phylogeny and Evolution of Holothuroids."

On behalf of the Geological Society of America, of the four paleontological societies that sponsor the *Treatise*, and of paleontologists and zoologists everywhere who will benefit from reference to Echinodermata 3, I express thanks and warm appreciation to the authors of the volume. Not least in deserving praise are Mrs. LAVON MCCORMICK and ROGER B. WILLIAMS, of my staff, for their prolonged, painstaking, able work on typescript, illustrations, and proofs, as well as work in libraries. Also, C. K. HYDER, editor of the University of Kansas Press, has continuously furnished very valuable aid.

ASTEROZOANS

By W. K. Spencer† and C. W. Wright
[†deceased; London]

CONTENTS

INTRODUCTION

In 1950 the late W. K. Spencer accepted responsibility for the section of the *Treatise* on Asterozoa. He invited me to prepare a contribution on the Mesozoic and Cenozoic members of the subclass Asteroidea. By 1953 he and I had, with great help from the Editor, R. C. Moore, virtually completed final typescript and illustrations on the basis of the then accepted classification, under which, in particular, Paleozoic Asteroidea were grouped in one ordinal arrangement and post-Paleozoic Asteroidea in a different and unrelated one.

In that year we both concluded that we should fail in our responsibility to the *Treatise* if we could not relate and integrate these two classifications. Much discussion and thought was devoted to this end. Shortly before Dr. Spencer's death in 1954 we had reached firm conclusions on the outline of a single classification and on the place therein of nearly all families of Asteroidea. Unfortunately the reorganization of the existing typescript could not be completed before Dr. Spencer died, although he left copious notes with both the Editor and myself. I have consequently undertaken reorganization of the whole text.

While this revision was proceeding Dr. H. Barraclough Fell told me of his exciting discovery that *Platasterias* is a living somasteroid and of some of the phylogenetic conclusions that flowed from his study of this genus. His subsequent work on the phylogeny of the sea stars, based on a thorough re-examination of many living forms of Asteroidea and Ophiuroidea, has led to a new appraisal, fully in accord with the paleontological evidence, of the fundamentals of the classification and evolution of Asterozoa. I have endeavored to take this into account so far as possible throughout the text.

In the original draft Dr. Spencer was responsible for most of the introductory matter and for the systematic description of the Paleozoic Asteroidea and all Ophiuroidea. Since his death I have rewritten most of the general matter and revised all the systematic part in the light of later information. Moreover, all extant genera of Asterozoa are now listed. Consequently the whole text is attributed to us jointly.

A number of new names of higher categories are introduced in the text. The need for this arises largely from the fundamental reclassification referred to above.

I must express my deep gratitude to the Editor, as I am sure Dr. Spencer would have wished to do, for his important contribution to our work in the repeated organization of our material, to Dr. H. B. Fell for timely information and much help, to Dr. Hans Hess, of Basel, for advance information about his work on fossil Ophiuroidea and to Miss Ailsa Clark, of the British Museum (Natural History), for ready assistance over Recent Asterozoa.

GENERAL CHARACTERS

DIAGNOSTIC FEATURES

The asterozoans are echinoderms distinguished by radial extensions from a central disc or body; the extensions are supported by calcified axes which are developed as sheaths around radial water vessels. Generally, these extensions have a distinct individuality and are called **arms**

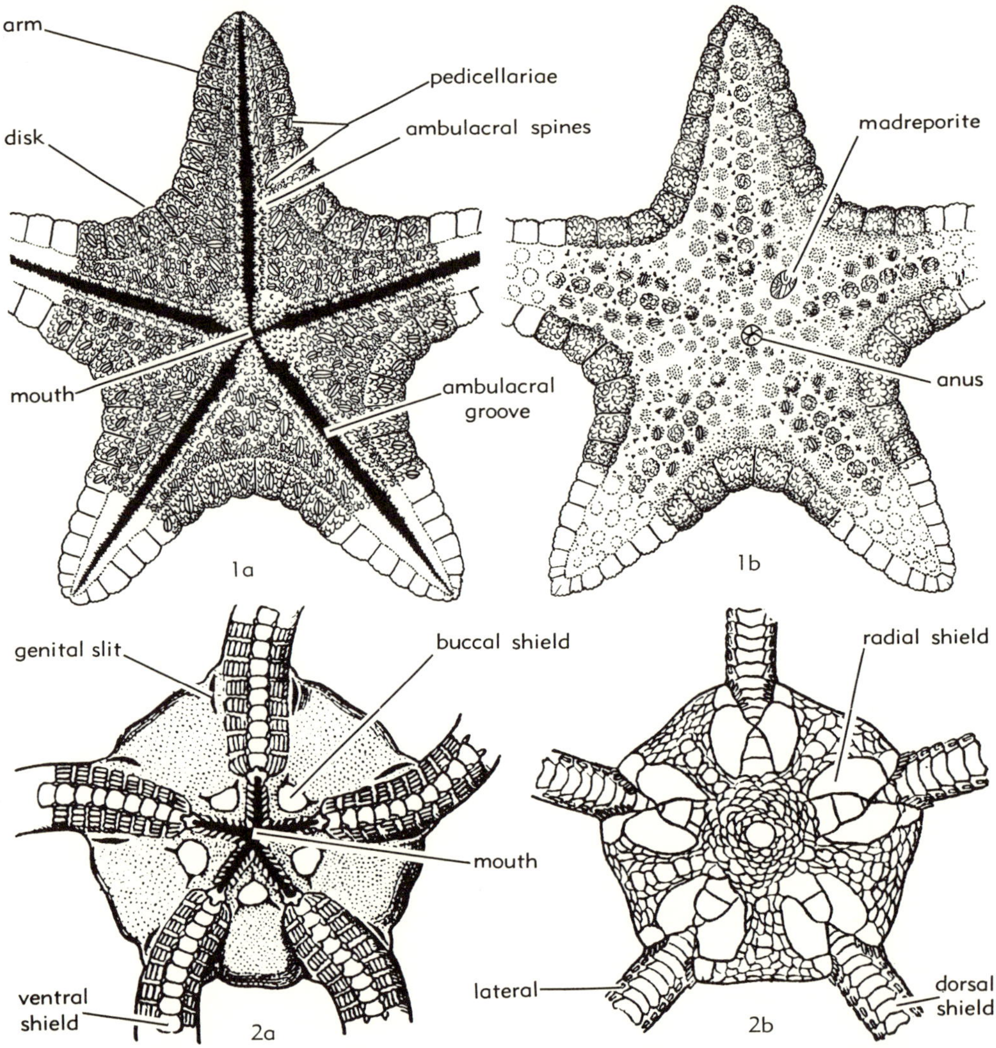

FIG. 1. Recent asterozoans, illustrating characteristic distinctions in form of asteroids and ophiuroids.

1. Starfish, *Anthenea flavescens,* common in shallow waters of the southwestern Pacific, showing features of its nearly flat underside (*1a,* oral surface) and more convex upper side (*1b,* aboral surface), ×1. The central disc is relatively large and the arms short. Narrow ambulacral grooves bordered by blunt spines radiate from the mouth. Pedicellariae and tubercles occur on both oral and aboral surfaces; they are arranged in rows of clusters on the upper surface (130).

2. Brittle-star, *Ophiura,* showing central disc and proximal (inner) parts of slender arms (*2a,* oral surface of one species; *2b,* aboral surface of another species), approx. ×2. Relatively large buccal shields are located interradially on the oral surface and pairs of large radial shields on the aboral surface; otherwise the disc is mainly covered by scalelike plates and granular ossicles. The arms bear rows of ventral (oral side), dorsal (aboral side), and lateral arm plates (137).

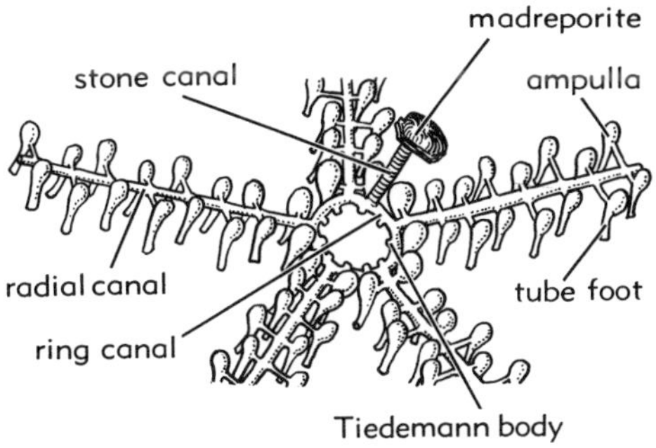

FIG. 2. Diagram of parts of starfish water-vessel system (104).

(Fig. 1). The mouth faces downward toward the sea floor and this side of the animal is called **oral.** Extending radially from the mouth along the oral surface to the extremities of the arms are rows of **tube feet,** which are mobile projections of the water-vessel system walled with soft tissue (Fig. 2). The surface opposite that bearing the mouth and tube feet is called **aboral.** In most asteroids it is characterized by the presence of a porous plate **(madreporite)** which acts as inlet to the water-vessel system; in most ophiuroids this plate is on the oral surface. Asterozoans are free-moving animals in the adult stage. Full-grown individuals range from a few millimeters to more than 50 cm. in diameter, measured from tip to tip of the outstretched arms.

Like other echinoderms, the asterozoans are exclusively marine. Living kinds in the main belong to two groups: asteroids, or starfishes, characterized in most taxa by lack of sharp separation of disc from arms; and ophiuroids, or brittle stars, distinguished by strong differentiation of the slender arms and rounded central disc. A group of correlative rank with primitive skeletal structures, the somasteroids, is represented by a few Paleozoic genera and a single extant species.

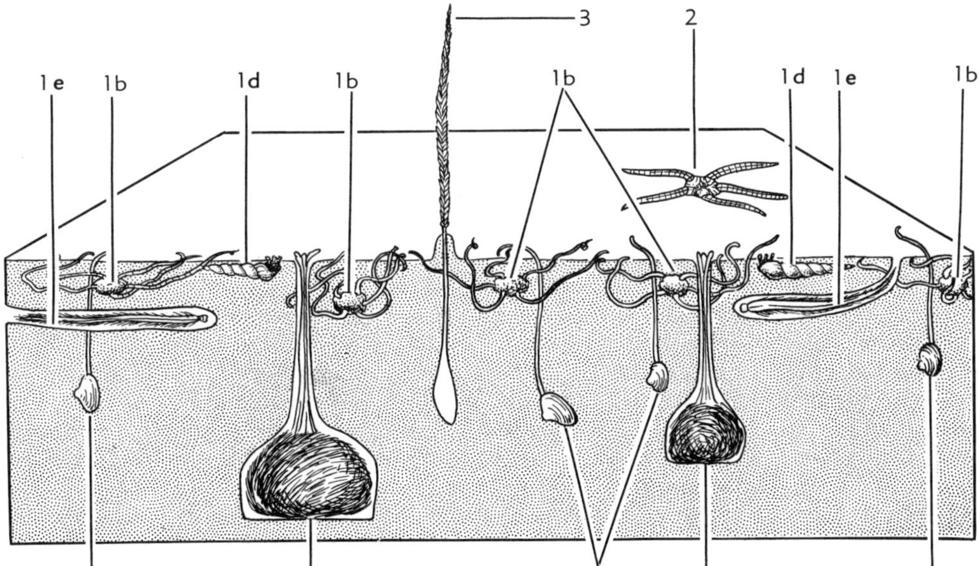

FIG. 3. Cross section of modern *Echinocardium-Turritella* community found in shallow seas, showing component groups of animals (101).

1. Animals living below surface of the sea floor, some in well-defined burrows; *1a,* sea urchin, *Echinocardium cordatum; 1b,* ophiuroid, *Amphiura filiformis; 1c,* pelecypod, *Abra nitida; 1d,* gastropod, *Turritella terebra; 1e,* worm.

2. Animals living on surface of sea floor; ophiuroid, *Ophiura texturata.*

3. Animals feeding in clear water above sea floor; anthozoan, *Virgularia mirabilis,* a suspension-feeder.

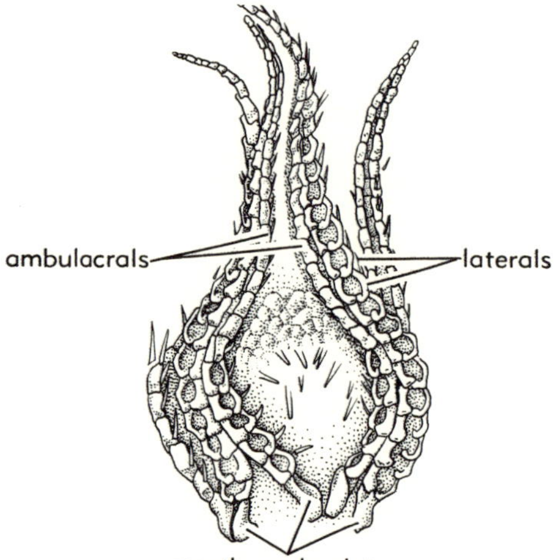

FIG. 4. Small Middle Ordovician ophiuroid *(Taeniaster spinosus)*, Trenton, Ohio, with arms upraised, indicating withdrawal of animal into its burrow before death, ×5 (133).

MODE OF OCCURRENCE AS FOSSILS

The skeletal elements of Asterozoa consist of discrete ossicles which readily separate from each other after death, and accordingly the greater part of our knowledge of fossil forms is derived from specimens preserved under exceptional conditions that prevented the dissociation of interrelated hard parts. In the Paleozoic such specimens have been found particularly in lower Arenigian rocks of southern France, upper Arenigian of Czechoslovakia, Upper Ordovician deposits of Girvan, in Scotland, and in the Early Devonian Bundenbach Slates of Western Germany. These beds comprise sediments of ancient sea floors on and in which vigorous animal communities lived, with starfishes an important element. A corresponding community of the present day is illustrated in Figure 3. Such communities are divisible into three components: animals living under the surface of the sea floor, animals living on the surface, and animals feeding in clear water above the sea floor (2).

Preservation of intact skeletal parts requires rapid burial after death. It is evident that animals living below the sea floor are best suited for preservation, for the surrounding mud prevents scattering of discrete hard parts and tends both to seal the remains and to retard somewhat the decay of soft parts; some fossil asterozoans show that fine mud had time to infiltrate internal vessels, so that the original shape and course of these soft parts can be traced in the form of casts. The surrounding matrix also retained the animal in postures comparable to those assumed in life, as is well shown by some fossil ophiuroids (Fig. 4).

Specimens consisting of external molds, formed by solution of the calcite skeleton, are very useful for study by means of rubber casts which reveal the finest detail of surface characters and by gentle bending may show important features of ossicle junctions. Fossils preserved as calcite or pyrites may be less valuable because recrystallization tends to obliterate surface and structural characters.

In the Mesozoic, an excellent source of material, especially of Asteroidea, is the chalk of the Upper Cretaceous. Whole specimens occur, but are rare, while isolated ossicles or groups of ossicles, representing dissociated remains of single individuals, are abundant. The soft chalk in which they are embedded readily weathers away or can be cleaned, exposing the ossicles in perfect condition for examination. Many studies and reconstructions have been made from such ossicles (55, 74, 89). Since the chalk was laid down as a continuous deposit during a very long time, it is possible to follow in great detail the stages of evolution of several asteroid lineages.

Many other occurrences are known in Paleozoic and later rocks, in addition to those already mentioned. Occasionally, "starfish beds" containing abundant individuals of one or more species are observed, but, generally speaking, Asterozoa are rare fossils and well-preserved specimens are exceptional. In many marine formations no individuals have yet been discovered. Inevitably, therefore, the detailed history of the subphylum is rather poorly known and a good deal of speculation is necessary to make a connected story.

MORPHOLOGY AND FUNCTION

SKELETAL STRUCTURES

MAIN ELEMENTS

The skeletal structures in Asterozoa develop in the outer layers of the body and thus outline its shape. Component elements, or **ossicles,** may be classified in three groups: **axial elements,** comprising ossicles formed in the sheath of water vessels; **adaxial elements,** consisting of ossicles which adjoin and are in series with the axial skeleton; and **extraxial elements,** comprising all other ossicles. The distinction between these groups is well shown in the early starfish *Villebrunaster,* in which the whole oral surface is occupied by strongly contrasted ossicles of the axial and adaxial skeletons (Fig. 5). The components so closely follow in similar series that they may be said to be arranged in metameres.

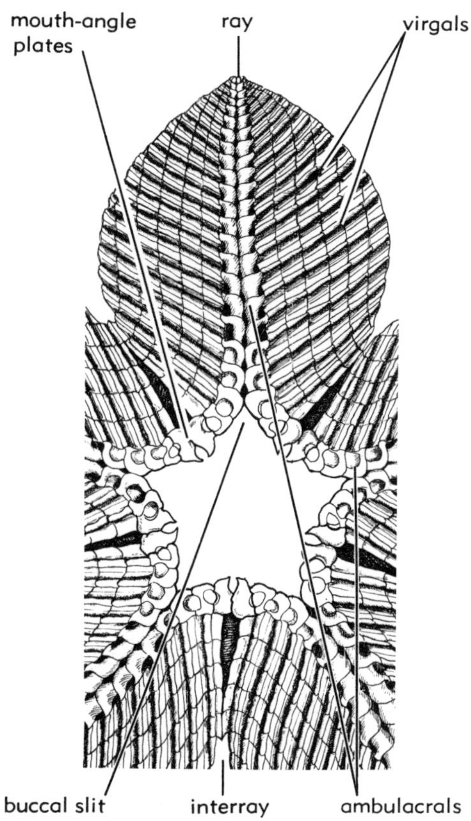

FIG. 5. Metameric segmentation of the skeleton of a primitive starfish, as shown by the oral surface of *Villebrunaster thorali* (reconstr.), ×5. The relatively large ossicles of the central disc and median part of the arms belong to the axial skeleton; others shown are adaxial elements. The double row of axial ossicles along each arm are ambulacrals, with basins for tube feet along their outer edges. Near the central opening the ambulacrals diverge to form V's of the mouth frame. In series with the ambulacrals are rod-shaped adaxial ossicles (virgals). Mouth-angle plates are located interradially, projecting into the central opening (131).

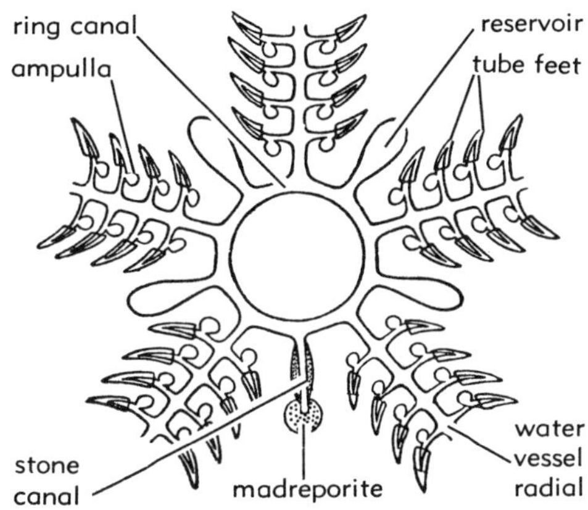

FIG. 6. Arrangement of water vessels in an extant asteroid (122).

The extraxial skeleton in *Villebrunaster* comprises triradiate ossicles making a wide-meshed network on the aboral surface, and not in series with the axial skeleton (see Fig. 20,*1*). A special extraxial plate is the madreporite which forms a sievelike opening from the outside into the water circulatory system. If present, it is located interradially; primitively, it was lateral in position but is on the aboral surface in most asteroids and on the oral surface in most ophiuroids and some early asteroids. The tube from the madreporite to the main water vessels is commonly calcified and is termed the **stone canal.**

AXIAL STRUCTURES

Arrangement. The pattern of axial skeletal elements exactly follows that of the main water vessels, one of which runs along the mid-line of each arm; a canal which encircles the mouth communicates between

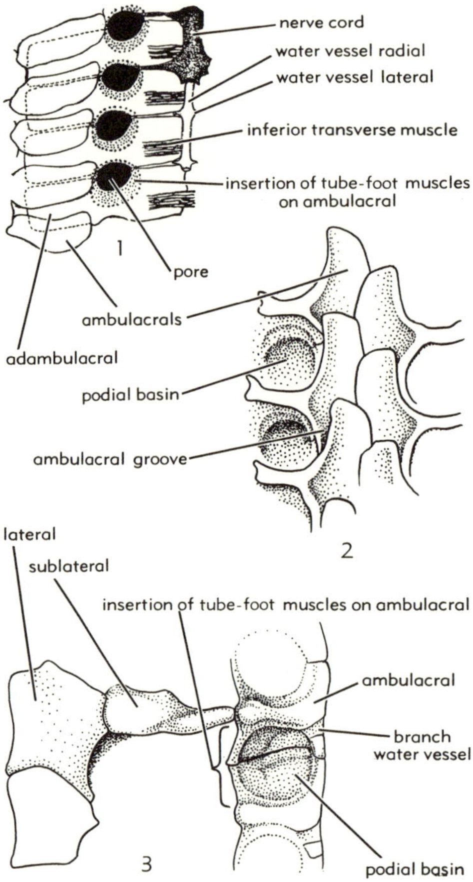

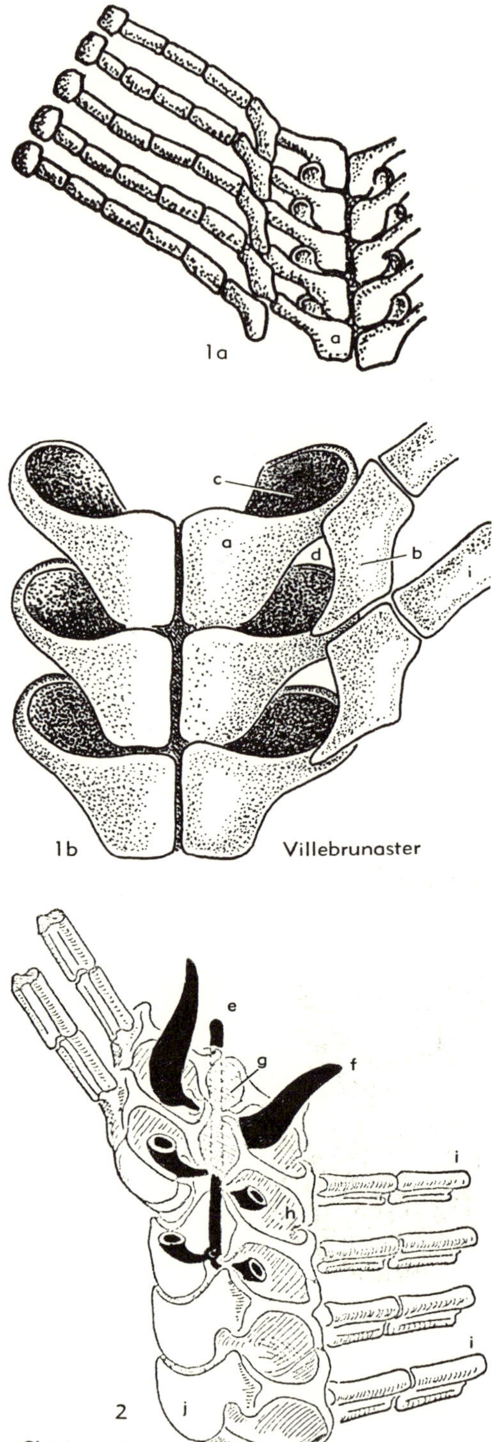

Fig. 7. Relationship between axial skeletal elements and seating of tube feet.——*1.* Structures visible along part of one side of ambulacrum of an extant starfish (*Astropecten irregularis*) with tube feet and spines attached to adambulacral and ambulacral ossicles removed and with parts of radial nerve cord and water vessel also cut away to show underlying structures (132).——*2.* Axial ossicles of primitive starfish (*Archegonaster*, U. Arenig.), showing structures corresponding with those in *1* (133).——*3.* Axial and adaxial ossicles of one side of arm of an early ophiuroid (*Eophiura*, U.Arenig.), showing basins for tube feet (133).

Fig. 8. Axial structures and tube feet of primitive somasteroids (108).——*1. Villebrunaster thorali* Spencer, L.Ord., Fr.; *1a,* arm skeleton near tip, ×6; *1b,* interpretation of mid-arm structure, ×10.——*2. Chinianaster levyi* Thoral, L.Ord., Fr. [Explanation: *a,* capitulum of ambulacral; *b,* adambulacral virgal; *c,* shelf of cupule (basin for tube foot); *d,* passage from cupule to arm interior; *e,* radial water vessel; *f,* tube foot; *g,* adductor muscle; *h,* cupule; *i,* virgal; *j,* ambulacral.]

the radial vessels of the arms and also connects with the stone canal (Fig. 6). The axial skeleton thus comprises the ossicles located along the water vessels of the arm and also the elements that form the **mouth frame.**

The axial skeleton is divided into segments, each of which corresponds with a branch water vessel approximately at right angles to the main canal. The branches are almost uniformly spaced. Each branch vessel emerges in a basin or depression readily recognizable in the skeleton. A tube foot is situated in each basin. Each of the ossicles that carry the basins is called an **ambulacral** (abbrev., *Amb,* pl. *Ambb*), the whole series along each arm forming an **ambulacrum.**

Relation of tube feet to their seatings. Fossil asterozoans commonly show in great detail the nature of the basins for the tube feet. The course of the water vessels and associated soft tissues can be determined by comparison with living starfishes (Fig. 7). The general relationship between water vessels and the axial skeleton has changed little since the time of the earliest known asterozoans. In all these primitive forms, whether somasteroids, asteroids, or ophiuroids, the seating of each tube foot was shared unequally between two ambulacrals, the larger part of the basin being located on the distal ossicles of the pair (Fig. 7,*2,3*). In Recent asteroids the basin is very shallow but still situated mostly on the distal ossicle (Fig. 7,*1,* shaded area, round pores).

The passage from the branch water vessel to the tube foot is distinguishable in many fossils as a break in the wall of the basin, just behind its transverse wall and situated on the proximal ambulacral, precisely as in extant asteroids. The derivation of this passage from a gap between adjacent ambulacrals can be seen from the situation in primitive somasteroids (Fig. 8).

Relation of structure to activities of tube feet. Minute details of the structure of the basins, as indicated by comparison with extant asteroids (71), are correlated with various activities of the tube feet. The chief of these are defined as protrusion and pointing. Protrusion is vigorous and extensive in extant forms. It is made possible by division of the tube foot into an internal sac, the **ampulla,** and an outward-directed closed tube, the **podium.** The tube foot is protruded by squeezing water from the ampulla into the podium and retracted by the reverse process (Fig. 9,*1a,b*).

It used to be thought that no Paleozoic asterozoans possessed passages between adjacent ambulacrals of the sort that imply the division of the tube foot into two parts, such as characterize extant asteroids (Fig. 10,*1*). It was necessary to suppose that if ampullae existed in Paleozoic forms they were situated outside the body cavity. JAEKEL (38) suggested that the deeply hollowed basins in many Paleozoic asterozoans held external ampullae (Fig. 10,*2*) and that this provided a mechanism for protrusion

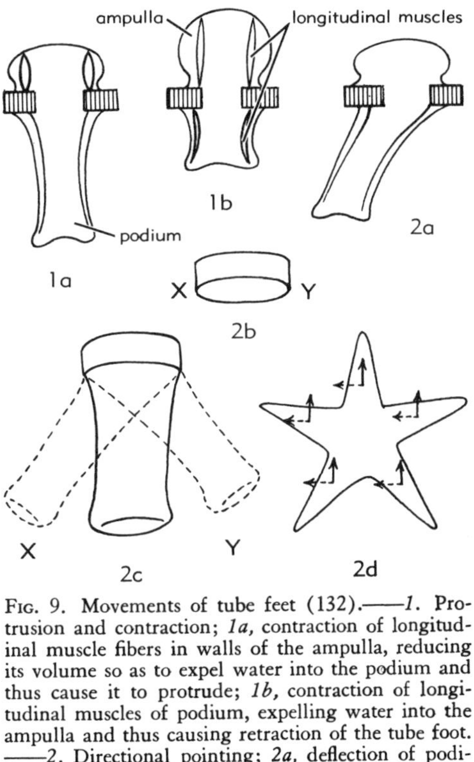

FIG. 9. Movements of tube feet (132).——*1.* Protrusion and contraction; *1a,* contraction of longitudinal muscle fibers in walls of the ampulla, reducing its volume so as to expel water into the podium and thus cause it to protrude; *1b,* contraction of longitudinal muscles of podium, expelling water into the ampulla and thus causing retraction of the tube foot. ——*2.* Directional pointing; *2a,* deflection of podium accomplished by contraction of muscles on one side and simultaneous relaxation of those on the other; *2b,* muscle ring at base of podium in oblique view, contraction of fibers at X being accompanied by relaxation of those at Y and similarly for any other opposite parts of the ring; *2c,* diagram showing tube foot expanded in three directions; *2d,* changes in direction of locomotion of a starfish accompanied by corresponding changes in direction of pointing of podia, as indicated by the two sets of arrows.

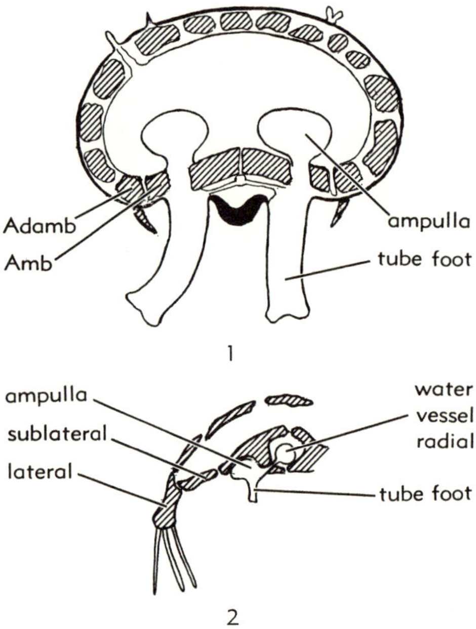

FIG. 10. Position of ampullae in relation to the body cavity of asterozoans.——*1.* Internal ampullae shown in cross section of arm of an extant asteroid (132).——*2.* External ampulla of an early ophiuroid *(Eophiura)* shown in cross section of arm, ampulla lodged in hollow of the ambulacral basin (114).

and retraction of the tube feet comparable with that of extant forms. In fact, most somasteroids and probably all Paleozoic asteroids had internal ampullae; the external ampullae are confined to Chinianasteridae (Fig. 8,*1b*) among somasteroids and to ophiuroids.

Directional protrusion or pointing of the tube feet is characteristic of living asterozoans. For example, in crawling, the podia are pushed outward in the direction of movement and in burrowing a sideways scooping action may be seen. SMITH (71) has shown that pointing of the podia is achieved by reciprocal contraction and relaxation of fibers on opposite sides of a ring of muscles between the wall of the upper part of the podium and the adjoining ambulacral (Fig. 9,*2a-d*). Thus, the angle of protrusion and the direction of pointing can be modified. Ancient asterozoans were probably able to point their tube feet like extant ones, for a well-defined narrow groove can be seen just inside the ambulacral basin in various early fossil asteroids and ophiuroids. This groove is plausibly interpreted as the line of insertion of the ring of muscle that controlled pointing.

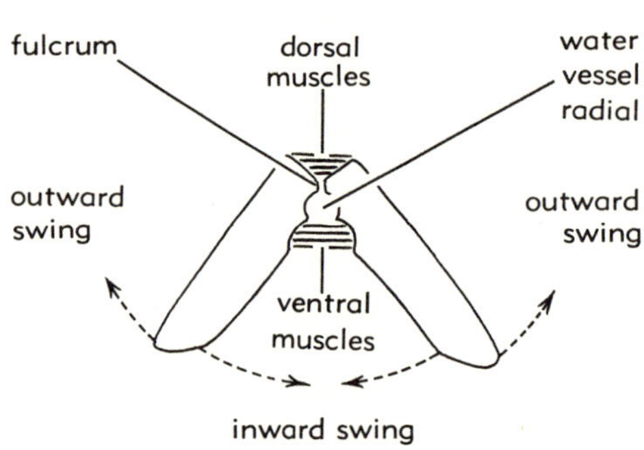

FIG. 11. Lever action of ambulacral ossicles (134).

Lever action of ambulacral ossicles. Asteroids are characterized by an arrangement of muscles attached to ambulacrals

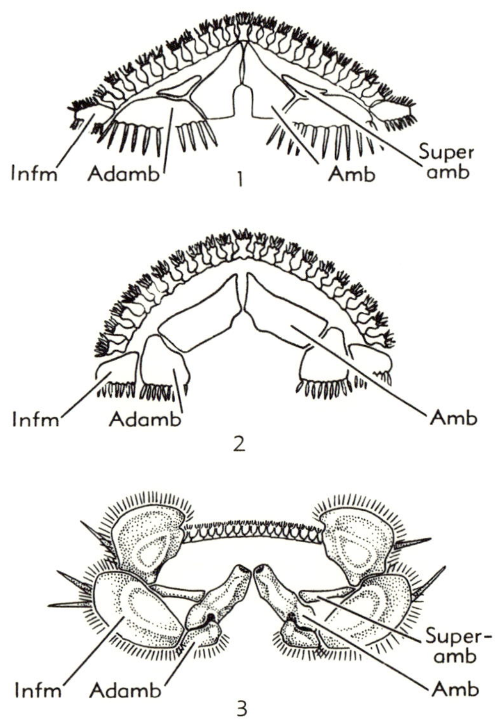

FIG. 12. Cross section of asteroid arms showing evolution of the ambulacral groove (133).——*1. Platanaster* (Ord.), with ambulacrals and adambulacrals at the same level.——*2. Schuchertia* (Ord.-Sil.), with ambulacrals arched above adambulacrals and rising well into the arm cavity.——*3. Astropecten* (Rec.), with steeply arched ambulacrals underlain by adambulacrals.

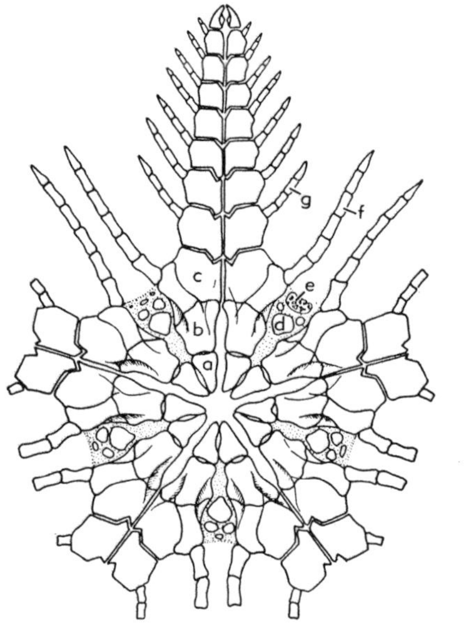

FIG. 13. *Chinianaster levyi* THORAL, L.Ord., Fr.; oral side of juvenile specimen, ×6.3 (108). [Explanation: *a-c*, 1st, 2nd, 3rd ambulacrals, *b* and *c* united syzygially; *d*, tegminal; *e*, inferred madreporite; *f*, oral pinnule; *g*, 4th metapinnule.]

that by a lever action allows opening or closing of the ambulacral groove. A pair of levers is formed by opposed bar-shaped ambulacrals arranged in an inverted V along the ambulacral groove on the oral surface of the arm; the fulcrum is near the tip of the V (Fig. 11). Excavations above and below the fulcrum provide for the insertion of dorsal and ventral muscles. Contraction of the dorsal muscles produces an outward swing of the lower ends of the bars, thus widening the ambulacral groove and assisting the protrusion of the tube feet. When the ventral muscles contract, the bars swing inward, narrowing the groove and sheltering the retracted tube feet.

The mouth-angle plates of primitive ophiuroids have a similar lever action, which was doubtless useful in digging burrows.

Ambulacral groove. The ambulacral groove with its arch formed by the ambulacrals is characteristic of asteroids. In this subclass the ambulacrals are invariably opposite one another, never alternating, as in somasteroids and early ophiuroids. In the earliest asteroid stage, exemplified by *Platanaster* (Fig. 12,*1*), the underside of the arm is flat and in substance of the ambulacrals interrupted by a shallow groove that carried the radial water vessel and radial nerve. A later stage, exemplified by *Schuchertia* (Fig. 12,*2*), shows a broad trough formed by transversely elongated ambulacrals, making an almost level roof of the groove which is bounded on either side by a wall of adambulacrals as innermost part of the adaxial skeleton. Subsequent deepening of the groove is accompanied by inward bending of these walls to make a nearly enclosed tube (Fig. 12,*3*).

Mouth frame. The mouth frame consists of the proximal ends of the rows of axial ossicles, more or less modified into an independent structure. There has been much argument about the origins and homologies of parts of the mouth frame, but a trustworthy conclusion appears to have been reached as a result of the latest studies by FELL (13) on primitive somasteroids. The following account is based on this work and therefore differs from standard descriptions.

Juvenile *Chinianaster* (Fig. 13) and *Villebrunaster* show that the proximal member of each ambulacral series is an acutely

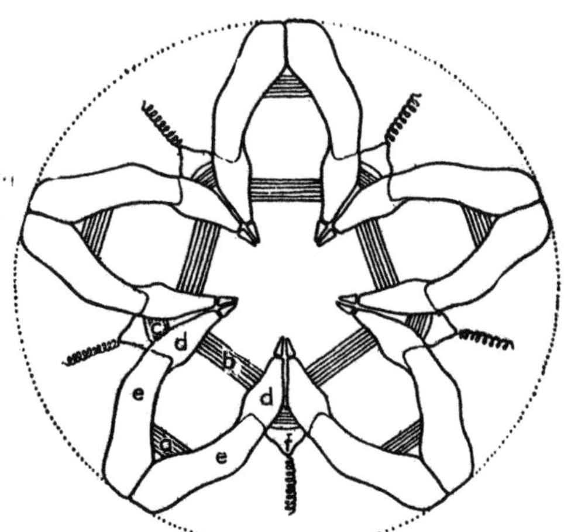

FIG. 14. Diagram of mouth frame of asteroid showing system of muscles that allows mouth-angle plates to be moved in and out (133). [Explanation: *a*, radial muscles between 1st ambulacrals; *b*, *c*, radial and interradial muscles attached to apophyses of mouth-angle plates; *d*, mouth-angle plate; *e*, 1st ambulacral; *f*, odontophore.]

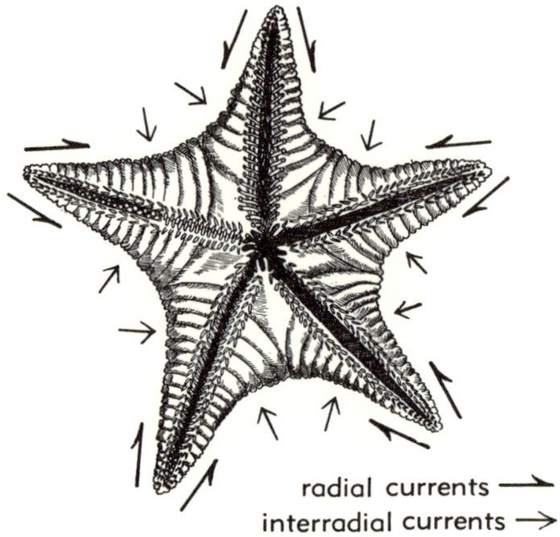

FIG. 15. Oral surface of recent asteroid *Porania*, showing large interradii and ciliate grooves; arrows indicate direction of flow of water currents (111).

triangular ossicle bordering the **buccal slit** and touching the adjacent ossicle of the neighboring axial series. These triangular ossicles are termed **mouth-angle plates.** The next two ambulacrals are united by syzygy into a massive ossicle which carries a complete cupule for a tube foot, as well as half cupules at either end. The differentiation of these elements of the mouth frame seems to be less distinct in adult *Chinianaster* and *Villebrunaster*. In the Recent *Platasterias*, the mouth-angle plates of each axial series have moved apart and superficially appear to be enlarged adambulacrals; the mouth frame thus appears to consist of powerful interradial "jaws" (pairs of mouth-angle plates of adjoining axial series) projecting between ends of the ambulacra. This is the typical appearance of the mouth frame of most Asteroidea, known as the **adambulacral type;** the mouth-angle plates have the appearance of specialized adambulacrals, whereas the apparent first ambulacrals are, in origin, fused second and third ambulacrals (Fig. 13).

A second type of mouth frame, termed **ambulacral type,** confined to the asteroid order Forcipulatida, is known. In this the mouth-angle plates are insignificant and most of the frame consists of the proximal ambulacrals, either straight or projecting into the mouth (see Fig. 63).

In some early Asteroidea with adambulacral mouth frames, an additional plate, the **torus,** is mounted on each pair of mouth-angle plates, projecting inward to the center of the mouth. Normally it carries several long spines. In *Cnemidactis,* for example, the five tori completely close the aperture. Forms without tori may have large spines that serve the same function.

In many forms the mouth frame was clearly more or less flexible, so that the mouth-angle plates could be moved in and out. The muscular system which allows this to be done (Fig. 14) involves an additional ossicle in each interradius, the **odontophore,** prominent in many early groups. Presumably this ossicle originally was an inframarginal which in the course of phylogeny became occluded from the margin and adapted as part of the "jaw" system. An analogous T-shaped plate is seen in somasteroids and some asteroids, but this may have a different origin.

The earliest Ophiuroidea (Stenurida) have deep radial V's, forming buccal slits, as in Somasteroidea, and somewhat similar mouth-angle plates. The "jaws" resemble those of Asteroidea of adambulacral type, but always have a torus.

ADAXIAL STRUCTURES

The adaxial skeleton of the somasteroid Chinianasteridae provides broad interradial tracts of grooves between parallel rows of narrow plates termed **virgals.** The rows are termed **metapinnules** and the whole structure is referred to as **metapinnular.** This arrangement of grooves persists in the extant somasteroid Platasteriidae but tends to be reduced and finally lost in the Paleozoic somasteroids. The sides of the grooves

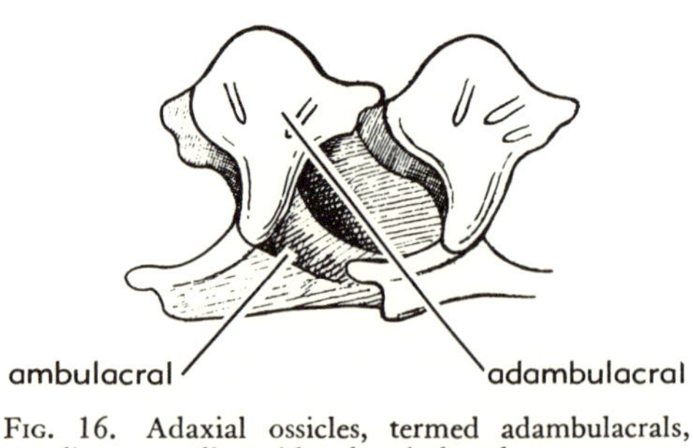

FIG. 16. Adaxial ossicles, termed adambulacrals, standing as wall at side of ambulacral groove; cross section of arm of *Archegonaster* (U.Arenig.), a somasteroid. Shelter for tube feet is provided by hollows between overhanging adambulacrals (133).

were ciliated, and thus water currents were produced from the periphery to the center of the arm. Many fossil and extant asteroids have analogous, if not homologous, systems for the production of such water currents. In principle, the currents serve two purposes—they bring small food particles to the ambulacral channel and thence to the mouth, and they also bring respiratory water to the tube feet and associated nerves, without which the tube feet become flaccid and do not function (70). The pattern of the oral surface and the arrangement of ciliary currents in the living asteroid *Porania* is much like that in *Chinianaster* (Fig. 15).

In later somasteroids and all asteroids some specialization of the primitive adaxial skeleton is observed. The row of ossicles next to the ambulacrals becomes continuous in a radial direction and the ossicles thicken to form a wall overhanging the ambulacral groove (Fig. 16); these ossicles are termed **adambulacrals.** They increase the depth of the ambulacral groove and shelter the podia when retracted; they are commonly armed with prominent spines.

The outermost row of virgals may similarly become continuous and form a row of **inferomarginals** (*Infm*, pl. *Infmm*). The intermediate rows may form a mosaic of ossicles, which in the most primitive asteroids (Platyasterida) retains the transverse series; but in most asteroids this pattern has been replaced by one composed of longitudinal (i.e., radial) gradients. The virgals of the row next to the adambulacrals, however, have become occluded in *Platasterias*, the living somasteroid, and rest across the internal surfaces of ambulacrals and adambulacrals; they are known as **superambulacral** ossicles. These persist in Platyasterida and Paxillosida but have disappeared in all but a very few other asteroids.

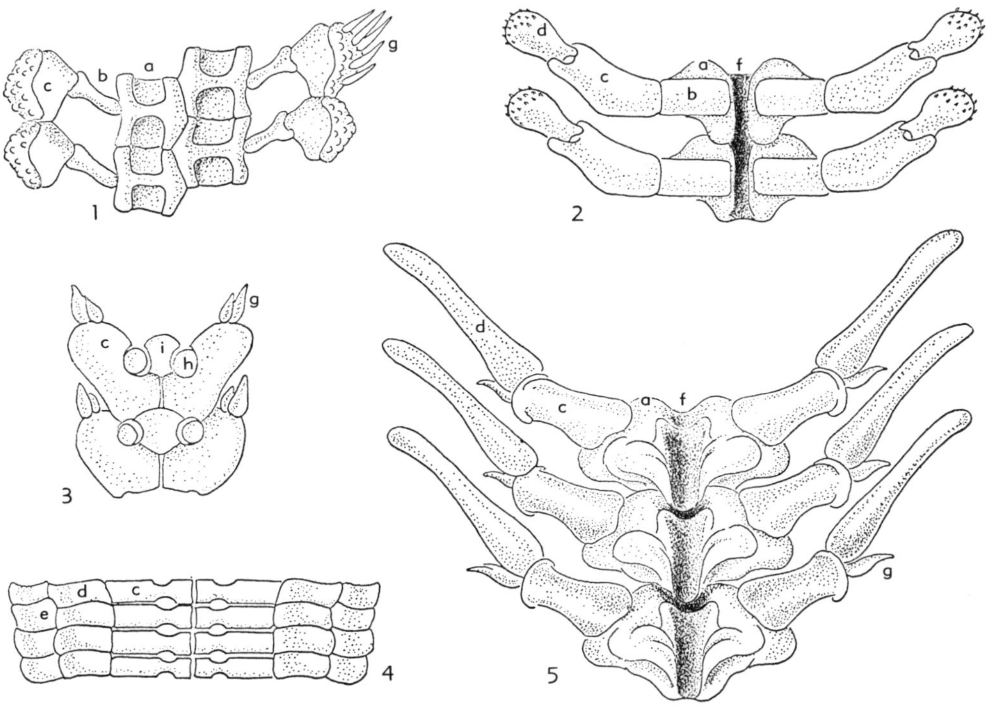

FIG. 17. Pinnate structure in Ophiuroidea (Fell).

1. Eophiura bohemica, ×2.3 (133).
2. Trichaster palmiferus, ×4 (108).
3. Ophiuraster symmetricus, ×13.5 (108).
4. Astrophyton sp., arm base, ×2 (119).
5. Asteronyx loveni, adoral view of arm skeleton dissection, ×6.7 (108).

[Explanations: *a*. ambulacral; *b*, virgal (sublateral); *c*, virgal (lateral); *d*, 3rd virgal; *e*, 4th virgal; *f*, hyponeural groove; *g*, lateral (secondary) spine; *h*, tentacle scale; *i*, ventral arm plate.]

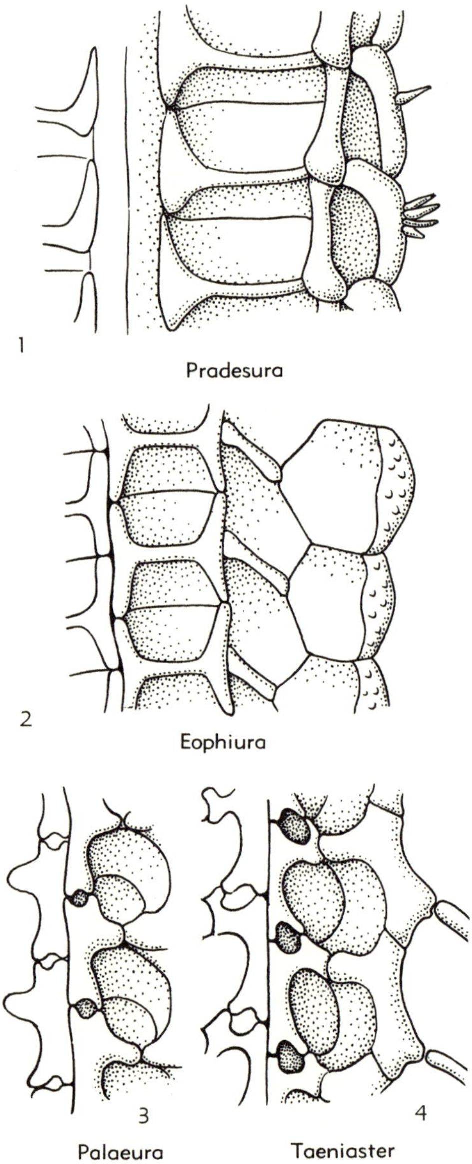

FIG. 18. Evolution of ambulacrals into vertebrae in arms of Ophiuroidea (133).——*1. Pradesura* (L. Arenig.), oldest known ophiuroid, has the basins for tube feet shared subequally by two ambulacrals with L-shaped ridges between, and an open ambulacral groove.——*2. Eophiura* (U.Arenig) has more equally shared basins, T-shaped ridges between, and closed ambulacral groove.——*3. Palaeura* (U. Arenig.) has the basins mostly on a single ambulacral, intervening ridges boot-shaped, and groove closed.——*4. Taeniaster,* much like *Palaeura,* has deeply excavated ridges for attachment of strong ventral muscles, and basins confined to single ambulacral.

OPHIUROID ARM (AXIAL AND ADAXIAL)

The most primitive ophiuroids show close resemblance to the contemporary somasteroids and, in particular, have a pinnate arm structure that is clearly derived from an ancestral metapinnular type. FELL has shown that several extant ophiuroids also preserve a distinctly pinnate arm structure, and he has demonstrated the homologies with the somasteroid arm (Fig. 17) (13). Even in a typical modern ophiuroid, the homology of lateral shield with a virgal can be made out (Fig. 17,5).

The critical development in the ophiuroid arm comprises fusion of the opposite ambulacrals to form single pieces that occupy most of the interior of the arm; these pieces, called **vertebrae,** articulated by a ball-and-socket joint. Associated with this step is transformation of the two inner rows of adaxial ossicles (homologues of first and second virgals) into side plates hinged to

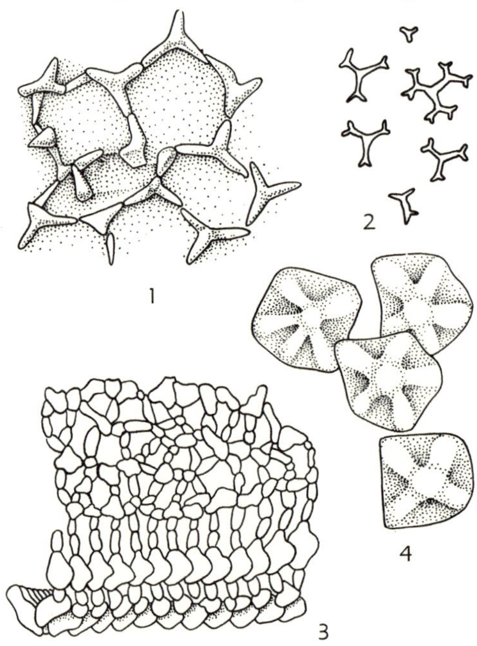

FIG. 19. Ossicles of extraxial skeleton of asterozoans.——*1.* Triradiate ossicles of aboral skeleton of an early somasteroid, *Chinianaster* (Arenig.) (133).——*2.* The developing network for comparison with an echinoid plate, which at this stage consists of discrete radiate ossicles (112).——*3.* Network of ossicles on sides and aboral surface of an extant starfish, *Asterias rubens* (139).——*4.* Ossicles from aboral surface of an early ophiuroid, *Encrinaster,* showing stellate ridges (133).

the vertebrae to act as cover plates; the side plates are known as **sublaterals** (*SubL*, pl. *SubLL*) and **laterals** (*L*, pl. *LL*). The stages of alteration of ambulacrals into vertebrae are shown within the order Stenurida (Fig. 18). (One suborder, Proturina, shows no sign of the change. For convenience the arm segments of all Stenurida are referred to as ambulacrals.)

The vertebral type of axial skeleton permits increased mobility, for with it the arm can twist and turn. Each arm segment becomes an independent unit, since seating of the tube foot is no longer shared by two ossicles. Ability to twist is provided by muscle bands which border the ball-and-socket joints.

EXTRAXIAL STRUCTURES

The extraxial skeleton comprises ossicles not associated with the tube feet, mainly ossicles of the aboral surface. In their simplest form these are spicules with three narrow rays diverging from a small center, together forming a wide-meshed net (Fig. 19). The holes of the network provide for respiratory exchange between the body fluids and the surrounding water. Many asteroids have a network similar to that of *Chinianaster* (Fig. 19,*1*) but more substantially built, in which the nodal points of the ossicles commonly carry a small knob surmounted by a spine or group of spinelets **(paxilla)**. The knobs may be arranged in diagonal rows marking the course of channels on the aboral surface, along which water currents are driven by cilia; these currents flow over holes in the network, which generally are occupied by external gills (**papulae**) (Fig. 20), consisting of projecting folds of the body wall.

This general type of respiratory arrangement characterizes the Paxillosida, Spinulosida, and Forcipulatida of all periods. In the Valvatida, however, from Paleozoic Hudsonasteridae to living Goniasteridae, the aboral surface tends to be more or less completely covered by large solid ossicles. The papulae emerge as a rule from specific papular pores between them and these may be grouped into special papular areas. There may also be other arrangements to compensate for the reduction of the respiratory area. In Hudsonasteridae a loosely built group

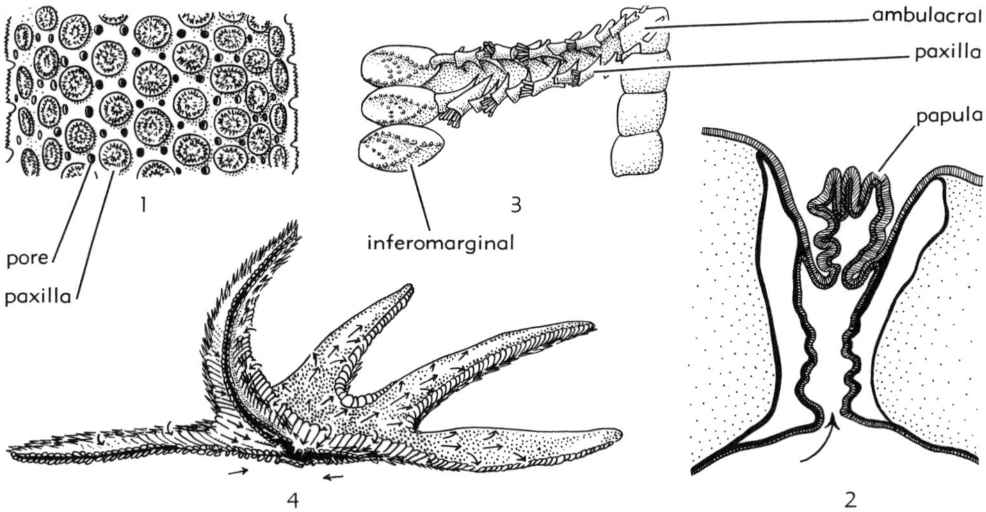

FIG. 20. Extraxial skeletal elements of the aboral surface of asteroids in relation to respiratory currents.

1. Diagonal channels on aboral surface of the extant *Chaetaster*, showing paxillae along borders of the channels and pores for papulae along the floors (118).

2. Cross section through a papula which has been withdrawn (118).

3. Diagonal channels on surface of the early Paleozoic *Platanaster* with the bordering paxillae fallen on their sides (133).

4. Diagram showing the way in which water collected on the aboral surface of a recent *Astropecten* is carried to the oral surface by intermarginal channels (111).

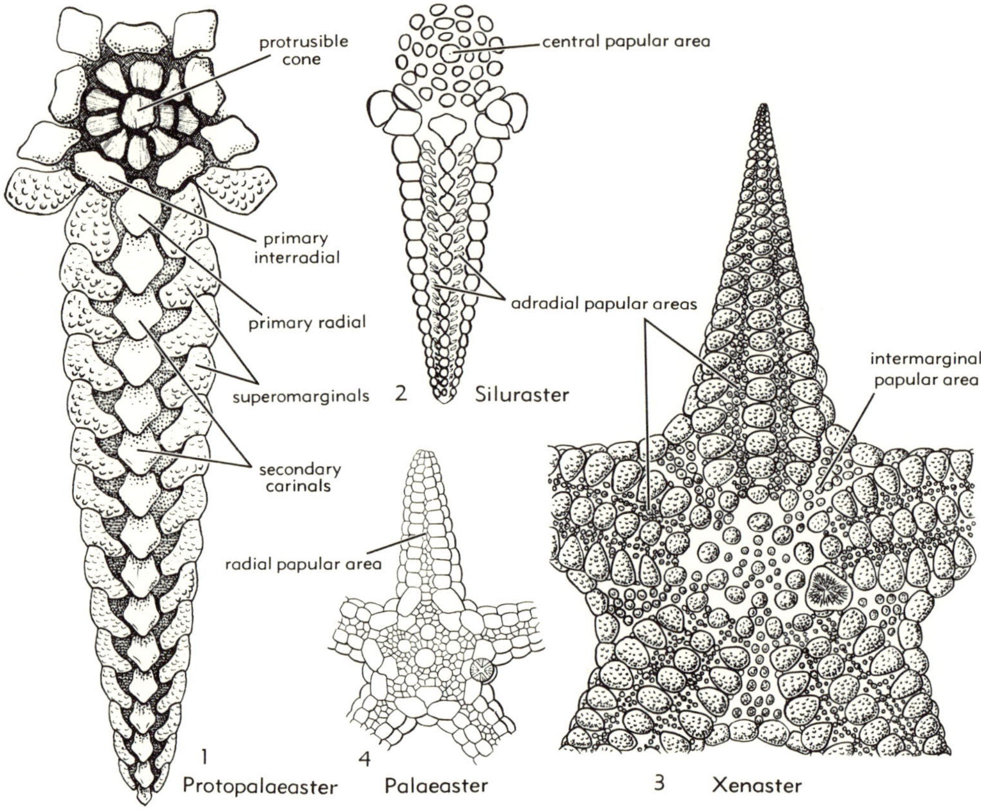

FIG. 21. Respiratory structures on aboral surface of early Valvatida (129).

1. Protopalaeaster (Hudsonasteridae), showing protrusible cone in center of disc, hinged to superomarginals through primary interradials (109).
2. Siluraster (Hudsonasteridae), showing central and adradial papular areas (133).
3. Xenaster (Xenasteridae) showing the addition of intermarginal papular areas (128).
4. Palaeaster (Palaeasteridae), showing radial papular areas consisting of numerous small ossicles (129).

of large ossicles in the center of the disc forms a **protrusible cone,** apparently for respiratory purposes; its ossicles are hinged to the superomarginals through the primary interradials (Fig. 21,*1*).

In *Siluraster* an adradial papular area occurs between carinals and superomarginals (Fig. 21,*2*). *Xenaster* has interradial intermarginal areas additionally (Fig. 21,*3*). Some *Mesopalaeaster* have an intermarginal area and in *Devonaster* they are of considerable extent; the abundant growth material of all ages found at Saugerties (57) shows that in ontogeny the young have an aboral surface like that of Hudsonasteridae and that older individuals develop successively adradial and intermarginal papular areas. In *Palaeaster* (Fig. 21,*4*) and *Neopalaeaster*, however, the whole of the middle of the upper surface of the arms becomes a respiratory area, being filled with small irregular ossicles. Generally, the appearance of additional respiratory areas seems to be correlated with a phylogenetic increase in size.

The extraxial skeleton of ophiuroids generally consists of overlapping scales which may show traces of their origin as components of a spicular web; scales on the aboral surface of *Encrinaster* (Ord.-Dev.) show the spicular rays surviving as stellate ridges with a shallow infilling between (Fig. 19,*4*). In many modern ophiuroids extraxial scales occupy considerable areas of the oral

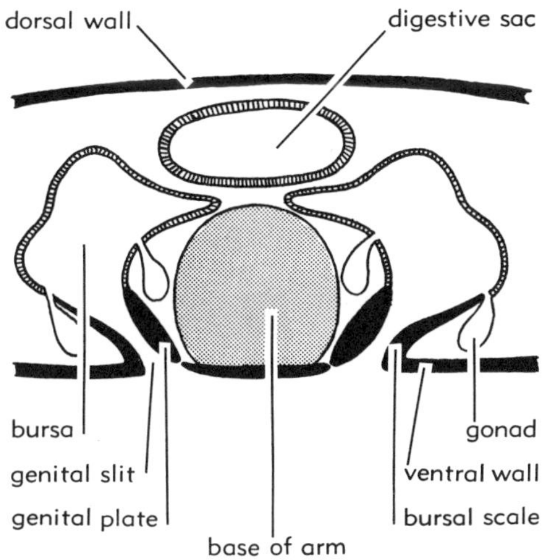

FIG. 22. Cross section of an ophiuroid disc showing gills and associated structures (115).

surface; this is mainly the consequence of extensions of the aboral skeleton downward to form pockets or pouches (**bursae**) for the internal gills. These constitute one of the main differences between ophiuroids and asteroids and presumably arose to meet the requirements of life in burrows. Damage to the delicate membrane of the respiratory surface is avoided by sinking it within the body wall (Fig. 22). Narrow slits allow the entry of water which is circulated within the cavity by ciliary action. The space for these internal gills is provided by secondary enlargement of the disc during ontogeny and by downgrowth of the aboral surface referred to above; during this process the madreporite migrates from a near-marginal aboral position to one very near the mouth. Since these characters are seen in the oldest known ophiuroid, *Pradesura,* they must be very stable features (Fig. 23,*3*).

Whereas early ophiuroids are characterized by a uniform covering of small scales, over a presumably flexible disc, a majority of later ophiuroids have a relatively stout and rigid covering (Fig. 24).

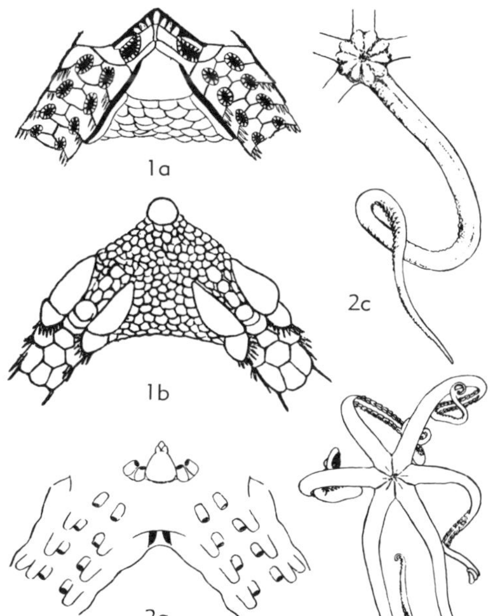

FIG. 23. Respiratory structures in ophiuroids.——*1*. Part of oral surface of the disc of *Ophiura* (Rec.); *1a,* arms covered ventrally by side shields (arm plates) and hexagonal small ventral shields, with pores for emergence of tube feet (or tentacles) surrounded by circlets of tentacle scales; gill slits along each side of arms with long genital bars at their outer edges; *1b,* radial shields on either side of base of arm, in aboral view (141).——*2*. *Asteroschema glutinosum* (Rec.); *2a,* part of disc and arms showing small openings of gill slits (shaded dark) in interrays; *2b,* entire animal, ×0.5; *2c,* disc and arm in aboral view, showing pairs of large radial shields, ×0.7 (120).——*3*. Oral surface of disc of *Pradesura,* oldest known ophiuroid, showing downgrowth of aboral surface of disc and accompanying movement of madreporite from lateral position, associated with internal position of gills, ×7 (133). [Explanation: *Amb,* ambulacrum; *bs,* buccal slit; *L,* lateral; *Mad,* madreporite; *MAP,* mouth-angle plate; *Subl,* sublateral.]

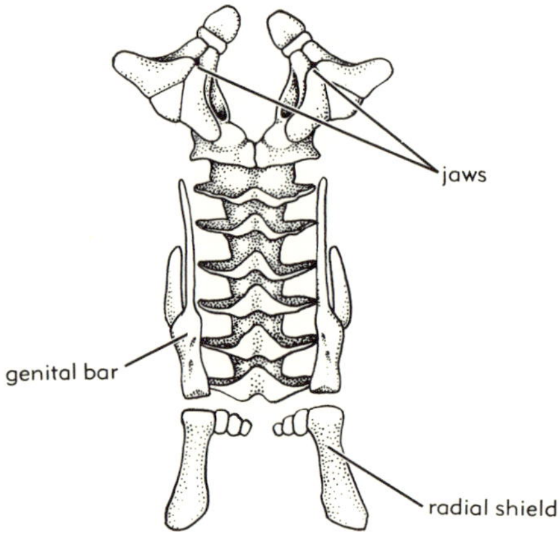

FIG. 24. Part of disc and arm base of *Ophiomyxa anisacantha* (Rec.); aboral view with radial shields (appearing as large marginal scales) upturned, showing simple articulation of the shields and genital bars and primitive jaws with feeble interradial musculature (120).

Scales persist but tend to become fused into larger ossicles. These normally include **radial shields** on the aboral surface of the disc and **genital bars** adjoining gill openings on the oral side. Each radial shield articulates with a genital bar, the shields being raised and lowered by muscles in aid of respiration; the shields are directly above the gill pouches, and by pressing against their flexible walls provide for the emptying and refilling of the pouches with water. Both paleontological and embryological evidence indicates that the radial shields are derived from fusion of scales at the edge of the disc. The marginal frame of enlarged scales in some Ordovician ophiuroids (e.g., *Euzonosoma*) was probably associated with pulse movement of the disc to circulate water in the burrow. Genital bars first appear in the Devonian.

In some early forms (e.g., *Stenaster* and *Stuertzaster*) and in Euryalina, the oral interrays of the disc are much reduced and the shape of the body and even some of the internal organs may resemble those of asteroids (14). In these forms the madreporite tends to retain its primitive lateral position and the gill pouches are concentrated about the center of the disc.

The density of calcification of the aboral skeleton varies considerably. In early members of the asteroid orders Paxillosida and Spinulosida the calcification of the central area of the disc is so weak that the mouth frame in the fossils is exposed in aboral view. Similar views of the mouth frame are commonly seen in early ophiuroids. Various extant asteroids and ophiuroids have the aboral surface covered by skin without any distinct ossicles in or below it.

Madreporite. As already mentioned, the madreporite was originally lateral but it migrated to the aboral or oral surface in various lineages. In early Spinulosida and Paxillosida, the madreporite, when recognizable, is a medium-sized rigid ossicle closely associated with a primary interradial but overlying neighboring ossicles (Fig. 25). In early Valvatida and Forcipulatida, however, it is a large thin, apparently flexible, plate situated in most genera on the oral edge of the side. This peculiar type of madreporite may have been concerned with both the water circulation and the hemal circulation. In modern Asteroidea

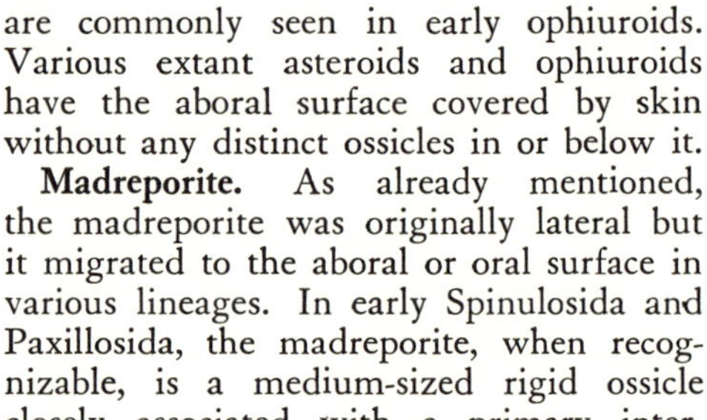

FIG. 25. Part of interradial aboral surface of *Palasterina primeva* (Sil.) showing flat madreporite overlying other ossicles (133).

[*Explanation of Figure 26*]

1. Comatulid crinoid, *Promachocrinus kerguelenensis*, pentacrinoid stage, ×12; *1a,b*, lateral and dorsal views.

2. Ophiuroid, *Ophiopyrgus wyvillethomsoni*, ×10; *2a,b*, lateral and aboral views.

3. Comatulid crinoid, *Eumorphometra aurora*, late pentacrinoid stage, ×12; *3a,b*, lateral and dorsal views.

4. Ophiuroid, *Ophiomastus stellamaris*, aboral view, ×30.

5. Ophiuroid, *Ophiomyxa* sp., aboral view of juvenile, ×15.

6. Asteroid, *Asterina* sp., ×15; *6a,b*, aboral views of larval stages.

7. Comatulid crinoid in early pentacrinoid stage, dorsal view showing circlet of infrabasals.

8. Ophiuroid, *Ophiosteira echinulata*, immature stage with basals still conspicuous, ×6.

[Explanation: *b*, basal; *c*, centrodorsal; *i*, infrabasal; *r*, radial; *t*, terminal.]

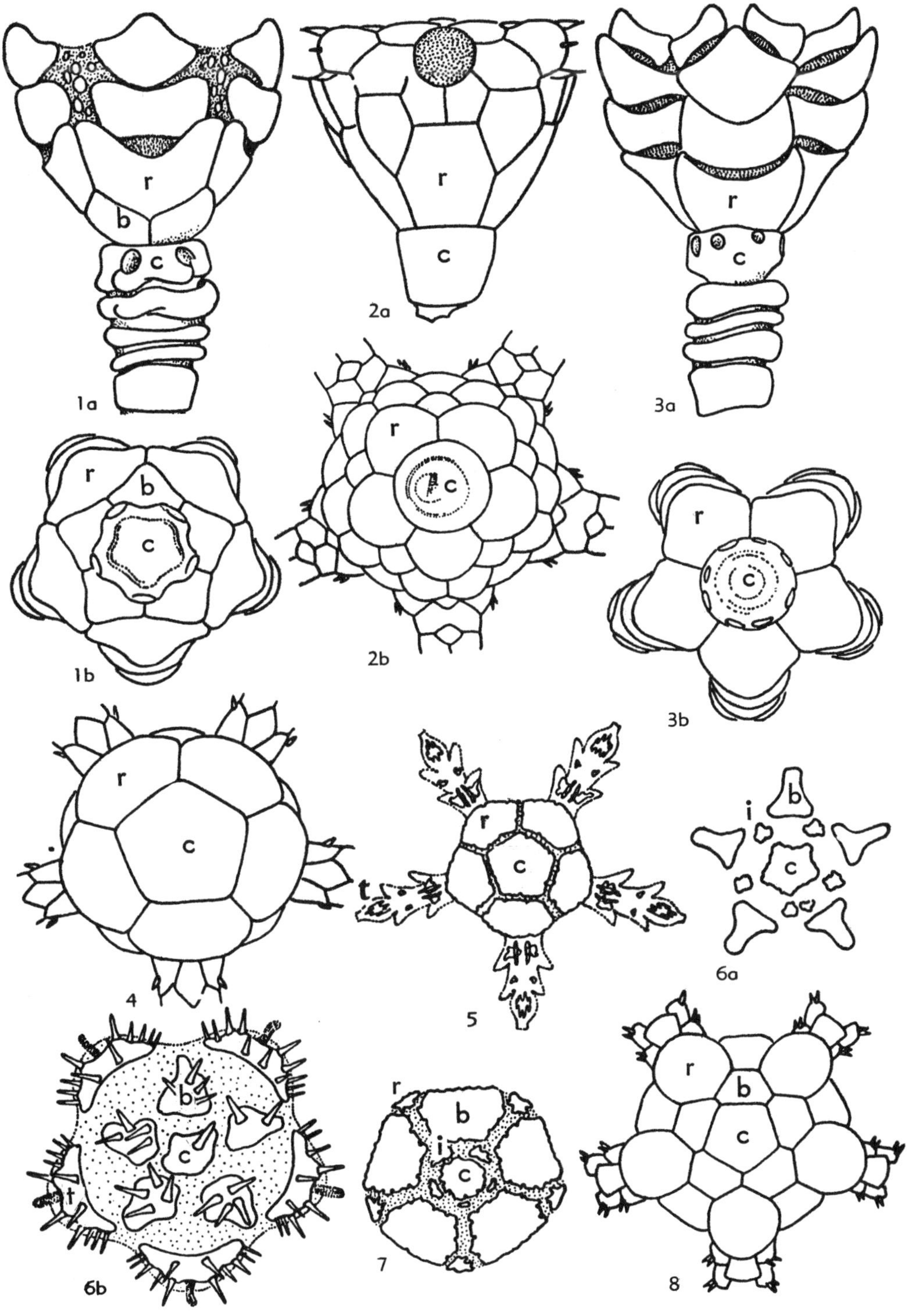

FIG. 26. Calyces of astroradiate echinoderms (108).
[*Explanation on facing page*]

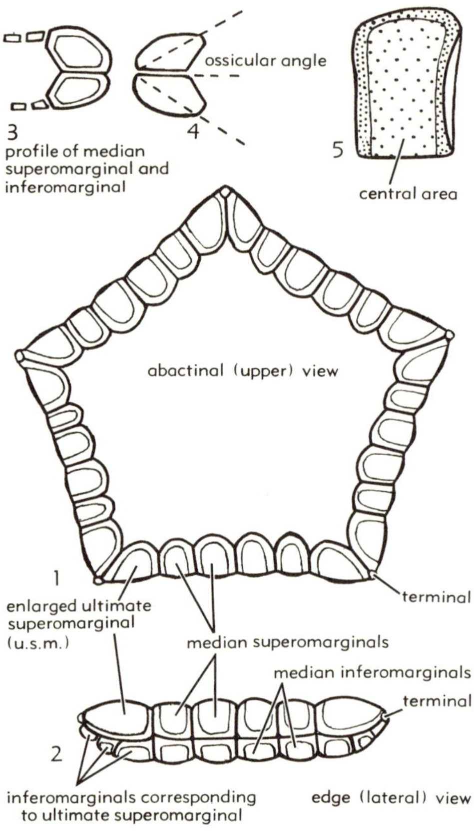

FIG. 27. Morphology of marginals in *Metopaster* (134).

it is almost solely concerned with the former function but in ontogeny a stage with both functions is seen; the very young asteroid has a heart, the **dorsal sac,** which lies near the opening of the water canal. In later life the activity of this heart is much diminished and it becomes embedded in the substance of the madreporite. The heart is very like that of *Balanoglossus.* The large flexible madreporite referred to above could well have capped a pulsating vesicle which retained its activity well into adult life. The primary interradial in juvenile asteroids forms such a cap, and this presumably accounts for the association of madreporite and a primary interradial in certain early forms.

Primary circlets. In early growth stages of both Asteroidea and Ophiuroidea the first-formed plates of the aboral skeleton are a few relatively large ossicles arranged in a pattern reminiscent of the structure of the calyx of Crinoidea. A centrale is surrounded by a circlet of five interradial plates and one of five radial plates. The homologies of these **primary circlets,** which in many groups persist in one form or another into the adult (Fig. 21), has been much discussed. It is clear that radial series **(carinals)** in adult asteroids are not homologous with the radials of crinoids or ophiuroids, but the basic calycinal system seems to be common to all three groups (Fig. 26) (13).

Marginals. At the edge of the body of many asteroids is a frame of enlarged ossicles called **marginals** (*M,* pl., *MM*). There may be only a single row or a lower row, homologous with the single row, and an upper one; the ossicles of the lower row are called **inferomarginals** (*InfM,* pl., *InfMM*) and of the upper **superomarginals** (*SupM,* pl., *SupMM*). The marginals may be rounded and sloping or square, forming a vertical wall at the edge of the body. Generally the marginals form a continuous frame, but in some early forms the arms are not fused together at their bases, so that the marginal frame is broken. In many Paxillosida, a central ridge occurs on each marginal, so that channels are developed over the edge of the frame, connecting with ciliated channels on the oral surface; in Cribellina similar grooves between marginals are occupied by papillated skin folds called **cribriform organs.** In some Goniasteridae (Valvatida) several distal marginals may be united in single enlarged ossicles, called **ultimate superomarginals** or **inferomarginals** (Fig. 27).

Superomarginals are normally placed directly over corresponding inferomarginals, but distally the correspondence may be less exact; in a few forms (e.g., *Trichasteropsis,* Paxillosida), the number of marginals differs markedly between upper and lower series. There is normally an equal number of marginals on either side of an interradial mid-line, but in a few families an odd **interradial** or **axillary marginal** is seen in either or both series.

In Notomyota the distal marginals are imbricate, thus allowing considerable flexure of the arms, by which means it is presumed that the animal swims.

Phylogenetically inferomarginals are derived from virgals and they are therefore part of the adaxial skeleton, whereas superomarginals are differentiated members of the extraxial skeleton of the aboral surface.

Terminals. Many and probably all asteroids have a single **terminal ossicle** at the tip of the arms, which in living and presumably fossil forms covers the base of an unpaired ocular tentacle. In some cases the terminals are large and have a characteristic shape.

Spines. Apart from protuberances that are one with the ossicles, most asterozoans bear spines. These originate, like ossicles of the oral and aboral skeleton, from radiate spicules, but they are formed at a higher level in the body wall than the ossicles. Paleontological evidence suggests that the spines are later-evolved skeletal elements that are secondarily attached to the older primary skeleton. A form of attachment common to all asterozoans is the ball-and-socket joint, very similar to that of echinoids (Fig. 28).

An important function of asterozoan spines is to clean the surface, especially in areas concerned with respiration. This is done by cilia on the spines that produce water currents from the base to tip of the spines, debris being thus lifted from the animal's surface. It is thus not surprising that spines and respiratory channels are commonly associated, particularly in asterozoans that have spines on paxillar shafts.

Various modifications of spines characterize different groups in Asteroidea and are of great classificatory importance. Paxillosida, many Valvatida, and some other groups carry, on the aboral surface or the margins, small ossicles with a ridged or pillar-like protuberance crowned by a tuft of spinelets or granules which are linked by muscles; these ossicles with their spines are known as **paxillae.** The tuft of spines can be opened or closed; when open, the network of paxillar spines provides effective protection for the papulae. In Pterasteridae (Spinulosida) the spines support a continuous supradorsal membrane, within which young are brooded.

Many Asteroidea carry **pedicellariae,** minute stalked, sessile, or sunken pincers. The stalked pedicellariae are formed of straight (forceps-like) or crossed (scissor-like) members embedded in tissue and attached to ossicles or spines; they occur in Forcipulatida (Fig. 29). Sessile pedicellariae consist of two or more opposed movable spines. Particularly in Valvatida, groups of such spines may be fused to form bivalved pedicellariae. Bivalved and other types occur in depressions in the aboral and marginal ossicles of Valvatida; such alveolate or foraminate pedicellariae are the only types commonly found in fossils.

FIG. 28. Spines of *Promopalaeaster bellulus* showing perforated ball-and-socket attachment, like that of echinoids (129).

SOFT PARTS

Following is a brief account of the soft parts of asterozoans other than those already dealt with in discussion of skeletal structures.

The **mouth** of asterozoans is in the center of the lower, oral, surface and opens into a sac-shaped stomach, which may be divided by a constriction. Paired extensions of the stomach reach into the arms in Asteroidea, where they form a liver. Most Asteroidea have an anus on the aboral surface, either in the center or close to it in an interradius, but some lack it, as do all Ophiuroidea.

In asteroids a **ring nerve** around the mouth gives rise to radial nerve cords that run along each ambulacral groove. The radial nerve is continuous with a general plexus of nerve fibers just under the epi-

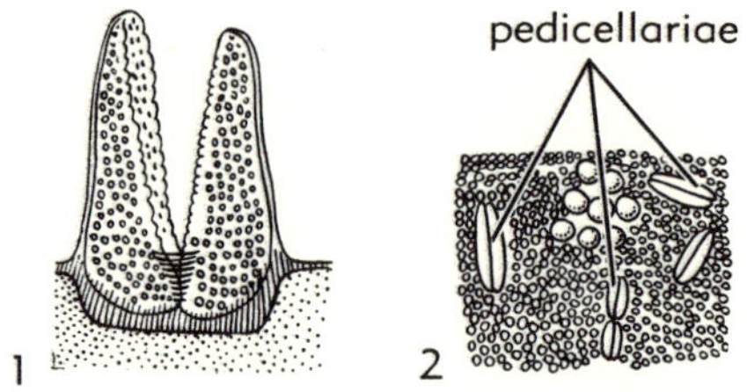

FIG. 29. Pedicellariae of Recent asteroids; *1*, forceps-like (forcipulate) pedicellaria of *Oreaster*, enlarged; *2*, bivalved (valvate) pedicellariae of *Culcita*, ×1 (118).

dermis of the whole body wall, including the podia; the plexus is thickened to form a marginal nerve cord along each side of the arms. In ophiuroids the nervous system is basically similar, but in forms with vertebrae the radial nerve has a ganglionic swelling in each vertebra; it gives off a branch into each podium and separate branches to the body wall and spines.

The asterozoan epidermis is liberally supplied with sensory cells, which are highly sensitive to touch or chemical stimuli. The ophiuroids have no special sense organs, but asteroids have a light-sensitive organ, generally pigmented, at the base of the terminal tentacle of each arm.

Asterozoa have a rather simple ring-and-radial hemal system.

ORIENTATION

Traces of bilateral symmetry are singularly few in Asterozoa. Several families of asteroids and all ophiuroids lack an anus; in most asteroids it is central or nearly so. There is no preferred direction of movement. Formally, orientation could be based on the position of the madreporite, save in the species or genera of asteroids with more than one, for example, *Acanthaster,* which has many. However, it is impossible to be certain that the position of the madreporite is homologous with that in other echinoderms. In any case, orientation in Asterozoa is of no practical and little theoretical importance.

FEEDING

Primitive somasteroids have a system of grooves on the oral side between the metapinnules, which are covered by small spines or plates and lead to the radial groove. These grooves clearly conduct water from the upper, aboral surface, primarily for feeding. Detrital particles falling on or near the sea star are thus swept by ciliary action, probably entangled in threads of mucus, to the mouth by way of the gaps between the aboral paxillae, along the interpinnular food grooves to the radial food groove and then to the mouth (Fig. 30). The living somasteroid *Platasterias* retains this method of suspension feeding and, in addition, captures relatively large food with its tube feet and passes it along the radial grooves to the mouth. The petaloid shape of the arms of somasteroids is probably associated with this system of food capture (11).

Similar methods of ciliary suspension-feeding persist in some living asteroids (e.g., *Porania,* Fig. 15). The frequent multi-armed forms among Paleozoic asteroids may also have been suspension-feeders, since the arms could constitute an effective net to catch drifting particles of food.

A development of this ciliary feeding is seen in mud-eating asteroids; *Ctenodiscus* (Goniopectinidae) plows through the upper layer of mud of the sea floor, entangling material in threads of mucus, which are then swept by ciliary action along channels on the oral surface to the radial groove and then to the mouth.

Typically, however, asteroids eat large food. Two main methods are observed in living forms. In the first, prey is passed to the mouth by the tube feet or the sea star positions itself with the mouth directly over the prey and the lips of the stomach

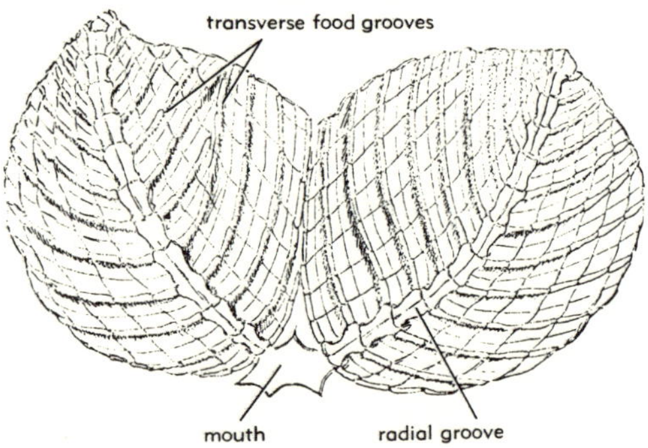

FIG. 30. Side view of *Villebrunaster thorali* (L.Ord., Fr.) with body flexed, showing transverse food grooves leading to radial groove and thence to mouth (133).

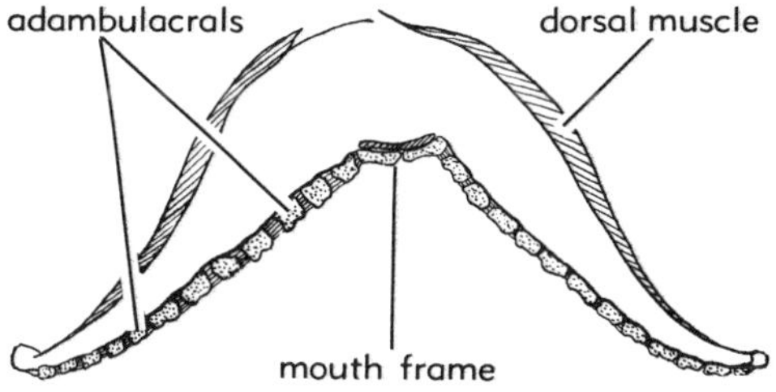

FIG. 31. Diagrammatic cross section of asteroid in position of feeding on large food, illustrating forces and resistances involved. The body is in the form of a dome supported by ends of the arms which give a firm grip on the sea floor. Resistance to collapse of the dome is provided by strong muscles which run from tips of the arms to the center of the dome, where they are firmly attached to each other, and by the chains of adambulacrals joined to each other by muscles, which run from a strong mouth frame below the center of the dome to tips of the arms (133).

are then everted through the mouth and draw the prey into the stomach. Living *Astropecten* has been observed to fill itself so full of small mollusks by this method that the upper surface is distended (42). This type of food capture is as old as the Ordovician, for specimens of *Girvanaster* have been found (U.Ord., W.Scot.) similarly dilated with small gastropods. Many modern asteroids eat other asteroids, ophiuroids, and echinoids. The predator digests the soft parts and ejects the hard remains. Fossil "pellets" consisting of such remains have been found in Upper Cretaceous chalks of western Europe; these are probably attributable to carnivorous asteroids. One specimen included identifiable remains of nine species of asteroids, two ophiuroids, and one echinoid (89).

The second method is that employed by *Asterias* and other members of its family. The sea star straddles a pelecypod and pulls open the upper valve with its tube feet; the lips of the stomach are then everted and inserted between the valves of the prey; the soft parts are then digested outside the mouth of the sea star, and the product is sucked in. This form of external digestion is applied to a variety of other prey, such as brachiopods, sickly fish, or coral polyps. An example of this system of feeding is supplied by a find of more than 400 *Devonaster* individuals associated with a bed of pelecypods (M.Dev., N.Am.).

How the strains involved in opening such large food are met is explained by Figure 31. The body, raised into a dome, is supported on the tips of the arms and prevented from collapse partly by long and strong dorsal muscles and partly by the chains of adambulacrals united by short muscles. The firm mouth frame supports the center of the dome. A specimen of

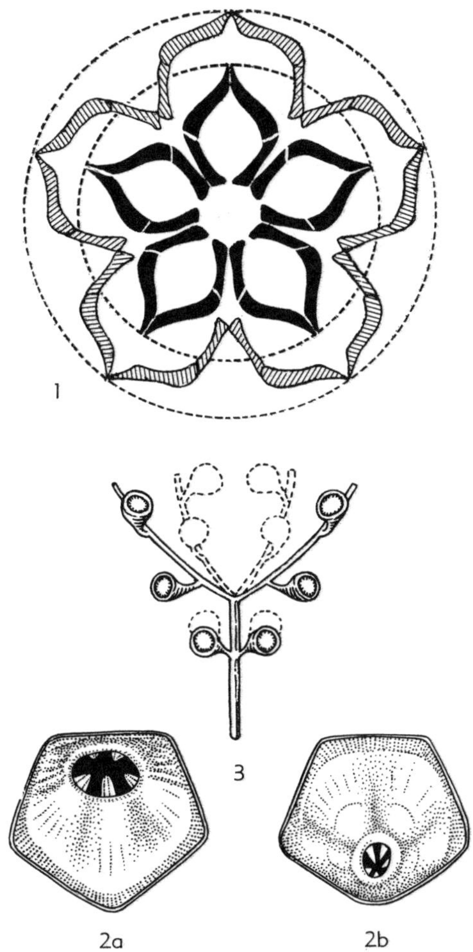

FIG. 32. Movements of echinoderm mouth parts useful for digging.——*1*. Ophiuroid mouth-frame movements associated with digging action of mouth-angle plates (133).——*2*. Movements of an echinoid's teeth showing position protruded *(2a)* and partly retracted *(2b)* (111).——*3*. Movements of second pair of buccal tentacles of an ophiuroid during digging action of mouth-angle plates; the tentacles are placed favorably for lateral scooping of loosened sediment and in some forms they are enlarged considerably (133).

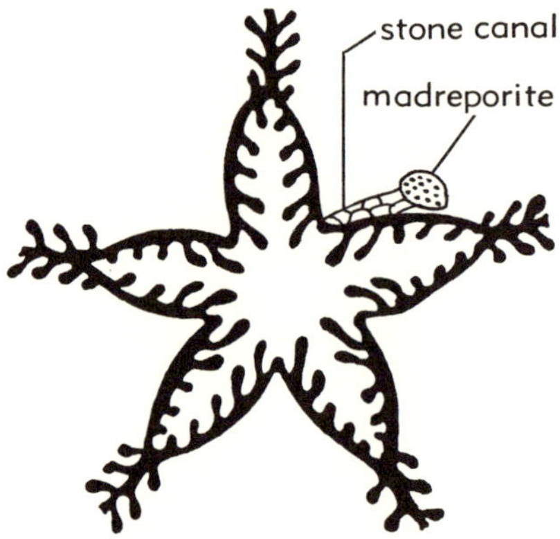

FIG. 33. Arrangement of buccal tentacles in *Eophiura;* these consist of tube feet disposed along edges of the buccal slits (133).

Ordovician *Salteraster* has been found in approximately this position.

Divergence of the first ophiuroids from the common stock of somasteroids may be associated with their living in burrows in the sea floor. The earliest ophiuroids (Arenig.) are found in nodules which show molds of some of the interior soft parts, indicating that the animals may have been buried in sediment at the time of death; moreover, central parts of the animal are preserved while distal parts of the arms are missing. The Recent *Amphiura* lives in a burrow with tips of the arms above or very near the sea bottom (Fig. 3). Where

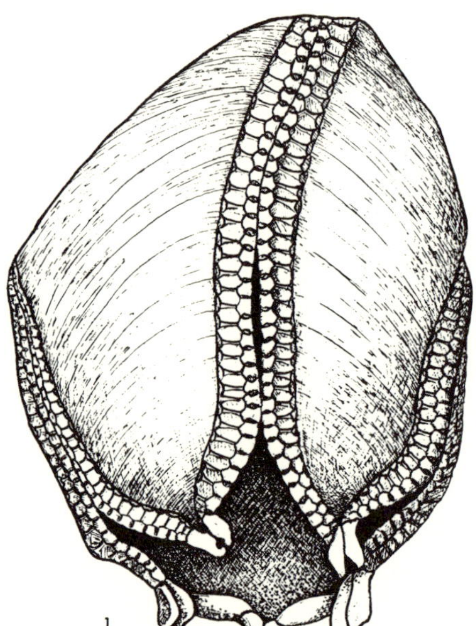

FIG. 34. *Cheiropteraster* (L.Dev.), showing downward projection of small mouth-angle plates for anchorage and diverging proximal ambulacrals next to the deep buccal slits, ×0.6 (133).

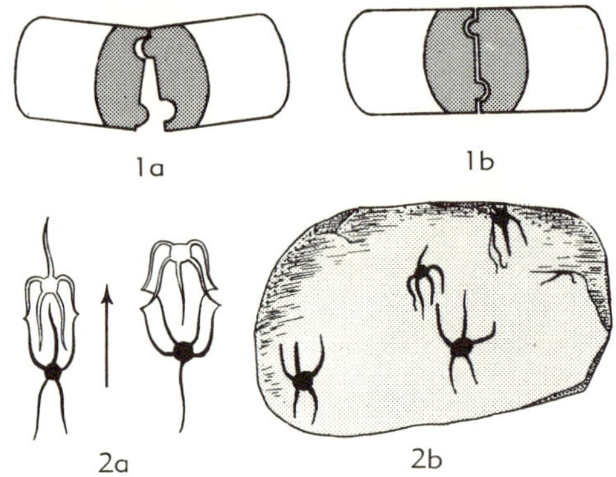

FIG. 35. Arm structure and movement in active bottom-dwelling ophiuroids (135).——*1.* Intervertebral joints of zygophiuroid vertebrae showing *(1a,b)* dorsal ball-and-socket and ventral peg-and-socket elements, latter maintaining linkage of vertebrae when the arm is used to push against the sea floor.——*2.* Position of arms in crawling; *2a, Ophiura* (Rec.) in first position (black) and at end of arm stroke (outline), showing inactive arm in front or rear (arrow shows direction of movement); *2b, Ophiaulax* (Dev.) in similar positions.

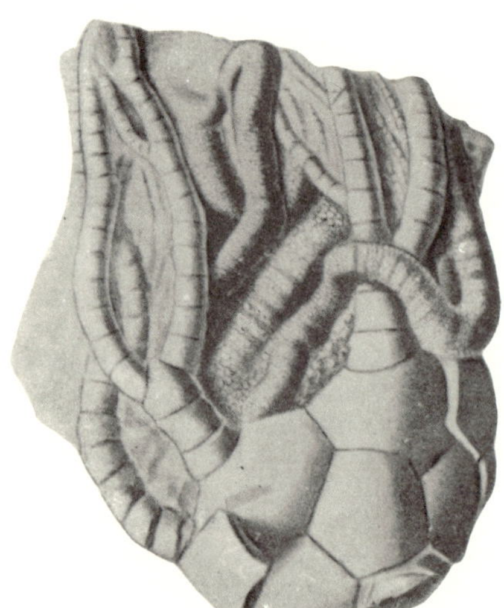

1

FIG. 36. *Onychaster flexilis* MEEK & WORTHEN (L.Carb.), intertwined in the arms of the crinoid *Barycrinus hoveyi* (103).

arm tips of early ophiuroids are preserved, they are generally pressed closely to the aboral surface of the disc, as if the arms had been withdrawn into the burrow before death (Fig. 4). As in living *Astropecten* (70), the tube feet may have been used as scoops in excavating, but the sediment was perhaps loosened by thrusts of the mouth-angle plates (Fig. 32,*1*) operating as do echinoids' teeth (Fig. 32,*2*); withdrawal of the cone of mouth-angle plates would swing the buccal tentacles into place to scoop away the loosened sediment (Fig. 32,*3*). While digging, the arms are bent upward and the disc is elongated to accommodate the animal to its narrow burrow.

Once in its burrow an ophiuroid can use only the tube feet of the tips of the arms for the capture of food; this is then passed along the series of tube feet protected by the lateral and sublateral plates and by rows of spines, until it reaches the mouth which is well provided with **buccal tentacles** (Fig. 33). The tunnel thus formed also serves for the passage of respiratory water. Arenigian ophiuroids (e.g., *Eophiura, Palaeura*) have been found in this position of feeding.

Other early ophiuroids, such as *Stenaster* (M.Ord.-U.Ord), had large soft slightly calcified discs providing a large food-collecting surface on the aboral side, which was connected by channels with the rows of tube feet on the oral surface. Presumably, they were sessile bottom-dwellers, living either as suspension-feeders or gathering organic detritus with their buccal tentacles, or both. *Cheiropteraster* (L.Dev., W.Ger.) had a large swollen disc anchored to the bottom by the mouth-angle plates (Fig. 34). The widely open buccal slits were bordered by buccal tentacles (only tube feet present in the genus) in a position to grasp food from debris floating near the bottom.

By Devonian time two groups had developed, with feeding habits associated with considerable structural modifications of the arms that allowed for active movement on the sea floor in one group and for a commensal life attached to crinoids and other hosts in the second. In the first of these (Ophiurida), the zygophiuroid joint is developed with a peg and a socket on the lower half of the vertebra, which allow the arm to swing downward rapidly (Fig. 35, *1*); the vertebral segments interlock during the swinging movement of the arm tip so as to push the animal along the sea floor. The arms are held in a characteristic posture during this process, two pairs of arms being used for propulsion and the fifth being inactive (Fig. 35,*2*). An abrupt change of direction is accomplished by using the inactive unpaired arm and the adjacent one of a pair, leaving the other arm of the pair to become inactive. *Ophiura*

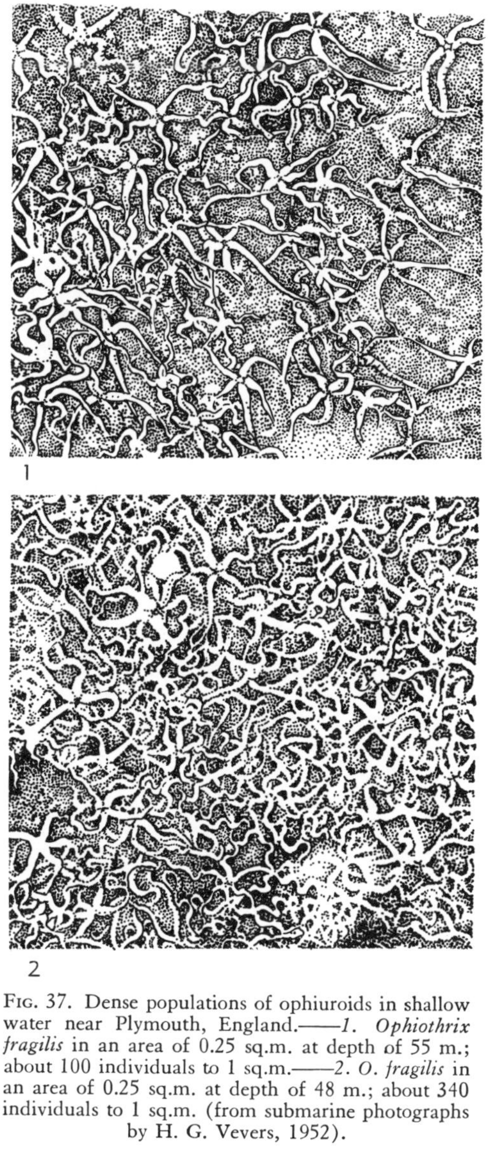

1

2

FIG. 37. Dense populations of ophiuroids in shallow water near Plymouth, England.——*1. Ophiothrix fragilis* in an area of 0.25 sq.m. at depth of 55 m.; about 100 individuals to 1 sq.m.——*2. O. fragilis* in an area of 0.25 sq.m. at depth of 48 m.; about 340 individuals to 1 sq.m. (from submarine photographs by H. G. Vevers, 1952).

can attain a speed of about 6 feet a minute, many times faster than an asteroid of comparable size using only its tube feet. Specimens of *Ophiaulax* (Dev.) have been found fossil in this "hunting" position (Fig. 35,*2b*). The system of vertebrae also allows the tips of the arms to coil around large prey and bring it to the mouth, like an elephant's trunk; in these forms the mouth-angle plates, used in earlier forms for digging burrows, now become jaws for crushing and tearing prey. Another important modification is the development of the laterals, which wrap around and are fixed to the vertebral axis. The laterals carry spines which may aid the grip of the arm on the bottom during movement; normally they do not entirely encase the arms, the gaps being filled by small dorsal and ventral plates.

The second of the two new groups in the Devonian (Phrynophiurida) had arms that could climb up and grip onto hosts such as crinoids by means of the vertical rolling of the arms and small hooked spines (Fig. 36). This group consists of suspension-feeders, collecting organic particles by their own ciliary action, aided by that of their hosts. Some extant members of the order, the "basket stars" (Gorgonocephalidae), have complex branched arms forming a tangle; they do not need the support of stalked animals but can support their food-gathering apparatus above the sea floor themselves.

Many living members of the order Ophiurida also retain the habit of suspension-feeding. Submarine photographs (Fig. 37) have shown very high densities of *Ophiothrix* in layers one above the other, apparently forming nets to capture food brought by tidal streams.

GLOSSARY OF MORPHOLOGICAL TERMS APPLIED TO ASTEROZOANS

The complex structures of asterozoans have given rise to many special terms. Variation in nomenclature and usage by specialists has greatly increased the number likely to be met in the literature. In the *Treatise* the number of such terms is kept to a minimum and in the following glossary many terms are mentioned merely as synonyms. Classification of terms is indicated typographically, boldface for most commonly used terms and italics for terms not recommended.

Many authors treat the names of particular ossicles ending in "al" as Latin nouns with a plural ending in "-ia" (e.g., ambulacral, pl. ambulacralia). In the *Treatise* such terms are treated as English words and the plurals formed by adding "-s" (e.g., ambulacrals).

abactinal. See aboral.

aboral. Applied to surface (or structures on it) opposite that bearing mouth and ambulacral grooves, or to direction away from mouth (*syn.*, abactinal, apical, dorsal).

accessory. Applied to ossicles of oral or aboral surface other than ossicles of primary circlet, carinals, ambulacrals, marginals, or terminal.

actinal. See oral.

actinostomial ring. See mouth frame.

adambulacral. Ossicle of series on oral surface of ray, next to ambulacrals (abbrev., *Adamb.*, pl., *Adambb*); derived from first virgal of primitive somasteroids.

adaxial. Applied to ossicles actually or in origin in transverse series wtih axial ossicles (i.e., with ambulacrals).

adoral. Directed toward mouth.

adradial. Ossicle of series on aboral surface of ray, between carinals and marginals (*syn.*, accessory radial, dorsolateral); also, directed toward axis of ray.

ambital. Pertaining to edge of body in plan view.

ambulacral. Ossicle of axial skeleton, one of double series of opposite or alternate ossicles formed along radial water vessel that constitutes axis of ray or arm (abbrev., *Amb*, pl., *Ambb*); also, pertaining to series of ambulacral ossicles.

ambulacral groove. Axial depression along oral surface of ray that is roofed by series of ambulacral ossicles.

ambulacral channel. Median channel between ambulacrals that houses radial water vessel and accompanying soft tissues.

ampulla (pl., **ampullae**). Dorsal saclike part of tube foot, either seated externally in cupule or internally and connecting with podium through podial pore; ampullae may be single or double.

anus. Vent of digestive tract, present only in some asteroids in which it is an inconspicuous pore near middle of aboral surface of disc.

apical. See aboral.

arc. Curved part of margin of asteroids in which arms are more or less distinct from disc in plan view; generally as interbrachial arc.

arm. Radial extension of body surrounding axis consisting of ambulacra; arms may be distinct from disc or not.

axial. Pertaining to axis formed by ambulacrals in sheath of radial water vessel in ray.

axil. Angle formed by junction of rays or arms in asteroids that have straight-sided arms and no interbrachial arcs.

axillary. In axil; generally applied to single large ossicle in axils of certain asteroids.

bivium. Part of asterozoan containing madreporite and ray on each side of it.

body wall. Integument, with any included calcareous skeleton, that encloses disc and arms.

brachial. See carinal.

buccal shield. Large, more or less triangular ossicle in interradial position adjoining mouth in ophiuroids (*syn.,* buccal plate).

buccal slit. Extension of mouth along axis of ray, bordered by single row of ambulacrals on each side (*syn.,* oral slit).

buccal tentacle. Tube foot on border of buccal slit.

bursa (pl., **bursae**). Internal gill pouch in ophiuroids, entered by gill slit.

caecal pore. See papular pore.

carinal. Ossicle of series along mid-line of aboral surface of ray, in line with primary radial if present (*syn.,* brachial, median dorsal, radial).

central plate. See centrale.

centrale. Prominent plate at center of aboral surface of disc in many asterozoans, center of primary circlet (*syn.,* central plate, centrodorsal).

centrodorsal. See centrale.

covering plate. Sometimes applied to laterals in certain primitive ophiuroids in which they can swing over ambulacral groove or to adambulacrals in certain asteroids.

cryptozonate. Referring to asteroids in which marginals are not conspicuously larger than other ossicles.

cupule. Cup-shaped depression on oral surface of ambulacrals in which tube foot is seated.

dental papilla. Scalelike ossicle projecting from jaw in ophiuroids.

disc. Central part of body, more or less distinctly separable from arms.

disc ambital. See intermarginal.

distal. Situated relatively farther away from mouth or center of disc; opposite of proximal.

dorsal. Same as aboral.

dorsal shield. Ossicles of series along mid-line of aboral surface of arm in ophiuroids (*syn.,* dorsal arm plate).

dorsolateral. See adradial.

fasciole. A specialized, heavily ciliated tract; applied to the intermarginal channel in some asteroids.

flooring plate. See ambulacral.

genital bar. Elongate ossicle along oral edge of gill slit at base of arm in ophiuroids.

genital papilla. Minute scalelike process adjoining gill slit.

genital slit. See gill slit.

gill slit. Fissure in disc of ophiuroids along side of base of arm, leading into bursa.

granules. Minute, more or less spherical skeletal elements situated on surface of ossicles, generally in pits or distributed in covering skin.

groove spine. Short blunt spines, generally recumbent, in clusters or rows bordering ambulacral grooves in many asteroids.

inferomarginal. Ossicle of a series along the oral edge of arms or disc or both (abbrev., *InfM,* pl., *InfMM*); in origin part of the adaxial skeleton, i.e., a virgal.

interactinal. See oral intermediate.

interbrachial. Between arms; applied to margin or surface of disc or to internal structures.

intermarginal. Small ossicle between rows of inferomarginals and superomarginals in some asteroids; also as epithet, applied to surface of one marginal adjoining next marginal.

intermediate. Apart from its common usage, applied specifically to ossicles of oral surface between adambulacrals and inferomarginals (*see* oral intermediate).

interradial. Indicating position midway between axis of adjacent rays or area between such rays.

interray. Area between a pair of adjacent rays.

jaw. Compound ossicle projecting into the mouth cavity in ophiuroids.

lateral. Ossicle of a series along the side of the arm in ophiuroids (abbrev., *L,* pl., *LL*) (*syn.,* side shield); in origin a virgal.

madreporite. Spongy or sievelike ossicle that serves as inlet to the water vascular system (abbrev., *Mad*); it is located interradially, lateral in some primitive asterozoans, but on the aboral surface in most asteroids and a few ophiuroids and on the oral surface in other ophiuroids.

marginal. Ossicle of a series along the ambitus (abbrev., *M,* pl., *MM*); either of inferomarginal or superomarginal series.

median dorsal. See carinal.

metapinnule. Structure running transversely outward from ambulacral and composed of series of more or less cylindrical ossicles called virgals; metapinnules constitute adaxial skeleton in most somasteroids and persist more or less modified in many asteroids and ophiuroids.

mouth. Entrance to digestive tract, invariably located in center of oral, or under, side of animal.

mouth-angle plate. More or less prominent ossicle projecting into mouth from proximal end of interray, forming pair with adjacent mouth-angle plate (abbrev., *MAP,* pl., *MAPP*); in origin part of series of ambulacrals.

mouth frame. Angulated girdle of ossicles surrounding mouth (*syn.,* actinostomial ring).

mouth shield. See buccal shield.

odontophore. Single axillary on distal edge of pair of mouth-angle plates in asteroids.

oral. Applied to surface of animal that contains mouth and in asterozoans is directed downward (*syn.,* actinal).

oral intermediate. Applied to ossicles of oral surface between adambulacrals and inferomarginals in asteroids, constituting part of adaxial skeleton (*syn.,* interactinal, ventrolateral).

oral papilla. Minute scalelike projection near mouth in ophiuroids.

oral slit. See buccal slit.

ossicle. Any individual calcified element of skeleton, but normally used for larger of such elements.

papilla (pl., **papillae**). Scalelike minute ossicle in ophiuroids.

papula (pl., **papulae**). Short protuberance of integument between ossicles of aboral or oral surface of asteroids that functions as external gill.

papular pore. Gap between ossicles for protrusion of papula (*syn.,* caecal, respiratory pore).

paxilla (pl., **paxillae**). Ossicle of extraxial skeleton with shaft surmounted by tuft of spinelets.

paxillose. With paxillae.

pedicellaria (pl., **pedicellariae**). Minute forceps- or pincer-like or valvate calcareous appendage borne on or in skin, ossicles, or spines of asteroids.

phanerozonate. With marginals conspicuously larger than other ossicles.

plate. See ossicle.

podial opening, podial pore. Passage between ambulacrals for emergence of tube foot.

podium (pl., **podia**). Cyclindrical outer part of tube foot.

primary circlet. Ring of prominent ossicles on aboral surface, typically consisting of five radials and five interradials surrounding centrale.

proximal. Nearer mouth or center of disc; opposite of distal.

pustule. Minute boss on ossicle with central depression in which spine articulates.

radial. Prominent ossicle on aboral surface of asteroids, in line with mid-line of arm, forming part of primary circlet; commonly used in older literature for any ossicle in series with primary radial (*see* carinal); also applied to organs (e.g., canal, nerve) extending along arms.

radial shield. Relatively large ossicle comprising one of pair adjacent to base of arm on aboral surface of disc in many ophiuroids.

ray. Segment of body that includes one ambulacral axis.

respiratory pore. See papular pore.

ring canal. Part of water vascular system that forms canal around mouth, from which radial canals radiate.

side shield. *See* lateral.

spicule. Very minute irregular, cylindrical, or radiate skeletal element.

spine. Sharp or blunt, short or long skeletal element, attached to ossicle by muscle.

spine pit. Coarse or fine depression in ossicle in which spine or granule articulates.

stone canal. Calcified tube leading from madreporite to ring canal.

streptospondyline. Type of articulation between vertebrae in some ophiuroids, in which there is simple ball-and-socket joint.

sublateral. Small ossicle between ambulacral and lateral on side of arm of some primitive ophiuroids, homologous with adambulacral of asteroids (abbrev., *SubL,* pl., *SubLL*).

superambulacral. Internal ossicle lying across the inner junction of ambulacral and adambulacral in some asteroids; originating from occluded virgal.

superomarginal. Ossicle of series along edge of disc or arms or both, above series of inferomarginals (abbrev., *SupM,* pl., *SupMM*); modified ossicle of extraxial skeleton in origin.

supramarginal. See superomarginal.

tentacle. May be used for tube feet in general or specialized one.

tentacle pore. Same as podial pore, but more commonly used than that term in describing ophiuroids.

terminal. Single ossicle at tip of arm, appearing very early in ontogeny; in asteroids it protects ocular tentacle (*syn.,* ocular plate).

tooth papilla. *See* dental papilla.

torus (pl., **tori**). Flat ossicle, commonly carrying spines, projecting into mouth from mouth-angle plate in some asteroids and from jaws of all ophiuroids.

trivium. Part of body containing three rays, excluding bivium.

tube foot. Extensible water-filled organ consisting of cylindrical podium and sac-shaped ampulla, connected with radial water canal; tube feet form two or four rows along ambulacral axis.

ventral. See oral.

ventral shield. Ossicle of secondary origin on oral side of arm in ophiuroids.

ventrolateral. See oral intermediate.

vertebra (pl., **vertebrae**). Fused pair of opposite ambulacrals, articulating with neighboring vertebrae by ball-and-socket joints.

virgal. More or less rod-shaped ossicle of metapinnule.

water-vascular system. Assemblage of water-filled canals comprising stone canal, ring canal, radial canals, and tube feet.

zygophiuroid. Type of articulation of vertebrae in some ophiuroids in which are several peg-and-socket joints that limit movement in horizontal plane between vertebrae (*syn.,* zygospondyline).

zygospondyline. See zygophiuroid.

REPRODUCTION AND ONTOGENY

Little is yet known of reproduction and ontogeny of somasteroids, but FELL has published a remarkable figure of a juvenile stage of the most primitive genus, *Chinianaster,* which shows many features resembling those of comatulid crinoids (Fig. 13).

Most Asteroidea are bisexual, but hermaphroditic individuals occur and some species are always hermaphrodites. Genital organs *(gonads)* vary in number and position from a pair in each interradius to large numbers arranged in two rows along each arm. In bisexual forms these gonads normally discharge ova in very large numbers or sperm into the water where fertilization occurs. Some asteroids, however, brood their young and these generally have rather few eggs; the method of fertilization is unknown. Certain cold-water species hatch the eggs under their arched body; some Astropectinidae brood young among the paxillae of the aboral surface; some Brisingidae brood them in cages of long spines between the bases of the arms; a few species even brood young in their stomachs. In Pterasteridae, a **supradorsal membrane** supported on tops of the paxillae forms the roof of a brood chamber.

Normally the embryo, when it escapes from the egg, develops a ciliated band and projections known as **larval arms**; this distinguishes the **bipinnaria stage.** Later, three short arms appear with a sucker between them **(brachiolaria stage).** The embryo attaches itself to some object by the sucker and metamorphoses into a minute star at one end, which then breaks free from or absorbs the remainder. The bipinnaria or the brachiolaria stages may be omitted in certain genera.

The skeleton begins to form as rods or spicules that expand to form flat plates with holes, typically 11 at first on the aboral surface (five terminals, five interradials, and one centrale). Later ossicles form between these original ones and push out the terminals so that they remain at the tips of the arms. In Goniasteridae the marginals are formed immediately proximal to the terminals. A series of very young Upper Cretaceous *Metopaster* is known from Denmark and England (55).

Many starfishes reproduce by fission and regeneration of the missing parts. *Linckia* (Ophidiasteridae) normally casts off single arms, which then regenerate the whole of the rest of the body.

Ophiuroidea are typically bisexual but some species are hermaphrodites. A few bisexual species have minute males permanently attached to the much larger females. Stenurida and Oegophiurida have gonads arranged serially inside the arms. More advanced forms have few to many gonads attached to the inner wall of the bursae; when ripe, they discharge into the bursae and thence through the bursal slits into the water. Some species brood their young in the bursae or the ovaries.

In forms in which the eggs are discharged into the sea, the embryo escapes from the egg in the **blastula stage,** much earlier than in Asteroidea. The free larva gradually develops into a **pluteus** similar to that of echinoids, with arms supported by skeletal rods. The hard and soft parts of the final stage gradually develop within the pluteus, the larval arms are absorbed, and the larva falls finally to the sea floor.

Reproduction by fission also occurs in some ophiuroids.

PHYLOGENY AND EVOLUTION

Recent work has demonstrated that the major subdivisions of sea stars appeared very early in geological history and that most of the higher taxa were extraordinarily long-lived (Fig. 38). Somasteroids and platyasterid asteroids persist to the present day, while the earliest known asteroids and ophiuroids are almost contemporary with the earliest somasteroids. One cannot therefore rule out the possibility that Stelleroidea had a long history in the Cambrian. If, however, as seems to be true, somasteroids are derived directly from crinoids, which are not yet known before the Ordovician, it is natural to assume that somasteroids originated no earlier than the Late Cambrian and that divergence from them of asteroids and ophiuroids was relatively rapid.

Only in the case of Platyasterida can we trace the course of this divergence in any detail. *Platanaster* differs little from some somasteroids, and even in its asteroid ambulacral furrow it is not really far from the condition seen in the living somasteroid *Platasterias*. Thereafter, Platyasterida are known only from the Devonian *Palasteriscus,* which seems to be merely a decalcifying offshoot of *Platanaster,* and by the Recent family Luidiidae. *Luidia* has diverged more fundamentally, by the development of strap-shaped arms and by telescoping and reduction of the metapinnules to form a pavement of squarish plates on the oral surface of the arms.

Paxillosida, Valvatida, and Forcipulatida are all present by the end of the Ordovician and at their first appearance they are very distinct from one another and from Somasteroidea. We do not yet know enough even to speculate usefully on the steps by which they diverged from Somasteroidea or its derivative, Platyasterida. There are undoubtedly whole groups of Late Cambrian and Early Ordovician sea stars of which so far we know nothing. However, the presence of superambulacral ossicles in many Paxillosida suggests their derivation from *Luidia*-like Platyasterida, a supposition supported by the regular occurrence in both groups of intermarginal fascioles. So far as morphology is concerned, it is conceivable that the other two suborders were derived from the same source, but there is nothing in geological occurrence to support this.

The Spinulosida include some Recent forms that retain suggestive somasteroid features, such as traces of metapinnular structure and interradial slots, and the suborder presumably, therefore, was derived from primitive Platyasterida or directly from the Somasteroidea.

The earliest known asteroid is, in fact, a member of the Paxillosida (Hemizonina)—namely, the Lower Ordovician *Petraster*. The Petrasteridae persisted into the Silurian, occurring in Europe, North America, and Australia. From Early Silurian to Triassic there is a widespread group, Palasterinidae, whose latest member, the Triassic *Trichasteropsis,* may by paedomorphy have given rise to the important stock of the Diplozonina, which first appeared in the Early Jurassic; the morphological gap, however, is considerable. Although the numerous Jurassic members can be distinguished generically from Recent forms, they all seem to belong to the Recent family Astropectinidae, which is abundant and widespread at the present day.

In living faunas two distinct groups generally are associated taxonomically with the Astropectinidae. They are here separated, following Fisher (17), as the suborders Cribellina and Notomyotina. The Cribellina show a number of apparently primitive characters, which in the Porcellanasteridae are combined with a highly specialized appearance. All members of the suborder are characterized by cribriform organs, a type of specialized fasciole, between some or all of the marginals. *Ctenodiscus* (Goniopectinidae) has channels between all marginals, with simple cribriform organs that consist of webbed spinelets, and the channels continue across the oral surface to the ambulacra. *Ctenodiscus* also has superambulacrals and true paxillae. These features suggest derivation from early Astropectinidae by development of the marginal fascioles in the direction of cribriform organs. Indeed, *Craspidaster* (Astropectinidae) differs from other members of its family in having the marginal channels covered by webbed spinelets; it thus neatly represents a transitional type between the Astropectinidae and Goniopectinidae. The Porcellanasteridae would then be derived from Ctenodiscinae by further specialization and localization of the cribriform organs, accompanying a change of ecology that allowed the general disappearance of spinulation of the body surface. If the Cribellina were thus derived from Diplozonia, one must assume that they reverted secondarily to having single ampullae. The only alternative possibility, an unlikely one, would be that they originated independently from early Somasteroidea and never passed through a stage with double ampullae. However, until early fossil representatives are found, one cannot speculate further about the suborder's history.

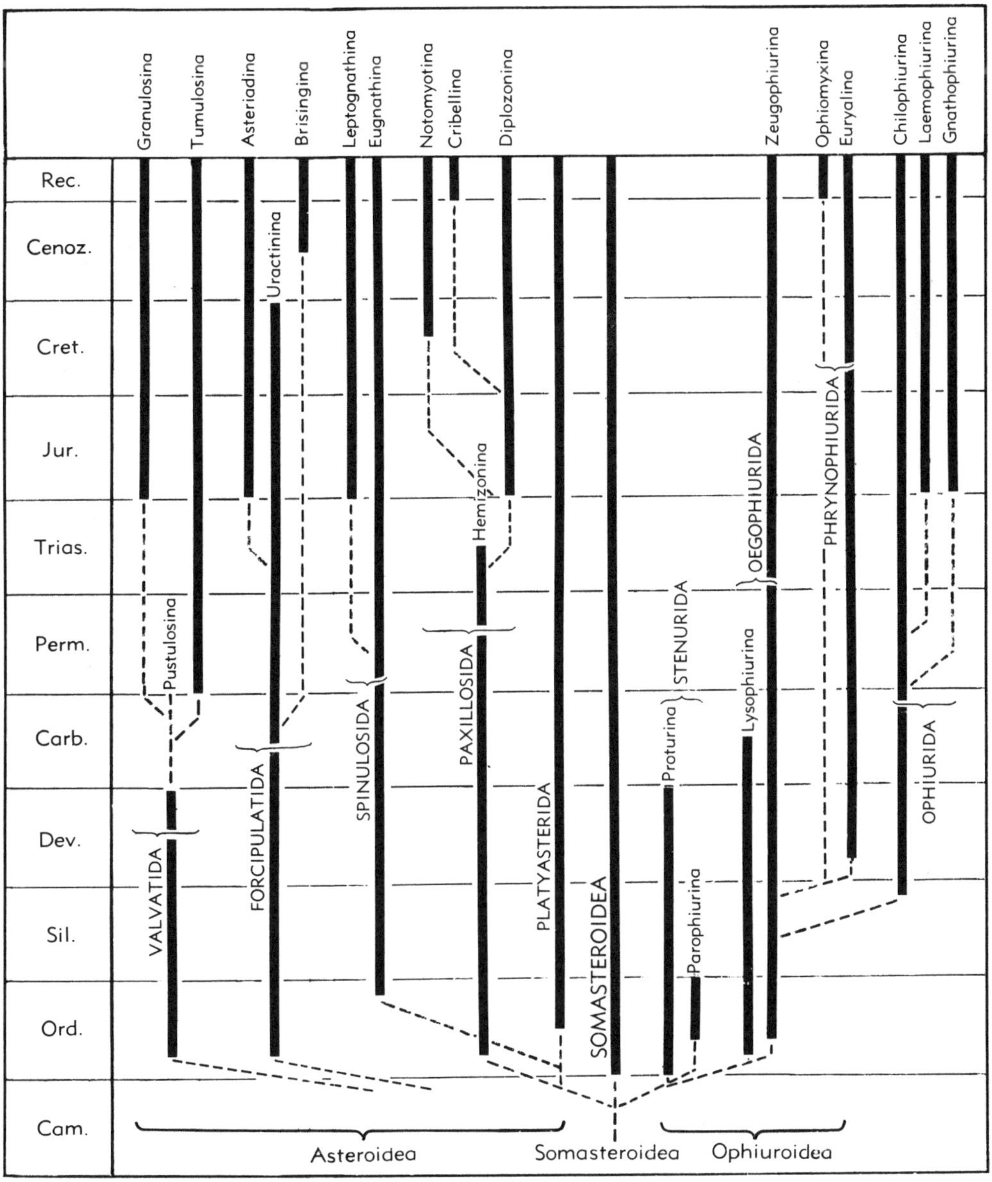

FIG. 38. Phylogeny of Asterozoa (134).

Notomyotina are characterized chiefly by longitudinal muscles in the arms; these with associated imbricate marginals are presumably an adaptation for swimming. The only known fossils consist of dissociated ossicles from the Cretaceous. They show close resemblance to certain living species. Thus, no useful discussion of phylogeny is possible, but it is likely that the suborder had a long history.

Species assigned to the order Valvatida provide a large proportion of known fossil Asteroidea, but, even so, our knowledge of the succession of species and genera is very limited. The first Valvatida (Palaeasteraceae) generally are small, having short wedge-shaped arms with inferomarginal frame extending to the arm tips; the aboral surface is composed of rather few large ossicles and is thus markedly paedomorphic. Such forms characterize the Middle and Upper Ordovician. Later forms tend to be larger and to lose the paedomorphic character of the aboral surface by introduction of accessory ossicles between the initial large ones; in addition, the bases of adjoining arms fuse and the large interbrachial axillary ossicles are occluded from the margin.

Promopalaeasteraceae have the inferomarginal frame limited and not extending to the arm tips; the arm becomes rounded in section, with the ambulacral segments compressed and the proximal ambulacrals in four rows, as in Forcipulatida. Presumably, this superfamily represents a continuation of the trends seen already in Palaeasteraceae. The Monasteridae, however, found only in the Permocarboniferous of Australia, have diverged in a different direction; here the adambulacrals are exceptionally wide and occupy most of the oral surface of the arms.

The Mesozoic Stauranderasteridae closely resemble *Monaster,* except that they have adambulacrals of more normal width. It would seem that the Stauranderasteridae are less aberrant descendants of the same group from which *Monaster* had diverged.

Early in the Jurassic the family Sphaerasteridae appeared, including forms with a closely fitting armature of ossicles and with no produced arms. The Jurassic *Sphaeraster* was hemispherical but an almost spherical genus survives to the present day. This group is probably derived from early Stauranderasteridae.

The Goniasteridae include many Mesozoic and Recent genera, whose general similarity is marked. They presumably originated in Palaeasteraceae, but there is a large gap in the late Paleozoic. The abundant Cretaceous forms allow recognition of some phyletic series which include links with a few Recent genera, but the detailed phylogeny of most of the family remains unknown. It was among Upper Cretaceous Goniasteridae that parallel evolution in several species series led to an orthogenetic interpretation. Many features that supported this interpretation are now known to be consequences of allometric growth, and there is no reason to postulate any process other than normal selection operating on a number of separate stocks in various niches in one broad environment.

The Oreasteridae are characterized by a high swollen disc, some with large tubercles or stout spines. They may have a superficial resemblance to certain Stauranderasteridae, but their young stages are flat and have inferomarginals and superomarginals like those of Goniasteriadae. They are thus probably derived from that family, but their coarsely reticulate aboral skeleton has diverged considerably from the tessellate one of the Goniasteridae.

The Ophidiasteridae are common in Recent seas but are known as fossils only from the Upper Cretaceous and Cenozoic. With their long cylindrical arms, they much resemble some Paleozoic forms, such as *Promopalaeaster,* but it is impossible to say if there is any direct connection.

The Spinulosida include a wide range of living forms that mostly have not evolved pedicellariae but tend to have groups of unmodified spinelets on their surface. This is a primitive feature, and the order seems to include a variety of stocks that separated at different times from the Platyasterida or even perhaps Somasteroidea. The Recent Tremasterinae retain metapinnular structure, and the Lower Jurassic Tropidasteridae show considerable resemblance to Palasteriscidae (Platyasterida).

Most of the known fossil Spinulosida belong to a suborder characterized by large spade-shaped mouth-angle plates, which persists from Ordovician to the present day. It includes several multiarmed genera, Recent and fossil, and already by the Silurian it had produced forms with reduced skeleton that apparently lived anchored by the jaws as suspension-feeders. The other suborder, characterized by small mouth-angle plates, includes some Recent forms of very primitive type but is virtually unrepresented by fossils.

The source of the Forcipulatida, all members of which have an "ambulacral" mouth frame, is unknown. They appeared early in the Ordovician. The Paleozoic suborder Uractinina mostly have very small discs and long narrow arms. They seem to be closely related to each other and to form in general a single evolving series. The only exception is the family Compsasteridae, which has spindle-shaped arms and somewhat resembles the Asteriidae (Asteriadina). However, there is no evidence among Compsasteridae of the pedicellariae which characterize the later suborder. The Compsasteridae and Asteriidae overlap in the Early Jurassic, and it may therefore be reasonable to derive the latter from the former. One stock of Uractinina, the Arthrasteridae, persisted until late in the Late Cretaceous.

The Brisingina, virtually unknown as fossils, comprise most peculiar forms. With their long narrow arms, small disc, and high axillary ossicle, they certainly resemble the Uractinina, and it is likely that they originated before the end of the Paleozoic.

The earliest order of Ophiuroidea, the Stenurida, includes forms that show a distinct metapinnular structure reminiscent of the Somasteroidea, but even the Pradesuridae (Early Ordovician) are quite distinct from Somasteroidea in their typically ophiuroid disc, covered with scales, and long slender arms. The order also includes forms that are losing the metapinnular aspect of the arms and tending toward typical ophiuroids in their arms structure.

The development of vertebral type of axial structure is carried further in Oegophiurida as early as the Ordovician and this order has recently been shown to persist to the present day (12). A single known Recent species retains a number of primitive "nonophiuroid" features.

Various specialized offshoots in both these early orders occur, particularly forms with globose or bag-shaped discs that seem to have been sessile suspension-feeders.

The remaining ophiuroids fall into two orders, Phrynophiurida and Ophiurida, with fully developed vertebrae of various types. Both these orders first appeared at the beginning of the Devonian and have persisted in large variety to the present day. They were derived presumably from different stocks within the Oegophiurida, but the detailed course of their evolution is not yet known. The Euryalina (Phrynophiurida) developed a type of vertebra that allowed vertical coiling of the arms and thus the ability to cling to other sessile organisms, particularly crinoids; this attribute is first seen in the Carboniferous. Ophiurida, on the other hand, became adapted for free movement on the sea floor, though many forms are actually sessile suspension-feeders. Although considerable evolutionary radiation of a minor sort has occurred in both groups, the earliest forms are much like modern ones, and there have been no changes of fundamental importance. Moreover, the fossil record is poor compared with the abundant and varied Recent fauna, most members of which probably have a long history.

CLASSIFICATION

LINNÉ in 1758 grouped all sea stars known to him in a single genus, *Asterias,* which was divided into three sections, of which the second, Stellatae, was equivalent to Asteroidea as now known, and the third, Radiatae, to Ophiuroidea plus Comatulae. The Stellatae corresponded to LINCK's (1733) Stellae fissae and the Radiatae to his Stellae integrae. LAMARCK (1801) recognized a "family" called les Stellérides, coordinate with his les Échinides; he established *Ophiura*, leaving asteroids and euryalids in *Asterias. Euryale* was set up by OKEN in 1815. LAMARCK in 1816 within an order termed radiaires échinodermes distinguished a section (les Stellérides) that included comatules, euryales, ophiures, and asteries. He thus finally separated Asteroidea and Ophiuroidea. By 1835 AGASSIZ had proposed Stellérides as an order of echinoderms, including in it the two families Asteriadae and Ophiuridae. Two years later BURMEISTER named the combined group Asteroidea, of which Hypostoma GRAY, 1840 (as a class), is a synonym. FORBES (1840), first in post-Linnean times, named Ophiuroidea at the suprafamilial level.

ZITTEL (1879), largely following BRONN (1860), divided a class Asteroidea into two orders, Ophiuridae, with suborders Euryaleae and Ophiureae, and Stelleridae, with suborders Encrinasteriae and "Asteriae verae" (91). The Encrinasteriae included a group of fossil forms, mainly Devonian, with marginals and stout petaloid arms. SCHÖNDORF (62) divided this group into two parts, one including forms with opposite ambulacrals, which he referred to Asteroidea, and the other with alternate ambulacrals, which he thought could not be referred either to Asteroidea or Ophiuroidea and therefore assigned to a new subclass, Auluroidea. Subsequently SPENCER (1930) showed that these forms had good ophiuroid characters and so he abandoned SCHÖNDORF's subclass. ZITTEL's "Asteriae verae" were long known as Euasteriae (BRONN's term) or Euasteroidea, in contrast with Encrinasteriae.

LUDWIG demonstrated that the Asteroidea and Ophiuroidea were built on a common plan, and he found in the ontogeny of Ophiuroidea what he regarded as an asteroid stage (42).

GREGORY (1899) established a class Stelleroidea with subclasses Asteroidea and Ophiuroidea: he recalled (25, p. 238) that these taxa are usually ranked as distinct classes but stated correctly that "no definite line of separation can be drawn between them" and that they are "constructed upon the same fundamental plan," "contain the same variations from the typical arrangement," and have "not a single constant difference between them." This is the view adopted in the *Treatise* and generally accepted by palaeontologists. Some neontologists, however, have argued for a wider separation of Asteroidea and Ophiuroidea, primarily on embryological grounds. The facts of palaeontology and also the occurrence among living forms of certain Ophiuroidea with asteroid characters outweigh the embryological evidence. That ophiuroid larvae at certain stages differ from asteroid larvae and resemble those of echinoids is presumably due to separate evolution of the pelagic larvae of Ophiuroidea (on which the pressures of selection operate just as much as on the adults) and convergence with echinoid larvae. Some biochemical evidence suggests closer affinity between ophiuroids and echinoids than between ophiuroids and asteroids (36, p. 700); but even if this is found to be valid in wider investigation it still would not outweigh the morphological and paleontological evidence.

Finally, SPENCER (77) based a third subclass, Somasteroidea, on certain Lower Ordovician sea stars and demonstrated that they were ancestral both to Asteroidea and to Ophiuroidea. FELL subsequently recognized a living species of somasteroid (11) and put the relationship of the three subclasses on a firm basis (13).

Subdivision of Asteroidea above family level began in 1875, when PERRIER established two unnamed sections, one with stalked and straight or crossed pedicellariae and quadriserial tube feet (family Asteriidae), the other with sessile pincer-shaped or valvate pedicellariae and biserial tube feet (six other families).

VIGUIER (1878) had two subclasses of a class Asteroidea based on nature of the mouth ring (86). The first, "Asteries ambulacraires," was characterized by predominance of the proximal functional ambulacral plates in the mouth ring and feebleness of the mouth-angle plates, by stalked, straight, or crossed pedicellariae, and by quadriserial tube feet (families Asteriidae, Heliasteridae, Brisingidae), the second by the predominance of adambulacral plates, to which the mouth-angle plates are assimilated, in the mouth ring, by sessile pincer-shaped or valvate pedicellariae and by biserial tube feet (seven families).

PERRIER in 1884 regarded the pedicellariae, although on fallacious grounds, as more important than other characters for classification and amplified his previous scheme (53). He divided the Asteroidea into four orders according to characters of the pedicellariae as follows: Forcipulatae (families Brisingidae, Pedicellasteridae, Asteriidae, Heliasteridae), Spinulosae (Echinasteridae, Pterasteridae, Asterinidae), Valvatae (Linckiidae, Goniasteridae, Asteropsidae), and Paxillosae (Archasteridae, Astropectinidae).

SLADEN (1889) rejected PERRIER's classification and established only two orders: Phanerozonia characterized by conspicuous marginals, and Cryptozonia, with marginals reduced or absent in the adult. Since some Cryptozonia have a phanerozonate stage in ontogeny, he regarded Phanerozonia as the more primitive order (67). A number of families cannot be definitely assigned to the Phanerozonia or Cryptozonia and in practice SLADEN's classification has been combined in various ways with PERRIER's. FISHER, for example, maintained the Phanerozonia to include PERRIER's Paxillosae and Valvatae and ranked it with Spinulosae and Forcipulatae (17).

The classification adopted for Asteroidea in the *Treatise* is essentially a combination of that of PERRIER and VIGUIER, modified in the light of paleontological evidence and the work of FELL. The four Recent orders of PERRIER are seen on a variety of evidence, including characters of the mouth ring and mouth-angle plates, to have their roots in Paleozoic families. The Luidiidae, however, are Recent survivors of the Paleozoic order Platyasterida, which represent the simplest modification of Somasteroidea to asteroid status. The Asteroidea are thus divided into five orders, Platyasterida, Paxillosida, Valvatida, Spinulosida, and Forcipulatida.

Above family level, the Ophiuroidea were split by most 19th-century authors into two divisions, ophiurids and euryalids, variously graded. BELL (1892), however, was struck by the importance of the system of articulation of the vertebrae. Accordingly, he divided the Ophiuroidea into three groups, Streptophiurae, with simple ball-and-socket articulation; Cladophiurae, with hour-glass-shaped articulatory surfaces; and Zygophiurae, in which free lateral movement of the arms was limited by processes and pits at sides of the vertebrae (1). GREGORY (1897) added Lysophiurae for Paleozoic form, with a double series of alternating ambulacrals instead of vertebrae (24).

JAEKEL, in 1923, erected a class "Brachioidea," divided into two subclasses, Parophiura, for certain early Paleozoic forms, and Ophiura (38). However, MATSUMOTO (1915) proposed an entirely new classification based on internal skeletal structures (45). Within the class Ophiuroidea he established two subclasses, Oegophiuroidea, for a group of Paleozoic forms, and Myophiuroidea, for the remainder. The latter subclass contained four orders, Phrynophiurida, Laemophiurida, Gnathophiurida, and Chilophiurida.

SPENCER (1951) recognized the subclass Ophiuroidea containing an order, Stenurida, with two suborders, co-ordinate with an order Ophiurida for the rest; within the Ophiurida he included as suborders MATSUMOTO's Oegophiuroidea and Myophiuroidea (77).

FELL (1962) has shown that MATSUMOTO's Oegophiuroidea and Phrynophiuroidea should stand as orders co-ordinate with the Stenurida, while MATSUMOTO's other orders could, if accepted, best be regarded as suborders of the Ophiurida (of which MATSUMOTO's Myophiuroidea is really a synonym, 12). This is the classification adopted in the *Treatise*. However, MURAKAMI has recently (1963) published a classification of extant ophiuroids derived from that of MATSUMOTO but based primarily on details of the jaw structure (52). He has distinguished the following orders and suborders: Phrynophiurida, Laemophiurida, Gnathophiurida, and Chilophiurida, with three suborders set up by himself in 1947, Trematophiurina, Holophiurina, and Agmatophiurina. Whether this rearrangement will find general acceptance remains to be seen.

OUTLINE OF CLASSIFICATION

The figures in parentheses indicate numbers of included genera. Where there is no oblique stroke, the figure represents fossil genera, some of which may have Recent species. Where there is an oblique stroke, the figure before it represents genera known as fossils (possibly including Recent species), while the figure after the oblique stroke represents Recent genera with no known fossil species. Figures after a colon (:) indicate numbers of subgenera exclusive of nominotypical subgenera.

Asterozoa *(subphylum)* (182/556:14). *L.Ord.-Rec.*
- Stelleroidea *(class)* (182/556:14). *L.Ord.-Rec.*
  - Somasteroidea *(subclass)* (7/1). *L.Ord.-Rec.*
    - Goniactinida *(order)* (7/1). *L.Ord.-Rec.*
      - Chinianasteridae (1). *L.Ord.*
      - Villebrunasteridae (2). *L.Ord.*
      - Platasteriidae (/1). *Rec.*
      - Archegonasteridae (1). *L.Ord.*
      - Archophiactinidae (3). *U.Ord.-U.Dev.*
  - Asteroidea *(subclass)* (111/288:12). *L.Ord.-Rec.*
    - Platyasterida *(order)* (3). *M.Ord-Rec.*
      - Palasteriscidae (2). *M.Ord.-L.Dev.*
      - Luidiidae (1). *Mio.-Rec.*
    - Paxillosida *(order)* (13/41:2). *L.Ord.-Rec.*
      - Hemizonina *(suborder)* (6). *L.Ord.-Trias.*
        - Petrasteridae (1). *L.Ord.-Sil.*
        - Lepidasteridae (2). *M.Sil.-U.Dev.*
        - Palasterinidae (3). *Sil.-Trias.*

Diplozonina *(suborder)* (7/20:2). *L.Jur.-Rec.*
Astropectinidae (7/20:2). *L.Jur.-Rec.*
Astropectininae (7/19:2). *L.Jur.-Rec.*
Craspidasterinae (/1). *Rec.*
Cribellina *(suborder)* (/13). *Rec.*
Goniopectinidae (/4). *Rec.*
Goniopectininae (/2). *Rec.*
Ctenodiscinae (/2). *Rec.*
Porcellanasteridae (/9). *Rec.*
Notomyotina *(suborder)* (1/8). *L.Cret.-Rec.*
Benthopectinidae (1/8). *L.Cret.-Rec.*
Valvatida *(order)* (69/93:2). *L.Ord.-Rec.*
Pustulosina *(suborder)* (27). *L.Ord.-L.Carb.*
Palaeasteraceae *(superfamily)* (22). *L.Ord.-L. Carb., ?Permocarb.*
Palaeasteridae (2). *M.Sil., ?Permocarb.*
Hudsonasteridae (6). *L.Ord.-U.Sil.*
Hudsonasterinae (4). *L.Ord.-U.Ord.*
Coccasterinae (1). *M.Sil.-U.Sil.*
Silurasterinae (1). *M.Ord.*
Neopalaesteridae (1). *L.Carb.*
Mesopalaesteridae (6). *U.Ord.-U.Dev.*
Mesopalaeasterinae (3). *U.Ord.-U.Dev.*
Lepidactininae (2). *M.Sil.-L.Dev.*
Clarkeasterinae (1). *U.Dev.*
Xenasteridae (5). *L.Dev.*
Family Uncertain (2).
Promopalaeasteraceae *(superfamily)* (5). *M. Ord.-Sil.*
Promopalaeasteridae (3). *M.Ord.-Sil.*
Eoactinidae (2). *Sil.*
Tumulosina *(suborder)* (6/1). *Permocarb.-Rec.*
Monasteridae (1). *Permocarb.*
Stauranderasteridae (3). *M.Jur.-U.Cret.*
Sphaerasteridae (2/1). *M.Jur.-Rec.*
Granulosina *(suborder)* (36/92:2). *L.Jur.-Rec.*
Odontasteridae (1/5). *M.Jur.-Rec.*
Chaetasteridae (/1). *Rec.*
Archasteridae (1). *?Mio., Rec.*
Goniasteridae (30/44:2). *L.Jur.-Rec.*
Goniasterinae (5/11:1). *U.Cret.-Rec.*
Chitonasterinae (/1). *Rec.*
Athenoidinae (/5). *Rec.*
Hippasteriinae (1/2:1). *U.Cret.-Rec.*
Nectriinae (/2). *Rec.*
Pseudarchasterinae (2/2). *U.Jur.-Rec.*
Pycinasterinae (2). *L.Jur.-Mio.*
Subfamily Uncertain (20/22).
Oreasteridae (/20). *Rec.*
Ophidiasteridae (4/19). *U.Cret.-Rec.*
Radiasteridae (/2). *Rec.*
Spinulosida *(order)* (15/72:2). *M.Ord.-Rec.*
Eugnathina *(suborder)* (13/23:2). *M.Ord.-Rec.*
Taeniactinidae (4). *U.Ord.-L.Carb.*
Lepyriactinidae (1). *L.Sil.*
Schuchertiidae (1). *M.Ord.-Sil.*
Helianthasteridae (3). *Dev.*
Solasteridae (2/7). *L.Jur.-Rec.*
Tropidasteridae (1). *L.Jur.*
Korethrasteridae (/4). *Rec.*
Pythonasteridae (/3). *Rec.*
Pythonasterinae (/1). *Rec.*
Myxasterinae (/2). *Rec.*
Pterasteridae (/9:2). *Rec.*
Leptognathina *(suborder)* (2/49). *L.Jur.-Rec.*
Asterinidae (/17). *M.Jur.-Rec.*
Asterininae (/12). *Rec.*
Anseropodinae (/3). *Rec.*
Tremasterinae (/2). *M.Jur., Rec.*
Ganeriidae (/8). *Rec.*
Poraniidae (/13). *Rec.*
Echinasteridae (1/1). *?U.Cret., Rec.*
Valvasteridae (1/1). *L.Jur.-Rec.*
Acanthasteridae (/1). *Rec.*
Mithrodiidae (/1). *Rec.*
Metrodiridae (/1). *Rec.*
Forcipulatida *(order)* (11/82:6). *L.Ord.-Rec.*
Uractinina *(suborder)* (10). *L.Ord.-U.Cret.*
Cnemidactinidae (1). *Ord.*
Urasterellidae (5). *L.Ord.-Permocarb.*
Calliasterellidae (3). *L.Carb.-U.Cret.*
Protarthrasterinae (1). *L.Carb.*
Calliasterellinae (1). *U.Carb.*
Arthrasterinae (1). *U.Cret.*
Compsasteridae (1). *L.Dev.-L.Jur.*
Asteriadina *(suborder)* (1/65:6). *L.Jur.-Rec.*
Heliasteridae (/1). *Rec.*
Zoroasteridae (/7). *Rec.*
Asteriidae (1/57:6). *?L.Jur., M.Jur.-Rec.*
Asteriinae (1/45:6). *?L.Jur., M.Jur.-Rec.*
Pedicellasterinae (/5). *Rec.*
Labidiasterinae (/4). *Rec.*
Pycnopodiinae (/2). *Rec.*
Neomorphasterinae (/1). *Rec.*
Brisingina *(suborder)* (/17). *L.Oligo.-Rec.*
Brisingidae (/17). *L.Oligo.-Rec.*
Ophiuroidea *(subclass)* (63/266:2). *L.Ord.-Rec.*
Stenurida *(order)* (10). *L.Ord.-U.Dev.*
Proturina *(suborder)* (6). *L.Ord.-U.Dev.*
Pradesuridae (2). *L.Ord.-L.Dev.*
Phragmactinidae (1). *U.Ord.*
Rhopalocomidae (2). *U.Sil.-U.Dev.*
Bdellacomidae (1). *U.Sil.-L.Dev.*
Parophiurina *(suborder)* (4). *L.Ord.-L.Dev.*
Eophiuridae (1). *L.Ord.*
Palaeuridae (2). *L.Ord.-L.Dev.*
Stenasteridae (1). *M.Ord.-U.Ord.*
Oegophiurida *(order)* (21/1). *L.Ord.-Rec.*
Lysophiurina *(suborder)* (15). *M.Ord.-L.Carb.*
Encrinasteridae (7). *U.Ord.-L.Carb.*
Protasteridae (8). *M.Ord.-L.Carb.*
Zeugophiurina *(suborder)* (6/1). *L.Ord.-Rec.*
Lapworthuridae (3). *L.Ord.-L.Dev.*
Furcasteridae (2). *U.Ord.-L.Carb.*
Klasmuridae (1). *U.Dev.*
Ophiocanopidae (1). *Rec.*
Phrynophiurida *(order)* (4/69). *L.Dev.-Rec.*
Ophiomyxina *(suborder)* (/23). *Rec.*
Ophiomyxidae (/23). *Rec.*
Ophiomyxinae (/16). *Rec.*
Ophiobyrsinae (/7). *Rec.*
Euryalina *(suborder)* (4/46). *L.Dev.-Rec.*
Eospondylidae (2). *L.Dev.*
Onychasteridae (1). *Miss.*
Asteronychidae (1/1). *?U.Cret., Rec.*
Asteroschematidae (/6). *Rec.*
Gorgonocephalidae (/33). *Oligo.-Rec.*
Euryalidae (/6). *Rec.*
Ophiurida *(order)* (26/194:2). *Sil.-Rec.*
Chilophiurina *(suborder)* (18/90:2). *Sil.-Rec.*
Ophiurinidae (5). *Sil.-L.Carb.*
Ophiuridae (10/48). *L.Carb.-Rec.*
Aganasterinae (1). *L.Carb.*
Ophiurinae (7/30). *L.Carb.-Rec.*
Ophiolepidinae (2/18). *?Perm., Rec.*

Ophioleucidae (/8). *Rec.*
Ophiocomidae (1/6). *?U.Cret., Rec.*
Ophionereididae (/5:2). *Rec.*
Ophiodermatidae (2/23). *L.Jur.-Rec.*
Laemophiurina *(suborder)* (4/45). *L.Jur.-Rec.*
Ophiacanthidae (3/37). *L.Jur.-Rec.*
Hemieuryalidae (2/8). *?L.Jur., Rec.*
Gnathophiurina *(suborder)* (4/59). *?L.Jur., Rec.*
Amphilepididae (/2). *Rec.*
Ophiactidae (/5). *Rec.*
Amphiuridae (3/36). *U.Cret.-Rec.*
Ophiothricidae (1/16). *?L.Jur., Rec.*
Suborder and family Uncertain (1).

SYSTEMATIC DESCRIPTIONS

Subphylum ASTEROZOA Zittel, 1895

Echinoderms characterized by generally depressed star-shaped body, composed of central disc with mouth on underside and symmetrical radiating arms; axial skeleton along arms protects radial water vessels and nerves; tube feet normally confined to lower side of body. *L.Ord.-Rec.*

In addition to genera known as fossils, Recent genera are listed with author, date, type-species, and synonymy, but generally diagnoses of genera not yet found fossil are omitted, despite the probability that most Recent asterozoan genera are of considerable geological age. Many fossil asterozoans have been referred to extant genera. Not all of these attributions are justified and some have been omitted here.

Class STELLEROIDEA Lamarck, 1816

[*nom. transl. et correct.* Gregory, 1900 (*pro* les Stellérides Lamarck, 1816)]

Characters of subphylum. *L.Ord.-Rec.*

Subclass SOMASTEROIDEA Spencer, 1951

Asterozoans with oral surface bearing shallow radial channels formed by recumbent ambulacrals, which in at least some forms can be raised to form temporary ambulacral furrows; tube feet seated in broad basins, which may or may not communicate with body cavity; radial water vessel enclosed to varying extent between ambulacrals. Axial skeleton consisting of ambulacrals in double series, generally in opposite pairs but apparently alternating in some forms, each ambulacral typically giving rise to transverse series of ossicles (**metapinnules**), consisting of more or less rod-like elements (**virgals**), which may be undifferentiated (Chinianasteridae) or differentiated into adambulacral, intermediate, and marginal ossicles; between metapinnules are food-groove channels on oral surface covered with small plates or protected by spines; jaws composed of proximal pairs of ambulacrals and their reduced metapinnules; aboral surface typically bearing paxillae with tetraradiate bases (11, 13, 77). *L.Ord.(Tremadoc.)-Rec.*

Order GONIACTINIDA Spencer, 1951

Characters of subclass. *L.Ord.-Rec.*

Family CHINIANASTERIDAE Spencer, 1951

Ambulacral skeleton of stout barrel-shaped ossicles, not forming any groove; tube feet exclusively external, set in cupules, pointed and covered with minute plates (as in Ophiocistioidea). Arms petaloid, formed from simple metapinnules which end in free marginal radiole; undifferentiated virgals flanged, carrying row of small plates on either side which cover channels. Mouth-angle plates subtriangular, elongate. Buccal slits extending into arm bases. Aboral surface with widely spaced paxillae. *L.Ord. (Tremadoc.)-L.Ord.(Arenig.).*

Chinianaster Thoral, 1935 [**C. levyi*; OD]. Characters of family. *L.Ord.*, S.Fr.——Fig. 39,*4*. **C. levyi*; oral surface (reconstr.), ×3.5 (133). (See also Fig. 8,*2*; 13.)

Family VILLEBRUNASTERIDAE Fell, 1963

Tube feet not plated, with internal ampulla; cupules communicating with internal cavity between wings on ambulacrals; water tube enclosed by capitula of ambulacrals; virgals differentiated into adambulacral, oral intermediate, and marginal elements; metapinnules of distal part of arms ending

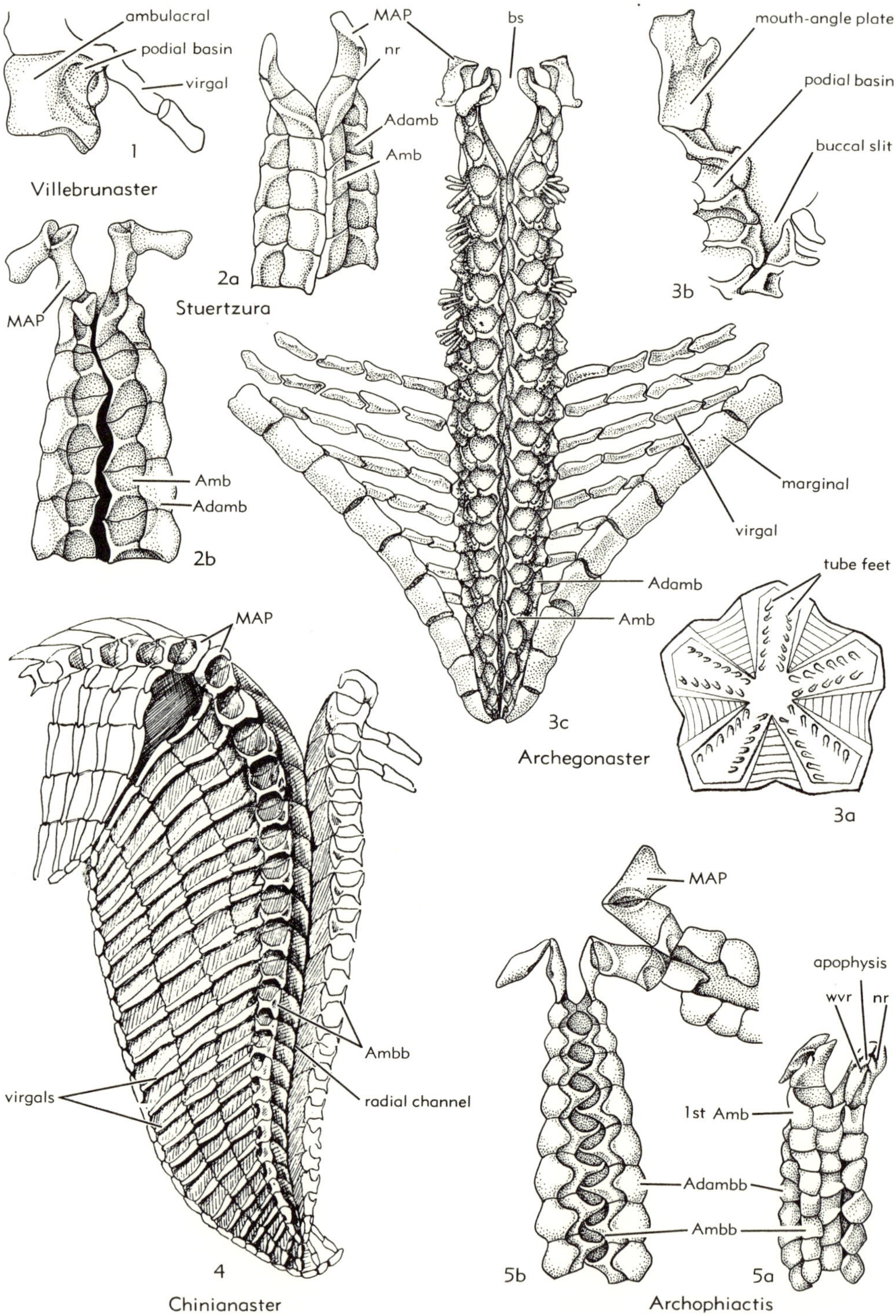

FIG. 39. Chiniasteridae *(4)*; Villebrunasteridae *(1)*; Archegonasteridae *(3)*, Archophiactinidae *(2,5)*. [Explanation: *Adamb*, adambulacrum; *Amb(b)*, ambulacrum; *bs*, buccal slit; *MAP*, mouth-angle plate; *nr*, nerve-ring groove, *wvr*, water-vascular-ring groove.] (p. U39, U41-U42).

in distinct marginals without terminal radioles. Intermediate virgals forming walls of food grooves, apparently without cover plates. *L.Ord.*

Villebrunaster SPENCER, 1951 [**V. thorali*; OD]. Arms petaloid. Metapinnules of slender virgals, undifferentiated except for marginals on distal part of arms and adambulacrals; ambulacrals changing in shape along arms. *L.Ord.*, S.Fr.——FIG. 39,*1*. **V. thorali*; ambulacrals showing podial basin and virgals, enl. (133). (See also Fig. 8.)

Ampullaster FELL [**A. ubaghsi*; OD]. Body more or less pentagonal, with rhombic arms separated by deep, narrow interradial clefts. Proximal metapinnules of slender virgals except for adambulacrals; distal ones of few strong virgals of which outermost are marginals. *L.Ord.*, S.Fr.——FIG. 40,*1*. **A. ubaghsi*; oral view, ×5 (108).

Family PLATASTERIIDAE Caso, 1945

Arms petaloid, separated by deep interradial fissures. Metapinnules consisting of virgals differentiated into adambulacral, occluded superambulacral, marginal, and terminal elements; adambulacrals and marginals forming walls of interpinnular grooves, covered by erectile series of small plates on either side; ambulacrals with adambulacrals erectable to form temporary ambulacral furrow of asteroid type. *Rec.*

Platasterias GRAY, 1871 [**P. latiradiata*; OD]. Characters of family. *Rec.*, Nicaragua.

Family ARCHEGONASTERIDAE Spencer, 1951

More or less pentagonal in outline with no interradial clefts; continuous series of marginals running along edge of body and adambulacrals along shallow radial grooves but metapinnules otherwise reduced to few rows of virgals near distal ambulacrals. Capitula of ambulacrals imbricating. *L.Ord. (U.Arenig.).*

A few specimens are preserved as horizontally flattened pentagons but most are elongated vertically, with distal ambulacrals flexed over the aboral surface so that upward stretched tube feet could grasp food (Fig. 39,*3a*). The mouth frame, as in primitive ophiuroids, has deep clefts at summit of mouth-angle plates, which indicate position of muscles for digging; the first pair of buccal tube feet projected directly into the mouth. The reduction of ossicles denotes transition to the Archophiactinidae.

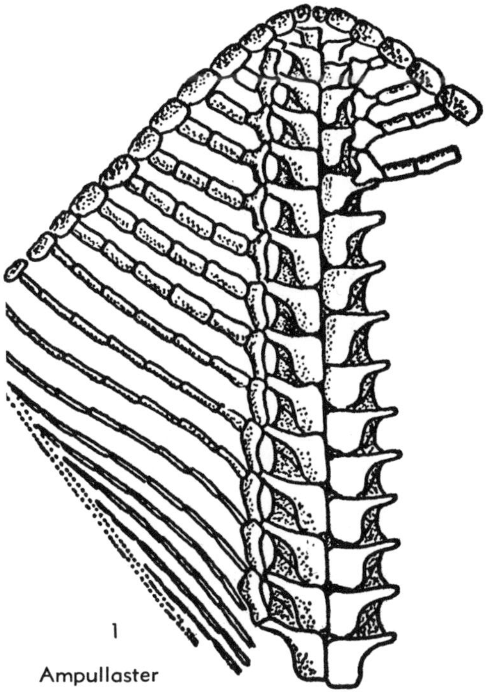

FIG. 40. Villebrunasteridae (p. *U*41).

Ball-and-socket joints between axial ossicles allowed flexing when the animal assumed a feeding posture.

Archegonaster JAEKEL, 1923 [**A. pentagonus* SPENCER, 1951; SM]. *L.Ord.(U.Arenig.)*, Czech.——FIG. 39,*3*. **A. pentagonus*; *3a*, diagram of arms flexed over aboral surface in feeding posture; *3b*, part of mouth frame from inner side, ×3; *3c*, part of oral surface reconstructed, ×3 (133).

Family ARCHOPHIACTINIDAE Spencer, 1927

Adaxial skeleton reduced to adambulacrals only. No buccal slits. *U.Ord.-U.Dev.*

The aboral surface was probably much swollen and nearly devoid of skeleton. The family was probably sessile.

Archophiactis SPENCER, 1925 [**A. grayae*; OD]. Adambulacrals broad; aboral surface of ambulacrals rounded; mouth-angle plates elongate, stout. *U.Ord.*, Scot.——FIG. 39,*5*. **A. grayae 5a,b*, aboral and oral surfaces of proximal part of arm and part of mouth frame, ×5 (133).

Stuertzura GREGORY, 1897 [**Protaster brisingoides* GERGORY, 1889; OD] [=*Stürtzura* GREGORY, 1897]. Adambulacrals narrow; aboral surface of ambulacrals ridged; mouth-angle plates long, narrow. *Sil.*, Australia.——FIG. 39,2. **S. brisingoides* (GREGORY); *2a,b*, aboral and oral surfaces of proximal part of arm and part of mouth frame, ×5 (133).

Lepidasterina RUEDEMANN, 1916 [**L. gracilis*; OD]. Multiarmed, with ossicles of radial groove as in *Stuertzura*. *U.Dev.*, USA(N.Y.).

Subclass ASTEROIDEA de Blainville, 1830

[*nom. transl.* GREGORY, 1900, p. 239 (*ex* order Asteroidea BURMEISTER, 1837, p. 467, *nom. correct. pro* order Asteridea DE BLAINVILLE, 1830, p. 216] [=order Astroides DE BLAINVILLE, 1822; order Asteriodea FLEMING, 1822, order Cirrhigrada FORBES, 1841; class Asterioidea BRONN, 1860, p. 240] [Diagnosis prepared by W. K. SPENCER & C. W. WRIGHT. Research on authorship and synonymy of subclass by H. B. FELL & J. WYATT DURHAM]

Asterozoans having relatively broad arms with considerable hollow space within ossicular frame; arms normally not separated from central disc; oral side bearing open ambulacral grooves which carry rows of tube feet, proximal tube feet in some early stocks forming buccal tentacles. Respiration through skin of aboral surface, which is commonly folded into external gills (papulae). Spines or granules generally well distributed over surface both of skin and of bare ossicles. Later genera commonly carrying pedicellariae. *L.Ord.-Rec.*

Order PLATYASTERIDA Spencer, 1951

Arms 5 to many; ossicles in transverse gradients emerging from axial ossicles; ambulacrals erect and ambulacra furrowed; single row of marginals, when present, channeled so that marginal grooves link with vestigial food grooves on oral surface, which persist as respiratory fascioles (13). *M.Ord.-Rec.*

Family PALASTERISCIDAE Gregory, 1900

[=Palaechinasteridae STÜRTZ, 1890 (invalid because not founded on generic name); Platanasteridae SPENCER, 1919]

Arms 5, adambulacrals very broad, with flat spines on long transverse ridge; ambulacral grooves very shallow, aboral surface swollen, with many parallel rows of paxillae; axillary broad, breastplate-shaped. *M.Ord.-L.Dev.*

Platanaster SPENCER, 1919 [**P. ordovicus*; OD]. Single row of marginals present. *M.Ord.*, Eng. (Shrops.).——FIG. 41,*1*; 42,*1*. **P. ordovicus*; 41,*1a*, oral side of arm, ×2; 41,*1b*, part of same, ×6; 42,*1*, aboral surface, ×1 (133). (See also Fig. 12,*1*.)

Palasteriscus STÜRTZ, 1886 [**P. devonicus*; OD]. Like *Platanaster* but lacking marginals. *L.Dev.*, Ger.——FIG. 41,*2*; 42,*2*. **P. devonicus*; 41,*2*, apical view of ambulacrals showing pegs, ×5; 42,*2*, aboral surface, ×0.5 (133).

Family LUIDIIDAE Verrill, 1899

Arms 5 to many, normally strap-shaped, adambulacrals, oral intermediate ossicles, and inferomarginals in regular transverse series as in *Platanaster*, recalling metapinnules; superambulacrals present. Aboral surface covered with paxillae not in wholly regular rows. *Mio.-Rec.*

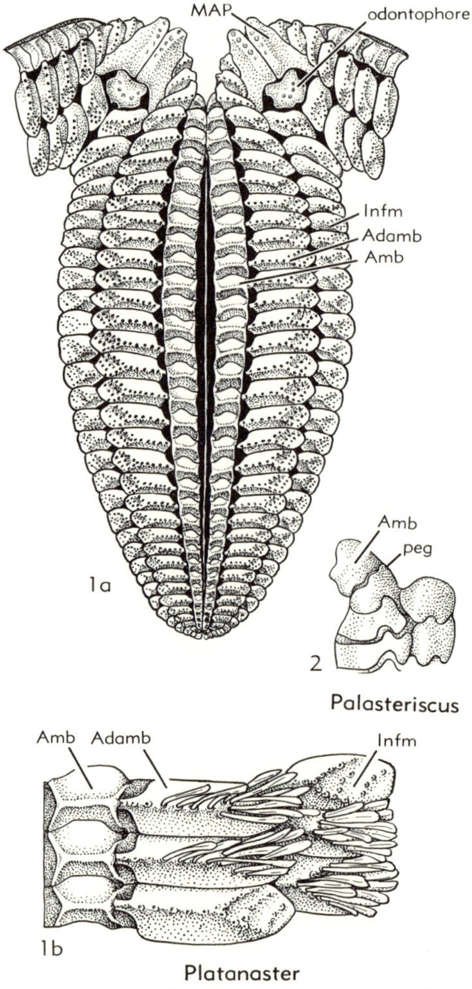

FIG. 41. Palasteriscidae. [Explanation: *Adamb*, adambulacrum; *Amb*, ambulacrum; *Infm*, inferomarginal; *MAP*, mouth-angle plate.] (p. *U*42).

Luidia Forbes, 1839 [**L. fragillissima* (=**Asterias ciliaris* Philippi, 1837); OD ICZN Opin. 129] [=*Hemicnemis* Müller & Troschel, 1840; *Petalaster* Gray, 1840; *Luydia* Düben & Koren, 1847; *Astrella* Perrier, 1882; *Integraster, Quinaster, Penangaster, Denudaster, Senegaster, Alternaster, Armaster, Maculaster* Döderlein, 1920]. Except for *Hemicnemis, Luydia,* and *Armaster,* synonyms may all be justifiable names for subgenera. *Mio.,* Hung.; *Rec.*——Fig. 42,*3*. *L. hungarica* Rakusj; oral surface of arm showing transverse gradients of ambulacrals, adambulacrals, oral intermediate ossicles, and inferomarginals, ×5 (124).

Order PAXILLOSIDA Perrier, 1884

[*no. correct.* Spencer & Wright, herein (*pro* Paxillosa)]

Mouth frame adambulacral, mouth-angle plates prominent, in many forms with keel and median furrow, marginal frame (when present) separated from mouth frame by interradial area with small ossicles. No transverse gradients. Interradial arc even, without axillary. Ambulacral areas never compressed. Tube feet in 2 rows. *L.Ord.-Rec.*

Many of this order have marginals channeled to form more or less specialized intermarginal fascioles, normally connecting with furrows on the oral side, presumably to conduct respiratory water from aboral to oral surface.

Suborder HEMIZONINA Spencer, 1951

[*nom. transl. et correct.* Spencer & Wright, herein (*pro* Hemizonida)] [=Gnathasterina Spencer, 1951]

No superomarginals or only on arms. Aboral surface generally with well-developed stellate ossicles. *L.Ord.-Trias.*

Family PETRASTERIDAE Spencer, 1951

[=Uranasteridae Spencer, 1916]

Arms 5; inferomarginals well developed; axillary area on oral surface with small ossicles or granules. *L.Ord.-Sil.*

Petraster Billings, 1858 [**Palasterina rigidus* Billings, 1857; OD] [=*Uranaster* Gregory, 1899]. Characters of family. *L.Ord.-Sil.,* Wales-Eire-Czech.-Australia.——Fig. 43,*4*. *P. kinahani* (Baily), L.Ord., Wales; *4a,* part of aboral surface of arm, ×4; *4b,* oral surface of arm and mouth region, ×2; *4c,* part of aboral surface, including aboral view of inferomarginals, adambulacrals, and mouth-angle plates, ×4 (133).

Family LEPIDASTERIDAE Gregory, 1899

Arms many, inferomarginals well devel-

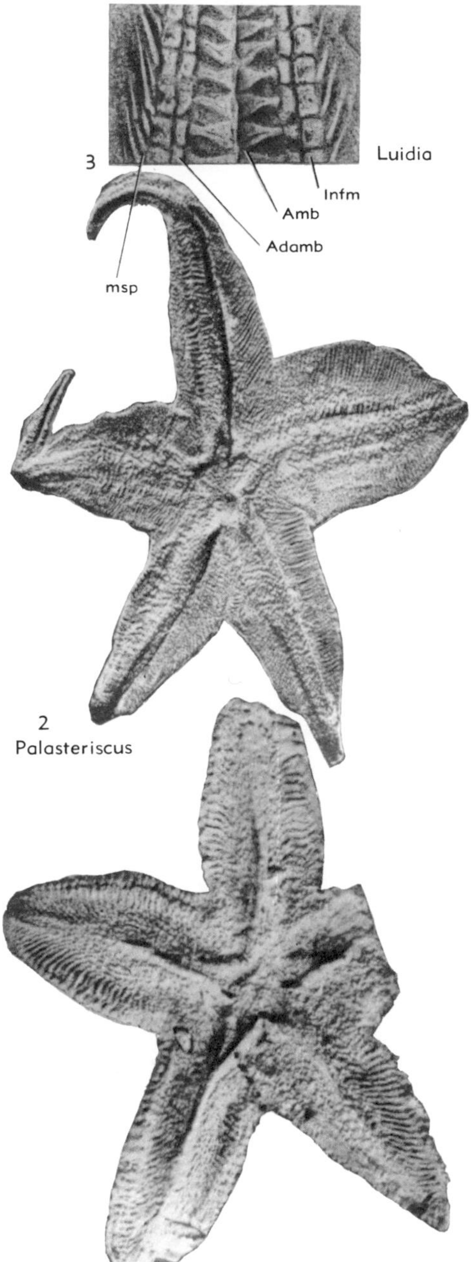

Fig. 42. Palasteriscidae *(1,2)*; Luidiidae *(3)*. [Explanation: *Adamb*, adambulacrum; *Amb*, ambulacrum; *Infm*, inferomarginal; *MAP*, mouth-angle plate; *msp,* marginal spine.] (p. *U*42-*U*43).

oped; oral interradial areas with large or small ossicles. *M.Sil.-U.Dev.*

Lepidaster Forbes, 1850 [**L. grayi*; OD]. Interradial areas with small ossicles. *M.Sil.*, Eng.——Fig. 43,6. **L. grayi*; part of oral surface, ×2 (133).

Devonistella Spencer, 1927 [**Helianthaster filiciformis* Woodward, 1874; OD]. Interradial areas

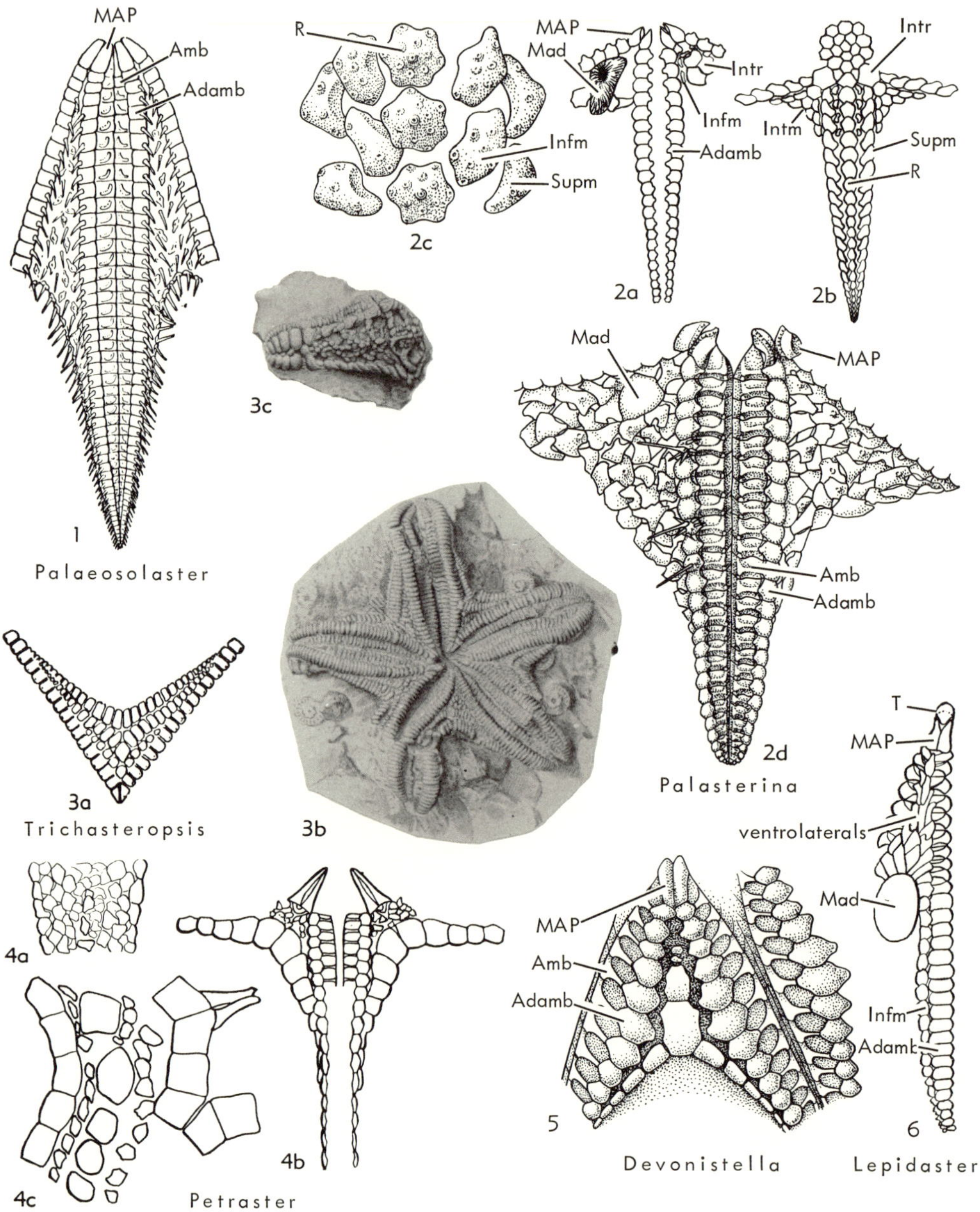

Fig. 43. Petrasteridae *(4)*; Lepidasteridae *(5,6)*; Palasterinidae *(1-3)*. [Explanation: *Adamb,* adambulacral; *Amb,* ambulacral; *Infm,* inferomarginal; *Intm,* intermarginal; *Intr,* interradial; *Mad,* madreporite, *MAP,* mouth-angle plate; *R,* radial; *Supm,* superomarginal; *T,* torus.] (p. *U43-U45*).

with few large ossicles. *U.Dev.*, Eng.——FIG. 43, *5.* **D. filiciformis* (WOODWARD); part of oral surface, ×3 (133).

Family PALASTERINIDAE Gregory, 1899

[=Palaesterinidae STÜRTZ, 1890 (not founded on included genus); Lindstroemasterininae GREGORY, 1899; Palaeosolasteridae SCHUCHERT, 1914; Protactininae SPENCER, 1926; Palaeosolasteridae LEHMANN, 1957]

Arms 5 to many; interradial arcs without differentiated marginals. *Sil.-Trias.*

Palasterina M'COY, 1851 [**Uraster primaevus* FORBES, 1848; OD] [=*Palaeasterina* ETHERIDGE, 1881; *Lindstroemaster* GREGORY, 1899; *Hisingeraster, Pseudopalasterina* STÜRTZ, 1900; *?Protactis* SPENCER, 1927; *?Archasterina* LEHMANN, 1957]. Arms 5, more or less cylindrical, bounded distally by adambulacrals; interradial areas aborally with few to many well-calcified ossicles. Odontophore weak to strong. Aboral surface of arms with rather large hexagonal marginals and carinals, in some forms with small intermediate ossicles. Larger ossicles may be pustulose and bear various spines. *L. Sil.-L. Dev.*, Ger.-Eng.-Sweden.——FIG. 43,2. **Palasterina primaeva* (FORBES), U.Sil., Eng.; *2a,b,* oral and aboral surfaces of arm and part of disc, ×3; *2c,* ossicles of aboral surface, ×15 (133).——FIG. 43,*2d. P. antiqua* (HISINGER), U.Sil., Eng.; oral surface, ×5 (133).

Palaeosolaster STÜRTZ, 1899 [**P. gregoryi*; OD] [=*Echinasterias, Echinostella* STÜRTZ, 1899; *Echinodiscus* STÜRTZ, 1899 (*non* WORTHEN & MILLER, 1883); *Echinodiscaster* DELAGE & HÉROUARD, 1904; *Echinodiscites* SCHUCHERT, 1914]. Arms many. Oral and aboral intermediate ossicles subequal, each carrying single spine. *L.Dev.*, Ger.——FIG. 43,*1.* **P. gregoryi*; part of oral surface, ×0.5 (133).

Trichasteropsis ECK, 1879 [*pro Trichaster* QUENSTEDT, 1875 (*non* AGASSIZ, 1836)] [**Asterias cilicia* QUENSTEDT, 1852 (=**Asterias weissmanni* MÜNSTER, 1843); OD]. Superomarginals much longer than inferomarginals, row of minute intermarginals present proximally. Ossicles in axillary areas in regular rows. *Trias.*, Ger.——FIG. 43, *3.* **T. weissmanni* (MÜNSTER); *3a,* part of oral surface, diagrammatic; *3b,c,* oral surface and aboral view of arm, ×1 (128).

Suborder DIPLOZONINA Spencer & Wright, new suborder

Regular double rows of marginals; aboral skeleton typically of true paxillae but early forms may have granular tessellate plates; superambulacral plates present. Tube feet pointed, without sucking discs; ampullae double. *L.Jur.-Rec.*

Family ASTROPECTINIDAE Gray, 1840

[incl. Plutonasteridae SLADEN, 1889; Priamasterinae KOEHLER, 1912 (*nom. transl. et correct.* FISHER, 1917 (*ex* Priamastérídées KOEHLER, 1912)]

Disc generally rather small; arms long and pointed, normally straight-sided, rarely petaloid; contact facets between marginals smaller in most genera than sides of these plates, leaving ridges on them and narrow channels between adjacent marginals for marginal fascioles; aboral surface covered with tessellate ossicles in some Mesozoic genera but otherwise with true paxillae; oral interradial areas with flat ossicles extending greater or less distance into arms; superambulacral ossicles present. Tube feet pointed, without sucking discs; ampullae double (26). *L.Jur.-Rec.*

Subfamily ASTROPECTININAE Gray, 1840

[*nom. transl.* SLADEN, 1899 (*ex* Astropectinidae GRAY, 1840)]

Marginal fascioles not webbed. *L.Jur.-Rec.*

Genera known as fossils are described first in alphabetical order, after the type-genus; those known only as extant genera are then listed in alphabetical order.

Astropecten GRAY, 1840 [**Asterias aranciaca* LINNÉ, 1758; SD FISHER, 1908] [=*Stellaria* NARDO, 1834 (*non* SCHMIDT in MÖLLER, 1832); *Astropectinides* VERRILL, 1914]. Intermarginal facet small, not angular; inferomarginals with irregularly distributed horseshoe-shaped tubercles, which bear long spines of varying size. Typical paxillae. *U. Mio.(Torton.)-Rec.*——FIG. 12,*3. A.* sp., Rec.; cross section of arm, enl. (133).

Advenaster HESS, 1955 [**A. inermis*; OD]. Lateral facets of marginals large so that intermarginal fasciole is very small; outer face only of marginals and all other ossicles except ambulacrals with pustules bearing fine spines. *L.Jur.(Bajoc.)*, Switz. ——FIG. 44,*1.* **A. inermis*; oral surface (reconstr.), ×1 (113).

Cuneaster HESS, 1955 [**C. hauteriviensis*; OD]. Intermarginal facets small, more or less quadrangular; ridge on marginals high and narrow so that intermarginal channel is wide. *L.Cret. (Hauteriv.)-L.Eoc.(Ypres.)*, Eu.——FIG. 44,*2.* **C. hauteriviensis*, L.Hauteriv., Switz.; side views of superomarginal and inferomarginal, ×4 (113).

Lambertella MERCIER, 1935 [**L. Valettei*; OD]. Ridge on marginals projecting laterally in club-shaped prominence with narrow neck, few or no spines or granules. [Known only from isolated marginals.] *M.Jur.(Bathon.)*, *?U.Cret.(Turon.)*, Fr.-?Eng.——FIG. 44,*5.* **L. valettei*, Bathon., Fr.; *5a,b,* aboral and profile views of ?superomarginal, ×3 (121).

Lophidiaster SPENCER, 1913 [**L. ornatus*; OD]. Intermarginal facet small, rounded; inferomarginals with rugosities but no horseshoe-shaped tubercles; typical paxillae. *L.Cret.(L.Alb.)-L.Mio.(Burdigal.),* Eu.-Can.

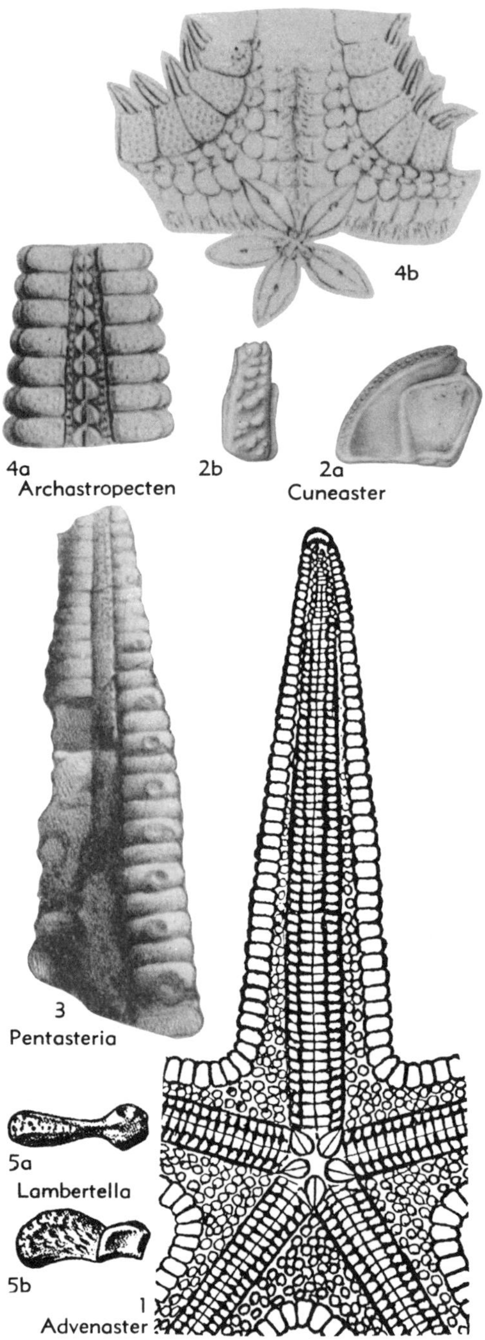

FIG. 44. Astropectinidae (Astropectininae) (p. *U45-U46*).

Pentasteria VALETTE, 1929 [*non* DE BLAINVILLE, 1834, in syn. in D'ORBIGNY, 1852 (*ex Pentasterias* AGASSIZ, 1842, *nom. correct. pro Pentastéries* DE BLAINVILLE, 1830)] [**P. boisteli*; OD] [=*Crenaster* D'ORBIGNY, 1850; *?Triboletia* DE LORIOL, 1908; ICZN pend.]. Lateral facets of marginals large; inferomarginals with rugosities and irregular row of large horseshoe-shaped tubercles; superomarginals with fine spine pits and (in some forms) large socket for stout spine; aboral surface covered by tessellated plates with pits; no paxillae. *L.Jur.-L.Eoc.,* Eu.

P. (Pentasteria). Some though not all superomarginals with short stout spine in large socket. *M.Jur.(Oxford.)-L.Cret.(Valangin.),* Eu.——FIG. 44,*3*. *P. (P.) rectus* M'COY, Oxford., Eng.; aboral view of part of arm showing large sockets for spines on superomarginals, ×1 (139).

P. (Archastropecten) HESS, 1955 [**Astropecten huxleyi* T. WRIGHT, 1862; OD]. No stout spines on superomarginals. *L.Jur.(?Pliensbach., Aalen.)-L.Eoc.(Ypres.).*——FIG. 44,*4*. *P. (A.) cotteswoldiae* (BUCKMAN), Bathon., Eng. (Oxfords.); *4a,* part of aboral, surface of arm showing superomarginals without large sockets, ×2; *4b,* oral surface of part of disc and arm, ×3 (139).

Plesiastropecten PEYER, 1944 [**P. hallovensis*; OD]. Only specimen too badly preserved for characters to be ascertained; marginals with long spines; aboral paxillae having stellate bases with 4 or 6 points. *L.Jur.(Hettang.),* Switz.

Astromesites FISHER, 1913 [**A. compactus*; OD]. *Rec.*

Bathybiaster DANIELSSON & KOREN, 1883 [**Astropecten pallidus* DANIELSSON & KOREN, 1877 (=**Archaster vexillifer* WYVILLE-THOMSON, 1873); OD] [=*Phoxaster* SLADEN, 1885 *(nom. nud.)*; *Phoxaster* SLADEN, 1889; *?Ilyaster* DANIELSSON & KOREN, 1883]. *Rec.*

Blakiaster PERRIER, 1881 [**B. conicus*; OD] [=*?Bunodaster* VERRILL, 1909]. *Rec.*

Ctenophoraster FISHER, 1906 [**C. hawaiiensis*; OD]. *Rec.*

Ctenopleura FISHER, 1913 [**C. astropectinides*; OD]. *Rec.*

Dipsacaster ALCOCK, 1893 [**D. sladeni*; OD]. *Rec.*

Dytaster SLADEN, 1889 [**D. nobilis*; SD FISHER, 1919] [=*Crenaster* PERRIER, 1885 (*non* D'ORBIGNY, 1850); *Dytaster* SLADEN, 1885 *(nom. nud.)*]. *Rec.*

Koremaster FISHER, 1913 [**Dytaster (Koremaster) evaulus*; OD]. *Rec.*

Leptychaster SMITH, 1876 [**L. kerguelenensis*; OD] [=*Leptoptychaster* SMITH, 1879; *Priamaster* KOEHLER, 1912]. *Rec.*

L. (Leptychaster). *Rec.*

L. (Parastropecten) LUDWIG, 1905 [**Parastropecten inermis*; OD] [=*Glyphaster* VERRILL, 1909]. *Rec.*

L. (Trophodiscus) FISHER [**Trophodiscus almus*; OD]. *Rec.*

Lonchotaster SLADEN, 1889 [**L. tartareus*; SD FISHER, 1919] [=*Lonchotaster* SLADEN, 1885 *(nom. nud.)*]. *Rec.*

Macroptychaster H.E.S. CLARK, 1962 [**Leptoptychaster accrescens* KOEHLER, 1920; OD]. *Rec.*

Mimastrella FISHER, 1916 [**Mimaster cognatus* SLADEN, 1889; OD]. *Rec.*

Patagiaster FISHER, 1906 [**P. nuttingi*; OD]. *Rec.*

Persephonaster WOOD-MASON & ALCOCK, 1891 [**P. croceus*; OD] [=*Psilasteropsis* FISHER, 1906]. *Rec.*

Plutonaster SLADEN, 1885 [**Archaster bifrons* WYVILLE-THOMSON, 1873; OD]. *Rec.*

Psilaster SLADEN, 1885 [**Astropecten andromeda* MÜLLER & TROSCHEL, 1842; OD] [=*Ripaster* KOEHLER, 1906; *Phidiaster* KOEHLER, 1909]. *Rec.*

Tethyaster SLADEN, 1889 [**Asterias subinermis* PHILIPPI, 1837; SD A. M. CLARK & A. H. CLARK, 1954] [=*Moiraster* SLADEN, 1889; *Sideriaster* VERRILL, 1899; *Anthosticte* FISHER, 1911]. *Rec.*

Thrissacanthias FISHER, 1916 [**Persephonaster penicillatus* FISHER, 1904; OD]. *Rec.*

Tritonaster FISHER, 1906 [**T. craspedotus*; OD]. *Rec.*

Subfamily CRASPIDASTERINAE Fisher, 1916

Marginal and oral fascioles webbed. *Rec.*

Craspidaster SLADEN, 1889 [**Archaster hesperus* MÜLLER & TROSCHEL, 1840; OD] [?=*Nauricia* GRAY, 1840 *(nom. dub.)*]. *Rec.*

Suborder CRIBELLINA Fisher, 1911

Arms five, short or long, disc large; marginals normally thin and lamelliform, high, naked or covered with membrane, smooth or with few large spines; cribriform organs between all or some marginals. No intestine or anus in most forms; tube feet pointed, without sucking disc; ampullae single. *Rec.*

Family GONIOPECTINIDAE Verrill, 1889

Abactinal surface paxillose; cribriform organs between all marginals; oral surface with transverse rows of ossicles separated by channels covered by webbed spinelets, continuous from cribriform organs to ambulacrals; superambulacral ossicles present. *Rec.*

Subfamily CTENODISCINAE Sladen, 1889

Marginals moderately solid; cribriform organs consisting solely of webbed spinelets. No intestine. *Rec.*

Ctenodiscus MÜLLER & TROSCHEL, 1842 [**Asterias polaris* SABINE, 1821 (=**A. crispata* RETZIUS, 1805); OD] [=*Anodiscus* PERRIER, 1869]. *Rec.*

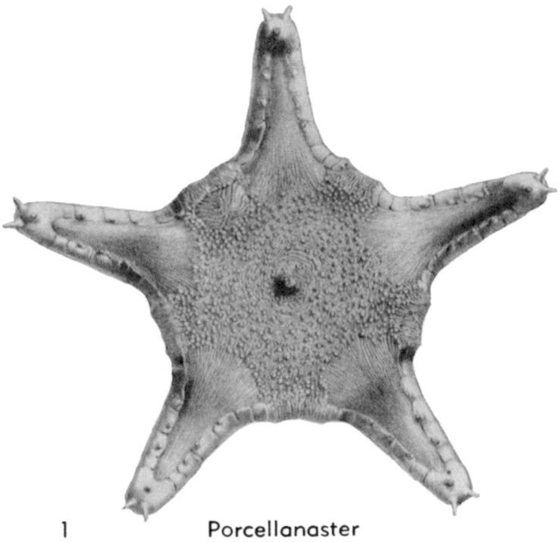

1 Porcellanaster

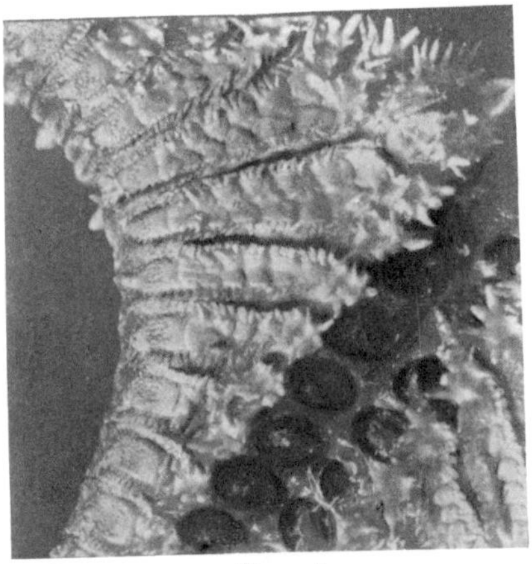

2 Ctenodiscus

FIG. 45. Goniopectinidae (Ctenodiscinae) *(2)*; Porcellanasteridae *(1)* (p. *U47-U48*).

——FIG. 45,*2*. **C. crispatus* (RETZIUS); part of aboral surface, enl. (110).

Pectinidiscus LUDWIG, 1900 [**P. annae*; OD]. *Rec.*

Subfamily GONIOPECTININAE Verrill, 1889

Cribriform organs composed of discrete spinelets covered by single-webbed series. Well-developed intestine and intestinal caecum. *Rec.*

Goniopecten PERRIER, 1881 [**G. demonstrans*; OD]. *Rec.*

Prionaster VERRILL, 1889 [**P. elegans*; OD]. *Rec.*

Family PORCELLANASTERIDAE Sladen, 1883

Marginals very thin; cribriform organs highly developed but (with single excep-

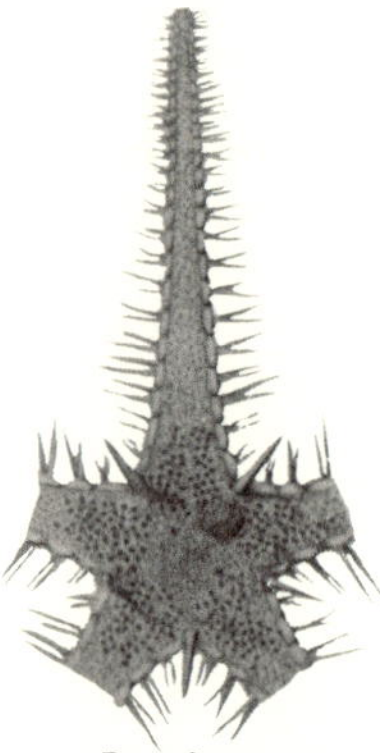

1

Benthopecten

FIG. 46. Benthopectinidae (p. U48).

tion) present only between some marginals; oral interradial areas without channels. *Rec.*

Porcellanaster WYVILLE-THOMSON, 1877 [**P. ceruleus*; OD] [=*Caulaster* PERRIER, 1882; *Albatrossaster* LUDWIG, 1907 (*pro Albatrossia* LUDWIG, 1905, *non* JORDAN & EVERMANN, 1898)]. *Rec.*——FIG. 45,*1*. **P. ceruleus*; aboral surface, ×2 (130).

Abyssaster MADSEN, 1961 [**Hyphalaster tara* WOOD-MASON & ALCOCK, 1891; OD]. *Rec.*

Benthogenia FISHER, 1911 [**B. cribellosa*; OD]. *Rec.*

Eremicaster FISHER, 1905 [**Porcellanaster tenebrarius* FISHER, 1905 (=**P. gracilis* SLADEN, 1883); OD]. *Rec.*

Hyphalaster SLADEN, 1883 [**Hyphalaster inermis* SLADEN, 1883; SD MADSEN, 1961]. *Rec.*

Lysaster BELL, 1909 [**L. lorioli*; OD]. *Rec.*

Sidonaster KOEHLER, 1909 [**S. vaneyi*; SD MADSEN, 1961]. *Rec.*

Styracaster SLADEN, 1883 [**S. horridus*; SD MADSEN, 1961] [=*Machairaster* PERRIER, 1884 (*nom. nud.*); *Chunaster* LUDWIG, 1907]. *Rec.*

Thoracaster SLADEN, 1883 [**T. cylindratus*; OD]. *Rec.*

Suborder NOTOMYOTINA Ludwig, 1910

[=Myonota VERRILL, 1914]

Arms flexible, with pair of dorsal muscle bands, perhaps allowing swimming; marginals alternate, imbricating with long spines; no superambulacral plates. Pedicellariae pectinate; tube feet with sucking discs, ampullae double. *L.Cret.-Rec.*

Family BENTHOPECTINIDAE Verrill, 1894

[*nom. transl.* VERRILL, 1899 (*ex* Benthopectininae VERRILL, 1894)] [=Pararchasterinae SLADEN, 1889]; Pontasterinae VERRILL, 1894; Cheirasteridae LUDWIG, 1910]

Disc small, arms long and slender; odd interradial marginal in each series present in some genera. *L.Cret.-Rec.*

Benthopecten VERRILL, 1884 [**B. spinosus*; OD] [=*Pararchaster* SLADEN, 1885]. Odd interradial superomarginal more prominent than others; superomarginals bearing 1 primary spine and inferomarginals 1 or 2. [An undescribed species probably belonging to this genus occurs in Albian rocks of England.] *?L.Cret.(U.Alb.)*, Eng., *Rec.*——FIG. 46,*1*. *B. armatus* (SLADEN), Rec.; aboral view, ×1 (130).

Acontiaster DÖDERLEIN, 1921 [**A. bandanus*; OD]. *Rec.*

Cheiraster STUDER, 1883 [**C. gazellae*; OD]. *Rec.*

Gaussaster LUDWIG, 1910 [**G. vanhoffeni*; OD]. *Rec.*

Luidiaster STUDER, 1883 [**L. hirsutus*; OD] [=*Acantharchaster* VERRILL, 1894; *Marcellaster* KOEHLER, 1907; *Marcelaster* KOEHLER, 1908]. *Rec.*

Myonotus FISHER, 1911 [**Acantharchaster intermedius* FISHER, 1900; OD]. *Rec.*

Nearchaster FISHER, 1911 [**Acantharchaster aciculosus* FISHER, 1910; OD]. *Rec.*

Pectinaster PERRIER, 1885 [**P. filholi*; OD]. *Rec.*

Pontaster SLADEN, 1885 [**Astropecten tenuispinus* DÜBEN & KOREN, 1846; OD]. *Rec.*

Order VALVATIDA Perrier, 1884

[*nom. correct.* SPENCER & WRIGHT, herein (*pro* "Valvata" PERRIER, 1884)]

Mouth frame of adambulacral type; mouth-angle plates relatively inconspicuous and normally only distinguishable from succeeding adambulacrals by their subtrigonal outline; infero- and superomarginals, if present, normally equal in number and without intermarginal channels; adambulacrals without transverse ridge. Ambulacrals normally in 2, rarely in 4 rows. Pedicellariae, when present, generally valvate and with bases sunk into substance of ossicles. *L.Ord.-Rec.*

Suborder PUSTULOSINA Spencer, 1951

[*nom. transl. et correct.* SPENCER & WRIGHT, herein (*ex* Pustulosa SPENCER, 1951)]

Marginals with many small undifferentiated spines elevated on small tubercles (pustules); superomarginals generally within frame of inferomarginals and loosely connected. Pedicellariae unknown. *L.Ord.-L. Carb.*

Superfamily PALAEASTERACEAE S. A. Miller, 1889

[*nom. transl.* SPENCER & WRIGHT, herein (*ex* Palaeasteridae S. A. MILLER, 1889)]

Ambulacral furrow generally closed by

short quadrangular adambulacrals. Inferomarginal frame reaches to ends of the arms. *L.Ord.-L.Carb., ?Permocarb.*

Family PALAEASTERIDAE S. A. Miller, 1889

Single axillary in each arc dividing row of inferomarginals that borders arms, which are thus not fully fused at base; superomarginals lying within inferomarginals; no carinals present but irregular small plates occurring along mid-line of arms; aboral surface of disc with distinct primary circlets, ossicles of which are separated by many

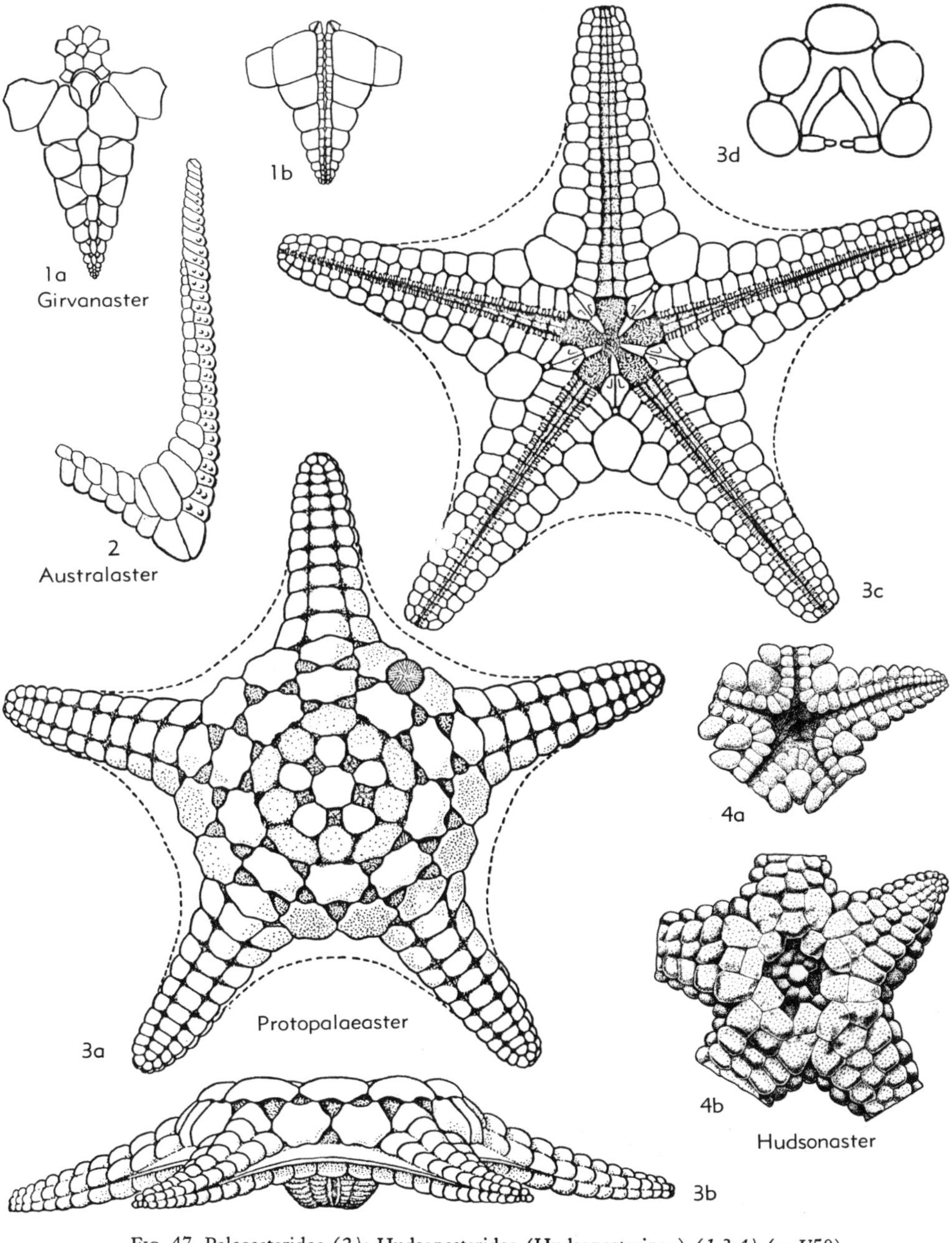

FIG. 47. Palaeasteridae *(2)*; Hudsonasteridae (Hudsonasterinae) *(1,3-4)* (p. *U*50).

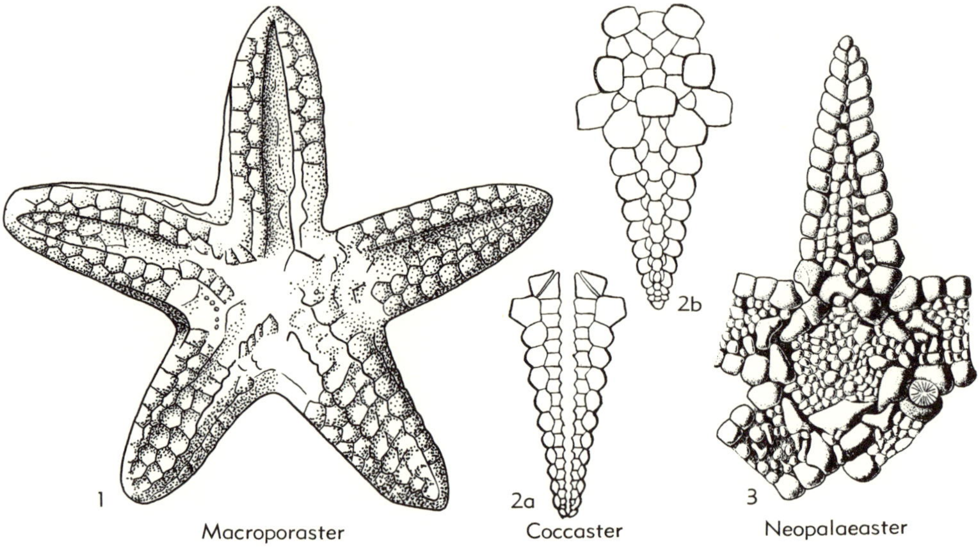

FIG. 48. Hudsonasteridae (Hudsonasterinae) *(1)*, (Coccasterinae) *(2)*, Neopaleasteridae *(3)* (p. *U51*).

small plates, probably forming rigid structure. *M.Sil., ?Permocarb.*

Palaeaster HALL, 1852 [**P. niagarensis*; OD]. Characters of family. *M.Sil.*, N.Am.——FIG. 21,*4*. **P. niagarensis*, USA(N.Y.); aboral side, ×2 (129).

?Australaster SCHUCHERT, 1914 [**Palaeaster (Monaster) giganteus* ETHERIDGE, 1892; OD] [=*Monaster* GREGORY, 1899 (*non* ETHERIDGE, 1892; ICZN pend.)]. Large, with slender arms; adambulacrals increasing in width and inferomarginals decreasing distally; aboral surface unknown. *Permocarb.*, New S. Wales.——FIG. 47,*2*. **A. giganteus* (ETHERIDGE); oral view, ×1 (129).

Family HUDSONASTERIDAE Schuchert, 1914

[=Protopalaeasteridae RASMUSSEN, 1962]

Single axillary with free distal edge dividing strong frame of inferomarginals bordering arms which generally are unfused but tend to join at base in some genera; superomarginals within frame of inferomarginals; aboral surface of arms generally consisting of superomarginals and row of carinals only, but with additional rows of ossicles in Silurasterinae; aboral surface of disc with centrale and primary circlets which may form protrusible cap, presumably for respiratory purposes; ambulacrals rarely exposed; in some genera apparently blocky, without large basins for tube feet, in others with strong T-shaped ridges and basins between; ampullar pores present in some genera but minute and may occur generally despite emphasis in descriptions on their absence. *L.Ord.-U.Sil.*

The strongly calycinal aboral skeleton of hudsonasterids is probably due to paedomorphic evolution from the young of an ancestral form and is not in itself necessarily a primitive feature.

Subfamily HUDSONASTERINAE Schuchert, 1914

Aboral arm skeleton composed of superomarginals and carinals only; protrusible cap present on central disc. *L.Ord.-U.Ord.*

Hudsonaster STÜRTZ, 1900 [**Palasterina rugosa* BILLINGS, 1857; OD]. Arms subtriangular, with broad base, clearly not fused. *U.Ord.*, N.Am.——FIG. 47,*4*. *H. incomptus* (MEEK), U.Ord.(Richmond.); Ohio; *4a,b*, oral and aboral sides, ×2 (129).

Girvanaster SPENCER, 1916 [**G. sculptus*; OD]. Axillaries very large; proximal superomarginals large; protrusible cap narrow. *U.Ord.*, Scot.——FIG. 47,*1*. **G. sculptus*; *1a,b*, oral and aboral sides, ×6 (133).

Protopalaeaster HUDSON, 1912 [**P. narrawayi*; OD] [=*Belaster* SPENCER, 1916; *Ordoviciaster* FEDOTOV, 1936]. Proximal superomarginals not conspicuously large; arms tending to fuse; ambulacrals apparently without T-shaped ridge. *L.Ord.-U.Ord.*, N.Am.-Eng.-Turkestan.——FIG. 47,*3*. **P. narrawayi*, L.Ord., USA(Minn.); *3a-c*, aboral, lat. view, oral view, ×2; *3d*, cross section of arm, ×4 (40). (See also Fig. 21,*1*.)

?**Macroporaster** RAYMOND, 1921 [*Asterias matutinus HALL, 1847; OD]. Similar to *Hudsonaster* but ambulacrals with T-shaped ridges forming wide basins for tube feet. *M.Ord.*, N.Am.——FIG. 48,*1*. **M. matutinus* (HALL); oral surface, ×1.5 (126).

Subfamily COCCASTERINAE Spencer & Wright, new subfam.

Like Hudsonasterinae but no protrusible cap on disc. *M.Sil.-U.Sil.*

Coccaster SPENCER, 1916 [**C. bulbiferus*; OD]. Primary radials much swollen. *M.Sil.-U.Sil.*, Eng. ——FIG. 48,*2*. **C. bulbiferus*, Herefords.; *2a,b*, aboral and oral sides, ×5 (133).

Subfamily SILURASTERINAE Spencer & Wright, new subfam.

Aboral arm skeleton with rows of small intermediate ossicles between carinals and superomarginals. *M.Ord.*

Siluraster JAEKEL, 1903 [**S. perfectus*; OD] [=*Caractacaster* SPENCER, 1916]. Characters of subfamily. *M.Ord.*, W.Eng.-Czech.——FIG. 21,*2*. *S. caractaci* (SPENCER), Eng.(Heref.); aboral side, ×3 (133).

Family NEOPALAEASTERIDAE Schuchert, 1915

Very like Palaeasteridae except that superomarginals overlie inferomarginals, but it is not certain whether axillary has free edge or is enclosed by inferomarginals. [A doubtful family. Lower Silurian species from Sweden ascribed to *Neopalaeaster* probably belongs to Palaeasteridae.] *L.Carb.*

Neopalaeaster SCHUCHERT, 1915 [**Palaeaster crawfordsvillensis* S. A. MILLER, 1880; OD]. Characters of family. *Miss.*, N.Am.——FIG. 48,*3*. **N. crawfordsvillensis* (MILLER), Miss., USA(Ind.); aboral side, ×2 (129).

Family MESOPALAEASTERIDAE Schuchert, 1914

[*nom. transl.* SPENCER & WRIGHT, herein (*ex* Mesopalaeasterinae SCHUCHERT, 1915)]

Axillaries enclosed by first pair of inferomarginals and arms fused at base; aboral skeleton with rows of small intermediate ossicles separating carinals from superomarginals. *U.Ord.-U.Dev.*

Large numbers of *Devonaster* found at Saugerties, N.Y. (RUEDEMANN, 1915) show various stages of development. The earliest has stellate aboral ossicles, as in *Petraster*. In the next, the aboral surface passes through a *Hudsonaster* stage. Finally, the large ossicles become separated by lightly calcified areas, which allowed the extrusion of papulae.

Subfamily MESOPALAEASTERINAE Schuchert, 1915

With papular areas on aboral surface at base of arms on either side of primary radial. *U.Ord.-U.Dev.*

Mesopalaeaster SCHUCHERT, 1914 [**Palaeaster shafferi* HALL, 1868; OD] [=?*Argaster* HALL, 1868].Disc compact, small; arms narrow, straight-sided. [FOERSTE (1919) distinguished *Hemipalaeaster* as a subgenus (type, *H. schucherti*, OD), since its row of carinals is not continuous but interrupted proximally. This seems to be of only specific importance.] *U.Ord.*, N.Am.-Scot.——FIG. 49,*2a*. *M. primus* SPENCER, U.Ord., Scot.; part of oral surface, ×6 (133).——FIG. 49,*2b,c*. **M. shafferi* (HALL), U.Ord., USA(Ohio); *2b*, oral side of young individual, ×5; *2c*, aboral side of adult, ×2 (133).

Arisaigaster SPENCER, 1953 [**Palaeaster parviusculus* BILLINGS, 1860; OD]. Disc large; arms short and broad. *U.Ord.-U.Sil.*, E.Can.-Scot.-Eng.——FIG. 49,*3a*. *A. leintwardinensis* SPENCER, U.Sil., Eng.; aboral side, ×9 (133).——FIG. 49,*3b*. **A. parviusculus* (BILLINGS), L.Sil., N.Scotia; oral side, ×4 (129).

Devonaster SCHUCHERT, 1914 [**Palaeaster eucharis* HALL, 1868; OD]. Disc large, arms broad and slightly petaloid; aboral surface of disc covered with small irregular plates which also extend between carinals and superomarginals which are well within frame of inferomarginals. *M.Dev.-U.Dev.*, N.Am.——FIG. 49,*4*. **D. eucharis* (HALL), M.Dev., USA(N.Y.); *4a,b*, aboral and oral sides, ×1 (129).

Subfamily LEPIDACTININAE Spencer, 1918

[*nom. transl.* SPENCER & WRIGHT, herein (*ex* Lepidactinidae SPENCER, 1918)]

Differs from Mesopalaeasterinae in having no papular areas at base of arms. *M.Sil-L.Dev.*

Lepidactis SPENCER, 1918 [**L. wenlocki*; OD]. Inferomarginals not extending to end of arms. *M. Sil.*, Eng.——FIG. 49,*1*. **L. wenlocki*; *1a,b*, oral and aboral sides of arm, ×2 (133).

Spaniaster SCHÖNDORF, 1907 [*pro Coelaster* SANDBERGER, 1855 (*non* AGASSIZ, 1835)] [**Coelaster latiscutatus* SANDBERGER, 1855; OD] [=?*Miomaster* SCHÖNDORF, 1909]. Inferomarginals extending to end of arms. *L.Dev.*, W.Eu.

Subfamily CLARKEASTERINAE Spencer & Wright, new subfamily

Like Mesopalaeasterinae but with double row of narrow carinals; primary radials swollen as in *Coccaster*. *U.Dev.*

Clarkeaster RUEDEMANN, 1916 [**Palaeaster clarki* CLARKE & SWARTZ, 1913; OD]. Characters of subfamily. *U.Dev.*, E.N.Am.——FIG. 49,*5*. **C. clarki* (CLARKE & SWARTZ); aboral side, ×2 (127).

Family XENASTERIDAE Gregory, 1899

[=Palaeogoniasteridae STÜRTZ, 1890 (invalid because not founded on generic name); Palaeostellidae LEHMANN, 1957]

Like Mesopalaeasteridae but with more than single ossicle in each axillary area; arms wedge-shaped, flattened orally and with narrow ambulacrals. *L.Dev.*

Xenaster SIMONOVITSCH, 1871 [**X. margaritatus*; OD] [=?*Archaeasterias* MÜLLER, 1855; *Archasterias* SIMONOWITSCH, 1871]. Two pairs of inferomarginals incorporated within marginal frame; superomarginals subordinate to inferomarginals; few intermediate ossicles on oral side. *L.Dev.*, Ger.——FIG. 21,*3*. **X. margaritatus*; part of aboral surface showing intermarginals between rows of superomarginals and small arc of inferomarginals; also lightly calcified space between carinals and superomarginals, ×1.5 (128).

Agalmaster SCHÖNDORF, 1909 [**A. miellensis*; SD SCHUCHERT, 1914]. Two pairs of inferomarginals

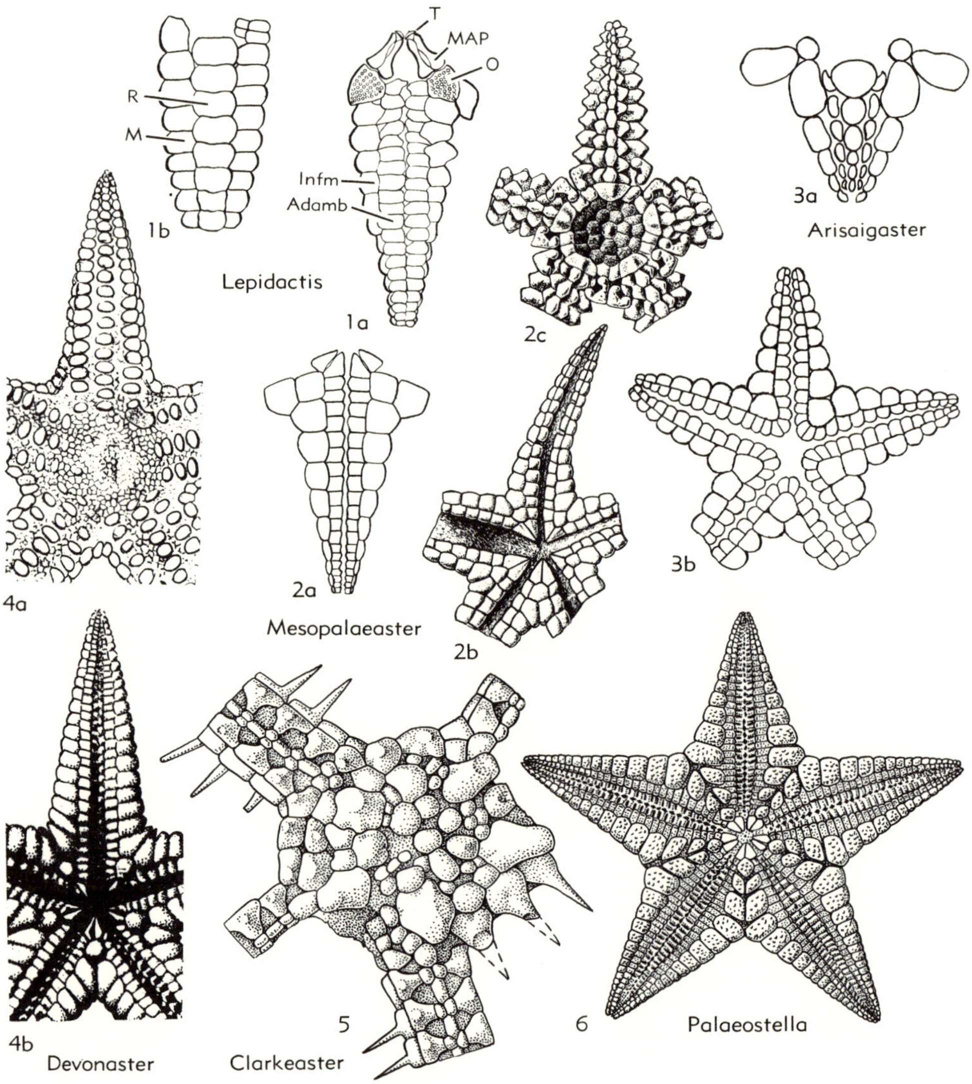

FIG. 49. Mesopalaeasteridae (Mesopalaeasterinae) *(2-4)*, (Lepidactininae) *(1)*, (Clarkeasterinae) *(5)*; Xenasteridae *(6)*. [Explanation: *Adamb*, adambulacral; *Infm*, inferomarginal; *M*, marginal; *MAP*, mouth-angle plate; *O*, odontophore; *R*, radial *T*, torus.] (p. *U51-U53*).

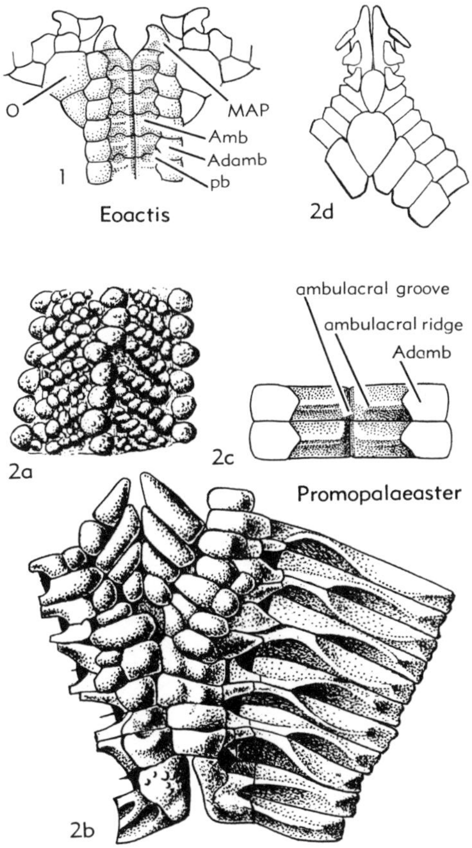

FIG. 50. Promopaleasteridae *(2)*; Eoactinidae *(1)*. [Explanation: *Adamb*, adambulacral; *Amb*, ambulacral; *MAP*, mouth-angle plate; *O*, odontophore; *pb*, podial basin.] (p. *U*53).

inside marginal frame; superomarginals prominent; oral intermediate plates numerous. *L.Dev.*, Ger.

Palaeostella STÜRTZ, 1890 [**P. solida*; OD] [=*Palaenectria* STÜRTZ, 1893; *Eifelaster* SCHÖNDORF, 1909]. Like *Agalmaster* but with no oral intermediate plates. *L.Dev.*, Ger.——FIG. 49,*6*. *P. follmanni* (SCHÖNDORF); oral side, ×1.5 (128). (128).

Rhenaster SCHÖNDORF, 1909 [**R. schwerdi*; OD]. No interray accessory plates on oral or aboral side. *L.Dev.*, Ger.

Trimeraster SCHÖNDORF, 1909 [**T. parvulus*; OD]. Only single pair of inferomarginals inside marginal frame. *L.Dev.*, Ger.

Family and Subfamily UNCERTAIN

Eostella LEHMANN, 1957 [**E. hunsrueckiana*; OD]. Marginals apparently T-shaped. [The single known specimen is too badly preserved for its characters and affinities to be made out.] *L.Dev.*, Ger.

Hunsrueckaster LEHMANN, 1957 [**H. peregrinus*; OD]. [The single known specimen is too badly preserved for its characters and affinities to be made out.] *L.Dev.*, Ger.

Superfamily PROMOPALAEASTERACEAE Schuchert, 1914

[*nom. transl.* SPENCER & WRIGHT, herein (*ex* Promopalaeasteridae SCHUCHERT, 1914)]

Marginal frame not reaching end of arms; adambulacrals broad; ambulacrals exposed and proximally may form 4 rows; arms well produced. *M.Ord.-Sil.*

Family PROMOPALAEASTERIDAE Schuchert, 1914

[=Anorthasterinae SCHUCHERT, 1915]

Arms fused at base and generally more or less cylindrical; ambulacrals compressed and proximally in 4 rows; adambulacrals broad. *M.Ord.-Sil.*

Promopalaeaster SCHUCHERT, 1914 [**Palaeaster granulosus* MEEK, 1872 (*non* HALL, 1868) (=**Palaeaster speciosus* MEEK, 1872); OD]. Characters of family. *M.Ord.-Sil.*, N.Am.-Scot.-Australia.——FIG. 50,*2a,b*. *P. magnificus* (MILLER), U.Ord.(Richmond.), USA(Ohio); *2a*, aboral side of distal part of arm, ×3; *2b*, oral side of interradial area, ×3 (129).——FIG. 50,*2c,d*. *P. elizae* SPENCER, U.Ord., Scot.; *2c*, oral view of ambulacrals and adambulacrals; *2d*, part of oral side showing inferomarginals enclosing 2 axillary ossicles between interradial arc and mouth frame, ×6 (133).

Anorthaster SCHUCHERT, 1914 [**Palaeaster miamiensis* S. A. MILLER, 1880]. Original diagnosis states "aberrant Promopalaeasteridae with the axillary and interbrachial areas composed entirely of adambulacral pieces" but may be founded on a pathological or damaged individual. *U.Ord.*, USA(Ohio).

?Kyraster LEHMANN, 1957 [**K. inermis*; OD]. Single known specimen badly preserved, may belong in this family. *L.Dev.*, Ger.

Family EOACTINIDAE Spencer, 1919

Marginal frame confined to few ossicles in arm axils. *Sil.*

Eoactis SPENCER, 1914 [**E. simplex*; OD]. Axillary large, hexagonal. *L.Sil.*, Eng.-N.Am.——FIG. 50,*1*. **E. simplex*, part of oral surface near mouth, ×8 (133).

Yarravaster SPENCER, 1950 [**Caractacaster yarraensis* WITHERS & KEEBLE, 1934; OD]. Axillary rounded, swollen. *Sil.*, Australia.

Suborder TUMULOSINA Spencer & Wright, new suborder

This suborder is erected for a presumed phyletic assembly of peculiar forms characterized by a high swollen disc covered with rather large ossicles notched in one way or another to allow for extrusion of papulae. The Monasteridae in many ways resemble early Palaeasteraceae, from which presumably they were derived. The Stauranderasteridae, though lacking the characteristic wide adambulacrals of Monasteridae, closely resemble the family in other important characters. The Sphaerasteridae probably were derived from Early Jurassic Stauranderasteridae and tended to a closely plated spherical form. *Permocarb.-Rec.*

Family MONASTERIDAE Schuchert, 1915

Disc large, tumid; arms club-shaped, with steep lateral borders formed by inferomarginals, which are visible in aboral aspect and

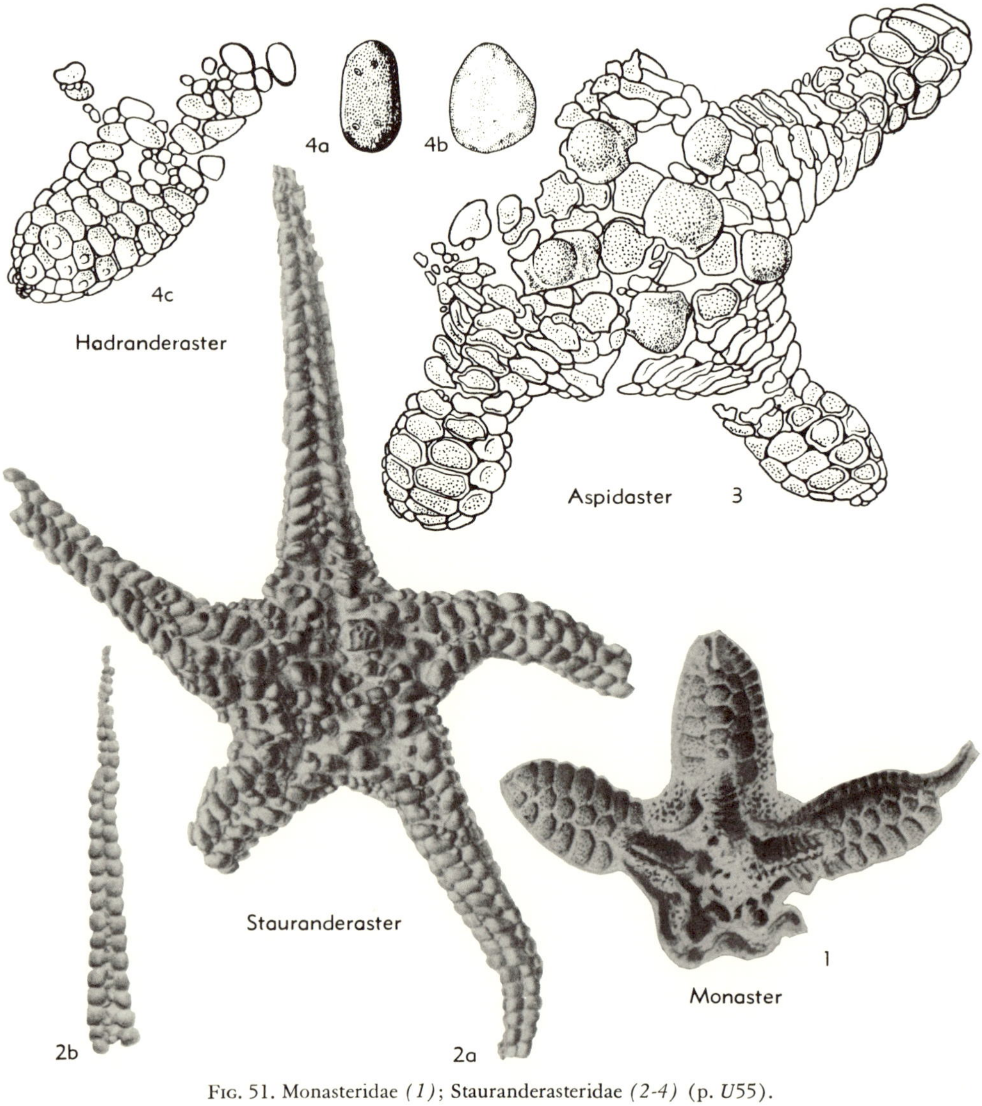

FIG. 51. Monasteridae *(1)*; Stauranderasteridae *(2-4)* (p. *U*55).

are twice as numerous as superomarginals, a series of which adjoins row of swollen carinals; adambulacrals exceptionally wide, occupying most of oral surface of arms; disc with centrale and primary interradials larger than remaining ossicles, most of which are spindle-shaped. *Permocarb.*

Monaster Etheridge, 1892 [**Palaeaster clarkei* de Koninck, 1877; SD Schuchert, 1914; ICZN pend.] [=*Etheridgaster* Gregory, 1899]. Characters of family. *Permo-Carb.*, Australia(New S. Wales).——Fig. 51,*1*. **M. clarkei* (de Koninck); aboral surface, ×1 (107).

Family STAURANDERASTERIDAE Spencer, 1913

Some forms closely resembling Monasteridae except that adambulacrals are square and inferomarginals are no more numerous than superomarginals; other forms with long narrow arms and intermarginals may occur; aboral skeleton of disc consisting of massive spindle- or breastplate-shaped ossicles notched at corners to allow protrusion of papulae and including prominent centrale and primary interradials. *M.Jur.-U.Cret.*

Stauranderaster Spencer, 1907 [**Oreaster boysii* Forbes, 1848; OD]. Arms long, narrow, and straight-sided; carinals weak or absent. *U.Jur. (Kimmeridg.)-U. Cret.(Maastricht.)*, Eu. - N. Am. (Tex.).——Fig. 51,*2*. *S. coronatus* (Forbes), U. Cret.(Cenoman.), Eng.; *2a*, aboral surface; *2b*, lat. view of arm showing intermarginals, ×0.75 (131).

Aspidaster de Loriol, 1884 [**A. delgadoi*; OD]. Disc high and swollen; arms club-shaped with large carinals; marginals and ossicles of disc generally with smooth rabbet edge. *M.Jur. (Bathon.)-U.Cret.(Dan.)*, Eu.——Fig. 51,*3*. *A. bulbiferus* (Forbes), U.Cret.(Santon.), Eng., aboral side, ×1 (131).

Hadranderaster Spencer, 1907 [**Pentaceros abbreviatus* Spencer, 1905 (=**Oreaster simplex* Geinitz, 1871); OD] [=?*Stauraster* Valette, 1928]. Marginals hexagonal or rounded, extremely thick, lacking smooth rabbet edge. *?L.Jur.(Charmouth.), M. Jur.(Bathon.) - U. Cret.(Campan.)*, W. Eu.——Fig. 51,*4*. **H. simplex* (Geinitz), U.Cret.(Santon.), Eng.; *4a,b*, lat. and profile views of marginal, ×2 (133); *4c*, aboral side of arm, ×1 (131).

Family SPHAERASTERIDAE Schöndorf, 1906

Body high, domed, and slightly pentagonal to globular, without produced arms, covered with close-fitting plates (3, 58). *M.Jur.-Rec.*

Sphaeraster Quenstedt, 1875 [*pro Sphaerites* Quenstedt, 1852 (*non* Duftschmid, 1805)] [**Sphaerites punctatus* Quenstedt, 1852; SD A. M. Clark, 1962]. Domed, with flat or slightly concave base, square inferomarginals and high, short, wide superomarginals forming ambital margin, broken only by ends of ambulacra; aboral surface covered by large thin hexagonal plates with pores for papulae along their sutures; oral interradial areas with close-fitting small plates. *U.Jur.(Oxford.)*, Ger.——Fig. 52,*2*. **S. punctatus* (Quenstedt); *2a,b*, upper and lower views of fragment, ×1 (123).

Podosphaeraster A. M. Clark, 1962 [**P. polyplax*; OD]. Spherical, with no ambital margin or distinct marginals, ambulacra reaching equator; more abactinal plates than in *Sphaeraster*, covered with thin skin containing granules. *Rec.*——Fig. 52,*3*. **P. polyplax*; dorsolateral view, ×2.5 (102).

Valettaster Lambert, 1914 [*pro Tholaster* Spencer, 1913 (*non* Seunes, 1896)] [**Oreaster ocellatus* Forbes; SD Rasmussen, 1950; ICZN Opin. 331] [=?*Asteriaceros* Valette, 1934]. Apparently like *Sphaeraster* in shape but aboral ossicles thick, irregular, generally low cones. *M.Jur.(Bathon.)-U.Cret.(Maastricht.)*, Eu.——Fig. 52,*1*. **V. ocellatus* (Forbes), U.Cret.(Santon.), Eng.; *1a*, oral side, ×1; *1b*, ossicle, ×2 (131).

Suborder GRANULOSINA Perrier, 1894

[*nom. transl. et correct.* Spencer & Wright, herein (*ex* Granulosa Perrier, 1894)]

Marginals conspicuous, invariably fewer than adambulacrals, in two series, opposite; aboral ossicles arranged in calycinal system in young, which generally cannot be distinguished in adults, ossicles generally covered with spines or granules in shallow sockets. Pedicellariae, if present, generally valvate and sunk in ossicles. *L.Jur.-Rec.*

Family ODONTASTERIDAE Verrill, 1889

[=Gnathasterinae Perrier, 1894]

Pentagonal or broadly stellate, with odd interradial marginal in each series, more paxillose than most Goniasteridae; mouth-angle plates triangular, generally bearing recurved spines with glassy tips. *M.Jur.-Rec.*

Odontaster Verrill, 1880 [**O. hispidus*; OD] [=*Gnathaster* Sladen, 1889; *Gnathodon* Verrill, 1899; *Peridontaster* Koehler, 1920; *Epidontaster* Koehler, 1921; *Gymnognathaster* Döderlein, 1928]. One spine common to each pair of mouth-angle plates. [A Jurassic species doubtfully belongs here.] *?M.Jur.(Bajoc.* or *Bathon.)*, N.Z.; *Rec.*

Acodontaster Verrill, 1899 [**Gnathaster elongatus*

Sladen, 1889; OD] [=*Heuresaster* Bell, 1908; *Pseudodontaster* Koehler, 1912; *Tridontaster* Koehler, 1920; *Metadontaster* Koehler, 1921]. *Rec.*

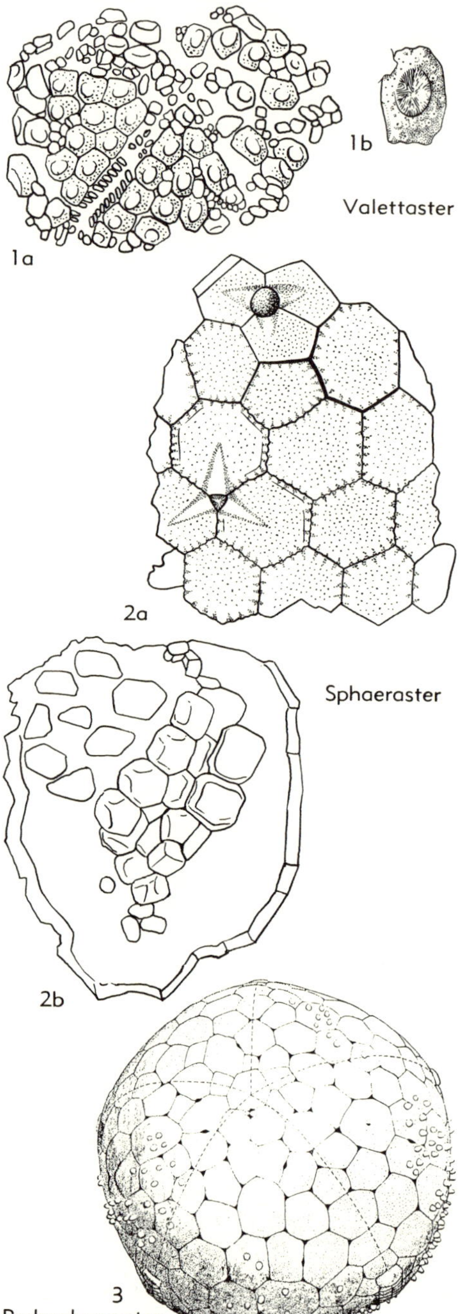

Fig. 52. Sphaerasteridae (p. U55).

Asterodon Perrier, 1891 [**Goniodiscus singularis* Müller & Troschel, 1842; SD Fell, 1953] [=*Diplodontias* Fisher, 1908 (*pro Goniodon* Perrier, 1894, *non* Herrick, 1888)]. *Rec.*

Eurygonias Farquar, 1913 [**E. hylacanthus*; OD]. *Rec.*

Hoplaster Perrier, 1882 [**H. spinosus*; OD]. *Rec.*

Family CHAETASTERIDAE Sladen, 1889

[*nom. transl.* Ludwig, 1897 (*ex* Chaetasterinae Sladen, 1889)]

Marginals small, with odd interradial marginal in each series; spinelets of paxillae slender and glassy; superambulacral plates present; calcareous interbrachial septa. Ampullae single. *Rec.*

The position of this family is altogether doubtful.

Chaetaster Müller & Troschel, 1840 [**Asterias subulata* Lamarck, 1816 (=**Asterias longipes* Retzius, 1805); OD]. *Rec.*

Family ARCHASTERIDAE Viguier, 1878

Superficially like Astropectinidae but tube feet with sucking discs; aboral ossicles tabulate and paxilliform, with internal imbricating ridges, arranged in oblique transverse rows on either side of conspicuous radial series. *?Mio.,Rec.*

Archaster Müller & Troschel, 1840 [**A. typicus*; OD]. Characters of family. Interradial arcs pointed. *?Mio.*, S.Afr., *Rec.*

Family GONIASTERIDAE Forbes, 1841

Pentagonal to narrowly stellate forms generally with large disc; marginals prominent, opposite, normally with no odd interradial marginal; plates on both oral and aboral surfaces in close contact; aboral plates flat, tabulate or paxilliform, with or without spines or granules but in several genera they may be tumid. Pedicellariae most commonly alveolate or valvate. *L.Jur.-Rec.*

Many Mesozoic genera belong here, but their phylogeny and therefore detailed taxonomy are determined only for a few groups. Few Cenozoic forms are known and it is therefore difficult to link the Mesozoic with abundant Recent genera. Hence, organization into subfamilies can only be provisional; certain groups of genera are distinct but there remain a large number of genera that must still be assigned to Goniasteridae, *incertae sedis*, but which probably ought to be grouped into several subfamilies.

Subfamily GONIASTERINAE Forbes, 1841

Shape varying from pentagonal to rather long-armed but all genera agree in having rather few and large marginals, normally with slightly sunken edge ("rabbet edge" of SPENCER) and raised smooth or distantly granulate central area; aboral plates may be flat and smooth, granulate or papillate, or may be raised into round tubercles. There is a strong tendency to specialization of distal superomarginals which may represent up to 7 marginals fused into an enlarged ultimate superomarginal. The subfamily comprises a series of genera radiating or secondarily derived from the Cretaceous and Cenozoic *Metopaster. U.Cret.-Rec.*

Goniaster AGASSIZ, 1835 [**Asterias tesselatus* LAMARCK, 1816; OD] [=*Phaneraster* PERRIER, 1894]. Prominent rounded or bluntly pointed tubercles on aboral plates, particularly on primary circlet and carinals; enlarged ultimate superomarginals. *Rec.*

Ceramaster VERRILL, 1899 [**Asterias granularis* RETZIUS, 1783; OD] [=?*Petalastrum* DE GREGORIO, 1895; *Philonaster* KOEHLER, 1909]. Like *Metopaster* in shape and marginal ornament but with no enlarged ultimate superomarginals and many more marginals (up to 18 in half arc) and with tabulate (not flat) aboral ossicles; arms not upturned strongly as in *Recurvaster. ?U. Cret.(Maastricht.),* Eu., *Rec.*——FIG. 53,*1*. *C.? dividuus* (RASMUSSEN), Maastricht., Denm.; *1a,b,* side and aboral views, ×1 (125).

Metopaster SLADEN, 1893 [**Goniaster (Goniodiscus) parkinsoni* FORBES, 1848; SD RASMUSSEN, 1950 (ICZN Opin. 331)] [=*Mitraster* SLADEN, 1893; *Ravniaster* BRÜNNICH-NIELSEN, 1943; ?*Dictydaster* MERCIER, 1935]. Pentagonal, with sharp-pointed arms; rarely with arms slightly produced; marginals very large and few (2 to 5 superomarginals in half arc); enlarged ultimate superomarginals corresponding to 2 to 7 inferomarginals; central area of marginals with fine pits for granules or smooth, surrounded by distinct narrow area with several rows of setae. *U. Cret.(Cenoman.)-Mio.,* Eu.-N.Am.-N.Z. —— FIG. 53,*3*. **M. parkinsoni* (FORBES), Santon., Eng.; *3a,b,* aboral and lat. views, ×1 (131); *3c,* profile of superomarginals and inferomarginals, ×2 (133). (See also Fig. 27.)

Recurvaster BRÜNNICH-NIELSEN, 1943 [**R. stevensensis* (=**Metopaster tumidus radiatus* SPENCER, 1913); OD]. Differs from *Metopaster* in having no enlarged ultimate superomarginals, more marginals, and arms produced and upturned, making distal marginals skew. *U.Cret.(Campan.)-Eoc.,* NW.Eu.——FIG. 53,*2*. **R. radiatus* (SPENCER), Campan., Denm.; *2a,b,* aboral and lat. views of arm, ×1; *2c,* profile of superomarginal and inferomarginal, ×1 (125).

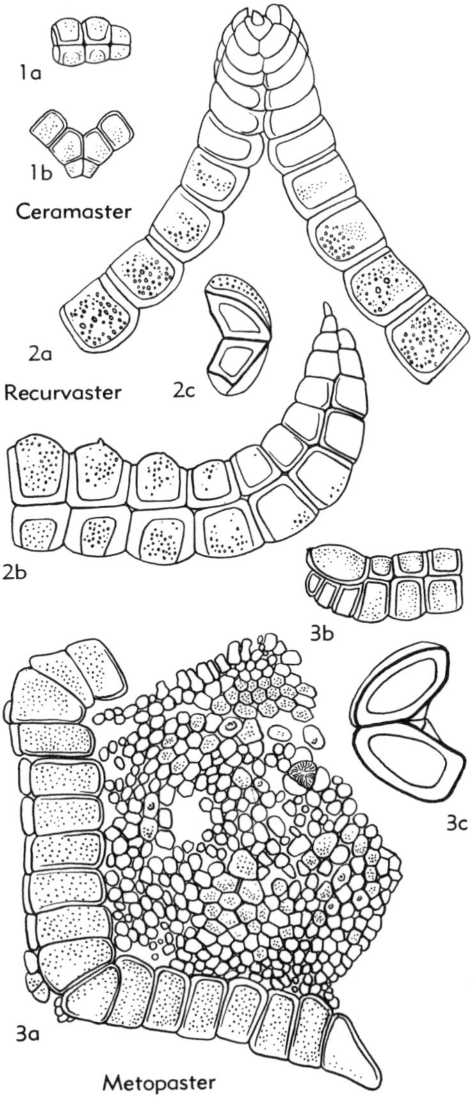

FIG. 53. Goniasteridae (Goniasterinae) (p. *U57*).

Spenceria FOURTEAU, 1914 [**Metopaster teilhardi* DE LORIOL, 1908; OD]. Ultimate inferomarginals seemingly enlarged like superomarginals. [Probably a young *Metopaster.*] *U.Cret.(Santon.),* Eng.

Sphaeriodiscus FISHER, 1910 [**Stephanaster bourgueti* PERRIER, 1894]. Only differs from *Metopaster* in that last few superomarginals are not united in single ossicle, penultimate superomarginals normally larger than median ones. *U.Cret. (Campan.),* Eng., *Rec.*

Cladaster VERRILL, 1899 [**C. rudis*; OD]. *Rec.*

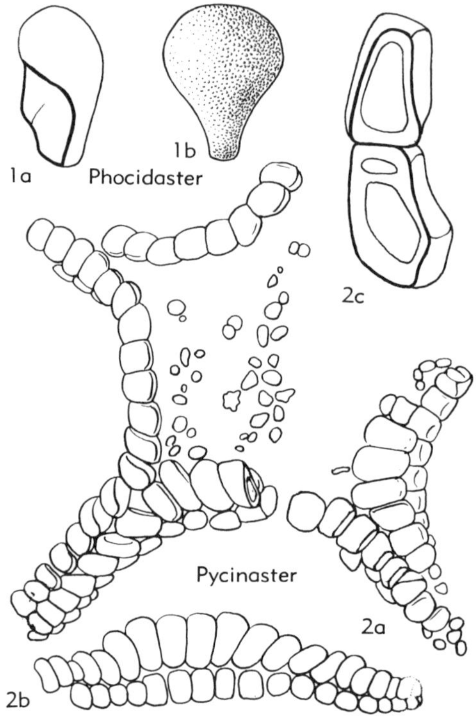

FIG. 54. Goniasteridae (Pycinasterinae) (p. U59).

Iconaster SLADEN, 1889 [**Astrogonium longimanum* MÖBIUS, 1859; OD]. *Rec.*

I. (Iconaster). *Rec.*

I. (Glyphodiscus) FISHER, 1917 [**Iconaster periectus* FISHER, 1913; OD]. *Rec.*

Pentagonaster GRAY, 1840 [**P. pulchellus*; OD] [=*Astrogonium* MÜLLER & TROSCHEL, 1842; *Stephanaster* AYRES, 1851]. Pentagonal or with moderately long arms, which have broad rounded ends; marginals with smooth central area; distalmost or one proximal to it in upper and lower series may be enlarged; inferomarginals corresponding with superomarginals except at extreme tip of arm. [There is very little difference between *Pentagonaster, Tosia,* and *Metopaster.*] *Rec.*

Pergamaster KOEHLER, 1920 [**P. tessellatus* (=?*Pentagonaster incertus* BELL, 1908)]. *Rec.*

Plinthaster VERRILL, 1899 [**Pentagonaster perrieri* SLADEN, 1889; OD] [=*Pyrenaster* VERRILL, 1889]. *Rec.*

Pontioceramus FISHER, 1911 [**P. grandis*; OD]. *Rec.*

Tesselaster H. L. CLARK, 1941 [**T. notabilis*; OD]. *Rec.*

Toraster A. M. CLARK, 1952 [**Astrogonium tuberculatum* GRAY, 1847]. *Rec.*

Tosia GRAY, 1840 [**T. australis*; OD]. *Rec.*

Tosiaster VERRILL, 1914 [**Tosia arcticus* VERRILL, 1909; OD]. *Rec.*

Subfamily CHITONASTERINAE Fisher, 1911

Disc small, arms moderately long, straight-sided, covered by membrane; aboral ossicles each with stout spine, like those present in vertical series on marginals; 3 similar spines on each adambulacral. *Rec.*

Chitonaster SLADEN, 1889 [**C. cataphractus*; OD] [=*Chitonaster* SLADEN, 1885 *(nom. nud.)*]. *Rec.*

Subfamily ANTHENOIDINAE Fisher, 1919

[=Leptogonasterinae PERRIER, 18--]

Body enclosed by thin membrane, generally covering or covered by granules; plates of aboral surface tending to be stellate; secondary aboral plates generally present. *Rec.*

Anthenoides PERRIER, 1881 [**A. peircei*; OD] [=*Leptogonaster* SLADEN, 1889; *Antheniaster* VERRILL, 1899]. *Rec.*

Atelorias FISHER, 1911 [**A. anacanthus*; OD]. *Rec.*

Ogmaster VON MARTENS, 1865 [**Goniodiscus capella* MÜLLER & TROSCHEL, 1842; OD] [=*Dorigona* GRAY, 1866]. *Rec.*

Siraster H. L. CLARK, 1915 [**S. tuberculatus*; OD]. *Rec.*

Stellaster GRAY, 1840 [**S. childreni* (=**Asterias equestris* RETZIUS, 1805)]. *Rec.*

Subfamily HIPPASTERIINAE Verrill, 1899

Aboral surface covered by well-spaced larger ossicles packed with intercalated smaller ones, in internal view forming coarse network; marginals with conical tubercles or stout spines. *U.Cret.-Rec.*

Hippasteria GRAY, 1840 [**H. europaea* (=**Asterias phrygiana* PARELIUS, 1768); OD (other included species is synonym)]. Disc large, arms short; marginals bare except for few large tubercle-like spines and granules around edge; larger aboral ossicles tumid and smooth except for marginal granules. *U.Cret.*, N.Z., *Rec.*

H. (Hippasteria) [=*Euhippasteria* DONS, 1938]. *U.Cret.*, N.Z., *Rec.*

H. (Nehippasteria) DONS, 1938 [**H. (N.) insignis*; OD]. *Rec.*

Cryptopeltaster FISHER, 1904 [**C. lepidonotus*; OD]. *Rec.*

Evoplosoma FISHER, 1906 [**E. forcipifera*; OD]. *Rec.*

Subfamily NECTRIINAE Perrier, 1894

Superambulacral plates present; with intermarginal as well as aboral papulae. *Rec.*

Nectria GRAY, 1840 [**Asterias ocellifera* LAMARCK, 1816; OD]. *Rec.*

Nectriaster H. L. CLARK, 1946 [**Mediaster monacanthus* H. L. CLARK, 1916; OD]. *Rec.*

Subfamily PSEUDARCHASTERINAE Sladen, 1889

Abactinal plates paxilliform or tabulate; superambulacra present, at least in rudimentary form. Pedicellariae, if present, spiniform, fasciolar, or incipiently bivalved. *U.Jur.-Rec.*

Pseudarchaster SLADEN, 1889 [**P. discus*; SD FISHER, 1911] [=*Pseudarchaster* SLADEN, 1885 *(nom. nud.)*]. Aboral plates paxilliform, in radial rows of which several extend along arm; marginals thick, with fasciolated grooves between; oral intermediate plates in transverse and longitudinal series; mouth-angle plates large and prominent. *U.Eoc.-L.Mio.*, N.Z., *Rec.*

Aphroditaster SLADEN, 1889 [**A. gracilis*; OD] [=*Aphroditaster* SLADEN, 1885 *(nom. nud.)*]. Doubtfully distinct from *Pseudarchaster. Rec.*

Paragonaster SLADEN, 1889 [**P. ctenipes*; SD FISHER, 1919]. Single series of flat granulose plates extending along aboral surface of arm between superomarginals. *U.Jur.-Mio.*, N.Z., *Rec.*

Perissogonaster FISHER, 1913 [**P. insignis*; OD]. Only differs from *Paragonaster* in having odd interradial marginal in each series. *Rec.*

Subfamily PYCINASTERINAE Spencer & Wright, n. subfamily

Very robust forms with rather small disc and long arms; superomarginals high, swollen, with rounded profile, and with large hollows for intermarginal muscles; spines almost absent; marginals and aboral ossicles may have feeble rugosities. Alveolar pedicellariae. *L.Jur.-Mio.*

This group seems to have no close relationship with any other Mesozoic Goniasteridae.

Pycinaster SPENCER, 1907 [*pro Pycnaster* SLADEN, 1891 (*non* POMEL, 1883)] [**Goniaster (Goniodiscus) angustatus* FORBES, 1848; OD]. Characters of subfamily. *L.Jur.-Mio.*, W.Eu.——FIG. 54,*2*. **P. angustatus* (FORBES), U.Cret.(Santon.), Eng. (Kent); *2a,b,* aboral and lat. views, ×1; 2c, profile of superomarginal and inferomarginal, ×2 (131).

?**Phocidaster** SPENCER, 1913 [**P. grandis*; OD]. Known only from large interradial superomarginals, which are high, short, club-shaped, with swollen aboral end; surface consisting of fine shallow spine pits separated by rugosities. *Cret. (U.Alb.-Cenoman.)*, Eng.——FIG. 54,*1*. **P grandis,* U.Alb., Devon; *1a,b,* profile and lat. views of superomarginal, ×2 (139).

Subfamily UNCERTAIN

Without thorough revision of the family the remaining fossil and Recent members cannot be satisfactorily placed in subfamilies. Some of the following genera might perhaps be placed in Goniasterinae, but most of them should obviously be assigned to one of several unnamed subfamilies. The subfamilial name Mediasterinae VERRILL, 1914, is available for one group. In the following account genera known as fossils are listed first in alphabetical order, then those known only as living forms.

Calliaster GRAY, 1840 [**C. childreni*; OD]. Arms rather long; marginals large, with large bosses that carry short stout spines; radial lines of stout spines on aboral surface and still larger spines on ossicles of primary circlet. *U.Eoc.-U.Oligo.*, N.Z., *Rec.*

Calliderma GRAY, 1847 [**C. emma*; OD] [=*Tomidaster* SLADEN, 1891]. Disc large, with short arms passing evenly into wide interradial arcs; marginals short, wide, relatively larger than in *Comptonia* and wider than in *Tylasteria,* with fine hexagonal spine pits and, irregularly, large shallow depressions; tessellate close-fitting oral and aboral plates. Valvate pedicellariae may be abundant. *U.Cret.(Cenoman.)-Oligo.*, Eu., *Rec.* ——FIG. 55,*6*. *C. smithiae* (FORBES); *6a,b,* oral and aboral sides, ×1 (Cenoman., Eng., Sussex) (131); *6c,* profile of superomarginal and inferomarginal, ×2 (Turon., Eng., Devon) (133).

Cenomanaster WRIGHT, 1951 [*pro Jacobella* MERCIER, 1935 (*non* JEANNET, 1908)] [**Jacobella cenomanensis* MERCIER, 1935; OD]. Disc rather large; arms narrow at base, long, straight-sided, tapering very gradually, with superomarginals not in contact; marginals short, wide, with granules; single large tubercle on aboral side of superomarginals; aboral ossicles granulose, irregularly rounded; some with conical tubercle. *U.Cret. (Cenoman.)*, Fr.——FIG. 56,*4*. **C. cenomanensis* (MERCIER), Sarthe; aboral side, ×1 (121).

Chomataster SPENCER, 1913 [**C. acules*; OD] [=*?Huraeaster* VALETTE, 1915]. Long slender arms sharply demarcated from disc; interradial margins generally straight, with wedge-shaped superomarginal at junction with arm; marginals tall, those of arm and interray differing in profile, generally with fine close spine pits. *U.Cret. (Santon.-Maastricht.)*, NW.Eu.——FIG. 55,*3*. *C.* sp., Campan., Eng.(Norfolk); aboral side, ×1 (139).

Comptonia GRAY, 1840 [**C. elegans*; OD]. Arms long, slender, distinct from slightly curved or straight interradial arcs; marginals square and rather small on arms, short and wide in interrays, with fine, close spine pits. Large valvate pedicellariae common. *L.Cret.(Apt.-U.Alb.), ?U.Cret. (Santon.)*, Eng.-?Egypt-?Can.——FIG. 55,*5*. *C. comptoni* (FORBES), U.Alb., Eng.; *5a,b,* aboral and lat. views, ×1 (131).

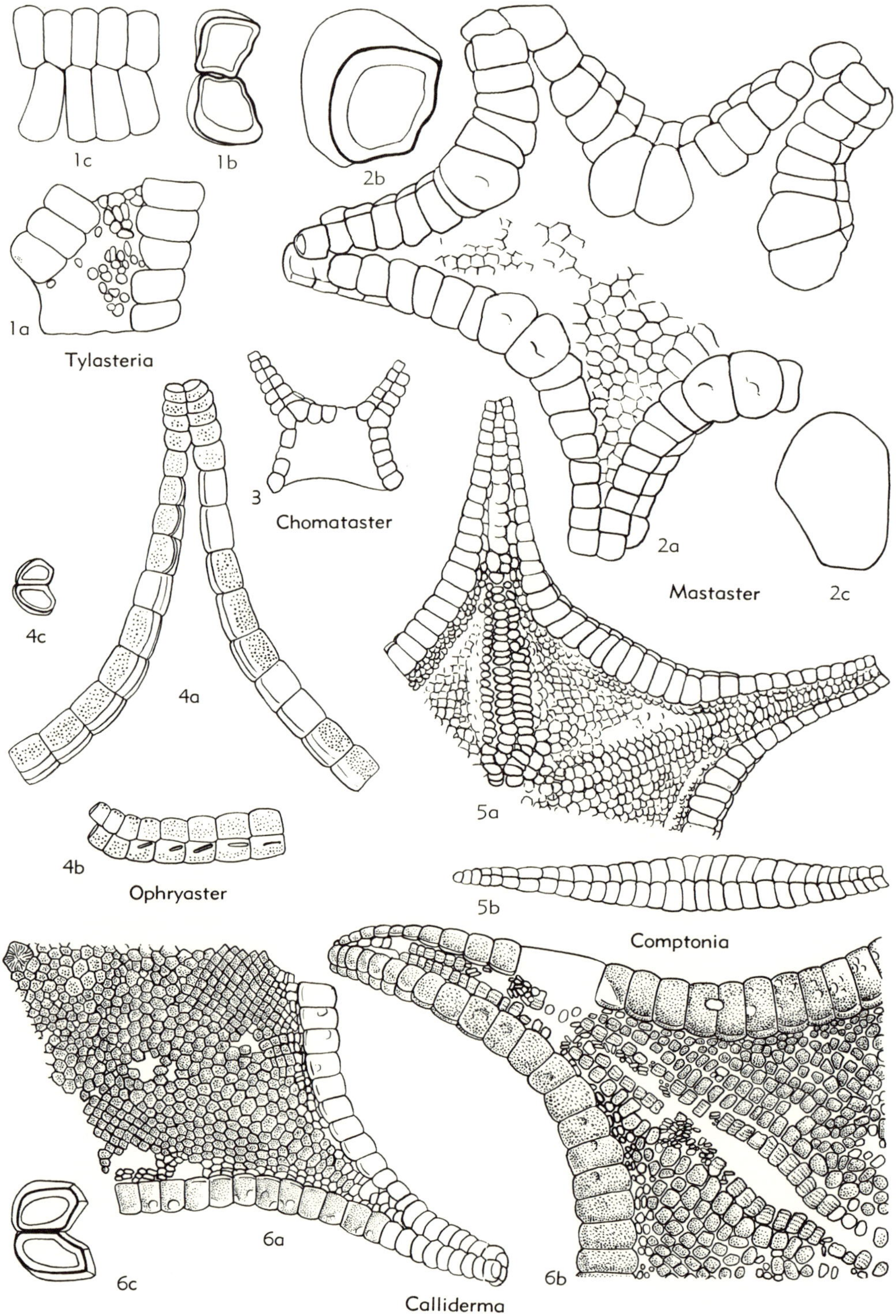

FIG. 55. Goniasteridae (subfamily uncertain) (p. *U*59, *U*62-*U*63).

Cottreauaster Wright, 1951 [*pro Spenceraster* Cottreau, 1929 (*non* Lambert, 1913)] [**Spenceraster lamarei* Cottreau, 1929; OD]. Disc small, arms long, narrow, straight-sided, flexible; marginals small, tumid with fine granules; superomarginals with tubercles or spines, aboral plates irregular. *M.Jur.(Bathon.)*, Fr.——Fig. 56,*3*. **C. lamarei* (Cottreau), Orne; aboral side, ×2 (105).

Crateraster Spencer, 1913 [**Asterias quinqueloba* Goldfuss, 1822; OD, ICZN Opin. 331] [=*Austinaster* Adkins, 1928]. Pentagonal to stellate, with arms slightly produced; marginals large, rather few (4 to 7 in half arc), more or less op-

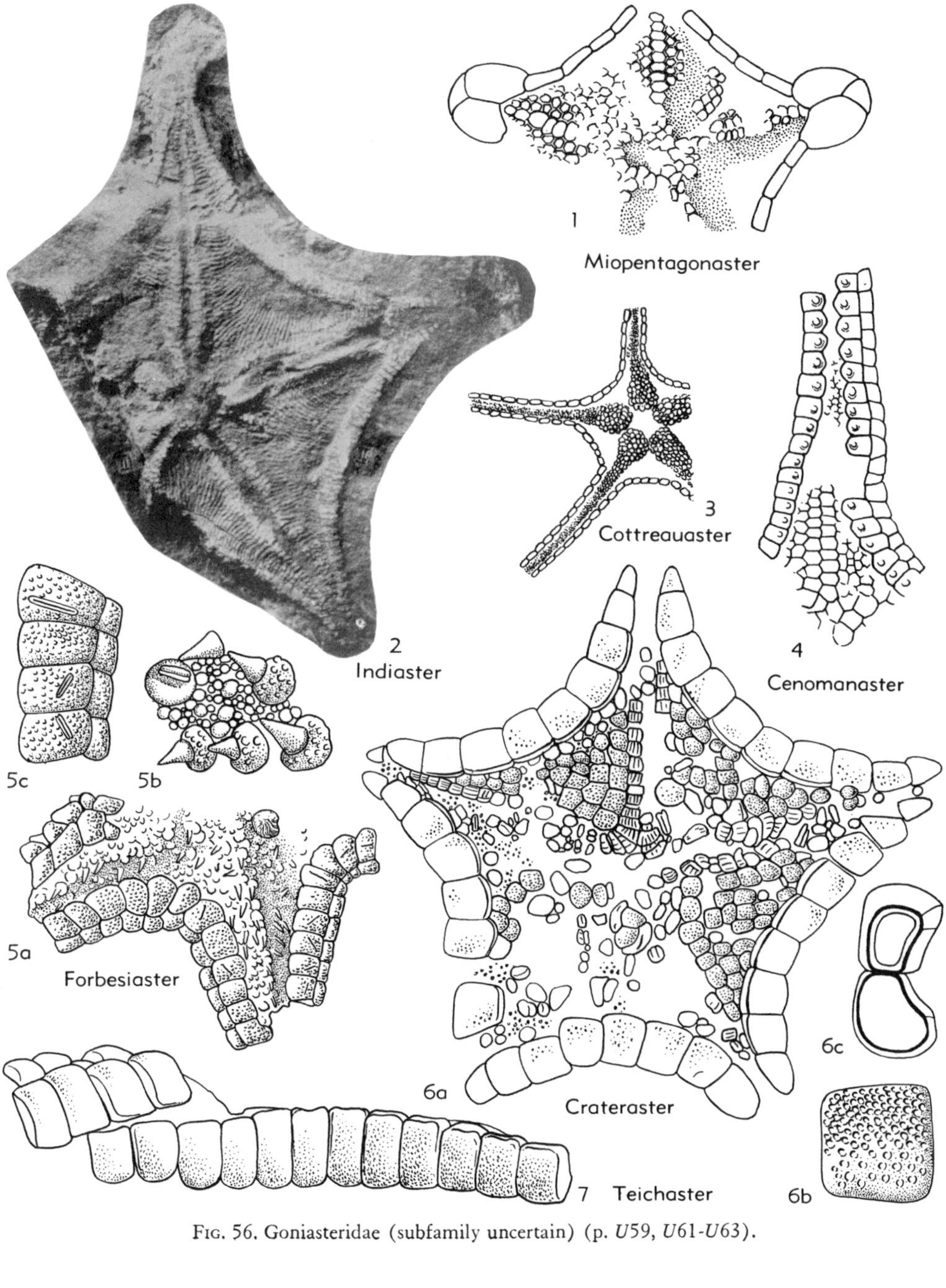

Fig. 56. Goniasteridae (subfamily uncertain) (p. *U*59, *U*61-*U*63).

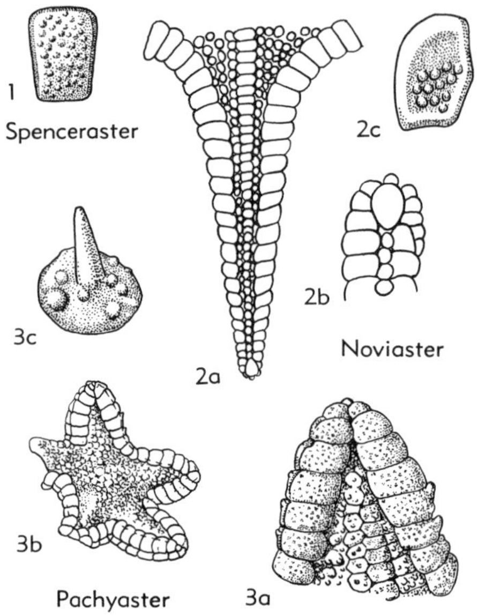

FIG. 57. Goniasteridae (subfamily uncertain) (p. U62-U63).

posite, last few superomarginals in contact on mid-line of arm; marginals with distinct lateral and oral or aboral faces, lateral faces at least with shallow crater-like pits; profile of marginals like that of some Jur. *Tylasteria;* oral and aboral ossicles large, tessellate. *Cret.(U.Alb.-Campan.)*, Eu.-N.Am.——FIG. 56,6. **C. quinqueloba* (GOLDFUSS), Santon., Eng.(Kent); *6a,* oral side, ×1 (131); *6b,* lat. face of marginal, ×2 (131); *6c,* profile of superomarginal and inferomarginal, ×2 (133).

Forbesiaster DE LORIOL, 1909 [**F. wrighti*; OD]. Arms wide at base, rounded at tip; no distinct interradial arc; marginals twice as wide as long, with widely spaced pits for granules and short spines around edges; aboral ossicles irregular, rounded or tumid, largest with granules and round conical spines. Large bivalved pedicellariae on most superomarginals. *U.Cret.(Santon.)*, Egypt.——FIG. 56,5. **F. wrighti*; *5a,* aboral side, ×1; *5b,* part of aboral surface, ×4; *5c,* superomarginals, ×2 (117).

Indiaster RAO, 1957 [**I. krishna*; OD]. Small, very short marginals; interradial areas on oral surface with rows of rod-shaped ossicles that simulate metapinnules. *M.Jur.(U.Bathon.)*, India(Cutch).——FIG. 56,2. **I. krishma*; oral surface, ×6 (Rao).

Leptogonium POMEL, 1887 [**L. mauritanicum*; OD]. Figure only; not recognizable beyond family. *Plio.(Sahelian)*, N.Afr.(Alg.).

Mastaster MERCIER, 1935 [**M. villersensis*; OD]. Like *Tylasteria* but with fewer, bulkier marginals and 2 very large highly swollen superomarginals in center of each interray. *U.Cret.(Cenoman.-Turon.)*, Eng.-Fr.——FIG. 55,2. **M. villersensis*; *2a,* aboral side, ×1 (Cenoman., Fr.); *2b,c,* profile and lat. views of interradial superomarginal, ×2 (Cenoman., Eng.) (139).

Mediaster STIMPSON, 1857 [**M. aequalis*; OD] [=*Isaster* VERRILL, 1894 (*non* DESOR, 1858)]. Long slender arms with several series of aboral ossicles separating superomarginals, one series generally reaching tip of arm; aboral ossicles tabulate. Rudimentary superambulacral ossicles present. *L.Mio.*, N.Z., *Rec.*

Miopentagonaster MERCIER, 1935 [**M. calloviensis*; OD]. Small, nearly straight interradial margin formed by 4 long, narrow, low granulose marginals in each series; distally much smaller marginal is followed by large terminal; aboral ossicles hexagonal, granulose, with marked spaces for papulae; carinals slightly raised and larger than other ossicles. *U.Jur.(Callov.)*, Fr.——FIG. 56,1. **M. calloviensis,* Calvados, aboral side, ×2 (121).

Noviaster VALETTE, 1929 [**N. lissajousi*; OD]. Arms long, straight-sided, moderately wide at base, tips blunt; interradial arcs rounded; superomarginals regular, distinctly skew, tumid, with large tubercles on faces between adjoining ossicles, surface with fine hexagonal pits; carinals prominent, 3 rows, reduced to 1 distally, along arms ending in large tumid oval ossicle. *M.Jur.(Bathon.)*, Eng.-Fr.——FIG. 57,2. **N. lissajousi,* Fr.; *2a,* aboral side, ×1; *2b,* tip of arm, ×4; *2c,* profile of superomarginal, ×4 (136).

Nymphaster SLADEN, 1889 [**Nymphaster protentus* SLADEN, 1889 (=**Pentagonaster arenatus* PERRIER, 1881 (obj.); SD FISHER, 1917] [=*Nymphaster* SLADEN, 1885 *(nom. nud.)*]. Arms long, slender, sharply distinct from disc; superomarginals in contact for whole length of arms; aboral plates tessellate. *L.Mio.*, N.Z.-Cuba, *Rec.*

Ophryaster SPENCER, 1913 [**Nymphaster oligoplax* SLADEN, 1891; OD]. Long slender arms passing into evenly rounded interradial arcs, marginals large, long, narrow, blocky, slightly tumid, with close or distant hexagonal or circular spine pits, which are generally absent from edge of superomarginals next to aboral ossicles; superomarginals in contact along distal part only of arms; no space for papulae between aboral ossicles. Long low bivalved pedicellariae common. *U.Cret.(Turon.-Campan.)-Mio.*, NW.Eu.——FIG. 55,4. *O. magnus* SPENCER, Campan., Denm.; *4a,b,* aboral and lat. views of arm, ×1; *4c,* profile of superomarginal and inferomarginal, ×1 (125).

Pachyaster DE LORIOL, 1909 [**P. aegyptiacus*; OD]. Aboral side resembling *Forbesiaster,* of which it may be synonym, based on juvenile.——FIG. 57,3. **P. aegyptiacus*; *3a,* arm, ×3; *3b,* aboral side, ×1; *3c,* aboral ossicle, ×20 (117).

Spenceraster LAMBERT, 1913 · [*pro Trachyaster* SPENCER, 1913 (*non* POMEL, 1883)] [**Nymphaster rugosus* SPENCER, 1907; SD SPENCER &

WRIGHT, herein]. Small, with short narrow arms and rounded interradial arcs; superomarginals in contact along mid-line of arms; marginals few, low, nearly as long as wide, with evenly curved profile, surface covered with rugosities but generally with narrow smooth band around edge. *Cret.(U.Alb.-Cenoman.)*, Eng.——FIG. 57,*1*. **S. rugosus* (SPENCER), aboral side of superomarginal, ×4 (133).

Teichaster SPENCER, 1913 [**T. favosus*; OD]. Arms more produced than in *Crateraster*, from which it is derived; body high but flat; marginals with high vertical lateral face; spine pits large, shallow, close, hexagonal or circular. *U.Cret. (Campan.)-Mio.*, Eu.——FIG. 56,*7*. **T. favosus*, U.Cret.(Maastricht.), Eng.(Norfolk); lat. view of fragment, ×1 (139).

Tylasteria VALETTE, 1930 [*pro Tylaster* SPENCER, 1913 (*non* DANIELSSON & KOREN, 1881)] [**Asterias jurensis* GOLDFUSS, 1822; OD]. Robust, with large disc and tapering arms, moderately broad at base; interradial arc well rounded; marginals wider than long, slightly tumid, profile evenly curved or square or undercut, densely covered by generally hexagonal spine pits; aboral plates large, flat, with hexagonal pits. *M.Jur.(Bajoc.)-L.Cret.(Alb.)*, Eu.——FIG. 55,*1*. **T. jurensis* (GOLDFUSS), U.Jur.(Oxford.), Ger.; *1a,c*, oral and lat. views; *1b*, profile of superomarginal and inferomarginal (123).

Amphiaster VERRILL, 1868 [**A. insignis*; OD]. *Rec.*

Astroceramus FISHER, 1906 [**A. callimorphus*; OD]. *Rec.*

Astrothauma FISHER, 1913 [**A. euphylacteum*; OD]. *Rec.*

Circeaster KOEHLER, 1909 [**Circeaster marcelli* KOEHLER, 1909; SD SPENCER & WRIGHT, herein].

Eugoniaster VERRILL, 1899 [**Pentagonaster investigatoris* ALCOCK, 1893; OD]. *Rec.*

Gigantaster DÖDERLEIN, 1924 [**G. weberi*; OD]. *Rec.*

Gilbertaster FISHER, 1906 (**G. anacanthus*; OD]. *Rec.*

Johannaster KOEHLER, 1909 [**J. superbus*; OD]. *Rec.*

Lithosoma FISHER, 1911 [**L. actinometra*; OD]. *Rec.*

Litonotaster VERRILL, 1889 [**Pentagonaster intermedius* PERRIER, 1884; OD]. *Rec.*

Lydiaster KOEHLER, 1909 [**L. johannae*; OD]. *Rec.*

Mahabissaster MACAN, 1938 [**M. zengi*; OD]. *Rec.*

Mariaster A. H. CLARK, 1916 [**Johannaster giganteus* GOTO, 1914; OD]. *Rec.*

Milteliphaster ALCOCK, 1893 [**M. woodmasoni*; OD]. *Rec.*

Notioceramus FISHER, 1940 [**N. anomalus*; OD]. *Rec.*

Peltaster VERRILL, 1899 [**P. hebes* =(**Goniaster nidarosiensis* STORM, 1881); OD]. *Rec.*

Progoniaster DÖDERLEIN, 1924 [**P. atavus*; OD]. *Rec.*

Pseudogoniodiscaster LIVINGSTONE, 1930 [**P. wardi*; OD]. *Rec.*

Rosaster PERRIER, 1894 [**Pentagonaster alexandri* PERRIER, 1881; OD] [=*Nereidaster* VERRILL, 1899]. *Rec.*

Sibogaster DÖDERLEIN, 1924 [**S. digitatus*; OD]. *Rec.*

Styphlaster H. L. CLARK, 1938 [**S. notabilis*; OD]. *Rec.*

Family OREASTERIDAE Fisher, 1911

[=Pentacerotidae GRAY, 1841; Antheneinae FISHER, 1911]

Disc large, generally high and swollen, even cushion-like in adult, with robust arms or none; younger stages generally resembling Goniasteridae; body normally covered with thick granulose membrane; marginals large; intermarginals may be present; abactinal skeleton reticulate, composed of stellate plates, in many forms bearing stout spines. Papulae numerous, in special areas; calcareous interbrachial septa. *Rec.*

Oreaster MÜLLER & TROSCHEL, 1842 [**Asterias reticulatus* LINNÉ, 1758; OD] [=*Pentaceros* (SCHULZE, 1760 *(non. binom.)*) GRAY, 1840 *(non* CUVIER & VALENCIENNES, 1829)]. *Rec.*

Anthaster DÖDERLEIN, 1915 [**Oreaster valvulatus* MÜLLER & TROSCHEL, 1843; OD]. *Rec.*

Anthenea GRAY [**A. chinensis* (=**Asterias pentagonula* LAMARCK, 1816); OD] [=*Hosia* GRAY, 1840; *Hosea* GRAY, 1866]. *Rec.*——FIG. 1,*1*. *A. flavescens* (GRAY), Rec.; *1a,b*, oral and aboral surfaces, ×1 (130).

Asterodiscus GRAY, 1847 [**A. elegans*; OD]. *Rec.*

Bothriaster DÖDERLEIN, 1916 [**B. primigenius*; OD]. *Rec.*

Choriaster LÜTKEN, 1869 [**C. granulatus*; OD]. *Rec.*

Culcita AGASSIZ, 1836 [**Asterias discoidea* LAMARCK, 1816 (=**Asterias schmideliana* RETZIUS, 1805); OD] [=*Randasia* GRAY, 1840; *Goniodiscus* MÜLLER & TROSCHEL, 1842; *Goniodiscoides* FISHER, 1906]. *Rec.*

Goniodiscaster H. L. CLARK, 1909 [**Asterias pleyadella* LAMARCK, 1816; OD]. *Rec.*

Gymnanthenea H. L. CLARK, 1938 [**Anthenea globigera* DÖDERLEIN, 1916; OD]. *Rec.*

Halityle FISHER, 1913 [**H. regularis*; OD] [=*Culcitaster* H. L. CLARK, 1915]. *Rec.*

Monachaster DÖDERLEIN, 1916 [**Goniodiscus sanderi* MEISSNER, 1892; OD]. *Rec.*

Nidorellia GRAY, 1840 [**Pentaceros (Nidorellia) armatus*; OD]. *Rec.*

Paulia GRAY, 1840 [**P. horrida*; OD] [=*Pauliella* LUDWIG, 1905]. *Rec.*

Pentaceraster DÖDERLEIN, 1916 [**Asterias mammillatus* AUDOUIN, 1827; OD]. *Rec.*

Pentaster DÖDERLEIN, 1935 [*pro Pentaceropsis* SLADEN, 1889 (*non* STEINDACHNER & DÖDERLEIN,

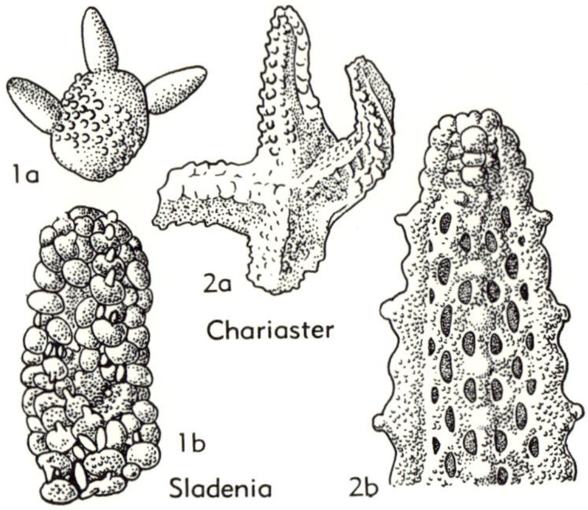

FIG. 58. Ophidiasteridae (p. U64).

1884)] [**Asterias obtusata* BORY DE SAINT VINCENT, 1827; OD]. *Rec.*

Poraster DÖDERLEIN, 1916 [**Oreaster productus* BELL, 1884 (=**Oreaster superbus* MOÖBIUS, 1859); OD]. *Rec.*

Protoreaster DÖDERLEIN, 1916 [**Asterias nodosa* LINNÉ, 1758; OD]. *Rec.*

Pseudanthenea DÖDERLEIN, 1916 [**Anthenea grayi* PERRIER, 1876; OD]. *Rec.*

Pseudoreaster VERRILL, 1899 [**Asterias obtusangulus* LAMARCK, 1816; OD]. *Rec.*

Stellasteropsis DOLLFUS, 1936 [**S. fouadi*; OD]. *Rec.*

Family OPHIDIASTERIDAE Verrill, 1867

[=Linckiidae PERRIER, 1875]

Disc small, arms long and slender, generally more or less cylindrical; body normally covered by granulose membrane; marginals small; aboral skeleton tessellate; small superambulacral plates generally present. Pedicellariae foraminate or excavate, or lacking. *U.Cret.-Rec.*

Ophidiaster AGASSIZ, 1835 [**Asterias ophidianus* LAMARCK, 1816; OD] [=?*Tamaria* GRAY, 1840; *Chione* GISTL, 1847]. *Rec.*

Austrofromia H. L. CLARK, 1921 [**Fromia polypora* H. L. CLARK, 1916; OD]. *Rec.*

Bunaster DÖDERLEIN, 1896 [**B. ritteri*; OD]. *Rec.*

Certonardoa H. L. CLARK, 1921 [**Scytaster semiregularis* MÜLLER & TROSCHEL, 1842; OD]. Like *Nardoa,* but oral surface of arms flat, not convex, and cross section of arms triangular at base; aboral ossicles in regular radial series on proximal part of arms. Papulae in groups, as in *Nardoa,* but none on oral surface. *Mio.,* Formosa, *Rec.*

Chariaster DE LORIOL, 1909 [**C. elegans*; OD]. Marginals in 2 prominent rows, with large mammillate tubercles generally on alternate ossicles; aboral surface flat but with median row of tubercles on distal part of arm formed by swollen intersections of long ossicles; adambulacrals with 2 rows of spines. *U.Cret.(Santon.),* Egypt.——FIG. 58,*2.* **C. elegans*; *2a,* aboral side, ×1; *2b,* tip of arm, ×4 (117).

Cistina GRAY, 1840 [**C. columbiae*; OD]. *Rec.*

Copidaster A. H. CLARK, 1948 [**C. lymani*; OD]. *Rec.*

Dactylosaster GRAY, 1840 [**Asterias cylindrica* LAMARCK, 1816; SD H. L. CLARK, 1921]. *Rec.*

Dissogenes FISHER, 1913 [**D. styracia*; OD]. *Rec.*

Ferdina GRAY, 1840 [**F. flavescens* GRAY, 1840; SD FISHER, 1919]. *Rec.*

Fromia GRAY, 1840 [**Asterias milleporella* LAMARCK, 1816; OD]. *Rec.*

Gomophia GRAY, 1840 [**G. egyptiaca*; OD]. *Rec.*

Hacelia GRAY, 1840 [**Ophidiaster (Hacelia) attenuatus*; OD]. *Rec.*

Leiaster PETERS, 1852 [**L. coriaceus*; SD FISHER, 1919] [=*Lepidaster* VERRILL, 1871 (*non* FORBES, 1850)]. *Rec.*

Linckia NARDO, 1834 [**L. typus* (=**Asterias laevigatus* LINNÉ, 1758); OD] [=*Cribella* AGASSIZ, 1835 (*non* FORBES, 1841); *Acalia* GRAY, 1840; *Catantes, Undina* GISTL, 1847]. Arms cylindrical; aboral ossicles not in regular longitudinal series; adambulacrals with 2 or 3 rows of granules. No pedicellariae. *U.Cret.,* Eng., *Rec.*

Nardoa GRAY, 1840 [**Asterias variolata* RETZIUS, 1805; SD H. L. CLARK, 1921] [=*Melia* GISTL, 1847]. *Rec.*

Narcissia GRAY, 1840 [**N. teneriffae* (=**Asterias canariensis* D'ORBIGNY, 1839); OD]. *Rec.*

Neoferdina LIVINGSTONE, 1931 [**Ferdina cumingi* GRAY, 1840; OD]. *Rec.*

Pharia GRAY, 1840 [**Ophidiaster (Pharia) pyramidatus*; OD]. *Rec.*

Phataria GRAY, 1840 [**Linckia (Phataria) unifascialis*; OD]. *Rec.*

Plenardoa H. L. CLARK, 1921 [**Linckia semiseriata* MARTENS, 1865; OD]. *Rec.*

Pseudophidiaster H. L. CLARK, 1916 [**P. rhysus*; OD] [=*Pseudolinckia* H. L. CLARK, 1916, *lapsus*]. *Rec.*

Sladenia DE LORIOL, 1909 [**Nardoa? fourteaui* DE LORIOL, 1904; OD]. Arms broadly flattened, rounded at tip; marginals rather large, mostly bearing short, stout spines; aboral ossicles oval, tumid, spinose; adambulacrals with ?2 rows of spines. *U.Cret.(Santon.),* Egypt.——FIG. 58,*1.* **S. fourteaui* (DE LORIOL); *1a,* aboral ossicle, enl.; *1b,* aboral side of arm, ×1 (Loriol).

Family RADIASTERIDAE Fisher, 1916

[=Mimasterinae SLADEN, 1889]

Marginals small and subpaxilliform; membranous interbrachial septa and superambulacral plates present; aboral plates consisting of penicillate paxillae; oral intermediate plates imbricated in transverse series. *Rec.*

Radiaster PERRIER, 1881 [**R. elegans*] [=*Mimaster* SLADEN, 1882]. *Rec.*

Gephyriaster FISHER, 1910 [**Mimaster swifti* FISHER, 1905]. *Rec.*

Order SPINULOSIDA Perrier, 1884

[*nom. correct.* SPENCER & WRIGHT, herein (*pro* Spinulosa PERRIER, 1884)] [=Velata PERRIER, 1894]

Mouth frame adambulacral; mouth-angle plates prominent, not keeled; marginal frame only rarely present except in juveniles; mouth-angle plates placed on prominent axillary if marginals are present. Pedicellariae, if present, generally consisting of grouped spines. Aboral skeleton reticulate, imbricate or absent, in many forms consisting of regular rows of paxillae, but in early family Taeniactinidae consisting of 3 rows of rather large plates in each radius. *M.Ord.-Rec.*

Since so few fossil forms are yet known, classification of this order must be provisional.

Suborder EUGNATHINA Spencer & Wright, new suborder

Mouth-angle plates large, spade- or plowshare-shaped, with conspicuous marginal and suboral spines; ambulacral furrows wide; adambulacral spines pectinate. *M. Ord.-Rec.*

Family TAENIACTINIDAE Spencer, 1927

[=Calyptactininae SPENCER, 1930]

Aboral surface of arms with 3 rows of prominent ossicles in each radius. *U.Ord.-Miss.*

Taeniactis SPENCER, 1927 [**T. wenlocki*; OD]. Oral interrays with few scattered ossicles; aboral skeleton confined to disc and bases of arms. *L.Sil.*, Scot.——FIG. 59,*3*. **T. wenlocki*; *3a*, individual with arms flexed upward, ×3; *3b,c*, oral and aboral surfaces, ×7.5 (133).

Baliactis SPENCER, 1922 [**B. ordovicus*; OD] [=*Leioactis*, ?*Palaeactis* LEHMANN, 1957]. Oral interrays bearing large broad axillary. *U.Ord.-Dev.*, Eng.-Ger.——FIG. 59,*2a,b*. *B. devonicus* SPENCER, L.Dev., Ger.; oral side of proximal part of arm, ×1 (133).

Calyptactis SPENCER, 1930 [**C. spinosus*; OD]. Aboral skeleton of closely fitting ossicles; arms apparently normally enrolled in life. *L.Carb. (Miss.)*, Eng.-N.Am.——FIG. 59,*4*. *C. demissus* (MILLER), Miss., USA; aboral side of enrolled specimen, ×1 (133).

Lepidasterella SCHUCHERT, 1914 [**L. babcocki* (=**Helianthaster gyalum* CLARKE, 1908); OD]. Arms 24 or more, with 3 rows of carinal and superomarginal ossicles on aboral surface. *U.Dev.*, N.Am.——FIG. 59,*5*. **L. gyalum* (CLARKE), USA (N.Y.); aboral side, ×1 (129).

Family LEPYRIACTINIDAE Spencer & Wright, new family

Aboral skeleton reduced, none preserved in fossils; adambulacrals very narrow, mouth-angle plates much elongated. *L.Sil.*

Lepyriactis SPENCER, 1927 [**L. nudus*; OD]. Arms 5. *L.Sil.*, Scot.——FIG. 59,*6*. **L. nudus; 6a,b*, oral and aboral views of proximal part of arm, ×3; *6c*, aboral view of 2 neighboring mouth-angle plates showing grooves for water vascular ring and neural ring, ×3 (133).

Family SCHUCHERTIIDAE Schuchert, 1915

Aboral skeleton not differentiated and all ossicles alike; inferomarginals present, with large axillaries abutting mouth-angle plates. *M.Ord.-Sil.*

Schuchertia GREGORY, 1899 [**Palasterina stellata* BILLINGS, 1858] [=*Trentonaster* STÜRTZ, 1900]. Characters of family. *M.Ord.-Sil.*, N.Am.-Scot.-Australia.——FIG. 59,*1*. *S. wenlocki* SPENCER, Sil., Scot.; oral view of arm, ×3 (133).

Family HELIANTHASTERIDAE Gregory, 1899

[*nom. transl.* SPENCER & WRIGHT, herein (ex Helianthasterinae GREGORY, 1899)] [=Palaechinasteridae STÜRTZ, 1890 (invalid because not based on included genus)]

Adambulacrals narrow, with single large spine or several at outer edge; aboral surface reticulate or granular. *Dev.*

Helianthaster ROEMER, 1863 [**H. rhenanus*; OD]. Arms 14 to 16, rather rigid; disc moderately large; aboral surface granular. *L.Dev.*, Ger.——FIG. 59,*7*. **H. rhenanus*; oral surface of arm, ×1 (133).

Echinasterella STÜRTZ, 1890 [**E. sladeni*; OD]. Arms 5, rather long and slender; adambulacrals with single spine; aboral surface reticulate, with small spines. *L.Dev.-U.Dev.*, Ger.——FIG. 60,*2*. **E. sladeni*, L.Dev.; X-ray view, ×0.5 (116).

Hystrigaster LEHMANN, 1957 [**H. horridus*; OD]. Arms 5, rather short, broad at base; long spines on aboral surface and in 2's or 3's on outer edge of adambulacrals. *L.Dev.*, Ger.——FIG. 60,*1*. **H. horridus*; X-ray view, ×0.5 (116).

Family SOLASTERIDAE Perrier, 1884

Disc rather large, arms long, 5 to many; aboral skeleton normally open and irregularly reticulate, rarely with large imbricating paxillate plates; oral intermediate plates present; single or double row of marginal paxillae. *L.Jur.-Rec.*

Solaster Forbes, 1839 [**Asterias endeca* Linné,

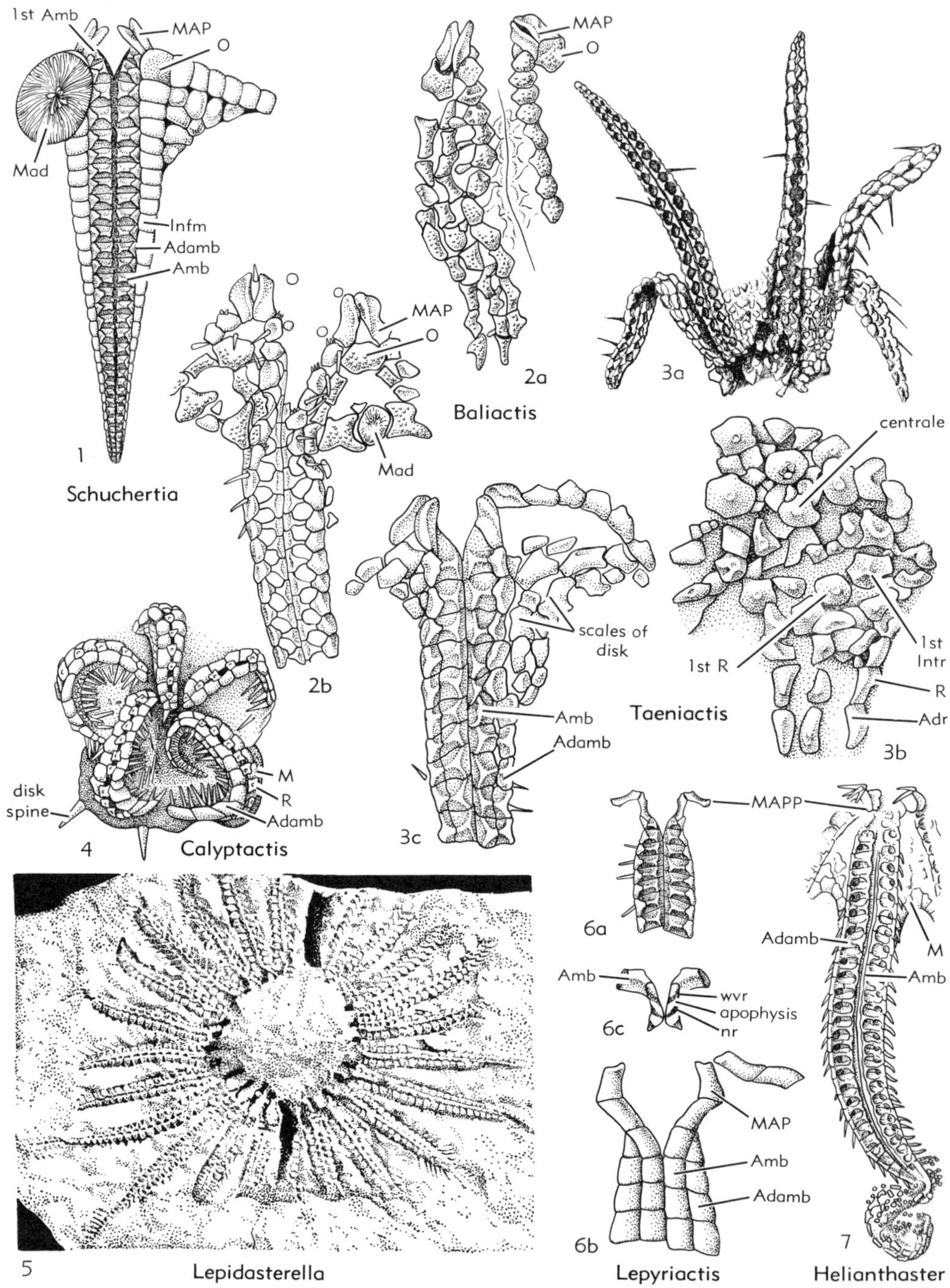

Fig. 59. Taeniactinidae *(2-5)*; Lepyriactinidae *(6)*; Schuchertiidae *(1)*; Helianthasteridae *(7)*. [Explanation: *Adamb*, adambulacral; *Adr*, adradial; *Amb*, ambulacral; *Infm*, inferomarginal; *Intr*, interradial; *M*, marginal; *Mad*, madreporite; *MAP*, mouth-angle plate; *nr*, nerve-ring groove; *O*, odontophore; *R*, radial; *wvr*, water-vessel-ring groove.] (p. *U65*).

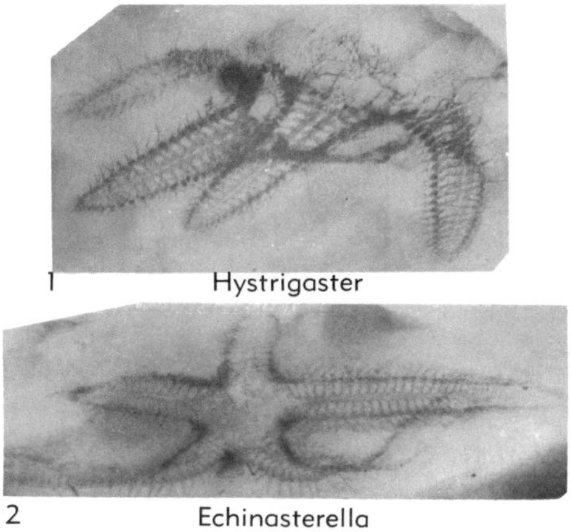

FIG. 60. Helianthasteridae (p. *U65*).

1771; OD] [=*Endeca, Polyaster* GRAY, 1840]. Arms 7 to 17 in Recent species; series of large marginal paxillae, with or without second smaller series. [Jurassic specimens rare, one with abnormally small disc and 33 arms.] *L.Jur. (Pliensbach.)-M.Jur.(Bathon.),* Eu.; *Rec.,* cosmop.——FIG. 61,*2*. *S.? moretonis* FORBES, Bathon., Eng.(Glos.); part of oral surface, ×1 (139).

Crossaster MÜLLER & TROSCHEL, 1840 [**Asterias papposus* LINNÉ, 1767; OD]. *Rec.*

Cuenotaster THIÉRY, 1920 [*pro Leucaster* KOEHLER, 1912 (*non* GAUTHIER, 1877)] [**Leucaster involutus* KOEHLER, 1912; OD]. *Rec.*

Heterozonias FISHER, 1910 [**Crossaster alternatus* FISHER, 1906; OD]. *Rec.*

Laetmaster FISHER, 1908 [*pro Ctenaster* PERRIER, 1881 (*non* AGASSIZ, 1836)] [**Ctenaster spectabilis* PERRIER, 1881; OD]. *Rec.*

Lophaster VERRILL, 1878 [**Solaster furcifer* DÜBEN & KOREN, 1884; OD] [=*Sarkaster* LUDWIG, 1905]. Arms 5; marginal paxillae in 2 well-developed rows. *Plio.,* Eng.; *Rec.*

Paralophaster FISHER, 1940 [**Solaster godefroyi* KOEHLER, 1912; OD] [=*Myoraster* FISHER, 1940]. *Rec.*

Rhipidaster SLADEN, 1889 [**R. vannipes*; OD]. *Rec.*

Xenorias FISHER, 1913 [**Rhipidaster (Xenorias) polyctemis*; OD]. *Rec.*

Family TROPIDASTERIDAE Wright, 1880

Arms 5 to many; adambulacrals broad,

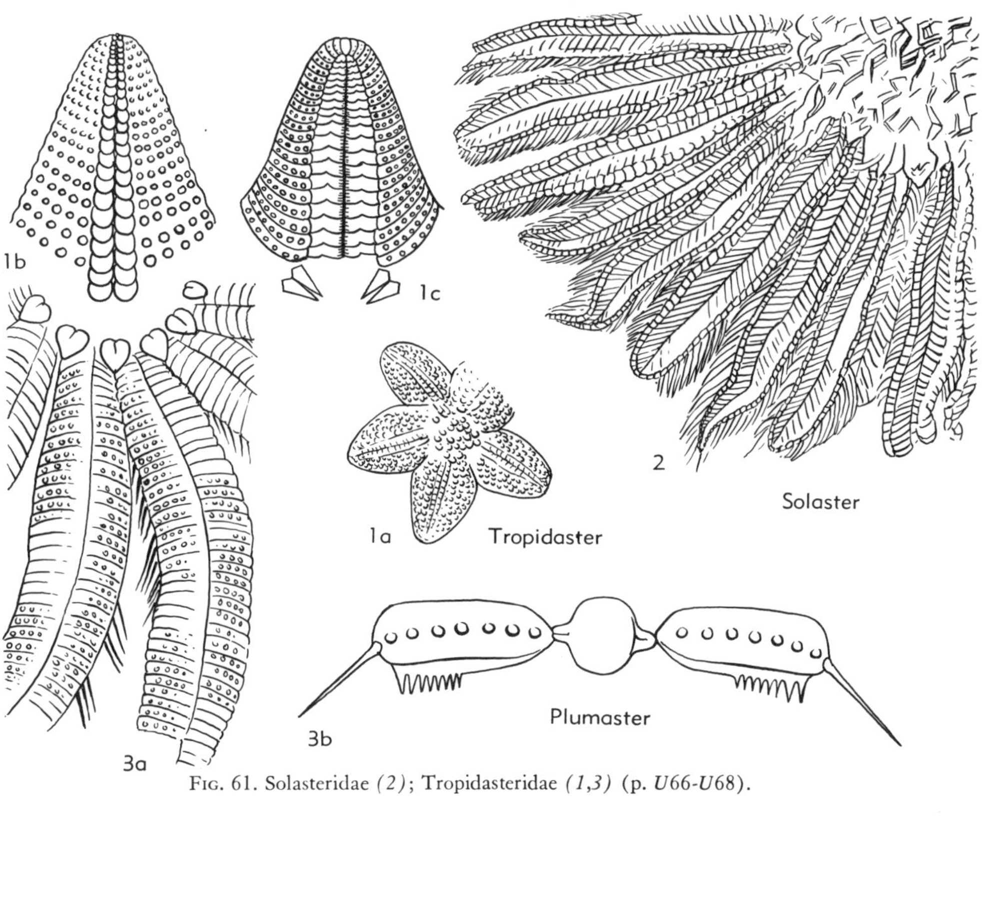

FIG. 61. Solasteridae *(2)*; Tropidasteridae *(1,3)* (p. *U66-U68*).

with transverse ridge carrying single row of spines; aboral surface with rows of spines. *L.Jur.*

Tropidaster FORBES, 1850 [**T. pectinatus*; OD]. Small, with 5 bluntly petaloid arms; ambulacral grooves wide, bordered by wide adambulacrals bent in middle and bearing 5 small spines which project into ambulacral furrow; continuous with adambulacrals are short wide inferomarginals with raised ridge and 8 or more long spines covering intermarginal grooves; mouth-angle plates prominent, plowshare-shaped; aboral surface with radial double row of overlapping rounded plates, otherwise covered with transverse rows of blunt conical spines. *L.Jur.(Pliensbach.)*, Eng.——FIG. 61,*1*. **T. pectinatus*, Eng.(Glos.); *1a*, aboral side, ×1; *1b,c*, diagrammatic aboral and oral views, ×2 (139).

?**Plumaster** WRIGHT, 1861 [**P. ophiuroides*; OD]. Arms many, narrow at base, widest at mid-length; adambulacrals short, wide, with row of 8 long slender spines and row of small spines on distal margins; mouth-angle plates prominent. *L.Jur. (Pliensbach.)*, Eng.——FIG. 61,*3*. **P. ophiuroides*, Eng.(Yorks.); *3a*, oral side, ×1; *3b*, adambulacrals, enl. (139).

Family KORETHRASTERIDAE Danielsson & Koren, 1884

Superficially like Solasteridae but without oral intermediate plates and with spines of adambulacrals and inferomarginals forming single series; aboral skeleton formed of rounded plates or wide mesh of isolated tufts of spinelets; marginals not paxilliform. *Rec.*

Korethraster THOMSON, 1873 [**K. hispidus*; OD]. *Rec.*

Anareaster FELL & H. E. S. CLARK, 1959 [**A. ganymede*; OD]. *Rec.*

Peribolaster SLADEN, 1889 [**P. folliculatus*; OD] [=*Peribolaster* SLADEN, 1885 (*nom. nud.*)]. *Rec.*

Remaster PERRIER, 1894 [**Korethraster (R.) palmatus*; OD]. *Rec.*

Family PYTHONASTERIDAE Sladen, 1889

[*nom. transl.* PERRIER, 1894 (*ex* Pythonasterinae SLADEN, 1889)]

Arms 5 to 10, long, cylindrical; aboral surface with bundles of long spinelets in webbed or ensacculated groups; mouth-angle plates and adambulacrals with combs of webbed spinelets. *Rec.*

Subfamily PYTHONASTERINAE Sladen, 1889

Pythonaster SLADEN, 1889 [**P. murrayi*; OD] [=*Pythonaster* SLADEN, 1885 (*nom. nud.*)]. *Rec.*

Subfamily MYXASTERINAE Perrier, 1894

Myxaster PERRIER, 1885 [**M. sol*; OD]. *Rec.*

Asthenactis FISHER, 1906 [**A. papyraceus*; OD]. *Rec.*

Family PTERASTERIDAE Perrier, 1875

Cross-shaped or lobed aboral plates bearing groups of spinelets which support membrane, distinct from aboral surface, forming cavity for young which escape by central valved aperture, termed osculum; lateral spines on adambulacrals either supporting oral web or merging in oral surface; no oral intermediate plates; mouth-angle plates broad and plowshare-shaped; internal septa membranous. *Rec.*

Pteraster MÜLLER & TROSCHEL, 1842 [**Asterias militaris* O. F. MÜLLER, 1776; OD]. *Rec.*

P. (Pteraster) [=*Pterasterides* VERRILL, 1909; ?*Lophopteraster* VERRILL, 1895]. *Rec.*

P. (Retaster) PERRIER, 1878 [**Pteraster capensis* GRAY, 1847; SD FISHER, 1911] [=*Hexaster* PERRIER, 1891; *Temnaster* VERRILL, 1894]. *Rec.*

P. (Apterodon) FISHER, 1940 [**Pteraster stellifer* SLADEN, 1882; OD]. *Rec.*

Benthaster SLADEN, 1882 [**Benthaster wyvillethomsoni* SLADEN, 1882; SD SPENCER & WRIGHT, herein]. *Rec.*

Calyptraster SLADEN, 1882 [**C. coa*; OD]. *Rec.*

Cryptaster PERRIER, 1885 [**C. personatus*; OD]. *Rec.*

Diplopteraster VERRILL, 1880 [**Pteraster multipes* SARS, 1877; OD]. *Rec.*

Euretaster FISHER, 1940 [**Retaster insignis* SLADEN, 1882; OD]. *Rec.*

Hymenaster THOMSON, 1873 [**H. pellucidus*; OD]. *Rec.*

Hymenasterides FISHER, 1911 [**H. zenognathus*; OD]. *Rec.*

Marsipaster SLADEN, 1882 [**M. spinosissimus* SLADEN, 1882; SD SPENCER & WRIGHT, herein]. *Rec.*

Suborder LEPTOGNATHINA Spencer & Wright, new suborder

Mouth-angle plates small, triangular; ambulacral furrows narrow. *L.Jur.-Rec.*

This suborder includes a variety of isolated families, of which some are clearly very primitive, but their geological record is poor.

Family ASTERINIDAE Gray, 1840

Minute marginals normally present; aboral skeleton composed of imbricated plates bearing grouped or single spinelets or granules; oral intermediate plates in regular transverse series, in some species consisting of virgals; interradial slits or dorsal depressions present in Tremasterinae. *M. Jur.-Rec.*

Known genera of this family are Recent except for a yet undescribed Middle Jurassic form from Switzerland.

Subfamily ASTERININAE Gray, 1840

Papulae widely distributed. Aboral ossicles thick. *Rec.*

Asterina NARDO, 1834 [**Asterias minuta* (=**Asterias gibbosa* PENNANT, 1777); OD] [=*Ctenaster* AGASSIZ, 1835; *Asteriscus* MÜLLER & TROSCHEL, 1840; *Asterinides* VERRILL, 1914]. *Rec.*

Allopatiria VERRILL, 1913 [**Patiria ocellifera* GRAY, 1840; OD]. *Rec.*

Asterinopsis VERRILL, 1914 [**Asterias penicillaris* LAMARCK, 1816; OD]. *Rec.*

Desmopatiria VERRILL, 1914 [**D. flexilis*; OD]. *Rec.*

Disasterina PERRIER, 1875 [**D. abnormalis*; OD] [=*Habroporina* H. L. CLARK, 1921]. *Rec.*

Manasterina H. L. CLARK, 1938 [**M. longispina*; OD]. *Rec.*

Nepanthia GRAY, 1840 [**Nepanthia maculata* GRAY, 1840; SD VERRILL, 1914] [=*Parasterina* FISHER, 1908]. *Rec.*

Paranepanthia FISHER, 1917 [**Nepanthia platydisca* FISHER, 1913; OD]. *Rec.*

Patiria GRAY, 1840 [**P. coccinea* GRAY, 1840 (=**Asterias miniata* BRANDT, 1835); OD] [=*Callopatiria, Enoplopatiria* VERRILL, 1914]. *Rec.*

Patiriella VERRILL, 1914 [**Asterina (Asteriscus) regularis* VERRILL, 1867; OD]. *Rec.*

Pseudonepanthia A. H. CLARK, 1916 [**P. gotoi*; OD]. *Rec.*

Tegulaster LIVINGSTONE, 1933 [**T. emburyi*; OD]. *Rec.*

?**Socomia** GRAY, 1840 *(nom. dub.). Rec.*

Subfamily ANSEROPODINAE Fisher, 1903

[=Palmipedinae SLADEN, 1889]

Papulae in narrow radial band. Ossicles of papular area stellate. Other aboral ossicles thin, scalelike, with a downward projection that meets an upward process from an oral ossicle, forming or supporting pillar for disc. *Rec.*

Anseropoda NARDO, 1834 [**Asterias membranacea* RETZIUS, 1805; (=**Asterias placenta* PENNANT, 1777); OD] [=*Palmipes* AGASSIZ, 1835; *Carna* GISTL, 1848]. *Rec.*

Kampylaster KOEHLER, 1920 [**K. incurvatus*; OD]. *Rec.*

Mirastrella FISHER, 1940 [**M. biradialis; OD*]. *Rec.*

Subfamily TREMASTERINAE Fisher, 1903

[*ex* SLADEN, 1889, provisional]

Papulae in radial areas, wider than in Anseropodinae. With interradial slits or aboral depressions. *M.Jur., Rec.*

Tremaster VERRILL, 1879 [**T. mirabilis*; OD]. *Rec.*

Stegnaster SLADEN, 1889 [**Pteraster inflatus* HUTTON, 1872; OD]. *Rec.*

Genus undescribed assigned to this subfamily. *M. Jur.(Bajoc.),* Switz.

Family GANERIIDAE Sladen, 1889

[incl. Cryasteridae KOEHLER, 1906 (as Cryasteridés, *nom. correct.* FISHER, 1911)]

Large marginals normally present but may be small and confined to interrays; aboral skeleton imbricate, reticulate, or reduced to minute plates in skin. *Rec.*

Ganeria GRAY, 1847 [**G. falklandica*; OD]. *Rec.*

Aleutiaster A. H. CLARK, 1939 [**A. schafferi*; OD]. *Rec.*

Cycethra BELL, 1881 [**C. simplex* (=**Goniodiscus verrucosus* PHILIPPI, 1857); OD] [=*Lebrunaster* PERRIER, 1891]. *Rec.*

Leilaster A. H. CLARK, 1938 [**Korethraster radians* PERRIER, 1881; OD]. *Rec.*

Magdalenaster KOEHLER, 1907 [**M. arcticus*; OD]. *Rec.*

Perknaster SLADEN, 1889 [**P. fuscus* SLADEN, 1889; SD FISHER, 1940] [=*Cribraster* PERRIER, 1888; *Cryaster* KOEHLER, 1906; *Cribellopsis* KOEHLER, 1917]. *Rec.*

Scotiaster KOEHLER, 1908 [**S. inornatus*; OD]. *Rec.*

Tarachaster FISHER, 1913 [**T. tenuis*; OD]. *Rec.*

Family PORANIIDAE Perrier, 1894

[=Gymnasteriidae SLADEN, 1889; Asteropidae FISHER, 1908]

Marginals varying from prominent, more or less overlapping, smooth, or with few spines, to absent; aboral skeleton loosely tessellate or reticulate, covered by skin which is smooth, granulose, or spinulose; extreme forms may have no solid skeleton except some axial elements. *Rec.*

Porania GRAY, 1840 [**Asterias gibbosus* LEACH, 1817 (=**Asterias pulvillus* O. F. MÜLLER, 1788); OD] [=*Glabraster* A. H. CLARK, 1916].*Rec.*——FIG. 15,*1*. *P.* sp., Rec.; oral surface (111).

Asteropsis MÜLLER & TROSCHEL, 1840 [*pro Asterope* MÜLLER & TROSCHEL, 1840 (*non* HÜBNER, 1819)] [**Asterias carinifera* LAMARCK, 1816; OD] [=*Gymnasteria* GRAY, 1840, December (*non* PHILIPPI, 1840, June)]. *Rec.*

Chondraster VERRILL, 1895 [**Porania grandis* VERRILL, 1878; OD]. *Rec.*

Dermasterias PERRIER, 1875 [**D. inermis* (=**Asteropsis imbricata* GRUBE, 1857); OD]. *Rec.*

Marginaster PERRIER, 1881 [**M. pectinatus* PERRIER, 1881; SD SLADEN, 1889] [=*Cheilaster* BELL, 1892]. *Rec.*

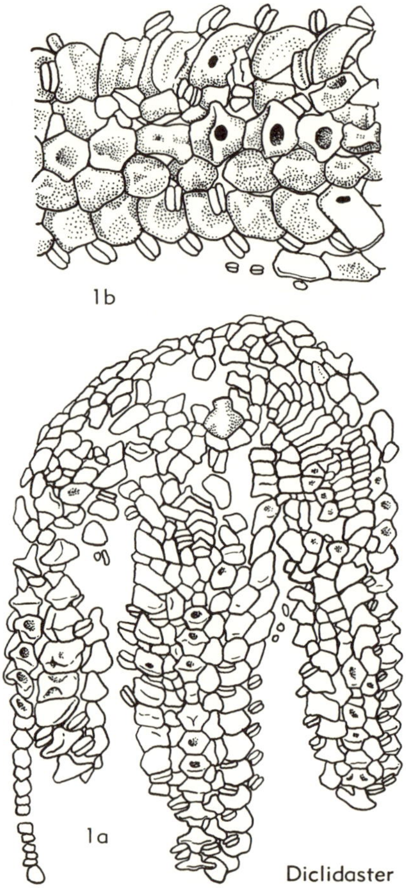

FIG. 62. Valvasteridae (p. U70).

Petricia GRAY, 1847 [*P. *punctata* (=*Asterias vernicina* LAMARCK, 1816); OD]. *Rec.*

Poraniella VERRILL, 1914 [*P. *regularis*; OD]. *Rec.*

Poraniomorpha DANIELSSON & KOREN, 1881 [*P. *rosea*; OD] [=*Rhegaster* SLADEN, 1883; *Lasiaster* SLADEN, 1889; *Culcitopsis* VERRILL, 1914]. *Rec.*

Poranisca VERRILL, 1914 [*P. *lepida*; OD]. *Rec.*

Pseudoporania DONS, 1936 [*P. *stormi*; OD]. *Rec.*

Sphaeriaster DONS, 1939 [*pro Sphaeraster* DONS, 1938 (*non* QUENSTEDT, 1875)] [**Sphaeraster berthae* DONS, 1938; OD]. *Rec.*

Spoladaster FISHER, 1940 [**Cryaster brachyactis* H. L. CLARK, 1923; OD]. *Rec.*

Tylaster DANIELSSON & KOREN, 1881 [*T. *willei*; OD]. *Rec.*

Family ECHINASTERIDAE Verrill, 1867

[*non* Echinasterina GRAY, 1840, based on *Echinaster* GRAY, 1840 (*non* MÜLLER & TROSCHEL, 1840)]

Disc small, arms long and slender; aboral ossicles reticulate, with spines single or in small groups. Pedicellariae lacking; ampullae single. *?U.Cret., Rec.*

Echinaster MÜLLER & TROSCHEL, 1840 [**Asterias seposita* LAMARCK, 1816 (*non* RETZIUS, 1783) (=**Asterias sagena* RETZIUS, 1805); SD FISHER, 1913] [*non Echinaster* GRAY, 1840 (see *Acanthaster*)] [=*Othilia, Rhopia* GRAY, 1840; *Henricides* VERRILL, 1914]. *Rec.*

Cribraster PERRIER, 1891 [*C. *sladeni*; OD]. *Rec.*

Dictyaster WOOD-MASON & ALCOCK, 1891 [*D. *xenophilus*; OD]. *Rec.*

Henricia GRAY, 1840 [*H. *oculata* (=**Asterias sanguinolenta* O. F. MÜLLER, 1776); OD] [=*Cribella* FORBES, 1841 (*non* AGASSIZ, 1835); *Cribrella* LÜTKEN, 1857 (*non* AGASSIZ, 1835); *Magdalenaster* KOEHLER, 1907; *Cyllaster* A. H. CLARK, 1916; *Spinohenricia* HEDING, 1936]. Marginals more or less distinguishable; aboral surface with many spinelets in groups or on ridges; single doubtful specimen found fossil (7). *?U.Cret.* (Calif.), *Rec.*

Plectaster SLADEN, 1889 [**Echinaster decanus* MÜLLER & TROSCHEL, 1843; OD]. *Rec.*

Poraniopsis PERRIER, 1891 [*P. *echinaster*; OD (other included species is synonym)] [=*Poraniopsis* PERRIER, 1888 (*nom. nud.*); *Lahillea* DE LORIOL, 1904; *Alexandraster* LUDWIG, 1905; *Ortmannia* DE LORIOL, 1906]. *Rec.*

Rhopiella FISHER, 1940 [*R. *koehleri*; OD]. *Rec.*

Thyraster IVES, 1890 [**Echinaster serpentarius* MÜLLER & TROSCHEL, 1842; OD]. *Rec.*

Family VALVASTERIDAE Viguier, 1878

[*nom. correct.* FISHER, 1911 (*pro Valvasteridés* VIGUIER, 1878)] [=Valvasterinae KOEHLER, 1910]

Marginals conspicuous; aboral ossicles regularly arranged, with small triangular papular areas between, with isolated spinelets. Pedicellariae large, low, bivalved on superomarginals and small, 2-jawed on aboral surface; ampullae double. *L.Jur.-Rec.*

Valvaster PERRIER, 1876 [**Asterias striata* LAMARCK, 1816; OD]. *Rec.*

Diclidaster DE LORIOL, 1897 [*D. *gevreyi*; OD]. Short wide imbricating plates at base of arms; some aboral ossicles bearing perforations (?for stalked pedicellariae). *L.Jur.(Hettang.)*, Fr.——FIG. 62,*1*. **D. gevreyi*, Ardêche; *1a*, oblique aboral view, arms turned down, ×2; *1b*, part of arm, ×4 (117).

Family ACANTHASTERIDAE Sladen, 1889

Many arms; madreporites numerous. Upright, 2-jawed pedicellariae; well-developed interbrachial septa; ampullae double. *Rec.*

Acanthaster GERVAIS, 1841 [*pro Echinaster* GRAY, 1840 (*non* MÜLLER & TROSCHEL, 1840)] [*A.

echinus (=**Asterias planci* LINNÉ, 1758); OD]. *Rec.*

Family MITHRODIIDAE Viguier, 1879

[*nom. transl.* PERRIER, 1894 (*ex* Mithrodiinae VIGUIER, 1879)]

Whole surface, including spines, overlaid with rough granules, tubercles or spinelets; no interbrachial septa; ampullae double. *Rec.*

Mithrodia GRAY, 1840 [**M. spinulosa* (=**Asterias clavigera* LAMARCK, 1816); OD] [=*Heresaster* MICHELIN, 1844]. *Rec.*

Family METRODIRIDAE Sladen, 1889

[*nom. transl.* FISHER, 1917 (*ex* Metrodirinae SLADEN, 1889)]

Abactinal surface and marginals covered with skin, bearing isolated skin-covered spines. *Rec.*

Metrodira GRAY, 1840 [**M. subulata*; OD] [=*Scaphaster* DE LORIOL, 1899]. *Rec.*

Order FORCIPULATIDA Perrier, 1884

[*nom. correct.* SPENCER & WRIGHT, herein (*pro* Forcipulatae PERRIER, 1884)]

Mouth of ambulacral type; madreporite always on aboral surface. Pedicellariae, when present, always straight or crossed. *L.Ord.-Rec.*

Suborder URACTININA Spencer & Wright, new suborder

[=Urasterina SPENCER, 1951 (name misleading because *Uraster* is synonym of *Asterias* of suborder Asteriadina)]

Arms typically narrow and well produced with parallel sides; aboral ossicles with numerous paxillae set on shafts in diagonal rows; in Arthrasterinae aboral ossicles are reduced to 5 rows and paxillar shafts broadened to form ridges; ambulacrals, except in some late Devonian and Carboniferous forms, not compressed; adambulacrals typically with median ridge carrying row of stout spines; single primary interradial present in earlier genera; odontophore high and wedge-shaped. Pedicellariae unknown. [For comparison of mouth frame with that of Asteriadina see Fig. 63, 64. Most Uractinina are Paleozoic but Compsasteridae linger into L. Jur. and a genus of Calliasteridae is rather common in U.Cret.] *L.Ord.-U.Cret.*

Family CNEMIDACTINIDAE Spencer, 1918

Arms steep-sided, with upper row of close ossicles (?inferomarginals) knit with row of adambulacrals which have flat oral surface; aboral surface with rows of small paxillae; oral side with wide mouth and several divergent ambulacrals arranged as if bordering buccal slits, aborally fused into closed girdle. Most apertures closed orally by 5 flaps, in same position as tori in ophiuroids but in horizontal plane. [The only described genus is M.Ord.-U.Ord. but undescribed material is known from L.Ord. of Czech.] *Ord.*

Cnemidactis SPENCER, 1918 [**Urasterella girvanensis* SCHUCHERT, 1914; OD]. Characters of family. *M.Ord.-U.Ord.*, Can.-Scot.——FIG. 63,*4*; 64,*1*. **C. girvanensis* (SCHUCHERT), U.Ord., Scot.; 63,*4*, part of mouth frame; 64,*1*, cross section of arm, oral surface of interray, ×5 (133).

Family URASTERELLIDAE Schuchert, 1914

[=Roemerasterinae GREGORY, 1900]

Arms rounded in section; adambulacra disc-shaped, with transverse ridge carrying stout spines, aboral ossicles paxilliform, subequal, in diagonal series, each corresponding with segment of ambulacral skeleton; ambulacrals not compressed. *L.Ord.-Permocarb.*

Urasterella M'COY, 1851 [**Uraster ruthveni* FORBES, 1858; OD] [=*Roemeraster, Palasteracanthion, Protasteracanthion* STÜRTZ, 1886]. Aboral side of arms flat, many paxillae with unequal-sized bases. *M. Ord. - Permocarb.*, Can.-Scot.-Ger.-USSR. ——FIG. 64,*6a,b*. **U. ruthveni* (FORBES), U.Sil., Eng.; *6a*, oral side of arm, ×2; *6b*, aboral side, ×5 (133).——FIG. 64,*6c,d*. *U. thraivensis* SPENCER, U.Ord., Scot.; *6c*, oral side, ×2; *6d*, aboral side, ×3 (133).——FIG. 64,*6e*. *U. verruculosa* LEHMANN, L.Dev., Ger.; aboral side, ×1 (116). (See Fig. 63,2.)

Bohemaster JAEKEL, 1923 [**B. primula*; OD]. Unrecognizable member of family. *L.Ord.*, Czech.

Phillipsaster SPENCER, 1950 [**Palaeaster coronella* SALTER, 1857; OD]. Like *Salteraster* but with only slightly swollen arms. *Sil.*, Eng.-E.Can.-Australia. ——FIG. 64,*3*. **P. coronella* (SALTER), L.Sil., Eng., part of aboral surface, ×5 (133).

Salteraster STÜRTZ, 1893 [**Palaeaster asperrima* SALTER, 1857; OD]. Arms strongly swollen, with single row of carinals separated from marginals by many paxillae. *M.Ord.-Sil.*, N.Am.-Eng.-Australia.——FIG. 64,*4a,b*. **S. asperrimus* (SALTER), U.Ord., Eng.(Heref.); *4a*, cross section of arm, ×5; *4b*, oral side of arm, ×2 (133).——FIG. 64,*4c*. *S. grandis* (MEEK), U.Ord.(Richmond.), USA(Ohio); specimen with arms folded together, ×1 (129).

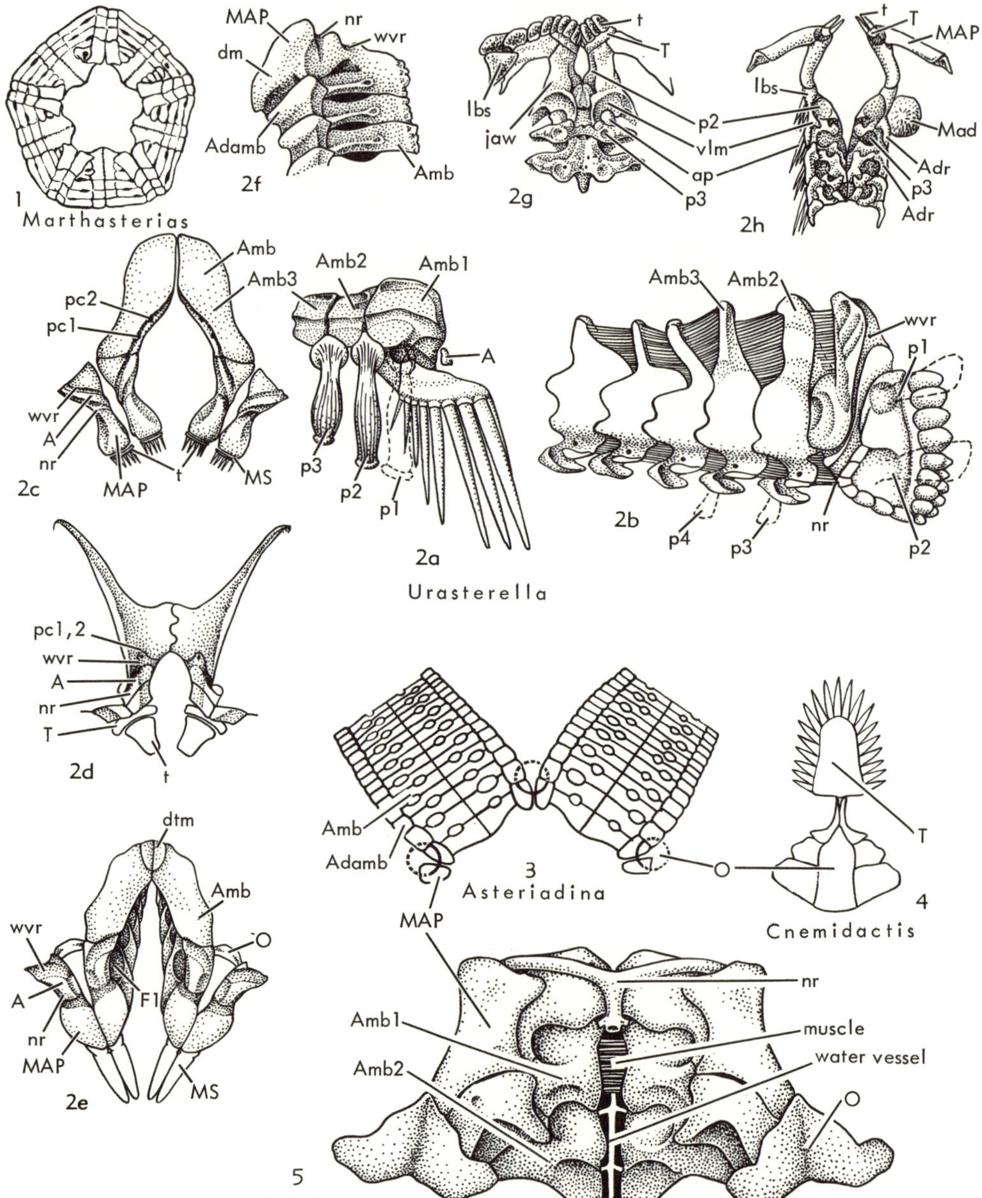

FIG. 63. Mouth frames of ambulacral type.——*1.* Aboral aspect of frame of *Marthasterias glacialis*, Rec. (86).——*2a-h.* Frame of *Urasterella* in oral aspect (133).——*3.* Diagram of ambulacral type of mouth frame as in Asteriadina, oral aspect (Spencer, n).——*4.* Part of frame of *Cnemidactis*, Ord., showing flap-like torus (29).——*5.* Part of frame of *Brisinga*, Rec., oral aspect (133). [Explanation: *A*, apophysis; *Adamb*, adambulacral; *Adr*, adradial; *Amb*, ambulacral; *ap*, articular peg (lateral hinge); *dm*, dental muscle attachment; *dtm*, dorsal transverse muscle attachment; *lbs*, lateral buccal shield; *Mad*, madreporite; *MAP*, mouth-angle plate; *MS*, mouth spine; *nr*, nerve-ring groove; *O*, odontophore; *p*, podial (tube foot) basin; *pc*, podial canal (to tube foot); *T*, torus; *t*, tooth (teeth); *vlm*, ventral longitudinal muscle attachment; *wvr*, water-vessel-ring groove.] (p. U71, U76-U77).

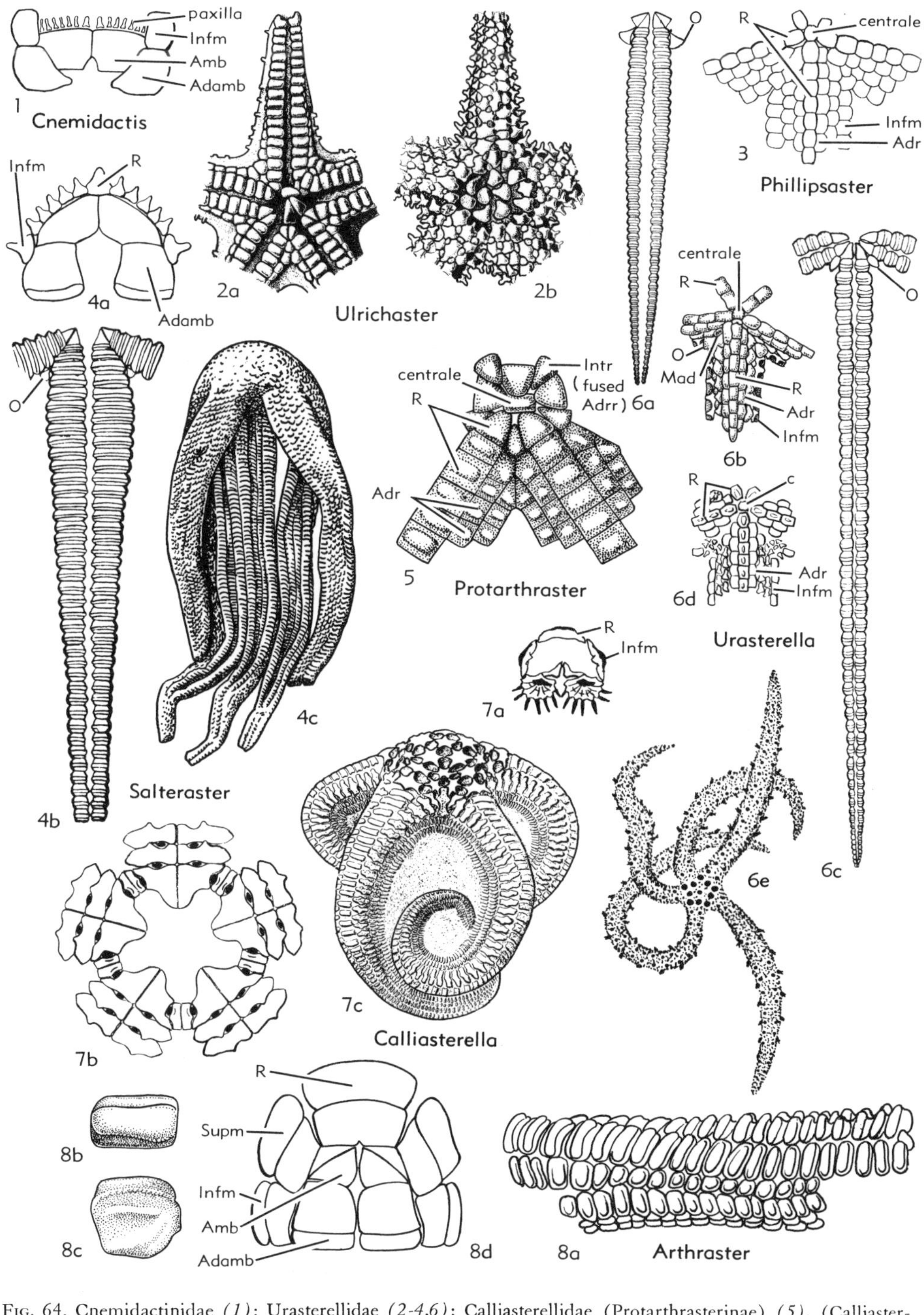

FIG. 64. Cnemidactinidae *(1)*; Urasterellidae *(2-4,6)*; Calliasterellidae (Protarthrasterinae) *(5)*, (Calliasterellinae) *(7)*, Arthrasterinae *(8)*. [Explanation: *Adamb*, adambulacral; *Adr*, adradial; *Amb*, ambulacral; *Infm*, inferomarginal; *Intr*, interradial; *Mad*, madreporite; *O*, odontophore; *R*, radial; *Supm*, superomarginal.] (*U*71, *U*74).

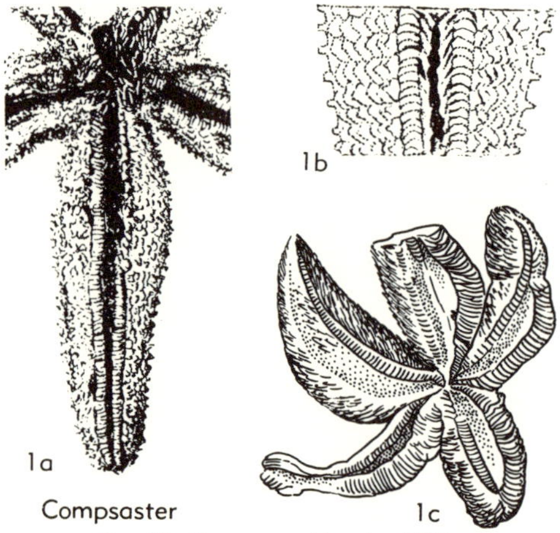

FIG. 65. Compsasteridae (p. U74).

Ulrichaster SPENCER, 1950 [**Urasterella ulrichi* SCHUCHERT, 1915]. Median oral surface of arms with 2 rows of ossicles, highly swollen in adults. *M. Ord.-L. Sil.,* N. Am.-Scot.——FIG. 64,*2.* **U. ulrichi* (SCHUCHERT), M.Ord.(Blackriv.), USA (Minn.); *2a,* oral side, ×3; *2b,* aboral side, ×2 (129).

Family CALLIASTERELLIDAE Schöndorf, 1910

[=Arthrasteridae SPENCER, 1918]

Disc small; arms long and straight-sided; oral face of adambulacrals with prominent transverse ridge; aboral ossicles of arms transversely elongate and bearing median ridge; with 5 primary radials. *L.Carb.-U. Cret.*

Subfamily PROTARTHRASTERINAE Spencer, 1918

Aboral ossicles in several rows, those along mid-line of arms with transverse ridge, others with paxillae. *L.Carb.*

Protarthraster SPENCER, 1918 [**Palaeaster longimanus* WHIDBORNE, 1896; OD]. Arms well rounded in section, inferomarginals on oral surface. *L.Carb.,* Eng.——FIG. 64,*5.* **P. longimanus* (WHIDBORNE); aboral surface, ×5 (133).

Subfamily CALLIASTERELLINAE Schöndorf, 1910

Aboral ossicles of arm reduced to 3 rows, no superomarginals present. *U. Carb.*

Calliasterella SCHUCHERT, 1914 [*pro Calliaster* TRAUTSCHOLD, 1879 (*non* GRAY, 1840)] [**Calliaster mirus* TRAUTSCHOLD, 1879; OD]. Arms enrolled toward oral side, as in *Calyptactis. U.Carb.,* Eu.(USSR).——FIG. 64,*7.* **C. mira* (TRAUTSCHOLD); *7a,* section of arm, ×1; *7b,* ossicles of mouth region from oral side, ×2; *7c,* reconstruction, ×0.5 (128).

Subfamily ARTHRASTERINAE Spencer, 1918

Aboral ossicles of arm in 5 rows of inferomarginals, superomarginals, and carinals. *U.Cret.*

Arthraster FORBES, 1848 [**A. dixoni*; OD]. Arms semicircular in section; arm ossicles more or less equal, short and wide, with tumid surface rising to smooth or rugose or pitted rounded transverse ridge. *U.Cret.(Turon.-Maastricht.),* Eng.-Fr.——FIG. 64,*8.* **A. dixoni,* Turon., Eng.; *8a,* side of arm, ×1; *8b,c,* top and side of marginal, ×2; *8d,* cross section of arm, ×2 (133).

Family COMPSASTERIDAE Schuchert, 1914

Arms swollen and fusiform; ambulacrals and adambulacrals numerous, compressed; ambulacrals transversely elongate, almost linear; aboral surface bearing many paxillae, with tall shafts. *L.Dev.-L.Jur.*

Compsaster WORTHEN & MILLER, 1883 [**C. formosus*; OD] [=*Jaekelaster* STÜRTZ, 1900; *Schlueteraster* LEHMANN, 1957]. Characters of family. *L. Dev.-L. Jur. (Bathon.),* Eng.-Ger.-N. Am.——FIG. 65,*1a,b.* **C. formosus,* U.Miss.(Chester.), USA (Ill.); *1a,* oral side, ×1; *1b,* part of same, ×2 (129).——FIG. 65,*1c.* *C. spiniger* (WRIGHT), Bathon., Eng.; *1c,* oral side, ×2 (139).

Suborder ASTERIADINA Fisher, 1928

Many ossicles of aboral surface modified to form crossed pedicellariae, basal plate representing spine, pincers on its summit comprising modified spinelets. *L.Jur.-Rec.*

Family HELIASTERIDAE Viguier, 1878

Disc large, with many (up to 50) short arms, divided internally from disc by wall. *Rec.*

Heliaster GRAY, 1840 [**Asterias helianthus* LAMARCK, 1816; SD H. L. CLARK, 1909]. *Rec.*

Family ZOROASTERIDAE Sladen, 1889

Disc small, arms normally long, subcylindrical, with ossicles in close radial series. Pelicellariae straight only. *Rec.*

Zoroaster WYVILLE THOMSON, 1873 [**Z. fulgens*; OD]. *Rec.*

Bythiolophus FISHER, 1916 [**B. acanthinus*; OD]. *Rec.*

Cnemidaster SLADEN, 1889 [**C. wyvillei*; OD]. *Rec.*

Mammaster Perrier, 1894 [**Zoroaster sigsbeei* Perrier, 1881; OD]. *Rec.*

Myxoderma Fisher, 1905 [**Zoroaster (Myxoderma) sacculatus*; OD]. *Rec.*

Pholidaster Sladen, 1889 [**P. squamatus*; SD Fisher, 1919]. *Rec.*

Prognaster Perrier, 1891 [**P. grimaldii*; OD]. *Rec.*

Family ASTERIIDAE Gray, 1840

Arms 5 to many; body swollen; ambulacrals and adambulacrals numerous, compressed; adambulacrals transversely elongate, ambulacrals narrow. Tube feet normally in 4 rows; pedicellariae straight and crossed. Aboral skeleton generally reticulate. Division into subfamilies is difficult. Neomorphasterinae and Pedicellasterinae are distinct groups, but Labidiasterinae and Pycnopodiinae perhaps less so. Attempts to divide the remainder have not been wholly successful, and therefore here they are all left in Asteriinae. *?L.Jur., M.Jur.-Rec.*

Subfamily ASTERIINAE Gray, 1840

[=Stichasteridae Perrier, 1885;. Coscinasteriinae, Notasteriinae Fisher, 1923]

Characters most closely similar to those of *Asterias*. [Includes a wide range of genera which lack features of other subfamilies.] *?L.Jur., M.Jur.-Rec.*

Asterias Linné, 1758 [**A. rubens*; SD Norman, 1865] [=*Stellonia* Nardo, 1834; *Uraster* Agassiz, 1835; *Asteracanthion* Müller & Troschel, 1840; *Allasterias* Verrill, 1909; *Parasterias* Verrill, 1914]. Disc fairly large, arms 5 or 6, rather broad and short; aboral skeleton open network with spines in nearly regular series along midline and margin of arms but elsewhere spaced irregularly. [Fossils can only be provisionally placed in the restricted genus.] *?L.Jur.(Pliensbach.), M.Jur.(Oxford.), Rec.*, cosmop.——Fig. 66,*1*. *A.? gaveyi* (Forbes), L.Jur., Pliensbach., Eng. (Glos.); *1a,* oral side of arm, ×1; *1b,* ambulacrals and adambulacrals, ×3 (139).

Adelasterias Koehler, 1914 [**Diplasterias papillosa* Koehler, 1906; OD]. *Rec.*

Allostichaster Verrill, 1914 [**Asteracanthion polyplax* Müller & Troschel, 1844; OD]. *Rec.*

Anasterias Perrier, 1875 [**A. minuta*; OD] [=*Asteroderma* Perrier, 1888; *Parastichaster* Koehler, 1920; *Eremasterias* Fisher, 1930]. *Rec.*

A. (Anasterias). *Rec.*

A. (Sporasterias) Perrier, 1894 [**Asterias rugispina* Stimpson, 1860; (=**Asteracanthion antarcticum* Lütken, 1856); OD]. *Rec.*

A. (Kalyptasterias) Koehler, 1923 [**K. conferta*; OD]. *Rec.*

Aphanasterias Fisher, 1923 [**A. pycnopodia*; OD]. *Rec.*

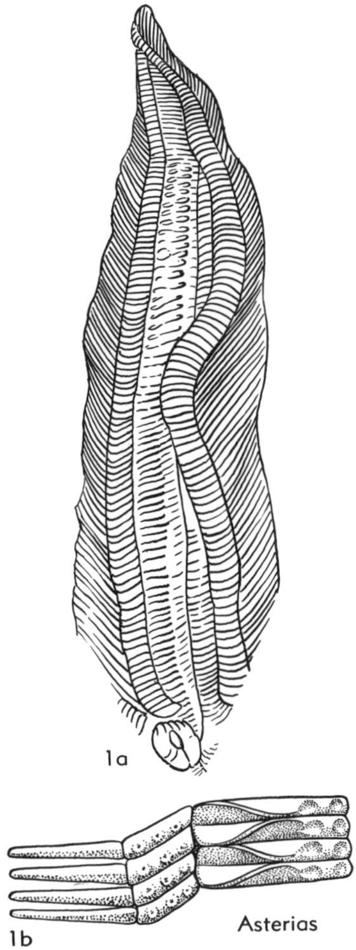

Fig. 66. Asteriidae (Asteriinae) (p. *U75*).

Aphelasterias Fisher, 1923 [**Asterias japonica* Bell, 1881; OD]. *Rec.*

Astrometis Fisher, 1923 [**Asterias sertulifera* Xantus, 1860; OD]. *Rec.*

Astrostole Fisher, 1923 [**Margaraster? scaber* Hutton, 1872; OD]. *Rec.*

Australiaster Fisher, 1923 [**Coscinasterias dubia* H. L. Clark, 1909; OD]. *Rec.*

Caimanaster A. M. Clark, 1962 [**C. acutus*; OD]. *Rec.*

Carlasterias da Costa, 1952 [**Mortensenia lusitanica* da Costa, 1941; OD] [=*Mortensenia* da Costa, 1941 (*non* Döderlein, 1905)]. *Rec.*

Calvasterias Perrier, 1875 [**C. asterinoides*; OD] [=*Stichorella* Koehler, 1920]. *Rec.*

Coscinasterias Verrill, 1867 [**C. muricata*; OD]. *Rec.*

C. (Coscinasterias). *Rec.*

C. (Stolasterias) Sladen, 1889 [**Asterias tenui-*

spina LAMARCK, 1816; SD FISHER, 1923] [=*Lytaster, Polyasterias* PERRIER, 1894]. *Rec.*

Cosmasterias SLADEN, 1889 [**Asteracanthion sulcifer* PERRIER, 1869 (=**Asteracanthion luridum* PHILIPPI, 1858; SD FISHER, 1930)] [=*Comasterias* PERRIER, 1891; *Quadraster* PERRIER, 1896]. *Rec.*

Cryptasterias VERRILL, 1914 [**Diplasterias turqueti* KOEHLER, 1906; OD]. *Rec.*

Displasterias PERRIER, 1888 [**Asterias brandti* BELL, 1881; ICZN pend.] [=*Podasterias* PERRIER, 1894; *Koehleraster* FISHER, 1922; *Bathyasterias* FISHER, 1930]. *Rec.*

Distolasterias PERRIER, 1896 [**Asterias (Stolasterias) stichantha* SLADEN, 1889; OD]. *Rec.*

Evasterias VERRILL, 1914 [**Asterias troscheli* STIMPSON, 1862; OD]. *Rec.*

Gastraster PERRIER, 1894 [**Pedicellaster margaritaceus* PERRIER, 1882; OD]. *Rec.*

Granaster PERRIER, 1894 [**Stichaster nutrix* STUDER, 1885; OD] [=*Hemiasterias* VERRILL, 1914]. *Rec.*

Icasterias FISHER, 1923 [**Asterias panopla* STUXBERG, 1878; OD]. *Rec.*

Kenrickaster A. M. CLARK, 1962 [**K. pedicellaris*; OD]. *Rec.*

Leptasterias VERRILL, 1866 [**Asteracanthion muelleri* SARS, 1844; OD] [=*Ctenasterias* VERRILL, 1914]. *Rec.*

L. (Leptasterias). *Rec.*

L. (Endogenasterias) DJAKONOV, 1938 [**Asteracanthion groenlandicum* STEENSTRUP, 1857; OD]. *Rec.*

L. (Eoleptasterias) DJAKONOV, 1938 [**Asteracanthion ochetense* BRANDT, 1835; OD]. *Rec.*

L. (Hexasterias) FISHER, 1930 [**Asteracanthion polaris* MÜLLER & TROSCHEL, 1842; OD]. *Rec.*

L. (Nesasterias) FISHER, 1930 [*L. (N.) stolacantha*; OD]. *Rec.*

Lethasterias FISHER, 1923 [**Asterias nanimensis* VERRILL, 1914; OD]. *Rec.*

Lysasterias FISHER, 1908 [**Anasterias perrieri* STUDER, 1885; OD] [=*Anasterias* LUDWIG, 1903 (*non* PERRIER, 1885); *Paedasterias* VERRILL, 1914]. *Rec.*

Marthasterias JULLIEN, 1878 [**M. foliacea* (=**Asterias glacialis* LINNÉ, 1758); OD]. *Rec.*

Meyenaster VERRILL, 1913 [**Asterias gelatinosus* MEYEN, 1834; OD]. *Rec.*

Neosmilaster FISHER, 1930 [**Asterias georgianus* STUDER, 1885; OD]. *Rec.*

Notasterias KOEHLER, 1911 [**N. armata*; OD] [=*Autasterias* KOEHLER, 1911]. *Rec.*

Orthasterias VERRILL, 1914 [**O. columbiana* (=**Asterias koehleri* DE LORIOL, 1897); OD]. *Rec.*

Perissasterias H. L. CLARK, 1923 [**P. polyacantha* H. L. CLARK, 1923; SD FISHER, 1926]. *Rec.*

Pisaster MÜLLER & TROSCHEL, 1840 [**Asteracanthion margaritifer* (=**Asterias ochraceus* BRANDT, 1835); OD] [=*Calliasterias* FEWKES, 1889]. *Rec.*

Psalidaster FISHER, 1940 [**P. mordax*; OD]. *Rec.*

Pseudechinaster H. E. S. CLARK, 1962 [**P. rubens*; OD]. *Rec.*

Saliasterias KOEHLER, 1920 [**S. bracheata*; OD]. *Rec.*

Sclerasterias PERRIER, 1891 [**S. guernei*; OD] [=*Eustolasterias* FISHER, 1923]. *Rec.*

Smilasterias SLADEN, 1889 [**Asterias (S.) scalprifera* SLADEN, 1889; SD FISHER, 1930] [=*Nanaster* PERRIER, 1894]. *Rec.*

Stenasterias VERRILL, 1914 [**Asterias (Leptasterias) macropora* VERRILL, 1909; OD]. *Rec.*

Stephanasterias VERRILL, 1871 [**Asteracanthion albulus* STIMPSON, 1853]. *Rec.*

Stichaster MÜLLER & TROSCHEL, 1840 [**S. striatus* (=**Asterias aurantiaca* MEYEN, 1834 (not invalidated by *A. aranciaca* LINNÉ, 1758); OD] [=*Tonia* GRAY, 1840; *Coelasterias* VERRILL, 1867 (*nom. nud.*); *Coelasterias* VERRILL, 1871]. *Rec.*

Stichastrella VERRILL, 1914 [**Asterias rosea* O. F. MÜLLER, 1776; OD]. *Rec.*

Stylasterias VERRILL, 1914 [**Asterias forreri* DE LORIOL, 1887; OD]. *Rec.*

Tarsastrocles FISHER, 1923 [**Hydrasterias verrilli* FISHER, 1903; OD]. *Rec.*

Triplasterias ENGELS & SCHROEVERS, 1961 [**T. mercatoris*; OD]. *Rec.*

Uniophora GRAY, 1840 [**U. globifera* (=**Asterias granifera* LAMARCK, 1816); OD]. *Rec.*

Urasterias VERRILL, 1909 [**Asteracanthion linckii* MÜLLER & TROSCHEL, 1842; OD]. *Rec.*

Subfamily PEDICELLASTERINAE Fisher, 1918

Alone in family with first proximal adambulacrals wholly or partly separated interradially. Tube feet may be biserial throughout. *Rec.*

Pedicellaster SARS, 1861 [**P. typicus*; OD]. *Rec.*

Ampheraster FISHER, 1923 [**Sporasterias marianus* LUDWIG, 1905; OD]. *Rec.*

Anteliaster FISHER, 1923 [**A. coscinactis*; OD]. *Rec.*

Hydrasterias SLADEN, 1889 [**Asterias (H.) ophidion*; OD]. *Rec.*

Peranaster FISHER, 1923 [**Pedicellaster chirophorus* FISHER, 1917; OD]. *Rec.*

Tarsaster SLADEN, 1889 [**T. stoichoides*; OD]. *Rec.*

Subfamily LABIDIASTERINAE Verrill, 1914

Arms many, long and slender. One spine on each inferomarginal, wreathed with crossed pedicellariae. No oral intermediate plates. Aboral skeleton very open or obsolescent. *Rec.*

Labidiaster LÜTKEN, 1871 [**L. radiosus*; OD] [=?*Gymnobrisinga* STUDER, 1884; *Labidiastrella* VERRILL, 1914]. *Rec.*

Coronaster PERRIER, 1885 [**C. parfaiti*; OD] [=*Heterasterias* VERRILL, 1914]. *Rec.*

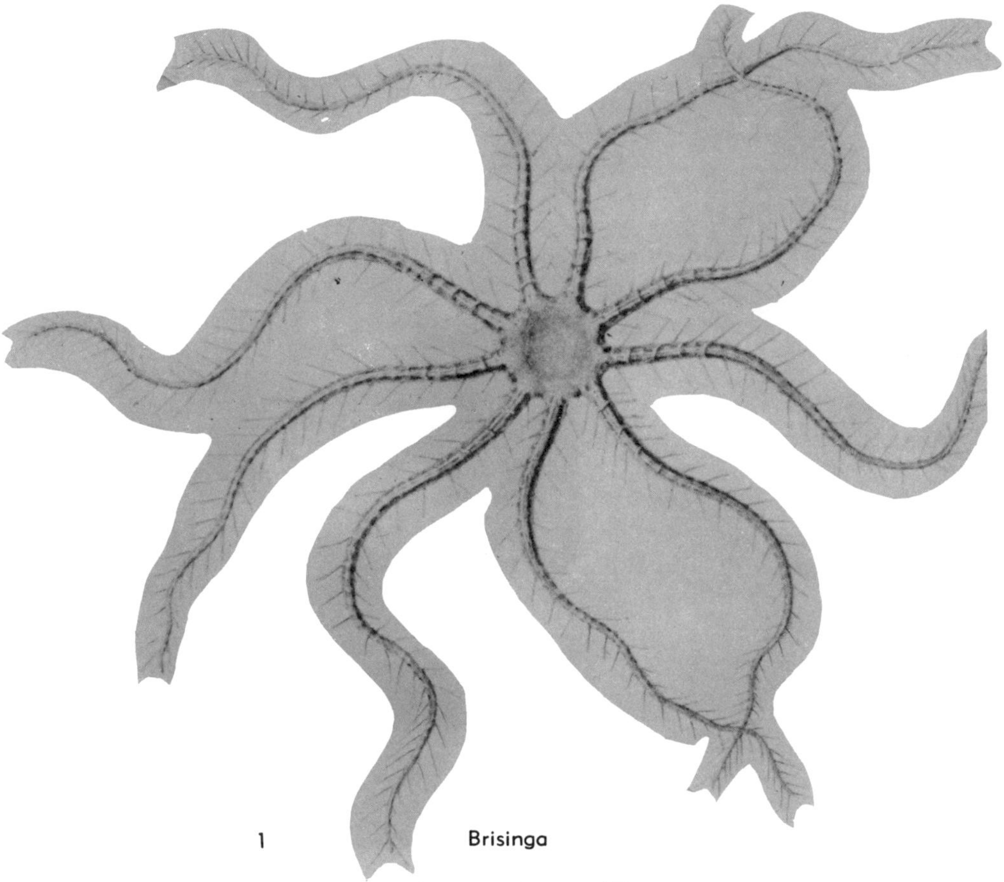

FIG. 67. Brisingidae (p. *U77*).

Plazaster FISHER, 1941 [**Labidiaster borealis* UCHIDA, 1938; OD]. *Rec.*

Rathbunaster FISHER, 1906 [**R. californicus*; OD]. *Rec.*

Subfamily PYCNOPODIINAE Stimpson, 1862

[*nom. transl.* VERRILL, 1914 (*ex* Pycnopodiidae STIMPSON, 1862)]

Inferomarginals with two spines, heavily wreathed with pedicellariae. No oral intermediate plates. Aboral skeleton obsolescent. *Rec.*

Pycnopodia STIMPSON, 1862 [**Asterias helianthoides* BRANDT, 1835; OD]. *Rec.*

Lysastrosoma FISHER, 1922 [**L. anthosticta*; OD]. *Rec.*

Subfamily NEOMORPHASTERINAE Fisher, 1923

Primary aboral plates conspicuously enlarged. Aboral skeleton of closely imbricated, sparsely granulated ossicles in regular radial series. *Rec.*

Neomorphaster SLADEN, 1889 [*pro Glyptaster* SLADEN, 1885 (*non* HALL, 1852)] [**N. eustichus* (=**Stichaster talismani* PERRIER, 1891); OD] [=*Calycaster* PERRIER, 1891]. *Rec.*

Suborder BRISINGINA Fisher, 1928

[=Euclasteroidea TORTONESE, 1958]

Arms many, sharply distinct and readily separated from very small disc; odontophore visible on edge of disc; ambulacrals and adambulacrals not compressed; pairs of ambulacrals articulated end to end; aboral skeleton weak. Crossed pedicellariae abundant; papulae in many species lacking. *L. Oligo.-Rec.*

Family BRISINGIDAE Sars, 1875

Characters of suborder. [A single fossil of unidentified genus is known from California.] *L.Oligo.-Rec.*

Brisinga ASBJORNSEN, 1856 [**B. endecacnemos*; OD].*Rec.*——FIG. 67,*1*. *B. mediterranea* PERRIER; aboral surface, slightly enlarged (Perrier). *Rec.*

Astrocles FISHER, 1917 [**A. actinodetus*; OD]. *Rec.*

Astrolirus FISHER, 1917 [**Brisinga panamensis* LUDWIG, 1905; OD]. *Rec.*

Astrostephane FISHER, 1917 [**Brisinga moluccana* FISHER, 1916; OD]. *Rec.*

Belgicella LUDWIG, 1903 [**B. racowitzana*; OD]. *Rec.*

Brisingaster DE LORIOL, 1883 [**B. robillardi*; OD]. *Rec.*

Brisingella FISHER, 1917 [**Brisinga fragilis* FISHER, 1906; OD]. *Rec.*

Brisingenes FISHER, 1917 [**Brisinga mimica* FISHER, 1916; OD]. *Rec.*

Colpaster SLADEN, 1889 [**C. scutigerulus*; OD]. *Rec.*

Craterobrisinga FISCHER, 1916 [**Brisinga panopla* FISHER, 1906; OD]. *Rec.*

Freyella PERRIER, 1885 [**Freyella spinosa* PERRIER, 1885; SD FISHER, 1917]. *Rec.*

Freyellaster FISHER, 1918 [**Freyella fecunda* FISHER, 1905; OD]. *Rec.*

Hymenodiscus PERRIER, 1884 [**H. agassizi*; OD]. *Rec.*

Odinia PERRIER, 1885 [**Odinia semicoronata* PERRIER, 1885; SD FISHER, 1917]. *Rec.*

Odinella FISHER, 1940 [**O. nutrix*; OD]. *Rec.*

Parabrisinga HAYASHI, 1948 [**P. pellucida*; OD]. *Rec.*

Stegnobrisinga FISHER, 1916 [**Brisinga (Stegnobrisinga) placoderma* FISHER, 1916; OD]. *Rec.*

Genus? Brisingid species. *Oligo.*, USA(Calif.).

GENERIC NAMES OF INDETERMINATE OR UNRECOGNIZABLE STATUS APPLIED TO FOSSIL ASTEROIDEA

Coelaster AGASSIZ, 1836 [**C. couloni*; OD] Not figured. Unrecognizable. *L.Cret.(Neocom.)*, Switz.

Cribellites TATE, 1864 [**C. carbonarius*; OD]. No species figured. Possibly a urasterellid. *Carb.*, Eire.

Cupulaster FRITSCH, 1893 [**C. pauper*; OD] Unidentifiable juvenile. *U.Cret.(Turon.)*, Czech.

Palmasterias SAVI & MENEGHINI, 1851 [*non* BLAINVILLE in GERVAIS, 1842, unrecognized asteroid]. Stated by NEAVE to be a crinoid.

Rumanaster POPESCU-VOITESTI, 1911 [**R. uhligi*; OD]. Unidentifiable terminals and ?marginals of a phanerozonate form. *Eoc.*, Rumania.

Subclass OPHIUROIDEA Gray, 1840

[*nom. transl.* GREGORY, 1900, p. 259 (*ex* order Ophiuroidea D'ORBIGNY, 1852, p. 132, *nom. correct. pro* order Ophiurida GRAY, 1840, p. 132] [=order Ophiuridae ZITTEL, 1880, p. 439] [Diagnosis prepared by W. K. SPENCER & C. W. WRIGHT. Research on authorship and synonymy by H. B. FELL]

Asterozoa with disc in almost all forms sharply distinct from slender elongate arms; most primitive forms retaining traces of metapinnular structures in arms, derived from Somasteroidea, but in most of subclass bulk of arm cavity filled with complex ossicles of axial skeleton; ossicles of adaxial skeleton forming side plates, primitively movable but in advanced forms firmly fixed to axial ossicles. Respiration by means of gills which typically are placed in enlarged interrays. Spines inconspicuous or absent except on lateral edges of arms and jaws. *L.Ord.(Arenig.)-Rec.*

Some ophiuroid stocks contain forms in which the internal gills are concentrated near the center of the body, the interrays being then much reduced and the general body shape stellate. Some Paleozoic fossils apparently of this type have often been classified as asteroids. Although the general shape of most ophiuroids has been remarkably constant since the time of their earliest appearance (*Pradesura,* L.Arenig.) to the present, profound changes have affected the skeletal, particularly axial, structure of the arms, resulting in production of the so-called vertebrae, which permit the snake-like movements of the arms that give the subclass its name.

Order STENURIDA Spencer, 1951

Basins for seating tube feet shared, usually subequally, by 2 ambulacrals; arm joints and musculature allowing only simple movements; buccal slits present in many genera. *L.Ord.(L.Arenig.)-U.Dev.*

In one of the two suborders, Proturina, the ambulacrals remain in a primitive condition; in the other, Parophiurina, some stocks have ambulacrals that approach the condition of vertebrae.

Suborder PROTURINA Spencer & Wright, new suborder

Tube enclosing radial water vessel not strengthened along its adradial edges; ambulacral basins shallow. *L.Ord.(L.Arenig.)-U.Dev.*

The Pradesuridae are the oldest known Ophiuroidea. When first introduced they exhibited a typical ophiuroid disc and long slender arms; these were burrowing forms. They were then absent from known faunas

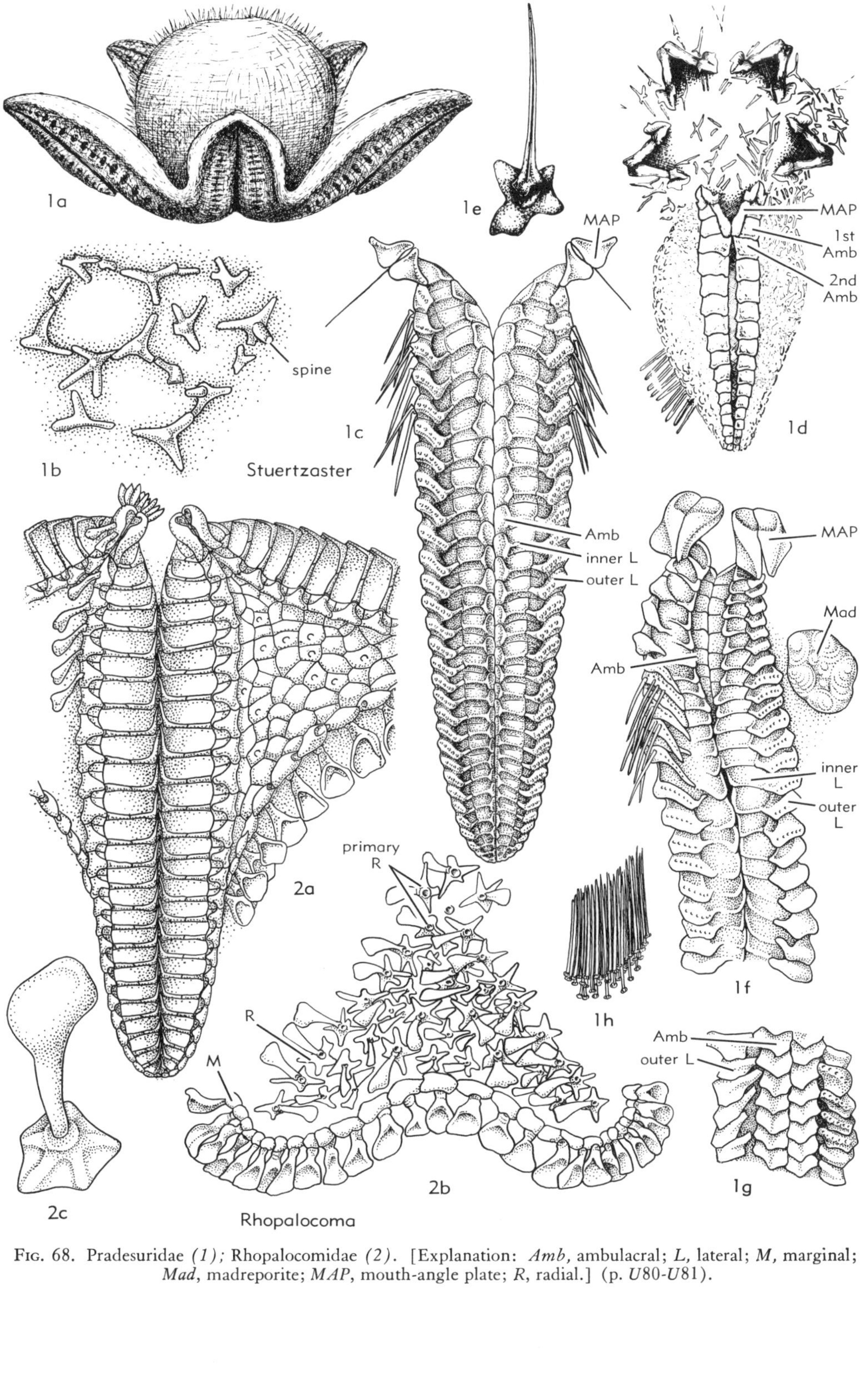

FIG. 68. Pradesuridae *(1)*; Rhopalocomidae *(2)*. [Explanation: *Amb*, ambulacral; *L*, lateral; *M*, marginal; *Mad*, madreporite; *MAP*, mouth-angle plate; *R*, radial.] (p. *U*80-*U*81).

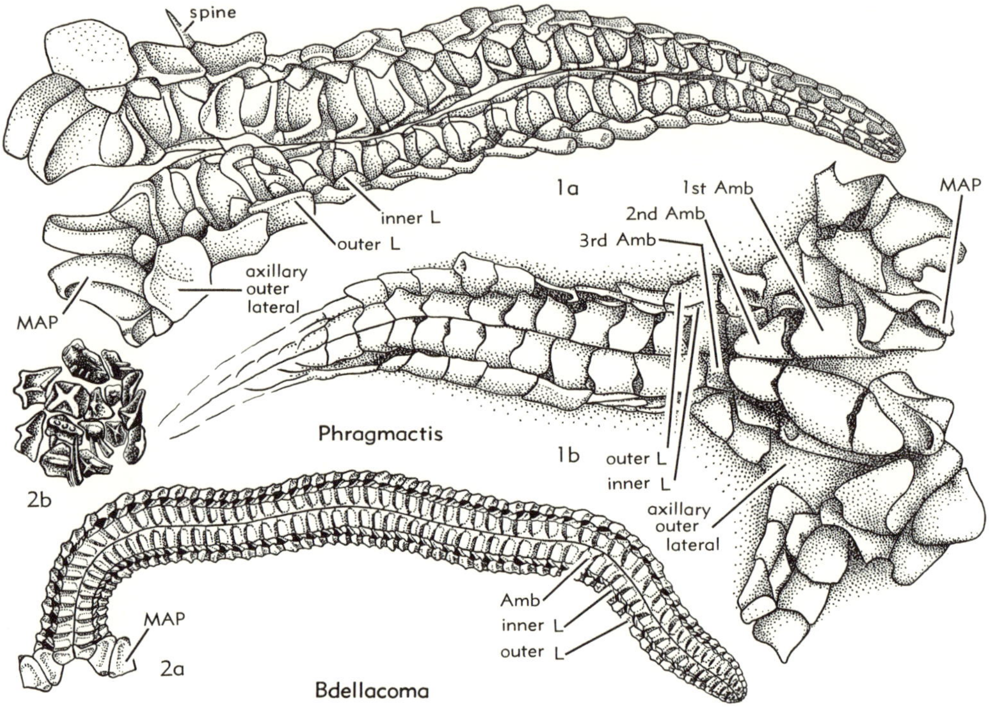

FIG. 69. Phragmactinidae *(1)*; Bdellacomidae *(2)*. [Explanation: *Amb*, ambulacral; *L*, lateral; *MAP*, mouth-angle plate.] (p. *U*81).

until the Late Silurian when they appeared in the lagoonal fauna at Leintwardine (Shropshire). Their general form was completely changed, for they exhibited high swollen aboral surface and short swollen arms. Evidently they had emerged to live on the sea floor. Several fossil specimens may be found crowded together on one slab, an indication of suspension feeding. The Rhopalocomidae, Phragmactinidae, and Bdellacomidae, whose ancestry is still unknown, are found in the same fauna as Pradesuridae, as well as later.

Family PRADESURIDAE Spencer, 1951

Laterals subventral, with only narrow swing. *L.Ord.-L.Dev.*

Earliest forms have a typical ophiuroid disc, undifferentiated proximal ambulacrals and laterals with narrow oral edges, whereas later ones have a swollen disc with reduced oral interrays, differentiated proximal ambulacrals and laterals with broad oral edges. All are assumed to have been sessile but not living in burrows.

Pradesura SPENCER, 1951 [**Palaeura jacobi* THORAL, 1935; OD]. Mouth-angle plates subtriangular; oral interrays large; aboral surface of disc covered with overlapping scales, each with central spine; madreporite small and thick with few grooves; proximal ambulacrals undifferentiated, buccal slits elongate; laterals with narrow oral edge. *L.Ord.(L.Arenig.)*, S.Fr.——FIG. 18,*1*, 23,*3*. **P. jacobi* (THORAL); 18,*1*, basins for tube feet; 23,*3*, oral surface of part of disc and arm, ×7 (133).

Stuertzaster ETHERIDGE, 1899 [*pro Palaeocoma* SALTER, 1857 (*non* D'ORBIGNY, 1850)] [**Palaeocoma marstoni* SALTER, 1857; SD SCHUCHERT, 1914] [=*Erinaceaster* LEHMANN, 1957]. Arms rather short, with rounded ends; aboral surface of disc highly swollen, with wide-meshed network of radiate spicules; oral interrays slight; mouth-angle plates subtriangular to elongate; proximal ambulacrals differentiated; laterals with broad oral edge. *U.Sil.-L.Dev.*, Eng.-Ger.——FIG. 68,*1a-d*. **S. marstoni* (SALTER), U.Sil., Eng. (Heref.); *1a*, profile, ×2; *1b*, part of aboral skeleton, ×8; *1c*, oral surface of arm; *1d*, ambulacrals and mouth frame in aboral view, ×3 (133).——FIG. 68,*1e*. *S. spinosissimus* (ROEMER), L.Dev., Ger.; ossicles of aboral surface, ×5 (133).——FIG. 68,*1f-h*. *S. colvini* (SALTER), U.Sil., Eng. (Heref.); *1f*, oral surface of arm, ×1; *1g*, aboral surface of part of arm, ×5; *1h*, aboral ossicles (133).

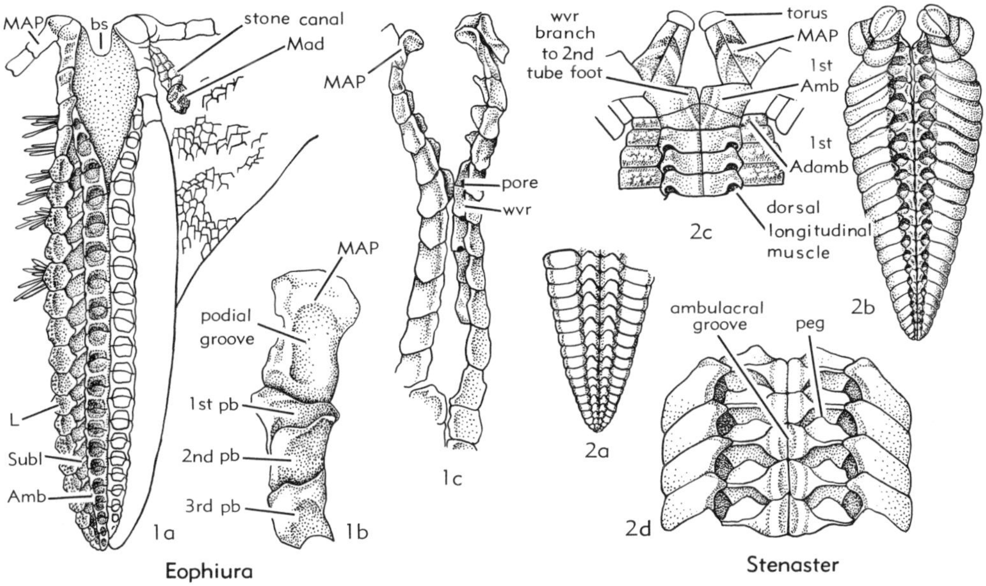

FIG. 70. Eophiuridae *(1)*, Stenasteridae *(2)*. [Explanation: *Adamb,* adambulacral; *Amb,* ambulacral; *bs,* buccal slit; *L,* lateral; *Mad,* madreporite; *MAP,* mouth-angle plate; *pb,* podial basin; *Subl,* sublateral: *wvr,* water-vessel-ring groove.] (p. *U*82).

Family PHRAGMACTINIDAE Spencer, 1951

[*nom. correct.* SPENCER & WRIGHT, herein (*ex* Phragmactidae SPENCER, 1951)]

Laterals embracing sides of arms; proximal buccal tentacles well developed; no aboral skeleton. *U.Ord.*

Phragmactis SPENCER, 1940 [**P. grayae*; OD]. Laterals with spines on low ridge; single interradial ossicles in position of buccal shields; mouth-angle plates short, deeply excavated for first buccal tentacles; proximal ambulacrals differentiated. *U. Ord.*, Scot.——FIG. 69,*1*. **P. grayae*, Girvan; *1a,b,* oral and aboral surfaces of arm, ×10 (133).

Family RHOPALOCOMIDAE Spencer & Wright, new family

Laterals subventral, with wide swing, bearing row of spines along their oral edge; ambulacral grooves wide; sublaterals well exposed. *U.Sil.-U.Dev.*

Rhopalocoma SALTER, 1857 [**Palaeocoma (R.) pyrotechnica*; OD]. Arms 5, short and blunt; disc with large oral interrays bordered by more or less cylindrical marginals, each bearing large club-shaped spine; proximal ambulacrals not differentiated; mouth-angle plates elongate; laterals articulating with sublaterals by ball-and-socket joints. *U.Sil.*, Eng.(Heref.).——FIG. 68,*2*. **R. pyrotechnica*; *2a,* oral surface of arm, ×6; *2b,c,* aboral ossicles, ×6, ×8 (133).

Ptilonaster HALL, 1868 [**P. princeps*; OD]. Arms 10; conical spine on each lateral; disc domed, with spicular skeleton; madreporite well developed near edge; mouth-angle plates elongate, wedge-shaped. *U.Dev.*, USA(N.Y.).

Family BDELLACOMIDAE Spencer & Wright, new family

Laterals embracing sides of very long narrow arms and bearing long spines; aboral skeleton composed of stout ossicles. *U.Sil.-L.Dev.*

Bdellacoma SALTER, 1857 [**Palaeocoma (B.) vermiformis*; OD]. Characters of family. *U.Sil.-L. Dev.*, Eng.-Ger.——FIG. 69,*2*. **B. vermiformis,* U.Sil., Eng.(Heref.); *2a,* oral surface of arm, ×4; *2b,* aboral ossicles, ×4 (133).

Suborder PAROPHIURINA Jaekel, 1923

[*nom. transl.* SPENCER & WRIGHT, herein (*ex* subclass Parophiura JAEKEL, 1923)]

Ambulacrals with strong median (adradial) ridges, completely enclosing radial channel; their distal ends not sufficiently elongated to form complete cups for seating tube feet. *L.Ord.-L.Dev.*

Members of this suborder have undifferentiated ambulacrals adjoining the buccal slits and thus all these plates resemble one

1 Medusaster

FIG. 71. Palaeuridae (p. U82).

another, as in *Pradesura* and early somasteroids. The canal leading to the madreporite is calcified, simulating the stone canal of asteroids; it is most readily seen in *Eophiura.*

Rather than coin a new subordinal name, it seems best to revive, redefine, and translate JAEKEL's name for a subclass that was erected primarily for *Eophiura* and *Palaeura.*

Family EOPHIURIDAE Schöndorf, 1910

[*non* Eophiuridae STÜRTZ, 1900 (*recte* Eophiuritidae) =Furcasteridae STÜRTZ, 1900]

Extremities of arms blunt; ambulaterals alternating; laterals and sublaterals in independent series. *L.Ord.*

Eophiura JAEKEL, 1903 [**E. bohemica* SCHUCHERT, 1914; SM]. Well-developed ball-and-socket joints between laterals and sublaterals, giving them wide swing; spines on laterals forming continuous border except at ends of arms where they are broadly scattered; buccal slits very deep, bordered by 5 or 6 ambulacrals; basins for tube feet deep, rounded. *L.Ord.(U.Arenig.),* Czech.——FIG. 70,*1*. **E. bohemica*; *1a,* part of oral surface, ×1; *1b,* angle of mouth frame from side, ×2.5; *1c,* mouth frame and ambulacrals from above, ×1 (133). (See Figs. 7,*3*; 10,*2*; 18,*2*.)

Family PALAEURIDAE Spencer, 1951

Like Eophiuridae but arm extremities tapering and no sublaterals visible; ambulacrals with incipient boot-shaped median ridge; disc rounded, with conspicuous interradial areas and well-defined scales. *L. Ord.-L.Dev.*

Palaeura JAEKEL, 1903 [**P. neglecta* SCHUCHERT, 1914; SM]. Arms 5. *L.Ord.(U. Arenig.),* Czech. ——FIG. 18,*3*. **P. neglecta* SCHUCHERT; ambulacrals (133).

Medusaster STÜRTZ, 1890 [**M. rhenanus*; OD]. Arms many. *L.Dev.,* Ger.——FIG. 71,*1*. **M. rhenanus*; aboral surface (116).

Family STENASTERIDAE Schuchert, 1914

Ambulacrals oposite, not alternating; laterals broad, each with several rows of pustules; disc with swollen aboral surface and reduced oral interrays. *M.Ord-U.Ord.*

Stenaster BILLINGS, 1858 [**S. salteri* (=**Uranaster obtusus* FORBES, 1848; SD SCHUCHERT, 1914] [=*Tetraster* NICHOLSON & ETHERIDGE, 1880]. *M.Ord.-U.Ord.,* Can.-Eu.-W.Asia.——FIG. 70,*2*. **S. obtusus* (FORBES), U.Ord., Scot.; *2a,b,* aboral and oral surface of arm, ×2.5; *2c,* mouth frame from aboral side, ×5; *2d,* aboral surface of arm, ×5 (133).

Order OEGOPHIURIDA Matsumoto, 1915

[*nom. transl. et correct.* FELL, 1962 (*ex* Oegophiuroidea MATSUMOTO, 1915)]

Hyponeural groove covered by soft skin, forming canal not closed over by ventral arm plates; disc covered by skin with or without granules or by imbricating scales; in Encrinasteridae scales at margin may fuse to form narrow frame of stout ossicles; madreporite lateral in early forms, though not so heavily calcified as in Stenurida, tending to move to oral side; no oral or radial shields, dorsal or ventral arm plates, genital plates or bursae; each ray with only 2 buccal tentacles; paired serial gonads extending along proximal part of arms. Gastric caeca entering arms (12). *L.Ord.-Rec.*

The laterals are fused with the sublaterals but junction of the components can be recognized in some cases (Fig. 73,*3*). The inner ends of the fused ossicles are articulated to the axial ossicles and can rotate to act as cover plates; as in Stenurida, they may be subventral in some families and wrapped round the side of the arms in others.

FELL's recognition of *Ophiocanops* as a living member of this largely Paleozoic order allows the soft-part characters to be diagnosed.

Large-scale models of vertebrae of fossil oegophiurids demonstrate their general resemblance to those of most modern ophiuroids (Fig. 72). An important functional difference lies in the very weak development of the ventral peg-and-socket joint; if any fossil oegophiurids had emerged from living

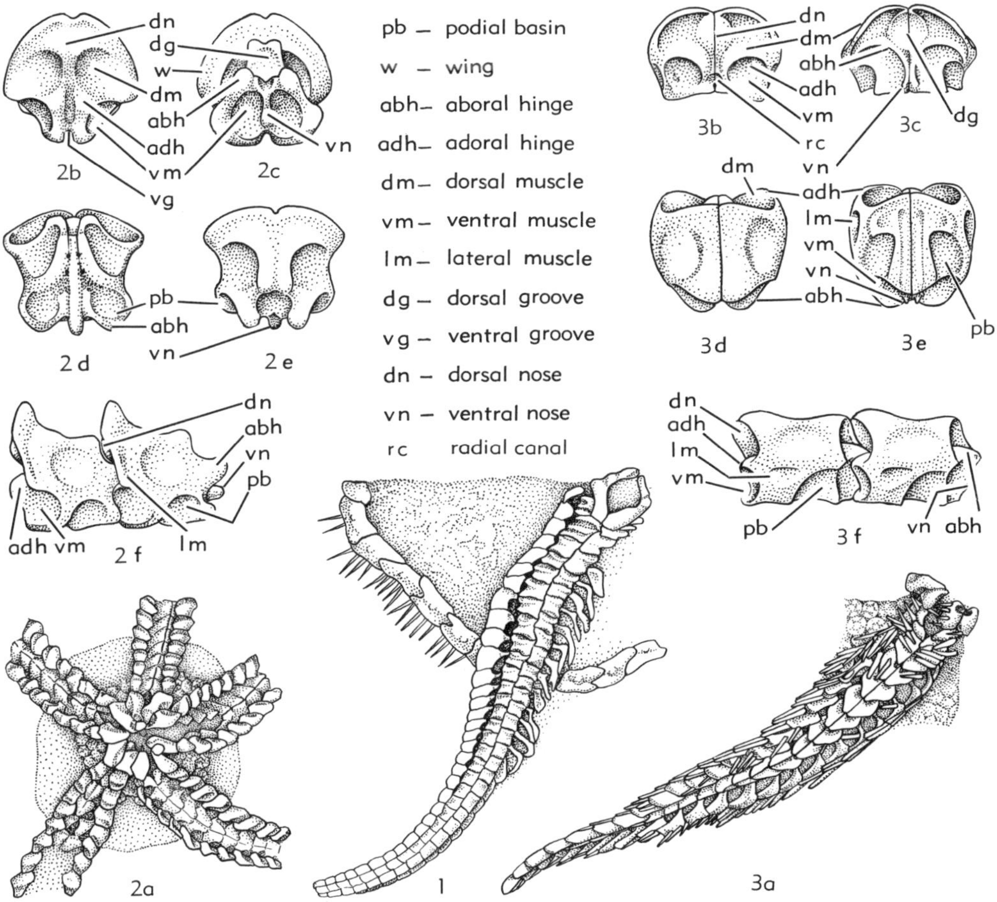

FIG. 72. Structural features of Oegophiurida (133).

1. Encrinaster grayae, U.Ord., Scot.; aboral surface showing marginal frame with slight spicular skeleton within frame, also showing ambulacrals with vertebrae and mouth frame from aboral side, ×2.

2. Vertebrae of *Hallaster* sp., Sil.. N.Am.; *2a,* disc and arm bases from oral side, ×2.5; *2b-f,* vertebra in adoral, aboral, apical, oral, and lateral views, enlarged.

3. Ophiura sp., Rec.; *3a,* oral surface of arm and part of disc, ×5; *3b-f,* vertebra in adoral, aboral, apical, oral, and lateral views, enlarged.

in burrows, they could not move speedily on the sea floor. The vertebrae have characteristic boot-shaped median ridges, under which lies the radial water vessel. As in Stenurida, the structure and muscles of the mouth frame allowed biting movements by the interradially placed mouth-angle plates, but the radial components, which remain passive in the bite, are different; in Stenurida, they comprise several ossicles in each ray but in Oegophiurida only a single pair. They move backwards in the initial stages of the bite, overriding the vertebrae (Fig. 73,*1,2*). The single row of long horizontal spines on the torus are quite unlike the vertically pointed teeth of typical Ophiurida.

Suborder LYSOPHIURINA Gregory, 1896

[*nom. correct.* SPENCER & WRIGHT, herein (*pro* Lysophiurae GREGORY, 1896)]

Halves of vertebrae alternating. *M.Ord.-L.Carb.*

Family ENCRINASTERIDAE Schuchert, 1914

[=Palaeobrisingidae STÜRTZ, 1890 (not founded on generic name) (*nom. nud.*); Aspidosomatidae GREGORY, 1899 (*Aspidosoma* is junior homonym); Schoenasteridae SCHUCHERT, 1915; Euzonosomatidae SPENCER, 1930; Cheiropterasteridae SPENCER, 1930]

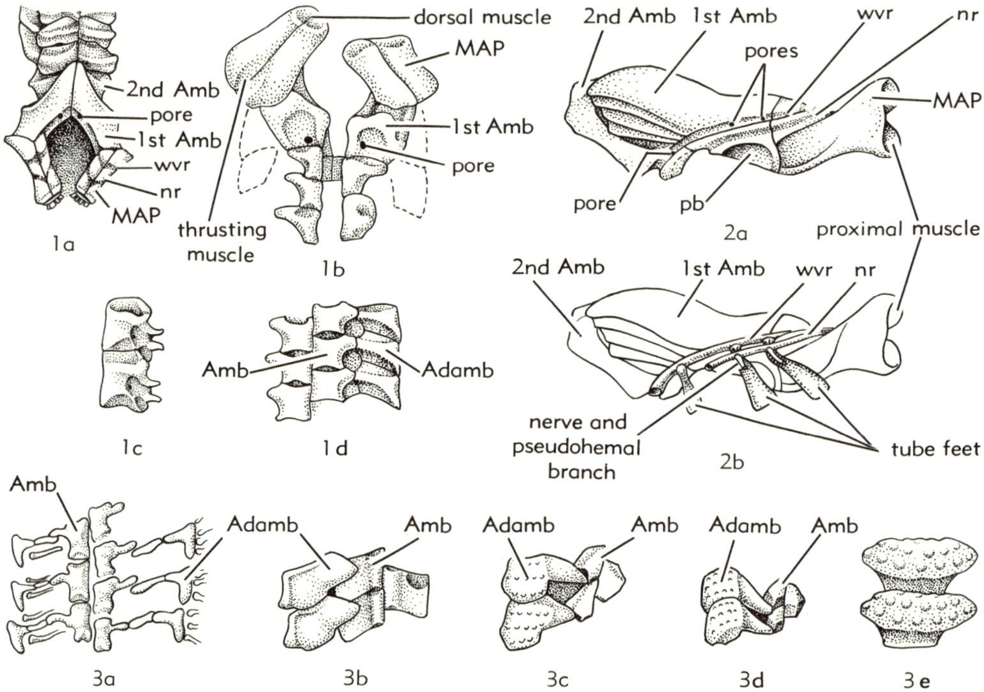

FIG. 73. Mouth frame, ambulacrals, and laterals of fossil Oegophiurida (133).

1. *Encrinaster; 1a,b, E. grayae,* U.Ord., Scot., apical and oral views of mouth frame showing low first pair of ambulacrals, reduced second ambulacrals, large cups for second buccal tentacles, and perforations in each cup for branch water vessel, ×10; *1c,* ambulacrals of *E. grayae* showing cups for tube feet and groove for outward swing of laterals, ×10; *1d, E. eifelensis,* L.Dev., Ger., showing laterals in outward position.

2. *Lapworthura* sp.; *2a,b,* inner views indicating large first two buccal tentacles and first ambulacral overriding next one, ×10.

3. *Euzonosoma; 3a,* laterals of *E. tischbeinianum* formed by fusion of two components, oral view, ×6; *3b-d,* laterals of *E. orbitoides,* apical and two oral views, ×10; *3e,* rows of pustules on laterals and grooves between adjacent ossicles of young *E. orbitoides,* ×20. [Explanation: *Adamb,* adambulacral; *Amb,* ambulacral; *MAP,* mouth-angle plate; *nr,* nerve-ring groove; *pb,* podial basin; *wvr,* water-vessel-ring groove.]

Laterals subventral, commonly with broad oral face, elongate transversely and with curved sutures, producing appearance of rope twists; oral interrays well developed; margin commonly bounded by frame of ossicles. *U.Ord.-L.Carb.*

Typical forms first discovered in the Lower Devonian of Germany were assigned to *Aspidosoma* or *Encrinaster* and thought to have affinities with Asteroidea. SCHÖNDORF (62) showed that they differed from modern Asteroidea and Ophiuroidea by having the radial water vessel enclosed as a canal within the ambulacrals but he gave too much importance to this feature in erecting a new suborder, Auluroidea; most early Ophiuroidea have such a feature and it persists in modern Euryalidae; the Encrinasteridae also have typical ophiuroid vertebrae.

Laterals of most genera of the Encrinasteridae have broad oral faces bearing rows of pustules, superficially resembling adambulacrals of Asteroidea (Fig. 73, *1c*). The stenurid *Stenaster* has laterals with similar oral face and with typical ophiuroid attachment to ambulacrals, so that they operate as cover plates with wide lateral swing.

A tendency for the disc to become swollen is carried to an extreme in *Cheiropteraster,* which also has tube feet confined to proximal parts of the arms and alongside deep buccal slits.

Encrinaster HAECKEL, 1866 [**Aspidosoma arnoldi* GOLDFUSS, 1848; SD SCHUCHERT, 1914] [=*Aspidosoma* GOLDFUSS, 1848 (*non* FITZINGER, 1845)]. Arms with many axial and adaxial elements; strong musculature between ambulacrals; marginal frame well marked. *U.Ord.-L.Carb.*, Eng.-Scot.-Ger.——FIG. 73,*1a-c*; 74,*3a-c*. *E. grayae* SPENCER, U.Ord., Scot.(Girvan); 73,*1a-c*, mouth frame and ambulacrals, ×10; 74,*3a*, oral side of arm and part of disc, ×3.3; *3b*, part of oral surface of arm, ×10; *3c*, adambulacrals and ambulacrals in wide part of arm, ×10 (133).——FIG. 73,*1d*. *E. eifelensis* SCHÖNDORF, L.Dev., Ger.; laterals (128).

Cheiropteraster STÜRTZ, 1890 [**C. giganteus*; OD].

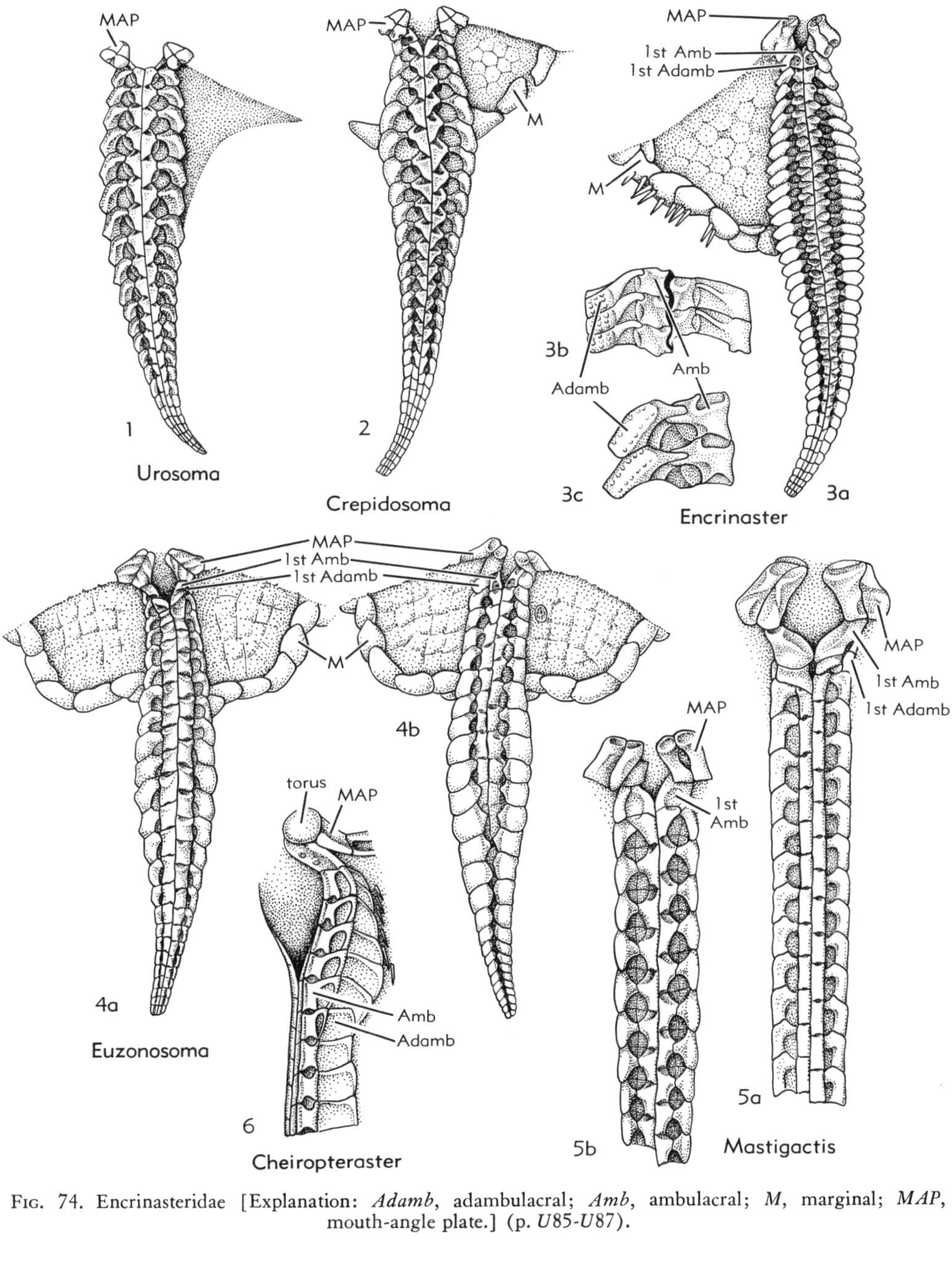

FIG. 74. Encrinasteridae [Explanation: *Adamb*, adambulacral; *Amb*, ambulacral; *M*, marginal; *MAP*, mouth-angle plate.] (p. *U85-U87*).

Vertebral boots elongate; lateral T-shaped, with prominent small spines at edge; proximal ambulacrals divergent, barely differentiated, disc highly swollen, covered by thick skin with granules but no visible scales; tube feet confined to bases of arms. *L.Dev.*, Ger.——FIG. 74,*6*. **C. giganteus*; area near mouth showing ambulacrals and laterals, ×1 (133). (See also Fig. 34.)

Crepidosoma SPENCER, 1930 [**C. wenlocki*; OD]. Like *Euzonosoma* but mouth frame weaker. *L.Sil., ?L.Dev.*, Scot.-Ger.——FIG. 74,2. **C. wenlocki*, L.Sil., Scot.; oral surface, ×7.5 (133).

Euzonosoma SPENCER, 1930 [**E. orbitoides*; OD]. [=*?Schoenaster* MEEK & WORTHEN, 1860 (virtually unrecognizable); *Jovaster* KEYES & BEANE, 1934; *Hymenosoma* LEHMANN, 1957]. Arms distinctly petaloid; laterals widest at middle of arm length; proximal ambulacrals well developed; fewer axial and adaxial elements in arms than in *Encrinaster*; mouth frame strong; marginal frame strong. *U. Ord.-U. Dev.*, Scot.-Ger.-USA (N.Y.-Ill.).——FIG. 73,*3*; 74,*4*. **E. orbitoides*, U.Ord., Scot.(Girvan); 73,*3*, laterals; 74,*4a,b*, aboral and oral surfaces of arm and part of disc, ×5 (133).

Loriolaster STÜRTZ, 1886 [**L. mirabilis*; OD]. Like *Cheiropteraster* but vertebral boots short; laterals deep, with short spines at edge; proximal ambulacrals in form of bow. *L.Dev.*, Ger.

Mastigactis SPENCER, 1930 [**Eugasterella aranea* RUEDEMANN, 1916; OD]. Arms long, narrow, straight-sided; vertebral boots elongate; mouth-

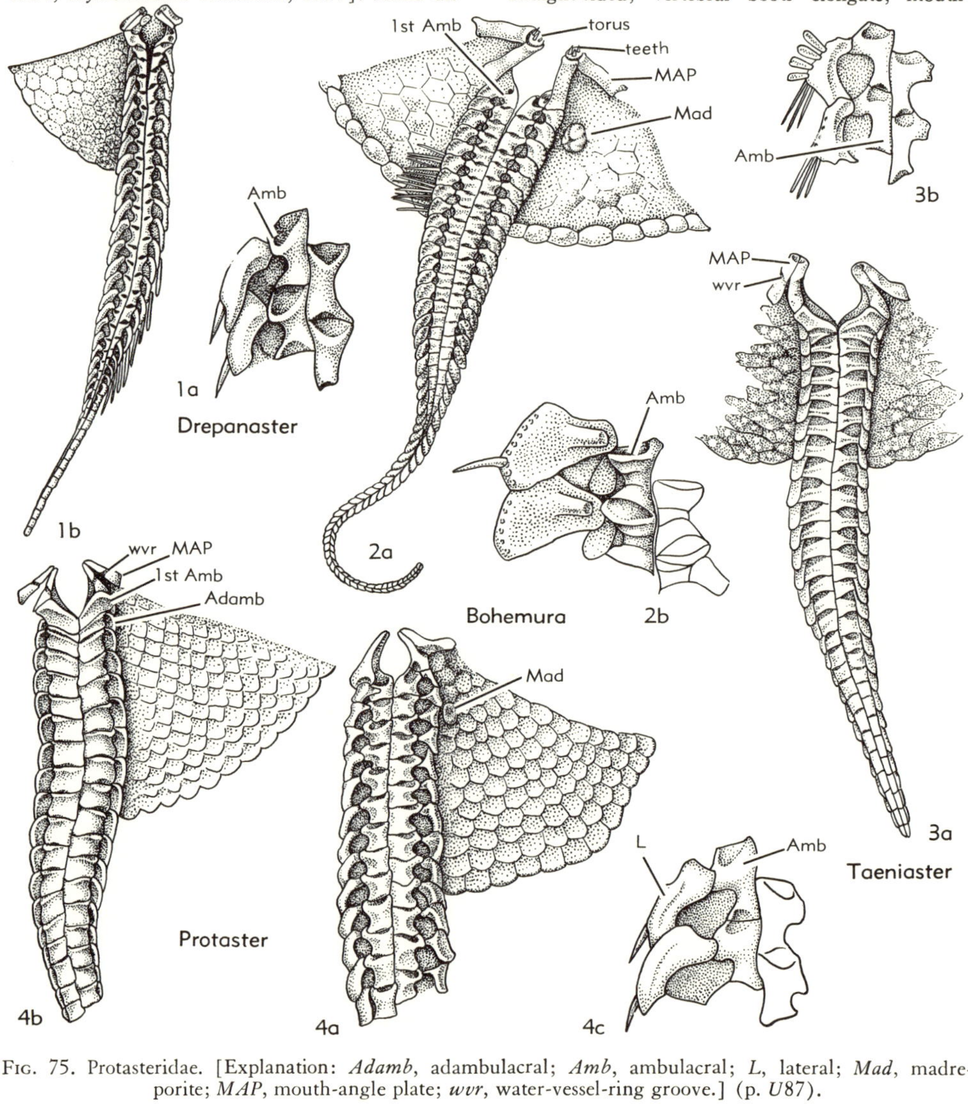

FIG. 75. Protasteridae. [Explanation: *Adamb*, adambulacral; *Amb*, ambulacral; *L*, lateral; *Mad*, madreporite; *MAP*, mouth-angle plate; *wvr*, water-vessel-ring groove.] (p. *U*87).

angle plates stout. *U.Ord.-U.Dev.*, Scot.-N.Am. ——FIG. 74,5. **M. aranea* (RUEDEMANN), U.Ord., Scot.(Girvan); *5a,b,* aboral and adoral surfaces of arm, ×4 (133).

Urosoma SPENCER, 1930 [**Uraster hirudo* FORBES, 1848; OD]. Like *Crepidosoma* but no marginal frame. *M.Ord.-U.Dev.*, Eu.-N.Am.——FIG. 74,*1*. **U. hirudo* (FORBES), U.Sil., Eng.; part of oral surface, ×7.5 (133).

Family PROTASTERIDAE S. A. Miller, 1889

[=Palaeophiuridae GREGORY, 1897; Taeniasteridae GREGORY, 1899; Palaeophyomyxidae STÜRTZ, 1900]

Laterals wrapped around sides of arms, forming side shields; edges of disc may be thickened but have no well-developed marginal frame; oral edges of ambulacrals narrow; laterals with vertical ridge bearing short or long spines; groove spines commonly present. *M.Ord.-L.Carb.*

Protaster FORBES, 1849 [**P. sedgwickii*; OD] [=*Eugaster* HALL, 1860; *Eugasterella* SCHUCHERT, 1914]. Depressions for attachment of dorsal arm muscles weak; laterals with articulating nose near oral edge. *M.Ord.-L.Carb.*, Eng.-USA(N.Y.).——FIG. 75,*4a,b*. **P. sedgwickii*, U.Sil., Eng.; *4a,b,* oral and aboral sides of arm and part of disc, ×0.6 (133).——FIG. 75,*4c*. *P. salteri* (FORBES), M.Ord., Wales; part of arm showing ambulacrals and laterals, ×20 (133).

Aulactis SPENCER, 1930 [**A. orthopaeda*; OD]. Vertebrae with wide shallow median groove on aboral side. *M.Ord.*, Wales.

Bohemura JAEKEL, 1903 [**B. jahni*; OD]. Musculature like that of *Protaster* but articulating nose of laterals plain, distant from oral edge. *U.Ord.-L.Carb.*, Eu.-W.Asia.——FIG. 75,*2a*. **B. jahni*, U.Ord., Czech.; oral side of arm and part of disc, ×1 (133).——FIG. 75,*2b*. *B. groomi* SPENCER, U.Ord., Wales; part of arm showing ambulacrals and laterals, ×10 (133).

Drepanaster WHIDBORNE, 1898 [**Protaster scabrosus* WHIDBORNE, 1896; OD]. Arms very long and narrow; muscle depressions as in *Taeniaster*. *U.Ord.-L.Carb.*, N.Am.-Eng.-Scot.——FIG. 75,*1a*. **D. scabrosus*, L.Carb., Eng.(Devon.); part of arm showing ambulacrals and laterals, ×10 (133). ——FIG. 75,*1b*. *D. grayae* SPENCER, U.Ord., Scot.; oral side of arm and part of disc, ×2 (133).

?**Inyoaster** PHLEGER, 1936 [**I. bradleyi*; OD]. Unrecognizable. *Ord.*, Calif.

Mastigophiura LEHMANN, 1957 [**M. grandis*; OD]. Differs from *Taeniaster* only in having large spines on disc, perhaps only a specific difference. *L.Dev.*, Ger.——FIG. 76,*1*. **M. grandis*; oral side, ×0.5 (116).

Palaeophiura STÜRTZ, 1890 [**P. simplex*; OD]. Vertebral boots long, narrow; laterals with spine-like ridge projecting considerably outward. *L. Dev.*, Ger.

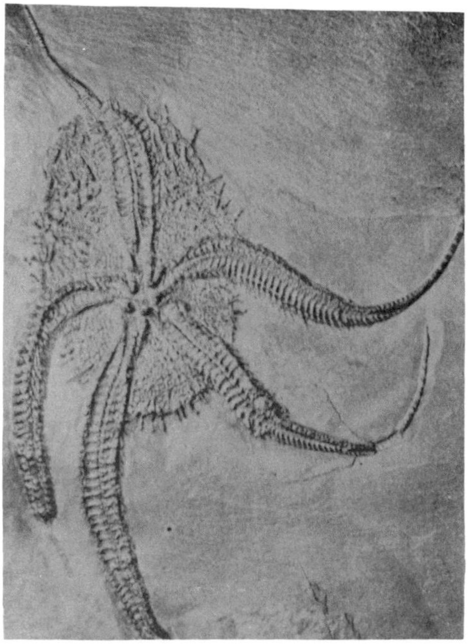

1 Mastigophiura

FIG. 76. Protasteridae (p. *U87*).

Taeniaster BILLINGS, 1858 [**Palaeocoma spinosa* BILLINGS, 1857; SD SCHUCHERT, 1914] [=*Alepidaster* MEEK, 1872; *Protasterina* ULRICH, 1878; *Bundenbachia* STÜRTZ, 1886; *Palaeophiomyxa* STÜRTZ, 1890]. Arms not conspicuously narrow; depressions for aboral muscles deep. *M.Ord.-L. Dev.*, N.Am.-Ger.——FIGS. 75,*3*. **T. spinosus* (BILLINGS), M.Ord.(Trenton.), Ont.; *3a,* aboral side and part of disc, ×2; *3b,* part of arm showing ambulacrals and laterals, ×20 (133). (See also Fig. 4, 18,*4*.)

Suborder ZEUGOPHIURINA Matsumoto, 1929

[*nom. transl. et correct.* FELL, 1963 (*ex* Zeugophiuroidea MATSUMOTO, 1929)]

Halves of vertebrae opposite, separate or fused. *L.Ord.-Rec.*

Members of this suborder in general structure closely resemble Protasteridae except in position of the vertebral halves. They are distinguished from early Ophiurida only by position of the laterals; in Zeugophiurina they are separated by a wide groove that exposes the oral surface of the vertebrae, except for a covering of soft skin,

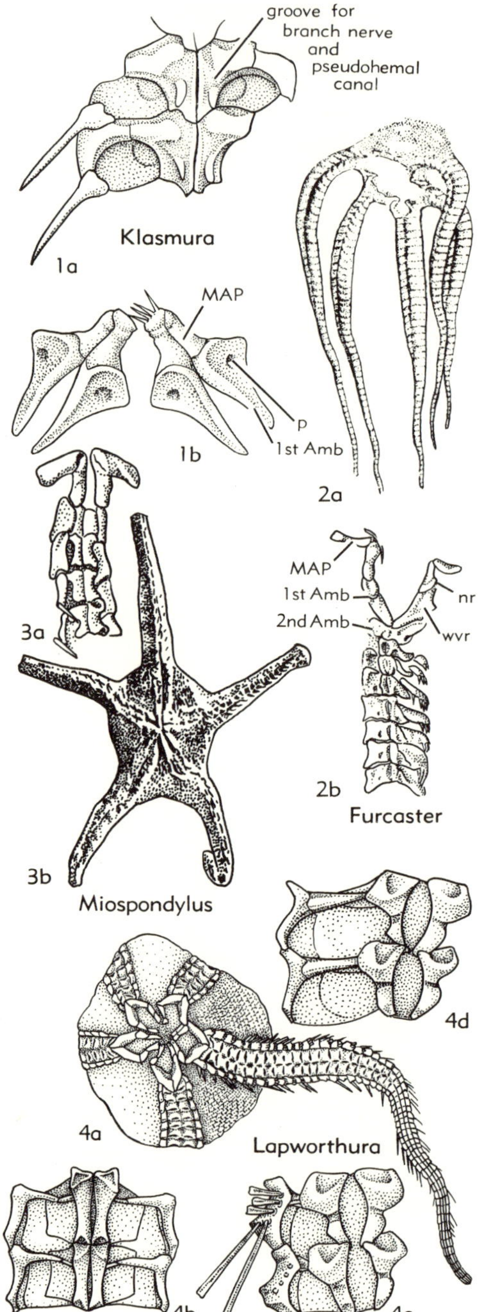

FIG. 77. Lapworthuridae *(3-4)*; Furcasteridae *(2)*; Klasmuridae *(1)*. [Explanation: *Amb*, ambulacral; *MAP*, mouth-angle plate; *nr*, nerve-ring groove; *p*, podial basin; *wvr*, water-vessel-ring groove.] (p. U88-U89).

whereas in Ophiurida the laterals meet and cover the oral surface of the vertebrae.

Specimens of the earliest genus, *Hallaster*, are found with arms upflexed, indicating a burrowing habit. Later genera (e.g., *Lapworthura, Klasmura*) were probably emergent. *Klasmura* seems to have lived commensally with crinoids.

Family LAPWORTHURIDAE Gregory, 1897

[=Hallasteridae SPENCER, 1925]

Disc large, arms robust, with short or long conical spines, generally set in row not parallel to arm axis. *L.Ord.-L.Dev.*

Lapworthura GREGORY, 1897 [**Protaster miltoni* SALTER, 1857; OD]. Arms low, broad; basins for tube feet large; laterals with elongate noses, vertical spines long, in rows at wide angle to arm axis. *U.Ord.-Sil.*, Eng.-Scot.-Australia.——FIG. 77, *4*. **L. miltoni* (SALTER), M.Sil., Eng.(Heref.), *4a*, aboral side of disc and arm, ×1.5; *4b-d*, arm showing ambulacrals and laterals, enl. (133).

Hallaster STÜRTZ, 1886 [**Protaster forbesi* HALL, 1861 (=**Palaeocoma cylindrica* BILLINGS, 1857); OD] [=*Taeniura* GREGORY, 1897 (*non* MÜLLER & HEULE, 1837); *Hypophiura* JAEKEL, 1903]. Vertical spines longer than arm segment; basins for tube feet narrow, boots with blunt toe; laterals with short noses. *L.Ord.-L.Dev.*, N.Am.-Scot.——FIG. 72,2. **H. cylindricus* (BILLINGS), M.Ord.(Trenton.), Ont.; *2a*, oral side of disc and arm bases; *2b-f*, vertebrae (133).

Miospondylus GREGORY, 1897 [**Ophiura rhenana* STÜRTZ, 1893; OD]. Arms moderately high; vertical spines mostly short and unequal, long spines lying across ambulacral groove. *L.Dev.*, Ger.——FIG. 77,*3*. **M. rhenanus* (STÜRTZ), Bundenbach; *3a*, side view of arm, ×3; *3b*, oral side, ×0.7 (133).

Family FURCASTERIDAE Stürtz, 1900

[=Eoluidiidae GREGORY, 1897; Eophiuridae (*recte* Eophiuritidae), Palaeospondylidae STÜRTZ, 1900]

Like Lapworthuridae but with subequal needle-shaped spines in rows close to sides of arms and parallel to arm axes. *U.Ord.-Miss.*

Furcaster STÜRTZ, 1886 [**F. palaeozoicus* (=**Protaster leptosoma* SLATER, 1857); OD] [=*Palastropecten, Eoluidia* STÜRTZ, 1886; *Squamaster* RINGUEBERG, 1886; *Eophiurites* STÜRTZ, 1900; *Palaespondylus, Palaeospondylus* STÜRTZ, 1900 (*non* TRAQUAIR, 1890); *Sympterura* BATHER, 1905; *Gregoriura* CHAPMAN, 1907; *Rhodostoma* SOLLAS & SOLLAS, 1912]. Interior of laterals with long vertical ridges normal to adoral edge; vertebrae with median hollow; mouth frame petaloid;

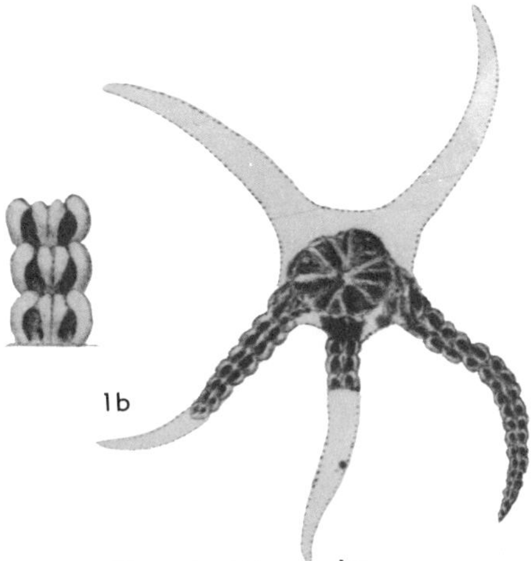

FIG. 78. Furcasteridae (p. U89).

fossils showing arms stiffly upright. *U.Ord.-Miss.*, Eu.-N.Am.-Australia.——FIG. 77,*2*. **F. leptosoma* (SALTER), L.Dev., Ger.; *2a*, specimen with raised arms, ×1; *2b*, part of aboral surface of arm, ×2.5 (133).

Tremataster WORTHEN & MILLER, 1883 [**T. difficilis*; OD]. Imperfectly known but apparently like *Furcaster* except that arms are flexuous. *Miss.*, USA(Ill.).——FIG. 78,*1*. **T. difficilis*; *1a*, oral surface; *1b*, part of arm, enl. (138).

Family KLASMURIDAE Spencer, 1925

Laterals with single long flat hollow spine; disc and arms covered by thick skin and long spines; mouth frame stout; madreporite ventral. *U.Dev.*

Klasmura RUEDEMANN, 1916 [**K. mirabilis*; OD]. Arms generally enrolled; specimens found in association with crinoids, with which they probably lived commensally. *U.Dev.*, USA(N.Y.).——FIG. 77,*1*. **K. mirabilis*; *1a*, part of oral surface, ×10; *1b*, oral side of mouth frame, ×10 (133).

Family OPHIOCANOPIDAE Mortensen, 1933

Laterals with rather few stout spines more or less parallel to arm axis, those on oral surface hooked and with course serrations, as in Euryalina; madreporite marginal; articulation of vertebrae tending to euryaline type (streptospondylous) (12). *Rec.*

Ophiocanops KOEHLER, 1922 [**O. fugiens*; OD]. *Rec.*

Order PHRYNOPHIURIDA Matsumoto, 1915

Disc and arms covered with skin; radial shields and genital plates articulating by simple facet or transverse ridge on each plate; peristomial plates large, entire or double or triple; oral frames entire, without well-developed lateral wings; dorsal arm plates absent or rudimentary; lateral arm plates ventral or subventral. *L.Dev.-Rec.*

Suborder OPHIOMYXINA Fell, 1962

Disc and arms covered by thick soft skin overlying plates and scales. *Rec.*

Family OPHIOMYXIDAE Ljungman, 1866

Characters of suborder. *Rec.*

Subfamily OPHIOMYXINAE Ljungman, 1866

[*nom. transl.* MATSUMOTO, 1915 (*ex* Ophiomyxidae LJUNGMAN, 1866)]

Oral shields small; adoral plates long and slender; vertebrae long and slender, articular peg well developed. *Rec.*

Ophiomyxa MÜLLER & TROSCHEL, 1840 [**Ophiura pentagona* LAMARCK, 1816; OD]. *Rec.*

Astrogeron VERRILL, 1899 [**Ophiogeron supinus* LYMAN, 1883; OD]. *Rec.*

Neoplax BELL, 1884 [**N. ophiodes*; OD]. *Rec.*

Ophiodera VERRILL, 1899 [**Ophiomyxa serpentaria* LYMAN, 1883; OD]. *Rec.*

Ophiogeron LYMAN, 1878 [**O. edentulus*; OD]. *Rec.*

Ophiohelus LYMAN, 1880 [**Ophiohelus umbella* LYMAN, 1880; SD H. L. CLARK, 1915]. *Rec.*

Ophiohyalus MATSUMOTO, 1915 [**O. gotoi*; OD]. *Rec.*

Ophiohymen H. L. CLARK, 1911 [**O. gymnodiscus*; OD]. *Rec.*

Ophioleptoplax H. L. CLARK, 1911 [**O. megapora*; OD]. *Rec.*

Ophiolycus MORTENSEN, 1933 [**O. inermis*; OD]. *Rec.*

Ophiomora KOEHLER, 1907 [**O. elegans*; OD]. *Rec.*

Ophiosciasma LYMAN, 1878 [**O. attenuatum*; OD]. *Rec.*

Ophioscolex MÜLLER & TROSCHEL, 1842 [**O. glacialis*; OD] [=*Ophiocynodus* H. L. CLARK, 1911]. *Rec.*

Ophiostiba MATSUMOTO, 1915 [**O. hidekii*; OD]. *Rec.*

Ophiostyracium H. L. CLARK, 1911 [**O. trachyacanthum*; OD]. *Rec.*

Ophiosyzygus H. L. CLARK, 1911 [**O. disacanthus*; OD]. *Rec.*

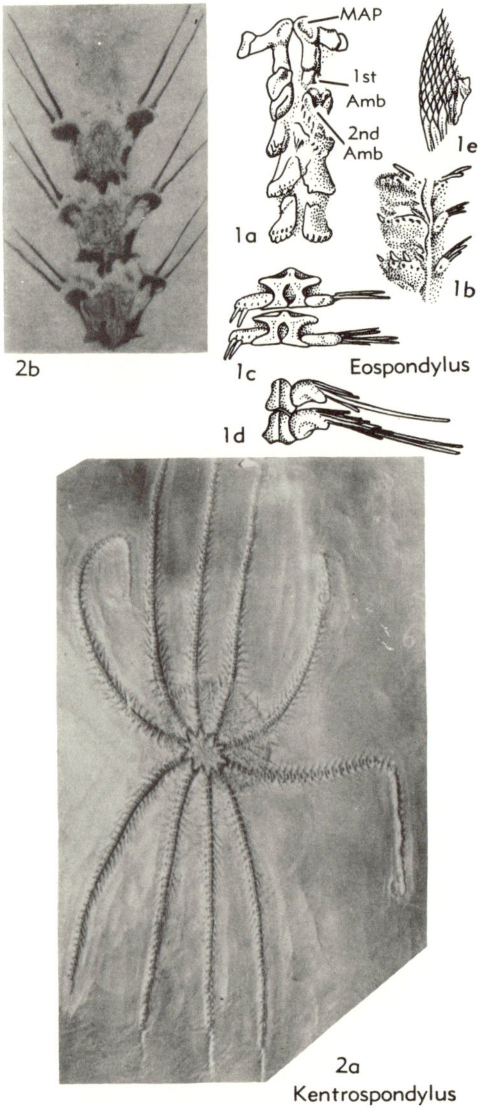

FIG. 79. Eospondylidae. [Explanation: *Amb*, ambulacral; *MAP*, mouth-angle plate.] (p. *U*90).

Subfamily OPHIOBYRSINAE Matsumoto, 1915

Oral shields and adoral plates fused together, massive; vertebrae short and thick, articular peg rudimentary or lacking. *Rec.*

Ophiobyrsa LYMAN, 1878 [**O. rudis*; OD]. *Rec.*

Astrogymnotes H. L. CLARK, 1914 [**A. catasticta*; OD] [=*Ophiovesta* KOEHLER, 1931]. *Rec.*

Ophiobrachion LYMAN, 1883 [**O. uncinatus*; OD]. *Rec.*

Ophiobyrsella VERRILL, 1899 [**Ophiobyrsa serpens* LYMAN, 1883; OD]. *Rec.*

Ophiophrixus H. L. CLARK, 1911 [**O. acanthinus*; OD]. *Rec.*

Ophioschiza H. L. CLARK, 1911 [**O. monacantha*; OD]. *Rec.*

Ophiosmilax MATSUMOTO, 1915 [**O. mirabilis*; OD]. *Rec.*

Suborder EURYALINA Lamarck, 1816

[*nom. correct.* FELL, 1962 (*pro* Euryalae MÜLLER & TROSCHEL, 1840, *nom. transl. ex* euryales LAMARCK, 1816)]

Disc small, with no plates or, in later forms, scales; disc and arms covered by thick skin, with or without granules; metapinnular structure in arms persisting in some genera; vertebrae typically articulating by broad hourglass-shaped surfaces but Onychasteridae retaining reduced zygophiuroid peg; arms coiling vertically and may branch. *L.Dev.-Rec.*

Family EOSPONDYLIDAE Spencer & Wright, new family

Arms 5 or 10; laterals well separated on aboral surface but closely approximated on oral surface of arms; laterals large and sickle-shaped. *L.Dev.*

Eospondylus GREGORY, 1897 [*pro Ophiurella* STÜRTZ, 1886 (*non* AGASSIZ, 1834] [**Ophiurella primigenia* STÜRTZ, 1886; OD]. Arms 5; vertical spines unequal, some very long; disc covered with smooth overlapping scales. *L.Dev.*, Ger.——FIG. 79,*1*. **E. primigenius* (STÜRTZ); *1a,b,* proximal part and side of arm; *1c,d,* oral and aboral surface of arm; *1e,* part of disc, ×5 (133).

Kentrospondylus LEHMANN, 1957 [**K. decadactylus;* OD]. Arms 10, very long and slender, round in section; vertical spines more or less equal, very long; disc with granules, some bearing long slender spines. *L.Dev.*, Ger.——FIG. 79,*2*. **K. decadactylus*; *2a,* aboral surface, ×0.5; *2b,* arm, enl. (116).

Family ONYCHASTERIDAE Miller, 1889

Arms 5, but may branch; laterals small. *L.Carb.*

This family closely resembles Recent Euryalina in the narrow high vertebrae with intervertebral articulation concentrated in the center of the ossicles, small laterals, small basins for the tube feet, and branching of the arms. The movement of the arms in *Onychaster*, however, was restricted by the small zygophiuroid peg (Fig. 80,2).

Onychaster MEEK & WORTHEN, 1868 [**O. flexilis*; OD]. Characters of family. *L.Carb.(Miss.),* N. Am.-Eng.-Scot.——FIG. 80,*1*. *O. barrisi* (HALL),

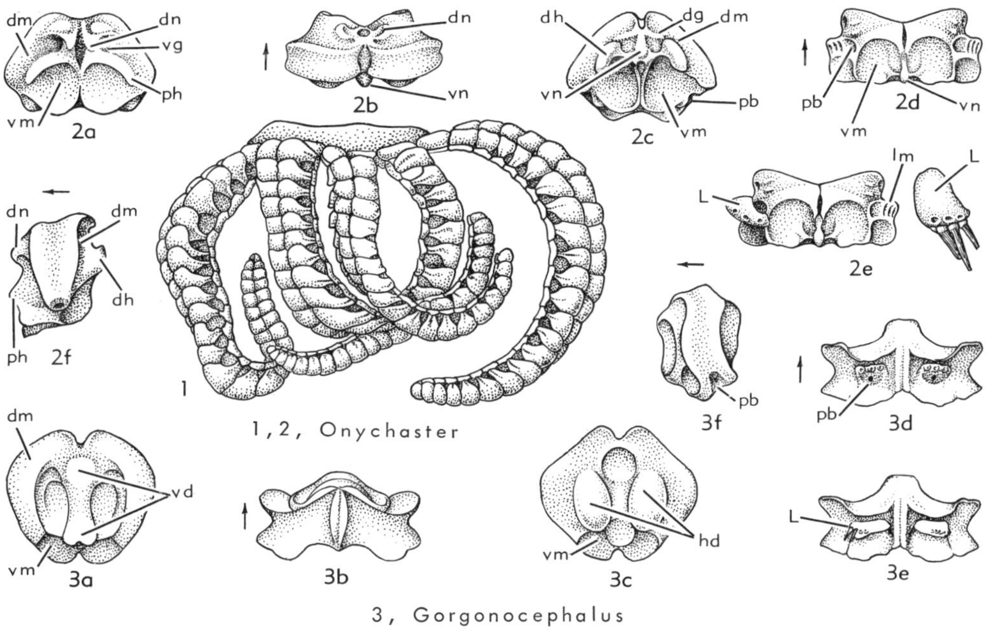

FIG. 80. Onychasteridae *(1-2)*; Gorgonocephalidae (3) (p. *U*90-*U*91).

Eng.(Devon.); side view, ×2 (133).——FIG. 80, *2,3.* **O. flexilis,* USA(Ind.); vertebrae *(2a-f)* compared with *Gorgonocephalus* (Rec.) *(3a-f)*, enlarged: *a,* proximal face; *b,* aboral side (arrow toward mouth); *c,* distal face; *d,* oral side, with laterals removed; *e,* oral side with laterals in position; *f,* lateral view (arrow toward mouth) (133).

Family ASTERONYCHIDAE Müller & Troschel, 1842

Arms not branched; distinct metapinnular structure; vertebrae with ventral furrow; distal arm joints not long and slender; distally lateral arms spines may be transformed into hooklets that do not have perforated lamina. Gonads restricted to disc. *?U.Cret., Rec.*

Asteronyx MÜLLER & TROSCHEL, 1842 [**A. loveni*; OD]. Disc and arms covered aborally by naked skin; more than 3 arm spines, outer ones modified as hooklets. [Isolated ossicles from Upper Cretaceous may belong here.] *?U.Cret.(Senon.)*, W.Eu., *Rec.*

Astrodia VERRILL, 1899 [**A. tenuispina*; OD]. *Rec.*

Family ASTEROSCHEMATIDAE Verrill, 1899

Similar to Asteronychidae but gonads extending at least midway along arms. *Rec.*

Asteroschema OERSTED & LÜTKEN, 1856 [**Asterias oligactes* PALLAS, 1788; OD] [=*Laspalia* LJUNGMAN, 1872 (*non* GRAY, 1840)]. *Rec.*

Astrobrachion DÖDERLEIN, 1927 [**Ophiocreas constrictus* FARQUAR, 1900; OD]. *Rec.*

Astrocharis KOEHLER, 1904 [**A. virgo*; OD]. *Rec.*

Astroscolex MORTENSEN, 1933 [**Ophiocreas adhaerens* STUDER, 1884; OD]. *Rec.*

Ophiocreas LYMAN, 1879 [**O. lumbricus*; OD]. *Rec.*

Ophiuropsis STUDER, 1884 [**O. lymani*; OD]. *Rec.*

Family GORGONOCEPHALIDAE Ljungman, 1867

[*incl.* Astrotominae MATSUMOTO, 1915]

Arms simple or branching; ventral furrow open; dorsal surface of arms bearing hooks without lamina of regularly arranged holes. [An undescribed genus occurs in the Oligocene of New Zealand (FELL, *in litt.*).] *Oligo.-Rec.*

Gorgonocephalus LEACH, 1815 [**Asterias caput-medusae* LINNÉ, 1758; SD H. L. CLARK, 1915]. *Rec.*——FIG. 80,*3*. *G.* sp.; *3a-f*, vertebrae (see under *Onychaster*, p. *U*90 for details).

Asteroporpa OERSTED & LÜTKEN, 1856 [**A. annulata* LÜTKEN, 1856; OD]. *Rec.*

Astracme DÖDERLEIN, 1927 [**Astrophyton mucronatum* LYMAN, 1869; OD]. *Rec.*

Astroboa DÖDERLEIN, 1911 [**Astrophyton clavatum* LYMAN, 1861; OD] [=*Astrorhaphis* DÖDERLEIN, 1911]. *Rec.*

Astrocaneum DÖDERLEIN, 1911 [**Astrophyton spinosum* LYMAN, 1875; OD] [=*Astrocynodus* A. H. CLARK, 1918]. *Rec.*

Astrochalcis KOEHLER, 1905 [**A. tuberculosus*; OD]. *Rec.*

Astrochele VERRILL, 1878 [**A. lymani*; OD]. *Rec.*

Astrochlamys KOEHLER, 1911 [**A. bruneus*; OD]. *Rec.*

Astrocladus VERRILL, 1899 [**Euryale verrucosum* LAMARCK, 1816 (=**Asterias euryale* RETZIUS, 1783); OD]. *Rec.*

Astroclon LYMAN, 1879 [**A. propugnatoris*; OD]. *Rec.*

Astrocnida LYMAN, 1872 [**Trichaster isidis* DUCHASSAING, 1850; OD]. *Rec.*

Astroconus DÖDERLEIN, 1911 [**Astrophyton australe* VERRILL, 1876; OD]. *Rec.*

Astrocrius DÖDERLEIN, 1927 [**Astrotoma sobrinus* MATSUMOTO, 1912; OD]. *Rec.*

Astrocyclus DÖDERLEIN, 1911 [**Astrophyton caecilia* LÜTKEN, 1856; OD]. *Rec.*

Astrodendrum DÖDERLEIN, 1911 [**Gorgonocephalus sagaminus* DÖDERLEIN, 1902; OD]. *Rec.*

Astrodictyum DÖDERLEIN, 1927 [**Astrophyton panamense* VERRILL, 1867; OD]. *Rec.*

Astroglymna DÖDERLEIN, 1927 [*pro Astrodactylus* DÖDERLEIN, 1911 (*non* HOGG, 1839)] [**Astrophyton sculptum* DÖDERLEIN, 1896; OD]. *Rec.*

Astrogomphus LYMAN, 1869 [**A. vallatus*; OD]. *Rec.*

Astrogordius DÖDERLEIN, 1911 [**Astrophyton cacaoticum* LYMAN, 1874; OD]. *Rec.*

Astrohamma DÖDERLEIN, 1930 [**Astrothamnus tuberculatus* KOEHLER, 1923; OD]. *Rec.*

Astrohelix DÖDERLEIN, 1930 [**Astrotoma bellator* KOEHLER, 1904; OD]. *Rec.*

Astrophyton FLEMING, 1828 [**Euryale muricatum* LAMARCK, 1816; SD H. L. CLARK, 1915]. *Rec.*

Astroplegma DÖDERLEIN, 1928 [**A. expansum*, OD]. *Rec.*

Astrospartus DÖDERLEIN, 1911 [**Euryale mediterraneus* RISSO, 1826; OD]. *Rec.*

Astrostephanus DÖDERLEIN, 1930 [**Astrotoma vecors* KOEHLER, 1904; OD]. *Rec.*

Astrothamnus MATSUMOTO, 1915 [**Astrotoma echinacea* MATSUMOTO, 1912; OD]. *Rec.*

Astrothorax DÖDERLEIN, 1911 [**A. misakiensis*; OD]. *Rec.*

Astrothrombus H. L. CLARK, 1909 [**A. rugosus*; OD]. *Rec.*

Astrotoma LYMAN, 1875 [**A. agassizii*; OD]. *Rec.*

Astrozona DÖDERLEIN, 1930 [**Astrogomphus munitus* KOEHLER, 1904; OD]. *Rec.*

Conocladus H. L. CLARK, 1909 [**C. oxyconus*; OD]. *Rec.*

Ophiocrene BELL, 1894 [**O. oenigma*; OD]. *Rec.*

Schizostella A. H. CLARK, 1952 [**S. bifurcata*; OD]. *Rec.*

Family EURYALIDAE Gray, 1840

[=Trichasteridae DÖDERLEIN, 1911]

Metapinnular structure may survive; *Trichaster*, for example, has metapinnules consisting of 3 virgals; ventral groove closed and radial canal and nerve enclosed within vertebrae; distal arm joints long and slender; no dorsal hooks on arms but distally lateral spines may be modified into hooklets, which have a perforated lamina. Gonads extending into arms. *Rec.*

Euryale? liasica QUENSTEDT, 1876, of which the figures suggest *Trichaster*, is reported by SEILACHER (1953) to consist of casts of resting places of normal ophiuroids, the traces of moving arm tips simulating branching arms.

Euryale OKEN, 1815 [**Euryale asperum* LAMARCK, 1816; SD H. L. CLARK, 1915]. *Rec.*

Asteromorpha LÜTKEN, 1869 [**A. steenstrupii* (=**Asteroschema rousseaui* MICHELIN, 1862); OD]. *Rec.*

Asterostegus MORTENSEN, 1933 [**A. tuberculatus*; OD]. *Rec.*

Astroceras LYMAN, 1879 [**A. pergamena*; OD]. *Rec.*

Sthenocephalus KOEHLER, 1898 [**S. indicus*; OD]. *Rec.*

Trichaster AGASSIZ, 1836 [**Euryale palmiferum* LAMARCK, 1816; OD]. *Rec.*

Order OPHIURIDA Müller & Troschel, 1840

[*nom. correct.* FELL, 1960 (*pro* Ophiureae MÜLLER & TROSCHEL, 1840)] [=Myophiuroidea MATSUMOTO, 1917]

Ambulacral grooves closed by growth of laterals on oral side toward mid-line of arms. Vertebrae subcylindrical, generally with zygophiuroid joints; vertebral halves opposite and united in pairs; dorsal and ventral shields (arm plates) present in all except most primitive forms. Radial shields, genital plates and buccal shields also generally present; no independent madreporite; stone canal opens on buccal shield. *Sil.-Rec.*

Earliest Ophiurida occur in Silurian rocks of the Argentine, followed by those of Devonian rocks in Belgium and western Germany. Minute details visible in Belgian specimens (35) show some features transitional between Oegophiurida and Ophiurida and others typically ophiurid. All stages in differentiation of aboral skeletal elements of disc into large units are found; the process begins at the margin and continues inward. *Ophiurina* (L.Dev.) has merely a thickened margin, whereas *Ophiaulax* (U. Dev.) and *Stephanoura* (U.Dev.) have a strong marginal frame not unlike that of *Euzonostoma*. Ossicles in the marginal frame of *Stephanoura* are incipient radial shields, for they articulate with a genital bar. In *Aganaster* (Miss.) the radial shields are fused in pairs and cover most of the disc except the center, which is occupied by a centrale and a primary circlet, as in some early Mesozoic ophiurids. *Stephanoura* also has small buccal shields.

Generic names were proposed for many of the early Mesozoic ophiurids by AGASSIZ (1835) and D'ORBIGNY (1850). They were founded on somewhat generalized characters. Later it was recognized that these groups had a distinctly modern aspect and T. WRIGHT and others placed them in Recent genera. With stricter standards of generic diagnosis, these identifications cannot all be maintained and in many cases essential diagnostic characters are not sufficiently exposed or preserved in the fossils. Attribution to Recent genera, or even families, is therefore often doubtful.

Suborder CHILOPHIURINA Matsumoto, 1915

[*nom. transl. et correct.* SPENCER & WRIGHT herein (*ex* Chilophiurida MATSUMOTO, 1915)]

Radial shield and genital plate articulate by 2 condyles and 1 pit on each plate. Genital plates and scales barlike. Peristomial plates large or small, normally double or triple. Oral frames with or without well-developed lateral wings. Oral papillae very well developed. *Sil.-Rec.*

The above is MATSUMOTO's diagnosis, applicable to the Recent forms. As here arranged, the suborder includes also the primitive transitional Paleozoic genera which lack fully developed radial shields.

Family OPHIURINIDAE Gregory, 1897

Disc tending to have marginals lengthened in some genera to form incipient short radial shields. Lateral shields wrapped well around arms but narrow dorsal and ventral shields may be present. Spines short, parallel or at slight angle to arm axis. *Sil.-Miss.*

Ophiurina STÜRTZ, 1890 [**O. lymani*; OD]. Disc with thickened margin but no plates visible; covered by granulose skin. No dorsal or ventral shields. Arm spines on low ridge. *L.Dev.*, Ger.——FIG. 81,*2*. **O. lymani*; *2a,b*, parts of aboral and oral surfaces, ×1 (133).

Argentinaster RUEDEMANN, 1916 [**A. bodenbenderi*; OD]. Marginal frame narrow. Lateral shields high, swollen. *Sil.*, Arg.

Ophiaulax UBAGHS, 1941 [**Protaster decheni* DEWALQUE, 1881; OD]. Like *Ophiurina* but disc bordered by well-developed marginals, dorsal shields present and arm spines on distinct ridge. *U.Dev.*, Belg.-Fr.——FIG. 81,*1*. **O. decheni* (DEWALQUE), Belg.; *1a,b*, oral and aboral surfaces, ×1 (135).

Silesiaster SCHWARZBACH & ZIMMERMANN, 1936 [**S. longivertebralis*; OD]. Only poor material known, but close to *Ophiurina*. *L.Carb.*, Ger.

Stephanoura UBAGHS, 1941 [**S. belgica*; OD]. Disc covered by slight skeleton, with incipient radial shields and interradial plates at margin; center with weak skeleton. Arms with ventral shields, arm spines on strong ridge, tentacle pores large. *U.Dev.*, Belg.——FIG. 81,*3*. **S. belgica*; *3a,b*, oral and aboral surfaces, ×1 (135).

Family OPHIURIDAE Lyman, 1865

Disc covered with thick scales or plates; primary circlet commonly prominent. No granulation. Radial shields normally stout. Genital papillae commonly present; oral papillae few; no dental papillae; an unpaired infradental papilla at apex of each jaw. Arms inserted laterally in and fused with disc. Arms short or moderately long, stout, widest at base, tapering rapidly. Arm plates all well developed. Arm spines short, flat on arm. *L.Carb.-Rec.*

Subfamily AGANASTERINAE Stürtz, 1900

[*nom. transl.* SPENCER & WRIGHT, herein (*ex* Aganasteridae STÜRTZ, 1900)]

Disc with large radial shields, united in pairs. *Miss.*

Aganaster MILLER & GURLEY, 1890 [**Protaster? gregarius* MEEK & WORTHEN, 1869; OD] [=*Ophiopege* BÖHM, 1893 (obj.)]. Characters of subfamily. *Miss.*, N.Am.-Scot.——FIG. 81,*4*. **A. gregarius* (MEEK & WORTHEN), L.Miss., USA (Ind.); *4a,b*, aboral surface; *4c*, oral surface of disc with exposed jaws; *4d*, side of arm; *4e*, oral surface of arm; *4f,g*, part of oral and aboral surfaces; all ×1 (133).

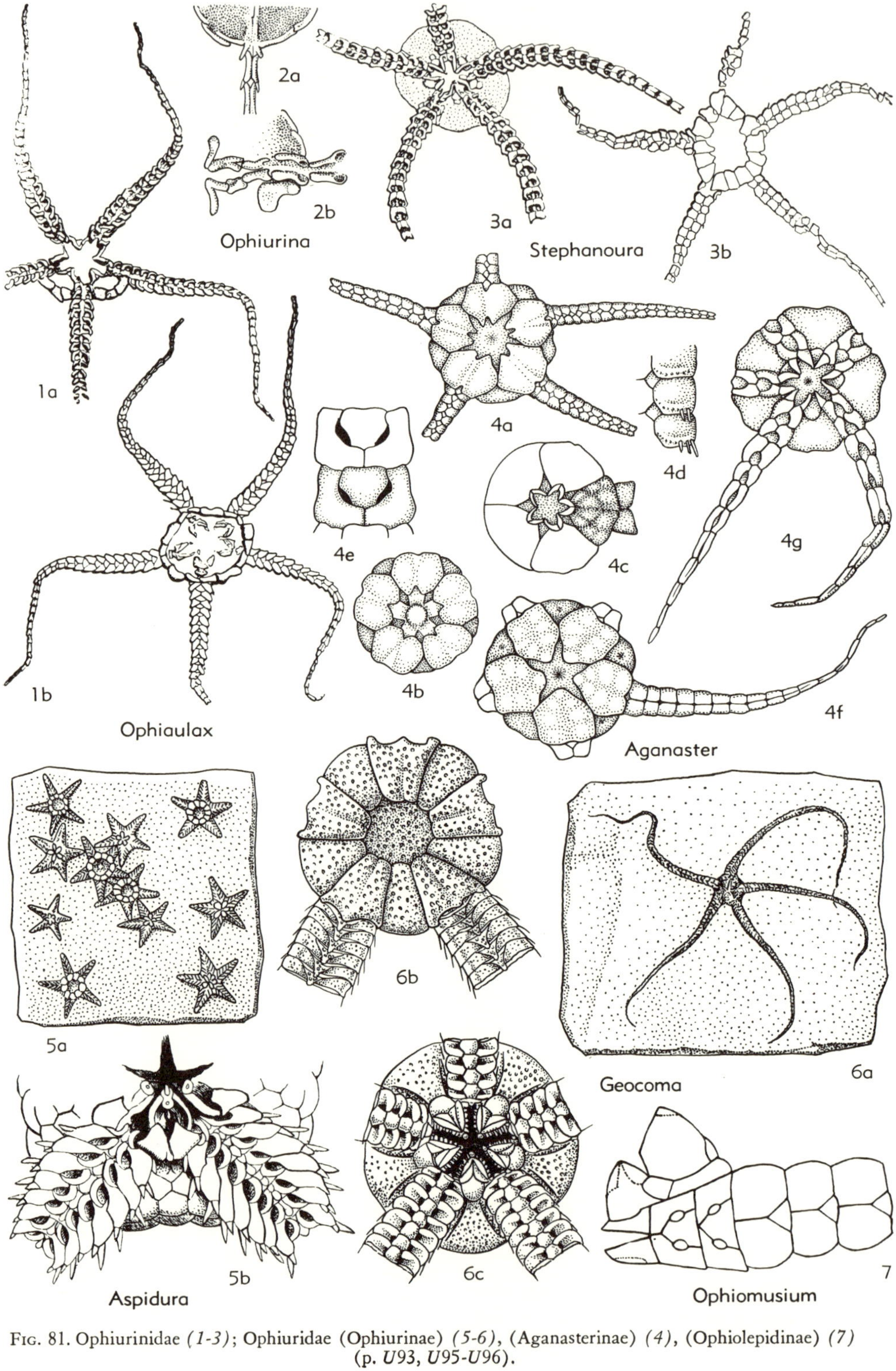

Fig. 81. Ophiurinidae *(1-3)*; Ophiuridae (Ophiurinae) *(5-6)*, (Aganasterinae) *(4)*, (Ophiolepidinae) *(7)* (p. *U*93, *U*95-*U*96).

Subfamily OPHIURINAE Lyman, 1865

[*nom. transl.* FELL, 1960 (*ex* Ophiuridae LYMAN, 1865)] [=Cholasteridae WORTHEN & MILLER, 1883; Ophiomastinae MATSUMOTO, 1915]

Second oral tentacle pore opening more or less entirely outside oral slit. *U.Miss.-Rec.*

MATSUMOTO divided Recent genera into two groups, the first with several proximal lateral shields extraordinarily wide and the disc merely with primaries and radial shields, the second without any wide lateral shields and with secondary scales on the disc.

Ophiura LAMARCK, 1801 [**Asterias ophiura* LINNÉ, 1758; OD] [=*Ophioglypha* LYMAN, 1860; *Ophioglyphina* LUDWIG, 1886; *Ophiozea* A. H. CLARK, 1920]. Disc covered with scales; primary plates inconspicuous. Edge of disc notched at base of arms, notch being filled with rudimentary dorsal arm plates. Ventral shields generally triangular, broader than long, not touching each other. Arm comb normally present but not continuous across arm base. Arm spines vestigial. [Many little-known Mesozoic forms have been referred to this genus.] *?Jur., U.Cret.-Rec.,* cosmop.——FIG. 1,*2*. *Ophiura* spp., Rec.; *2a,b,* oral and aboral surfaces, ×2 (137).

Amphiophiura MATSUMOTO, 1915 [**Ophioglypha bullata* WYVILLE-THOMSON, 1873; OD]. Arms gradually tapering with blunt end. Arm spines in single row, well spaced. Oral shield large. *Oligo.,* N.Am.; *U.Mio.,* Sakhalin; *Rec.*

Anophiura H. L. CLARK, 1939 [**A. simplex*; OD]. *Rec.*

Anthophiura H. L. CLARK, 1911 [**A. axiologa*; OD]. *Rec.*

Aplocoma D'ORBIGNY, 1852 [**Acroura agassizi* MÜNSTER, 1831; OD]. Arms moderately long. Dorsal and ventral shields small and well separated. [A doubtful genus.] *L.Jur.,* Eng.-Ger.——FIG. 82,*1a*. *A. leckenbyi* (FORBES), Pliensbach., Eng.; aboral side, ×1 (139).——FIG. 82,*1b,c*. *A. murravii* (FORBES), Pliensbach., Eng.; oral and aboral surfaces, ×1 (139).

Aspidophiura MATSUMOTO, 1915 [**A. watasei*; OD]. *Rec.*

Aspidura AGASSIZ, 1835 [**Ophiura loricata* GOLDFUSS, 1826; OD]. Arms short, very broad at base. Disc surface consisting solely of centrale, primary circlet and radial shields. *M.Trias.,* Bulg.-Ger.——FIG. 81,*5*. **A. loricata* (GOLDFUSS); *5a,* slab with specimens showing aboral surface, ×1; *5b,* oral side, ×10 (141).

Astrophiura SLADEN, 1879 [**A. permira*; OD]. *Rec.*

Cholaster WORTHEN & MILLER, 1883 [**C. peculiaris*; OD]. Disc apparently much as in *Aspidura* except that secondary scales occur. Arms abruptly truncated, ending in enlarged ossicles in only known specimen, but this may be pathological. *U.Miss.,* USA(Ill.).——FIG. 83,*1*. **C. peculiaris*; *1a,* aboral view, ×1; *1b,* part of arm, enl. (138).

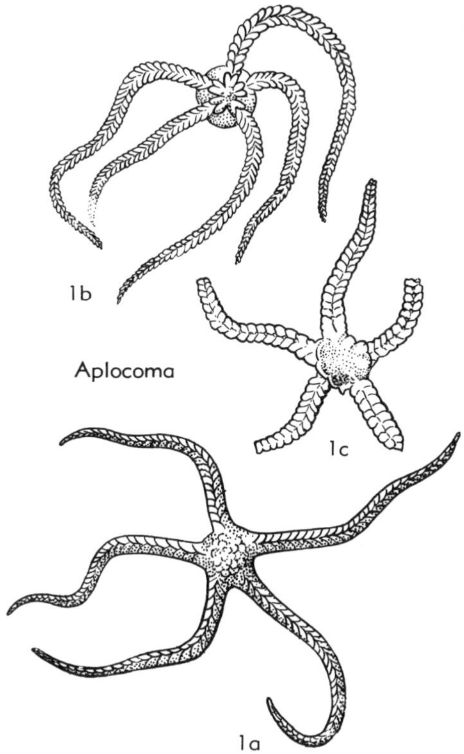

FIG. 82. Ophiuridae (Ophiurinae) (p. *U*95).

Dictenophiura H. L. CLARK, 1923 [**Ophiura carnea* LÜTKEN, 1858; OD]. *Rec.*

Euvondria FELL, 1961 [**E. floretta*; OD]. *Rec.*

Geocoma D'ORBIGNY, 1850 [**Ophiura carinata* MÜNSTER in GOLDFUSS, 1833; OD]. Radial shields large, reaching almost to center of disc. *L.Jur.,* ——FIG. 81,*6*. **G. carinata* (MÜNSTER); *6a,* aboral side, ×1; *6b,* aboral surface of disc, ×7; *6c,* oral surface of arm, ×7 (141).

Gymnophiura LÜTKEN & MORTENSEN, 1897 [**G. mollis*; OD]. *Rec.*

Haplophiura MATSUMOTO, 1915 [**Ophiozona gymnopora,* H. L. CLARK, 1909; OD]. *Rec.*

Homalophiura H. L. CLARK, 1915 [**Ophioglypha inornata* LYMAN, 1878; OD]. *Rec.*

Ophiochalcis KOEHLER, 1931 [**O. aspera*; OD]. *Rec.*

Ophiochorus H. L. CLARK, 1939 [**O. granulatus*; OD]. *Rec.*

Ophiochrysis KOEHLER, 1904 [**O. ornata*; OD]. *Rec.*

Ophiocrossota H. L. CLARK, 1928 [**Ophioglypha multispina* LJUNGMAN, 1867; OD]. *Rec.*

Ophiocten LÜTKEN, 1955 [**O. kroyeri* (=**Ophiura*

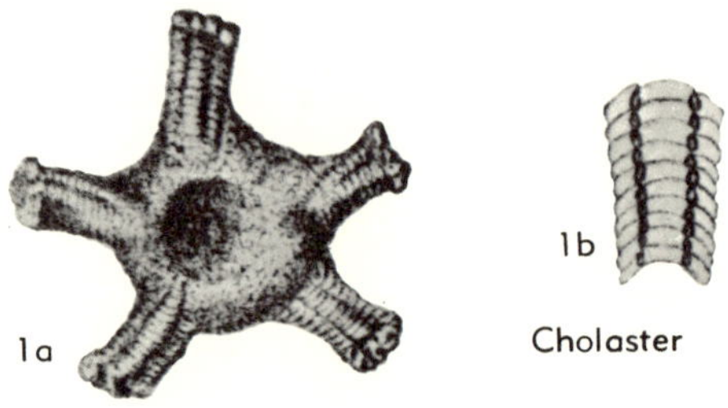

FIG. 83. Ophiuridae (Ophiurinae) (p. *U95*).

sericea FORBES, 1852); OD]. Primary plates more or less conspicuous. Disc not notched at arm base. Arm comb normally continuous across arm base. *Jur.*, Ger.-Eng.; *Mio.*, USSR; *Rec.*

Ophiogona STUDER, 1876 [*O. *laevigata*; OD] [=*Ophiagona* LÜTKEN, 1877; *Ophiomaria* A. H. CLARK, 1916]. *Rec.*

Ophiomastus LYMAN, 1878 [*O. *tegulitius*; OD]. *Rec.*

Ophiomisidium KOEHLER, 1914 [*O. *speciosum*; OD]. *Rec.*

Ophionotus BELL, 1902 [*O. *victoriae*; OD]. *Rec.*

Ophiophycis KOEHLER, 1901 [*O. *mirabilis*; OD]. *Rec.*

Ophiopleura DANIELSSON & KOREN, 1877 [*O. *borealis*; OD] [=*Luetkenia* DUNCAN, 1878 (*non* CLAUS, 1864)]. *Rec.*

Ophioplinthus LYMAN, 1878 [**Ophioplinthus medusa* LYMAN, 1878; SD H. L. CLARK, 1915]. *Rec.*

Ophiopyrgoides H. L. CLARK, 1939 [**Ophiopyrgus trispinosus* KOEHLER, 1904; OD]. *Rec.*

Ophiopyrgus LYMAN, 1878 [*O. *wyvillethomsoni*; OD]. *Rec.*

Ophiosteira BELL, 1902 [*O. *antarctica*; OD] [=*Ophiomages* KOEHLER, 1923]. *Rec.*

Ophiotjalfa MORTENSEN, 1915 [*O. *vivipara*; OD]. *Rec.*

Ophiotypa KOEHLER, 1897 [*O. *simplex*; OD]. *Rec.*

Ophiuraster H. L. CLARK, 1939 [*O. *perissus*; OD]. *Rec.*

Ophiuroglypha HERTZ, 1926 [**Ophioglypha lymani* LJUNGMAN, 1870; OD]. *Rec.*

Ophiurolepis MATSUMOTO, 1915 [**Ophiolepis carinata* STUDER, 1876; OD]. *Rec.*

Stegophiura MATSUMOTO, 1915 [**Ophiura nodosa* LÜTKEN, 1855; OD]. *Rec.*

Theodoria FELL, 1961 [**Amphiophiura relegata* KOEHLER, 1922; OD]. *Rec.*

Subfamily OPHIOLEPIDINAE Ljungman, 1867

[*nom. transl.* MATSUMOTO, 1915 (*ex* Ophiolepididae LJUNGMAN, 1867)]

Second oral tentacle pores opening inside oral slits. *?Perm.,Rec.*

Ophiolepis MÜLLER & TROSCHEL, 1840 [**Ophiura annulosa* DE BLAINVILLE, 1834 (*non* LAMARCK, 1816); SD LYMAN, 1865 (=**Ophiolepis superba* H. L. CLARK, 1915)]. *?U.Trias.(Rhaet.)*, Fr.; *Rec.*

Amphipholizona H. L. CLARK, 1915 [**A. delicata*; OD]. *Rec.*

Ophioceramis LYMAN, 1865 [**Ophiolepis januarii* LÜTKEN, 1856; OD]. *Rec.*

Ophioceres KOEHLER, 1922 [*O. *incipiens*; OD]. *Rec.*

Ophiocrates KOEHLER, 1904 [*O. *lenta*; OD]. *Rec.*

Ophiocypris KOEHLER, 1931 [*O. *tuberculosus*; OD]. *Rec.*

Ophiolebella MORTENSEN, 1936 [**Ophiolebes biscutifer* E. A. SMITH, 1879; OD]. *Rec.*

Ophiolipus LYMAN, 1878 [*O. *agassizii*; OD]. *Rec.*

Ophiomidas KOEHLER, 1904 [*O. *alatus* KOEHLER, 1904; SD H. L. CLARK, 1915]. *Rec.*

Ophiomusium LYMAN, 1869 [*O. *eburneum*; OD] [=*Ophiomusa* HERTZ, 1927; *?Ophiuraster* MILLER, 1958 (*non* H. L. CLARK, 1939)]. Disc and arm plates not obscured by skin, disc covered by regular porcelaneous plates and radial shields. Dorsal and ventral arm plates minute, not developed in distal part of arm where lateral arm plates meet on dorsal and ventral mid-lines. Tentacle pores 2 to 5. Continuous ridge of fused oral papillae round edge of jaw. *?Perm.-?Trias., L.Jur.-Rec.*, cosmop.——FIG. 81,7. *O. granulosum* (ROEMER), U.Cret.(Senon.), Eng.; oral surface of proximal part of arm, ×5 (125).

Ophiopenia H. L. CLARK, 1911 [*O. *disacantha*; OD]. *Rec.*

Ophiophyllum LYMAN, 1878 [*O. *petilum*; OD]. *Rec.*

Ophioplocus LYMAN, 1861 [**Ophiolepis imbricata* MÜLLER & TROSCHEL, 1842; OD]. *Rec.*

Ophiosphalma H. L. CLARK, 1941 [**Ophiomusium planum* LYMAN, 1878; OD]. *Rec.*

Ophioteichus H. L. CLARK, 1938 [*O. *parvispinum*; OD]. *Rec.*

Ophiothyreus LJUNGMAN, 1871 [*O. *goesi*; OD]. *Rec.*

Ophiotitanos SPENCER, 1907 [*O. *tenuis*; OD]. Disc heavily granulate, except for feebly swollen radial shields. Dorsal arm plates distinctly swollen. Five short arm spines. Tentacle pores along whole length of arm. *U.Cret.*, Eng.——FIG. 84,*1*. *O. *tenuis*, Cenoman., Eng.(Kent); aboral view, ×2 (113).

Ophiozona LYMAN, 1865 [**Ophiolepis impressa* LÜTKEN, 1859; SD H. L. CLARK, 1915]. *Rec.*

Ophiozonella MATSUMOTO, 1915 [**Ophiozona longispina* H. L. CLARK, 1908; OD]. *Rec.*

Ophiozonoida H. L. CLARK, 1915 [*O. *picta*; OD] [=*Ophiotylos* MURAKAMI, 1943]. *Rec.*

Family OPHIOLEUCIDAE Matsumoto, 1915

Arms long and slender, commonly constricted at nodes, inserted ventrally below disc and not firmly fused with it. Arm spines few, small and adpressed. Disc with granules and spinules. Continuous series of oral papillae along free margin of jaws. *Rec.*

Ophioleuce KOEHLER, 1904 [**Ophioleuce seminudum* KOEHLER, 1904; SD H. L. CLARK, 1915]. *Rec.*

Amphitarsus H. L. CLARK, 1941 [**A. mirabilis*; OD]. *Rec.*

Ophiocirce KOEHLER, 1904 [**O. inutilis*; OD]. *Rec.*

Ophiernus LYMAN, 1878 [**O. vallincola*; OD]. *Rec.*

Ophiopallas KOEHLER, 1904 [**O. paradoxa*; OD]. *Rec.*

Ophioperla KOEHLER, 1912 [**O. ludwigi* (=**Ophiura koehleri* BELL, 1908); OD]. *Rec.*

Ophiopyren LYMAN, 1878 [**Ophipyren longispinus* LYMAN, 1878; SD H. L. CLARK, 1915]. *Rec.*

Ophiotrochus LYMAN, 1878 [**O. panniculus*; OD]. *Rec.*

Family OPHIOCOMIDAE Ljungman, 1867

[=Ophiospilinae MATSUMOTO, 1915]

Arms stout, widest at some distance from base. Arm spines long, at angle to arm. Oral frame with well-developed lateral wings. Teeth stout, quadrangular. Oral papillae border each jaw. Dental papillae in clump at apex of each jaw. *?U.Cret.,Rec.*

Ophiocoma AGASSIZ, 1836 [**Ophiura echinata* LAMARCK, 1816; SD H. L. CLARK, 1915]. Disc granulate. Arm spines solid. tentacle scales short, leaflike. *?U.Cret.(Cenoman.)*, Eng.; *Rec.*

Ophiarthrum PETERS, 1851 [**O. elegans*; OD]. *Rec.*

Ophiocomella A. H. CLARK, 1939 [**O. caribbaea*; OD]. *Rec.*

Ophiocomina KOEHLER, 1920 [**Asterias nigra* ABILDGAARD, 1789; OD]. *Rec.*

Ophiomastix MÜLLER & TROSCHEL, 1842 [**Ophiura annulosa* LAMARCK, 1816; OD] [=*Acantharachna* E. A. SMITH, 1877]. *Rec.*

Ophiopsila FORBES, 1843 [**O. aranea*; OD] [=*Ophianoplus* M. SARS, 1857]. *Rec.*

Ophiopteris E. A. SMITH, 1877 [**O. antipodum*; OD]. *Rec.*

Family OPHIONEREIDIDAE Ljungman, 1867

[*nom. transl.* SPENCER & WRIGHT, herein (*ex* Ophionereidinae LJUNGMAN, 1867)] [=Ophiochitonidae MATSUMOTO, 1915]

Arms robust, not constricted at nodes, widest some distance from base. Keel on mid-line of ventral and commonly also on dorsal shields. Arms inserted ventrally below disc and not fused with it. Arm spines long, at angle to arm. Disc large and flat, with no granules or spines. *Rec.*

Ophionereis LÜTKEN, 1859 [**Ophiura reticulata* SAY, 1825; SD LYMAN, 1865]. *Rec.*

O. (Ophionereis). *Rec.*

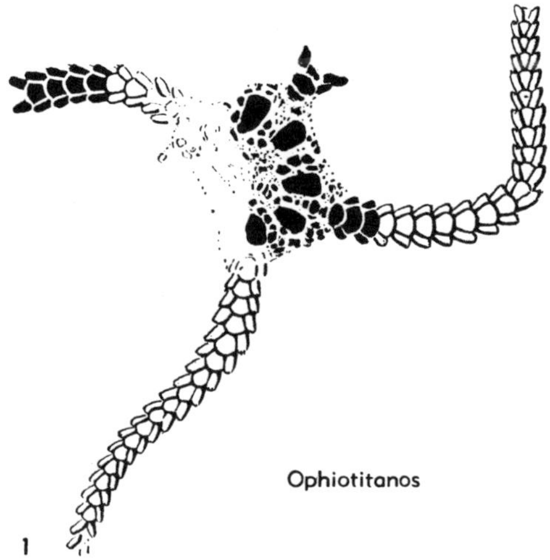

FIG. 84. Ophiuridae (Ophiolepidinae) (p. U96).

O. (Ophiotriton) DÖDERLEIN, 1896 [**O. semoni*; OD]. *Rec.*

O. (Ophiocrasis) H. L. CLARK, 1911 [**O. dictydisca*; OD]. *Rec.*

Ophiochiton LYMAN, 1878 [**O. fastigatus*; OD]. *Rec.*

Ophiodesmus ZIESENHENNE, 1940 [**O. amphilogus*; OD]. *Rec.*

Ophiodoris KOEHLER, 1904 [**Ophiodoris malignus* KOEHLER, 1904; SD H. L. CLARK, 1915]. *Rec.*

Ophioplax LYMAN, 1878 [**O. ljungmani*; OD]. *Rec.*

Family OPHIODERMATIDAE Ljungman, 1867

[=Ophiarachninae MATSUMOTO, 1915]

Arms inserted laterally and firmly fused with disc, moderately long, stout, widest at base. Granules cover disc scales of both surfaces and commonly jaws also. Unpaired infradental papillae at apex of each jaw. *L.Jur.-Rec.*

Ophioderma MÜLLER & TROSCHEL, 1840 [**Asterias longicauda* RETZIUS, 1805; SD H. L. CLARK, 1915]. *Rec.*

Bathypectinura H. L. CLARK, 1909 [**Pectinura lacertosa* LYMAN, 1883; OD]. *Rec.*

Cryptopelta H. L. CLARK, 1909 [**Ohiopeza aster* LYMAN, 1879; OD] [=*?Ophiodyscrita* H. L. CLARK, 1938]. *Rec.*

Diopederma H. L. CLARK, 1913 [**Ophiura daniana* VERRILL, 1867; OD]. *Rec.*

Distichophis ELY, 1942 [**D. clarki*; OD]. *Rec.*

Ophiarachna MÜLLER & TROSCHEL, 1842 [**Ophiura incrassata* LAMARCK, 1816; SD LÜTKEN, 1869]. *Rec.*

Ophiarachnella LJUNGMAN, 1872 [**Ophiarachna gorgonia* MÜLLER & TROSCHEL, 1842; SD H. L. CLARK, 1915]. *Rec.*

FIG. 85. Ophiodermatidae (p. U98).

Ophiochaeta LÜTKEN, 1869 [*O. hirsuta; OD]. *Rec.*

Ophiochasma GRUBE, 1868 [*O. adspersa (=*Ophiarachna stellata* LJUNGMAN, 1867); OD] [=*Ophiopinax* BELL, 1884]. *Rec.*

Ophioclastus MURAKAMI, 1943 [*O. hataii; OD]. *Rec.*

Ophioconis LÜTKEN, 1869 [*Pectinura forbesi HELLER, 1862; OD]. *Rec.*

Ophiocormus H. L. CLARK, 1915 [*O. notabilis; OD] [=*Ophiostegastus* MURAKAMI, 1944]. *Rec.*

Ophiocryptus H. L. CLARK, 1915 [*O. maculosus; OD]. *Rec.*

Ophioncus IVES, 1889 [*O. granulosus; OD]. *Rec.*

Ophiopaepale LJUNGMAN, 1871 [*O. goesiana; OD]. *Rec.*

?**Ophiopetra** HESS, 1962 [*O. lithographica; OD]. Arm spines 3. Ventral shields pentagonal, dorsal shields triangular. [Perhaps belongs to Ophionereididae.] *U.Jur.(Kimmeridg.)*, Fr.——FIG. 85,*1*. *O. *lithographica*; aboral surface, ×10 (Hess).

Ophiopeza PETERS, 1851 [*O. fallax; OD] [=*Ophiopsammus* LÜTKEN, 1869] *?M.Jur.(Bathon.)*, *Rec.*

Ophiopezella LJUNGMAN, 1871 [*Ophiarachna spinosa LJUNGMAN, 1867; SM LYMAN, 1882]. *Rec.*

Ophiurochaeta MATSUMOTO, 1915 [*Ophiochaeta mixta LYMAN, 1878; OD]. *Rec.*

Ophiuroconis MATSUMOTO, 1915 [*O. monolepis; OD]. *Rec.*

Ophiurodon MATSUMOTO, 1915 [*Ophioconis grandisquama KOEHLER, 1904; OD]. *Rec.*

Palaeocoma D'ORBIGNY, 1850 [*Ophiura milleri PHILLIPS, 1829; OD]. Arms cylindrical. Disc with or without granules. Radial shields large; other parts of disc lightly calcified. Dorsal shields broad, touching along most of length of arm. Ventral shields similar. Lateral shields high. Arm spines rudimentary. *L.Jur.*, Eu.——FIG. 86,*1*. *P. milleri* (PHILLIPS); Pliensbach., Eng.(Yorks.); *1a*, part of aboral surface of arm, ×3; *1b*, aboral surface of disc, ×1 (139).——FIG. 86,*1c*. *P. escheri* (HEER), Hettang., Switz.; oral side (reconstr.), ×3.6 (113).

Pectinura FORBES, 1843 [*P. vestita; OD]. *Rec.*

Schizoderma NIELSEN, 1932 [*S. diplax; OD]. *Rec.*

Toporkovia DJAKONOV, 1954 [*T. fragilis; OD]. *Rec.*

Suborder LAEMOPHIURINA Matsumoto, 1915

[*nom. transl. et correct.* SPENCER & WRIGHT, herein (*ex* Laemophiurida MATSUMOTO, 1915)]

Radial shields and genital plates articulate by means of transverse ridge or simple facet on either plate. Peristomial plates large, normally entire. Oral frames entire, without well-developed lateral wings. Dorsal arm plates commonly very small; lateral arm plates well developed, generally meeting in pairs dorsally and ventrally. *L.Jur.-Rec.*

Family OPHIACANTHIDAE Perrier, 1891

[=Ophiomycetidae VERRILL, 1899]

Arms slender, commonly constricted at nodes. Dorsal and ventral arm plates very small. Arm spines long, numerous, at angle to arm, commonly glassy and serrate. Disc with granules and spinules. Distal vertebrae may be partly divided longitudinally by series of pores. *L.Jur.-Rec.*

Ophiacantha MÜLLER & TROSCHEL, 1842 [*O. spinulosa 1842 (=*Asterias bidentata* RETZIUS, 1805); SD H. L. CLARK, 1915] [=*Ophiectodia, Ophientodia, Ophiopristis, Ophioscalus, Ophiotreta* VERRILL, 1899; *Ophiodiplax* KOEHLER, 1911]. Disc covered with thin skin bearing granules and stumpy spines. Arm spines hollow. *L.Jur.(Pliensbach.)*, Switz.; *Rec.*

Amphipsila VERRILL, 1899 [*A. maculata; OD]. *Rec.*

Glaciacantha FELL, 1961 [*G. jason; OD]. *Rec.*

Microphiura MORTENSEN, 1911 [*M. decipiens; OD]. *Rec.*

Ophiacanthella VERRILL, 1899 [*Ophiacantha troscheli LYMAN, 1878; OD]. **Rec.**

Ophialcaea VERRILL, 1899 [*Ophiacantha tuberculosa LYMAN, 1878; SD H. L. CLARK, 1915]. *Rec.*

Ophiambix LYMAN, 1880 [*O. aculeatus; OD]. *Rec.*

Ophientrema VERRILL, 1899 [*Ophiacantha scolopendrica LYMAN, 1883; OD]. *Rec.*

Ophioblenna LÜTKEN, 1859 [*O. antillensis; OD]. *Rec.*

Ophiocamax LYMAN, 1878 [*O. vitrea; OD]. *Rec.*

Ophiochondrella VERRILL, 1899 [*Ophiochondrus squamosus LYMAN, 1883; OD]. *Rec.*

Ophiocopa LYMAN, 1883 [*O. spatula; OD]. *Rec.*

Ophiocymbium LYMAN, 1880 [*O. cavernosum; OD]. *Rec.*

Ophiodaces KOEHLER, 1922 [*O. inanis; OD]. *Rec.*

Ophiodelos Koehler, 1931 [**O. insignis*; OD]. *Rec.*

Ophiodictys Koehler, 1922 [**O. uncinatus*; OD]. *Rec.*

Ophiogema Koehler, 1922 [**O. punctata*; OD]. *Rec.*

Ophioglyphoida Chapman, 1934 [**Ophiacantha (Ophioglyphoida) fosteri*; OD]. Doubtful, as aboral surface unknown. *L.Cret.(Alb.),* Australia.

Ophiolebes Lyman, 1878 [**Ophiolebes scorteus* Lyman, 1878; SD H. L. Clark, 1915]. *Rec.*

Ophiolimna Verrill, 1899 [**Ophiacantha bairdi* Lyman, 1883; OD]. *Rec.*

Ophiologimus H. L. Clark, 1911 [**O. hexactis*; OD]. *Rec.*

Ophiomedea Koehler, 1906 [**O. duplicata*; OD]. *Rec.*

Ophiomelina Koehler, 1922 [**Ophiomitrella placida* Koehler, 1904; OD]. *Rec.*

Ophiomitra Lyman, 1869 [**Ophiomitra valida* Lyman, 1869; SD Verrill, 1899]. *Rec.*

Ophiomitrella Verrill, 1899 [**Ophiacantha laevipellis* Lyman, 1883; OD]. *Rec.*

Ophiomyces Lyman, 1869 [**Ophiomyces frutectosus* Lyman, 1869; SD H. L. Clark, 1915]. *Rec.*

Ophiomytis Koehler, 1904 [**O. weberi*; OD]. *Rec.*

Ophiophrura H. L. Clark, 1911 [**O. liodisca*; OD]. *Rec.*

Ophiophthalmus Matsumoto, 1917 [**Ophiacantha cataleimmoida* H. L. Clark, 1911; OD] [=*Ophiosemnotes* Matsumoto, 1917]. *Rec.*

Ophiopinna Hess, 1960 [**Geocoma elegans* Heller, 1858; OD]. Disc covered with small thin scales. Base of arms with cuff of short wide plates. Teeth rather weak, skittle-shaped. Four square and 2 elongate mouth papillae. Proximal part of arm with about 10 arm spines, but on middle part dorsal arm plates are rudimentary or absent and some normal spines are transformed into tall feather-shaped spines, arranged in double dorsal row. Ventral arm plates keeled. Distal part of arm excessively long and thin. *?L.Jur. (Pliensbach.), M.Jur.(Callov.),* Fr.-Switz.——Fig. 88,*1*. **O. elegans* (Heller), Callov., Fr. (Ardéche); *1a,b,* aboral and oral surfaces of disc, ×5; *1c,d,* aboral and lateral views of arm, ×5; *1e,* growth stages in natural position, ×2 (Hess).

Ophioplinthaca Verrill, 1899 [**Ophiomitra dipsacos* Lyman, 1878; OD]. *Rec.*

Ophioprium H. L. Clark, 1915 [**Ophiacantha curvicornis* Lyman, 1883; OD]. *Rec.*

Ophioripa Koehler, 1922 [**O. marginata*; OD]. *Rec.*

Ophiosparte Koehler, 1922 [**O. gigas*; OD]. *Rec.*

Ophiothamnus Lyman, 1869 [**O. vicarius*; OD] [=*Ophioleda* Koehler, 1906]. *Rec.*

Ophiothauma H. L. Clark, 1938 [**O. heptactis*; OD]. *Rec.*

Ophiotholia Lyman, 1880 [**O. supplicans*; OD]. *Rec.*

Ophiotoma Lyman, 1883 [**Ophiotoma coriacea* Lyman, 1883 (=**Ophiotoma bartletti* Lyman, 1883); SD H. L. Clark, 1915] [=*Ophiopora* Verrill, 1899]. *Rec.*

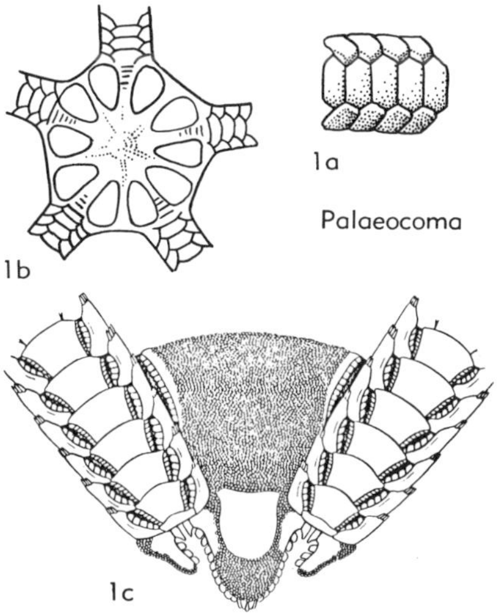

Fig. 86. Ophiodermatidae (p. *U*98).

Ophiotrema Koehler, 1896 [**O. alberti*; OD]. *Rec.*

Ophiurothamnus Matsumoto, 1917 [**Ophiomitra dicyla* H. L. Clark, 1911; OD]. *Rec.*

Family HEMIEURYALIDAE Verrill, 1899

[=Ophiochondrinae Verrill, 1899]

Disc and arm plates very stout. Vertebrae very stout, articulating as in Euryalina and arms coil in vertical plane. [Epizoic.] *?L.Jur.(Pliensbach.), Rec.*

Hemieuryale von Martens, 1867 [**H. pustulata*; OD]. Dorsal arm plates completely divided, forming mosaic. Arm spines 3, short, flat. [A Jurassic species known only from isolated ossicles may belong here.] *?L.Jur.(Pliensbach.),* Switz.; *Rec.*

Ophiochondrus Lyman, 1869 [**O. convolutus*; OD]. *Rec.*

Amphigyptis Nielsen, 1933 [**A. perplexa*; OD] [=*?Ophiocyclus* H. L. Clark, 1939]. *Rec.*

Ophiogyptis Koehler, 1905 [**O. nodosa*; OD]. *Rec.*

Ophioholcus H. L. Clark, 1915 [**Sigsbeia sexradiata* Koehler, 1914; OD]. *Rec.*

Ophioleila A. H. Clark, 1949 [**O. elegans*; OD]. *Rec.*

Ophiomoeris Koehler, 1904 [**Ophiomoeris spinosa* Koehler, 1904; SD H. L. Clark, 1915] [=*Ophiurases* H. L. Clark, 1911]. *Rec.*

Ophioplus Verrill, 1899 [**Hemieuryale tuberculosa* Lyman, 1833; OD]. *Rec.*

Quironia A. H. Clark, 1934 [**Q. johnsoni*; OD]. *Rec.*

Sigsbeia Lyman, 1878 [**S. murrhina*; OD]. Large supplementary plate present on either side of each

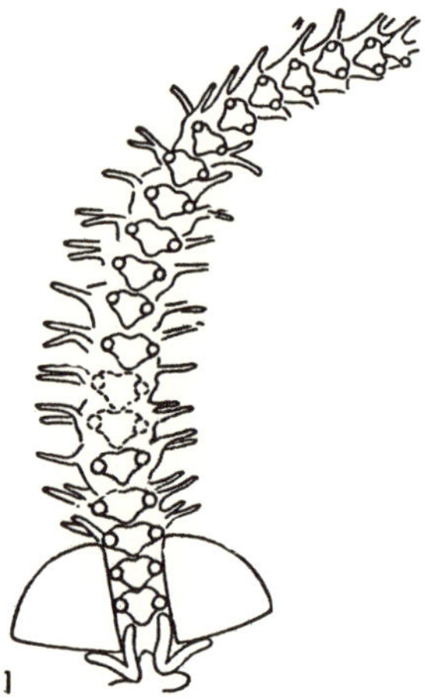

FIG. 87. Amphiuridae (p. U102).

dorsal arm plate; 2 genital clefts in each interradius. [Fossil doubtfully referred here.] *?Mio.*, Victoria; *Rec.*

Suborder GNATHOPHIURINA Matsumoto, 1915

[*nom. transl. et correct.* SPENCER & WRIGHT, herein (*ex* Gnathophiurida MATSUMOTO, 1915)]

Radial shield and genital plate articulating by conspicuous socket in former and large ball-like condyle on latter. Genital plates normally fixed firmly to vertebrae. Peristomial plates generally small and entire, rarely large or double. Oral frames normally with well-developed lateral wings. *?L.Jur., Rec.*

Family AMPHILEPIDIDAE Matsumoto, 1915

Arms long and slender, inserted ventrally below disc and not firmly fused with it. Vertebrae long and slender, commonly divided longitudinally by series of pores. Disc without granules or spines. *Rec.*

Amphilepis LJUNGMAN, 1866 [**Amphiura norvegica* LJUNGMAN, 1864; OD]. *Rec.*

Ophiochytra LYMAN, 1880 [**O. epigrus*; OD]. *Rec.*

Family OPHIACTIDAE Matsumoto, 1915

[*nom. transl.* FELL, 1960 (*ex* Ophiactinae MATSUMOTO, 1915)]

Arms slender, commonly constricted at nodes. Disc with granules or spinules. Jaws with lateral oral papillae separated by gap from dissimilar infradental papillae at apex. *Rec.*

Ophiactis LÜTKEN, 1856 [**O. krebsii* (=**Ophiolepis savignyi* MÜLLER & TROSCHEL, 1842); SD H. L. CLARK, 1915] [=*Amphiactis* MATSUMOTO, 1915]. *Rec.*

Hemipholis LYMAN, 1865 [**Ophiura elongata* SAY, 1825; OD]. *Rec.*

Ophiodaphne KOEHLER, 1931 [**O. materna*; OD]. *Rec.*

Ophiopholis MÜLLER & TROSCHEL, 1840 [**Ophiolepis scolopendrica* MÜLLER & TROSCHEL, 1840 (=**Asterias aculeatus* RETZIUS, 1783); SD H. L. CLARK, 1915]. *Rec.*

Ophiopus LJUNGMAN, 1866 [**O. arcticus*; OD] [=*Ophiaregma* SARS, 1872]. *Rec.*

Family AMPHIURIDAE Ljungman, 1867

Disc covered by fine imbricating scales or with minute spines or naked. Arms inserted ventrally in disc. Arm spines conical and stout. No dental papillae. Paired infradental papillae at apex of each jaw (10). *U.Cret.-Rec.*

Amphiura FORBES, 1843 [**A. chiajii*; SD VERRILL, 1899] [=*Hemilepis* LJUNGMAN, 1871]. Oral papillae not forming continuous row along jaw but having single infradental separated by gap from single outer papilla with internal one invisible above gap. Disc with fine, flat, imbricating scales. Tentacle scales 2. [Jurassic and Cretaceous fossils referred to this genus are all doubtful.] *Rec.*

Acrocnida GISLÉN, 1926 [**Asterias brachiata* MONTAGUE, 1804; OD] [=*Ophiocentrus* LJUNGMAN, 1867]. *Rec.*

Ailsaria FELL, 1962 [**Amphioplus echinulatus* MORTENSEN, 1940; OD]. *Rec.*

Amphiacantha MATSUMOTO, 1917 [**Amphioplus acanthinus* H. L. CLARK, 1911; OD]. *Rec.*

Amphichilus MATSUMOTO, 1917 [**A. trichoides*; OD]. *Rec.*

Amphichondrius NIELSEN, 1933 [**A. granulosus*; OD]. *Rec.*

Amphicontus HILL, 1940 [**A. minutus*; OD]. *Rec.*

Amphilimna VERRILL, 1899 [**Ophiocnida olivacea* LYMAN, 1869]. *Rec.*

Amphilycus MORTENSEN, 1933 [**A. androphorus*; OD]. *Rec.*

Amphinephthys FELL, 1962 [**Amphiura crossota* MURAKAMI, 1943; OD]. *Rec.*

Amphiocnida VERRILL, 1899 [**Ophiocnida putnami* LYMAN, 1871; OD]. *Rec.*

Amphiodia VERRILL, 1899 [**Amphiura pulchella* LYMAN, 1869; OD]. *Rec.*

Amphiomya H. L. CLARK, 1939 [**A. notabilis*; OD]. *Rec.*

Amphioncus H. L. CLARK, 1939 [**A. platydiscus*; OD]. *Rec.*

Amphipholis LJUNGMAN, 1866 [**A. januarii*

(=**Ophiolepis gracillima* STIMPSON, 1852); OD]. *Rec.*

Amphioplus VERRILL, 1899 [**Amphiura tumida* LYMAN, 1899; OD]. Four or 5 oral papillae, outermost on adoral shield, small. Radial shields divergent. *Mio.,* Venezuela; *Rec.*

Anamphiura H. L. CLARK, 1939 [**A. valida*; OD]. *Rec.*

Ctenamphiura VERRILL, 1899 [**Amphiura maxima* LYMAN, 1879; OD]. *Rec.*

Diamphiodia FELL, 1962 [**Amphiura violacea* LÜTKEN, 1856; OD]. *Rec.*

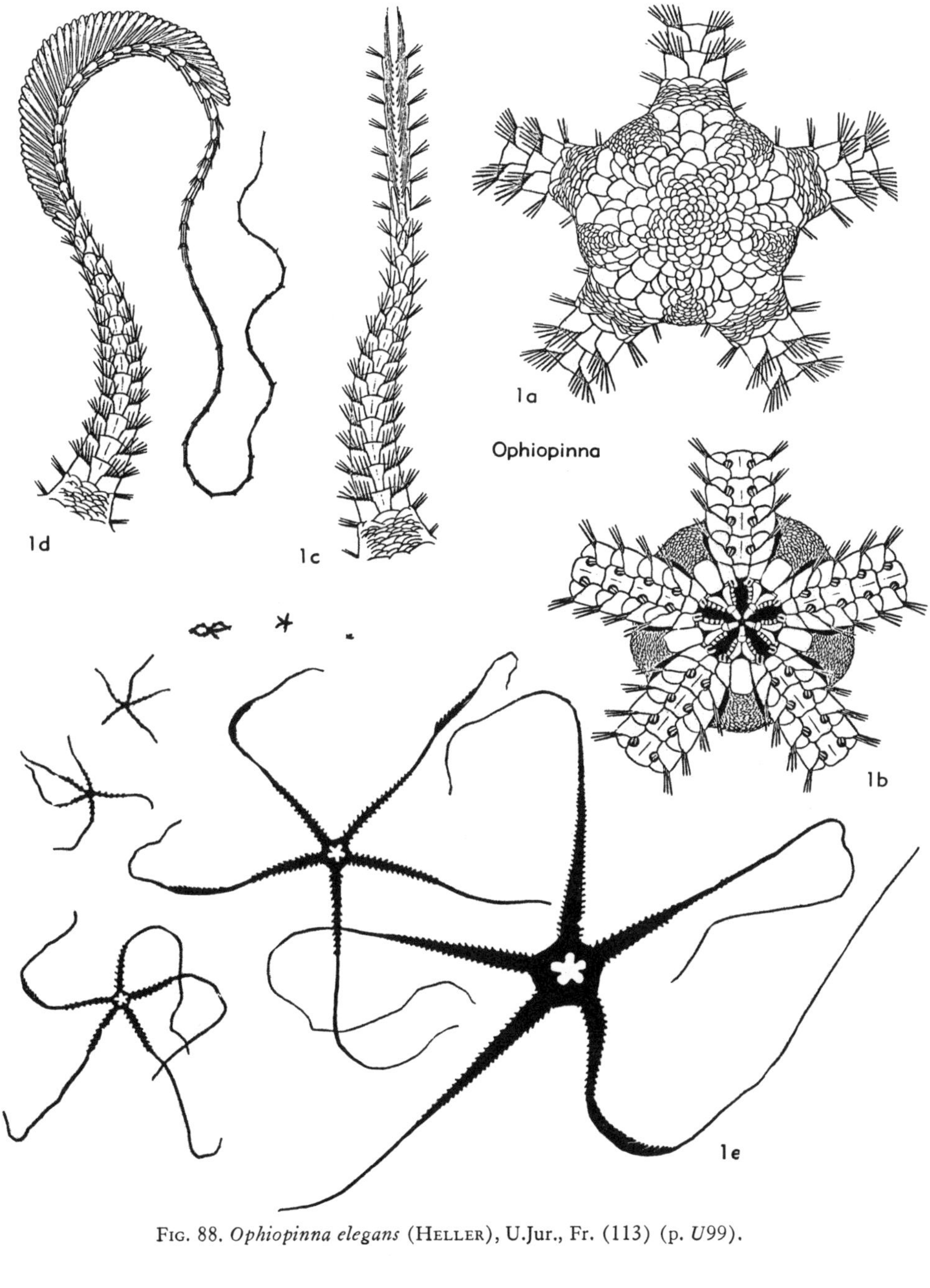

FIG. 88. *Ophiopinna elegans* (HELLER), U.Jur., Fr. (113) (p. *U*99).

Gymnodia FELL, 1962 [**Amphiodia tabogae* NIELSEN, 1932; OD]. *Rec.*

Icalia FELL, 1962 [**Amphiura denticulata* KOEHLER, 1896; OD]. *Rec.*

Monamphiura FELL, 1962 [**Amphiura alba* MORTENSEN, 1924; OD]. *Rec.*

Monopholis FELL, 1962 [**Amphiura vitax* KOEHLER, 1904; OD]. *Rec.*

Nannophiura MORTENSEN, 1933 [**N. lagani*; OD]. *Rec.*

Nullamphiura FELL, 1962 [**Amphiura psilopora* H. L. CLARK, 1911; OD]. Like *Amphiura* but no tentacle scales or only few rudimentary ones. *Cret.-Rec.*——FIG. 87,*1*. *N. felli* SKWARKO, Cenoman., Australia(Bathurst I.), oral view of arm, ×10 (129a).

Nullopholis FELL, 1962 [**Amphipholis nudipora* KOEHLER, 1944; OD]. *Rec.*

Ophiocentrus LJUNGMAN, 1867 [**O. aculeatus*; OD]. *Rec.*

Ophiocnida LYMAN, 1865 [**Ophiolepis hispida* LECONTE, 1851; SD VERRILL, 1899] [=*Ophiocnidella* LJUNGMAN, 1872]. Three or 4 subequal oral papillae. Disc with numerous scattered spines. *Pleist.*, Eng.; *Rec.*

Ophiomonas DJAKONOV, 1952 [**O. bathybia*; OD]. *Rec.*

Ophionema LÜTKEN, 1869 [**O. intricata*; OD]. *Rec.*

Ophionephthys LÜTKEN, 1869 [**O. limicola*; OD]. *Rec.*

Ophiophragmus LYMAN, 1865 [**Amphiura wurdemanii* LYMAN, 1860; SD H. L. CLARK, 1915] [=*Amphispina* NIELSEN, 1933]. *Rec.*

Ophiostigma LÜTKEN, 1856 [*O. *tenue*; OD] [=*?Amphistigma* H. L. CLARK, 1938]. *Rec.*

Pandelia FELL, 1962 [**Amphiura hinemoae* MORTENSEN, 1924; OD]. *Rec.*

Paracrocnida MORTENSEN, 1940 [**P. persica*; OD]. *Rec.*

Paramphiura KOEHLER, 1895 [**Ophiocoma punctata* FORBES, 1841; OD]. *Rec.*

Silax FELL, 1962 [**Aphiura verrilli* LYMAN, 1879; OD]. *Rec.*

Unioplus FELL, 1962 [**Amphioplus falcatus* MORTENSEN, 1933; OD]. *Rec.*

Family OPHIOTHRICIDAE Ljungman, 1866

Teeth stout, quadrangular, spiniform tooth papillae clustered at apex of each jaw. No oral papillae. *?L.Jur., Rec.*

Ophiothrix MÜLLER & TROSCHEL, 1840 [**Ophiura rosula* FLEMING, 1828 (=**Asterias pentaphylla* PENNANT, 1777); SD LYMAN, 1865] [=*Ophionyx* MÜLLER & TROSCHEL, 1840]. Both sides of disc with many plates, bearing spines; aboral surface granular. Radial shields small or partly concealed. Dorsal arm plates smooth. Arm spines 4 or more. *?L.Jur.(Sequan.)*, Fr.; *Rec.*

Amphiophiothrix H. L. CLARK, 1946 [**Ophiothrix demessa* LYMAN, 1861; OD]. *Rec.*

Gymnolophus BROCK, 1888 [**Ophiothela holdsworthi* E. A. SMITH, 1878; OD]. *Rec.*

Lissophiothrix H. L. CLARK, 1938 [**L. delicata*; OD]. *Rec.*

Macrophiothrix H. L. CLARK, 1938 [**Ophiura longipeda* LAMARCK, 1816; OD] [=*Placophiothrix* H. L. CLARK, 1938]. *Rec.*

Ophioaethiops BROCK, 1888 [**O. unicolor*; OD] [=*Ophiohelix* KOEHLER, 1895]. *Rec.*

Ophiocnemis MÜLLER & TROSCHEL, 1842 [**Ophiura marmorata* LAMARCK, 1816; OD]. *Rec.*

Ophiogymna LJUNGMAN, 1866 [**O. elegans*; OD] [=*Ophiocampsis* DUNCAN, 1887]. *Rec.*

Ophiolophus MARKTANNER-TURNERETSCHER, 1887 [**O. novarae*; OD]. *Rec.*

Ophiomaza LYMAN, 1871 [**O. cacaotica*; OD] [=*Luetkenia* BROCK, 1888]. *Rec.*

Ophiopeltis DÜBEN & KOREN, 1846 [**O. securigera*; OD]. *Rec.*

Ophiophthirius DÖDERLEIN, 1898 [**O. actinometrae*; OD]. *Rec.*

Ophiopsammium LYMAN, 1874 [**O. semperi*; OD]. *Rec.*

Ophiopteron LUDWIG, 1888 [**O. elegans*; OD]. *Rec.*

Ophiosphaera BROCK, 1888 [**O. insignis*; OD]. *Rec.*

Ophiothela VERRILL, 1867 [**O. mirabilis*; OD] [=*Ophioteresis* BELL, 1892]. *Rec.*

Ophiotrichoides LUDWIG, 1882 [**O. lymani*; OD]. *Rec.*

GENERIC NAMES OF INDETERMINATE OR UNRECOGNIZABLE STATUS APPLIED TO FOSSIL OPHIUROIDEA

Acroura AGASSIZ, 1836 [**Ophiura prisca* MÜNSTER in GOLDFUSS, 1831; OD]. The holotype of the type-species is quite indeterminable and the genus must therefore be treated as *nom. dub. Trias.*, Ger.

Ataxaster JAEKEL, 1903 [**A. pygmaeus*; OD]. Undescribed ophiuroid. *Ord.*, Czech.

Ephipiellum LOMNICKI, 1899 [**E. symmetricum*; OD]. Unidentifiable ophiuroid vertebrae. *Mio.*, Pol.-Crimea.

Dolicharthra BERRY, 1938 [**D. bemelenica*; OD]. Based on isolated ossicles of several genera, including vertebrae that may belong to *Ophiomusium*. *U.Cret.(Maastricht.)*.

Ophiaxina MÜLLER, 1950 [**O. intercarinata*; OD]. Based on vertebrae only, which somewhat resemble those of *Ophiomyxa*. *U.Cret.(Campan.)*, Rügen.

Ophioma POMEL, 1887 [**Ophioma juliensis*; OD]. No description, only a figure of unidentifiable arm fragment. *Plio.*, Algeria.

Ophiotrigonum HESS, 1960 [**O. oxfordiense*; OD]. Disc not known. Arms sharply triangular in section. Dorsal and lateral arm plates smooth. Spines rudimentary. *M.Jur.(Oxford.)*, Switz.——FIG. 89,

*1. *O. oxfordiense; 1a-d,* lat., dorsal, ventral views and cross section of arm, ×3 (Hess).

Ophiurella AGASSIZ, 1836 [**Ophiura speciosa* MÜNSTER in GOLDFUSS, 1831; OD]. Material is too poorly preserved for characters to be assessed. *M.Jur.-U.Jur.,* Ger.——FIG. 89,*2*. **O. speciosa,* Kimmeridg.; oral surface, ×1 (111a).

Ophiuriocoma VALETTE, 1929 [**O. mazenoti*; OD]. Description insufficient for affinities to be decided. *L.Jur.(Aalen.),* Fr.

Platyarthra BERRY, 1938 [**P. jekerica*; OD]. Based on ossicles of several genera, including lateral arm plates perhaps belonging to *Ophiomusium. U.Cret. (Maastricht.),* Neth.

Pseudaspidura KOLOSVÁRY, 1941 [**P. hungarica*; OD]. Unidentifiable. *Oligo.,* Hung.

Schizospondylus MÜLLER, 1950 [**S. jasmundiana*; OD]. Based on vertebrae only, which are similar to those of *Ophiocamax. U.Cret.(Campan.),* Rügen.

Transspondylus MÜLLER, 1950 [**T. bubnoffi*; OD]. Based on unidentifiable vertebrae. *U.Cret. (Campan.),* Rügen.

Xenura SCHÖNDORF, 1938 [**X. kobuldi*; OD]. Unrecognizable. *Dev.,* Ger.

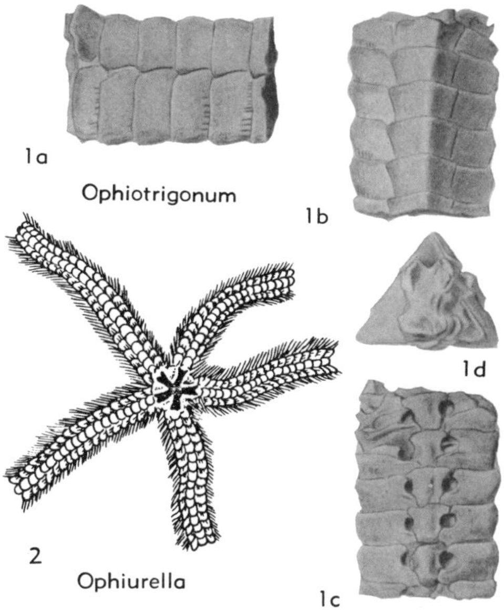

FIG. 89. *Ophiotrigonum* and *Ophiurella*; *1a-d, Ophiotrigonum oxfordiense* HESS, U.Jur., Switz. (113) (p. *U*102); *2, Ophiurella speciosa* (MÜNSTER), U.Jur., Ger. (113) (p. *U*102-*U*103).

GENERIC NAMES OF INDETERMINATE OR UNRECOGNIZABLE STATUS APPLIED TO FOSSIL ASTEROZOA

Asteriatites VON SCHLOTHEIM, 1813 [=*Asteriacites* VON SCHLOTHEIM, 1820; *lapsus*]. The name was first published (p. 68, *non* p. 109, for a foraminifer), with three included species, names for which apparently were based on nonbinominal names in KNORR, since the only description was an indication to KNORR's figures. These are of Solenhofen fossils (U.Jur., Kimmeridg.) all crinoids *(Saccocoma)* except for a single ophiuroid; the figures on the plate are not separately identified in SCHLOTHEIM's indication. In 1820 SCHLOTHEIM published *Asteriacites* with a single included species, *A. ophiurus,* which is an unidentifiable ophiuroid from the "Muschelkalk." If, as NEAVE maintains, *Asteriacites,* 1820, is a *lapsus* for *Asteriatites,* 1813, the type-species must be found among the 1813 species, if any of their names are available. Although SCHLOTHEIM's indications to KNORR's figures are in puzzling form, it seems that at least two of the names could be fixed among the crinoids and could therefore be treated as available from 1813. One of them, *pennatus,* was attributed by SCHLOTHEIM in 1820 to *Ophiurites,* with again an indication to KNORR's figures. It seems best to place *Asteriaties* SCHLOTHEIM, 1813 [=*Asteriacites* SCHLOTHEIM, 1820; *lapsus*] as a *nom. dub.* in the synonymy of *Saccocoma* AGASSIZ, 1836.

Stereoaster FOERSTE, 1919 [**S. squamatus*; OD]. Not asterozoan. *Sil.,* Ohio.

Trichotaster T. WRIGHT, 1873 [**T. plumiformis*; OD] [=*Trochitaster* WOODWARD, 1874; *lapsus*]. Specimen not traced. Probably not asterozoan. *Sil.,* Eng.

REFERENCES

The following list comprises only selected references; it is not intended to be a complete bibliography of fossil, let alone of extant, Asterozoa. Works are listed which are specifically referred to in the *Treatise,* together with a number of monographs and other publications of broad scope, containing lists of references to previous works, and a limited number of shorter papers of special interest. The original source for generic names generally can be found in NEAVE's *Nomenclator Zoologicus* or the *Zoological Record.*

Bell, F. J.

(1) 1892, *A contribution to the classification of ophiuroids, with descriptions of some new and little known forms:* Zool. Soc. London, Proc. (1892), p. 175-183.

Blegvad, Harald

(2) 1915, *Food and conditions of nourishment among the communities of invertebrate animals found on or in the sea-bottom in Danish waters:* Danish Biol. Sta., Rept. Bd. of Agric., v. 22, p. 41-78.

Clark, A. M., & Wright, C. W.

(3) 1962, *A new genus and species of recent starfishes belonging to the aberrant family Sphaerasteridae, with notes on the possible origin and affinities of the family:* Ann. & Mag. Nat. History, ser. 13, v. 5, p. 243-251, text-fig. 1-4.

Clark, H. L.

(4) 1915, *Catalogue of the recent ophiurans: based on the collection of the Museum of Comparative Zoology:* Harvard Univ. Museum Comp. Zoology, Mem., v. 25, no. 4, 376 p., 20 pl.

Clarke, J. M.

(5) 1921, *Organic dependence and disease:* N.Y. State Museum, Bull. 221, 222, 113 p., 115 text-fig.

Cottreau, Jean

(6) 1929, *Echinodermes du Bradfordien des environs d'Alençon (Orne):* Soc. Géol. France, Bull., ser. 4, v. 29, p. 463-469, pl. 36.

Durham, J. W., & Roberts, W. A.

(7) 1948, *Cretaceous asteroids from California:* Jour. Paleontology, v. 22, p. 432-439, pl. 65-66.

Etheridge, Robert

(8) 1892, *A monograph of the Carboniferous and Permo-Carboniferous Invertebrata of New South Wales. Part II. Echinodermata, Annelida, and Crustacea:* New South Wales Geol. Survey, Paleontology, Mem., no. 5, p. i-ix, 67-131, pl. 12-22.

Fedotov, D. M.

(9) 1926, *Die morphologie der Euryalen:* Zeitschr. Wiss. Zoologie, v. 127, p. 403-528, text-fig. 1-83.

(10) 1928, *Über die Beziehungen der Echinodermenklassen zueinander:* Zool. Lab. Sevastopolsk Biol. Sta., Akad. Nauk SSSR, ser. 2, no. 2, p. 1-94.

(11) 1936, *Zur Morphologie und Evolution der Seesterne und Ophiuren des unteren Silur:* Acad. Sci. URSS Inst. Paléozool., Travaux, v. 5, p. 3-33, 2 pl., 6 text-fig.

Fell, H. B.

(12) 1960, *Synoptic keys to the genera of Ophiuroidea:* Victoria Univ. Wellington, Zool. Publ., no. 26, 44 p., 6 text-fig.

(13) 1961, *A revision of the major genera of amphiurid Ophiuroidea:* Royal Soc. New Zealand, Trans., v. 2, no. 1, p. 1-26, pl. 1.

(14) 1962, *A living somasteroid, Platasterias latiradiata Gray:* Univ. Kansas Paleont. Contrib., Art. 6, 16 p., 4 pl., 8 text-fig.

(15) 1962, *Evidence for the validity of Matsumoto's classification of the Ophiuroidea:* Seto Marine Biol. Lab. Publ., v. 10, no. 2, p. 145-152, text-fig. 1-3.

(16) 1963, *The phylogeny of sea-stars:* Royal Soc. London, Philos. Trans., ser. B, v. 246, p. 381-435, pl. 50-51, 16 text-fig.

Fisher, W. K.

(17) 1911-30, *Asteroidea of the North Pacific and adjacent waters:* U. S. Natl. Museum, Bull. 76, pt. 1 (1911), 419 p., 122 pl.; pt. 2 (1928), 245 p., 81 pl.; pt. 3 (1930), 356 p., 93 pl.

(18) 1919, *Starfishes of the Philippine seas and adjacent waters:* Same, Bull. 100, v. 3, 712 p., 156 pl.

(19) 1940, *Asteroidea:* Discovery Rept., v. 20, p. 69-306, pl. 1-23.

Forbes, Edward

(20) 1848, *On the Asteriadae found fossil in British strata:* United Kingdom, Geol. Survey, Mem., v. 2, p. 457-482.

Gemmill, J. F.

(21) 1914, *The development and certain points in the adult structure of the starfish Asterias rubens:* Royal Soc. London, Philos. Trans., ser. B, v. 205, p. 213-294, pl. 18-24.

(22) 1915, *On the ciliation of Asterias and on the question of ciliary nutrition in certain species:* Zool. Soc. London, Proc. (1915), p. 1-19, pl. 1-3, text-fig. 1-2.

Gray, J. E.

(23) 1840, *A synopsis of the genera and species of the class Hypostoma (Asterias Linn.):* Ann. & Mag. Nat. History, v. 6, p. 175-184, 275-290.

Gregory, J. W.

(24) 1897, *The classification of the Palaeozoic echinoderms of the group Ophiuroidea:* Zool. Soc. London, Proc. (1896), p. 1028-1044.

(25) 1900, *The Stelleroidea:* in E. R. Lankester, A Treatise on Zoology: v. 3, p. 237-281, text-fig. 1-33, A. & C. Black (London).

Hess, Hans

(26) 1955, *Die fossilen Astropectiniden (Asteroidea); neue Beobachtungen und Übersicht über die bekannten Arten:* Schweiz. Palaeont. Abh., v. 71, p. 1-113, pl. 1-4, text-fig. 1-62.

(27) 1960, *Über die Abgrenzung der Astropectiniden-Gattungen Pentasteria Valette und Archastropecten Hess:* Eclogae Geol. Helv., v. 53, p. 329-331.

(28) 1960, *Neubeschreibung von Geocoma elegans (Ophiuroidea) aus dem unteren Callovien von La Voulte-sur-Rhone (Ardeche):* Same, v. 53, p. 335-385, text-fig. 1-37.

(29) 1960, *Ophiurenreste aus dem Malm des Schweizer Juras und des Departements Haut-Rhin:* Same, v. 53, p. 385-421, text-fig. 1-50.

(30) 1960, *Über zwei Ophiuren (Ophiocoma? rasmusseni n. sp. und Ophiotitanos tenuis Spencer) aus der Englischen Kreide:* Same, v. 53, p. 747-757, text-fig. 1-8.

(31) 1960, *Ophioderma escheri Heer aus dem unteren Lias der Schambelen (Kt. Aargau) und verwandte Lias-Ophiuren aus England und Deutschland:* Same, v. 53, p. 757-793, text-fig. 1-21.

(32) 1962, *Mikropaläontologische Untersuchungen an Ophiuren:* Same, v. 55, p. 595-656, text-fig. 1-142.

(33) 1963, *Mikropaläontologische Untersuchungen an Ophiuren. III. Die Ophiuren aus dem Callovien-Ton von Liesberg (Berner Jura):* Same, v. 56, p. 1141-1164, text-fig. 1-78.

———, & Enay, Raymond

(34) 1962, *Sur la découverte d'Ophiures (Ophiopetra lithographica, n. g., n. sp.) dans le Jurassique supérieur du Haut-Vabromey (Jura méridional):* Eclogae Geol. Helv., v. 55, p. 657-673, pl. 1-2, text-fig. 1-6.

Hunt, O. D.

(35) 1925, *The food of the bottom fauna of the Plymouth fishing grounds:* Marine Biol. Assoc., United Kingdom, Jour., new ser., v. 12, p. 500-598.

Hyman, L. H.

(36) 1955, *The invertebrates: Echinodermata:* 763 p., 280 text-fig., McGraw-Hill (New York).

Jaekel, O. M. J.

(37) 1903, *Asteriden und Ophiuriden aus dem Silur Böhmens:* Deutsch. geol. Gesell., Zeitschr., v. 55, p. 106-113, text-fig. 1-6.

(38) 1923, *Zur Morphogenie der Asterozoa:* Paläont. Zeitschr., v. 5, p. 344-350, 1 text-fig.

Jennings, H. S.

(39) 1907, *Behavior of the starfish Asterias ferreri de Loriol:* Calif. Univ. Publ. Zoology, v. 4, no. 2, p. 53-185.

Kesling, R. V.

(40) 1962, *Notes on Protopalaeaster narrawayi Hudson:* Jour. Paleontology, v. 36, p. 933-942, pl. 133-134, text-fig. 1-2.

Lehmann, W. M.

(41) 1957, *Die Asterozoen in den Dachschiefern des rheinischen Unterdevons:* Hess. Landesamt Bodenf., Abhandl., v. 21, 160 p., 55 pl., 31 text-fig.

Loriol, M. P. de

(42) 1909, *Note sur quelques stellérides du Santonien d'Abou-Roach:* Inst. Egypte, Bull., ser. 5, v. 2 (1908), p. 169-184, pl. 10-12.

Ludwig, Hubert

(43) 1899, *Echinodermen (Stachelhauter). II Klasse. Die Seesterne:* in Bronn, H. G., Klassen und Ordnungen des Thierreich: v. 2, pt. 3, no. 22, p. 621-623 (Leipzig).

MacBride, E. W.

(44) 1906, *Echinodermata:* in Cambridge Natural History, v. 1, p. 425-623, text-fig., Macmillan (London).

Matsumoto, Hikoshichirô

(45) 1915, *A new classification of the Ophiuroidea: with descriptions of new genera and species:* Acad. Nat. Sci. Philadelphia, Proc., v. 67, p. 43-92.

(46) 1917, *A monograph of Japanese Ophiuroidea arranged according to a new classification:* Tokyo Univ., Jour. Coll. Sci., v. 38, art. 2, 408 p., 7 pl.

(47) 1929, *Outline of the classification of the Echinodermata:* Tohoku Univ., Sci. Repts., ser. 2 (Geol.), v. 13, no. 2, p. 27-33.

Mercier, Jean

(48) 1935, *Les Stellérides mésozoïques du Bassin de Paris (bordure occidentale):* Soc. Linnéenne Normandie, Mém., new ser., Géol., v. 1, pt. 2, no. 3, p. 1-66, pl. 1-3.

Mortensen, Theodor

(49) 1927, *Handbook of the echinoderms of the British Isles:* 471 p., 269 text-fig., Oxford Univ. Press.

(50) 1932, *Studies of Indo-Pacific euryalids:* Vidensk. Medd., fra Dansk. Naturh. Foren., v. 96, no. 2, p. 1-75, pl. 1-5, text-fig. 1-57.

Müller, Johann, & Troschel, F. H.

(51) 1842, *System der Asteriden:* xx+134 p., 12 pl. (Braunschweig).

Murakami, Shiro

(52) 1963, *The dental and oral plates of Ophiuroidea:* Royal Soc. New Zealand, Trans., v. 4, no. 1, 48 p., 7 pl.

Perrier, J. O. E.

(53) 1884, *Mémoire sur les étoiles de mer recueillies dans la Mer d'Antilles et le Golf du Mexique:* Muséum Histoire Nat., Paris, Nouv. Arch., ser. 2, v. 6, p. 127-276, pl. 1-10.

(54) 1894, *Expéditions scientifiques du Travailleur et du Talisman:* 431 p., 26 pl., Masson (Paris).

Rasmussen, H. W.

(55) 1950, *Cretaceous Asteroidea and Ophiuroidea, with special reference to the species found in Denmark:* Danmarks Geol. Unders., ser. 2, no. 77, 134 p., 18 pl., 8 text-fig.

(56) 1951, *Cretaceous Ophiuroidea from Germany, Sweden, Spain and New Jersey:* Dansk. Geol. Foren., Medd., v. 12, p. 47-57, 1-5 text-fig.

Ruedemann, Rudolf

(57) 1916, *Account of some new or little known fossils, mostly from Palaeozoic rocks of New York:* N. Y. State Museum, Bull. 189, p. 7-97, pl. 1-31, text fig. 1-32.

Schöndorf, Friedrich

(58) 1906, *Die organisation und systematische Stellung der Sphaeriten:* Arch. Biont., v. 1, p. 245-306, pl. 24-25, text-fig. 1-36.

(59) 1909, *Palaeozoische Seesterne Deutschlands. I Teil. Die echten Asteriden der rheinischen Grauwacke:* Palaeontographica, v. 56, p. 37-112, pl. 7-11.

(60) 1909, *Die Asteriden des russischen Karbon:* Same, v. 56, p. 323-338, pl. 23-24.

(61) 1909, *Organisation und Aufbau der Armwirbel von Onychaster:* Nassau. Vereins Naturk., Jahrg. 62, p. 47-63, pl. 6.

(62) 1910, *Über einige Ophiuriden und Asteriden des englischen Silur und ihre Bedeutung für die Systematik paläozoischer Seesterne:* Same, Jahrg. 63, p. 206-256.

(63) 1910, *Die Asteriden der deutschen Trias:* Niedersächs. Geol. Vereins, Jahresbericht, v. 3, p. 90-116, text-fig.

(64) 1910, *Palaeozoische Seesterne Deutschlands II Teil. Die Aspidosomatiden des deutschen. Unterdevon:* Palaeontographica, v. 57, p. 1-66, pl. 1-3.

Schuchert, Charles

(65) 1914, *Fossilium Catalogus, I: Animalia, pars 3, Stelleroidea palaeozoica:* 53 p., W. Junk (Berlin).

(66) 1915, *Revision of Paleozoic Stelleroidea with special reference to North American Asteroidea:* U.S. Natl. Museum, Bull. 88, 311 p., 38 pl., 41 text-fig.

Sladen, W. P.

(67) 1889, *Report on the scientific results of the voyage of H.M.S. Challenger during the years 1873-76: Zoology—vol. 30, Report on the Asteroidea:* xlii+893 p., 117 pl., H. M. Stationery Office (London).

———, & Spencer, W. K.

(68) 1891-1908, *British fossil Echinodermata from the Cretaceous formations; vol. 2, Asteroidea and Ophiuroidea:* Palaeontograph. Soc. London, Mon., pt. 1, v. 1890 (1891), p. 1-28, pl. 1-8; pt. 2, v. 1893 (1893), p. 29-66, pl. 9-16; pt. 3, v. 1905 (1905), p. 67-90, pl. 17-26; v. 1907 (1907), p. 91-132, pl. 27-29, fig. 1-34; pt. 5, v. 1908 (1908), p. 133-138.

Smith, J. E.

(69) 1937, *On the nervous system of the starfish Marthasterias glacialis:* Royal Soc. London, Philos. Trans., ser. B, v. 227, p. 111-173, text-fig. 1-32.

(70) 1946, *The mechanics and innervation of the starfish tube-ampulla system:* Same, v. 232, p. 279-310, text-fig. 1-15.

(71) 1950, *Some observations on the nervous mechanisms underlying the behaviour of starfishes:* Soc. Experiment. Biol., Symposium no. 4 on Animal Behaviour, p. 196-220, text-fig. 1-32.

Sollas, I. B. J.

(72) 1913, *On Onychaster, a Carboniferous brittle-star:* Royal Soc. London, Philos. Trans., ser. B, v. 204, p. 51-62, pl. 8, 9.

Sollas, W. J., & Sollas, I. B. J.

(73) 1912, *Lapworthura: a typical brittle-star of the Silurian age:* Royal Soc. London, Philos. Trans., ser. B, v. 202, p. 213-232, pl. 9-10.

Spencer, W. K.

(74) 1913, *The evolution of the Cretaceous Asteroidea:* Royal Soc. London, Philos. Trans., ser. B, v. 204, p. 99-177, pl. 10-16.

(75) 1914-40, *British Palaeozoic Asterozoa:* Palaeontograph. Soc. London, Mon., pt. 1, v. 1913 (1914), p. 1-56, pl. 1, text-fig. 1-31; pt. 2, v. 1915 (1916), p. 57-108, pl. 2-5, text-fig. 32-63; pt. 3, v. 1916 (1918), p. 109-168, pl. 6-13, text-fig. 64-112; pt. 4, v. 1917 (1919), p. 169-196, text-fig. 113-138; pt. 5, v. 1920 (1922), p. 197-236, pl. 14-17, text-fig. 139-171; pt. 6, v. 1922 (1925), p. 237-324, pl. 18-22, text-fig. 172-212; pt. 7, v. 1926 (1927), p. 325-388, pl. 23-34, text-fig. 213-248; pt. 8, v. 1928 (1930), p. 389-436, pl. 25-28, text-fig. 249-280; pt. 9, v. 1933 (1934), p. 437-494, pl. 29-32, text-fig. 281-321; pt. 10, v. 1940 (1940), p. 495-540, pl. 33-37, text-fig. 322-348.

(75a) 1965 (by H. G. Owen), *Table of contents, supplement and index* [to 75], Same, p. 541-583.

(76) 1950, *Asterozoa and the study of Palaeozoic faunas:* Geol. Mag., v. 87, p. 393-408.

(77) 1951, *Early Palaeozoic starfish:* Royal Soc. London, Philos. Trans., ser. B, p. 235, p. 87-129, pl. 2-8, text-fig. 1-24.

Stürtz, Bernhard

(78) 1886, *Beitrag zur Kenntniss paläozoischer Seesterne:* Palaeontographica, v. 32, p. 75-98, pl. 8-14.

(79) 1890, *Neuer Beitrag zur Kenntniss paläozoischer Seesterne:* Same, v. 36, p. 203-247, pl. 26-31.

Thoral, Marcel

(80) 1935, *Contribution à l'étude paléontologique de l'Ordovicien inférieur de la Montagne Noire:* 363 p., 35 pl. (Montpellier).

Ubaghs, Georges

(81) 1941, *Description de quelques ophiures du Famennian de la Belgique:* Musée Royal Histoire Nat. Belgique, Bull., v. 17, no. 44, 31 p., 2 pl., 1 text-fig.

(82) 1942, *Bohemura constellata (Thorent) et Drepanaster sp., Ophiuroïdes du Devonien inférieur de la Belgique et du nord de la France:* Same, v. 18, no. 7, 19 p., 2 pl., 1 text-fig.

(83) 1953, *Classe des Stelléroïdes* in PIVETEAU, JEAN, Traité de Paléontologie, v. 3, p. 774-842, 64 text-fig., Masson et Cie (Paris).

Valette, Aurelien

(84) 1929, *Note sur quelques Stellérides jurassiques du Laboratoire de Géologie de la Faculté des Sciences de Lyon:* Lab. Géologie Univ. Lyon, Travaux, pt. 16, Mém. 13, p. 1-39, pl. 1-5.

Vevers, H. G.

(85) 1952, *A photographic survey of certain areas of sea floor near Plymouth:* Marine Biol. Assoc., United Kingdom, Jour., v. 31, p. 215-222.

Viguier, Camille

(86) 1878, *Anatomie comparée du squelette des stellérides:* Arch. Zoologie Expér. Gen., v. 7, p. 33-250, pl. 5-16.

Withers, R. P., & Keeble, R. A.

(87) 1934, *The Palaeozoic starfish of Victoria:* Royal Soc. Victoria, Proc., new ser., v. 46, p. 220-249, pl. 10-12.

(88) 1934, *The Palaeozoic brittle-stars of Victoria:* Same, v. 47, p. 196-212, 1 pl., 1 text-fig.

Wright, C. W., & Wright, E. V.

(89) 1940, *Notes on Cretaceous Asteroidea:* Geol. Soc. London, Quart. Jour., v. 94, p. 231-248, pl. 13-17.

Wright, Thomas

(90) 1863-1880, *British Fossil Echinodermata of the Oolitic formations: vol. 2, Asteroidea and Ophiuroidea:* Palaeontograph. Soc. London, Mon., pt. 1, v. 1861 (1863), p. 1-130, pl. 1-12; pt. 2, v. 1864 (1866), p. 131-154, pl. 13-18; pt. 3, v. 1880 (1880), p. 155-203, pl. 19-21.

Zittel, K. A. von

(91) 1880, *Handbuch der Paläontologie:* Abt. 1, *Paläzoologie:* v. 1, 765 p., 558 text-fig., R. Oldenbourg (München & Leipzig).

ADDITIONAL SOURCES OF ILLUSTRATIONS

(101) **Blegvad, Harald,** 1915
(101a) **Clark, A. M., & Wright, C. W.,** 1962
(102) **Clark, H. L.,** 1915
(103) **Clarke, J. M.,** 1921
(104) **Coe, —. —.**
(105) **Cottreau, Jean,** 1929
(106) **Eck, H.,** 1879
(107) **Etheridge, Robert,** 1892
(108) **Fedotov, D. M.,** 1926-36
(109) **Fell, H. B.,** 1960-63
(110) **Fisher, W. K.,** 1911-40
(111) **Gemmill, J. F.,** 1914-15
(111a) **Goldfuss, August,** 1826-33
(112) **Gordon, Isabella,** 1926
(113) **Hess, Hans,** 1955-62
(114) **Jaekel, O. M. J.,** 1903-23
(114a) **Kesling, R. V.,** 1962
(115) **Lang, A.,** 1896
(116) **Lehmann, W. M.,** 1957
(117) **Loriol, M. P. de,** 1909
(118) **Ludwig, H.,** 1899
(119) **Lyman, Theodore**
(120) **Matsumoto, Hikoshichirô,** 1915-29
(121) **Mercier, Jean,** 1935
(122) **Mortensen, Theodor,** 1927-32
(122a) **Perrier, J. O. E.,** 1894
(123) **Quenstedt, F. A.,** 1875
(124) **Rakusj, J.,** 1927
(124a) **Rao, Ragvendra,** 1957
(125) **Rasmussen, H. W.,** 1950-51
(126) **Raymond, P. E.,** 1921
(127) **Ruedemann, Rudolf,** 1916
(128) **Schöndorf, Friedrich,** 1906-1910
(129) **Schuchert, Charles,** 1914-15
(129a) **Skwarko, S. K.,** 1963
(130) **Sladen, W. P.,** 1889
(131) **Sladen, W. P., & Spencer, W. K.,** 1891-1908
(132) **Smith, J. E.,** 1937-50
(133) **Spencer, W. K.,** 1913-51
(134) **Spencer, W. K., & Wright, C. W.**
(135) **Ubaghs, Georges,** 1941-53
(136) **Valette, Aurelien,** 1929
(137) **Woods, Henry,** 1937
(138) **Worthen, A. H., & Miller, S. A.,** 1883
(139) **Wright, C. W., & Wright, E. V.,** 1940
(140) **Wright, Thomas,** 1863-80
(141) **Zittel, K. A. von,** 1880

ECHINOZOANS

By J. Wyatt Durham, K. E. Caster, Harriet Exline, H. B. Fell, A. G. Fischer, D. L. Frizzell, R. V. Kesling, P. M. Kier, R. V. Melville, R. C. Moore, D. L. Pawson, Gerhard Regnéll, Georges Ubaghs, and Carol D. Wagner

CONTENTS

Subphylum ECHINOZOA Haeckel in Zittel, 1895

[Diagnosis prepared by H. B. Fell]

Fundamentally globoid, including secondarily discoid and cylindroid echinoderm forms that entirely lack arms, brachioles, or outspread rays such as characterize most crinozoans and asterozoans. Earliest members with mouth and anus at opposite ends of body but these are secondarily displaced in many later forms. Meridional water vessels traversing body wall in direction of anus, such vessels originally lying on surface of theca but in later Paleozoic and all post-Paleozoic groups sinking into its substance. Skeleton, nervous system, reproductive organs, and muscular system tending to be differentiated into meridional systems, although underlying bilateral symmetry is discernible generally and in course of evolution may become expressed strongly. *L.Cam.(Olenellus Zone)-Rec.*

GENERAL FEATURES AND RELATIONSHIPS OF ECHINOZOANS

By H. Barraclough Fell and Raymond C. Moore

[Harvard University; University of Kansas]

INTRODUCTION

Echinozoans are fundamentally globoid, cylindroid, or discoid echinoderms which entirely lack outspread rays, such as characterize asterozoans, and arms or brachioles,[1] such as occur in most crinozoans.

Earliest echinozoan classes have the mouth and anus at opposite extremities of the body, thus defining an anteroposterior axis, but in some later classes these orifices of the alimentary tract are found to be secondarily displaced. Meridionally disposed water vessels traverse the body wall in the direction of the anus, such vessels originally lying on the surface of the theca (as interpreted by Fell, strongly doubted by Dur-

[1] The calcareous scale-covered podia of ophiocistioids cannot be compared closely with the arms of crinoids, supported by an internal skeleton of solid ossicles, or with similar structures of other crinozoans.

HAM) but in later Paleozoic and all post-Paleozoic groups sinking into its substance. Skeletal elements, reproductive organs, and the muscular and nervous systems tend to be differentiated in meridional patterns, although an underlying bilateral symmetry almost invariably is evident.

The echinozoans are defined as a subphylum and included classes are designated Helicoplacoidea (L.Cam.), Edrioasteroidea (L.Cam.-L.Carb.), Ophiocistioidea (L.Ord.-U.Ord.), Cyclocystoidea (M.Ord.-L.Sil.), Holothuroidea (?M.Ord., Dev.-Rec.), and Echinoidea (M.Ord.-Rec.).

BACKGROUND

Until recently, zoologists customarily have divided the phylum Echinodermata into two contrasted subphyla, respectively named Pelmatozoa and Eleutherozoa. The Pelmatozoa, predominantly represented by fossil forms, have been construed to include groups that throughout all or at least part of their postlarval life are attached in fixed manner to the substrate and that carry the oral and anal openings of the spirally twisted gut on the upwardly directed surface of the body or the anus may be located laterally on the theca (this side being defined as posterior). Crinoids, cystoids, blastoids, and the much less common fossil groups known as eocrinoids, paracrinoids, and edrioblastoids are typical pelmatozoans. The Eleutherozoa comprise almost exclusively free-living echinoderms in which the mouth is directed downward or anteriorly and the anus (if present) is usually placed on the upper surface or posteriorly at end of the body opposite the mouth. Best-known eleutherozoans are the sea urchins (echinoids), sea cucumbers (holothurians), starfishes (asteroids), and brittle stars (ophiuroids).

The various pelmatozoan and eleutherozoan groups differ from one another so widely that the task of elucidating their interrelationships and defining the nature of their presumed common ancestry has been extremely difficult. Recent morphological and paleontological studies by FELL (4-6) have led to definite conclusion that the star-shaped members of the so-called eleutherozoans are so interrelated as to comprise a single grouping classifiable as a subphylum, and for it the name Asterozoa is available. Similarly, other evidence implies that the globoid, cylindroid, and discoid so-called eleutherozoans are probably interrelated also, and these may be associated in another subphylum named Echinozoa. Consequently, the too-inclusive "Eleutherozoa" are an outmoded polyphyletic assemblage and the name should be abandoned for taxonomic purposes. On the other hand, if confined to characterization of life habit only, the descriptive noun and adjective "eleutherozoan" are conceded to have usefulness.

The discovery of a class of echinoderms named Helicoplacoidea by DURHAM & CASTER (3), found in oldest known (*Olenellus* Zone) fossil-bearing rocks of California, has shown that primitive, free-living members of the Echinodermata already had become differentiated in earliest Cambrian time. The morphological characters of the Helicoplacoidea partly resemble those of Echinoidea, Holothuroidea, and Edrioasteroidea, suggesting a relationship of all four classes to a common ancestral stock. This stock, then, is inferred to be the source of eleutherozoan helicoplacoids, holothurians, and echinoids on one hand and to prevailingly pelmatozoan edrioasteroids on the other. On the basis of common features of body form, however, especially absence of ray-like and armlike extensions from it, all are assignable to the subphylum Echinozoa.

The Echinozoa represent an ancient stock, modern representatives of which are the Holothuroidea and Echinoidea, whereas the Asterozoa are of later origin, interpreted by FELL (6) to have been derived from pinnulate pelmatozoans belonging or allied to the Crinoidea. Recent studies of dendrochirote holothurians (PAWSON & FELL, 12), partly discussed in the later section of this volume devoted to Holothuroidea (p. *U*641), indicate the essentially archaic nature of this group and point to possibly significant relationships with the Cambrian helicoplacoids. Also, some observed parallelism of the psolid dendrochirotes and edrioasteroids serves to reinforce judgment that the Edrioasteroidea belong with echinozoan echinoderms, rather than the subphylum Pelmatozoa, where previously they have been placed (4, 7).

Relationships of the early Paleozoic Homalozoa examined critically by UBAGHS

(13) are doubtful. These are asymmetrical echinoderms which in specialized forms show a tendency toward bilateral symmetry, though none fully attain it. They were evidently free-living (eleutherozoan) in habit (Fig. 90,*1*), for none fixed permanently to the substrate are known. Classification as homalozoans is not based on life habit of the animals.

In summation, FELL (7) has pointed out that four contrasted structural patterns are clearly discernible in echinoderms as follows (see also chapter by UBAGHS on "General Characters of Echinodermata," *Treatise,* Part S).

(1) A **homalozoan pattern,** represented by early Paleozoic forms with skeletal parts arranged asymmetrically or displaying some degree of bilateral symmetry, is seen in classes named Homostelea (Fig. 90,*1*), Stylophora, and Homoiostelea, formerly grouped together as "Carpoidea." They are assigned to the subphylum named Homalozoa WHITEHOUSE, 1941. Other classes named Cyamoidea, Cycloidea, and Machaeridia, have very doubtful status.

(2) A **crinozoan pattern** comprises prevailingly globoid forms with partial radiate meridional symmetry from which ambulacral feeding appendages (arms,

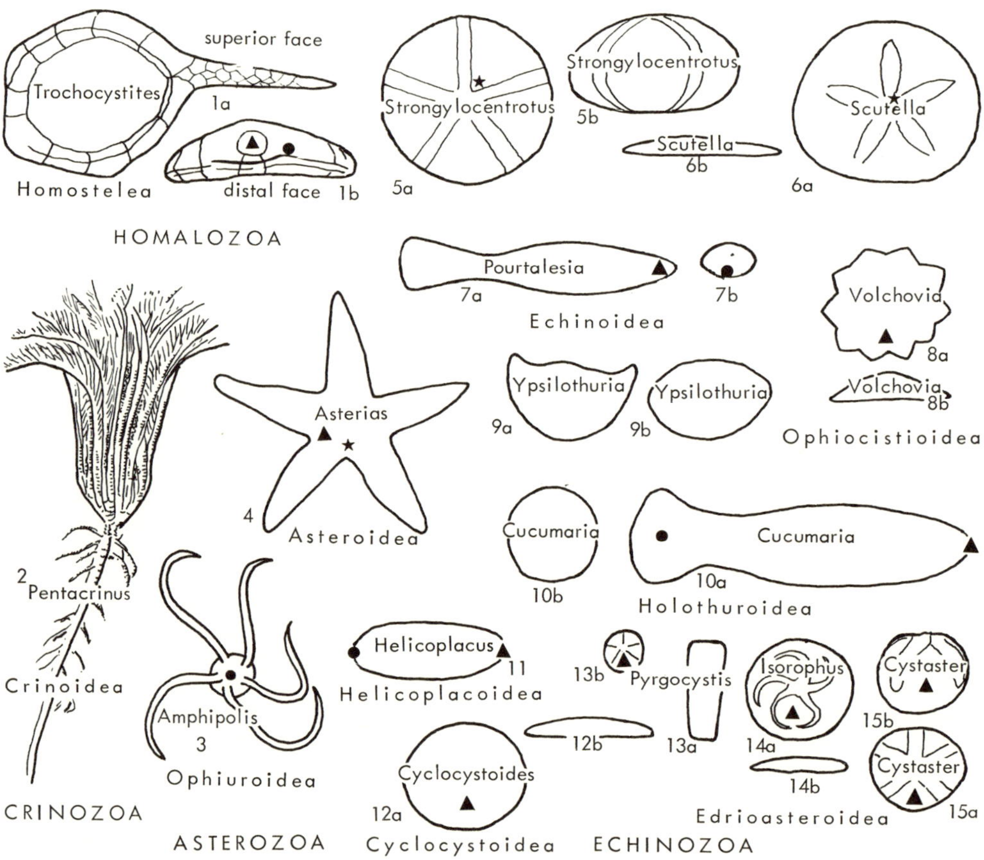

Explanation: ● mouth; ▲ anus; ★ madreporite

FIG. 90. Types of thecae representative of echinoderm subphyla (diagrammatic, not to scale).——*1*. Homalozoa (carpoid).——*2*. Crinozoa (crinoid).——*3,4*. Asterozoa (ophiuroid, asteroid).——*5-15*. Echinozoa (*5-7*, echinoids; *8*, ophiocistioid; *9,10*, holothurians; *11*, helicoplacoid; *12*, cyclocystoid; *13-15*, edrioasteroids).

brachioles) extend upward or outward. Most of them are attached to the substrate in fixed location throughout postlarval life, but some are free-moving (eleutherozoan) as adults. They include classes named Eocrinoidea, Paracrinoidea, Cystoidea, Crinoidea (Fig. 90,*2*), Edrioblastoidea, and Blastoidea. These are grouped in the subphylum Crinozoa MATSUMOTO, 1929 (=Pelmatozoa LEUCKART, 1848, *partim*).

(3) An **asterozoan pattern** is characterized by radially divergent axes of symmetry expressed by relatively broad to elongate narrow extensions of the body spread laterally outward. Commonly the central body surrounded by its rays has a star-shaped outline. Three subclasses named Somasteroidea, Asteroidea (Fig. 90,*4*), and Ophiuroidea (Fig. 90,*3*) are recognized, grouped together in the class Stelleroidea and the subphylum Asterozoa HAECKEL in ZITTEL, 1895.

(4) An **echinozoan pattern** consists of globoid, cylindroid, and discoid forms with well-marked radial meridional symmetry which entirely lack outspread extensions comparable to the arms or brachioles of crinozoans or the rays of asterozoans. The classes Helicoplacoidea (Fig. 90,*11*), Holothuroidea (Fig. 90,*9,10*), Ophiocistioidea (Fig. 90,*8*), Cyclocystoidea (Fig. 90, *12*), Edrioasteroidea (Fig. 90,*13-15*), and Echinoidea (Fig. 90,*5-7*) are divisions of the subphylum Echinozoa HAECKEL in ZITTEL, 1895. Most echinozoans are free-moving throughout life, but some edrioasteroids and dendrochirote holothurians are recognized as sessile animals.

In agreement with FELL (7), it seems evident that the mode of life adopted by various echinoderm assemblages is not acceptable as the governing criterion for classification of them in first-rank divisions (subphyla), even though this may affect morphological features importantly. Eleutherozoan tendencies exclusively characterize Homalozoa, Asterozoa, and nearly all Echinozoa, whereas they are confined to a minority of Crinozoa (comatulid and various other crinoids, possibly a few blastoids and cystoids). Clearly, the four patterns of symmetry which have been described briefly are entirely unrelated to these tendencies. Free-living echinoderms acquire locomotor mechanisms that facilitate quest for food by browsing on algae, by preying on other animals, and by swallowing large quantities of mud in order to extract its small organic content. This is accompanied generally by the development of jaws or special oral appendages suited to gross (macrophagous) feeding. The anus, if present, tends to be placed remote from the mouth, commonly on the opposite side of the body. Among sessile echinoderms locomotor organs are partly or entirely lost, forcing the animals to depend on such planktonic food sources as sea currents may provide. Feeding is of microphagous type, food particles being carried by some ciliary or comparable mechanism with aid of the tube feet along food grooves of appendages and the upper body surface to the mouth. The alimentary canal has a more or less contorted U-shape, with mouth and anus directed upward in locations not far apart. These features characterize most Crinozoa, in contrast to the Asterozoa and Echinozoa.

EVOLUTION OF ECHINOZOA

The oldest known organism classed as an echinozoan is *Helicoplacus* (L.Cam., *Olenellus* Zone, Calif.-Nev.), although a locally associated fossil (edrioasteroid, ?*Stromatocystites*) is also classed as an echinozoan (2, p. 52). Eocrinoids are other echinoderms found in this ancient fauna.

The mouth of *Helicoplacus* is located at the broadly rounded end of its fusiform body (therefore interpreted by DURHAM & CASTER, 1963, as anterior) and the anus probably occurs at the tapered opposite extremity (Fig. 91,*1*). The very numerous small plates of the theca are disposed in closely adjoined spiral rows twisted counterclockwise. A single narrow band of minute platelets (bifurcated in some individuals) which winds around the body is interpreted by FELL (7) to denote an external water vessel, although observations by DURHAM (personal communication) indicate a possibility that this water vessel may have been internal. The narrow band of platelets, identified as an ambulacrum, divides the thecal surface into halves that define a sort of bilateral symmetry greatly affected by torsion. It is combined wtih an apparent radial symmetry defined by arrangement of the thecal plates. FELL (7) has noted that simi-

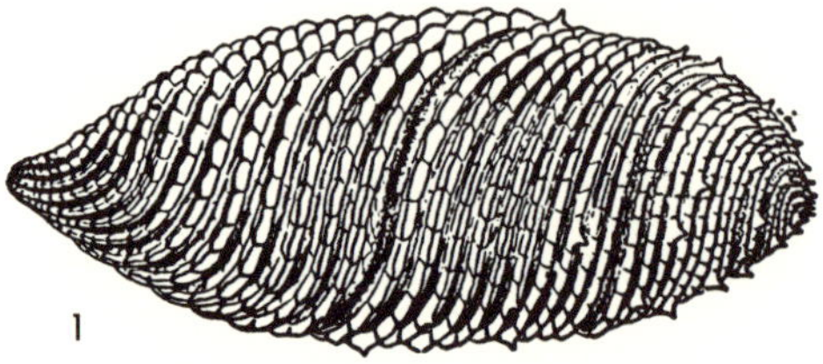

1

FIG. 91. *Helicoplacus*, L. Cam., USA (Calif.); lat. view of theca (reconstr.), ×0.33 (3).

lar torsion is observed in earliest echinoids (e.g., *Eothuria*) of Ordovician age, although in them the symmetry is overtly radial (more correctly meridional) as defined by the five twisted ambulacral meridians, regularly spaced at intervals of 72 degrees. Analogous torsion also is seen in the Edrioasteroidea, persisting to their extinction in the Lower Carboniferous, but it has not been reported in the Holothuroidea and Ophiocistioidea. The torsion soon was lost in the echinoid line. The oldest known edrioasteroid (*Stromatocystites,* L.Cam., Czech., ?Calif.; M.Cam., France, Sweden) had straight ambulacra, rather than curved ones. Also, the echinoid *Aulechinus,* a contemporary of *Eothuria,* and the somewhat older *Bothriocidaris* exhibit no signs of torsion.

Helicoplacus may have been an eleutherozoan bottom-feeding echinoderm rather similar to a plated dendrochirote holothurian. Its thecal plates formed a complete, robust, flexible test. The varying degrees of expansion and contraction observed in the fossils (3) imply existence of musculature capable of altering thecal shape in a manner comparable to operation of a concertina. Perhaps *Helicoplacus* crept over the sea floor like an annelid. The occurrence of the fossils in a very fine clastic matrix denotes a mud-bottom habitat and suggests that this echinoderm may have been a mud-swallower, like many nondendrochirote holothurians. The small size of the mouth has led DURHAM (personal communication) to guess, rather, that these echinoderms depended for food on small organic particles gathered from surrounding water by a mucus-secreting or ciliary mechanism, such food particles being then passed along the ambulacrum to the mouth. *Helicoplacus* may thus have lived much of the time in an upright position as a somewhat sedentary inhabitant of the muddy sea bottom, only occasionally assuming a horizontal attitude.

In the view of FELL (7), the primitive state of the lone ambulacrum of *Helicoplacus* implies that this structure carried rudimentary tube feet which may have been respiratory organs or possibly only sensory tentacles, like the dorsal tube feet of many holothurians. The completeness of thecal plating and probable rudimentary nature of the tube feet are evidence that no effective respiratory mechanism existed on the body surface. If this is correct, FELL has judged it reasonable to infer that rectal respiration was required, either of the pulsatory crinoid type or by means of respiratory trees, as in holothurians. The distribution of respiratory trees in various orders of holothurians suggests that these structures are related directly to habits of the animals and implies that the earliest holothuroids already had developed such trees. It seems likely, then, that rudimentary respiratory trees were present in the Helicoplacoidea, although the divergent speculations on the significance of nearly all observed morphological features leave much doubt.

The earliest Echinoidea, represented by such forms as *Eothuria,* resemble the helicoplacoids not only in torsion of the body wall, as previously noted, but in the flexible nature of the multiplated theca. Possibly these features indicate derivation of the echinoid line from helicoplacoids. The early Paleozoic echinoids possessed five well-developed ambulacra, on which (as interpreted by FELL) the meridional water vessels lay as external structures (though with internal ampullae for the tube feet). This is doubted by DURHAM and UBAGHS (personal communication) who note that in *Bothriocidaris, Ectinechinus* and *Eothuria* the radial water vessels clearly were internal; in *Aulechinus* they are susceptible of either interpretation.

Structural details of the ambulacral pores of early echinoids show that the tube feet were large and probably suctorial; certainly they were extensile and muscular. Thus they could serve the double function of locomotor and respiratory organs, as in modern echinoids. The fossils exhibit a moderately well-developed jaw mechanism, which shows that the early echinoids were capable of feeding in the manner of their extant endocyclic descendants, biting and grinding organisms in the substrate and chewing algae. Such features demonstrate an eleutherozoan habit and deny pelmatozoan tendencies, for echinoderms provided with feeding mechanisms of this sort rapidly would starve if they adopted a sessile existence.

The Ophiocistioidea developed a rigid skeleton by solid union of adjacent thecal plates in a manner comparable to that seen in modern echinoids. Locomotion was effected by use of the grossly enlarged and plated tube feet of the oral surface. These tube feet, which specially distinguish the class, doubtless also served the function of nutrition by sweeping up detrital material and cramming it into the downwardly directed mouth. The anus, as in endocyclic echinoids, was located on the upper surface next to the margin or midway between it and the apical pole. Habits of the ophiocistioids surely were eleutherozoan, as indicated by their morphology and by lack of any known sessile forms.

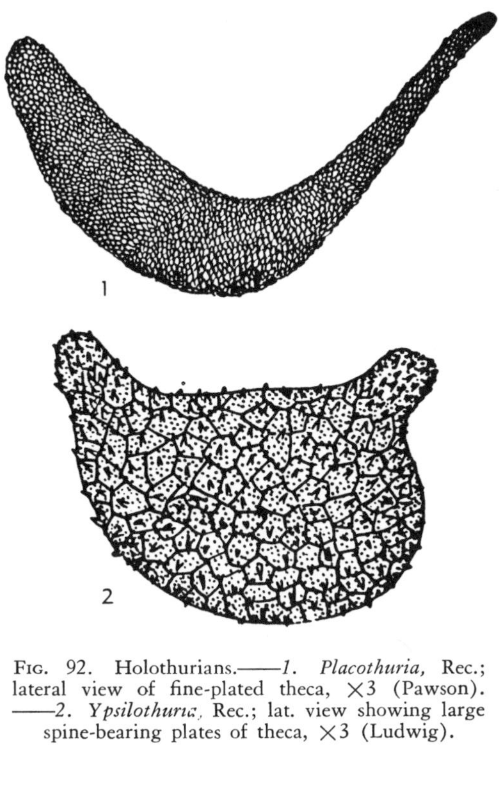

Fig. 92. Holothurians.——*1. Placothuria,* Rec.; lateral view of fine-plated theca, ×3 (Pawson).——*2. Ypsilothuria,* Rec.; lat. view showing large spine-bearing plates of theca, ×3 (Ludwig).

Oldest known fossil remains of Holothuroidea consist of isolated diminutive skeletal plates. However, recent studies (12; see also p. *U*646) indicate strong probability that the Ordovician and later Paleozoic holothurians closely resembled some modern members of the Dendrochirotida (e.g., *Placothuria*; Fig. 92,*1*) and of a new order named Dactylochirotida (12), the latter exemplified by *Ypsilothuria* (Fig. 92, *2*). These holothurians cited for comparison are all heavily plated forms with a complete test made up of large plates which commonly are provided with rigid spinous processes. The early holothurians are judged to be similar also to *Helicoplacus* and to the Ordovician echinoids with flexible theca (e.g., *Eothuria,* originally considered to be a holothurian). Suctorial tube feet probably were lacking in earliest holothurians, judging from their rudimentary state in extant plated genera, but this is uncertain. If such tube feet had not yet been developed, locomotory movement of the animals must have been effected by contraction and expansion of the body wall and its flexible test (7). When they were provided with suctorial tube feet, movement on the echinoid plan would have been possible. Morphological evidence indicates that some kind of jaw apparatus comparable to the echinoid lantern was developed early in evolution of the holothurians, but with development of the dendrochirote type of tentacles, apparently the apparatus was abandoned and its elements served the new purpose of providing attachments for the radial and retractor muscles. In this fashion the organ persisted in later holothurians as the calcareous ring surrounding the pharynx.

Probably the many-branched dendrochirote type of tentacle was evolved from initially simple tube feet of finger-like form in the oral region (12). Repeated dichotomy could lead to the very complex dendritic tentacles of the Dendrochirotida, which are efficient for collecting planktonic food and conveying it to the mouth by ciliary action accompanied by contractions of the tentacles and spooning movements of the two ventral tentacles. The dendrochirotids, whether motile or not, are able to trap sufficient nourishment by filtering of sea water, provided that currents replace the

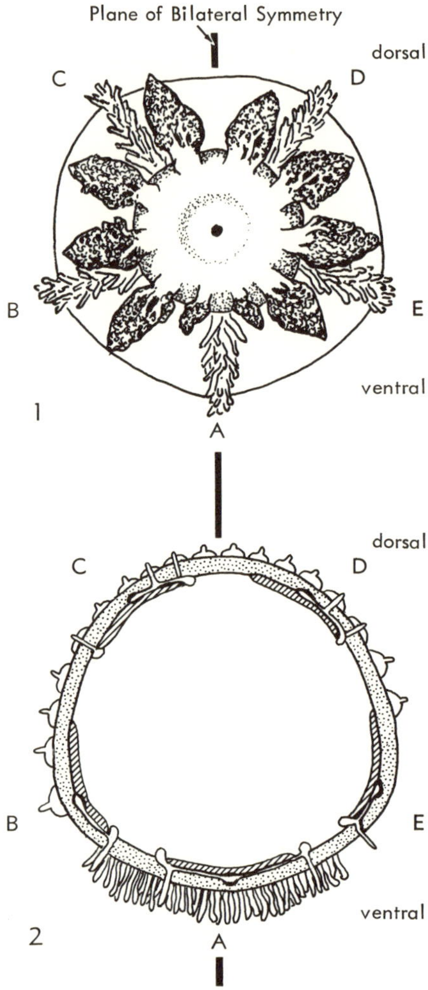

FIG. 93. Anterior and cross section views of holothurians showing differentiation of ventral and dorsal sides and vertical plane of bilateral symmetry. ——*1. Cucumaria,* Recent dendrochirotidan from front, showing pair of reduced ventral tentacles and crowded tube feet of five rays (*A-E,* designations of rays, Carpenter system) (diagram.).——*2.* Cross section (diagram.) of body of *Holothuria,* Rec. aspidochirotidan, showing ventral locomotory tube feet and paxillate tube feet of dorsal and lateral surfaces (*A-E,* Carpenter designations of rays).

surrounding water so as to bring fresh supplies of food particles. From the dendrochirotids more than one line of evolution is possible, for they have the means of adopting either eleutherozoan or pelmatozoan habits. If a locomotor system is retained, the oral tentacles can be adapted to serve for collection of food, operating in various ways. In the Cucumariidae, for example, the body may be held erect, attached only by the posterior tube feet, with tentacles around the upraised mouth spread outward in the manner of a sea anemone (Fig. 93,*1*). In holothurians that adopt a

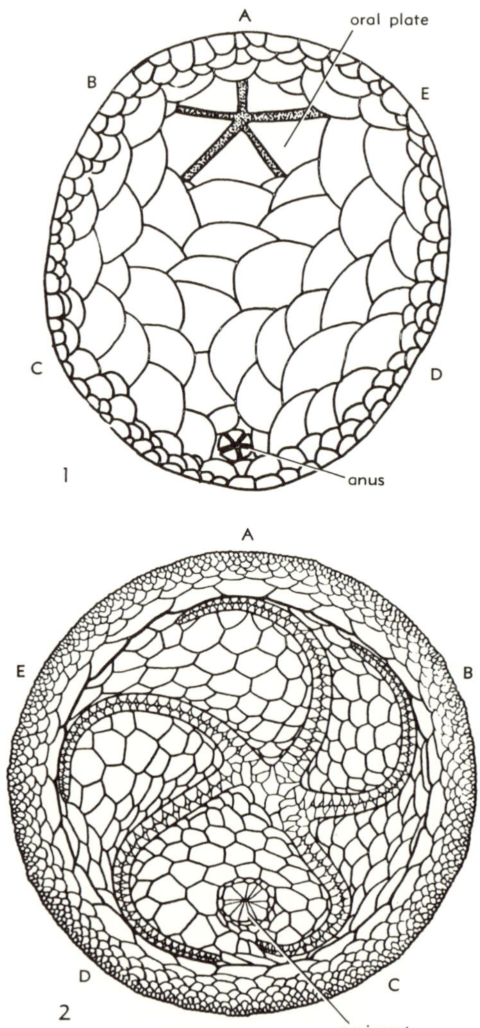

FIG. 94. Comparison of psolid holothurian with edrioasteroid (diagram.). ——*1.* Dorsal view of *Psolus,* a dendrochirotidan, showing oral and anal plates near opposite extremities and imbricated nature of other thecal plates (*A-E,* Carpenter designations of rays).——*2. Isorophus,* upper (oral) surface of typical edrioasteroid showing prominent curved ambulacra and periproct in posterior *(CD)* interray (10).

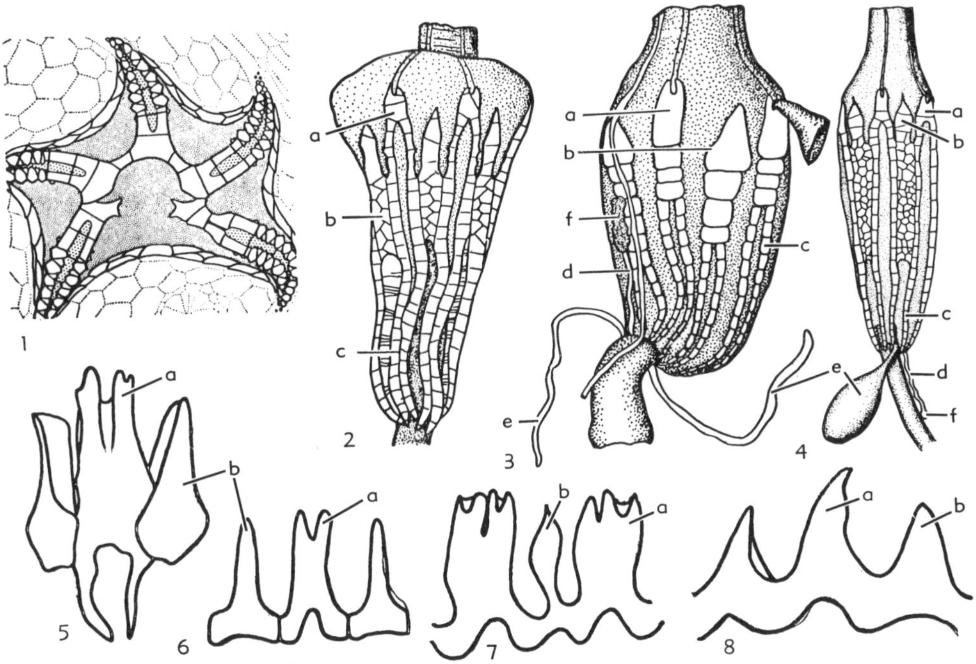

FIG. 95. Calcareous rings of edrioasteroid and holothurians.——*1.* Oral side of *Isorphus,* U.Ord. edrioasteroid, showing calcareous ring and associated ambulacral plates (10).——*2-4.* Side views of pharyngeal region of dendrochirotidan holothurians *Pentadactyla, Placothuria,* and *Neothyonidium,* showing calcareous rings and associated structures, all Rec. (Pawson).——*5-8.* Diagrammatic outlines of part of calcareous rings of *Euthyonidiella, Psolus, Thyonidium,* and *Mitsukuriella,* from side (all Rec. dendrochirotidan holothurians) (*5,7,8,* from Heding & Panning, 1954; *6,* Pawson). [Explanation: *a,* radial piece of calcareous ring; *b,* interradial piece; *c,* posterior process of radial; *d,* madreporic duct (or stone canal); *e,* Polian vesicle; *f,* madreporite.]

horizontal attitude the lower and upper sides are not determined at random or changed from time to time. The lower (ventral) side bears abundant locomotory tube feet, whereas the upper (dorsal) and lateral surfaces have much less numerous (paxillate) tube feet associated with low pimple-like elevations (Fig. 93,*2*). Aspidochirote forms can readily evolve from dendrochirotids, thus permitting gross mud-swallowing and a markedly eleutherozoan habit (7).

Another possibility is for the locomotory system to be converted to a purely adhesive role, thus leading to a sessile (pelmatozoan) habit, though no known holothurian fully attained this. It is illustrated by the psolid dendrochirotes, some genera of which exhibit a flattened limpet-like body which adheres by its broad ventral surface to a firm substrate (Fig. 94,*1*). They may be classed as statozoans (temporarily fixed), rather than true pelmatozoans (permanently fixed). The exposed dorsal and lateral surfaces are covered by a test composed of robust imbricated plates. The mouth and anus are placed on the upper surface and commonly protected by valvate plates similar to those of edrioasteroids and various cystoids.

The whole body, in fact, is somewhat comparable to that of an edrioasteroid, except for its lack of external ambulacral plates (Fig. 94,*2*). Morphological features of the psolids are closer to those of the Edrioasteroidea than to characters of many Holothuroidea and Echinoidea. They chiefly differ from edrioasteroids in the same way that distinguishes them from echinoids, namely, in the internal placement of their water vessels, as a result of which the psolid test lacks ambulacral plates. The

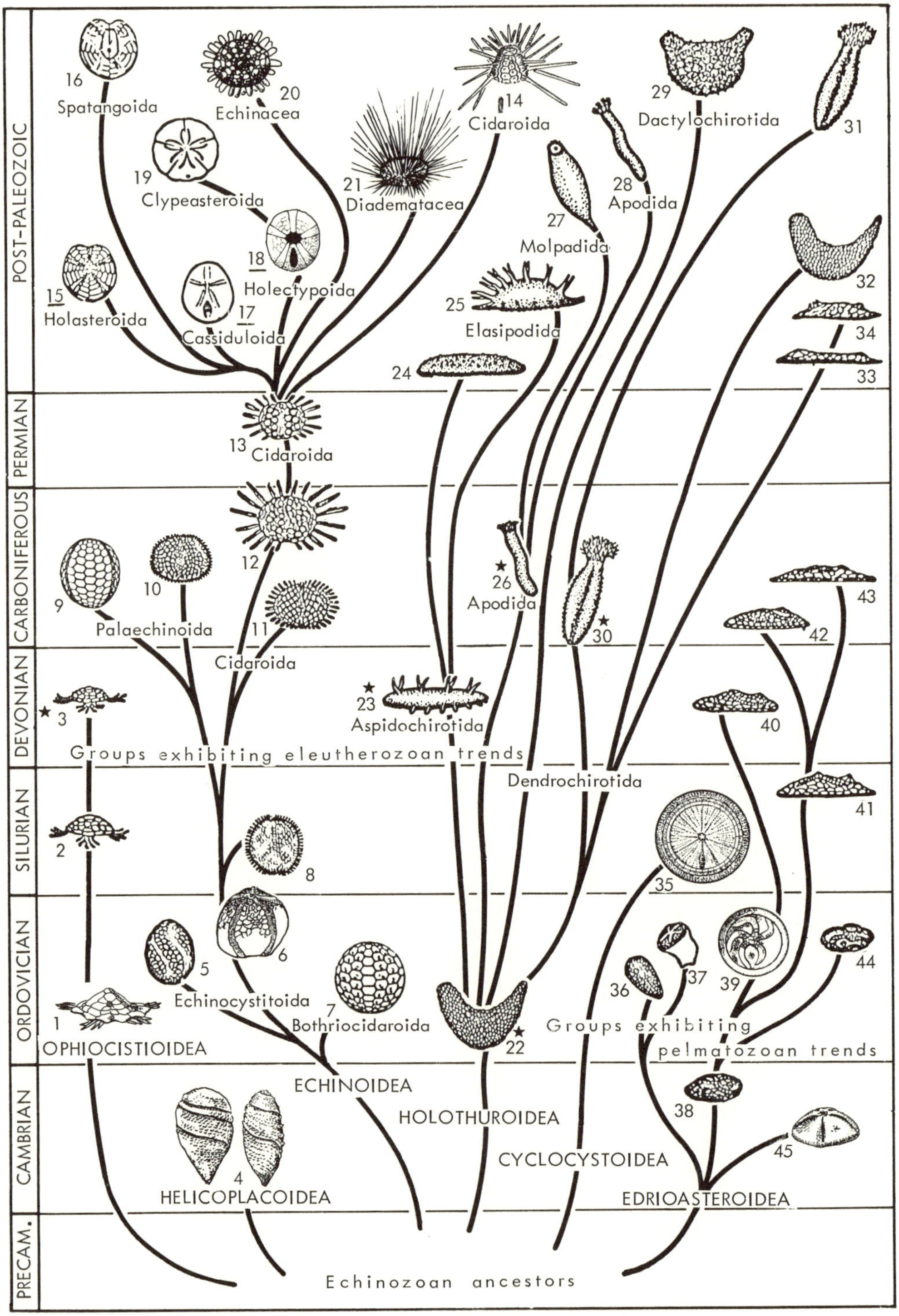

FIG. 96. Diagram representing postulated phylogenetic relationships of echinozoan classes and orders (modi-

(Continued on facing page.)

edrioasteroids and psolids are similar derivatives from an early echinozoan stock and both developed pelmatozoan characters in response to adopting a sessile mode of life (7). However, we must not overlook the fact that latest known edrioasteroids are some 300 million years older than earliest recorded psolids. This is a great hiatus in the fossil record.

Comparative studies of pharyngeal skeletal elements found in the dendrochirote holothuroids suggest that original representatives of the class must have possessed external ambulacra formed by modified thecal plates similar to those of edrioasteroids illustrated by Kesling & Mintz (10) (Fig. 95,*1*). Seemingly, the evolution of large dendrochirote tentacles required the existence of a protective mechanism suited to allow withdrawal of them into the body. This was achieved by the introvert, which comprises a telescoping of the anterior part of the body with retraction induced by muscles derived from the radial muscle group. Evolution of the introvert implies a conversion of the original external ambulacral areas of the test into internal structures surrounding the pharynx and these structures serve for insertion of the retractor muscles. In primitive dendrochirotes the pharyngeal skeleton is still recognizable (in the opinion of Pawson and Fell) as equivalent to the ambulacral plates of an edrioasteroid, but in most surviving holothurians the mechanism is very much reduced or vestigial (Fig. 95,*2-8*).

Although the edrioasteroids (Fig. 94,*2*; 95,*1*) adopted a habit similar to that of psolid holothurians (Fig. 92), as previously noted, their ambulacral tracts remained external, instead of sinking inward, and this permitted an alternative method of feeding suited to the pelmatozoan way of life. The feeding mechanism was provided by the whole complex of tube feet distributed along the five ambulacra which extended outward from the mouth as meridians on globoid forms (see Fig. 102,*2a,b*) but confined to the upper (oral) surface generally, as on discoid forms (Fig. 94,*2*). Each ambulacrum carried a median groove that was bordered on either side by tube feet which must have functioned in manner similar to those of crinoids, waving about in the water so as to entrap small organic particles on their mucous surface, sweeping this food inward to the mouth along food grooves of the ambulacra. Fell (7) has stressed that no introvert structure evolved in edrioasteroid stocks, and consequently tentacles of dendrochirote type doubtless never developed.

In the holothurian line, radial water vessels early became concentrated into internal canals and this occurred also in postechinocystitoid echinoids. Hence, among forms that adopted pelmatozoan habits a preexisting dendrochirote nutritive mechanism inevitably was demanded, and in holothurians external ambulacra never were involved. The fact pointed out by Bassler (1) that a fully plated ventral sole is retained by Cambrian edrioasteroids may be taken as evidence of an originally spherical form of the test, and the ambulacra of these early members of the class were more simple than in later forms. The development of such pelmatozoan features as cover plates along the ambulacra indicates specialization analogous to that seen in somasteroids and crinoids (7).

A comparison of dissections of psolids with edrioasteroids suggests some reasonable inferences concerning the internal anatomy of the latter. Lacking evidence to the contrary, Fell (7) assumed that edrioasteroids had a gonad placed in the posterior *(CD)* interray. In psolids the gonopore lies

---

(Fig. 96. *Continued from facing page.)*

fied from Fell). [Explanation: Highly conjectural reconstructions of some genera indicated by star. Numbers are for identification of genera selected to illustrate class and order groups, forms known only as fossils in post-Paleozoic part of chart marked by underlined numbers. 1, *Volchovia;* 2, *Sollasina;* 3, *Rhenosquama;* 4, *Helicoplacus;* 5, *Eothuria;* 6, *Aulechinus;* 7, *Bothriocidaris;* 8, *Echinocystites;* 9, *Palaechinus;* 10, *Pholidocidaris;* 11, *Lepidocidaris;* 12, *Archaeocidaris;* 13, *Miocidaris;* 14, *Cidaris;* 15, *Holaster;* 16, *Echinocardium;* 17, *Nucleolites;* 18, *Holectypus;* 19, *Arachnoides;* 20, *Colobocentrotus;* 21, *Diadema;* 22, *Thuroholia;* 23, *Protocaudina;* 24, *Stichopus;* 25, *Deima;* 26, *Thallatocanthus;* 27, *Molpadia;* 28, *Chiridota;* 29, *Ypsilothuria;* 30, *Calclamna;* 31, *Cucumaria;* 32, *Placothuria;* 33, *Lepidopsolus;* 34, *Psolus;* 35, *Cyclocystoides;* 36, *Cystaster;* 37, *Cyathocystis;* 38, *Walcottidiscus;* 39, *Isorophus;* 40, *Agelacrinites;* 41, *Hemicystites;* 42, *Lepidodiscus;* 43, *Hemicystites;* 44, *Edrioaster;* 45, *Stromatocystites.*]

on the introvert, just behind the mouth. The corresponding position in edrioasteroids is that in which a pore is known to occur, although hitherto this pore has been supposed to be a hydropore. Since psolids respire (at least in part) by means of respiratory trees, it seems likely that similar trees occurred in edrioasteroids, and it is probable that the hydropore was internal, as in dendrochirote holothurians. Irrespective of these inferences, in FELL's opinion the Edrioasteroidea should be recognized as bona fide members of the subphylum Echinozoa, exhibiting various pelmatozoan features no more fundamental than the same features in psolid holothurians, where undoubtedly they constitute purely secondary responses to demands of a sessile habit.

Inferred relationships of the echinozoan classes are illustrated approximately and very diagrammatically in Figure 96. The reconstructions of forms most highly subject to conjecture are prominently marked by asterisks.

Subsequent chapters in this volume of the *Treatise* contain discussions of varying length which express the views of their authors on distinctive features of the several classes of Echinozoa, with appropriate morphological comparisons within and between the classes. Also, more detailed consideration of evolution and phylogeny is presented. Important literature is cited in a composite list of references for Echinoidea and in separate lists for other classes.

REFERENCES

Bassler, R. S.

(1) 1935, *The classification of the Edrioasteroidea:* Smithsonian Misc. Coll., v. 93, no. 8, p. 1-11, pl. 1.

Durham, J. W.

(2) 1964, *Occurrence of the Helicoplacoid[e]a (Echinodermata):* Geol. Soc. America, Spec. Paper 76, p. 52.

———, & Caster, K. E.

(3) 1963, *Helicoplacoidea: a new class of echinoderms:* Science, v. 140, p. 820-822, text-fig. 1.

Fell, H. B.

(4) 1962, *A classification of echinoderms:* Tuatara, v. 10, p. 138-140.

(5) 1963, *The phylogeny of sea-stars:* Royal Soc. London, Philos. Trans., ser. B, v. 246, p. 381-435, pl. 50-51, text-fig. 1-16.

(6) 1963, *The evolution of the echinoderms:* Smithsonian Inst., Ann. Rept. 1962, p. 457-490, pl. 1-3.

(7) 1965, *Early evolution of the Echinozoa:* Harvard Univ., Museum Comp. Zoology, Breviora, v. 219, p. 1-17, text-fig. 1-13.

(8) 1965, *The Echinodermata:* in PARKER & HASWELL, Textbook of Zoology, 7th ed. (in press), Macmillan (London, New York).

Jaekel, Otto

(9) 1918, *Phylogenie und System der Pelmatozoen:* Paläont. Zeitschr., v. 3, no. 1, p. 1-128, text-fig. 1-114.

Kesling, R. V., & Mintz, L. W.

(10) 1960, *Internal structures in two edrioasteroid species, Isorophus cincinnatiensis (Roemer) and Carneyella pilea [pileus] (Hall):* Michigan Univ., Museum Paleontology, Contrib., v. 15, no. 14, p. 315-348, pl. 1-9, text-fig. 1-4.

Matsumoto, Hikoshichirô

(11) 1929, *Outline of the classification of the Echinodermata:* Tohoku Imper. Univ., Sci. Repts., ser. 2 (Geology), v. 13, no. 2, p. 27-33.

Pawson, D. L., & Fell, H. B.

(12) 1965, *A revised classification of the dendrochirote holothurians:* Harvard Univ., Museum Comp. Zoology, Breviora, v. 214, p. 1-7.

Ubaghs, Georges

(13) 1961, *Sur la nature de l'organe appeletige ou pedoncle chez les Carpoides Cornuta et Mitrata:* Acad. Sci. Paris, Comptes Rendus, v. 253, p. 2738-2740, text-fig. 1-5.

Whitehouse, F. W.

(14) 1941, *The Cambrian faunas of northeastern Australia; Part 4, Early Cambrian echinoderms similar to the larval stages of Recent forms:* Queensland Museum, Mem., v. 12, pt. 1, p. 1-28, pl. 1-4, text-fig. 1-9.

Zittel, K. A. von

(15) 1900 [1896-1900], *Text-book of palaeontology* (transl. and ed. C. R. Eastman), v. 1, ed. 1, 706 p., 1476 text-fig. (Echinodermata, p. 122-250, text-fig. 217-402, publ. 1896), Macmillan & Co., Ltd. (London, New York).

HOMOLOGY OF ECHINOZOAN RAYS

By RAYMOND C. MOORE and H. BARRACLOUGH FELL
[University of Kansas; Harvard University]

INTRODUCTION

An important subject in the study of Echinozoa relates to the homology of their five ambulacral rays as seen both in different classes of the subphylum and in representatives of other echinoderm subphyla. If corresponding parts of the theca in various groups can be distinguished reliably, morphological comparisons are facilitated and advances may be made in solving questions of phylogeny and evolution. Are uniform means of designating homologous skeletal parts possible, and if so, to what extent is it desirable to adopt them?

As a first step, consideration must be given to orientation. The oral face of most echinozoans is directed downward, as in the myriad kinds of echinoids and the small group of ophiocistioids. In others, such as the edrioasteroids and cyclocystoids, it is pointed upward, and in the holothurians it is directed sideward. The mode of life of helicoplacoids may have resembled that of holothurians, moving about with the long axis of the body parallel to the substrate, or alternatively the long axis of the body may have been subvertical most of the time. An anterior extremity is definable in the holothurians, irregular echinoids, and seemingly in the helicoplacoids, whereas anterior and posterior directions are ill-defined or determined somewhat arbitrarily by conventions in the regular echinoids, edrioasteroids, ophiocistioids, and cyclocystoids. Nevertheless, a plane of bilateral symmetry can be recognized in all, and its orientation is identifiable as normal to the substrate. The part of the plane that coincides with one of the ambulacral rays or that approximately marks its median position is considered anterior and the part that bisects an interray is classed as posterior, because in many, if not most, forms this interray contains the anus. Right and left sides then can be differentiated but with opposite signification in forms having the oral face upward as compared with those in which it is downward. Partly for this reason, right and left as directional terms are not favored, even though formerly much used for crinoids and other Crinozoa and though judged by FISCHER (1952) to be a preferred method in referring to parts of echinoids.

The Lovén system of marking the rays and interrays of echinoids, explained and illustrated in the subsequent chapter on echinoid morphology (p. *U*220), has been employed very widely by specialists and in our opinion wisely has been adopted in the *Treatise* for descriptions and figures of members of the Echinoidea. This is because its application to both regular and irregular echinoids is trustworthy and unambiguous (Fig. 97). In this system the plane of bilateral symmetry passes through the anterior ray (designated III) and posterior interray (designated 5), which in irregular echinoids contains the anus. In clockwise order on the oral surface interrays (Arabic numbers) and rays (Roman numbers) are 5 (posterior), I, 1, II, 2, III (anterior), 3, IV, 4, and V. The arrangement of rays in many irregular echinoids shows a well-defined grouping of the three anterior rays, forming a so-called trivium, and the two posterior rays, making an opposed bivium (Fig. 97,*D*,*E*). Between rays of the bivium is interamb 5 containing the anus. The plane of bilateral symmetry, which coincides with the Lovén plane, is emphasized, whereas this is much less readily discerned in the regular echinoids, among which it is positively determinable by the symmetrical pattern of large and small plates of rays at the peristomial margin and by location of the madreporite in interamb 2.

A trivium and bivium are defined by junction of the oral plates of many crinoids (e.g., *Haplocrinites*, Fig. 98,*1*) and by the pattern of ambulacral grooves or plate rows on the tegmen of numerous crinoids (e.g., *Cyathocrinites*, Fig. 98,*2*; modern *Antedon*, and others). In our view it is highly significant that the grouping of pelmatozoan rays in threesome and twosome is not by any means at random, for as in irregular echinoids the median ray of the trivium invariably coincides with the anterior part of the plane of bilateral symmetry and likewise the interray enclosed by the bivium

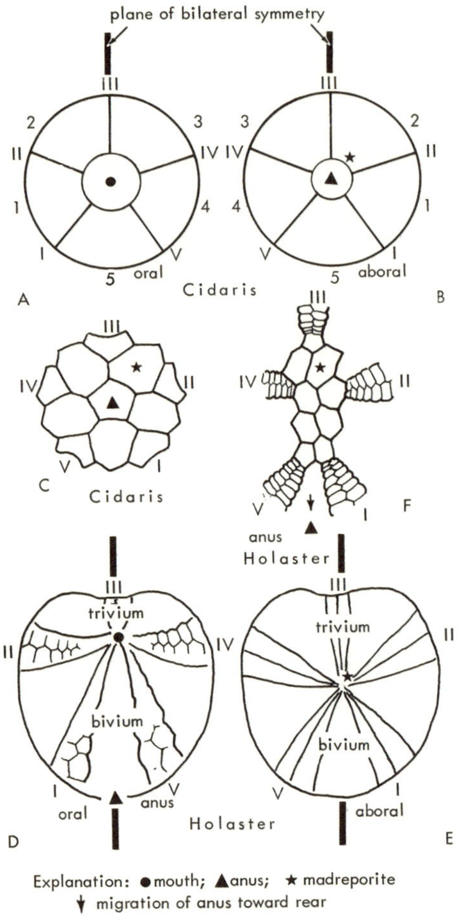

FIG. 97. Lovénian numerical notations for rays and interrays of regular echinoid *(Cidaris, A-C)* and irregular echinoid *(Holaster, D-F)*, latter showing trivium and bivium; enlargements of apical systems shown in *C* and *F*.

contains the anus. It may also contain other small openings identified as hydropore, gonopore, or hydrogonopore. If the Echinozoa, as well as Asterozoa, have been derived from ancestral echinoderms that gave rise also to the Crinozoa, it is not surprising to find in the different subphyla common features of bilateral symmetry, and this should aid in reaching trustworthy conclusions concerning homologies. Reasonable proof of the descent of members belonging to one subphylum from stocks classified in another is not demanded, for the purpose here is simply to find common denominators. The anteroposterior plane of bilateral symmetry is thought to be paramount among these.

A system for designating the rays and interrays of echinozoans, similar to that devised by LOVÉN in being somewhat arbitrarily defined, uses capital letters instead of numerals. This was introduced by P. H. CARPENTER (1884) for identifying main divisions of the thecal skeleton of crinoids and the procedure has been found equally well suited to other pelmatozoans. Various authors have extended the Carpenter system to echinozoans and even to asterozoans (e.g., CUÉNOT, 1948; HYMAN, 1955; AILSA CLARK, 1963), but in a manner decidedly open to question. As applied to crinoids (Fig. 98), the Carpenter letters are very simple, unambiguous, and convenient. The ray coinciding with the anterior part of the plane of bilateral symmetry (opposite to the interray containing the anus) is marked by *A*, and then on the oral surface other rays are designated in clockwise succession by *B*, *C*, *D*, and *E*. Interrays can be indicated in terms of their bordering rays, as *AB*, *BC*, etc. To echinoderm workers this is elementary; they do not need to be reminded that in aboral views of crinoids the sequence of Carpenter letters runs counterclockwise.

CARPENTER LETTERS APPLIED TO ECHINOIDS

As previously stated, the Lovén system of ray and interray designation has been adopted as "official" in the *Treatise* for application to the Echinoidea. Even so, correlation of the Lovén numerals with Carpenter letters needs to be considered in order to indicate homologies between members of this class and representatives of the Crinozoa, as well as other echinozoan classes and possibly the Asterozoa. Crinoids, blastoids, edrioasteroids, echinoids, holothurians, ophiocistioids, asteroids, and ophiuroids have been depicted with parts marked by letters of the Carpenter system, and wherever this is done erroneously by presumed authorities, they and others are led to false conclusions concerning homologies.

What guides are available for determining homologous parts of echinoderm tests? Among forms that display entirely perfect pentameral symmetry of the skeleton—fos-

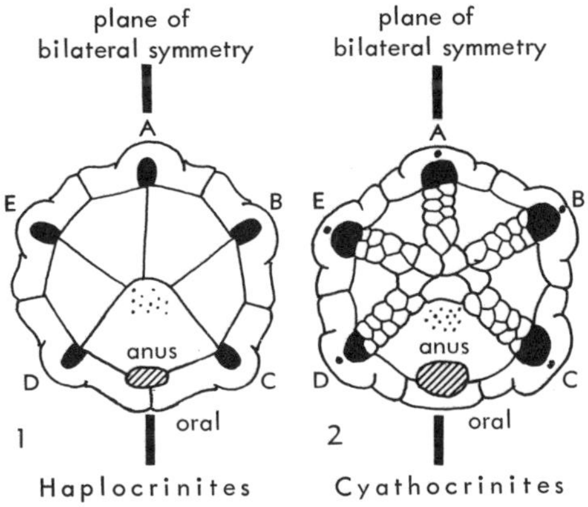

FIG. 98. Oral views of crinoids showing bilateral symmetry and designation of rays by letters of the Carpenter system.

sils rarely if ever provide evidence of soft parts—no clues are provided for distinguishing one ray or interray as different from others. The *A* ray cannot be discriminated and accordingly others are unidentifiable. Such perfect symmetry is found in some crinozoans that in them it is possible only to determine oral (ventral) and aboral (dorsal) sides of the test. The vast majority of echinoderms exhibit one or more asymmetrical features, which in turn may lead to recognition of bilateral symmetry superposed on the basic pentameral symmetry of the phylum. Here we may employ as guides for recognizing significant departures from perfect pentameral symmetry any single excentrically located structure (e.g., mouth, anus, hydropore, gonopore, madreporite, etc.), any distinctive skeletal element (e.g., posterior oral plate differing in shape and size from other orals), and any groups of such features (e.g, insert oculars on one side of periproct of echinoid, exsert oculars on other sides). Greatest in value is modification of thecal outline and the pattern of rays and interrays (e.g., trivium, bivium) in some manner that clearly defines overriding bilateral symmetry with respect to the anteroposterior plane perpendicular to the oral and aboral surfaces of the theca. This may be accompanied and accentuated by such surface features as fascioles, grouped areas of specialized spines, and tracts of particular kinds of pedicellariae. In varying degree this pattern is clearly evident in all of the subphyla, possibly excepting the Homalozoa, and it is discernible in nearly all echinoderm classes. Reliable recognition of ray homologies depends on correct correlation of the available guides.

In order to determine the correct application of Carpenter letters to echinozoans, attention may be directed first to regular echinoids and later extended to the irregular forms, inasmuch as all kinds of irregular echinoids indubitably constitute modifications derived from primordial regular ancestors, chiefly represented by the Paleozoic Cidaroida. Then, we will turn to the classes Edrioasteroidea, Cyclocystoidea, Ophiocistioidea, and Holothurioidea. The spirally twisted theca of the Helicoplacoidea lacks evidence of differentiated rays and therefore is not considered here.

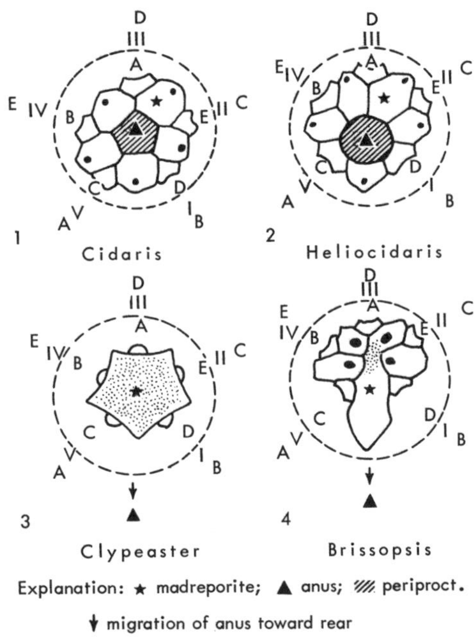

FIG. 99. Apical systems of regular echinoids (*Cidaris, Heliocidaris, 1,2*) and irregular echinoids (*Clypeaster, Brissopsis, 3,4*) showing Lovénian numerals for designation of rays accompanied on inner side by Carpenter letters recognized by *Treatise* and on outer side different placement of these letters according to common usage of authors. The diagrams show (in *1*) nearly perfect radial symmetry, (in *2*) incipient bilateral symmetry marked by exsert oculars II, III, IV and insert oculars I and V, (in *3*) central "monobasal" madreporite, and (in *4*) rearward migration of madreporite.

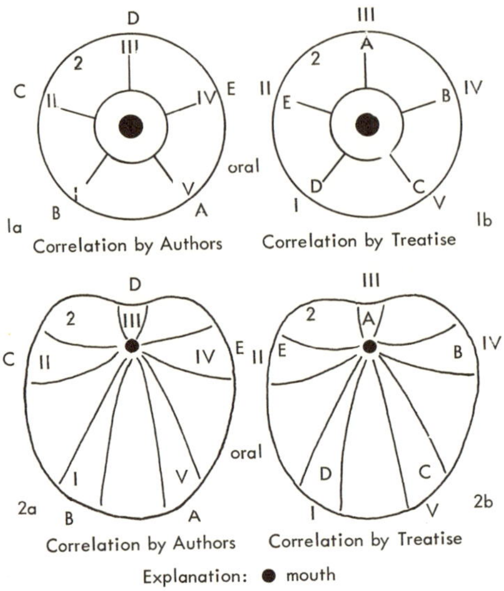

FIG. 100. Oral views of regular and irregular echinoids showing divergent correlation of Carpenter letters with Lovénian numerals by authors and by *Treatise*.

Regular echinoids have been regarded universally as a group that differs markedly from nearly all irregular echinoids. Their nearly perfect pentaradiate symmetry is disturbed only by differentiation of one of the genitals in the ocular ring as a madreporite. It lies next to the periproct in the Lovén interamb 2 (Fig. 97,*C*; 99,*A*). Regular urchins creep over the substrate with any ray forward and they can reverse the direction of their movement without turning around (HYMAN, 1955, p. 550), although some have a very slight preference for locomotion with the III ray in front. Application of the Lovén numerals to the rays and interrays of the regular echinoids has been accepted by all on the basis of the single clue furnished by position of the madreporite, for if this element of the theca prevailingly (not quite universally) occurs in interamb 2 of irregular echinoid tests, it is entirely reasonable to infer the same location for it in the tests of regular echinoids. Then, other Lovén numerals for the rays and interrays can be assigned with measurable confidence (Fig. 97,*A-C*). This is not the last word with respect to the regular echinoids, however, for in many of them signs of incipient bilateral symmetry corresponding to that marked by the anterolateral axis of irregular echinoids (ray III, interray 5) can be recognized, though it has been overlooked as a significant feature by specialists (Fig. 97,*D-F*; 99,*2*).

Granting that application of the Lovén numerical designations of rays and interrays in irregular and regular echinoids is the same, as agreed by all authors who have employed this system, substitution of Carpenter letters for the numerals is unacceptable—certainly so if the letters are placed in the manner adopted by such authors as CUÉNOT (1948), HYMAN (1955), and AILSA CLARK (1963). The disposition of letters for rays shown in Figure 100,*1a,2a* is based on the assumption that interamb 2 (Lovén), which typically contains the madreporite, corresponds to interray *CD* (Carpenter), which typically contains the excentric anus and may also have a hydropore, gonopore, or hydrogonopore. No account is taken of pervasive bilateral symmetry defined by the anteroposterior axial plane, expressed not only by the location of various excentric structures but commonly by characters of the whole skeleton. In our view correct placement of Carpenter letters, identical in crinozoans and echinozoans, is unequivocally indicated by relationships to the fundamental anteroposterior bilateral symmetry. Thus *A* (Carpenter) corresponds to III (Lovén), rather than to V, and the posterior interray *CD* (Carpenter) is equivalent to interamb 5 (Lovén), and not to interamb 2 (Fig. 100,*1b,2b*). The importance of these conclusions with respect to ray homologies is obvious, especially in connection with studies of echinoderm evolution and phylogeny.

With an initially skeptical approach, one of us (FELL, who is author of the *Treatise* chapters on cidaroids and other echinoid groups) undertook a special study of the regular echinoids aimed at finding out whether this group furnishes independent evidence that the axis passing through amb III and interamb 5 corresponds to the anteroposterior axis defined by ray *A* and interray *CD* of the Carpenter system. Some of the results were communicated to MOORE in a letter (7 January 1963), excerpts from which follow.

"Propositions that seem pertinent initial statements are the following. (1) Lovénian symmetry is a universal characteristic of echinoids, with only very minor exceptions. (2) *Bothriocidaris,* if not an echinoid in strictest sense, is so close to one that all writers who have referred to it in the past decade are unanimous in regarding it as best classified among echinoids and at least susceptible to analysis as an echinoid. [Evidence published by MYANNIL (1962) indicates beyond doubt that *Bothriocidaris* is an echinoid.] (3) *Bothriocidaris* exhibits full Lovénian symmetry. (4) Cidaroids, as the only known group of Paleozoic echinoids to cross into the Mesozoic, must be the ancestors of all post-Paleozoic echinoids. (5) The plane of bilateral symmetry of all post-cidaroid echinoids passes through amb III and interamb 5 (Lovénian notation). It is required to ascertain whether the post-cidaroid plane of symmetry presents a stable relationship to other structures in all known echinoids including Paleozoic forms in which the Lovénian plane has proved recognizable, and whether the madreporite (or hydropore) presents a similar stable relationship.

"My approach to this problem is outlined in succeeding paragraphs. (1) It is self-evident that if any morphological proof of the fundamental character of the plane of bilateral symmetry in post-cidaroid echinoids exists, such proof can only be sought in the apical system of regular echinoids, for it is in this region that the first observable signs of movement of the anus into an interambulacrum are found, and it is such movement that enables us to recognize the anteroposterior axis. (2) LOVÉN established his law on the basis of the plate arrangement of the ambulacra at the peristome in spatangoids, among which bilateral symmetry is conspicuous. He extrapolated backward on the basis of the position of the madreporite in spatangoids, showing that if a regular echinoid is oriented with the madreporite in the same relative position, then the only possible plane of symmetry yielding the Lovénian sequence is that passing through the apex, amb III, and interamb 5, which defines the anteroposterior axis and plane in spatangoids and also exists in regular echinoids. By extending his inquiry to the apical region, LOVÉN was able to show that a bilateral symmetry generally exists in adapical, as well as adoral, parts of the test, and that it corresponds to the plane derived by him from adoral plate arrangement. (3) Cidaroids generally have been regarded as exhibiting aborally an almost perfect radial symmetry, not susceptible to Lovénian analysis. Therefore, it has been inferred that the anteroposterior axis and plane is a feature evolved in post-cidaroid groups. Since the madreporite is necessarily to be regarded as a pre-cidaroid structure, its significance as a morphological "marker" has been given much importance, and as others are lacking, it has been used as the main reference point in attempting to correlate echinoid orientation with that used in other groups.

"I have investigated the symmetry of the apical system of regular echinoids as part of my *Treatise* assignment, paying special attention to signs of incipient bilateral symmetry. I have reached the judgment that the common method of describing the apical system of any regular echinoid by statements such as 'oculars I and V insert' actually puts the cart before the horse. As pointed out by MORTENSEN in the first volume of his monograph on Echinoidea, the embryonic cidaroid apical system is monocyclic, with all oculars inserted between genitals (one comprising the madreporite) so as to form a single ring. With increasing growth the genitals overtake the oculars, intruding upon the periproct so as to exclude the oculars effectively from this region, thus converting the apical system into a dicyclic type. The dicyclic type clearly is secondary to the monocyclic.

Post-cidaroid echinoids generally inherit a dicyclic type of apex as their juvenile pattern, but commonly they tend to depart from it (as a tertiary modification) by developing one or more insert oculars in adults. Why is this so? The answer, in my opinion, is found by referring to such groups as the pygasteroids, nucleolitids, and others in which the first signs of migration of the anus out of the apex can be observed. The postembryonic development of the surviving nucleolitid *Apatopygus* shows in the life history of the animal how the anus leaves the apex, beginning to descend interamb 5, with new plates introduced between the periproct and apical system. No one can doubt that this is an example of recapitulation and that it was the way in which the so-called irregular echinoids arose, with conspicuous bilateral symmetry coming to involve the whole test. I interpret the 'insert oculars I and V' to mean, not that these oculars moved back inward so as to adjoin the periproct, but that in virtually all regular echinoids a constant tendency exists for the periproct to *move out* toward oculars I and V. Resorption of the neighboring genitals occurs and the emargination of their adapical edges is the real reason why oculars I and V come into contact with the periproct. Hence, it is easy to understand why irregular echinoids arose from more than one stock of regular echinoids, as stressed by DURHAM & MELVILLE (1957); it was because the periproct for some reason tends constantly to move toward oculars I and V, which is rearward in spatangoids and all other irregular echinoids. Accordingly, any tendency in a regular echinoid for oculars I and V to become insert must be treated as prima-facie evidence of an anteroposterior axis.

"With these considerations in mind, I have reexamined the apical systems of all cidaroids available to me, studying also the photographic plates in MORTENSEN's monograph and plates in my own earlier papers. Soon, most decisive witness to the existence of an anteroposterior axis in the cidaroids became evident, despite my earlier belief that this

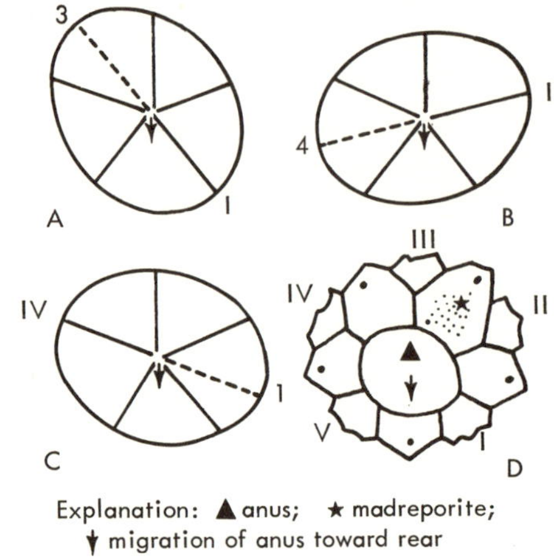

FIG. 101. "Misshapen" regular echinoids of the Echinometridae indicated by outlines of thecae in aboral views showing elongation in three different directions *(A-C)* and enlarged apical system belonging to these with exsert and insert oculars which denote bilateral symmetry with respect to the anteroposterior plane.

was not so. A recurrent tendency of oculars I and V toward *failure to become exsert* during growth constitutes evidence that this axis is present in many genera of such diverse subfamilies of the Cidaridae as the Histocidarinae (most primitive surviving group), Goniocidarinae, Stereocidarinae, Rhabdocidarinae, and Ctenocidarinae, as well as other families. Therefore, I do not doubt that it is a universal features of the cidaroids, even though expressed sporadically, some adults showing insert oculars I and V, others with all oculars exsert but I and V nearest to the periproct, and still others with evenly disposed insert or exsert oculars.

"If differences among the oculars are due to purely chance variations during growth, all oculars should be affected equally, but this is not the case. An unmistakable bias—almost an exclusive one—caused only oculars I and V to become insert, or rather, to fail in being exsert. Unavoidably, therefore, I must conclude that an anteroposterior axis of symmetry already exists in the cidaroids. It is defined by the recurrent tendency of oculars I and V to remain next to the periproct, or expressed otherwise, by the recurrent tendency of the cidaroid periproct to move rearward in the direction of interamb 5. That this anteroposterior axis of the cidaroids exactly matches the Lovénian axis on the oral surface cannot be emphasized too strongly, for it means that we are dealing here with a true plane of symmetry identical with the one found in all post-cidaroid echinoids [Fig. 99].

"A test now can be applied to the reasoning which has been outlined. Certain regular echinoids (e.g., Parasaleniidae, Echinometridae) are 'misshapen' in exhibiting a bilateral symmetry of the whole test expressed by markedly oval outlines in many species when viewed from the oral or aboral sides. In the Echinometridae (MORTENSEN, v. 3, pt. 3, p. 278, fig. 130) the axis of elongation plainly differs from that of spatangoids and other irregular echinoids, for in different forms it is observed to coincide with amb I and interamb 3, or with amb II and interamb 4, or with amb IV and interamb 1 [Fig. 101]. Now if the bilateral symmetry associated with elongation of the test in any of these directions is as fundamental as that postulated for symmetry with respect to the amb III and interamb 5 axis, the insert oculars ought then to be located in the appropriate ambulacra—not in I and V. *But this is not so* [Fig. 101,*D*]. Instead, the echinometrids are entirely characterized by insertion of oculars I and V, just as in other cidaroid groups. Hence, the slightly displaced periproct completely ignores the new symmetry along axes other than that marked by amb III and interamb 5, remaining faithful to this latter. Thus, I conclude that the amb III-interamb 5 plane of symmetry is a fundamental feature of all orders of echinoids from the Cidaroida onward. It was already present in at least one of the Paleozoic orders, for we know that cidaroids as defined in the strictest sense range back into the Permian (Miocidaridae), and the Archaeocidaridae, which generally are classed as cidaroids, extend back to the Lower Carboniferous and possibly to the Silurian.

"Including the Bothriocidaroida, three orders of Paleozoic echinoids in addition to the Cidaroida are discriminated. The Lovénian law seems to be observed exactly in *Bothriocidaris,* as stressed by JACKSON, HAWKINS, MORTENSEN, and DURHAM & MELVILLE. Fossils now available to show characters of the Paleozoic orders Echinocystitoida and Palaechinoida are too fragmentary to allow determination of the presence in them of Lovénian symmetry. The archaic *Bothriocidaris* (Ord.) at least already exhibited the same anteroposterior plane of symmetry that is manifested in the Cidaroida and in post-cidaroid groups, as indicated by the potential, incipient, or consummated migration of the anus into interamb 5.

"What about the madreporite, to which so far relatively little notice has been given? In *Bothriocidaris,* which we might reasonably expect to show such a structure in interamb 5, it is located rather in one of the ambulacra, not even in interamb 2, as customarily in echinoids. By definition, a posterior direction is indicated by the location of the anus, and observation of other echinoderms indicates that the hydropore tends to move into the interambulacrum containing the anus. The hydropore itself thus may be regarded as a marker of posterior direction, though of secondary value. When we trace the post-Paleozoic history of the echinoid hydropore (madreporite), we find that it fulfills expectations, albeit belatedly and in a most uncertain, devious manner, experimenting, so to speak, with various locations but ultimately yielding to interamb 5 after crossing the middle of the apical disc. It set out from genital 2 (seemingly reached as a post-bothriocidaroid translocation), toyed briefly with genital 4 (with pores developed on both genitals 2 and 4, as seen in juvenile *Apatopygus*, MORTENSEN, Vid. Medd. 73, p. 187, fig. 20), or invaded all of the genital plates, producing pores in each of them simultaneously (e.g., *Discoidea*, MORTENSEN, monograph) and even converted all of them into a single compound 'monobasal' plate at the apex [Fig. 99,*3*], or it sailed across the apex into interamb 5 with a long trailing lobe joined to its old host (genital 2) [Fig. 99,*4*]. The last-mentioned arrangement represents the so-called ethmolytic condition of spatangoids. Finally, the madreporite may swallow up all of the residual genitals, pushing a long process into the posterior interambulacrum (e.g., *Palaeopneustes, Heterobrissus*, and other spatangoids of more specialized nature).

"This brief sketch shows that the history of the hydropore (madreporite) in echinoids is one of continuing migration, with interamb 2 its longest occupied home and interamb 5 its ultimate destination. As a trigonometric landmark, the madreporite is hopeless. As an indicator of posterior direction and position, the anus is far superior in guidance to correct orientation, for its morphologic relationships are entirely consistent. By its influence on the behavior of oculars I and V, the anus very early indicated the tendency to migrate into interamb 5, enabling us to recognize the anteroposterior axis and plane of bilateral symmetry in many of the oldest regular echinoids. The great adventure in evolution of the echinoid anus was its slide from dead-center location on the aboral surface to a place on the underside of the test within sight of the mouth, where finally it halted futile pursuit of the mouth, which was seeking to escape forward.

"In summation, it seems to me that inescapably we must recognize the anteroposterior plane passing through amb III and interamb 5 of Lovénian notation as the fundamental plane of symmetry in all echinoids. Therefore, corresponding notations of the Carpenter system equate amb *A* with amb III, interamb *AB* with interamb 3, amb *B* with amb IV, and so on. Interamb *CD* is not the equivalent of interamb 2, as commonly supposed by authors, but corresponds to interamb 5."

CARPENTER LETTERS APPLIED TO EDRIOASTEROIDS

The ambulacral rays of some edrioasteroids, for example as seen on the globose test of *Cystaster* (Fig. 102,*2*). are straight and they diverge radially in nearly perfect pentamerous manner. Actually, the interray containing the prominent anal pyramid of *Cystaster* is wider than others. Bilateral symmetry with respect to the vertical plane that bisects this interamb and that coincides with the opposite ambulacrum is clearly evident. The oral surface is directed upward, and Carpenter letters are applicable without any question, beginning with *A* for the ray in the plane of anteroposterior bilateral symmetry and proceeding clockwise for designation of others in alphabetical sequence. An aperture next to peristomial plates at summit of the test is located in interray *CD*, like the anus. It has been distinguished as a hydropore by KESLING (1960) but designated noncommittally as "third aperture" by REGNÉLL (p. *U*150). If the supposition

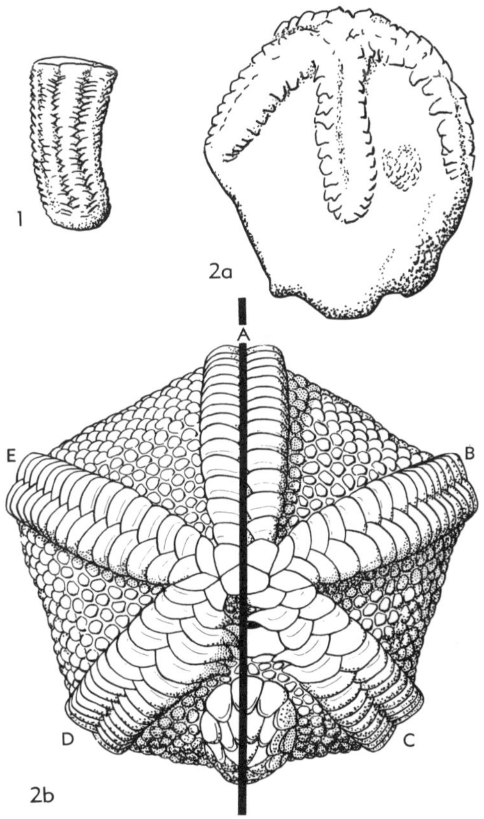

FIG. 102. Edrioasteroids, showing side view of cylindroid form *(1)* and *D*-ray side view and oral view *(2a,b)* of globose form, anteroposterior plane of bilateral symmetry strongly defined in all, not to scale (*1*, from Aurivillius; *2a*, from Jaekel; *2b*, from Kesling).

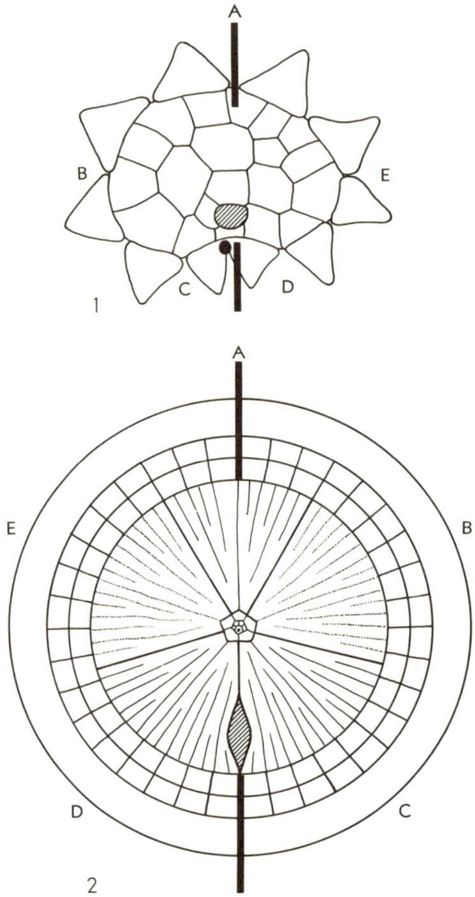

FIG. 103. Ophiocistioidea and Cyclocystoidea showing anterodorsal plane of bilateral symmetry.——*1. Volchovia* (ophiocistioid), aboral view showing application of Carpenter letters to rays, reconstr., ×0.5 (from Gekker).——*2. Cyclocystoides*, oral surface, reconstr., showing Carpenter letters, ×2.25 (from Kesling).

expressed by FELL & MOORE (p. *U*117) is correct, that respiration of the edrioasteroids probably resembled that of holothuroids, utilizing internal respiratory trees not preservable in fossils, this opening may very well be a gonopore. Whatever its physiological function may have been, this structure fits in with the *A-CD* plane of symmetry.

The same symmetry is definable on the oral surface of cylindroid edrioasteroids (e.g., *Pyrgocystis*, Fig. 102,*1*) and discoid forms, which include a majority of genera belonging to the class. The latter are characterized by more or less strongly curved ambulacra, as well known. Bilateral symmetry defined by the anteroposterior plane is less evident, but nonetheless uniformly marked by location of the anus in interray *CD*.

CARPENTER LETTERS APPLIED TO OPHIOCISTIOIDS

UBAGHS (p. *U*176) has reported that notation of rays and interrays of the ophiocistioids presents no difficulty or ambiguity. In *Volchovia* (Fig. 103,*1*), for example, the anteroposterior plane of bilateral symmetry is defined by the arrangement of plates on the aboral surface and position of the anus, supplemented by presence in the same interray of a small aperture doubtfully identified as a hydropore, gonopore, or hydrogonopore. Carpenter letters have been employed by UBAGHS, therefore, in describing fossils of this class.

CARPENTER LETTERS APPLIED TO CYCLOCYSTOIDS

Although most specimens of cyclocystoids are not very well preserved, enough is known concerning structure of their tests to establish definitely nearly perfect pentameral symmetry of the many-branched rays (Fig. 103,*2*). No system of notation for the rays has been adopted by authors, however, even though a vertical plane of bilateral symmetry transecting the nearly flat discoid test is recognized. This coincides with the mid-line of one of the rays and an opening in the opposite interray that must be the anus. Accordingly, the ray just mentioned is here defined by the letter *A* of the Carpenter system and other rays can then be distinguished in customary manner. Three cyclocystoid genera have been described, but in the view of KESLING (p. *U*188) they are synonymous and in the *Treatise* all are included in *Cyclocystoides*.

CARPENTER LETTERS APPLIED TO HOLOTHURIANS

The holothurians prevailingly comprise cylindroid echinozoans, as indicated by their common name sea cucumbers; some are decidedly wormlike in form and others thick discoid to globoid. They differ in mode of life from echinoids and eleutherozoan

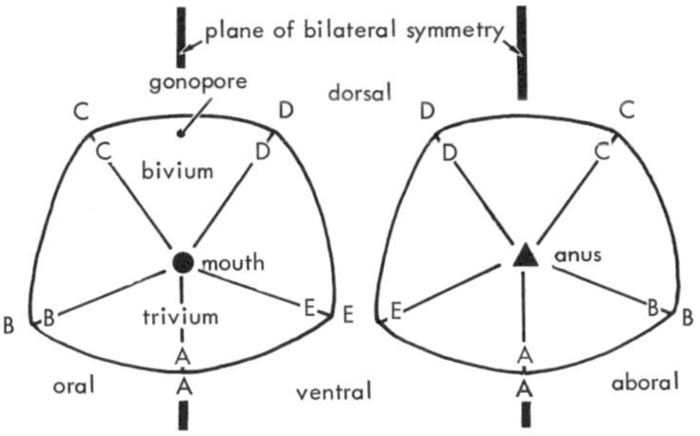

FIG. 104. Diagrammatic oral and aboral views of holothurian showing plane of bilateral symmetry and Carpenter letters for designation of rays belonging to trivium and bivium.

echinoderms generally in displaying a strongly marked fore and aft orientation, with the mouth at one extremity and the anus at or near the other. The animals crawl or burrow with one of their sides lowermost and the opposite side directed upward. Accordingly, the lower side is designated as ventral and the other as dorsal. Three rays on the lower side are differentiated as a trivium and the opposite two comprise a bivium, and between rays of the bivium a gonopore commonly is recognized near the anterior end of the body.

Authors who have employed Carpenter letters for the different rays of holothurians agree in adopting *A* for the median ventral ray, others then being identified in clockwise sequence around the mouth. This results in application of *C* and *D* to rays of the bivium (Fig. 104). The tenuous basis for this agreed application of Carpenter letters seems to be the premise that the gonopore corresponds to the genital plate of the echinoid apical disc which prevailingly (but not exclusively) includes the madreporite, and on the additional premise that the interray containing the echinoid madreporite corresponds to the Carpenter *CD* interray. The first-mentioned premise may be correct, whereas the second is judged by us to be incorrect. The truly significant, and therefore basic, consideration is that the obvious vertical plane of bilateral symmetry in the holothurian body which coincides with the median ventral ray (middle one of the trivium) and which bisects the bivium is identical in its relationships to the vertical anteroposterior plane of symmetry recognized in all echinoids (now including regulars as well as irregulars), in crinoids, and in nearly all other crinozoans. The placement of the holothurian gonopore is the same as in crinoids but not at all equivalent to the common location of the madreporite in echinoids which would call for finding the holothurian gonopore in interamb *AB* of the trivium. Fortuitously, it seems, because we have been guided by criteria entirely at variance with those accepted by other authors, the application of Carpenter letters to holothurians happens to be in complete agreement. Designation of the rays of holothurians adopted in the *Treatise* is as shown in Figure 104, and this is identical to designation given by CUÉNOT, HYMAN, and others.

CARPENTER LETTERS APPLIED TO ASTEROZOANS

So uncertain and insecure is identification of individual rays and interrays of somasteroids, asteroids, and ophiuroids that authors generally have declined to use any system of notations for them. In many of these echinoderms radial symmetry appears to be perfect, with no clue whatever for the adoption of orientation other than oral and aboral. Asteroids commonly possess an easily distinguished madreporite in one of the interrays on the aboral side of the body, and in addition, some show the presence of an anus, also located on the aboral surface in the interray at left (in aboral view) of the one containing the madreporite. Some asteroids (e.g., *Acanthaster*) carry several madreporites scattered about on the aboral side. Ophiuroids commonly appear to be perfectly symmetrical radially, but the presence of a madreporite in one of the interrays next to the mouth can be determined. Several genera of the suborder Euryalina (order Phrynophiurida), however, have five madreporites, one in each interray, or similarly disposed hydropores not associated with any skeletal element may be found (e.g., *Trichaster*) (HYMAN, 1955, p. 613).

In asteroids and ophiuroids having a single madreporite, authors (e.g., CUÉNOT, 1948; HYMAN, 1955; AILSA CLARK, 1963) have accepted this as basis for defining the interray containing it as *CD* (Fig. 105).

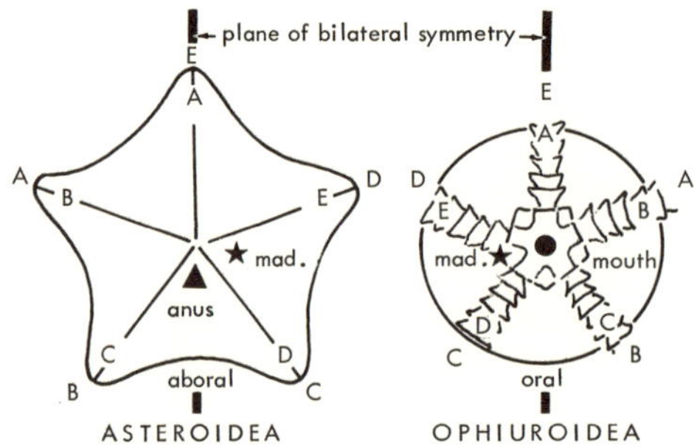

FIG. 105. Diagrammatic aboral view of asteroid and oral view of ophiuroid showing doubtfully identified plane of bilateral symmetry and application of Carpenter letters to rays, inner letters as here identified and outer letters according to authors.

Guidance for conclusions on homology is the same as accepted by these and other authors—firmly fixed location of the madreporite in all echinoderm groups—even though this has been shown by us to be untrustworthy, commonly serving to mislead rather than to point out homologous skeletal parts correctly. Since the rays of no asterozoans are grouped in trivia and bivia, and since a vertical plane of bilateral symmetry defined by such arrangement is unavailable to help us, only guesswork concerning orientation remains for use. When account is taken of earliest known asterozoans, which in all classes exhibit structural affinities with pinnulate crinoids (FELL, 1963) and which indicate interrelationships pointing to a common ancestry, judgment is reached that location of the anus outweighs that of the madreporite as marker for orientation. Thus we are led to apply Carpenter notations of rays as shown in Figure 105,*1* (letters distributed outside of the outline) for asteroids, and on assumption that the interray bearing a madreporite in ophiuroids corresponds to the madreporite-bearing interray in asteroids, Carpenter letters may be applied to ophiuroids as shown in Figure 105,*2*. Whereas great confidence can be expressed as to the correctness of ray and interray homologies for echinozoans as here presented, surely this cannot be extended to include the asterozoans, at least on the basis of present knowledge.

SUMMARY

A vertical plane of bilateral symmetry which is clearly defined in the Crinoidea and various other classes of the Crinozoa provides the basis for applying in uniform manner letter symbols introduced by CARPENTER for the different rays and interrays. The system is especially suited for description and illustration of echinoderms belonging to this subphylum, many of which display arrangement of the rays in a well-marked trivium and bivium. The anus is invariably located in the *CD* interray, defined as posterior, and commonly a hydropore or gonopore or both occur in the same interray (Fig. 98; 106,*1*).

An identical plane of bilateral symmetry is demonstrated to exist in echinoids, incipiently expressed in the regular echinoids (Fig. 97,*B,C*; 99,*1,2*; 101; 106,*2,3*) but strongly marked in the irregular echinoids (Fig. 97,*D-F*; 99,*4*; 106,*6,7*), most of which display grouping of the anterior three rays in a trivium and the posterior two rays in a bivium, the latter enclosing the anus. The Lovén system of notation, using Roman numerals for rays and Arabic numerals for interrays, is adopted in the *Treatise* (Fig. 97; 100; 106,*2,6*), but correlation of it with the Carpenter system is indicated (Fig. 100, *1b,2b*; 106,*2,6*). In addition, authors' application of Carpenter letters to echinoids in manner judged to be entirely erroneous is discussed and illustrated (Fig. 99; 100; 106,*3,7*). Reasons are presented for relegating the madreporite to a very subordinate status as a structure to be considered in studies of homology, and accordingly conclusions mainly or entirely based on this are rejected.

The application of Carpenter letters to edrioasteroids (Fig. 102), ophiocistioids (Fig. 103,*1*), and cyclocystoids (Fig. 103,*2*) offers no problems and is straightforward.

For the holothurians (Fig. 104; 106,*4*) recognition of far-reaching homologies and use of Carpenter ray notations to express them are curious in that identical conclusions have been reached in different ways, one being quite faulty and the other strongly supported by trustworthy evidence. The faulty approach is that generally accepted by authors, based on trust in the significance of gonopore placement, in our opinion erroneously correlated with the madreporite-bearing genital 2 element of the echinoid apical disc. The differentiation of holothurian rays into a ventrally oriented trivium and dorsal bivium defines a fundamental vertical plane of bilateral symmetry equivalent to that in echinoids, crinoids, and most other echinoderm classes. The presence of a gonopore in the *CD* interray corresponds to the gonopore in crinoids, for example, and not to genital 2 of echinoids.

Asterozoans are doubtfully oriented in terms of Carpenter ray notations, but phylo-

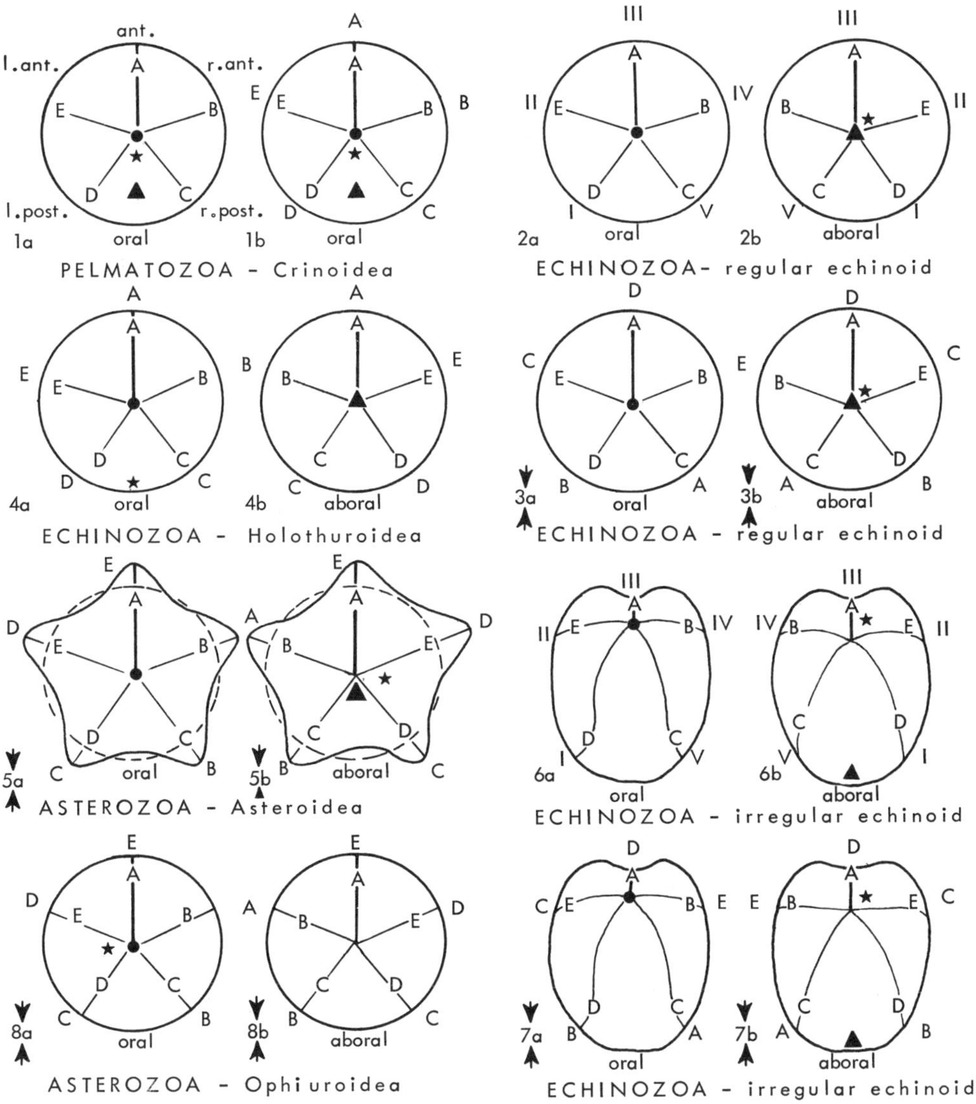

FIG. 106. Summary of ray designations applied to echinozoans and other echinoderms brought together for comparison. [Ray designations outside of test outlines indicate usage of authors; those inside of these outlines show designations adopted in the *Treatise,* except that for echinoids Carpenter letters merely indicate correlation with *Treatise*-adopted Lovénian numerals. Arrows pointing to some figures call attention to ray designations which are judged to be erroneous.]

genetic considerations support tentative identification of the anus-bearing interray of asteroids with the *CD* interray of crinozoans and interamb 5 of echinoids (Fig. 105,*1*; 106,*5*). Correlation of madreporite placement in ophiuroids with that observed in asteroids indicates that the interray bearing this structure is *DE* (not *CD*), granting the orientation of asteroids just stated.

REFERENCES

Agassiz, Alexander

(1) 1872-74, *Revision of the Echini:* Harvard Univ., Museum Comp. Zoology, Mem., v. 3, p. 1-762, pl. 1-38.

Bather, F. A.

(1a) 1900, *The Echinoderma:* in E. R. Lankester, ed., A Treatise on Zoology, pt. 3, 344 p., 309 text-fig., A. & C. Black (London).

Carpenter, P. H.

(2) 1884, *Report upon the Crinoidea collected during the voyage of H.M.S. Challenger during the years 1873-1876:* Rept. Scientific Results Explor. Voyage H.M.S. Challenger, Zoology, v. 11, xii+442 p., 62 pl., 21 text-fig.

Clark, Ailsa M.

(3) 1962, *Starfishes and their relations:* British Museum (Nat. History), 119 p., 16 pl., 30 text-fig. (London).

Clark, H. L.

(4) 1925, *A catalog of the Recent sea-urchins in the British Museum:* British Museum (Nat. History), 278 p., 12 pl. (London).

Cuénot, Lucien

(5) 1948, *Anatomie, éthologie, et systématique des échinodermes:* in Traité de zoologie, P.-P. Grassé (ed.), v. 11, p. 1-275, text-fig. 1-312, Masson (Paris).

Durham, J. W.

(6) 1955, *Classification of clypeasteroid echinoids:* California Univ., Publ. Geol. Sci., v. 31, p. 73-198, pl. 3-4, 38 text-fig.

———, & Melville, R. V.

(7) 1957, *A classification of echinoids:* Jour. Paleontology, v. 31, p. 242-272, text-fig. 1-9.

Fell, H. B.

(8) 1963, *The phylogeny of sea-stars:* Royal Soc. London, Philos. Trans., ser. B, v. 246, p. 381-435, pl. 50-51, text-fig. 1-16.

Gregory, J. W.

(9) 1900, *Echinoidea,* in E. R. Lankester, A. Treatise on Zoology, Part III, the Echinoderma: p. 282-332, text-fig. 1-47, A. & C. Black (London).

Hawkins, H. L.

(10) 1931, *The first echinoid:* Biol. Reviews, v. 6, p. 443-458.

Hyman, L. H.

(11) 1955, *The invertebrates: Echinodermata:* vii & 763 p., 280 text-fig., McGraw-Hill (New York).

Jackson, R. T.

(12) 1896, *Studies of Palaeechinoida:* Geol. Soc. America, Bull., v. 7, p. 171-254, pl. 2-9, 5 text-fig.

(13) 1912, *Phylogeny of the Echini, with a revision of Paleozoic species:* Boston Soc. Nat. History, Mem., v. 7, p. 1-490, pl. 1-76, 256 text-fig.

(14) 1929, *The status of Bothriocidaris:* Harvard Univ., Mus. Comp. Zoology, Bull, v. 69, p. 481-512.

Jaekel, Otto

(15) 1918, *Phylogenie und System der Pelmatozoen:* Paläont. Zeitschr., v. 3, no. 1, p. 1-128, text-fig. 1-114.

Kesling, R. V.

(16) 1960, *Hydropores in edrioasteroids:* Michigan Univ., Museum Paleontology, Contrib., v. 15, no. 8, p. 139-192, pl. 1-13, text-fig. 1-14.

Kier, P. M.

(17) 1962, *Revision of the cassiduloid echinoids:* Smithsonian Misc. Coll., v. 144, no. 3, 262 p., 44 pl., 184 text-fig.

Lovén, Sven

(18) 1874, *Etudes sur les Echinoidées:* K. Svenska Vetenskapsakad. Handl., v. 11, no. 7, p. 1-91, pl. 1-53.

Mortensen, Theodor

(19) 1928, *Bothriocidaris and the origin of echinoids:* Vidensk. Medd. Dansk Naturh. Foren., v. 86, p. 93-124.

(20) 1928-51, *A monograph of the Echinoidea:* v. 1, Cidaroidea, 551 p., 173 text-fig., atlas 24 pl. (1928); v. 2, Bothriocidaroida, Melonechinoida, Lepidocentroida, and Stirodonta, 647 p., 377 text-fig., atlas 89 pl. (1935); v. 3, no. 1, Aulodonta, 370 p., 197 text-fig., atlas 77 pl. (1940); v. 3, no. 2, Camarodonta, 533 p., 321 text-fig., atlas 56 pl. (1943); v. 3, no. 3, Camarodonta, 446 p., 215 text-fig., atlas 66 pl. (1943); v. 4, no. 1, Holectypoida, Cassiduloida, 363 p., 14 pl., 326 text-fig. (1948); v. 4, no. 2, Clypeasteroida, 471 p., 258 text-fig., atlas 72 pl. (1948); v. 5, no. 1, Spatangoida, 432 p., 25 pl., 315 text-fig. (1950); v. 5, no. 2, Spatangoida, 593 p., 286 text-fig., atlas 64 pl. (1951); index to v. 1-5, 63 p. (1951),

C. A. Reitzel (Copenhagen); Milford, Oxford Univ. Press (London).
(21) 1931. *Bothriocidaris and the ancestry of echinoids, a reply:* Vidensk. Medd. Dansk Naturh. Foren, v. 90, p. 313-352.

Nichols, David
(22) 1959, *Changes in the Chalk heart-urchin Micraster interpreted in relation to living forms:* Royal Soc. London, Philos. Trans., ser. B, v. 242, p. 347-437, 1 pl., 46 text-fig.
(23) 1962, *Echinoderms:* 200 p., 26 text-fig., Hutchinson (London).

Regnéll, Gerhard
(24) 1945, *Non-crinoid Pelmatozoa from the Paleozoic of Sweden:* Lunds Geol.-Mineral. Inst., Medd., v. 108, viii+255 p., 15 pl., 30 text-fig. (Lund).

Zittel, K. A. von
(25) 1900 [1896-1900], *Text-book of palaeontology* (transl. and ed. C. R. Eastman), v. 1, ed. 1, 706 p., 1476 text-fig. (Echinodermata, p. 122-250, text-fig. 217-402, publ. 1896), Macmillan & Co., Ltd. (London, New York).

HELICOPLACOIDS

By J. Wyatt Durham and Kenneth E. Caster
[University of California (Berkeley); University of Cincinnati]

Class HELICOPLACOIDEA Durham & Caster, 1963

[Helicoplacoidea Durham & Caster, 1963, p. 820]

Free-living, fusiform to pyriform placoid echinoderms with spirally pleated, expansible and flexible test; apical and oral poles at opposite extremities; columns of plates disposed spirally; ambulacra and "interambulacra" present, new "interambulacral" plates originating at apical pole and becoming more oral in position as subsequent plates are added, origin of ambulacral plates obscure, but possibly similar. *L.Cam.(Olenellus Zone).*

In the retracted state the known species of this specialized extinct group are pyriform, but when expanded they become fusiform. The plates are not firmly sutured to one another as in the echinoids and many pelmatozoans, and in the expanded state (Fig. 107) the body was flexible, much as in the holothurians. Because the plates were not sutured to one another the test usually became disassociated upon death. In consequence, the small isolated plates are much more abundant in the fossil record than partial or entire tests.

The "interambulacral" areas are composed of three columns of plates extending from the oral to the apical pole. In the retracted state (Fig. 108,*A*) the central column is external and the two lateral columns folded internally; in order to expand, the lateral column folded outward laterally and formed the floors of troughs adjacent to the ridgelike medial column (Fig. 108,*B*). The three columns of an area originate from a single center in the apical area. The minute plates when first recognizable appear in a multiserial (?triserial) column. As the apical pole becomes more distant and the plates grow larger, the single column differentiates into three columns, with the medial plates forming the central column. In each of the two described species there are 10 "interambulacral" areas.

The principal ambulacrum makes at least two full spiral turns, starting at the mouth, but does not reach to the apical pole. The secondary ambulacrum first appears about 180 degrees along the spiral from the mouth and then continues for approximately another 180 degrees, being separated from the first by two "interambulacral" areas throughout most of its length. In one specimen (Fig. 109,*B*) the two ambulacra clearly join, apparently adapically (although theoretical objections can be made to this interpretation), and the principal ambulacrum continues. In the retracted state the medial "interambulacral" columns imbricate adapically posterior to the branching of the ambulacrum, and slightly adorally anterior to this point. The ambulacra are composed

FIG. 107. *Helicoplacus gilberti* DURHAM & CASTER, L.Cam. (*Olenellus* Zone), USA (Calif.); well-preserved adoral portion of theca showing clockwise spiral torsion of plate rows interpreted as "interambulacral" and small part of ambulacrum, oral pole upper right at center of spiral, ×6.6 (2).

of four or more rows of small plates. As yet no podial pores have been recognized certainly, although possible grooves for tube feet seem to be present on adjacent ambulacral plates on one poorly preserved specimen.

Anal and genital orifices have not been recognized. The structure of the peristome is uncertain, but the mouth apparently was not more than 1 mm. in diameter. No supportive or masticatory structures have been recognized in the oral area. Likewise no tentacular or brachial appendages have been found, and no evidence of any attachment area for such structures has been discovered.

The symmetry of the test, at least as far as the "interambulacra" are concerned, is radial, modified by torsion to a spiral form. However, the single primary ambulacrum imposes a distinct bilaterality upon the

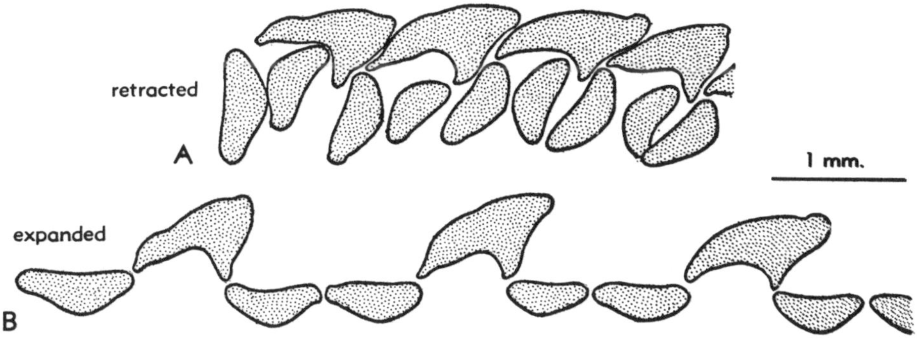

Fig. 108. Diagrammatic sections of thecal plates of *Helicoplacus gilberti* parallel to oral-aboral axis, showing *(A)* infolded "interambulacral" columns in retracted state of theca, and *(B)* spread-out lateral "interambulacral" columns in expanded state of theca, ×18 (1).

original radial pattern of the "interambulacra." Nevertheless, it should be noted that the two "interambulacral" areas separating the branch for most of its length from the primary ambulacrum are strongly suggestive of possible subsequent development of a pentameral pattern. If a similar branching were repeated three additional times, and if the appearance of branching were accelerated in ontogeny, the result would be the usual pentameral echinoderm pattern.

The small area of the ambulacra in comparison with the total bulk of the animal suggests that the ambulacra were primarily respiratory structures. Inasmuch as most living echinoderms have a ciliated epidermis, it seems probable that the integument was likewise ciliated in this group. The small mouth indicates that *Helicoplacus,* the type and presently only known genus of the class, was a small-particle-feeder. This seems likely in view of the lack of indication of other food-gathering organs, and it may be inferred that *Helicoplacus* was usually in the expanded state, with organic particles gathered by its cilia being passed along the spiral "interambulacral" grooves towards the mouth. Presumably, only in times of danger from predators or in encountering an unfavorable environment or in periods of inactivity would the retracted state be assumed.

The small size of the apical pole, the fact that it is the point of origination for new plates, the lack of any specialized structure for adhesion, and the small-particle size of the enclosing sediments indicate that the known helicoplacoids were free-living. Their mode of life is unknown. They may have been stationary, with the apical pole buried in the soft substrate upon which they lived, or they may have lain on the sea floor, crawling about like many holothurians. Another possibility is that they rested passively on the sea floor when in the retracted state but when expanded slowly pulsed through the water by rhythmic contractions of the accordion-like test. With respect to the last-suggested hypothesis it may be noted that in the completely expanded state the volume of the body was probably more than twice that of the retracted state and the specific gravity of individuals then may not have been much above that of sea water.

In California *Helicoplacus* occurs in the same beds as the eocrinoid *Eocystites,* as well as various trilobites, a few archaeocyathids, and inarticulate brachiopods. Stratigraphically, the shales in which it occurs are intercalated in a thick sequence of archaeocyathid-bearing beds, and the known occurrences are located at levels about one-third of the thickness of the sequence above the base of the olenellid-trilobite-bearing beds. In nearby western Nevada, a *Stromatocystites*-like edrioasteroid occurs in association with numerous disarticulated plates of *Helicoplacus.* Seemingly, these occurrences of Edrioasteroidea, Eocrinoidea, and Helicoplacoidea represent the oldest known records of the Echinodermata. The presence of these dissimilar echinoderm types close to the beginning of the good fossil record indicates that differentiation within the phylum must have been initiated before the beginning of the Cambrian. The free-living character of *Helicoplacus* suggests that, contrary to usual concepts, ancestral echinoderms may have been free-living. Whatever their character, the common ancestor must have had the potential to give rise to such divergent types as *Helicoplacus, Eocystites,* and the edrioasteroids.

Because of the highly specialized nature of the expansion-contraction mechanism which is judged to distinguish the Helicoplacoidea, this group probably represents a branch of the echinoderm stock that left no descendants. The imbrication of the test in the retracted state, however, is strongly reminiscent of such edrioasteroids as *Lepidodiscus* and *Agelacrinus,* suggesting that the Edrioasteroidea may be related to the Helicoplacoidea. Also, it seems possible that the holothurians might have been derived from the immediate, pretorsion ancestor of the helicoplacoids. The test of the adherent but flexible holothurian *Psolus,* with its heavy imbricating plates, is suggestive of the retracted test of *Helicoplacus.* The origination of new plates from the apical pole, the lack of circumoral appendages, and the probable body shape of the pretorsion ancestor also suggest such early echinoids as *Aulechinus* and the equivocal echinoid-holothurian *Eothuria.* If these similarities are significant, they suggest that the Edrioasteroidea, as well as the Helicoplacoidea, belong in the subphylum Echinozoa and that this group probably was derived from the immediate pretorsion ancestor of the Helicoplacoidea.

Family HELICOPLACIDAE
Durham & Caster, n. fam.

Characters of the class. *L.Cam.* (*Olenellus Zone*).

Helicoplacus Durham & Caster, 1963, p. 82c [**H. gilberti*; OD]. Test of 10 "interambulacra" and single ambulacrum with short branch; peristome

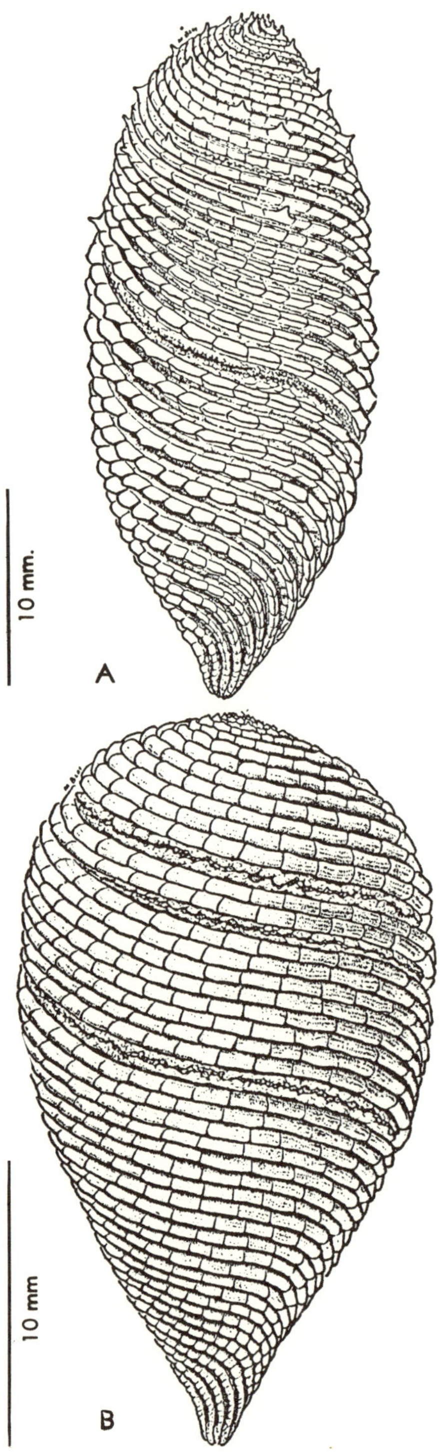

Fig. 109. Side views (reconstr.) of species of *Helicoplacus,* both L.Cam. (*Olenellus* Zone), USA (Calif.), showing strong torsion of theca, oral extremity at top, pointed aboral end directed downward.——*A. H. curtisi* Durham & Caster, partially expanded, a spinose species (1).——*B.* **H. gilberti* Durham & Caster, individual in retracted state (1).

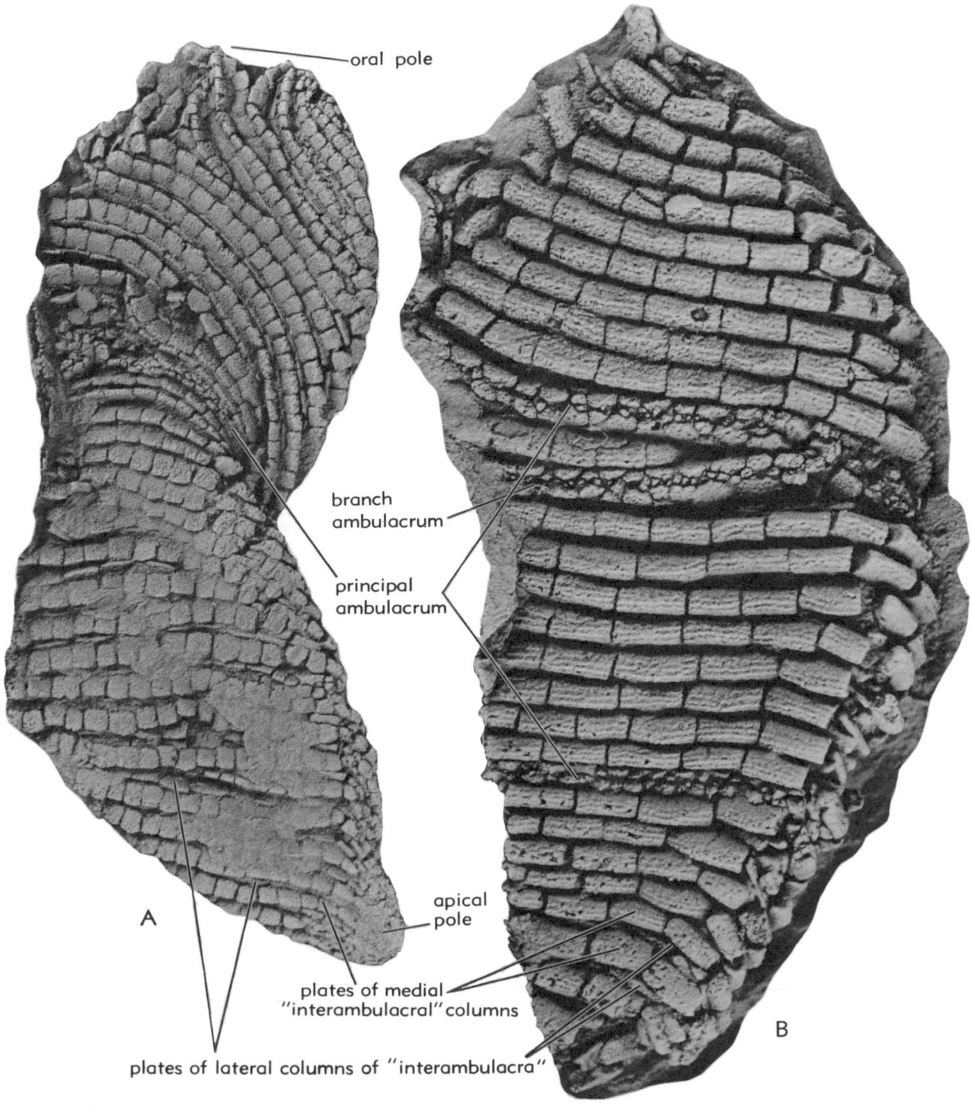

FIG. 110. Specimens of *Helicoplacus,* L.Cam. (*Olenellus* Zone), USA (Calif.).——*A*. Flattened, nearly complete specimen, *H.* sp., showing pointed apical pole and partly dissociated upper part of test with oral pole, ×4 (2).——*B.* **H. gilberti* DURHAM & CASTER, incomplete flattened specimen with branching ambulacrum, lowermost ambulacrum being continuation of this, ×7 (2).

small; type-species with longitudinal ribs on plates of medial "interambulacral" column; random nonarticulating spines on plates of medial interambulacral column, especially near peristome. *L. Cam. (Olenellus Zone),* USA (Calif.-W. Nev.). ——FIG. 107; 108, 109, 110,A. **H. gilberti*; 107, flattened and expanded oral pole of incomplete specimen, oral area at center of spiral, showing a few plates of incomplete ambulacrum in lower right, and spines on random plates of medial interambulacral column, ×6.6 (2); 108,*A,B,* sections of test parallel to oral-aboral axis, in retracted state *(A)* showing infolded "interambulacral" columns, and in expanded state *(B)* showing lateral "interambulacral" columns in expanded position, both ×18 (1); 109,*B,* restoration, retracted state ×3.75 (1); 110,*B*, incomplete, flattened specimen showing branching of ambulacrum, lowermost ambulacrum same as uppermost but on next volution, ×7 (2).——Fig. 109,*A*. *H. curtisi*; restoration of spinose species partially expanded, ×2.5 (1).——Fig. 110,*A*. *H.* sp., flattened, nearly complete specimen, showing apical pole and partially dissociated oral pole, ×4 (2).

REFERENCE

Durham, J. W., & Caster, K. E.

(1) 1963, *Helicoplacoidea, a new class of echinoderms:* Science, v. 140, no. 3568, p. 820-822, text-fig. 1, photo on cover.

ADDITIONAL SOURCE OF ILLUSTRATIONS

(2) **Kier, P. M.**, n.

EDRIOASTEROIDS

By Gerhard Regnéll

[Lunds Universitet, Lund, Sweden]

CONTENTS

Class EDRIOASTEROIDEA Billings, 1858

[as suborder Edrioasteridae] [=Thyroidea Chapman, 1860; Agelacrinoidea S. A. Miller, 1877; Cystasteroidea Steinmann, 1890; Agelacystida Haeckel, 1895 *(partim)*; Thecoidea Jaekel, 1895; Cystostelleroidea Steinmann, 1903] [excludes forms now referred to class Edrioblastoidea[1]]

[1] The stalked genus *Astrocystites* Whiteaves, 1897 (=*Steganoblastus* Whiteaves, 1897) has formerly been placed with the Edrioasteroidea as the single representative of the family Astrocystitidae Bassler, 1935. Since *Astrocystites* differs in several respects from typical Edrioasteroidea R. O. Fay (14) found reason to institute a new class, Edrioblastoidea Fay, 1962, to reecive the genus. This procedure will be followed here, though not without a certain reluctance.

Many-plated echinoderms with well-developed (normally) quinqueradiate endothecal ambulacral system; no thecal pores, but pores may be present between ambulacral elements (thus not piercing substance of plates); no arms or brachioles; anal opening in posterior interradius, generally covered by valvular pyramid; a third aperture, interpreted generally as a hydropore, may be recognizable between mouth and anus; unstalked (Fig. 111,*1*). *L.Cam.-L.Carb. (Miss.)*.

INTRODUCTION

The pelmatozoan nature of the Edrioasteroidea, although some of them show eleutherozoic tendencies in mode of life, is clearly demonstrated by the following observations. Even so, the class here is included in the dominantly eleutherozoan subphylum Echinozoa.

(1) The adoral surface, with mouth, anus, and a third opening, was directed upward. In some genera the theca was modified so as more or less to simulate a stem.

(2) The Edrioasteroidea fed as whirlers, according to Remane (Spencer, 38), that is, food was brought to the mouth by a subvective system of ciliated grooves protected by cover plates.

(3) Evidence is found in some genera

(e.g., *Edrioaster*) of an aboral motor nerve center, but it may be assumed *a priori* that this system is strongly reduced because it has little or no importance in sessile or almost sessile forms wanting a stem and movable arms (7, pt. 7).

The most significant features of the Edrioasteroidea, by which they differ from all other pelmatozoans, are the nature of ambulacral structures and absence of all exothecal appendages.

MORPHOLOGY

GENERAL FEATURES OF THECA AND AMBULACRA

As in other noncrinoid pelmatozoans, the viscera are enclosed in a capsule, termed **theca.** However, unlike the theca of the Rhombifera, this is not closed, for the ambulacra are lodged between the thecal skeletal elements and do not rest upon them (Fig. 111,*2*). It may be that, morphogenetically, the ambulacral skeleton is not different from the other thecal plates.

As mentioned already, the **adoral surface**

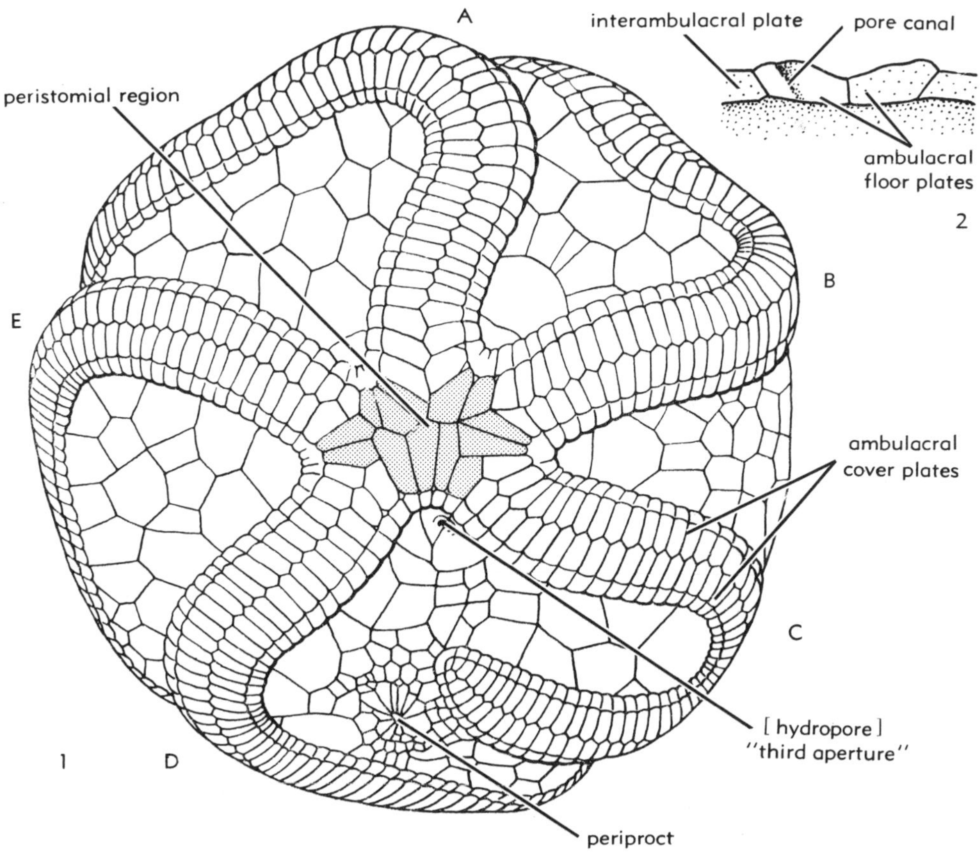

FIG. 111. Typical edrioasteroid, *Edrioaster bigsbyi* (BILLINGS), M.Ord., Ontario, illustrating some morphological features.——*1*. Adoral surface showing ambulacra with biserially arranged cover plates adjoined on each side by row of adambulacral plates, interambulacral areas distinguished by relatively large irregularly arranged plates; ambulacra marked by letters of Carpenter system; ill-differentiated plates of peristomial region unshaded; posterior interambulacrum with low anal pyramid (periproct) and near peristome with "third aperture" interpreted as hydropore; ×2.5 (after 24, modified from 7, pt. 4).——*2*. Transverse section of ambulacral floor plates showing their relation to adjoining interambulacrals; floor plate at left viewed on its sutural surface, showing pore canal, floor plate at right viewed on cut surface; ×5 (7, pt. 4).

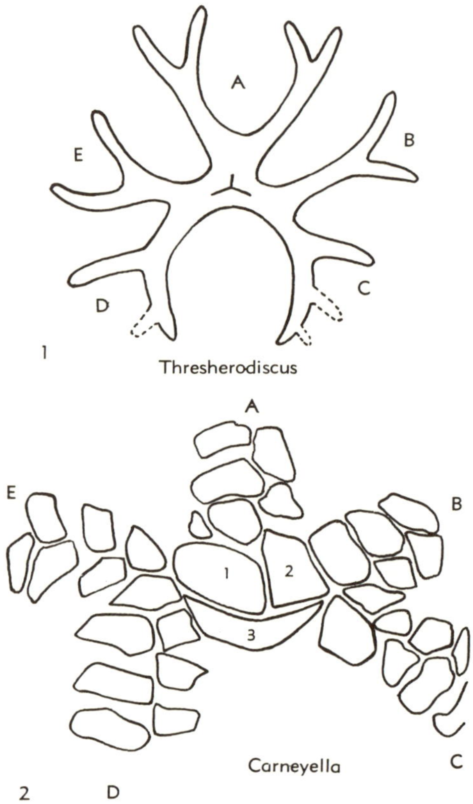

FIG. 112. Trimerous arrangement of ambulacra rarely exhibited by edrioasteroids.——*1.* Pattern of rays shown in diagram of adoral surface of *Thresherodiscus ramosus* FOERSTE (M.Ord., Ontario), ×4 (16).——*2.* Adoral part of theca of *Carneyella pileus* (HALL) (U.Ord., Ohio), with somewhat disguised trimerous disposition of rays owing to separation by large orotegminal plates (left and right anterior, *1,2;* posterior, *3*), ×8 (40).

(known also as oral, ventral, or actinal surface—"adoral" should be preferred to "oral" because the latter term refers primarily to the area occupied by the mouth and its skeleton, the peristome) contains the apertures of the thecal wall and was directed upward (hence sometimes referred to as "upper surface"), whereas the **aboral surface** (known also as apical, adapical, dorsal, or abactinal surface) was directed downward ("lower surface"). In most genera, the ambulacral grooves are restricted to the adoral surface and do not reach the aboral surface except in the family Edrioasteridae.

In the oldest known edrioasteroid, *Stromatocystites* (L.Cam.-M.Cam.), and in many later genera, the theca is depressed semiglobular, having the shape of a slightly convex disc. The theca developed variously, however, being either very thin and almost flat in some forms (e.g., *Agelacrinites*), or saclike in others (e.g., *Cystaster*). The tendency to elevate the adoral surface over the sea bottom has found its extreme expression in *Pyrgocystis.* In this genus the theca is transformed into a high turret, the height of which is many times the diameter of the adoral surface.

Pentamerous symmetry in these forms is demonstrated mainly in the disposition of the ambulacral grooves, which differentiate the theca into 5 ambulacral (radial) and 5 interambulacral (interradial) fields, called **ambulacra** and **interambulacra,** respectively. One or more extra rays may occur in several genera. The ambulacra will be referred to by the letters *A-E* in agreement with the system introduced by P. H. CARPENTER (1884). The ambulacrum opposite to the posterior interradius *(CD)* is designated by *A*; ambulacra *B-E* follow in clockwise direction when the oral surface is directed upward. The interambulacra have the symbols *AB, BC,* etc. (Fig. 111,*1*).

Pentamerism is not reflected generally by the arrangement of thecal plates. A trimerous disposition of the radial extensions, with one anterior ray and two forking lateral ones, is apparent in *Thresherodiscus* (M. Ord.) (Fig. 112,*1*) and it can be distinguished in *Carneyella* (M.Ord.-U.Ord.) (Fig. 112,*2*) *Dinocystis* (U.Dev.), and *Lepidodiscus squamosus* (Miss.). This has been interpreted by BATHER and others as an archaic pattern, but that view is not in accord with recent opinion which regards trimerism in echinoderms as a secondary feature. [*Tribrachidium* GLAESSNER, 1959, from the Precambrian of S. Australia, which is characterized by perfect threefold symmetry, has a superficial resemblance to disc-shaped edrioasteroids. However, there is no ground for assuming that *Tribrachidium* developed into some primitive echinoderm. Rather was it an aberrant coelenterate (cf. this *Treatise,* p. *W*228)].

All Edrioasteroidea are small or moderate in size. The thecal diameter of adult specimens varies generally between 5 and 60 mm.; in *Pyrgocystis* it may be less than 5 mm.; the largest of the two known specimens of *Timeischytes* is 4.5 mm.

THECAL SKELETON

As a rule, the interambulacral (interradial) sections are broad as compared with the ambulacral (radial) ones. Their structure therefore largely controls constitution of the theca. Where the ambulacra are exceptionally broad, as in *Hemicystites,* the interambulacra are narrow, in consequence. Very commonly, the posterior *(CD)* interambulacrum differs in shape from the others.

GENERAL CHARACTER

The plates of the interambulacra are called **interambulacrals** (interambulacralia, interradials, interradialia). They may be scale-like and more or less imbricating, which provides the theca with a certain degree of flexibility. In the oldest known family (Stromatocystitidae), flexibility is achieved by other means; as by weak calcification of the skeleton or by presence of stroma strands between the polygonal nonimbricate plates, which thereby admitted some mobility. Attachments of the stroma strands are marked generally by two or three diplopore-like depressions that extend across the sutures of adjoining plates. Because of the superficial resemblance of these structures to diplopores, *Stromatocystites* has been supposed to indicate relationship between edrioasteroids and diploporite Hydrophoridea. This interpretation is quite implausible in view of the fact that true diplopores only exceptionally cross sutures between thecal plates (e.g., *Glyptosphaerites*). It is not more convincing to compare the dumbbell-shaped depressions on edrioasteroid suture faces with the pore canals of the Rhombifera (as suggested by Cuénot, 10).

A weakly calcified theca, with minute plates, is also present in the saclike *Cystaster*.

It is noteworthy that flexibility of the theca among discoidal and hemispherical forms surely is not correlated with their temporary fixation. Otherwise, one might suppose that the purpose of flexibility was to allow the theca to act as a sucking disc, as indeed it may have been in *Stromatocystites.*

The turret-shaped *Pyrgocystis* cannot have been able to relinquish its attachment at an adult stage. In spite of this, its theca is formed by imbricating plates, which are not markedly different from those of the interradial areas of the adoral surface. The skeletal elements in this genus show a tendency, although not absolute, to develop phylogenetically toward greater plasticity without loss of firmness. This is effected by modification of the inosculating plates of early species into obliquely disposed plates arranged in distinct columns separated by grooves and in late forms by dense crowding of small plates without overlapping. Development of this sort seems to have occurred independently in Lower Ordovician and Silurian stocks of *Pyrgocystis.* The improved flexibility of the theca thus acquired may have served for adjustment of the theca to the substratum and for directing the oral region toward food-bearing water currents. It may be that flexibility of the theca had some importance also for the mechanism regulating the opening and closing of the ambulacral grooves.

In forms where the interambulacrals comprise a mosaic of polygonal plates, the theca generally is characterized by greater rigidity. In the Cyathocystidae, the lateral plates are fused into a solid saclike mass cemented to the substratum so as to form, with plates of the adoral surface, an extremely firm theca. The stalked *Astrocystites* has a very rigid theca composed of relatively few plates, which are extraordinarily large and thick as compared with the plates of all Edrioasteroidea (*Astrocystites* now assigned to Edrioblastoidea).

INTERAMBULACRALS

The interambulacrals are either polygonal plates arranged in a mosaic, or scalelike imbricating ossicles. Some genera, however, have interambulacrals intermediate between these types, with polygonal, slightly imbricate plates (e.g., *Walcottidiscus, Ulrichidiscus*). Species with mosaic plates and species with imbricate plates may belong to one and the same genus (e.g., *Lebetodiscus, Isorophus, Agelacrinites*).

In most genera, the interambulacrals are tolerably uniform, but in some they are clearly differentiated in size and shape. The interambulacral plates of *Hemicystites* and *Anglidiscus* have a tendency to become larger in a centrifugal direction, whereas in *Lebetodiscus* and *Thresherodiscus* the largest plates are found near to the center; in *Agelacrinites* they vary strongly in shape.

Minute plates bordering on the valvular plates of the periproct are found in several genera (e.g., *Anglidiscus, Isorophusella*). Not seldom (e.g., *Edrioaster*) an agglomeration of numerous small irregular plates on the right side of the anus marks the position of an expanded rectum.

Further, one or more rows of small adradial plates may line the ambulacra (e.g., *Anglidiscus, Isorophus, Lepidodiscus, Stromatocystites*).

A definite arrangement of the interradial plates is usually not recognizable. A noteworthy exception is found among members of the Cyathocystidae and *Timeischytes* among the Hemicystitidae. *Cyathotheca* and *Cyathocystis* have only one large, triangular plate in each interambulacrum. Whether these have developed from a number of ordinary interambulacrals fused into larger solid plates or are primary structures is not obvious from the fossil material, in which not the slightest indication of sutures can be traced. Yet it is easy to realize that an amalgamation of mosaic interambulacrals (as in *Stromatocystites*) would produce plates like those present in the Cyathocystidae, just as has been suggested to explain the sides of saclike thecae by fusion of lateral plates corresponding to those in *Cystaster*. This seems reasonable. Although corresponding morphologically and functionally to the orals of certain crinoids, the interambulacrals of the Cyathocystidae are not necessarily homologous with orals. *Cyathocystis* also possesses an inner circle of five interradially disposed plates which cover the oral field. These probably do not belong to the interambulacral series, however, for they seem to have originated by coalescence of the most proximal ambulacral cover plates of each two adjoining rays, like the oral cover plates of other Edrioasteroidea.

Timeischytes is remarkable in that all interambulacra save the posterior one are each covered by a single large sublunate or sublinguiform plate. In interambulacrum *CD* five differently shaped interambulacrals are disposed around the anal pyramid.

In forms with imbricating plates, the overlap is invariably in a proximal direction, and greater toward the periphery. A diagonal arrangement of the interambulacrals may be discerned in some species.

Advocating the view that the Edrioasteroidea were derived from some cystoidean ancestor, Foerste (16) observed that imbrication of thecal plates can hardly be considered a primitive feature, because the plates of cystoids have polygonal outlines and are arranged in a mosaic. He suggested that the change was due to "assuming of the sessile habit, together with the enormous shortening of the theca in a vertical direction. This caused the distal edge of one plate to collapse within the proximal edge of the adjoining plate." Without entering now on a discussion of the supposed phylogeny of the Edrioasteroidea, it should be pointed out, in objection to this theory, that an extraordinarily strong imbrication of the thecal plates is found in *Pyrgocystis,* in which the vertical axis of the theca is extremely long.

PERIPHERAL RING

Plates of the adoral surface (excluding those of the ambulacra) commonly are differentiated into interambulacrals proper and distal plates forming a peripheral ring. This is true of most discoidal to hemispherical forms. Naturally, a peripheral ring is lacking among the edrioasteroids in which the ambulacra pass on to the aboral surface.

Generally, plates of the peripheral ring decrease in size in a centrifugal direction. It is a common feature that those nearest to the central part of the adoral surface are even bigger than the interambulacrals and are extended tangentially, especially in *Agelacrinites* and *Timeischytes.* Plates of the border nearest to the periphery are minute.

The outer portion of the peripheral ring undoubtedly was mobile and thus capable of adjusting to the surface of the substratum. The larger plates forming an inner band of the peripheral ring in many Edrioasteroidea had much greater rigidity; they were rather firmly locked horizontally but capable of some vertical extension.

Most plates of the peripheral ring bear one or more processes on their aboral surface. These processes may have served for attachment of muscles of the muscular wall in which the plates were imbedded.

The peripheral ring of *Cyathocystis* is formed by a single row of subquadrate marginals. In *Cyathotheca* it is wanting entirely.

Not seldom the peripheral ring is raised over the central portion of the theca. This is a post-mortem feature due to sinking in of the central body after decay of the animal.

ABORAL SIDE

Among edrioasteroids which lack a definite peripheral ring, the aboral side of the theca is made up of plates more or less continuous with and similar to the interambulacrals.

In *Stromatocystites* the aboral center is occupied by a fairly distinct dorsocentral (not necessarily homologous with the dorsocentral of other echinoderms) surrounded by large polygonal plates; toward the periphery the plates decrease somewhat in size and tend to be more rounded.

The aboral surface of *Cooperidiscus* is bordered by a projecting periphery of more prominent plates. The surface enclosed in this ring is covered with squamose plates, the imbrication of which is centrifugal and thus continuous in direction with the imbrication of the oral face. This suggests that *Cooperidiscus* developed from some more or less globular body covered by plates overlapping in a direction from the base toward the oral pole.

The Edrioasteridae are characterized by differentiation of the aboral surface into three regions: a central area covered with more flexible integument, bearing smaller-than-average plates; a circular frame of relatively stout plates (corresponding to the peripheral ring?); and a peripheral area of plates serially homologous with the interambulacrals of the oral face but a little smaller than the majority of these (Fig. 113,*1*).

In *Pyrgocystis (Pyrgocystis)* the basalmost plates are closely amalgamated so as to form a sort of ferrule (Fig. 113,*2*), while in *P. (Rhenopyrgus)* numerous minute plates are scattered in a coriaceous skin forming a saccate base.

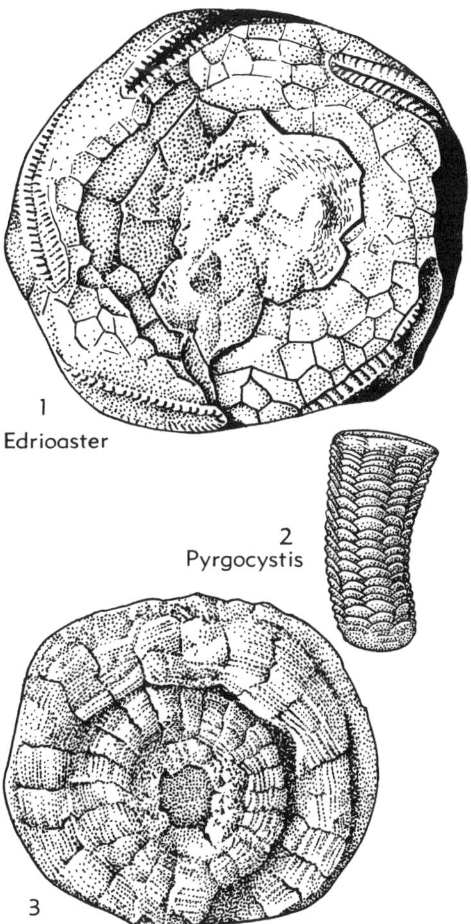

FIG. 113. Aboral surface of edrioasteroids.——*1. Edrioaster buchianus* FORBES, M.Ord., Wales, anal interradius at left, ×1.6 (7, pt. 2).——*2. Pyrgocystis (Pyrgocystis) sulcata* (AURIVILLIUS), U.Sil. (Wenlock.), Sweden (Gotl.), from side, showing smoothly rounded aboral extremity below, ×3 (2).——*3. Discocystis kaskaskiensis* (HALL), U.Miss. (Chester.), USA (Ala.), ×2 (4).

Little information is available about the structure of the aboral side of most other Edrioasteroidea. Conditions similar to those in the Edrioasteridae have been traced in *Lebetodiscus* (7). The lower side of *Discocystis kaskaskiensis* is remarkable in having been described originally as an echinoid, "*Echinodiscus optatus*" WORTHEN & MILLER, 1883. It is composed of many rows of fused imbricating plates (interambulacrals

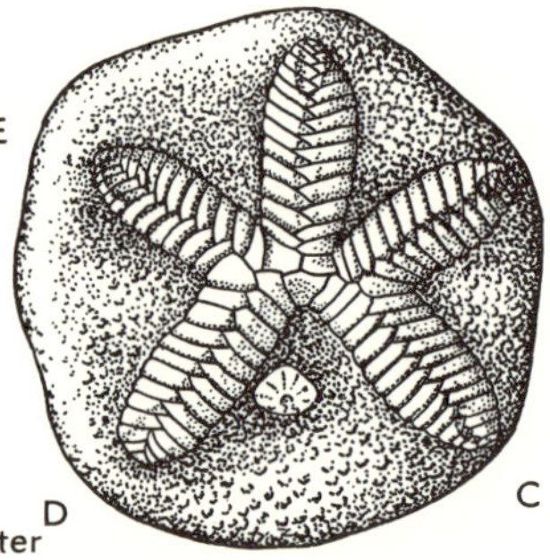

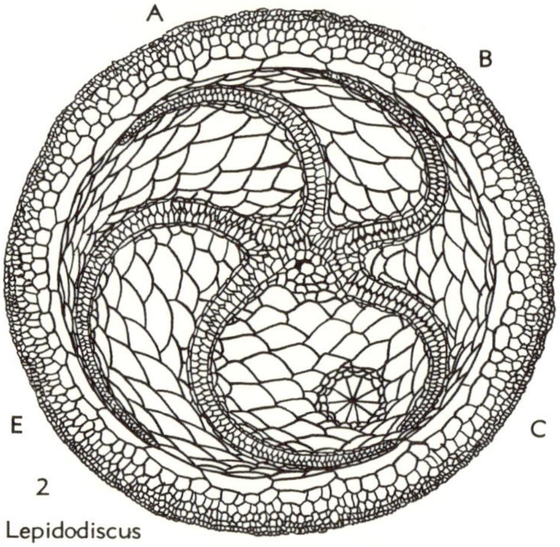

FIG. 114. Edrioasteroids with straight and curved ambulacra.——*1. Cystaster granulatus* HALL, U. Ord., USA (Ohio), a simple type characterized by relatively wide, short, and straight rays, ×4.5 (20). ——*2. Lepidodiscus ephraemovianus* (BOGOLUBOV), U.Dev. (Famenn.), USSR, showing contrasolar curvature of long, narrow ambulacra in all rays except *C,* which is solar, ×2.8 (18).

of the adoral surface mosaic) and a central area of attachment (Fig. 113,*3*).

In species attached by the entire aboral surface to some foreign object, the lower surface is, of course, not accessible for investigation in complete specimens. FOERSTE (16) has reported several specimens of *Carneyella cincinnatiensis* in which a few plates of the adoral surface of the theca were missing. On etching away the clay inside the theca, he found no trace of aboral plates, even though the finest details in sculpture of the shell of *Rafinesquina* supporting the theca were preserved. From this FOERSTE concluded that "it may be assumed that in those forms which assumed the sessile habit, the original plates of the aboral surface became obsolete, a fleshy surface, unprotected beneath, being much better adapted for attachment to an underlying surface."

AMBULACRA

GENERAL CHARACTER

It has been pointed out above that the thecal skeleton is not completely continuous but is interruped by the skeletal elements of the **ambulacra** enclosing the ambulacral groove (subvective groove, food groove). The ambulacral structures are intercalated between (not extended over) the interradial plates. It is likely, however, that from the outset the floor plates formed part of the thecal wall, having acquired later the appearance of a separate system. According to this view, the grooves were originally epithecal, very much as in the Diploporita.

The primitive condition of the ambulacra, as displayed by *Stromatocystites* and by immature stages of several species assigned to other genera, is characterized by a straight course. Straight ambulacra are further found in the Cyathocystidae, some Hemicystitidae (e.g., *Pyrgocystis, Cystaster, Cincinnatidiscus, Hemicystites,* and *Timeischytes*) (Fig. 114,*1*). In *Isorophus* the ambulacra may be almost straight; in *Thresherodiscus* they are straight and repeatedly branch dichotomously, which is quite unique among the Edrioasteroidea.

In all other genera, the ambulacra are more or less curved. The curve may be **contrasolar** (counter-clockwise, to the left) or **solar** (clockwise, to the right). All rays may curve in the same direction, or they may behave differently in this respect (Fig. 114,*2*). All ambulacra curving in a contrasolar direction are found in *Lebetodiscus, Lepidoconia, Streptaster, Ulrichidiscus,* and *Dinocystis.* There is evidence that contrasolar curvature is a primitive feature among species with curved rays, for in many forms (if not all) in which ambulacrum *B* has a solar curve, its proximal part has a distinct tendency to be directed contrasolarly.

Solar curvature of the rays is characteristic of *Foerstediscus, Cooperidiscus,* and certain species of *Edrioaster* (Fig. 115).

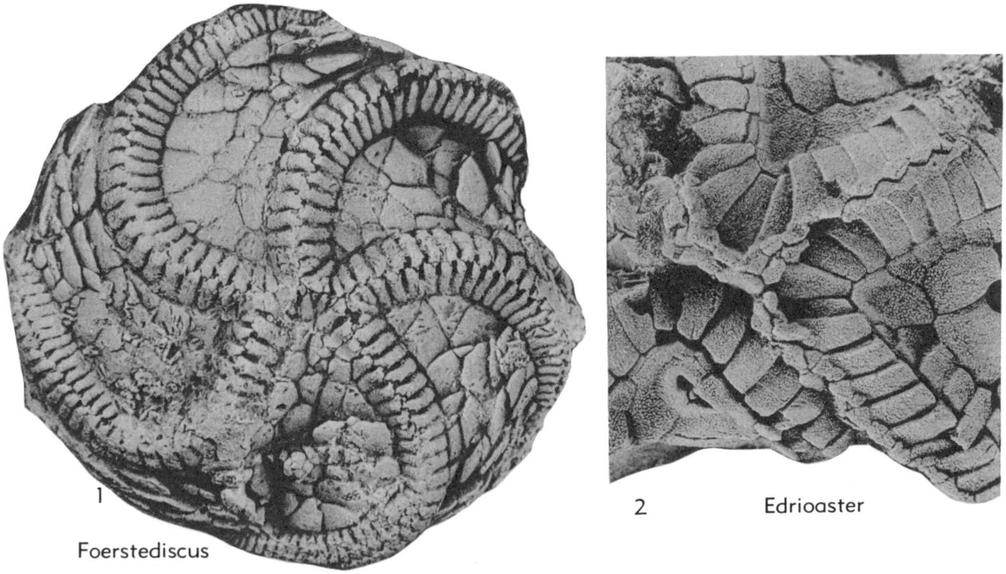

FIG. 115. Adoral views of Middle Ordovician edrioasteroids.——*1. Foerstediscus splendens* BASSLER, showing all ambulacra curved in solar (clockwise) direction, ×2.7 (24).——*2. Edrioaster levis* (BATHER), showing at upper right cover plates of ambulacrum *B* and at lower right those of ambulacrum *C;* relatively large "third aperture" near peristome in posterior interambulacrum at lower left, ×3.3 (24).

Where the rays curve in different directions, the most common case is that the rays *A, B, D,* and *E* are contrasolar and *C* solar (e.g., *Walcottidiscus, Bassleridiscus, Carneyella, Anglidiscus, Isorophus, Lepidodiscus, Discocystis, Edrioaster bigsbyi*). Two rays *(B, C)* that curve in a solar direction and others in a contrasolar direction occur in *Isorophusella* and *Agelacrinites* (and, maybe, *Xenocystites*).

Provided that ambulacra *C* and *D* curve in opposite directions and are long enough to approach each other, their tips are seen to meet approximately on the same level (e.g., *Carneyella* and some *Lepidodiscus*). The distal portion of ambulacrum *C* runs proximally to the distal part of ambulacrum *D* (thus nearest to the periproct) in *Anglidiscus, Discocystis, Edrioaster* and some *Lepidodiscus.* This is also true of some species of *Isorophus* and *Agelacrinites,* whereas in other members of these genera the distal part of ambulacrum *C* passes on the distal side of ambulacrum *D.* There may be a certain variation in this respect.

The diagnostic significance of mode of curvature of the ambulacra has been called in question, most recently by SINCLAIR (36). It is evident that direction of curvature is not constant in all species; nevertheless it seems legitimate to maintain that disposition of the ambulacra follows a pattern characteristic of each genus and species, although curvature is subject to a certain variation in the same way as number of ambulacra, for example.

The solar coiling of ambulacrum *C* has been postulated to originate from a differential pull of gravity on different rays of the growing animal in position attached to the sloping valve of a brachiopod with interradius *BC* assumed to be directed upward and with the anus at right of the mouth. Apart from theoretical considerations, such a position is indicated by the fact that the theca rather commonly exhibits a sag toward the left side.

By slight turning of the theca so that ambulacrum *B* became directed more to the right, this ray as well came under the pull of gravity of that side and acquired a solar curve, too. Once acquired, the mode of curving of the ambulacra might have remained unchanged, on principle, in later generations in the state fixed in ancestral forms and thus unaffected by the more or less at random position of the theca. The deeper explanation of the varying (but apparently specific) orientation of the theca on the sloping surface of the substratum as probably controlling the bend of the rays is not obvious, nor is it easy to comprehend the origin of solar or contrasolar curvature affecting all rays of a theca (7).

The ambulacra are grooves underlain by floor plates, and both walled in and roofed over by cover plates which could open or close at will.

AMBULACRAL FLOOR PLATES

The plates forming the ambulacral grooves are known as **floor plates** (ambulacrals, ambulacralia *auctorum,* adambulacrals, adambulacralia *auctorum*—terms that should be avoided because of their inconsistent usage —subambulacral plates).

In the primitive condition (e.g., *Stromatocystites*), these are arranged in a double series (two rows) of alternating plates meeting in a perradial (mid-radial) line. Biserial floor plates are further met within the Edrioasteridae and, doubtfully, in the Cyathocystidae. The perradial suture is more or less zigzag. In the Edrioasteridae a lateral pore occurs between successive floor plates on each side, thus not piercing the plates but situated in the tangential suture between them (Fig. 116,*1*). The Stromatocystitidae are probably devoid of such pores, whereas the Cyathocystidae are not known in this respect.

The Hemicystitidae and Agelacrinitidae, on the other hand, are characterized by a single row of concave floor plates (Fig. 116,*2*) which may partly have become fused but still separated at intervals. An apparent exception to the rule that no pores are present between the uniserial floor plates is offered by *Anglidiscus,* according to a statement by ANDERSON (1) (Fig. 116,*3*). BATHER (7) pointed out that the floor plates of *Lebetodiscus* were disposed in a double series of alternating plates with intervening

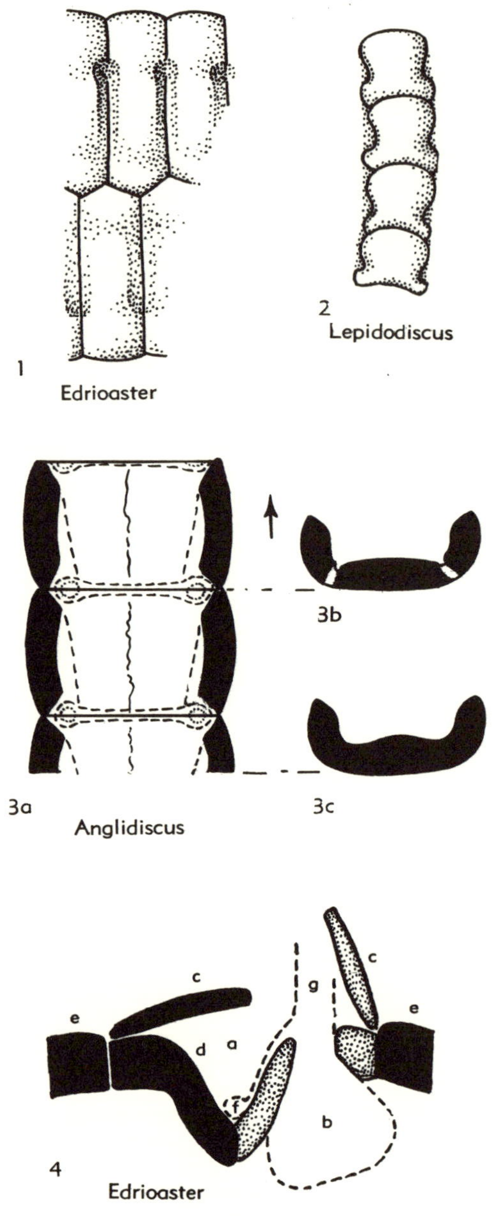

FIG. 116. Ambulacral floor plates of edrioasteroids.——*1. Edrioaster bigsbyi* (BILLINGS), M.Ord., Ontario; upper (adoral) surface of floor plates showing grooves that lead from pore depressions to perradial canal, ×10 (7, pt. 4).——*2. Lepidodiscus beecheri* (CLARKE), Miss., USA (Pa.); lower (aboral) side of floor plates, ×6 (9).——*3. Anglidiscus fistulosus* (ANDERSON), L.Carb., Eng., diagrams of floor plates, arrow pointing toward mouth, ×30 (1); *3a,* plan of mature floor plates from center of ambulacrum; *3b,c,* transverse sections of floor plate at suture and through center.——*4. Edrioaster* sp., diagrammatic transverse section across ambulacrum showing relations of cover plates, floor plates and inferred water vessel (ampulla) with its connection to tube foot through pore (natural suture surfaces stippled, cut surfaces solid black, supposed soft parts indicated by broken line) [*a,* ambulacral groove; *b,* ampulla; *c,* cover plate; *d,* floor plate; *e,* interambulacral plate; *f,* perradial water vessel; *g,* tube foot] (7, pt. 2).

pores. As first observed by RAYMOND (28), this proposal is due to misinterpretation of the ambulacral skeletal structure in this instance.

The floor plates of *Stromatocystites* are fairly polymorphous but invariably more or less elongated in a radial direction, alternating irregularly. These structures are more fully known in the Edrioasteridae, thanks to BATHER's thorough investigations. *Edrioaster bigsbyi* may be taken as an example (condensed from BATHER). In this species the floor plates are elongate at right angles to the perradius. Their outer margins are convex and their abutting margins straight. The plates rise sharply up from the interradial areas, forming a rounded margin on each side of the groove and then they dip almost straight down to the perradial suture. Proximally, the depression thus formed is more marked, whereas distally it becomes slighter and almost disappears. The suture between two floor plates on the same side of the groove is depressed from a point just within the rounded margin right down to the perradial suture. The depression is deepest at its outer end, where also it is slightly expanded in circular manner to form a pore under lateral margins of the cover plates. The perradial suture also is depressed. At the extreme distal end of a ray, the floor plates diminish considerably in size but continue to alternate and are arranged fanwise. The canals of the lateral pores passing into the thecal cavity run obliquely downward about parallel with the sutures between the floor plates and the adjacent interambulacrals, that is, sloping from the exterior inward toward the perradius. The pores must have been podial pores housing a tube foot connected with a perradial water vessel and an endothecal ampulla (Fig. 116,*4*).

Other members of the Edrioasteridae agree, on principle, with *Edrioaster bigsbyi* with regard to structure of the floor plates, though, of course, minor differences occur.

It is evident that the development of floor plates as described is largely influenced by the presence of a tube-foot ampulla system. Where such a system is absent, as in the Hemicystitidae (*Anglidiscus* apparently making a bewildering exception) and the Agelacrinitidae, the skeletal elements beneath the ambulacral groove are less differentiated. As pointed out above, the floor plates are uniserial in the families just referred to. There is hardly anything to indicate whether the uniseriality arose by a "straightening" of the biserial row or by fusion of a pair of more or less opposed plates of the biserial row.

In some forms the uniserial floor plates exhibit a considerable overlap in a proximal direction as seen from above. FOERSTE (16), who paid much attention to intimate structure of the Edrioasteroidea, observed that the "proximal overlapping suggests that the floor plates may be modified thecal plates belonging to the upper face of the theca—the food grooves extending over the thecal plates themselves, without intermediate flooring." However that may be, the ambulacral system has to be treated as a morphological unity.

Not uncommonly, the individual floor plates (speaking of the uniserial type only) were formed in fact by fusion of two primitive floor plates, one proximal and another distal. The floor plates are not invariably broad enough to underlie the entire width occupied by the cover plates and thus basal extensions of the cover plates may project beyond their margins (e.g., *Carneyella pileus*). The floor plates are quadrangular or are much wider than long (in distal and proximal regions of the ambulacra). The median portion of their adoral surface is excavated radially. The (ambulacral) groove thus formed may be fairly wide (e.g., *C. cincinnatiensis*) or extremely narrow (e.g., *Streptaster,* in which each whole ambulacrum is very narrow). A much fainter groove has been stated to run, in several species, along the border of the floor plates, on each side of the ambulacral groove. These lateral grooves had some connection with the fulcra of the cover plates, by means of which the latter were opened and closed over the ambulacral furrow.

AMBULACRAL COVER PLATES

Irrespective of the biseriality or uniseriality of the floor plates, the ambulacral **cover plates** (adambulacrals, adambulacralia *auctorum*, —cf. "Ambulacral Floor Plates" above—*Saumplättchen*) are invariably bi-

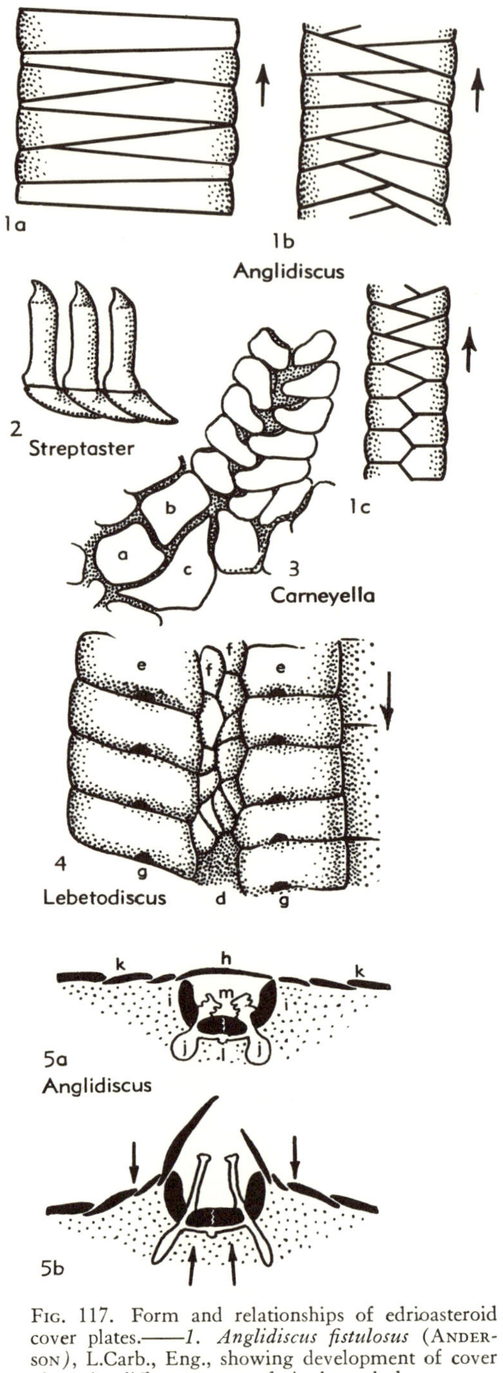

Fig. 117. Form and relationships of edrioasteroid cover plates.——*1. Anglidiscus fistulosus* (Anderson), L.Carb., Eng., showing development of cover plates in different parts of single ambulacrum, arrows pointing toward mouth, ×28 (1); *1a-c,* proximal, medial, distal.——*2. Streptaster septembrachiatus* Miller & Dyer, U.Ord., USA (Ohio), 3 vertically disposed palisade-like cover plates borne by ambulacral floor plates, seen from side, enlarged (16).——*3. Carneyella pileus* (Hall), U.Ord., USA (Ohio); peristomial region with orotegminals (*a, b,* left and right ant.; *c,* post.), proximal part of ambulacrum *B* showing boot-shaped lateral cover plates, ×12 (37).——*4. Lebetodiscus dicksoni* (Billings), M.Ord., Ontario, part of ambulacrum, arrow pointing toward mouth (*d,* ambulacral groove; *e,* lateral cover plate; *f,* median cover plate; *g,* pore between lateral cover plates), ×6 (7, pt. 3).——*5. Anglidiscus fistulosus;* hypothetical transverse sections of ambulacral groove and adjacent skeletal elements indicating supposed mechanism of cover-plate movements, enlarged; *5a,* animal deflated, with groove closed; *5b,* animal inflated, with groove open (*h,* ambulacral cover plate; *i,* ambulacral floor plate; *j,* ampulla; *k,* interambulacral plate; *l,* radial canal; *m,* tube foot) (1).

serial, although this is concealed in some very few special examples. These plates form the walls and roof of the canal made up of the ambulacral skeletal elements.

The cover plates stand on the upper margin or outer extension of the floor plates, which, more or less, form lateral shelves (Fig. 116,*4*). In various genera (e.g., *Edrioaster*), the cover plates have been reported to articulate on a beveled facet just within the edge of the radial groove. Primarily, a single cover plate matches each floor plate among those arranged biserially and a pair of cover plates corresponds to each floor plate among those arranged uniserially. The crowding and doubling of the cover plates seen in many edrioasteroid species is correlated with the aforementioned fusion in a radial direction of uniserial floor plates (40).

In *Anglidiscus* and, less pronounced, in *Lepidodiscus lebouri,* there is an apparent uniseriality of the cover plates in the proximal region of the ambulacrum (Fig. 117,*1*). However, as observed by Anderson (1) in *Anglidiscus,* structural details indicate that this is very likely a masked biserial arrangement. This interpretation is corroborated by the fact that, in *Anglidiscus,* a gradual transition is observed in a distal direction into very regular pentagonal plates interlocking along the perradius.

The type of cover plates last mentioned is characteristic of many early Edrioasteroidea, although the cover plates may, of course, vary a good deal in general outline according to width of the ambulacrum, number and length of the floor plates, and other characters. They are very regular and symmetrical in the Edrioasteridae as well. In *Stromatocystites* they are still little advanced, being minute and less regular.

However, in many species examined in this respect, the cover plates are differentiated further morphologically. They are more or less boot-shaped, with the "sole" of the boot proximal and the "toe" admedian (Fig. 117,*3*). In consequence, they cannot effect closure of the groove as completely as do the symmetrical cover plates of the Edrioasteridae, Cyathocystidae (e.g., *Cyathocystis*), and others. But this drawback is counteracted in many instances by the presence of additional cover plates along the median line (first and foremost in *Lebetodiscus, Carneyella,* and *Thresherodiscus,* further in *Isorophus, Isorophusella,* and *Edrioaster*) The ordinary cover plates have been interpreted by Foerste (16) to be distinguished as "lateral cover plates," in opposition to "median" or "intercalated cover plates" (Fig. 117,*4*). Those last mentioned are smaller, and only their triangular tips may be seen intercalated between the tips of the lateral cover plates, interlocking along the perradius. Some evidence indicates that the additional cover plates abutted against the floor plates by means of facets. However, a perfect closing mechanism could also be achieved in forms provided with both medial and lateral projections of the lateral cover plates proper, as described by Sinclair (36) in a new species of *Foerstediscus*. The medial projections interlock in a closely fitting perradial zigzag suture.

Streptaster is affected by a conspicuous elongation of its cover plates so as to cause a palisade-like effect, especially along the concave curvature of the distal parts of the rays, when observed in lateral view (Fig. 117,*2*). A similar tendency is obvious in *Bassleridiscus.*

In *Timeischytes* only four or five cover plates occur on each side of the perradial line.

The movements of the cover plates may have been brought about by the action of muscles. Small processes interpreted as points of muscular attachment have been recognized at the lateral end of the lower side of the cover plates in several species (36, 40). It has also been suggested, however, that the movements were less due to muscular activity of the cover plates than to contraction and relaxation of the thecal skeleton as a whole in forms with overlapping interambulacrals. The lateral edges of the cover plates appear to have been overlain, in some forms, by inner edges of the adradial interambulacrals. Inflation of the test would result in "pressure on the outer edge of the covering plates, causing them to rotate about the upper edge of the flooring plates, so that their inner ends become elevated and the ambulacral groove uncovered" (1) (Fig. 117,*5*). Whether the specifically developed cover plates in the distal part of the ambulacra in *Anglidiscus,* stretching over the entire width of the ambulacrum, were actually affected by the movement mechanism may be questioned.

Pores between the ambulacral cover plates appear in some few genera (e.g., *Lebetodiscus, Lepidoconia*). In *Lebetodiscus,* a single large pore occurs in the suture between each pair of lateral cover plates (Fig. 117,*4*), whereas in *Lepidoconia* there are 5 pores. In so far as can be ascertained, these are opposite on either side of each plate, but those of adjacent plates are alternate (41).

As opposed to the pores connected with the floor plates, those of the cover plates cannot possibly have been openings for tube feet, for obvious reasons. They may have played a role for circulation of water carrying food particles to the subvective groove. The most probable explanation may be that they were outlets for excessive water rather than intakes for water. But why the genera mentioned, and only those, were provided with such pores remains obscure.

PERISTOME

From a strictly morphological point of view, plates constituting the peristome should not be separated from the ambulacral and, in part, interambulacral skeleton. Yet it is convenient to deal separately with elements integrated in the peristomial (and substomial) structures. These have been described and discussed fully by Bather (7), following whom very little information has been added until 1960, when a study by Kesling & Mintz (26) was published.

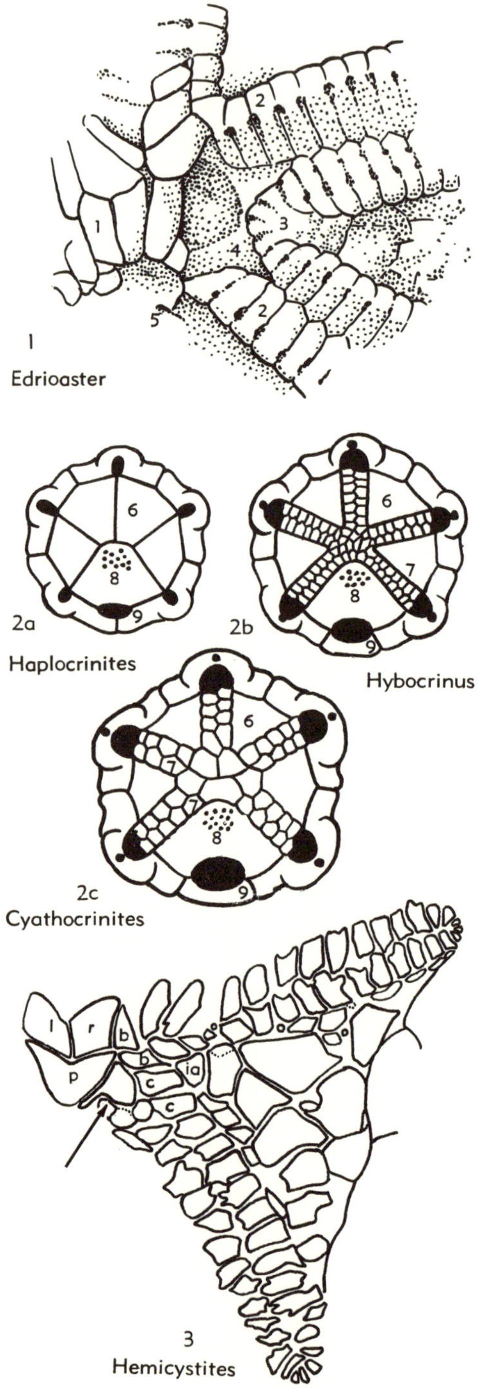

FIG. 118. Morphological features of peristomial region, viewed from above (adorally).——*1. Edrioaster bigsbyi* (BILLINGS), with proximal parts of *B* and *C* ambulacra at right (*1*, ambulacral cover plate; *2*, ambulacral floor plate; *3*, interradial peristomial element; *4*, radial peristomial element; *5*, "third aperture"), ×4 (7, pt. 4).——*2*. Stages in evolution of crinoid tegmen for comparison (diagrammatic); *2a, Haplocrinites; 2b, Hybocrinus; 2c, Cyathocrinites* (*6*, deltoid or oral plate; *7*, ambulacral cover plate; *8*, madreporite, *9*, periproct) (6).——*3. Hemicystites rectiradiatus* (SHIDELER), part of adoral surface, arrow indicating probable location of "third aperture" (inferred hydropore) (*l,p,r*, left, posterior, and right orotegminals); (*b,c*, ambulacral cover plates of *B* and *C* ambulacra, respectively; *i,a*, interambulacrals), ×9 (40).

The peristome is more or less similarly developed in all genera examined. The internal peristomial skeleton consists of a firm subcircular or subpentagonal mouth frame ("substomial chamber"), through which the gullet receiving the ambulacral grooves passed into the stomach. This was evidently attached to the inner adcentral border of the mouth frame. This frame is made up of five perradial plates, triangular as viewed from above (after removal of the plates covering the mouth) (Fig. 118,*1*), and five interradial plates, seen only in aboral aspect. Morphogenetically, all plates mentioned, apart from the elements in interradius *CD*, are undoubtedly fused portions of floor plates. In the interradius *CD* (posterior) the frame is integrated by part of the large interradial plate that is pierced by the canal interpreted generally as a hydropore canal (stone canal), and, contingently, by further peristomial interradial plates.

Also the peristomial cover plates forming a tegmen were derived mainly from the primitive ambulacral elements, namely cover plates which were modified along somewhat different lines in various groups of Edrioasteroidea.

BATHER (6, p. 100) has outlined three stages in the evolution of the tegmen in Crinoidea (Fig. 118,*2*): (1) tegmen composed of five large plates (deltoids or orals) covering the ambulacra and mouth (e.g., *Haplocrinites*); (2) deltoids as in the preceding but the ambulacra with alternating cover plates and the mouth covered by an indefinite number of modified ambulacral cover plates (e.g., *Hybocrinus*); (3) as in the preceding but with plates covering the mouth enlarged and reduced in number, usually five (e.g., *Cyathocrinites*).

A certain parallelism may be pointed out between development of the orotegminal plates in the Crinoidea and Edrioasteroidea, although the phylogenetic significance of this is not very evident in the latter. In the Stromatocystitidae the peristomial cover does not agree with the primitive type just indicated but consists of four large plates and an indefinite number of smaller ones. Whether the large plates held a definite position cannot be made out on the evidence available.

However, the first stage recognized by BATHER is apparent in *Cyathotheca.* The second stage is represented by the Agelacrinitidae, Edrioasteridae, and, perhaps, by the Stromatocystitidae. The third stage is found first and foremost in *Cyathocystis,* but also the Hemicystitidae can properly be included. Both in *Cyathocrinites* and in the Hemicystitidae the mouth slit opens transversely with respect to the sagittal plane (Fig. 118,*3*). The posterior orotegminal, located in interradius *CD,* is the largest one, its adcentral margin forming more or less a straight line transversely to the sagittal plane. In *Cyathocystis* five orotegminals are seen, as in *Cyathocrinites,* but in the Hemicystitidae the number is reduced to three, according to interpretation by BASSLER and others. The anterolateral orotegminals, which are thus smaller in size than the posterior one, were probably derived morphogenetically from ambulacral coverplate elements belonging to the ambulacra *E, A,* and *B.* They border on each other by a straight suture in the perradius of the anterior ambulacrum *(A).*

KESLING & EHLERS (25) and EHLERS & KESLING (13) have the opinion that, in *Carneyella pileus,* one additional plate, namely that designated by FOERSTE (16) as plate "X" and representing according to him one of the interambulacral plates in the posterior interambulacrum, is probably part of the peristomial region. This would be true also of *Hemicystites* and related genera. A still greater number of peristomial cover plates are present in *Timeischytes,* including two large, elongate, subpentagonal ones in the anterior part of the peristome and six diversiform plates in the posterior part of the peristome. However, these orotegminals are arranged in a definite pattern unique of its kind. Additional investigation is required before it can be stated that all genera included in the Hemicystitidae are characterized by having more than three plates covering the peristome.

PERIPROCT

In disc-shaped edrioasteroids the periproct is invariably located on the same surface as the peristome. Also in the saclike Cyathocystidae it lies on the adoral surface, whereas in *Cystaster* it has a lateral position. It is located in the posterior interambulacrum *(CD).* In many genera (the bulk of these belonging to the Hemicystitidae, apart from *Cystaster* and *Anglidiscus* and further in *Ulrichidiscus, Discocystis, Isorophusella, Agelacrinites, Dinocystis,* and others) it lies fairly central in this interambulacrum, whereas in others (e.g., *Stromatocystites, Lepidodiscus, Cooperidiscus, Edrioaster,* and others) it is more or less close to the posterior margin of the interambulacrum. In *Timeischytes* the pyramid is in contact with the peristomial region and occupies a good deal of the posterior interambulacrum. *Cyathotheca* differs from all other known Edrioasteroidea in that the periproct occupies a lunate area of the adoral surface behind the posterior orotegminal. In *Cyathocystis* the anal opening lies on the border of the posterior and adjacent marginals.

Not uncommonly, two sets of periproctal plates can be recognized, namely the circumanal plates proper, which form a more or less pyramidal cover of the anal opening **(anal pyramid),** and the distal periproctal plates surrounding the anal pyramid (e.g., *Lebetodiscus, Anglidiscus, Foerstediscus, Isorophusella,* and others). But in many genera the anal area is occupied entirely by the anal pyramid. On the other hand, no distinct anal pyramid may be present, the anus being covered by a number of irregular plates (e.g., *Edrioaster*) (Fig. 119,*1*). The construction of the anal cover in *Cyathotheca* is unknown, but the anal opening differs from that of all other Edrioasteroidea in being lunate. Otherwise, the periproct—whether consisting of a simple pyramid or of numerous plates—is more or less circular, oval, or polygonal in outline. The shape of the anal pyramid is dependent

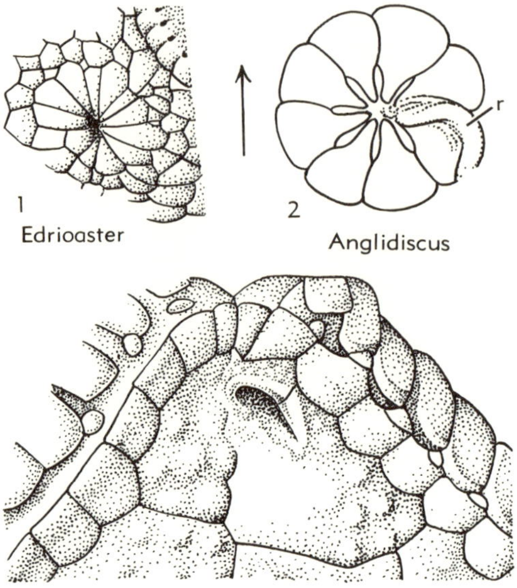

FIG. 119. Features of periproct and "third aperture."——*1.* Periproct of *Edrioaster bigsbyi* (BILLINGS), M.Ord., Ontario, ×6.5 (7, pt. 4).——*2.* Inner side of anal pyramid of *Anglidiscus fistulosus* (ANDERSON), L.Carb., Eng., showing at right impression (r) which may denote rectum, arrow pointing toward mouth); ×16 (1).——*3.* Part of posterior interambulacrum of *Edrioaster bigsbyi* adjacent to peristome, showing so-called third aperture, ×6 (7, pt. 4).

on the number of plates composing it. Mostly, these are regularly triangular pieces. In *Timeischytes* they are only four in number; in *Cyathocystis* they are five, but other genera have a pyramid made up by 6 to 15 plates. The distal periproctal plates are usually smaller and more irregular in outline than those of the pyramid, and this is true also of plates protecting the vent in genera devoid of a regular anal pyramid.

The structure of the anal pyramid warrants the conclusion that respiration through the anus played at least a certain role in edrioasteroids (Fig. 119,*2*).

THIRD APERTURE

The noncommittal designation "third aperture" (the other two apertures in the thecal wall of the adoral surface being the mouth and anus) has been preferred for the small opening which for a long time has been known to be present in the posterior interambulacrum, behind the mouth, notably in the Edrioasteridae (Fig. 111, 131). It is reasonable to assume that a third aperture was present in many other Edrioasteroidea, although not noticed—possibly obliterated—in the actual fossil material. Lately, KESLING (24) has published a very careful study of a number of edrioasteroids in which he was able to demonstrate the presence, in the right posterior part of the peristomial region or in the nearby part of the posterior interambulacrum, of a structure that he interpreted as a hydropore. Additional comments were presented by KESLING & MINTZ (26). However, in some genera an external pore may have been wanting throughout ontogeny. If so, we have to reckon with a stone-canal opening in the body cavity, perhaps differentiated into a number of weak tubes hanging down from the water ring, as in most holothurians.

The structure of the third aperture was well known only in *Edrioaster* prior to the appearance of KESLING's (24) 1960 paper. BATHER (7, pt. IV) reported the presence of a small, obliquely transverse, slightly curved slit that crossed at right angles the suture between two plates at the adoral end of the interradius *CD* (Fig. 119,*3*). A slight widening and deepening of the aperture in an adoral direction indicates that the canal for which it served as intake (or outlet, according to the function assigned to the pore) passed obliquely through the test from right to left.

To illustrate the nature of the third aperture in different edrioasteroids (Fig. 120) it may be well to quote in full the paragraphs of KESLING (24) in which he summarizes the six basic types of hydropores (the third aperture is referred to throughout as the "hydropore") that his study has led him to recognize.

> *Type I.*—Hydropore within the posterior interambulacrum, in the right proximal region, consisting of a large permanent opening shared by two plates. *Edrioaster* the only known genus having this type.
>
> *Type II.*—Hydropore within the posterior interambulacrum near ambulacrum V [*i.e.*, ambulacrum *C*], not a large opening. Subtype A, exemplified by *Thresherodiscus,* consists of two large "bordering plates," each semioval, with a long juncture between them. When submersed in xylol, the holotype of *T. ramosus* shows a dark area along the juncture, suggesting that a larger opening underlies the thin margins of these

plates. Subtype B, exemplified by *Foerstediscus splendens,* apparently involves three plates in the posterior interambulacrum around a short slot set perpendicular to the edge of ambulacrum V. The anterior of these plates is fused to another interambulacral plate and, judging from the appearance in xylol, is very thin. The arrangement is such that the larger of the two fused plates may have acted like a hinge.

Type III.—Hydropore a small opening along

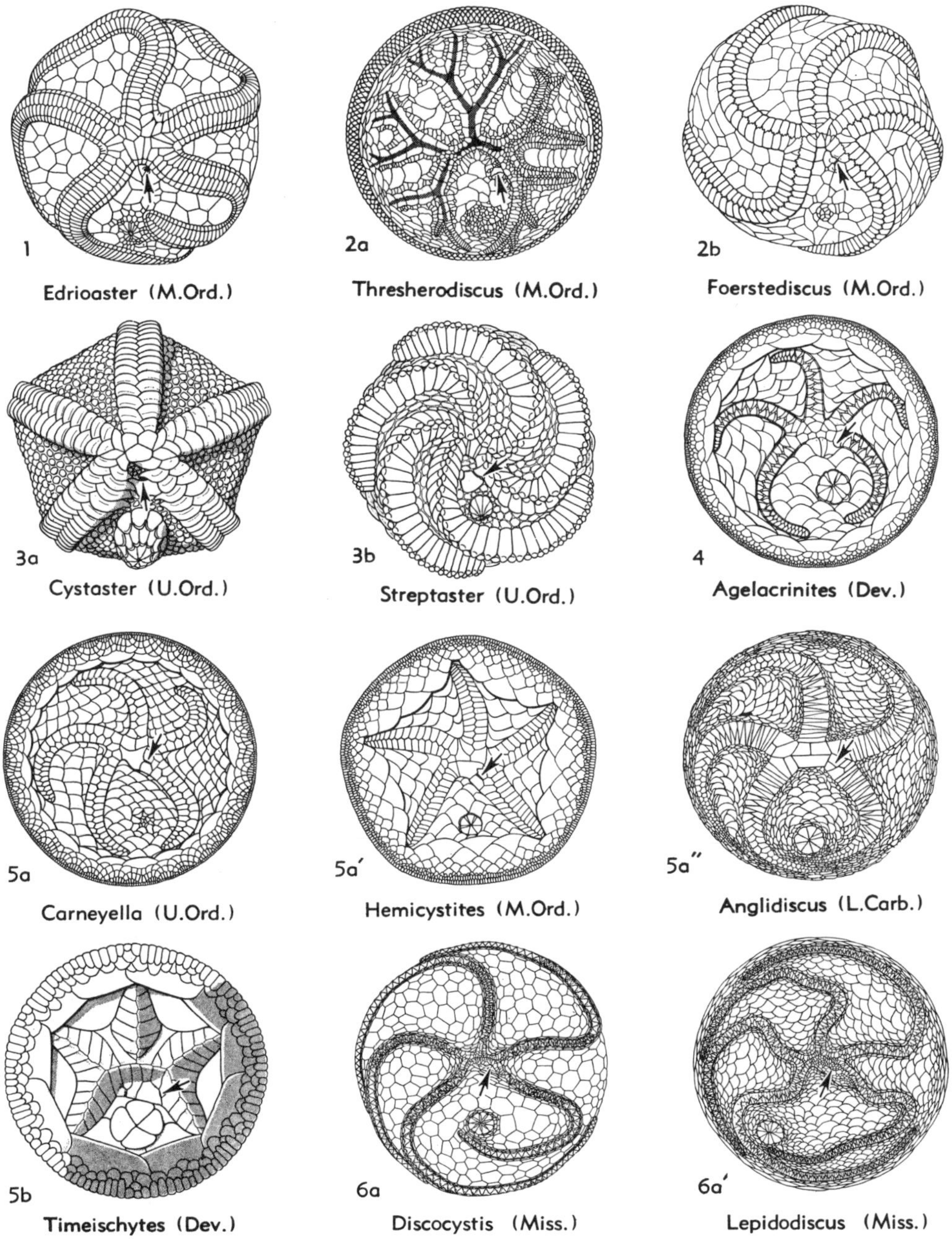

FIG. 120. Types of hydropores distinguished by KESLING, illustrated by restorations: I *(1)*; IIA *(2a)*, IIB *(2b)*; IIIA *(3a)*; IIIB *(3b)*; IV *(4)*; VA *(5a,a′,a″)*, VB *(5b)*; VI *(6a,a′)* (24).

the boundary between the posterior interambulacrum and the edge of ambulacrum V. In subtype A, found in *Cystaster granulatus,* it is bounded posteriorly by a very large interambulacral plate and anteriorly by a long lateral extension of an ambulacral plate. In subtype B, noted in *Streptaster,* the opening is much smaller and extends along the edge of ambulacrum V to the right of an unusually large interambulacral plate.

Type IV.—Hydropore along the juncture between an expanded proximal ambulacral plate and a large plate of the peristomial region. The posterior part of the peristomial region projects into the posterior interambulacrum, but is not distinctly set off from it, and includes two plates with raised edges that form a vaulted structure where they join. Possibly, one of these plates hinged on the other to expose more of the hydropore. *Agelacrinites* is the only genus in which this type occurs.

Type V.—Hydropore along the juncture between a short proximal ambulacral plate and the right posterior plate of the peristomial region. In subtype A, exemplified by *Carneyella pilea, Hemicystites chapmani, Isorophus cincinnatiensis,* and *Anglidiscus fistulosus,* the peristomial plate is aligned with the left series of covering plates of ambulacrum V, so that it appears to be the proximal left plate of the ambulacrum. However, if the plate arrangement is more carefully analyzed, it will be seen that the ambulacrum is sharply indented to accommodate this plate, and the plate must then be regarded as the right posterior extension of the peristomial region. In subtype B, represented by *Timeischytes megapinacotus* and *Hemicystites devonicus,* the right posterior plate is larger and more clearly associated with the peristomial region. Although the proximal plates on the left side of ambulacrum V are shorter than those on the right side of ambulacrum I [*i.e.,* ambulacrum *D*], directly opposite, they do not completely accommodate the anomalous unpaired plate, which does not appear to be a continuation of the ambulacrum.

Type VI.—Hydropore within the right posterior part of the peristomial region. Although this transverse structure is in part bordered by the covering plates of ambulacrum V, it is not closely related to the ambulacrum. *Discocystis laudoni* and *Lepidodiscus squamosus* furnish good examples, and *L. ephraemovianus* also seems to belong to this type (cf. Fig. 120 reproduced from KESLING, 24).

Various alternatives can be suggested in interpreting the nature of the third aperture. Either it was the intake for water feeding the water-vascular system (i.e. a **hydropore**), or it was the outlet for the sexual product (i.e., a **gonopore**). Contingently, it was a common opening for the stone canal and gonoduct (i.e., a **hydrogonopore**).

The hydropore alternative, which is the orthodox and at the same time the prevalent one, includes the possibility that the gametes, if produced in a single gonad, were discharged through the mouth or anus but, if produced in five gonads, through the ambulacral grooves. In this case there was only one stone canal.

The gonopore alternative implies that there was a single gonad, since we would hardly expect a common opening for, say, five gonads. The probable behavior of the stone canal in this case has already been touched upon above. A hydropore piercing the outer wall of the theca may also have been lacking on account of having migrated into the rectum (as probable in the Blastoidea), or into the mouth.

The hydrogonopore alternative, finally, would mean an organization very much the same as in certain cystoids in which the hydrocoelic and genital openings have joined together to form a common aperture lying between the mouth and anal opening, as shown by GISLÉN (19).

It is difficult to decide which alternative should be given preference; in fact, it seems impossible to give a definite answer. In support of interpretation of the third aperture as a hydropore, BATHER (7) pointed out that the presence of this structure correlates with the presence of pores between the ambulacral floor plates, whereas in the Agelacrinitidae the apparent absence of a hydropore is correlated with the absence of pores between the ambulacral floor plates. BATHER's reasoning does not hold good in full any longer because a third aperture is found in several Agelacrinitidae, as demonstrated by KESLING (24). On the other hand, the argument is not decisively weakened by the fact that the Edrioasteroidea very likely had only one gonad (as judged by CUÉNOT, 10, who, also, favors the gonopore hypothesis). Morphologically, the slitlike third aperture in *Edrioaster,* for example, recalls the aperture generally interpreted as a hydropore in the Hydrophoridea, which exhibit two openings between the mouth and anus, that nearest the mouth being considered as the hydropore (cf. GISLÉN, 19).

STEM

A stem of normal pelmatozoan character is not found in any edrioasteroid (*Astrocystites* having been removed to the class Edrioblastoidea). In a few genera, the vertical axis of the theca itself is prolonged very pronouncedly as compared with the disclike theca of most Edrioasteroidea. In the case mentioned, the theca, exclusive of the adoral surface, functionally served as a stem. Such conditions are met within the Cyathocystidae, in which this portion of the theca forms a solid mass, as well as in *Cystaster* and *Pyrgocystis* among the Hemicystitidae. The most extreme manifestation of the tendency toward thecal prolongation shows up in the turret-like form of *Pyrgocystis,* the aboral pole of which is covered with a ferrule-like structure.

ORNAMENT

The ornament of edrioasteroids, if present, commonly is found on all plates of the adoral face but it may be restricted mainly to the interambulacrals and marginals. Spines may have been present in species of various genera. Because of their fragility, however, such structures are not likely to be preserved in the fossil state. As a matter of fact, the only known example of genuine spines is afforded by *Pyrgocystis sardesoni,* in which the ambulacral cover plates were provided with spines articulating on tubercles. Their function as structures protective of the ambulacral groove is not as obvious as in the case of the adambulacral spines in Asteroidea, which have open grooves without cover plates. Spinules seem to have been present on the edges of the turret plates in this and other species of *Pyrgocystis.*

Pustules, produced more or less into papillae or spines, are found in *Carneyella* (especially in *C. vetusta* and *C. ulrichi,* in which the ornament has been interpreted erroneously as a coating of the stromatoporoid *Dermatostroma*) and in *Edrioaster.* Knobs, so small as to be classified as granules, occur on the plate surfaces in *Lepidodiscus.* In *Foerstediscus calderi* all plates are rather coarsely granulose.

Interambulacrals sculptured with heavy ribs, a type of ornament otherwise not typical of the Edrioasteroidea, occur in *Agelacrinites hamiltonensis* and a couple of other species assigned to this genus.

More or less minutely pitted plates are found in *Carneyella* and *Anglidiscus,* and in some species of *Hemicystites* (e.g., *H. paulianus*).

GLOSSARY OF MORPHOLOGICAL TERMS APPLIED TO EDRIOASTEROIDEA

[Terms of lesser importance are printed in italics.]

A ray (CARPENTER). Ray (anterior) located opposite interray designated *CD* (posterior) that contains the periproct; corresponds to ray III in system advocated by JAEKEL.

AB interray (CARPENTER). Interray (anterior right) next adjoining *A* ray in clockwise direction when edrioasteroid is viewed from adoral (ventral) side; between *A* and *B* rays; corresponds to interray III-IV in system advocated by JAEKEL.

abactinal. Applied to side of theca or plate opposite actinal (oral) surface of edrioasteroid (syn., aboral, adapical, apical, dorsal).

aboral. Applied to part of theca or plate directed away from mouth, surface of theca opposite that bearing mouth, in edrioasteroids directed downward (syn., abactinal, adapical, apical, dorsal).

actinal. See oral.

adambulacral (adambulacralium, pl. *adambulacralia). See* floor plate.

adapical. See aboral.

admedian. Applied to skeletal element located along median line of ray.

adoral. Applied to surface of theca or plate directed toward mouth.

adradial. Applied to small plates lining ambulacra in some edrioasteroid genera.

ambulacral (ambulacralium, pl., *ambulacralia). See* floor plate.

ambulacral elements. Plates forming ambulacral system.

ambulacral groove. Groove formed by double or single row of floor plates of ambulacra; served to convey food particles to mouth by means of ciliary currents (syn., food groove, subvective groove).

ambulacral system. Organ system peculiar to echinoderms, its main elements being a ring canal encircling the mouth, and 5 radial ambulacral vessels radiating from the ring canal and lodged in the ambulacral grooves.

ambulacrum (pl., **ambulacra**). Any of 5 straight or curved skeletal zones in rays (radii) of theca, enclosing ambulacral groove; adj., ambulacral.

ampulla. Vesicle protruding into perivisceral coelom and associated with a tube foot so that, on contraction of the ampulla, fluid can pass into the tube foot, or, reversely, on contraction of the tube foot, into the ampulla.

anal pyramid. Cover of anal opening composed of

several more or less triangular plates forming conical elevation.

apical. See aboral.

arm. Portion of ray extending from theca (not present in Edrioasteroidea).

B ray (CARPENTER). Ray (right anterior) next to *A* (anterior) ray in clockwise direction when edrioasteroid is viewed from adoral (ventral) side; corresponds to ray IV in system advocated by JAEKEL.

BC interray (CARPENTER). Interray (posterior right) next adjoining *B* ray in clockwise direction when edrioasteroid is viewed from adoral (ventral) side; between *B* and *C* rays; corresponds to interray IV-V in system advocated by JAEKEL.

brachiole. Biserial, nonpinnulate exothecal appendage springing independently from surface and containing no extension of the body systems; Edrioasteroidea are devoid of brachioles.

C ray (CARPENTER). Ray (right posterior) next to *B* ray in clockwise direction when edrioasteroid is viewed from adoral (ventral) side; corresponds to ray V in system advocated by JAEKEL.

CD interray (CARPENTER). Interray (posterior) next adjoining *C* ray in clockwise direction when edrioasteroid is viewed from oral (ventral) side; between *C* and *D* rays; differs frequently in shape from the other interrays and contains the periproct; corresponds to interray V-I in system advocated by JAEKEL.

cover plate. Any of biserially arranged plates forming walls and roof of ambulacral groove (syn., adambulacral, ambulacral, Saumplättchen).

D ray (CARPENTER). Ray (left posterior) next to *C* ray in clockwise direction when edrioasteroid is viewed from adoral (ventral) side; corresponds to ray I in system advocated by JAEKEL.

DE interray (CARPENTER). Interray (posterior left) next adjoining *D* ray in clockwise direction when edrioasteroid is viewed from adoral (ventral) side; between *D* and *E* rays; corresponds to interray I-II in system advocated by JAEKEL.

deltoid. *See* oral.

[**diplopore.** Any of double pores piercing thecal plates in Diploporita (Hydrophoridea); absent in Edrioasteroidea but employed for a morphologically remotely similar structure in *Stromatocystites.*]

dorsal. Referring to direction or side away from mouth, in edrioasteroids directed downward. *See* aboral.

dorsocentral. Plate in the aboral center of *Stromatocystites*; not necessarily homologous with dorsocentral of other echinoderms.

E ray (CARPENTER). Ray (left anterior) next to *D* ray in clockwise direction when edrioasteroid is viewed from adoral (ventral) side; corresponds to ray II in system advocated by JAEKEL.

EA interray (CARPENTER). Interray (anterior left) next to *E* ray in clockwise direction when edrioasteroid is viewed from adoral (ventral) side; between *E* and *A* rays; corresponds to interray II-III in system advocated by JAEKEL.

eleutherozoic. Applied to free-living echinoderm.

endothecal. Applied to ambulacral plates that pass between thecal plates.

[**epithecal.** Applied to ambulacral plates that *rest on* thecal plates (absent in edrioasteroids).]

[**exothecal.** Applied to structures, like brachioles, not incorporated in theca proper (does not occur in edrioasteroids).]

floor plate. Any of double or single row of plates forming an ambulacral groove (syn., adambulacral, ambulacral, subambulacral).

food groove. *See* ambulacral groove.

gonopore. Simple opening serving as exit from genital system.

hydrogonopore. Supposedly common opening for stone canal and gonoduct.

hydropore. Pore or slit serving as adit to water-vascular system.

interambulacral *(interambulacralium,* pl., *interambulacralia).* Any of thecal plates forming interambulacra (syn., interradial).

interambulacrum (pl., **interambulacra**). Any of 5 interradial sections of theca (syn., interradius, interray); adj., interambulacral.

interradial (interradialium). See interambulacral.

interradius. *See* interambulacrum (syn., interray); adj., interradial.

lateral groove. Faint groove running at each side along border of floor plates in many edrioasteroids; served articulation of cover plates on floor plates.

lateral pore. Any of pores in tangential suture of successive floor plates in Edrioasteridae and some other edrioasteroids; cf. podial pores.

marginal. Any plate in peripheral ring.

oral. Any of 5 interradially disposed plates forming circlet on tegmen surrounding or covering mouth (syn., deltoid); adj., surface of theca bearing mouth, in edrioasteroids directed upward (syn., actinal, adoral, ventral).

orotegminal (orotegminal plate). Any of peristomial cover plates forming tegmen.

pelmatozoan. Applied to echinoderm fixed to substratum, with or without stem.

pentameral. *See* quinqueradiate.

pentamerous. *See* quinqueradiate.

peripheral ring. Peripheral zone formed by distal interambulacral plates in most edrioasteroids.

periproct. Space in *CD* interray in which anal opening is located.

peristome. Space in which mouth opening is located.

perradius. Center line of ambulacrum; adj., perradial.

podial pore. Pore admitting passage of tube foot (podium); lateral pores must have been podial pores.

podium. *See* tube foot.

quinqueradiate. Applied to radial symmetry characterized by 5 rays extending from mouth (syn., pentameral, pentamerous).

radial, radius. Radial plates together with all structures borne thereupon (syn., ray); as adjective, belonging to, or in direction of radius.

ray. *See* radial.

sagittal plane. Plane extending anteroposteriorly and dorsoventrally in mid-line, dividing bilaterally symmetrical animal into two similar halves.

Saumplättchen. See cover plate.

sessile. Of animal fixed to substratum, sedentary.

stone canal. Canal leading from hydropore (madreporite) to ring canal of ambulacral system.

stroma. Endoskeletal mesenchyme.

subambulacral. See floor plate.

substomial chamber. Frame formed by internal peristomial skeleton, through which gullet receiving ambulacral grooves passed into stomach.

subvective groove. See ambulacral groove.

subvective system (HAECKEL). Applied to organ system serving transportation of food particles to mouth; from morphological point of view, subvective system cannot be separated from ambulacral system.

suture. Straight line along which adjacent plates meet.

tegmen. Term applied to oral cover of edrioasteroids; strictly roof in crinoid theca.

theca. Capsule of mesodermal skeleton enclosing intestine; in edrioasteroids, the theca is no closed structure, because the skeleton of ambulacra is lodged between the thecal skeletal elements.

thecal plate. Any of numerous plates, mosaic or imbricating, that form theca.

[**thecal pore.** Pore piercing substance of thecal plate (not present in Edrioasteroidea).]

"third aperture." Noncommittal designation for small opening in *CD* interray behind mouth; may represent a hydropore.

trimerous. Applied to radial symmetry characterized by three primary rays extending from mouth, each of two lateral rays giving off two branches.

tube foot. Muscular cylinder protruding through pore between ambulacral floor plates in some Edrioasteroidea; may have served locomotion, and may also have had other function (syn., podium).

valvular. Applied to anal pyramid composed of several more or less triangular plates (valvules).

ventral. *See* oral.

whirler (REMANE). Animal feeding by aid of cilia producing an eddy in which particles are eventually caught by other cilia, and, helped by a secretion of mucus, conducted into the place of digestion.

SOFT PARTS

A few points relating to the internal anatomy of the Edrioasteroidea have been touched upon cursorily, mainly in discussing the third aperture. For further information reference may be made to papers by ANDERSON (1), FOERSTE (16), JAEKEL (22), and WILLIAMS (40). Even if we assume that the internal organization was much the same, on principle, in all Edrioasteroidea, it is obvious that the location and extent of several organ systems must have varied considerably in the disc-shaped theca of, say, *Agelacrinites* and the theca of the turret-shaped *Pyrgocystis*.

ONTOGENY

Growth stages have been observed in a number of species, the smallest one measuring no more than 0.7 mm. *(Isorophus austini)* in diameter.

A most prominent feature in the ontogeny of forms with curved ambulacra is that the curvature is acquired gradually, which, in fact, is only what should be expected. WILLIAMS (40) studied a series of successively larger (and hence more fullgrown) individuals of *Isorophus austini* in abundant specimens, most of which exhibit the aboral side (Fig. 121). In the smallest larva (diam., 1.3 mm.) examined with regard to development of the ambulacral system, only the most proximal floor plate of each ambulacrum was present, encircling the so-called "substomial chamber." During larval development new floor plates are introduced in a distal direction, forming (in this case) a uniserial row. It may be assumed that a pair of ambulacral cover plates was developed simultaneously with addition of a floor plate. On reaching the peripheral limit of the theca, the growing ray, hitherto straight, was turned aside either in a solar or in a contrasolar direction in a way specific of each ambulacrum.

In forms characterized by straight ambulacra, the rays stopped growing at the moment of coming into contact with the peripheral ring or, as in *Stromatocystites,* when the margin of the adoral surface was reached. Contingently, the rays had ceased to extend in a radial direction even before reaching so far.

WILLIAMS (40) has noted that differentiation into radial and interradial skeletal elements cannot be recognized in very small specimens. Also, they cannot be oriented properly, because the anal pyramid is not found until in later growth stages.

It should be noted that the proportions between different parts of the theca may shift during development. Thus, in *Isorophus,* the oral area is proportionally broader in immature than in mature forms (4). In larval forms of some species (e.g., *Lepidodiscus buttsi, Agelacrinites hamiltonensis*), the marginal ring is proportionally wider than the central disc of the adoral surface, which is much contracted (9).

The ontogeny of Edrioasteroidea in comparison with other pelmatozoans has been discussed by Regnéll (31) elsewhere.

EVOLUTIONARY TRENDS

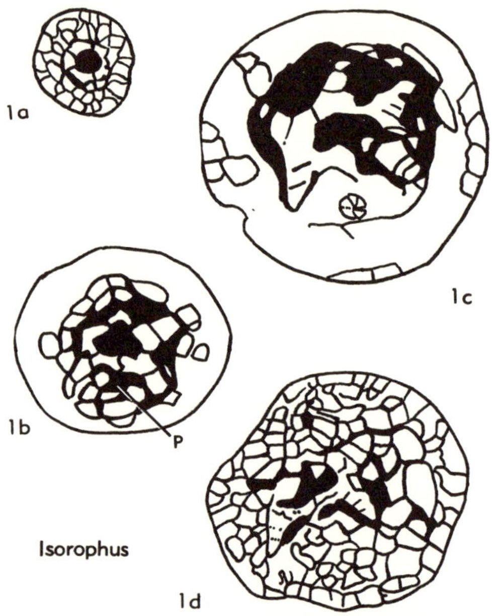

Fig. 121. Successive stages in development of *Isorophus austini* (Foerste) shown in aboral views, all ×12 (40).——*1a.* Substomial chamber and peristomial ring of modified floor plates distinct, no other floor plates separable.——*1b.* Stage with additional floor plate in ambulacra *C, D,* and *E,* location of anal pyramid *(p)* indistinct.——*1c.* Stage showing 2 floor plates in each ambulacrum and presence of peristomial ring.——*1d.* Stage with all plates distinguishable, 3 floor plates in ambulacra *C* and *D* but others less complete.

A general trend in respect of thecal construction is not obvious in phylogenetic development of Edrioasteroidea. A tendency towards prolongation of the vertical axis of the theca serves to elevate the adoral surface over the substratum. Contemporaneous with this tendency, however, most evident in *Pyrgocystis* (L.Ord.-L.Dev.), which clearly was a successful edrioasteroid well adapted to environmental requirements, the more or less discoidal type of theca was in existence from the very first appearance of Edrioasteroidea (e.g., *Stromatocystites*) in Early Cambrian time and it persisted throughout most of the Paleozoic (e.g., *Anglidiscus, Lepidodiscus,* U.Miss.). Essentially the same observation applies to hemispherical forms. Thus, once brought into shape, the various modifications of thecal development were remarkably persistent.

Nor is any positively discernible trend distinguished within each of the thecal types. This may be exemplified by *Pyrgocystis,* which does not consistently follow up the line of development operating at increased plasticity of the theca without loss of firmness (29).

A reduction of the ambulacral grooves, brought about in various ways, is a phenomenon met with in several groups of echinoderms. Not so in the Edrioasteroidea. On the contrary, the lengthening of the ambulacral grooves during ontogenetic development is paralleled by a similar trend in phylogeny. This may be correlated with the fact that in the Edrioasteroidea the ambulacral grooves retained their primary function of collecting nourishment and transporting food particles to the mouth. So an extension of the distal part of the ambulacra was an advantage in the economy of the animal, the more so since these organisms were mostly sedentary. In the Edrioasteridae, which were surely not permanently fixed, the ambulacra even passed on to the aboral surface. A similar effect was achieved by branching in the unique genus *Thresherodiscus.*

A limit for extension of the ambulacra in a radial direction was set by the inner zone of the peripheral ring, which in many genera was rigid. In consequence, the ambulacra had to deviate in a tangential direction, thus producing curved structures. The curvature of rays seen in adult specimens may have been initiated well before reaching maturity. An explanation possibly is found in the fact, referred to above, that the peripheral ring is proportionally much broader in young individuals of certain species than in fully grown individuals.

The trend toward enlargement of food-collecting areas is demonstrated by the predominance of curved ambulacra in geologically younger genera. It is true that forms with curved ambulacra are met with even in the Middle Cambrian (e.g., *Walcottidiscus*) and some Ordovician edrioasteroids developed extremely long ambulacra. But until Devonian time there is still a comparatively large number of genera in which the ambulacra are not affected by curving. The following figures support this statement, though available material provides only a narrow basis for generic judgment. Of known pre-Devonian genera, 9 have curved ambulacra out of a total of 17, i.e., 53 percent. Of known post-Silurian genera, 8 have curved ambulacra out of a total of 11, i.e., 73 percent. (No attention has been paid to the fact that, in *Hemicystites,* the ambulacra may tend to curve slightly, and *Thresherodiscus* is left out of account.)

The direction of curvature of the ambulacra is not governed by any apparent trend. It has been pointed out in discussion of ambulacral structures that the primitive condition of ray curvature conjecturally was a contrasolar one. To this one may object that in the oldest genus in which curving of the ambulacra has been observed (viz., *Walcottidiscus,* M.Cam.) one ray, ambulacrum *C,* curves in a solar direction. In view of the very scarce material of Edrioasteroidea obtained from Cambrian strata, however, it may be justified in this case to refer to the incompleteness of the fossil record available for study. In fact, the various types of ambulacral curving are distributed fairly equally, as known now, throughout the stratigraphic column, from Middle Cambrian to Upper Mississippian, as appears from the following survey. (1) All contrasolar: Middle Ordovician *(Lebetodiscus, Lepidoconia)*; Upper Ordovician *(Streptaster)*; Upper Devonian *(Dinocystis)*; Upper Mississippian *(Ulrichidiscus)*. (2) *A, B, D, E* contrasolar, *C* solar: Middle Cambrian *(Walcottidiscus)*; Middle Ordovician *(Bassleridiscus, Carneyella, Isorophus, *Edrioaster bigsbyi)*; Upper Ordovician *(Carneyella, Isorophus)*; Middle Devonian *(Lepidodiscus)*; Upper Devonian *(Lepidodiscus)*; Lower and Upper Mississippian *(Anglidiscus, Lepidodiscus, Discocystis)*. (3) *A, D, E* contrasolar, *B, C* solar: Middle Ordovician *(Isorophusella)*; Lower to Upper Devonian *(Agelacrinites, ?Xenocystites)*; Lower Mississippian *(Agelacrinites)* (4). All solar: Middle Ordovician (*Foerstediscus, Edrioaster* in part); Upper Ordovician (*Foerstediscus, Edrioaster* in part); Upper Devonian *(Cooperidiscus)*.

The structure of the ambulacra also seems not to have changed in a distinct direction. A tendency toward uniseriality of the ambulacral cover plates may have operated during phylogeny of Edrioasteroidea, however. As to the floor plates of the ambulacral grooves, it is difficult to trace a general line in development of their arrangement in the actual fossil material; however, they seem to have progressed from a biserial to uniserial plan, rather than the reverse, in conformity with the cover plates (30).

It should be remarked, finally, that a tendency toward a free-living mode of life is apparent in a few forms (e.g., Edrioasteridae), although this is met with even in some early types (e.g., Stromatocystitidae). CABIBEL, TERMIER & TERMIER (8) have suggested a line of evolution leading from *Stromatocystites* (which they consider an eleutherozoan) through *Eikosacystis* CABIBEL, TERMIER & TERMIER, 1959 *(incertae sedis)* to the Carpoidea Cincta (Trochocystitida), which, too, are thought to have been free-living.

ECOLOGY

Edrioasteroidea lived in various lithotopes, their fossil remains having been met within sandstones (more or less pure, micaceous, calcareous, etc.), shales (sandy, micaceous, calcareous, etc., black, brown, etc.), marls, and limestones. Optimal edaphic conditions apparently prevailed on a calcareous substratum. Sandy bottoms seem not to have appealed to most Edrioasteroidea, since relatively few forms have been found in sandstones (which in part may be due to poorer chances of preservation in this kind of rock). As far as known, only *Stromatocystites* seems to have been restricted to sandy bottoms. This genus may have lived in the intertidal zone. A few genera (e.g., *Pyrgocystis, Edrioaster*) were tolerant of varied lithological characters of the substratum, but the bulk of genera were fairly selective in this respect. Most, if not all, species of Edrioasteroidea had specific demands on the nature of the substratum.

Some edrioasteroids seemingly were attached directly to the indurated bottom surface (18, 36). *Pyrgocystis* mostly may have stuck in soft mud or ooze. In general, however, the Edrioasteroidea were attached to some hard foreign object, preferably the shell of some organism. Most commonly valves of brachiopods (especially *Rafinesquina alternata* where forms contemporaneous with that species are concerned, and in addition *Chonetes, Hebertella, Platystrophia, Spirifer,* etc.) were suited for this purpose. Other examples are afforded by sponges, corals, bryozoans, conulariids, pelecypods, cystoids, and crinoids. Most Maysville (Upper Ordovician) specimens from North America are found on *Rafinesquina alternata* (40), and most *Hemicystites* from Bohemia on conulariids (22). On the whole, all species seem to have been strictly selective in their choice of host. This raises question as to what sort of symbiotic relations existed between the two organisms. The edrioasteroid might be suspected of parasitism, but, as observed by Gekker (17) in the case of *Cyathocystis plautinae,* which he found both on species of *Heliocrinites* and on the bryozoan *Dianulites (Monticulipora) petropolitanus,* undoubtedly we have to do with commensalism. It is possible also that edrioasteroids were attached to exoskeletons of dead animals, but this probably was not the rule, because it would then have been reasonable to find Edrioasteroidea attached to, for example, trilobites, the exuviae of which were undoubtedly often abundantly available. That they did not attach themselves to living trilobites is easy to realize.

The associated fauna and sediments indicate clearly that the Edrioasteroidea lived in a littoral environment. From comparison with the habitat of pelmatozoans in general we may assume that the water was neither polluted nor turbid.

Most Edrioasteroidea were permanently attached. Others were evidently capable of restricted shifts in position. This must be true for the Edrioasteridae provided with pores between the ambulacral floor plates, in which the presence of a tube-foot-ampulla system is indicated and in which the ambulacra pass on to the marginal area of the aboral surface.

DISTRIBUTION

The Edrioasteroidea range from Lower Cambrian to Mississippian, where they disappear abruptly. They reached a very pronounced acme in the Middle Ordovician and in smaller degree in the Upper Ordovician, as is evident from the graph (Fig. 122) based on the number of genera represented in each division of the geological systems. Another high point, but far less extreme, was attained in the Upper Devonian. The edrioasteroid stock was subjected to severe crises in Late Cambrian and the Late Silurian time. The curves showing number of new genera appearing in each division and the rate of appearance of genera during the several periods mutually conform, on the whole, as shown by the diagram. It is noticeable that, just before becoming extinct, the Edrioasteroidea rose, in a way, to new activity. It should be observed that essentially coincident patterns are obtained in this instance, whether a graph of this kind is based on the stratigraphic range of families, genera, or species. If we construct a curve based on number of species and this curve be resolved into its North American and European components, we find that the maximum development of the Edrioasteroidea in the Middle Ordovician is due almost exclusively to the prolificity of the North American stock, mainly the Hemicystitidae. The second high point of the joint curve transpires as due to an accumulation of the two individual curves, so that for North America alone the high point appears in the Lower Mississippian (no European representatives being known) instead of in the Upper Devonian. Apart from the peaks mentioned, the two curves also show other, minor, differences which indicate that the Edrioasteroidea at various times contributed in a varying degree to the Paleozoic biota of North America on the one hand and Europe on the other.

It is a remarkable fact that Edrioasteroidea have not been recorded so far from any areas outside of North America and Europe[1]. Further, it is striking that forms with a more or less a turret-shaped theca are, on the whole, very scanty in North American deposits. Clearly, this group had an eastern Baltic origin. In spite of the great predominance of *Hemicystites* in North America, there is evidence that this type arose in central European seas in Early Ordovician time. Otherwise, the more or less discoidal forms are mainly characteristic of North America, which undoubtedly was the center of evolution and dispersal of the Agelacrinitidae and Stromatocystitidae, since most genera belonging to these families, as well as to Hemicystitidae, are confined to North America. The Edrioasteridae are less significant from a biogeographical point of view (30).

[1] Except a unique find in Victoria, Australia, of a form (Upper Silurian?) said to resemble *Cincinnatidiscus* (G. M. Philip, 1963. Austral. Jour. Sci., v. 26, p. 25).

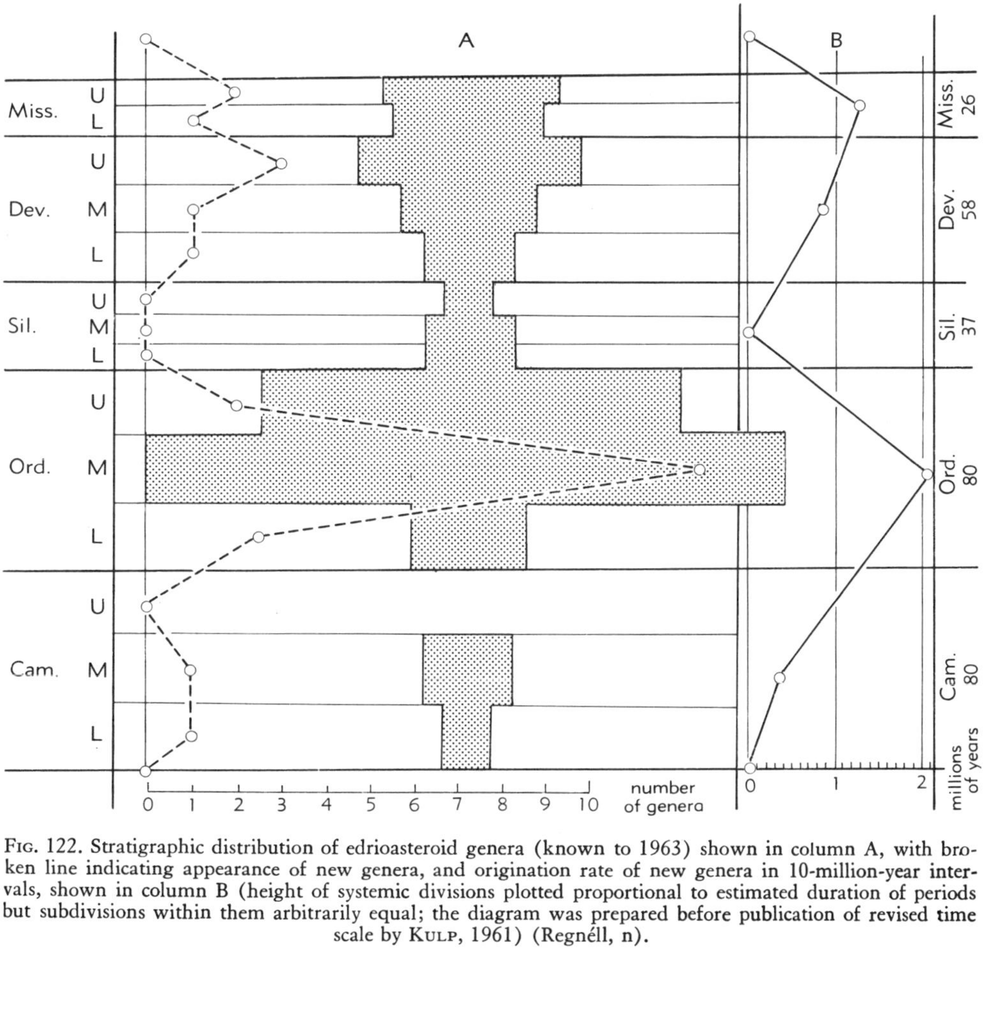

FIG. 122. Stratigraphic distribution of edrioasteroid genera (known to 1963) shown in column A, with broken line indicating appearance of new genera, and origination rate of new genera in 10-million-year intervals, shown in column B (height of systemic divisions plotted proportional to estimated duration of periods but subdivisions within them arbitrarily equal; the diagram was prepared before publication of revised time scale by KULP, 1961) (Regnéll, n).

CLASSIFICATION

In several textbooks and indices (e.g., BASSLER & MOODEY, 1943, 5), the Edrioasteroidea have been ranked as an order of the class Cystoidea. On the other hand, most specialists have credited them with a rank equal to that of the Cystoidea. Undoubtedly, this latter opinion is supported by many features in the structural organization of the Edrioasteroidea, especially the presence of pores between the ambulacral skeletal elements and the absence of skeletal subvective appendages (29).

The current classification of the Edrioasteroidea was established by BASSLER (3, 4) with due attention to proposals made by previous authors. This classification seems to be useful and therefore its essential features have been retained in the present *Treatise,* although other arrangements call for notice. For example, two main divisions might be recognized, one group having pores between the ambulacral floor plates, the other group having none. But how, then, should we classify the Stromatocystitidae and Cyathocystidae, which are insufficiently known in this respect? In my opinion, we lack evidence needed for assigning the families here recognized to units of suprafamilial rank.

For the definition of families, it would be preferable to make consistent use of some one morphological character, or, rather, set of characters, if such were available. Leaving out of account the Edrioasteridae which are known to be characterized, among other features, by pores intervening between the ambulacral floor plates, remaining families are in fact distinguished from each other, mainly on the basis of the structure of the mouth cover. There is no reason why this feature should not be as valid from a taxonomic point of view as any other expression of morphological differentiation. The nature of the oral tegmen demonstrates namely a modification of the ambulacral cover plates resulting in morphologically distinct types of orotegminal construction. One consequence of a classification on this basis is the recognition of two lines of development, parallel in many respects, represented by the Hemicystitidae, on the one hand, and the Agelacrinitidae, on the other.

The behavior of the ambulacra—whether straight or curved and if curved, in what direction—has been deemed to be a feature of importance for discrimination of species, as agreed by previous authors. A variation in the direction of curvature of the ambulacra does exist in some species, but this may be compared with the occasional excess of the normal number of rays, for example. As pointed out above, the ambulacra observed in early growth stages of forms having curved ambulacra in the adult state are not affected either by solar or contrasolar curvature. As noted by BATHER (7, pt. 8), "The distribution, however, being characteristic of species separated by other characters, cannot be fortuitous. There must have been some structures or habit in each species predisposing in turn of the coil in a solar or contrasolar direction."

Outline of Edrioasteroid Classification

[Figures enclosed by parentheses indicate numbers of known genera and subgenera]

Edrioasteroidea *(class)* (27; 1). *L.Cam.-L.Carb.(Miss.).*
- Stromatocystitidae (3). *L.Cam.-U.Dev.*
- Cyathocystidae (2). *?L.Ord., M.Ord.-L.Sil.*
- Hemicystitidae (12; 1). *Ord.-L.Carb.(Miss.).*
- Agelacrinitidae (8). *M.Ord.-L.Carb.(U.Miss.).*
- Edrioasteridae (2). *M.Ord.-U.Dev.*

SYSTEMATIC DESCRIPTIONS

Family STROMATOCYSTITIDAE Bassler, 1936

Mouth surrounded and covered by 4 large imperforate plates and many small ones; theca flexible, subpentagonal or circular, adoral side depressed convex, aboral side slightly concave, enclosing a shallow cavity; interambulacral areas and aboral side mostly composed of polygonal plates; ambulacra 5, narrow, straight or curved, confined to the upper surface; unstalked and unattached. *L.Cam.-U.Dev.*

Stromatocystites POMPECKJ, 1896, p. 505 [**S. pentangularis*; OD] [=*Stromatocystis* BATHER, 1900; *?Cambraster* JAEKEL, 1923]. Theca subpentagonal, reaching about 35 mm. in diam.; skeleton of numerous nonimbricate plates, usually 5- or 6-

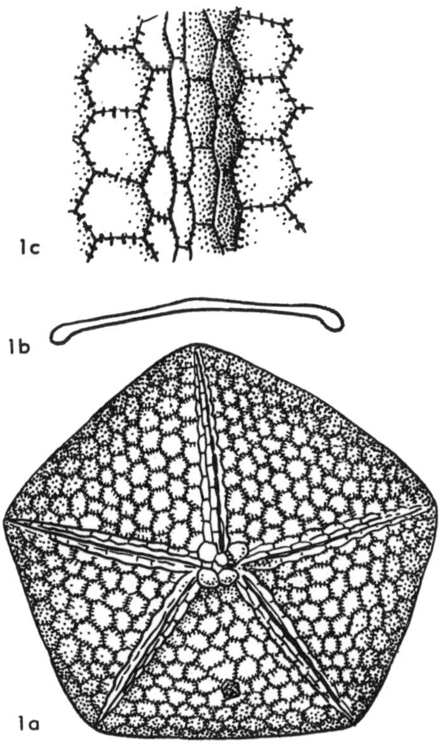

FIG. 123. Stromatocystitidae (p. *U*160-*U*161).

sided (on aboral side larger toward center, which is occupied by a more or less well-defined centrodorsal plate), or indistinctly differentiated and weakly calcified (*S. balticus* JAEKEL, 1899); sutures of interambulacrals crossed by 1 to 3 diploporoid structures, which undoubtedly are places of stroma strands uniting individual plates; ambulacra straight, terminating in angles of theca; ambulacral floor plates in 2 rows of elongate alternating plates; side plates long, about equal in number to floor plates and alternating with them; ambulacral grooves roofed over by small cover plates; anal opening pentagonal, covered by a pyramid of (about 9) small plates (35). *L.Cam.-M. Cam.,* N. Am.(Newf.)-Eu.(Baltic-Czech.-?Fr.). ——FIG. 123,*1*. **S. pentangularis,* M.Cam., Czech.; *1a,* adoral side (restored), ×1.5 (6); *1b,* detail of ambulacrum and adjacent plates, ×6; *1c,* median section of theca, ×1.1 (27).

[The name *Cambraster* was introduced by JAEKEL, 1923 (Palaeont. Zeitschr., vol. 5, p. 344), for a fossil from the Middle Cambrian of Hérault, south France, that was considered to be a primitive asterozoan. JAEKEL gave only a very generalized description and no figure. Therefore, the meaning of *Cambraster* remains obscure. Probably it must be looked upon as a *nomen nudum*. Only with great hesitation has it been entered here as a synonym of *Stromatocystites* (cf. REGNÉLL, 29, p. 199, and CABIBEL, TERMIER & TERMIER, 8, p. 284)].

Walcottidiscus BASSLER, 1935, p. 3 [**W. typicalis*; OD]. Theca as in *Stromatocystites* but larger (up to 60 mm.), its edge formed by small imbricate plates; interambulacrals and plates of aboral side polygonal, slightly imbricate; ambulacra more or less curved, *A, B, D, E,* contrasolar, *C,* solar. Insufficiently known (3, 4). *M.Cam.(Burgess Sh.),* Can.(B.C.).

Xenocystites BASSLER, 1936, p. 3 [**X. carteri*; OD]. Theca circular, about 45 mm. in diam.; interambulacral areas and aboral side covered by polygonal and mosaic plates; ambulacra slightly curved, 2 of them probably solar; ambulacral floor plates long, narrow, in single row. Insufficiently known (4). [G. W. SINCLAIR doubts assignment of this genus to Stromatocystitidae, suggesting it may actually belong to *Cooperidiscus* (personal communication).] *U.Dev.(Chemung.),* USA (N.Y.).

Family CYATHOCYSTIDAE Bather, 1899

[=Thecocystidae JAEKEL, 1918, emend. 1927; by error as Cyathothecidae JAEKEL, 1927, in REGNÉLL, 1945]

Adoral face with 5 short, straight ambulacra, which have either a triangular plate between each two ambulacra, adoral surface being surrounded by a border of marginal plates, or are covered by 5 triangular interradial plates; theca saclike, forming a solid mass permanently attached at aboral end to some foreign object; no third aperture observed. *?L.Ord., M.Ord.-L.Sil.*

Cyathocystis F. SCHMIDT, 1879, p. 2 [**C. plautinae*; SD HAECKEL, 1896, p. 115]. Theca up to about 15 mm. high, attached by broad or tapering aboral surface more or less drawn out into rootlike processes; deltoids as in *Cyathotheca* but not covering ambulacra, which are straight, short, broad or comparatively broad, and taper distally; 2 rows of alternating ambulacral cover plates; mouth covered by 5 large plates; anal pyramid of 5 triangular plates located in posterior interradius between deltoid and adjacent marginals; adoral surface bordered by frame of about 40 marginals (4). [A couple of the skeletal elements of the adoral surface recall strongly the conditions in *Stephanocrinus,* possibly affording an example of homeomorphism.] *M.Ord.-U.Ord.,* Eu.(Est.)-N. Am.(Okla.-Tenn.).——FIG. 124,*2*. **C. plautinae,* M.Ord., Est.; *2a,* adoral, ×3; *2b,* lat. view of two specimens attached to bryozoan, ×3 (22).

Cyathotheca JAEKEL, 1927, p. 4 [**C. suecica*; OD]. Theca irregular, goblet-shaped, height up to 8 mm., with irregular or pointed attachment surface; ambulacra apparently very narrow, practically hidden at surface, adoral face covered by very flat pyramid of 5 triangular deltoids (orals),

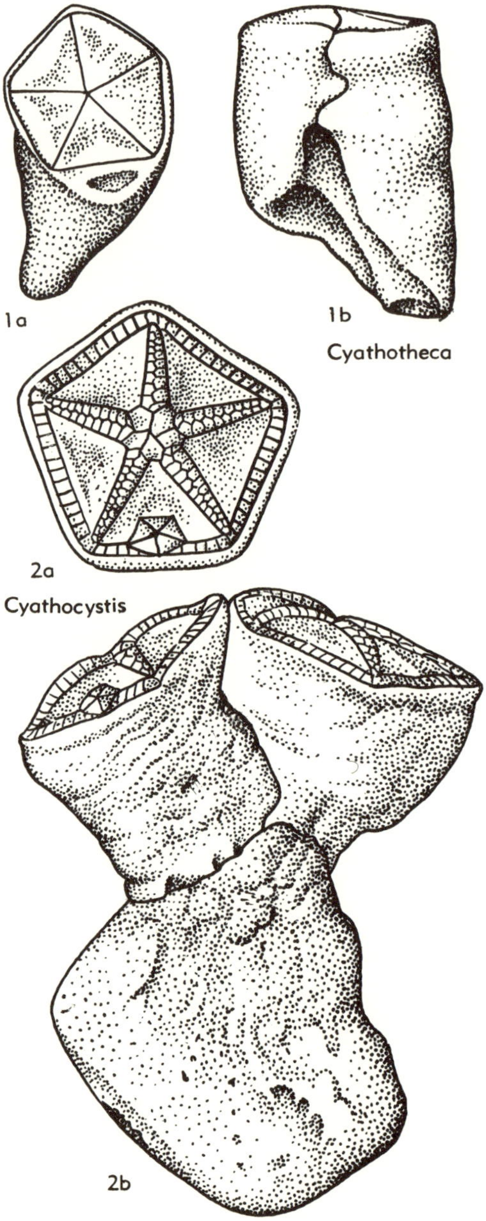

FIG. 124. Cyathocystidae (p. U161-U162).

behind which is lunate space with anal opening; no marginal plates (29). *?L.Ord., L.Sil.,* Eu. (Swed.-USSR).——FIG. 124,*1*. **C. suecica,* L.Sil., Sweden; *1a,b,* adoral, left lat. views, ×6 (23).

Family HEMICYSTITIDAE Bassler, 1936

Peristome covered by ?3, 4, or more plates, ?3, 4 of which have definite arrangement; 3 meeting in center of peristomial region, 1 lying over posterocentral part of peristome, other 2 with 1 on each side of ambulacrum *A*; theca flat, saclike or turret-shaped, composed of thin plates, attached permanently (or ?temporarily) to some foreign object by greater part of aboral surface; ambulacra normally 5, straight or curved; single row of ambulacral floor plates more or less overlapping proximally. *Ord.-L.Carb. (Miss.).*

Hemicystites HALL, 1852, p. 245 [**H. parasiticus*; OD] [=*Hemicystis* CARPENTER, 1891]. Theca subcircular, depressed or disclike, 5 to 20 mm. in diameter, attached preferably to shells of *Rafinesquina* and other brachiopods, or to cephalopods; thecal plates polygonal, squamose, imbricating, differentiated into larger interambulacrals (mainly increasing in size toward periphery) and zone of smaller marginals; ambulacra 5 or more *(H. multibrachiatus)*, in which rays *B*, *D*, and *E* bifurcate anomalously ("normal" specimens with 5 rays do occur), typically short, broad, in some forms tapering), and straight (or with a tendency to slight solar curvature of *C* and solar or contrasolar curvature of *A* especially); floor plates delicate, probably one for each pair of cover plates; anal pyramid elevated, of triangular plates arranged in regular circle; third aperture near the posterior oral plate (4). [Most prolific of all known edrioasteroid genera.] *Ord.-M.Dev., ?Miss.,* N.Am.-Eu.(Fr.-Czech.).——FIG. 125,*7*; 126,*1*. *H. chapmani* (RAYMOND), M.Ord.(Trenton.), Can. (Ont.); 125,*7*, adoral, view of holotype, ×3(4); 126,*1*, same (reconstr.) ×3(24). [See also Fig. 118,*3*; 120,*5a'*.]

Anglidiscus REGNÉLL, 1950, p. 226 [**Lepidodiscus fistulosus* ANDERSON, 1939, p. 68; OD]. Like *Carneyella*, but theca not attached (at least not permanently); ambulacral cover plates varying in different regions so that proximally only single row present but distally with 2 rows of interlocking plates, those near middle of ambulacrum being broad, wedge-shaped and relatively shorter than regular, pentagonal distal ones; ambulacral floor plates apparently pierced by lateral pores (for tube feet); periproct located farther back; peripheral ring less definite (30). *L.Carb.,* Eu.(Eng.).——FIG. 125,*4*; 126,*2*. **A. fistulosus* (ANDERSON); 125,*4*, adoral view (reconstr.), *ca.* ×5(1); 126,*2*, same, ×5 (24). [See also Fig. 116,*3*; 117, *1,5*; 119,*2*; 120,*5a"*.]

Bassleridiscus FISHER, 1951, p. 691 [**B. mohawkensis*; OD]. Intermediate between *Streptaster* and *Carneyella* in that ambulacra are strongly curved

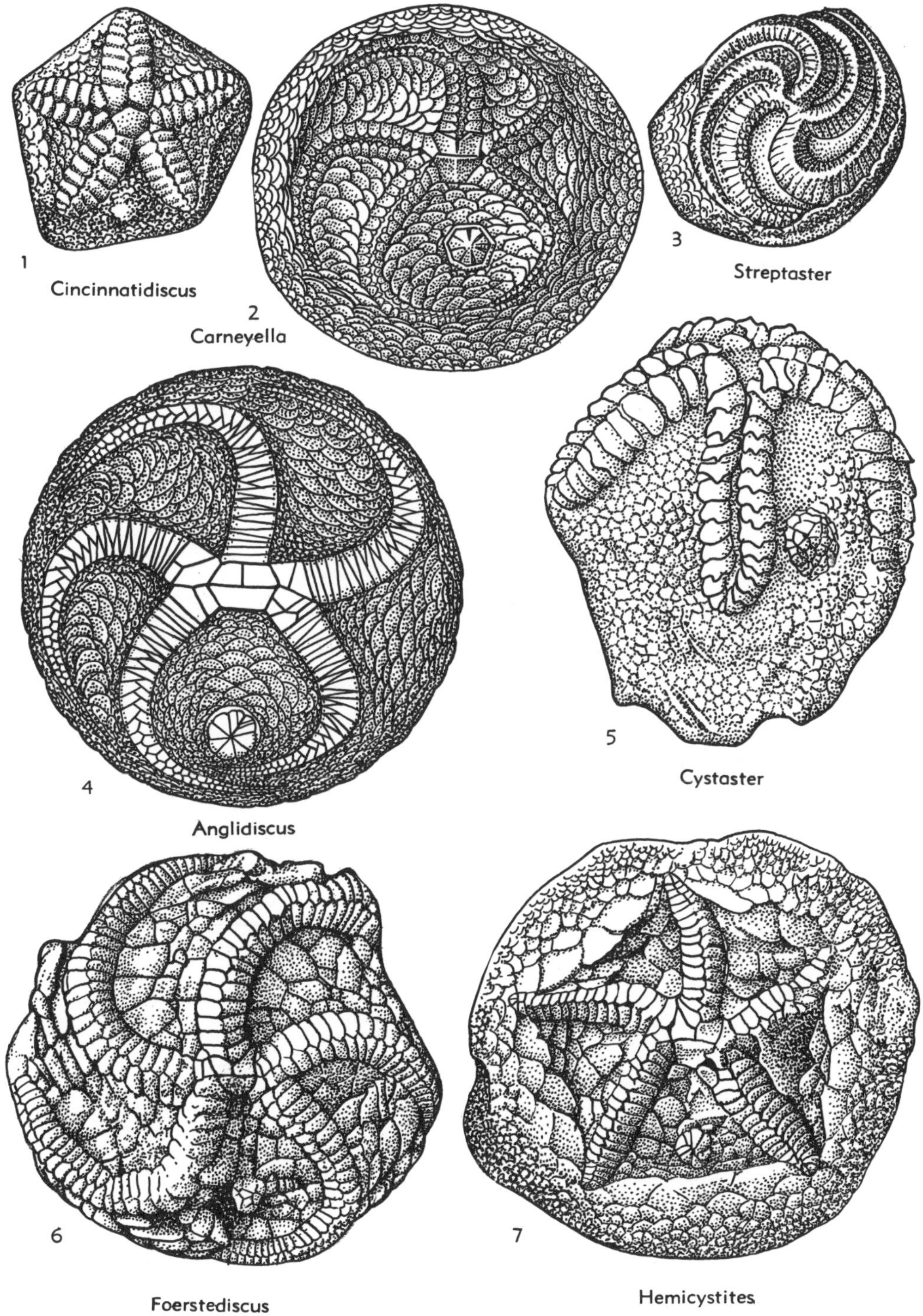

FIG. 125. Hemicystitidae (p. *U*162, *U*165, *U*167).

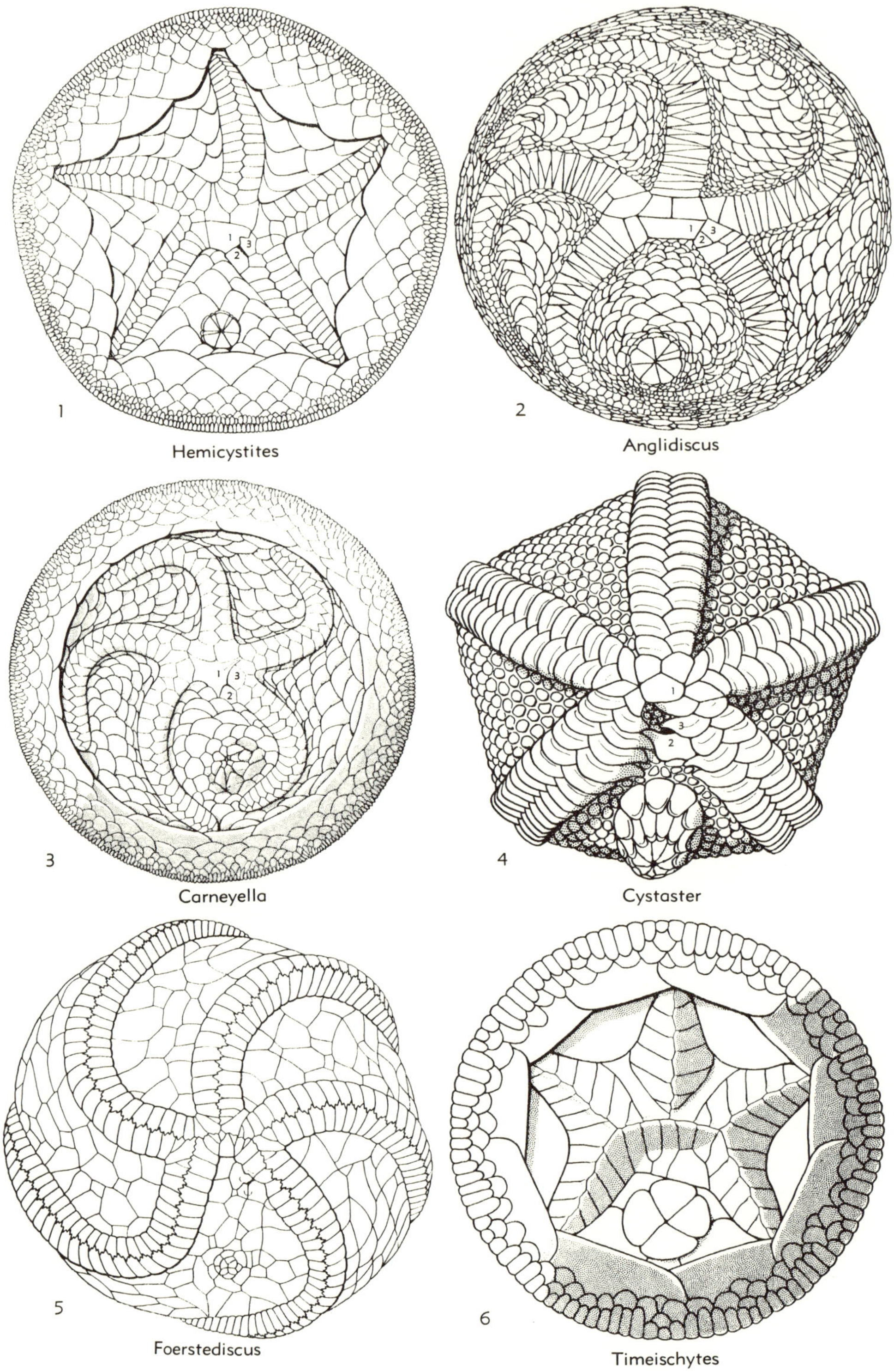
1
Hemicystites
2
Anglidiscus
3
Carneyella
4
Cystaster
5
Foerstediscus
6
Timeischytes

with sharply elevated ambulacral cover plates in 2 rows (as in *Streptaster*), ambulacrum *C* curving in solar direction, other ambulacra in contrasolar direction (as in *Carneyella*); theca depressed saclike, diam. about 15 mm.; interambulacrals small (0.25 to 1 mm.), slightly imbricating, polygonal to circular; ambulacra rather broad and very long, extending to aboral side (15). *M.Ord.(Trenton.),* USA(N.Y.).

Carneyella Foerste, 1916, p. 340 [**Agelacrinus pileus* Hall, 1866, p. 7; OD]. Theca subcircular, typically elevated, saclike, attached by broad basal part, but also forming depressed epizoic discs usually on brachiopods, diameter reaching 30 mm. but usually 15 to 20 mm.; thecal plates more or less imbricating, differentiated into interambulacral fields and marginal zone; ambulacra normally 5, varying in length, curved, ambulacrum *C* solar, others contrasolar, some (as in figured species) with small additional cover plates along mid-line of ambulacra; surface of plates minutely pitted and commonly ornamented with ridges and conspicuous nodes (spiniferous tubercles); anal pyramid of about 7 (or more) triangular pieces (4, 26). *M.Ord.-U.Ord.*, N.Am.——Fig. 125,*2*. *C. cincinnatiensis* (Hall), U.Ord.(Maysvill.), USA (Ohio); adoral view (reconstr.), ×3(20).——Fig. 126,*3*. **C. pileus* (Hall), U.Ord., USA (Ohio); adoral view (reconstr.), ×3 (26). [See also Fig. 112,*2*; 117,*3*; 120,*5a*; 129,*1*,*5*.]

Cincinnatidiscus Bassler, 1935, p. 3 [**Agelacrinus (Hemicystites) stellatus* Hall, 1866, p. 8; OD]. Like *Cystaster* except that the theca is depressed or flat, subpentagonal, up to 10 mm. in diameter, attached by the whole aboral surface, and that the interambulacral plates are squamose, imbricating distinctly, being surrounded by a marginal zone of small nodose plates; casually more than 5 ambulacra (4). *M.Ord.-U.Ord.,* USA(Ky.-Ohio).——Fig. 125,*1*. **C. stellatus* (Hall), U.Ord. (Maysvill.), USA(Ohio); adoral view, ×9 (20).

Cystaster Hall, 1871, p. opp. pl. 6 (as subgenus of *Hemicystites*) [**C. granulatus* (=*Thecocystis sacculus* Jaekel, 1899); OD] [=*Thecocystis* Jaekel, 1899]. Theca saclike (height and width about 6 mm.), attached by aboral surface to some foreign object (or ?free); interambulacrals minute (0.25 mm.), fused, rounded or polygonal, forming mosaic; ambulacra 5, short, straight, with rather coarse, alternating cover plates supported by small side plates; raised anal pyramid composed of about 10 small, elongate, nodose plates (22). *U.Ord.,* USA(Ohio-Ky.).——Fig. 125,*5*; 126,*4*. **C granulatus*, Maysvill., Ohio; 125,*5*, posterolateral view, ×9(22); 126,*4*, adoral view (reconstr.), ×7.5 (24). [See also Fig. 114,*1*; 120,*3a*.]

Foerstediscus Bassler, 1935, p. 6 [**F. grandis*; OD]. Theca depressed to flattened, diameter 6 to 30 mm.; interambulacrals relatively large, polygonal, only slightly imbricating in center, piled up on edge along margin; plates generally smooth; ambulacra 5, short or long, all strongly curved in solar direction, with elongate cover plates; anal pyramid a narrow elevated tube or flat broad area of small plates (4). *M.Ord.-U.Ord.*, USA(Minn.-Ky.)-Can.(Ont.).——Fig. 125,*6*; 126,*5*. *F. splendens* Bassler, M.Ord.(Blackriver.); USA(Minn.); 125,*6*, adoral view of holotype, ×4.5 (4); 126,*5*, same (reconstr.), ×2.2 (24). [See also Fig. 115, *1*; 120,*2b*.]

Lebetodiscus Bather, 1908, p. 543 [**Agelacrinites dicksoni* Billings, 1857, p. 294; OD]. Theca discoidal, up to 24 mm. in diameter, adoral surface convex, aboral surface attached to some foreign body; interambulacral plates large, with slight adoral imbrication or forming mosaic, decreasing in size toward marginal zone of very small plates; ambulacra 5, long, contrasolar, reaching periphery but not passing it; ambulacral cover plates in 2 sets, smaller median ones and larger lateral ones with single large pore between each pair; anal opening large, covered with small plates (28, 41). *M.Ord.*, N.Am.(Can.).——Fig. 127,*4*. **L. dicksoni* (Billings), (Trenton.), Ont.; adoral view, ×2.25 (28). [See also Fig. 117,*4*.]

Lepidoconia Wilson, 1946, p. 21 [**Lebetodiscus loriformis* Raymond, 1915, p. 56; OD] Similar to *Lebetodiscus* but with 5 pores on either side of ambulacral cover plates; theca about 23 mm. in diameter; marginal ring wide; ambulacra curving in a contrasolar direction, long and stout, covering large part of disc surface and rising considerably above it, tip of each ambulacrum touching or nearly touching coil of adjacent one so that interambulacrals have little or no contact with peripheral plates (41). *M.Ord.*; N.Am.(Can.).——Fig. 127,*1*. **L. loriformis*(Raymond), Trenton., Ont.; adoral view; ×2.25(4).

Pyrgocystis Bather, 1915, p. 5 [**P. sardesoni*; OD]. Theca elevated, subcylindrical or polygonal [*P. (Rhenopyrgus?) octogona* R. Richter, 1930], turret-shaped, with subparallel or aborally tapering sides which are curved or nearly straight, composed of scalelike plates that imbricate from below upward, disposed in spirals or columns which inosculate or are separated by grooves; ambulacra 5, restricted to adoral surface, broad and straight; plates of (incompletely known) oral face, as well as free borders of turret plates may bear spines (7, 29). *L.Ord.-M.Dev.*, Eu.-N.Am.

P. (Pyrgocystis). Theca of moderate size (ranging upward to 20 mm. in Ordovician species, 11 mm.

(Facing page.)

Fig. 126. Hemicystitidae, adoral views (reconstr., 24), *1-4* showing posterior orotegminals (numbered 1, 2) and proximal cover plate of *C* ambulacrum (numbered 3), with "third aperture" between them (p. *U*162, *U*165, *U*167).

in Silurian species); basal part of theca developed as sort of ferrule formed by small number of closely amalgamated plates. *L.Ord.-M.Dev.* Eu. (USSR-Norway-Czech.-Br.I.-Gotl.)-N.Am.(USA). ——FIG. 127,*2b*. *P.(P.)sulcata* (AURIVILLIUS), Sil.(U.Wenlock), Sweden(Gotl.); long. sec., ×4.7(29). [See also Fig. 113,*2*.]——FIG. 127, *2a*. **P. (P.) sardesoni,* M.Ord. (Blackriver.); USA (Minn.); adoral view, ×5.5(7).

P. (Rhenopyrgus) DEHM, 1961, p. 16 [**P. coronaeformis* RIEVERS, 1961, p. 10; OD]. Theca big (ranging to 95 mm.), widening in adoral

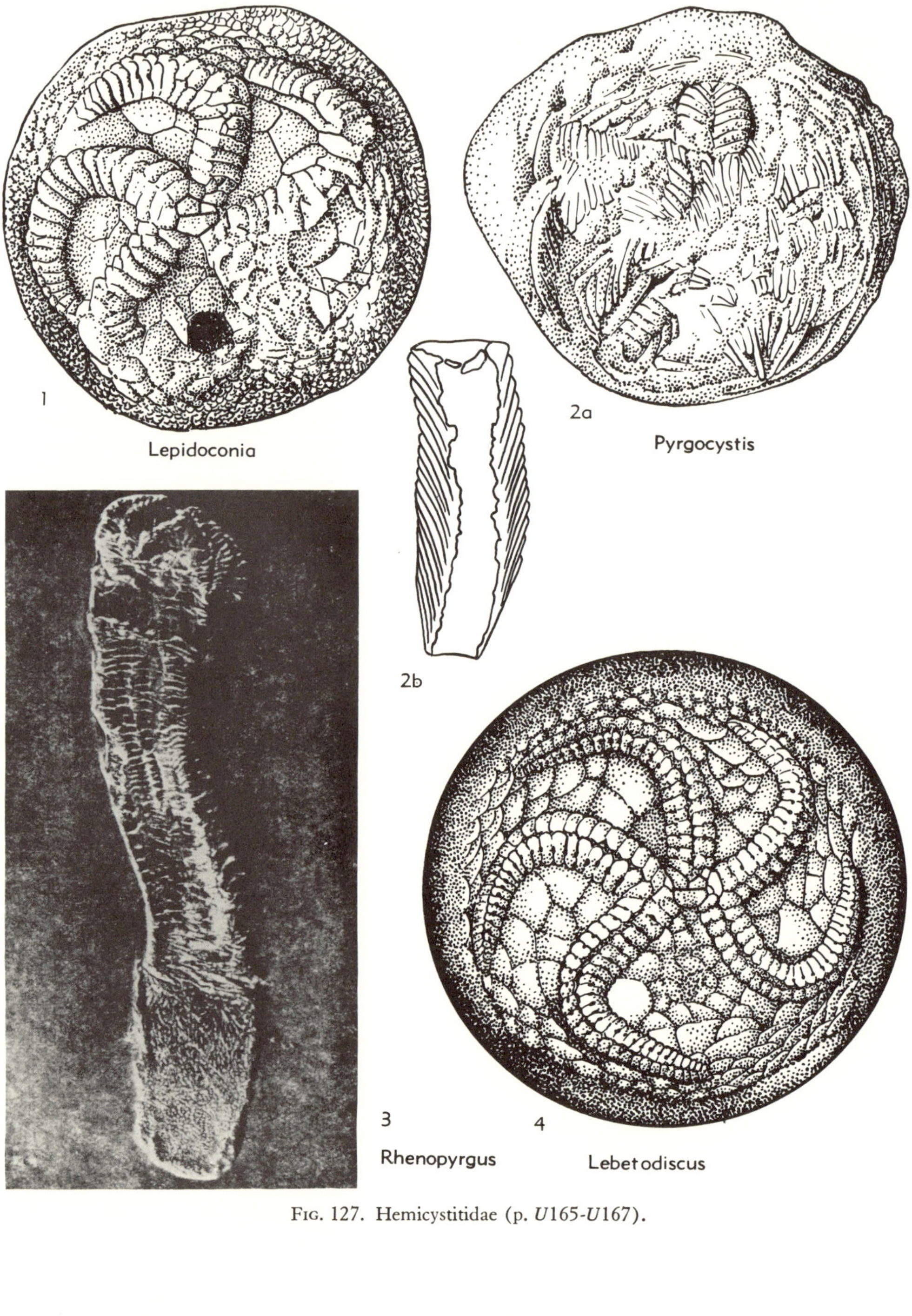

FIG. 127. Hemicystitidae (p. U165-U167).

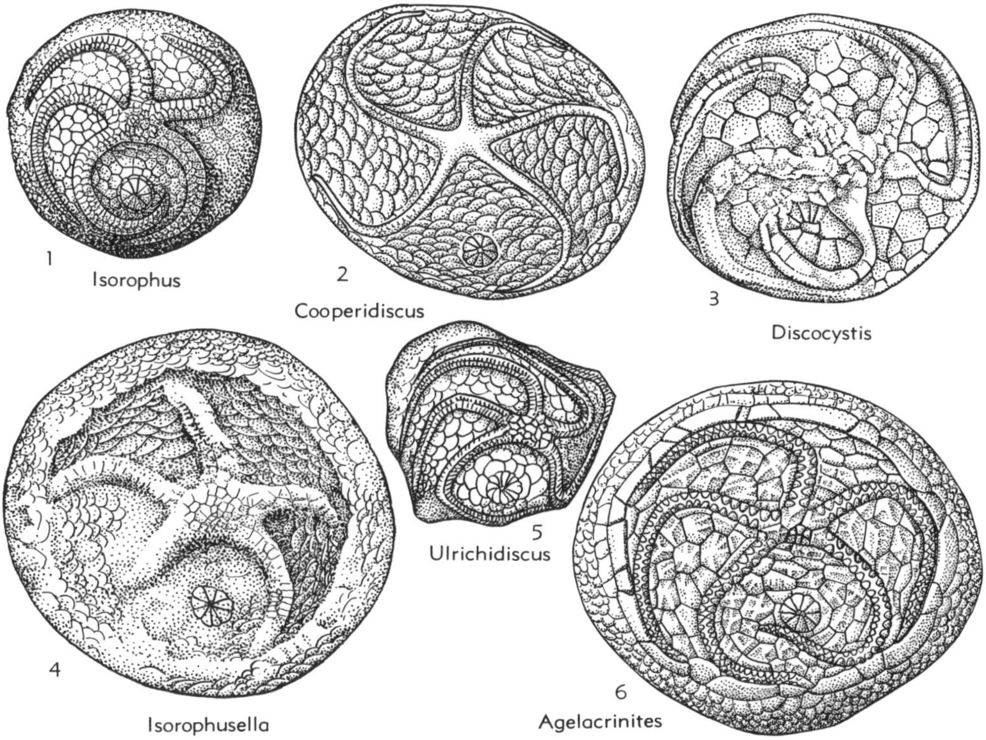

FIG. 128. Agelacrinitidae (p. *U*167-*U*169, *U*171).

direction; basal part of theca saccate, composed of numerous minute plates scattered in coriaceous skin. *L.Dev.*, Eu.(Ger.).——FIG. 127,*3*. *P. (Rhenopyrgus) coronaeformis*; lat. view, ×1 (32).

Streptaster HALL, 1872, p. following pl. 6 (as subgenus of *Agelacrinus*) [**Agelacrinus vorticellatus* HALL, 1866, p. 7; OD]. Like *Carneyella* but ambulacra revolving strongly in contrasolar direction (except solar ambulacrum *C* in *S. reversatus*, which may be a random variation) and composed of highly elevated plates; interambulacrals polygonal, small (0.5 to 1 mm. diameter), forming mosaic; ambulacra normally 5, but *S. septembrachiatus* (MILLER & DYER, 1878) having 7 (not constantly); single set of ambulacral cover plates (2 rows), which show pronounced palisade-like elongation in lateral view (16). *U.Ord.*, USA (Ohio-Tenn.-?Ky.).——FIG. 125,*3*. **S. vorticellatus* (HALL), Maysvill., Ohio; adoral view, ×3.75 (20).——FIG. 130,*2*. S. sp., *cf.* **S. vorticellatus* (HALL), Maysvill., Ohio; adoral view (reconstr.), ×6 (24). [See also Fig. 117,*2*; 120,*3b*.]

Timeischytes EHLERS & KESLING, 1958, p. 934 [**T. megapinacotus*; OD]. Similar to *Hemicystites* from which it differs in having several plates in the posterior half of peristome, and in having only one large sublunar or sublinguiform interambulacral in all interambulacra except the posterior one; theca very small (4 to 4.5 mm. in 2 specimens known); anal pyramid proportionally very large, quatrefoliated (13). *M.Dev.*, USA(Mich.).——FIG. 126,*6*. **T. megapinacotus*; adoral view (reconstr.), ×15.5(24). [See also Fig. 120,*5b*.]

Family AGELACRINITIDAE Clarke, 1901

[=Agelacrinidae auctt. *(partim)*]

Theca as in Hemicystitidae except that plates covering mouth are small, numerous, and without any definite order; ambulacra normally 5, more or less curved, rarely branching; ambulacral floor plates in single row, overlapping proximally. *M.Ord.-L. Carb.(U.Miss.)*.

Agelacrinites VANUXEM, 1842, p. 158 [**A. hamiltonensis*; OD] [=*Agelacrinus Roemer*, 1851; *Haplocystites* ROEMER, 1855; *Haplocystis, Agelacystis* HAECKEL, 1896]. Theca developed as thin epizoic disc attached by entire aboral surface, chiefly on *Rafinesquina*, subcircular to oval in outline; marginal zone formed by several rows of small plates bordering on ring of tangentially extended submarginal plates (diam., 10 to *ca.* 32 mm.) in single row; interambulacrals typically polygonal, with very slight overlap forming mosaic, sculptured, but in some species smooth and imbricating; ambulacra 5, long, usually narrow, much curved, *A,D,E* contrasolar, *B,C* solar; ambulacral cover plates in double row; anal area circular, covered by about 10 triangular plates regularly arranged in circle located approximately at center of posterior interambulacrum which may

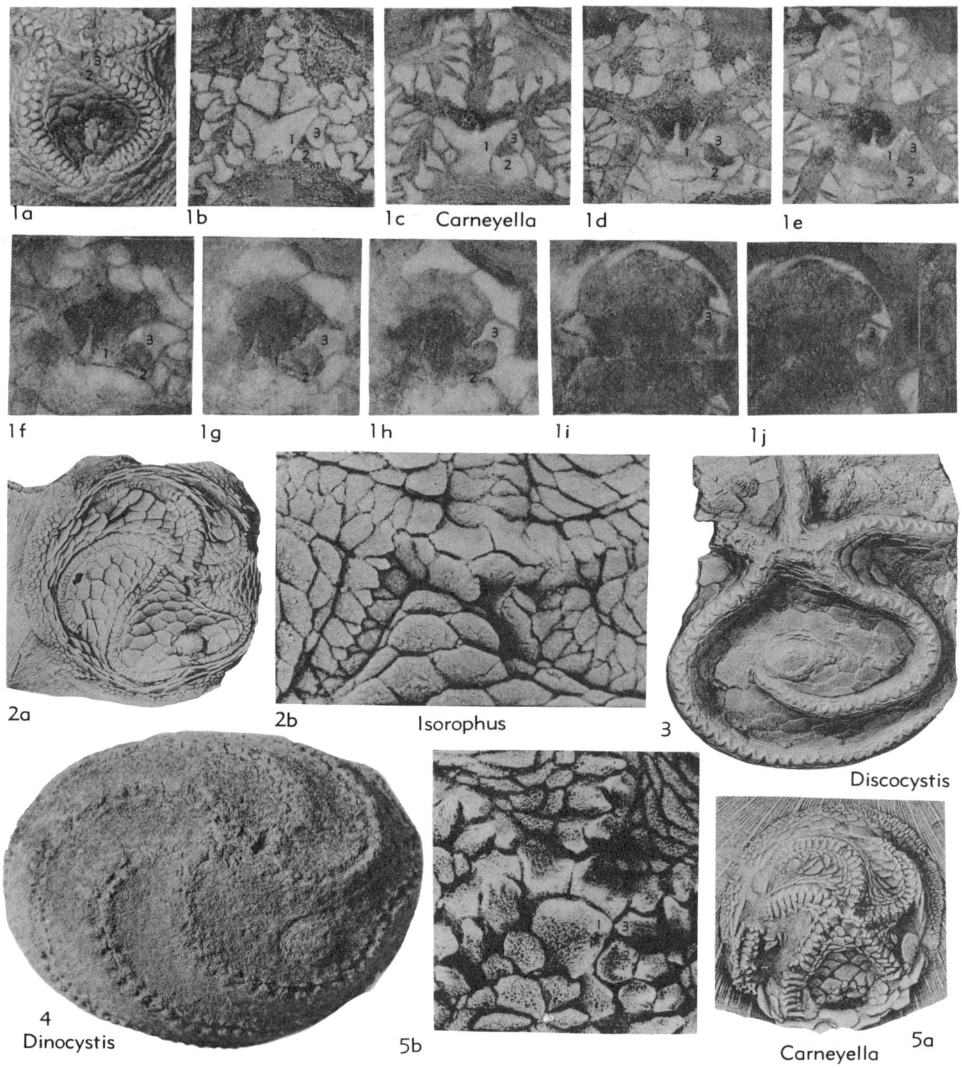

FIG. 129. Adoral views of edrioasteroids.——*1. Carneyella pileus* (HALL), U.Ord., USA(Ohio); *1a,* part of exterior surface, peristomial region (upper central) with posterior orotegminals indicated by numerals 1 and 2, proximal cover plate of *C* ambulacrum by 3, ×3; *1b-j,* photos of tangentially ground sections in succession toward aboral side of skeleton, showing "third aperture" (interpreted by KESLING as hydropore) between posterior orotegminals (numbered 1, 2) and proximal cover plate of *C* ambulacrum *(3),* ×5.5 (26).——*2. Isorophus cincinnatiensis* (ROEMER), U.Ord., USA(Ohio); *2a,* entire specimen, ×1.5; *2b,* peristomial region, ×6 (24).——*3. Discocystis laudoni* BASSLER, L.Miss., USA(Iowa); central and posterior parts of specimen, ×1.5 (24).——*4. Dinocystis barroisi* BATHER, U.Dev., Belg.; all ambulacra curved in contrasolar direction, ×1.5 (Regnéll, n).——*5. Carneyella pileus* (HALL), U.Ord., USA(Ohio); *5a,* entire specimen, ×1.5; *5b,* peristomial region with posterior orotegminals numbered 1,2, and proximal cover plate of *C* ambulacrum, 3, ×6 (24).

be entirely surrounded by ambulacra *C* and *D* (3,4). *L.Dev.-L.Miss.*, N.Am.(N.Y.-Mo.-Iowa-Ont.)-Eu.(Ger.).——FIG. 128,*6*. **A. hamiltonensis*, M.Dev.(Chenango), USA(N.Y.); adoral view, ×2.25(9).——FIG. 130,*6*. *A.* sp., M.Dev.(Traverse), USA(Mich.); adoral view (reconstr.), ×4.5(24). [See also Fig. 120,*4*.]

Cooperidiscus BASSLER, 1935, p. 8 [**Lepidodiscus alleganius* CLARKE, 1901, p. 194; OD]. Theca semiglobose, diameter about 40 mm., not attached, even temporarily; interambulacrals strongly imbricating; ambulacra 5, long, very narrow, strongly curved, all solar; ambulacral cover plates in double row; anal pyramid distinct, composed of 10 triangular plates in circle located near posterior margin (3). *U.Dev.*, USA(N.Y.-Pa.). —— FIG. 128,*2*. **C. alleganius*(CLARKE), Chemung., USA (N.Y.); adoral view, ×1.5 (9).

Discocystis GREGORY, 1897, p. 131 [**Agelacrinus kaskaskiensis* HALL, 1858, p. 696 (=*Echinodiscus optatus* WORTHEN & MILLER, 1883); OD] [=*Echinodiscus* WORTHEN & MILLER, 1883 (*non* LESKE, 1778, *nec* D'ORBIGNY, 1854, STÜRTZ, 1900); *Ageladiscus* S. A. MILLER, 1897]. Theca subcircular in outline, diameter to at least 30 mm., somewhat saclike, attached by central part of aboral side, with many rows of closely imbricated marginal plates; interambulacrals polygonal, forming mosaic; ambulacra normally 5, long, narrow, with strong curvature directed as in *Isorophus*; ambulacral cover plates in double row or forming cyclic pattern on each side of zigzag perradial line, with each cycle composed of 6 plates (e.g., *D. laudoni* BASSLER, 1936); periproct encircled by ambulacra *C* and *D*, anal pyramid distinct, composed of 7 or 8 triangular plates in circle located relatively near oral field (4, 12). [EHLERS & KESLING (12) have suggested that *Discocystis* might be a junior synonym of *Lepidodiscus* on evidence of the remarkable resemblance in shape and cyclic arrangement of the ambulacral cover plates in *D. laudoni* BASSLER, 1936, and *Lepidodiscus squamosus* (MEEK & WORTHEN, 1868).] *Miss.*, USA(Ala.-Ky.-Ill.-Mo.-Iowa).——FIG. 128,*3*. **D. kaskaskiensis*(HALL), U.Miss.(Chester), Ala.; adoral view, ×2(4).——FIG. 130,*5*. *D. laudoni* BASSLER, L.Miss. (Kinderhook.), USA(Iowa); adoral view (reconstr.), ×1.8 (24). [See also Fig. 113,*3*; 120,*6a*; 129,*3*.]

Isorophus FOERSTE, 1916, p. 340 [**Agelacrinus cincinnatiensis* ROEMER, 1851, p. 372; OD]. Theca circular in outline, diameter 8 to 40 mm., attached by entire aboral surface to *Rafinesquina* and other shells, disc-shaped to hemispherical, usually bordered by wide peripheral ring; interambulacrals scalelike, imbricating or nearly forming mosaic; ambulacra 5, almost straight to strongly curved, ambulacrum *C* solar, others contrasolar, relatively short and broad in most species; extra series of ambulacral cover plates commonly present between usual pair of rows; anal pyramid well defined, composed of 6 to 12 plates, in some specimens surrounded by zone of small plates, in species with curved rays periproct encircled by ambulacra *C* and *D* (4, 26, 30). [As illustrated by ROEMER (1851, pl. 5, fig. 3), curvature of the ambulacra is opposite to the true direction, owing to reversal of the image by the camera lucida used.] *M.Ord.-U.Ord.*, USA(N.Y.-Ky.-Tenn.-Ohio-Ind.).——FIG. 128,*1*; 130,*1*. **I. cincinnatiensis* (ROEMER), U. Ord. (Maysvill.), Ohio; 128,*1*, adoral view, ×1.5 (4); 130,*1*, same (reconstr.), ×2.4 (26). [See also Fig. 121; 129,*2*.]

Isorophusella BASSLER, 1935, p. 5 [**Lebetodiscus inconditus* RAYMOND, 1915, p. 61; OD]. Theca circular in outline, diameter up to 20 mm., slightly convex, resting on or attached by entire aboral surface to a hard substratum; with broad border of closely imbricating plates which decrease in size toward periphery; interambulacrals scalelike, strongly imbricating; ambulacra 5, short and stout, curving slightly in same direction as in *Agelacrinites*; median cover plates commonly present; oral area broad; anal pyramid distinct but almost flat, composed of about 10 elongate triangular plates arranged in circle located approximately at center of posterior interambulacrum (3, 4). *M.Ord.*, Can.(Ont.-Que.).——FIG. 128,*4*. **I. incondita* (RAYMOND), Trenton, Que.; adoral view, ×3 (3).

Lepidodiscus MEEK & WORTHEN, 1868, p. 357 [**Agelacrinites (Lepidodiscus) squamosus*; OD]. Theca subcircular or oval in outline, diameter 7 to 25 mm. in most species but 45 mm. in type-species, depressed, mostly with flat marginal rim by which it was attached to hard objects; interambulacrals strongly imbricating, with granular ornamentation in some species; ambulacra normally 5 (except for some specimens of *L. lebouri* SLADEN), long, narrow, much curved in same direction as in *Isorophus*, more seldom relatively short and broad; in *L. squamosus* ambulacral cover plates of one side are divided from those on other by prominent zigzag perradial line and are disposed in regular cycles of 6 throughout most of ambulacrum; ambulacral cover plates of other species arranged in double row, in some bordered by series of small side plates; anal pyramid distinct, composed of 9 to 15 triangular plates, usually not surrounded by ring of small plates, located near posterior margin and encircled by ambulacra *C* and *D* (4, 25, 30). *M.Dev.-U.Miss.*, N. Am. (N.Y.-Pa.-Mich.-Ind.)-Eu. (Eng.-USSR). [EHLERS & KESLING (12) have expressed doubt that many species assigned for the present to *Lepidodiscus* are correctly classified.]——FIG. 130, *4*. **L. squamosus* (MEEK & WORTHEN), L.Miss. (Keokuk), USA(Ind.); adoral view (reconstr.), ×1.5 (24). [See also Fig. 114,*2*; 120,*6a'*.]

Thresherodiscus FOERSTE, 1914, p. 433 [**T. ramosus*; OD]. Theca discoid, with gently convex adoral surface, diameter about 16 mm.; interambulacrals large, squamose, imbricating in central region, decreasing in size distally so as to

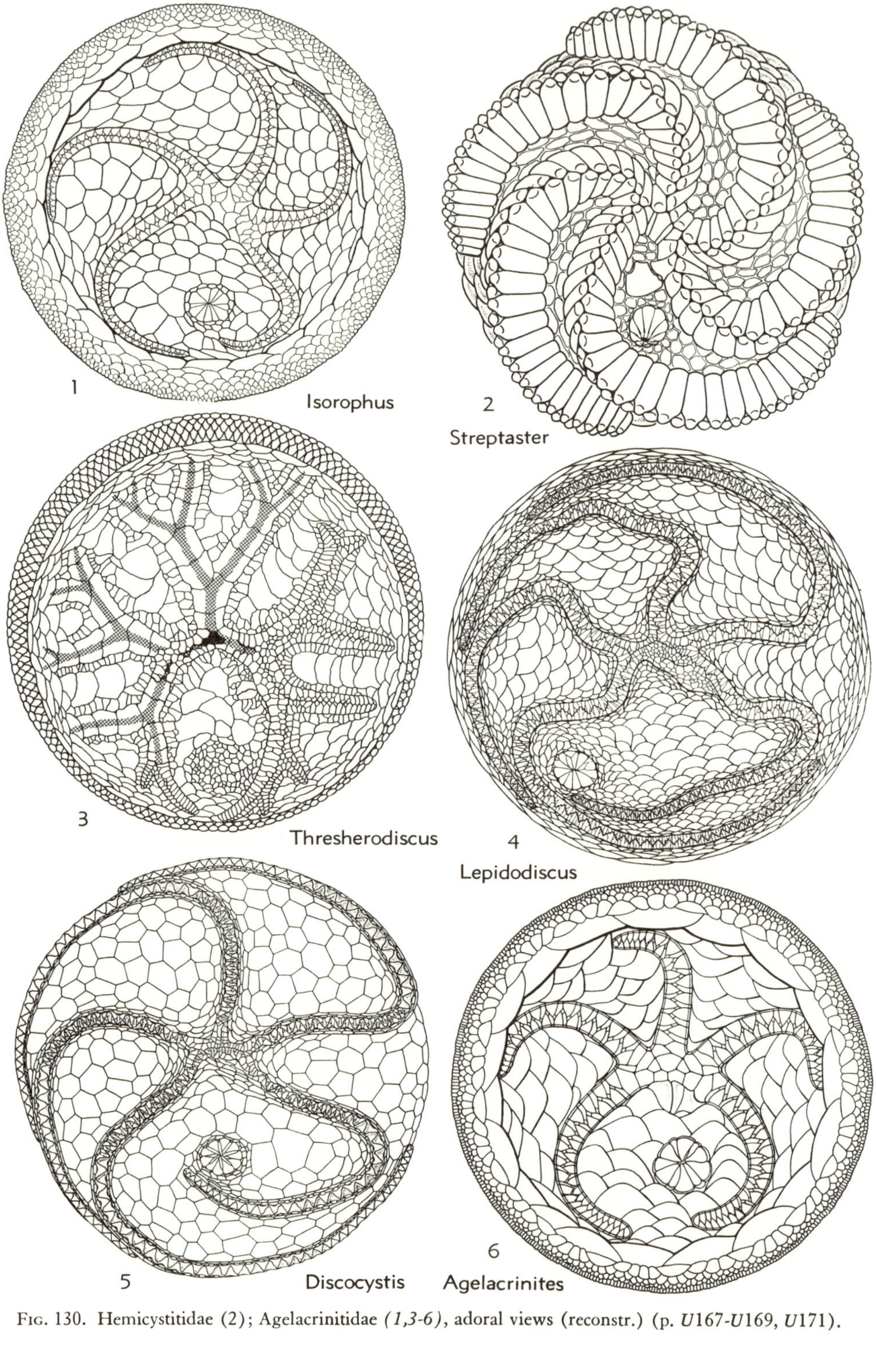

Fig. 130. Hemicystitidae (2); Agelacrinitidae *(1,3-6)*, adoral views (reconstr.) (p. *U*167-*U*169, *U*171).

merge with much smaller plates of marginal zone; ambulacra 5, rather broad, of pronounced trimerous origin, branching dichotomously with bifurcations of 1st, 2nd, and probably 3rd order; ambulacral cover plates in 2 sets consisting of very small median ones and much larger lateral ones; periproct unknown (16). *M.Ord.,* Can. (Ont.).——FIG. 130,*3*. **T. ramosus,* Trenton.; adoral view of holotype (reconstr.), ×4 (24). [See also Fig. 112,*1*; 120,*2a*.]

Ulrichidiscus BASSLER, 1935, p. 8 [**Agelacrinus pulaskiensis* MILLER & GURLEY, 1894, p. 16; OD]. Theca semiglobose, diameter about 26 mm.; interambulacrals relatively large, polygonal, slightly imbricating; ambulacra 5, narrow but well defined, very long, all strongly curving in contrasolar direction; anal pyramid well defined, composed of many long triangular plates in circle (3). *U.Miss.,* N.Am.(USA).——FIG. 128,*5*. **U. pulaskiensis* (MILLER & GURLEY), Chester., Ky.; adoral view, ×1 (3).

Family EDRIOASTERIDAE Bather, 1898

Theca shaped like tam-o'-shanter, not permanently attached; central part of aboral surface excavated, covered by flexible integument; interambulacrals polygonal, forming mosaic, continuous with those of aboral face; ambulacra 5, very long, strongly curved, passing on to aboral surface; ambulacral floor plates in double rows of alternating plates that meet along zigzag median suture, with pores between floor plates and each cover plate corresponding in position with a floor plate; third aperture located in posterior interambulacrum. *M.Ord.-U.Dev.*

Edrioaster BILLINGS, 1858, p. 82 [**Cyclaster bigsbyi* BILLINGS, 1857, p. 293; OD] [=*Cyclaster* BILLINGS, 1857 (*non* COTTEAU, 1856; *nec* MALLORY, 1904); *Aesiocystites* MILLER & GURLEY, 1894; *Edriocystis* HAECKEL, 1896; *Aesiocystis* BATHER, 1900]. Theca subcircular to subpentagonal, diameter 14 to 50 mm., height 0.25 to 0.5 of width; interambulacrals generally more or less pustulose or granulose, separated from central aboral region by frame of stouter plates; peripheral plates of aboral face variable in size but not minute but central plates minute and tending to imbricate; ambulacra comparatively broad, raised or not raised above the general surface, *A, B, D, E* curving in a contrasolar direction, *C* in solar direction (type-species only), or all curving solarly; small median cover plates may be present; peristome covered by solid tegmen of plates serially homologous with ambulacral cover plates; anus well defined, covered by variable number of irregular plates (7). [Forms in which all of the ambulacra

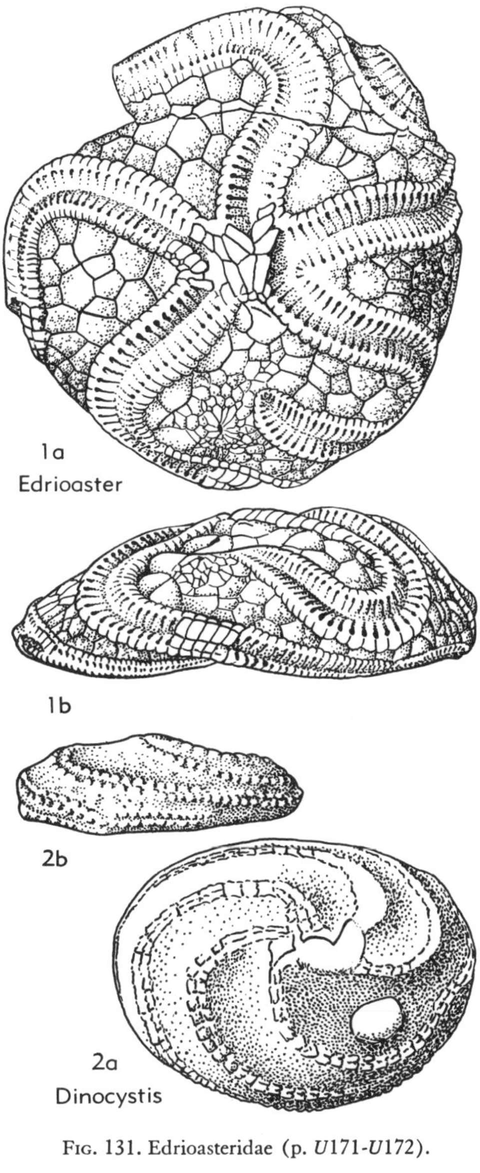

FIG. 131. Edrioasteridae (p. *U*171-*U*172).

curve in a solar direction possibly should be separated generically from *Edrioaster*.] *M.Ord., ?U. Ord.,* N. Am. (Ont.-Que.-Ky.-Mich.-Minn.-N. Y.)-Eu.(N.Wales).——FIG. 131,*1*. **E. bigsbyi* (BILLINGS), M.Ord.(Trenton.), Ont.; *1a,b,* adoral, post. views, ×1.7 (7, pt. 4). [See also Fig. 111; 113,*1*; 115,*2*; 116,*1,4*; 118,*1*; 119,*1,3*; 120,*1*.]

Dinocystis BATHER, 1898, p. 547 [**D. barroisi*; OD]. Theca elliptical in outline, diameter to about 40 mm., like *Edrioaster* but with frame on aboral surface slighter and surrounding region composed of thin flexible integument containing narrow imbricating plates; ambulacra narrow, all curving strongly in a contrasolar direction (7). *U.Dev.*, Eu.(Belg.).——FIG. 131,*2*. **D. barroisi*, Famenn.; 131,*2a*, oral, ×1 (Regnéll, n); 131,*2b*, ant., ×1.2 (7, pt. 1).

REFERENCES

Anderson, F. W.

(1) 1939, *Lepidodiscus fistulosus sp. nov. from Lower Carboniferous rocks, Northumberland:* Great Britain, Geol. Survey, Bull. 1, p. 67-81, pl. 5, text-fig. 1-10.

Aurivillius, C. W. S.

(2) 1892, *Über einige ober-silurische Cirripeden aus Gotland:* [K.] Svenska. Vetenskaps-akad., Handl., Bihang, ser. 4, v. 18, pt. 3, p. 1-24, pl. 1.

Bassler, R. S.

(3) 1935, *The classification of the Edrioasteroidea:* Smithsonian Misc. Coll., v. 93, pt. 8, p. 1-11, pl. 1.

(4) 1936, *New species of American Edrioasteroidea:* Same, v. 95, pt. 6, p. 1-33, pl. 1-7.

———, & Moodey, M. W.

(5) 1943, *Bibliographic and faunal index of Paleozoic pelmatozoan echinoderms:* Geol. Soc. America, Spec. Paper 45, vi+734 p.

Bather, F. A.

(6) 1900, *The Echinoderma* (with contributions by J. W. GREGORY and E. S. GOODRICH): in LANKESTER, E. R., A Treatise on Zoology, v. 3, viii+344 p., 309 text-fig., A. & C. Black (London).

(7) 1915, *Studies in Edrioasteroidea, 1-9:* reprinted, with additions, from Geol. Mag. for 1898, 1899, 1900, 1908, 1914, 1915, The Author (Wimbledon).

Cabibel, Jaques, Termier, Henri, & Termier, Geneviève

(8) 1959, *Les échinodermes mésocambriens de la Montagne Noire (Sud de la France):* Ann. Paléontologie, v. 44 (1958), p. 279-294, pl. 6-7, text-fig. 1-10.

Clarke, J. M.

(9) 1901, *New Agelacrinites:* New York State Museum, Ann. Rept. 55 (Bull. 49), p. 182-198, pl. 10, text-fig. 1-7.

Cuénot, Lucien

(10) 1948, *Anatomie, éthologie et systématique des échinodermes:* in Traité de zoologie, GRASSÉ, P.-P (ed.), v. 11, p. 3-272, text-fig. 1-312, Masson (Paris).

Dehm, Richard

(11) 1961, *Über Pyrgocystis (Rhenopyrgus nov. subgen.) coronaeformis Rievers aus dem rheinischen Unter-Devon:* Bayer. Staats-sammlung Paleont. Hist. Geol., Mitteil., v. 1, p. 13-17.

Ehlers, G. M., & Kesling, R. V.

(12) 1958, *Cyclic pattern of ambulacral covering plates in Discocystis laudoni and its taxonomic implication:* Michigan Univ., Museum Paleontology, Contrib., v. 14, no. 15, p. 265-276, pl. 1-3.

(13) 1958, *Timeischytes, a new genus of hemicystitid edrioasteroid from the Middle Devonian Four Mile Dam Limestone of Michigan:* Jour. Paleontology, v. 32, p. 933-936, pl. 121, text-fig. 1.

Fay, R. O.

(14) 1962, *Edrioblastoidea, a new class of Echinodermata:* Jour. Paleontology, v. 36, p. 201-205, pl. 34, text-fig. 1-3.

Fisher, D. W.

(15) 1951, *A new edrioasteroid from the Middle Ordovician of New York:* Jour. Paleontology, v. 25, p. 691-693, text-fig. 1-2.

Foerste, A. F.

(16) 1914, *Notes on Agelacrinidae and Lepadocystinae, with descriptions of Thresherodiscus and Brockocystis:* Denison Univ. Sci. Lab., Bull. 17, p. 399-487, pl. 1-6, text-fig. 1-8.

Gekker [Hecker], R. F.

(17) 1928, *Paläobiologische Betrachtungen an untersilurischen Invertebraten. I:* Ann. Soc. Paléont. Russie, v. 7, p. 47-86, pl. 3-4, text-fig. 1-3.

(18) 1940, *Agelacrinites aus dem Devon der russischen Tafel:* Acad. Sci. URSS, Inst. Paléont., Travaux, v. 9, pt. 4, p. 83-104, pl. 11-12, text-fig. 1-3.

Gislén, Torsten

(19) 1930, *Affinities between the Echinodermata Enteropneusta, and Chordonia:* Zool. Bidrag. Uppsala, 12, p. 199-304, text-fig. 1-46.

Hall, James

(20) 1872, *Description of new species of Crinoidea and other fossils from strata of the age of the Hudson River Group and Trenton Limestone:* New York State Museum, 24th Ann. Rept., p. 205-224, pl. 5-8, text-fig. 1-3.

Harker, Peter

(21) 1953, *A new edrioasteroid from the Carboniferous of Alberta:* Jour. Paleontology, v. 27, p. 288-289, text-fig. 1.

Jaekel, Otto

(22) 1899, *Stammesgeschichte der Pelmatozoen. 1. Thecoidea and Cystoidea:* x + 442 p., 18 pl., 88 text-fig.

(23) 1927, *Cyathotheca suecica n. g., n. sp., eine Thecoidee des schwedischen Ordoviciums:* K. Svenska. Vetenskapsakad., Arkiv Zool., v. 19A, pt. 5, p. 1-5, pl. 1.

Kesling, R. V.

(24) 1960, *Hydropores in edrioasteroids:* Michigan Univ., Museum Paleontology, Contrib., v. 15, no. 8, p. 139-192, pl. 1-13, text-fig. 1-14.

———, & Ehlers, G. M.

(25) 1958, *The edrioasteroid Lepidodiscus squamosus (Meek & Worthen):* Jour. Paleontology, v. 32, p. 923-932, pl. 119-120, text-fig. 1.

———, & Mintz, L. W.

(26) 1960, *Internal structures in two edrioasteroid species, Isorophus cincinnatiensis (Roemer) and Carneyella pilea (Hall):* Michigan Univ., Museum Paleontology, Contrib., v. 15, no. 14, p. 315-348, pl. 1-9, text-fig. 1-4.

Pompeckj, J. F.

(27) 1896, *Die Fauna des Cambrium von Tejřovic und Skrej in Böhmen:* K.-K. Geol. Reichsanst. Wien, Jahrb., v. 45 (1895), p. 495-614, pl. 13-17, text-fig. 1.

Raymond, P. E.

(28) 1921, *A contribution to the description of the fauna of the Trenton Group:* Canada, Geol. Survey, Dept. Min. & Res., Museum Bull. 31, p. 1-64, pl. 1-10.

Regnéll, Gerhard

(29) 1945, *Non-crinoid Pelmatozoa from the Paleozoic of Sweden:* Lunds Geol.-Min. Inst., Meddel. 108, p. i-viii+1-255, pl. 1-15, text-fig. 1-30.

(30) 1950, *"Agelacrinites" ephraemovianus Bogolubov and "Lepidodiscus" fistulosus Anderson:* K. Fysiogr. Sällsk. Lund Förhandl., v. 20, pt. 20, p. 218-237, text-fig. 1-2.

(31) 1960, *Données concernant le développement ontogénétique des pelmatozoaires du Paléozoïque (échinodermes):* Soc. Géol. France, Bull., ser. 7, v. 1 (1959), p. 773-783, text-fig. 1-6.

Rievers, Johannes

(32) 1961, *Eine neue Pyrgocystis (Echinod., Edrioasteroidea) aus den Bundenbacher Dachschiefern (Devon):* Bayer. Staatssammlung Paleont. Hist. Geol., Mitteil., v. 1, p. 9-11, pl. 2.

Rösler, Artur

(33) 1953, *Agelacrinites taunicus n. sp. (Edrioaster.) aus dem Unterdevon von Mensfelden (Blatt Limburg, Lahn). Mit Bemerkungen zu Agelacrinites? rhenanus (F. Roemer) und einer Übersicht über die Funde von nichtcrinoiden Pelmatozoen im rheinischen Unterdevon:* Hess. Landesamt Bodenf., Notizbl., v. 81, p. 69-74, pl. 3.

Schmidt, Friedrich

(34) 1879, *Ueber Cyathocystis Plautinae, eine neue Cystideenform aus Reval:* Russische-Kaiserl. Min. Gesell., Verhandl., ser. 2, v. 15, p. 1-7, text-fig. 1-3. [The entire volume was published in 1880, as given on the title-page, but separate of the present paper appeared in October 1879 (*fide* O. Jaekel, 1899, p. 439).]

Schuchert, Charles

(35) 1919, *A Lower Cambrian edrioasterid:* Smithsonian Misc. Coll., v. 70, pt. 1, p. 1-8, pl. 1, text-fig. 1.

Sinclair, G. W.

(36) 1951, *Three notes on Ordovician agelacrinitids:* (unpublished manuscript).

Spencer, W. K.

(37) 1904, *On the structure and affinities of Palaeodiscus and Agelacrinus:* Royal Soc. London, Proc., v. 74, p. 31-46, pl. 1, text-fig. 1-12.

(38) 1938, *Some aspects of evolution in Echinodermata:* Evolution, essays presented to E. S. Goodrich, p. 287-303, pl. 1-2.

Strimple, H. L., & Graffham, A. A.

(39) 1955, *A new species of Cyathocystis:* Washington Acad. Sci., Jour., v. 45, p. 353-355.

Williams, S. R.

(40) 1918, *Concerning the structure of Agelacrinites and Streptaster, Edrioasteroidea of the Richmond and Maysville divisions of the Ordovician:* Ohio Jour. Sci., v. 19, p. 59-86, pl. 1-9.

Wilson, Alice E.

(41) 1946, *Echinodermata of the Ottawa Formation of the Otawa-St. Lawrence Lowland:* Canada, Geol. Survey, Dept. Min. & Res., Bull. 4, p. i-v + 1-61, pl. 1-6, text-fig. 1-2.

OPHIOCISTIOIDS

By Georges Ubaghs

[Université de Liège]

[The author expresses sincere thanks to R. C. Moore for an excellent translation of his French typescript. Also, he acknowledges with great appreciation aid given to him by officers of the British Museum (Natural History) and the Oxford University Museum in making available for study by him various type-specimens and other ophiocistioid materials contained in their collections.]

CONTENTS

INTRODUCTION

Ophiocistioids, named from the Greek words *ophis* ("snake") and *kiste* ("box"), are a small group of pentaradiate free-moving echinoderms known only as fossils in Paleozoic strata. Their body may be entirely enclosed, except for the peristome, by a test composed of plates like that of echinoids, or it may be covered by plates on one side only, the other being protected by an integument that is little calcified if at all. A resemblance to echinoids does not include the mode of union of the test plates, for in the ophiocistioid *Sollasina,* at least, junction of the plates seems to differ from that observed in echinoids. Ophiocistioids have no arms comparable to those of asterozoans and crinoids or brachioles like those of cystoids. Instead, they bear hollow tubular organs with a cover of small imbricated plates, and despite their relatively large size, these organs are interpreted to be ambulacral tube feet.

The test has an oral surface, on which the animal rests or travels about, and an aboral surface. One is approximately equal to the other in area, but otherwise the two surfaces are quite dissimilar. Plates of the aboral surface are irregularly disposed in some ophiocistioids, whereas they exhibit a more or less definite radial arrangement in others. The anus is located on this surface or at its edge in genera represented by specimens with the anal vent preserved. At the center of the oral surface is the mouth, equipped with a masticatory apparatus consisting of five interradially placed jaws, each of which is composed of two pieces. The buccal apparatus is surrounded by a flexible peristomial membrane not unlike that of echinoids. Beyond the peristomial area, the oral surface is divided into five large ambulacral tracts and five narrow interambulacral ones. No ambulacral grooves are present. The ambulacra terminate at the border of the test without extending onto the aboral surface. Each ambulacral tract contains three columns of plates, one perradial and the other adradial. A single pore for each ambulacral tube foot is located at the intersection of adradial and perradial plate pairs. One of the oral interrays in two genera contains a madreporite associated with one or several orifices interpreted as gonopores. In another genus a single orifice on the aboral surface is possibly a gonopore, or hydropore, or combined gonopore-hydropore.

DEVELOPMENT OF KNOWLEDGE

Ophiocistioids were first differentiated from other echinoderms by Sollas (15) in 1899 when he proposed recognition of them as an order named Ophiocistia in the class

Ophiuroidea. In 1912 SOLLAS & SOLLAS (16) elevated the group to class rank, even though up to that time it included only three Silurian species from Great Britain. These are now judged to represent three monotypic genera: *Eucladia* WOODWARD, 1869; *Euthemon* SOLLAS, 1899; *Sollasina* FEDOTOV, 1926. In 1930 RICHTER (13) introduced the name *Rhenosquama* for organic remains from the Middle Devonian of Germany which he interpreted as ambulacral tube feet of an otherwise unknown ophiocistioid. The assignment of *Rhenosquama* to the ophiocistioids is by no means certain, however. In 1938 and 1940 GEKKER (6, 7) described certain fossils from the Lower Ordovician (Skiddavian) in the vicinity of Leningrad. Earlier, JAEKEL (1901, 1918) and BATHER (1913) had considered these fossils to be thecas of the fantastic *Rhipidocystis* JAEKEL, 1901. GEKKER concluded that in reality they are fragments of the test of a quite different organism, which he named *Volchovia* and classified among the ophiocistioids even though it differs greatly from typical ophiocistioids from the British Silurian. Subsequently, *Volchovia* has been discovered in Lower Ordovician rocks of Norway (REGNÉLL, 12) and in Upper Ordovician deposits of Ohio (POPE, 1960). Finally, a fossil from the Lower Devonian (Helderberg) of New York which SCHUCHERT (1915) described as *Eucladia? beecheri* and attributed (14) doubtfully to the ophiocistioids, is an enigmatic form impossible to classify in the state of present knowledge.

Owing to the strange features and rarity of ophiocistioids, interpretation of these echinoderms has been very difficult. Only a single specimen of *Eucladia* and one of *Euthemon* are known. Other genera are represented by a mere handful of specimens including fragments. First descriptions (WOODWARD, 18; GREGORY, 8) are partly erroneous. SOLLAS (15) may be credited with having shown accurately the structure of most typical genera and BATHER (1,2) with having been first to indicate that the tubular appendages of these forms are not arms but equivalents of the tube feet found in other echinoderms. More recently, FEDOTOV (4), GEKKER (7), and REGNÉLL (12) have contributed important new information concerning the morphology of ophiocistioids and discussion of their systematic position has been published, especially by FEDOTOV (4, 5), REGNÉLL (12), and UBAGHS (17).

For many years the ophiocistioids were placed among the ophiuroids, as suggested by their name, or in a separate group allied to the Ophiuroidea. At present they are accorded the rank of a distinct class, coordinate taxonomically with the Stelleroidea, Echinoidea, and Holothuroidea, for example. This is the view adopted by SOLLAS & SOLLAS (16), FEDOTOV (4, 5), RICHTER (13), GEKKER (6, 7), REGNÉLL (12) and a majority of modern treatises and manuals. Although they have been almost universally considered as Eleutherozoa, MATSUMOTO placed them in his subphylum Crinozoa, corresponding to Pelmatozoa less Edrioasteroidea. FELL (1963) has assigned them to the Echinozoa, and this classification is accepted in the *Treatise*.

MORPHOLOGY

GENERAL FORM AND SIZE

In outline the theca of ophiocistioids may be elliptical *(Sollasina),* suboval *(Volchovia),* subpentagonal *(Euthemon),* or faintly pentalobate *(Eucladia).* The border is sinuous and somewhat angulated in *Volchovia* but evenly regular in other genera.

The test bears no arms or rays corresponding to those of starfishes or crinoids, and, as previously noted, its downwardly directed oral surface with central mouth is very unlike the opposite aboral surface on the upper side. The oral surface of *Eucladia* (Fig. 132) and *Euthemon* (Fig. 133) is flat or gently convex medially but progressively curved upward near the border so as to grade into the unknown, probably arched aboral surface. In *Sollasina* (Fig. 134) both surfaces of all observed specimens are crushed, yet position of the appendages suggests that the theca in undistorted condition was low dome-shaped. The central part of the aboral side of *Volchovia* has the form of a moderately ele-

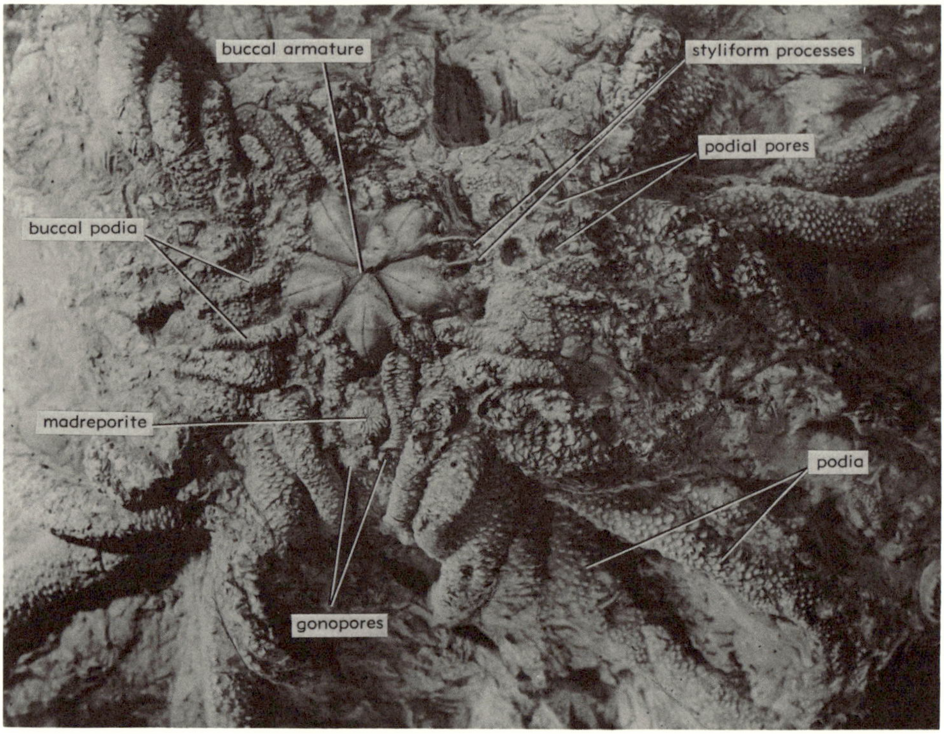

FIG. 132. *Eucladia johnsoni* WOODWARD, U.Sil.(L.Ludlov.), Eng.; part of oral surface showing morphological features, ×1.3 (Ubaghs, n).

vated mound which grades peripherally into very gently sloping, nearly flat marginal areas (Fig. 135). The sharply defined border of the theca is indented by angular embayments between the several pointed extensions of its circumference. The oral surface, not observed in this genus, is doubtless flat or concave. The theca of *Rhenosquama* is entirely unknown.

The size of tests belonging to ophiocistioids is small to modest, as shown by following measurements: *Euthemon,* diameter 7 or 8 mm.; *Sollasina,* length 30 mm., width 20 mm., *Eucladia,* length 50 mm., width 40 mm.; *Volchovia,* maximum diameter, 90 mm.

SYMMETRY

The oral surface of all genera in which it is known exhibits pentaradiate symmetry. This is marked by (1) distribution of plates in five radial and five interradial areas, (2) distribution of tube feet in five groups, and (3) presence of a buccal armature composed of five pairs of interradially disposed plates. A tendency toward pentaradiate symmetry appears also in the arrangement of aboral plates in *Volchovia,* but as noted later, this symmetry is far from perfect.

Orientation of the test can be determined in various ophiocistioids belonging to dif-

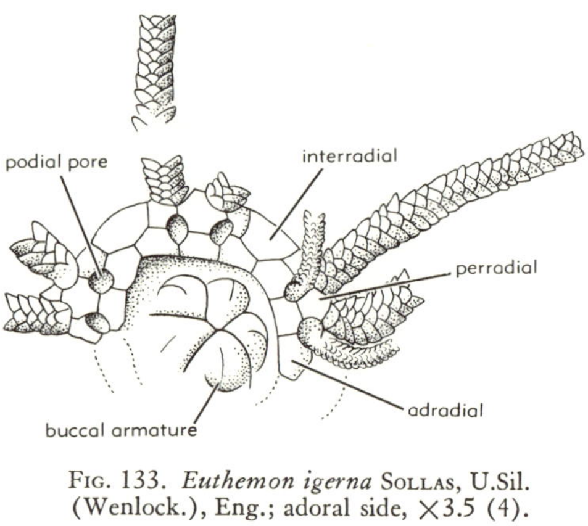

FIG. 133. *Euthemon igerna* SOLLAS, U.Sil. (Wenlock.), Eng.; adoral side, ×3.5 (4).

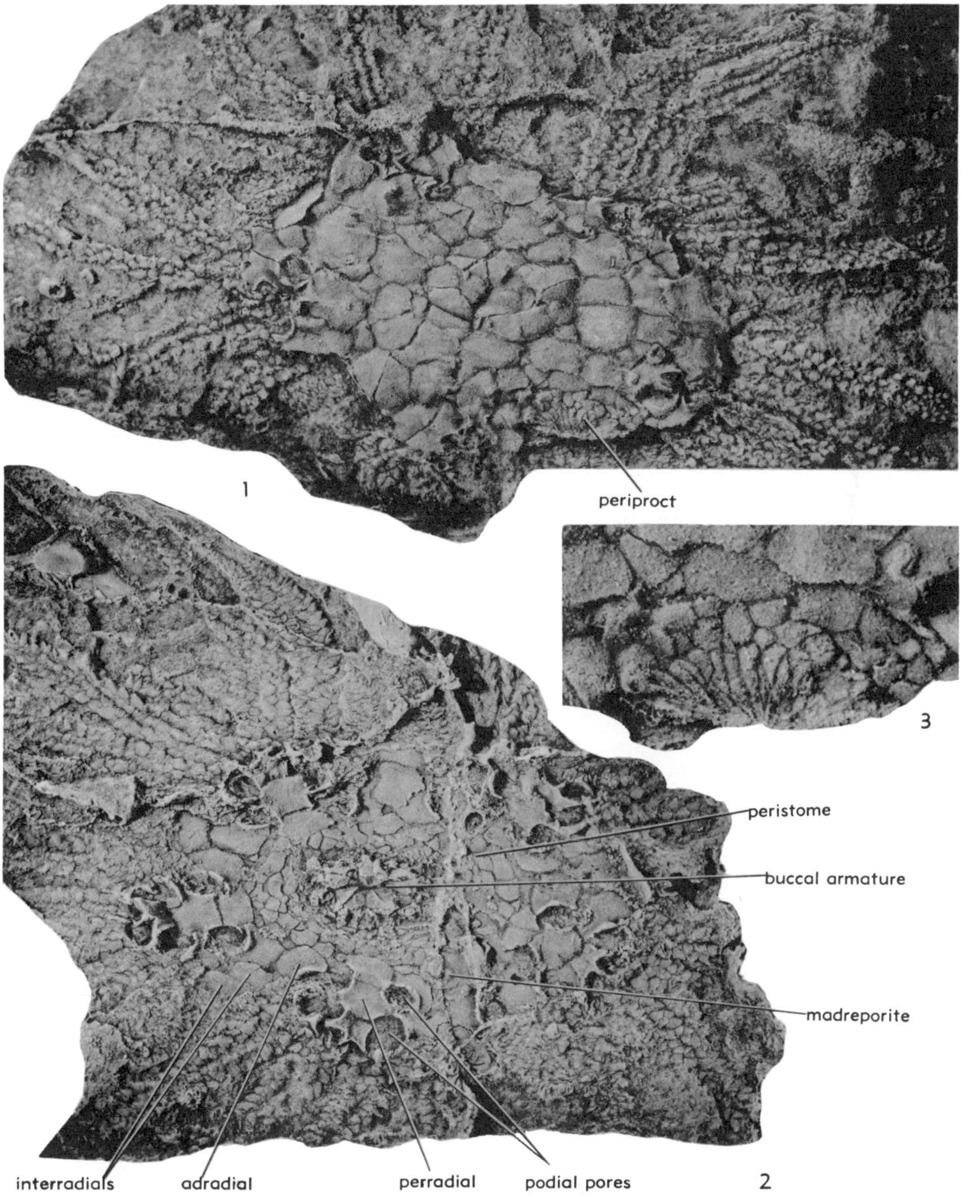

FIG. 134. *Sollasina woodwardi* (SOLLAS), U.Sil.(L.Ludlov.), Eng.; *1,2,* aboral and oral sides, ×2; *3,* periproct, ×15 (all Ubaghs, n).

ferent genera. In *Eucladia* (Fig. 132) and *Sollasina* (Fig. 134,*2*) the presence of a madreporite accompanied by one or more perforated tubercles (?gonopores) in an interray of the oral surface serves to identify this interray as *CD,* in terms of the Carpenter system. Moreover, some specimens of *Sollasina* (Fig. 134,*1,3*) show presence of the periproct on the aboral side of the test adjacent to the border and located in this interray or one next to it (decision as to which being difficult). The *CD* inter-

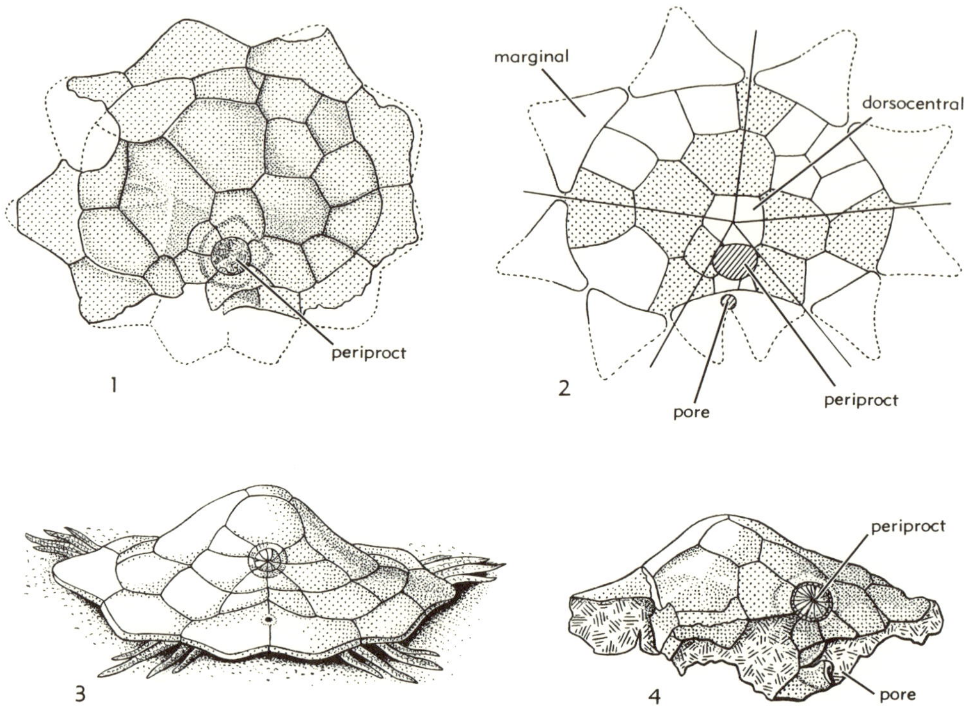

FIG. 135. *Volchovia,* L.Ord., Eu.; *1-3, V. mobilis* GEKKER, USSR (7); *4, V. norvegica* REGNÉLL, Norway (12); all ×1.

ray is identifiable also in *Volchovia* (Fig. 135) by the presence of an orifice previously mentioned as probably a gonopore or hydropore or a combined gonopore-hydropore; this opening occurs between two marginal plates on the aboral surface. The anus occurs in the same interray approximately midway between the supposed gonopore and the summit of the theca. The major axis of the test is perpendicular to the *CD* interradius in *Volchovia* and oblique to it in *Sollasina.*

ABORAL SURFACE

The upper surface of the theca is known only in *Sollasina* and *Volchovia.*

In *Sollasina* many irregular polygonal or rounded plates form the aboral surface (Fig. 134,*1*). The plates are thin and do not exceed 5 mm. in diameter. Their surface is smooth or covered by very minute scattered granules. Arrangement of the aboral plates seems to fit no determinable pattern, and no sign of an apical system resembling that of echinoids and many starfishes can be discerned. The small plates probably overlapped one another when the animals were alive, but the extent of their overlap clearly has been increased by post-mortem flattening of the test found in all specimens studied. An anal pyramid, described below, occurs between two groups of tube feet on a certain part of the margin.

In *Volchovia* the aboral surface, evidently rigid, is composed of large plates (Fig. 135). A flat framework formed by ten pentagonal or subtrigonal plates surrounds a central dome, composed of 20 to approximately 30 plates in different species. The dome plates are arranged in an outer, submarginal circle and an inner central area. REGNÉLL (12) has shown that the placement of aboral plates indicates a degree of pentaradiate symmetry, and evidence previously noted (location of periproct and ?gonopore) permits determination of the *CD* interray. Inasmuch as the two marginal plates adjoining the supposed gonopore belong to this interray, it is reasonable to conclude that the eight

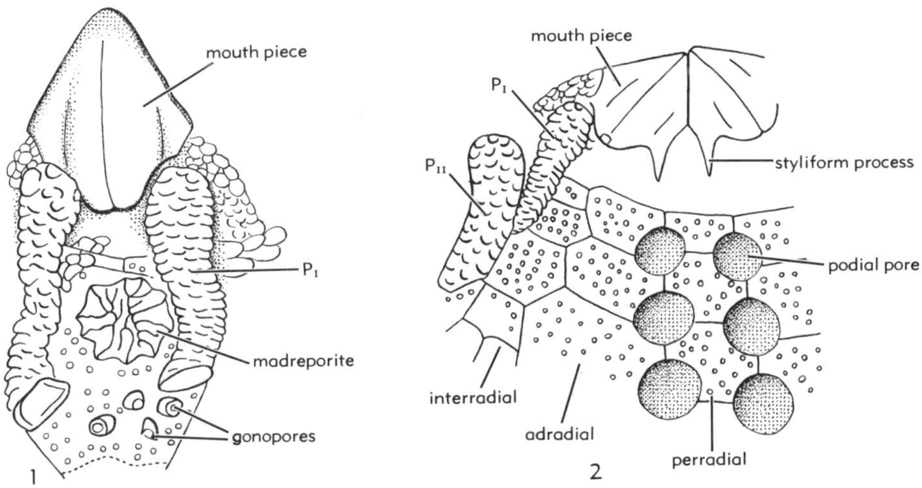

FIG. 136. *Eucladia johnsoni* WOODWARD, U.Sil.(L.Ludlov.), Eng.; *1*, part of *B* ray and *BC* interray, ×2 (4); *2*, part of *CD* interray, ×2.7 (4). [Explanation: P_I, P_{II}, 1st, 2nd pairs of podia.]

other marginal plates also are interradial, each interray having two such plates. Plates of the central region seemingly include (1) a "dorsocentral" located between the geometrical center of the test and the periproct, (2) a ring of five "radial" plates, which, in company with the first-mentioned element, form a kind of apical system, and (3) another circle of plates which in each ray contains a supplementary "radial" and in each interray one or more "interradials." The rays and interrays thus defined differ from one another, for plates of the *C* and *D* rays are smaller than those of other rays, which also are distinctly longer than the *C* and *D* rays. In *V. norvegica* the *D* ray contains only a single "radial," whereas in *V. mobilis* the *C* ray has three such plates. The "radials" and "interradials" are dissimilar in shape and size and the latter vary in number from one interray to another. These inequalities are distributed in such a way as to indicate a tendency toward acquisition of bilateral symmetry with respect to an anteroposterior axis (through *A* ray and *CD* interray).

ORAL SURFACE

The lower surface of the theca is known in *Eucladia, Sollasina,* and *Euthemon,* but unobserved in any specimen of *Volchovia,* doubtless owing to absence or very weak development of an oral skeleton in this genus.

The three genera in which the oral surface has been studied show that this part of the ophiocistioid theca consists of central peristome, described subsequently, and surrounding plates divisible into five radial or ambulacral tracts and five interradial or interambulacral tracts. The radial tracts are large, and as previously stated, contain a median column of perradial plates and two lateral columns of adradial plates (Fig. 134, *2,3*; 136,*1*). The interradial tracts are narrow, consisting of few plates disposed in single columns. The radial areas are pierced by double rows of simple circular orifices (ambulacral or tube-feet pores) with thickened borders, each pore being adjoined by two perradial and two adradial plates. In addition, an unpaired orifice occurs at the aboral extremity of each ambulacrum; it is adjoined by two plates, which in *Sollasina* are adradials but in other genera undetermined. In living ophiocistioids this orifice must have been located at the margin of the theca, but in fossils it is near the edge of the aboral side owing to compression which the specimens have undergone. From each orifice, including the unpaired terminal one of each ambulacrum, arises one of the tubular organs with scaly skeletal cover which is inferred to be a tube foot.

Around the peristome of *Sollasina,* and probably of *Euthemon* also, is a circle of 15 or 16 plates, of which five are perradial, ten adradial, and presumably one in addition with madreporite and gonopore. In *Eucladia* the 15 radial plates are accompanied by five interradial plates, making 20 in all. The radial tracts of *Eucladia* contain seven plates in each column, whereas in *Euthemon* these are reduced to three. *Sollasina* has three plates in each perradial column and four in each adradial column. Five or six plates occur in each interradial area of *Eucladia,* two in *Sollasina,* and a single one in *Euthemon.*

The pores of each ambulacral area are placed opposite one another or in slightly alternating position (e.g., *Eucladia,* Fig. 136,*2*). Those of the pair farthest from the peristome are closely adjoined, with only a thin calcareous partition between them, and accordingly, the perradial plate in this position has a cruciform outline.

TUBE FEET

The nature of relationships to the theca of appendages distributed along the ambulacral areas proves that these organs, in spite of their appearance and considerable size, are homologous with the ambulacral tube feet or podia of other echinoderms. Each is cylindrical, undivided, and hollow, and probably was flexible in the living animal. A skeletal cover consists of minute scalelike calcareous plates with imbricate arrangement, overlap being directed toward the distal extremity. Since the presence of such organs is unknown in *Volchovia,* its tube feet may have been naked or protected only by a very weak skeleton.

In *Eucladia,* the first pair of podial appendages, nearest to the mouth, are much smaller than others, with length of approximately 9 mm. and diameter of 2 mm. (Fig. 132; 136,*2*; 137,*3*). They are distinguished otherwise in being associated closely with the jaw plates and in forming part of the peristomial area; thus they have been termed buccal tentacles. Their imbricated cover of very minute plates extends to a distal flattened expansion that possibly denotes a sucker. These podia are comparable to the oral tube feet of stelleroids. Equivalent tube feet of *Sollasina* are too poorly preserved for description but their existence seems to be indicated by depressions corresponding to their insertions and by rows of tiny plates apparently belonging to them (Fig. 137,*1*). No trace of buccal tube feet is discernible in the single available specimen of *Euthemon.*

Other ambulacral tube feet are close together in slightly alternating position in *Eucladia* (Fig. 132), which has five pairs in each ray, though some rays show the presence of an additional single tube foot which may correspond to the unpaired tentacle of *Sollasina* noted subsequently. The diameter and length of the tube foot appendages increase toward the periphery of the theca, largest ones having a length of approximately 40 mm. and a diameter up to 6 mm. Their surface is completely covered by small shingled plates which overlap distally and are divisible into two groups. A set of larger, more salient cover plates ends distally in a conical point. Another set consists of much smaller, less salient plates which terminate in a gently rounded extremity. The distal end of these tube feet is not well preserved and therefore of an uncertain nature.

The tube feet of *Sollasina* other than those associated with the buccal apparatus number seven in each ray: three pairs in opposed positions and an unpaired terminal tentacle in perradial position at the extreme edge of the theca. The two proximal tube feet differ from others in their smaller size —length approximately 12 mm. and diameter 2 mm.—and in their extremely numerous, densely crowded, and strongly imbricated thin cover plates. Remaining tube feet and the unpaired tentacle resemble one another in consisting of tubes approximately 25 mm. in length and 3 mm. in diameter at the base. They end in points. Their skeleton consists of small imbricate plates arranged in more or less even longitudinal rows. Distally there are four rows: median aboral, median oral, and two lateral. Toward the base of the appendage the number of rows increases to six or more, with supplementary much smaller plates intercalated between them. All these plates have rhomboid outlines. The largest are patelliform, with an excentrically pointed distal apex. In the view of Nichols (11), pores probably existed between these plates, serving for passage to papillae like those of modern crinoids and ophiuroids.

In *Euthemon* only two pairs of tube feet are present in each ray (Fig. 133).

Those next to the mouth are smallest, with length not more than 5 mm., and they are covered with greatly crowded, very thin calcareous scales. The distal pair of tube feet in each ray attain a length of 12 or 13 mm. and diameter of 1.5 mm. or a little more. Their skeleton consists of four rows of plates, dorsal, ventral, two lateral, resembling in structure the arms of ophiuroids.

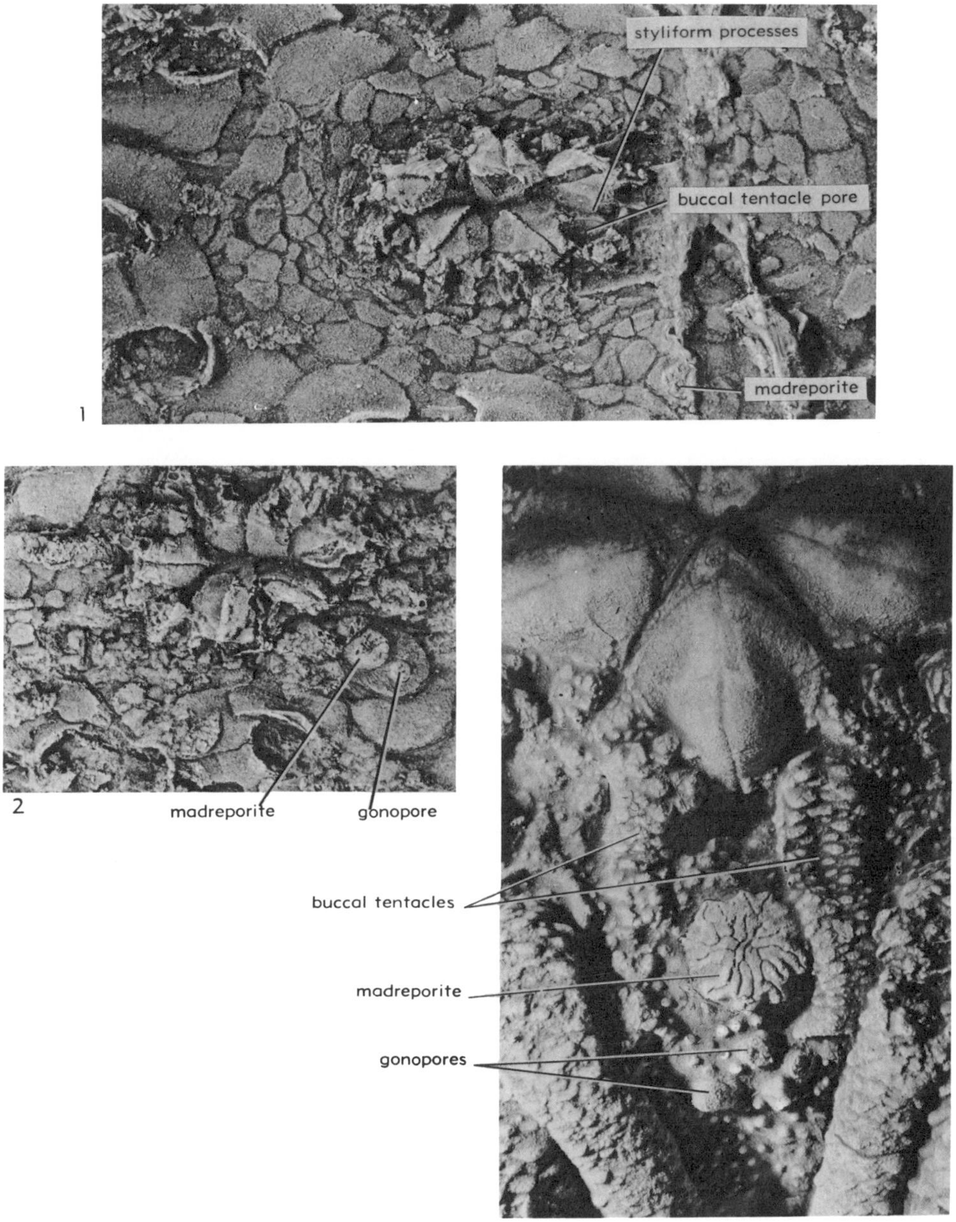

FIG. 137. *Sollasina* and *Eucladia,* U.Sil.(L.Ludlov.), Eng.; *1,2, S. woodwardi* (SOLLAS), features of oral side, ×5; *3, E. johnsoni* WOODWARD, part of oral side, ×5 (all Ubaghs, n).

Finally, the tube feet of *Rhenosquama*—assuming that this genus belongs with the ophiocistioids—are covered by imbricated plates which are all of similar nature. They are arranged in longitudinal and transverse rows, those of the latter alternating with one another in adjacent rows. Since four plates occur in each transverse row, their alternating placement produces eight longitudinal rows. The small arcuate plates have a small median point distally.

PERISTOME

The peristome occupies the center of the oral surface. It is circular or subcircular in outline as seen in *Eucladia* and *Euthemon* but is clearly elliptical in *Sollasina.* Its diameter equals 0.4 to 0.5 that of the test in *Eucladia* and *Euthemon,* respectively, and its longer axis in *Sollasina* is equal to, or slightly greater than, one-half that of the test.

A pentagonal rosette *(Eucladia, Euthemon)* or elliptical apparatus *(Sollasina)* at the center of the peristome is formed by five strong jaws, each of which is rhomboid in outline and located interradially (Fig. 132, 133, 136, 137). Each jaw consists of two pieces united along the interray axes. The adradial edge of each piece projects aborally in a styliform process, between which and the body of the jaw is a rather small podial pore. Whether the process is actually a separate ossicle is difficult to determine. The presence of the pore is firmly established in *Eucladia,* probable in *Sollasina* (Fig. 137,1), and possible but uncertain in *Euthemon.*

A flexible membrane extending from the buccal apparatus to borders of the peristome is very wide in *Sollasina* but much narrower in *Eucladia* and *Euthemon.* It is well preserved in the first-mentioned genus, where the membrane is reinforced by a pavement of many polygonal small plates arranged irregularly (Fig. 134,*2*; 137,*1*).

PERIPROCT

A periproct has been observed only in *Volchovia* and *Sollasina.*

In *Volchovia* (Fig. 135) the periproct consists of a cone composed of some 20 cuneiform plates surrounded by a circle of very small plates. It covers a rounded vent on the aboral surface located in the *CD* interray just behind the dorsocentral.

In *Sollasina* (Fig. 134,*1,3*) the periproct also is found on the aboral side but in marginal position close to one of the two extremities of the test. It is located in an interray doubtfully identified as *CD.* Narrow and elongate small plates form a pyramid, at the base of which irregularly polygonal small plates constitute a transition to other plates of the aboral surface.

MADREPORITE

One of the interrays of the oral surface bears a madreporite not far from the edge of the peristome, according to observations of *Eucladia* and *Sollasina.*

In *Eucladia* (Fig. 136,*2*, 137,*3*) it is found on the second plate of the *CD* interray, proceeding distally from the peristome. Edges of the madreporite, which has a diameter of 4 mm., are indented and its surface bears several irregularly forked grooves, reminiscent of the meandriform pattern of the madreporite in living asteroids.

The madreporite of *Sollasina* (Fig. 137,*2*) occurs on a special plate located between the peristome border and adjacent adradial plates of the *CD* interray. It consists of a small truncate cone, 1.2 mm. in diameter at the base, with a low, probably perforate summit.

GONOPORES

On the interradial plate which next follows that bearing the madreporite in *Eucladia* four prominent papillae with hollows filled by sediment may be observed (Fig. 136,*2*; 137,*3*). In *Sollasina* the cone interpreted as a madreporite is accompanied on the same plate by a large, seemingly perforated tubercle (Fig. 137,*2*). The papillae and tubercle probably represent genital pores. GEKKER (6, 7) has recorded the presence behind the periproct of *Volchovia* of a second orifice, located between the two posterior marginals (Fig. 135). This opening may be a gonopore, hydropore, or gonopore-hydropore (REGNÉLL, 12).

MODE OF LIFE

Ophiocistioids were free-moving echinoderms, as demonstrated by the absence of a stem or other means of fixation, by lack of ambulacral furrows, by the development of enormous tube feet, by the aboral location of the anus, and by the seemingly masticatory buccal apparatus. Almost certainly these animals rested and moved about on their oral surface. The extraordinary size of the ambulacral tube feet and their superficial resemblance, in *Euthemon* at least, to the arms of ophiuroids suggests that ophiocistioids belonged to the vagile benthos. This may not apply to *Volchovia,* however, for its form and weak development of an adoral skeleton denote a passive existence, possibly like that of patelloid gastropods or such ophiuroids as *Astrophiura,* which in resting position attach themselves by the ventral surface to hard surfaces so that the whole body acts as a cupping glass (Regnéll, 12).

The structural diversity of the podia even in a single individual evidently reflects differences of function. The buccal tentacles, smaller than others, possibly were sensory organs mainly, like the buccal podia of sea urchins used for exploring the sea bottom over which the animal crawls. Other tube feet may have served as locomotory organs, aiding also in grasping prey and in pushing it to the mouth. Fedotov (4) judged that the podia of *Euthemon* and *Sollasina* provided creeping movement like the arms of ophiuroids, whereas those of *Eucladia* behaved like the ambulacral tube feet of asteroids. That the podia were very mobile is indicated by the highly varied positions in which they are found in fossils. Perhaps the importance of their activities is related to dimensions of the madreporite (Regnéll, 12).

Finally, respiration of these animals may be explained very possibly by the presence of papillae which extended out between scales of the skeletal cover of their tube feet (Nichols, 11).

SYSTEMATIC POSITION

The body of ophiocistioids has often been compared, as by Bather (1, 2), to the central disc of ophiuroids. In line with this concept, developed by Fedotov (4, 5), the perradial and adradial plates of ophiocistioids are judged respectively homologous to orals and laterals in the ophiuroid arm, and interradial plates would correspond to the skeletal pavement of interbrachial areas in ophiuroids. Since the ophiocistioids lacked arms, however, comparison can be made only with proximal segments of the rays incorporated in the disc of ophiuroids.

These resemblances are balanced by countervailing differences. Whereas interradial areas are narrow and radial areas wide in ophiocistioids, the opposite is true of ophiuroids. The ambulacral pores of ophiocistioids are not found on a level with plates postulated as corresponding to the ventrals and laterals of ophiuroids, but instead, each occurs at the junction of two perradial and two adradial plates. Moreover, the perradials of ophiocistioids cannot be homologous with oral plates of ophiuroid arms, since the latter certainly are a secondary development produced late in the history of the group—consequently, without possible genetic relationship to ophiocistioid perradials. Such resemblance as appears between the disc of ophiuroids and the body of ophiocistioids therefore seems to be of superficial nature, pertaining to some aspects only, and even this disappears when comparison is extended to the archaic ophiuroids, which ought to be most similar to ophiocistioids if truly they were derived from ophiuroids.

Regnéll (12) is the one who has shown that the disposition of plates on the aboral surface of *Volchovia* offers analogies with the apical skeletal organization of very young ophiuroids and of some adult ophiuroids, such as *Astrophiura.* In each of these the apical system is predominantly composed of radial elements. However, in addition to question as to placement of *Volchovia* as an ophiocistioid (Nichols, 11), the nature of the apical system in this genus can hardly denote more than topographic similarities to the system of ophiuroids, since similar morphogenetic potentialities in this part of the body tend to be manifested in almost all echinoderm groups.

Furthermore, various other characters serve to distinguish the ophiocistioids from ophiuroids and from stelleroids generally. Among these are (1) entire absence of arms in ophiocistioids and of any structure allowing supposition that growth of the body at any stage was dominated by five divergent radial axes; (2) the presence in ophiocistioids of a well-developed aboral anus protected by a pyramid of small plates, which probably is a fundamental character of the group strikingly in contrast to the absence of such an orifice in ophiuroids, archaic asteroids, and somasteroids; (3) nature of the ophiocistioid buccal apparatus, which is very unlike that of stelleroids and differs from these in lacking any feature suggestive of its derivation by a transformation of primary ossicles belonging to the rays; and (4) the probable presence in ophiocistioids of an unpaired interradial gonad, a character that allies them with archaic pelmatozoans or holothuroids much more than with stelleroids.

If the ophiocistioids thus differ in numerous ways from asterozoans, resembling them only superficially, various characters typical of echinoids are not to be overlooked. Both assemblages possess a cap- or dome-shaped test formed of plates that enclose the entire body, except for the peristomial region, and this test, lacking arms, is divided on the oral side at least into five ambulacral areas and five intervening interambulacral areas. The buccal apparatus of ophiocistioids is closely analogous to that of echinoids in its seemingly different origin from the ray skeleton, its association with a peristomial membrane inlaid by small plates, and its structure and probable function as a masticatory mechanism. The presence of a well-developed periproct and its resemblance to the anal pyramid of archaic urchins such as *Aulechinus* and *Ectinechinus* is also an echinoid character, for the existence of a periproct on the test is one of the fundamental features of these echinoderms that dates from their origin. Interradial areas consisting of a single column of plates in each tract are restricted to the echinoid *Bothriocidaris*. Finally, if it is true that only a single gonad existed in ophiocistioids, this character perhaps is matched in certain archaic echinoids which possessed only one genital plate.

Evidently, such similarities do not imply direct parentage bonds, for they could have been acquired independently. As a matter of fact, other morphological aspects indicate that the ophiocistioids are far removed from echinoids. Especially important among these is division into ambulacral and interambulacral areas only on the lower side of the body, structure of the ophiocistioid ambulacrum, the gigantic size and distinctive construction of the ambulacral tube feet, the eccentric and even marginal location of the periproct, the oral placement of the madreporite and gonopores, the nature of the madreporite in closely resembling that of asteroids, the absence of any apical system, at least in most typical forms—all these characters suffice to show that ophiocistioids cannot be closely allied to echinoids.

The presence of certain features suggestive of archaic pelmatozoans also characterizes the morphology of ophiocistioids. Such features include structure and position of the anal pyramid, close association of the mouth with the madreporite and genital pores, and the probable existence of a single gonad.

Finally, FEDOTOV (4, 5) has emphasized general resemblances of the ophiocistioids to very young stelleroids and echinoids, in which the discoid body exhibits equally well-developed oral and aboral faces and bears some relatively very large podia. Ultimately, the stelleroids develop arms and acquire the star-shaped outline that characterizes them. Among echinoids, the oral part of the body undergoes enormous expansion, as result of which the aboral part becomes reduced to a very restricted area at the summit of the test. The interbrachial areas of ophiuroids expand mostly in adoral directions so as to modify profoundly the oral side of the disc. The ophiocistioids appear to have become fixed in their evolution at the stage when the disc, devoid of arms but provided with podia of considerable size in relation to the body, had developed subequal oral and aboral sides. Their unpaired gonad corresponds to the primary gonad of stelleroids and echinoids, which developed subsequently into five pairs of distinctive genital organs. Interpretable as a juvenile character are the large dimensions of the terminal tentacle, which are somewhat comparable to the primary podia of echinoderm larvae. The remarkable resemblance of ophiocistioids to some Recent echinoderms shortly after metamorphosis is perhaps the most typical feature of their organization.

In conclusion, it is obvious that ophiocistioids cannot be assigned to any class of living echinoderms. They comprise an extinct, entirely separate class of unknown origin, but on the whole seeming to offer greater similarities to echinoids than to stelleroids.

SYSTEMATIC DESCRIPTIONS

Class OPHIOCISTIOIDEA Sollas, 1899

[*nom. correct.* UBAGHS, 1953 (*pro* class Ophiocistia SOLLAS & SOLLAS, 1912, *nom. transl. ex* order Ophiocistia SOLLAS, 1899)] [=class Ophiocista MATSUMOTO, 1929 (*nom. null.*); Ophiocistioïdes LAMEERE, 1931 (*nom. neg.*); class Ophiocystia MOORE, LALICKER & FISCHER, 1952 (*nom. null.*)]

Unattached pentaradiate echinoderms with more or less depressed dome-shaped body entirely covered by plated test or with cover of plates on one side only, lacking arms or comparable projections; center of oral face occupied by peristome with buccal apparatus consisting of 5 interradially disposed jaws, each consisting of 2 pieces; remainder of oral surface divided into 5 large ambulacral tracts composed of 3 plate columns (1 perradial, 2 adradial) and 5 narrow interradial tracts composed of a single plate column; ambulacra confined to oral face; ambulacral pores simple, each located at junction of 2 perradial and 2 adradial plates; podia of relatively enormous size, covered by imbricated plates; periproct aboral, eccentric or marginal; madreporite and gonopores in same interray, typically on oral face. *L.Ord.-U.Sil., ?M.Dev.*

Family EUCLADIIDAE Gregory, 1896

[=Eucladidae SOLLAS, 1899 (*nom. null.*)]

Test subpentalobate; peristome rather restricted, surrounded by ring of 20 plates (5 perradial, 10 adradial, 5 interradial); ambulacra with at least 7 plates in perradial or adradial columns and 6 pairs of slightly alternating podia, plus unpaired terminal tentacle, most proximal pair of podia reduced in size and placed in peristomial area, others increasingly large toward periphery of test; interambulacra with 5 or 6 plates in single column; strong buccal apparatus occupying nearly entire peristome; madreporite well developed, with branched grooves, located in adoral interray that also contains several gonopores. *U.Sil.*

Eucladia WOODWARD, 1869 [**E. johnsoni*; OD]. Characters of family. *U.Sil.(L.Ludlov.)*, Eu.(Eng.).——FIG. 132; 136; 137,*3*. **E. johnsoni*; 132, part of oral surface, ×1.3 (Ubaghs, n); 136,*1*, part of *B* ray and *BC* interray, ×2 (4); 136,*2*, part of *CD* interray, ×2.7 (4); 137,*3*, part of oral side, ×5 (Ubaghs, n).

Family SOLLASINIDAE Fedotov, 1926

Well-developed peristome surrounded by ring of 15 (or 16) plates comprising 5 perradials, 10 adradials, and possibly additional plate bearing madreporite and gonopore; podia located opposite one another, ending in point and covered by small plates arranged typically in longitudinal rows, podia next to buccal tentacles (if present) unlike all others. *U.Sil.*

Sollasina FEDOTOV, 1926 [**Eucladia woodwardi* SOLLAS, 1899; OD]. Test elliptical in outline; aboral surface covered by irregular plates; oral surface with 3 perradials and 2 rows of 4 adradials in each ambulacrum, which bears pair of ?buccal podia, 3 other pairs, and terminal tentacle identical with adjacent podia; each interambulacral tract with 2 interradials; elliptical peristome with relatively small oval buccal apparatus surrounded by membrane reinforced by minute irregular plates; periproct with anal pyramid dorsomarginal; madreporite and gonopore on single plate in interray at border of peristome. *U.Sil.(L.Ludlov.)*, Eu.(Eng.).——FIG. 134; 137,*1,2*; 138. **S. woodwardi* (SOLLAS); 134,*1,2*, aboral and oral sides, ×2 (Ubaghs, n); 134,*3*, periproct, ×5 (Ubaghs, n); 137,*1,2*, buccal apparatus, peristome, madreporite, and gonopores, ×5 (Ubaghs, n); 138, ambulacral tube feet, ×5, ×10 (Ubaghs, n).

Euthemon SOLLAS, 1899 [**E. igerna*; OD]. Test rounded, each ambulacrum with 3 perradials and

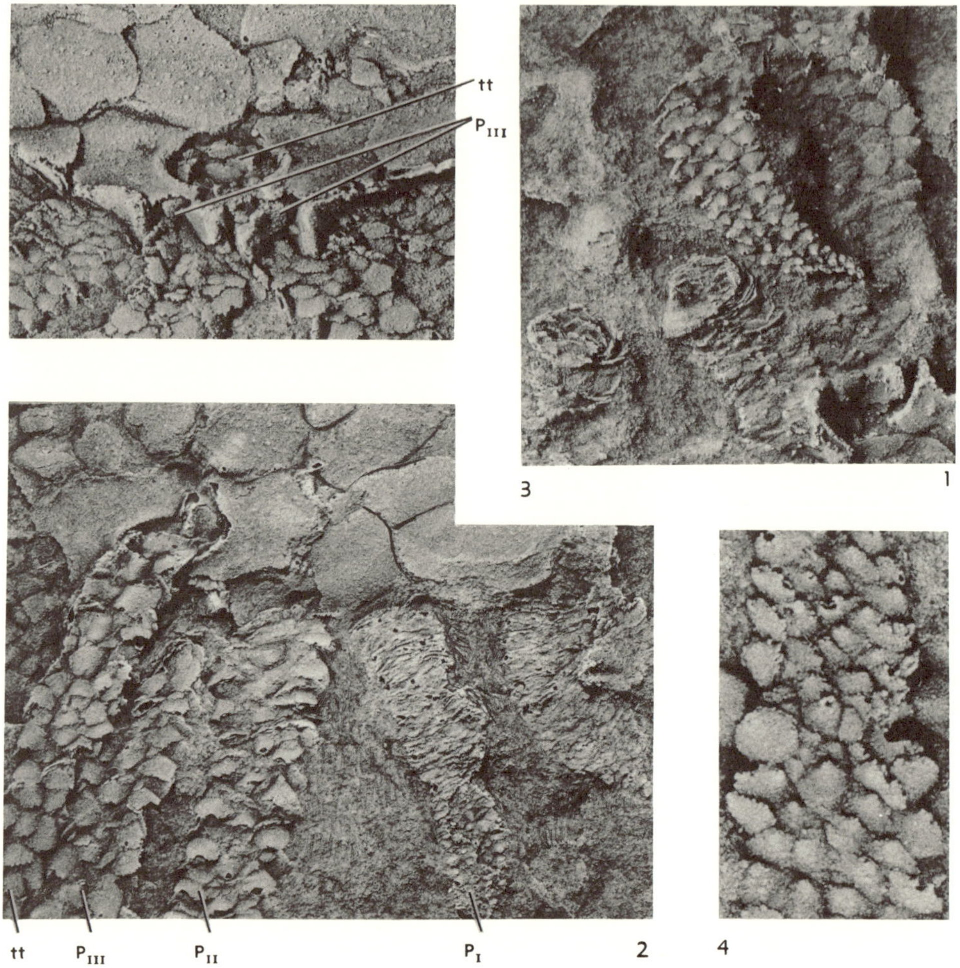

FIG. 138. *Sollasina woodwardi* (SOLLAS), U.Sil.(L.Ludlov.), Eng., ambulacral podia and pores; *1*, unpaired pore for terminal tentacle and paired pores for 3rd podial pair, ×5; *2*, part of surface showing terminal and associated paired podia, ×5; *3*, first pair of podia (at left) and one of following pair, ×5; *4*, part of podium, ×10 (all Ubaghs, n). [Explanation: PI, PII, PIII, 1st, 2nd, 3rd pairs of podia; *tt*, terminal tentacle.]

?3 adradials adjoined by 2 pairs of podia, small near peristome, other pair very large; interambulacral areas containing single interradial; buccal apparatus large, strong. *U.Sil.(Wenlock.)*, Eu.(Eng.).——FIG. 133. **E. igerna*; oral side, ×3.5 (4).

Family VOLCHOVIIDAE Gekker, 1938

Body oval dome-shaped; aboral face of test showing 10 subpentagonal or subtriangular marginals which form flattened border with indentations on outer side, central area composed of 20 to 30 plates well uparched medially; anal pyramid excentrically placed in *CD* interray with single pore interpreted as gonopore, hydropore, or gonopore-hydropore in same interray nearer margin; oral surface and podia unknown. *L.Ord.-U.Ord.*

Volchovia GEKKER, 1938 [**V. mobilis*; SD GEKKER, 1940]. Characters of family. [Genus differs from others referred to class much more than these differ from one another; therefore its assignment to the ophiocistioids is not entirely firm.] *L.Ord.-U. Ord.*, Eu. (USSR-Norway)-N. Am. (Ohio).——FIG. 135,*1-3*. **V. mobilis*, L.Ord., USSR; *1*, aboral view; *2*, plate diagram; *3*, oblique post. view (reconstr.); all ×1 (7).——FIG. 135,*4*. *V. norvegica* REGNÉLL, L.Ord., Norway; post. view from slightly left of *CD* interray, ×1 (12).

?Family RHENOSQUAMIDAE R. Richter, 1930

Organs doubtfully identified as ophiocistioid podia terminating distally in point, covered by mutually similar scales arranged in transverse rows of alternately placed plates so as to produce 8 longitudinal rows. *M.Dev.*

The inclusion of this taxon and its single presently known genus among the ophiocistioids is reasonable but decidedly open to question.

Rhenosquama R. RICHTER, 1930 [**R. westfalica*; OD, M]. Characters of family. *M.Dev.(Couvin.)*. Eu.(Ger.).——FIG. 139,*1*. **R. westfalica*, Eifel; impressions of ?podia, ×3 (13).

FIG. 139. *Rhenosquama westfalica* R. RICHTER, M. Dev., Ger.; impressions of ?podia, ×3 (13).

REFERENCES

Bather, F. A.

(1) 1907, *A guide to the fossil invertebrate animals in the department of geology and paleontology in the British Museum (Natural History)*: 182 p., 7 pl., 96 text-fig. (London).

(2) 1910, *Echinoderma*, in Encyclopaedia Britannica: Edit. 11, v. 8, p. 871-882, Univ. Press (Cambridge, Eng.).

Cuénot, Lucien

(3) 1948, *Ophiocistia*: in Traité de Zoologie (P.-P. Grassé, ed.), v. 11, p. 270-272, text-fig. 310-312, Masson (Paris).

Fedotov, D. M.

(4) 1926, *The plan of structure and systematic status of the Ophiocistia (Echinoderma)*: Zool. Soc. London, Proc., p. 1147-1157, pl. 1-2.

(5) 1928, *Ueber die Beziehungen der Echinodermenklassen zueinander*: Zool. Lab. Sevastopolsk. Biol. Sta., Akad. Nauk SSSR, Travaux, ser. 2, v. 12, p. 31-94.

Gekker [Hecker], R. F.

(6) 1938, *A new member of the class Ophiocistia Sollas (Volchovia n.g.) from the Ordovician of Leningrad province and changes in the diagnosis of this class*: Acad. Sci. URSS, Compte Rendus (Doklady), new ser., v. 19, p. 425-427, 1 text-fig.

(7) 1940, *Carpoidea, Eocrinoidea und Ophiocistia des Ordoviziums des Leningrader Gebietes und Estlands*: Acad. Sci. URSS, Inst. Paléont., Travaux, v. 9, p. 5-82, pl. 1-10.

Gregory, J. W.

(8) 1897, *On the classification of the Palaeozoic echinoderms of the group Ophiuroidea*: Zool. Soc. London, Proc. (1896), p. 1028-1044.

Lameere, Aug.

(9) 1931, *Précis de zoologie*: v. 2, 456 p., Univ. Inst. Geol. Torley Rousseau (Bruxelles).

Matsumoto, Hikoshichirô

(10) 1929, *Outline of a classification of Echinodermata*: Tohoku Univ., Sci. Rept., ser. 2 (Geol.), v. 13, no. 2, p. 27-33.

Nichols, David

(11) 1962, *Echinoderms*: 200 p., 26 text-fig., Hutchinson Univ. Library (London).

Regnéll, Gerhard

(12) 1948, *Echinoderms (Hydrophoridea, Ophiocistia) from the Ordovician (Upper Skiddavian, 3Cβ of the Oslo Region*: Norsk Geol. Tidsskrift, v. 27, p. 14-58, pl. 1-2.

Richter, Rudolf

(13) 1930, *Schuppenröhren als Anzeiger von zwei im deutschen Devon neuen Echinodermen-Gruppen (Edrioasteroidea Billings und Ophiocistia Sollas?)*: Senckenbergiana, v. 12, p. 279-304.

Schuchert, Charles

(14) 1915, *Revision of Paleozoic Stelleroidea with special reference to North American Asteroidea:* U.S. Natl. Museum, Bull. 88, 311 p., 38 pl., 41 text-fig.

Sollas, W. J.

(15) 1899, *Fossils in the University Museum, Oxford: I. On Silurian Echinoidea and Ophiuroidea:* Geol. Soc. London, Quart. Jour., v. 55, p. 692-715, text-fig.

———, & Sollas, I. B. J.

(16) 1912, *Lapworthura: a typical brittle-star of the Silurian age:* Royal Soc. London, Philos. Trans., ser. B, v. 202, p. 213-232, pl. 9-10. text-fig. 1-5.

Ubaghs, Georges

(17) 1953, *Classe des ophiocistioïdes (Ophiocistioidea):* in Traité de Paléontologie, Piveteau, Jean (Ed.), v. 3, p. 843-856, text-fig. 1-7, Masson & Cie (Paris).

Woodward, Henry

(18) 1869, *On Eucladia, a new genus of Ophiuridae, from the Upper Silurian, Dudley:* Geol. Mag., dec. 1, v. 6, p. 241-245, pl. 8.

CYCLOCYSTOIDS

By Robert V. Kesling

[Paleontological Museum, University of Michigan]

CONTENTS

INTRODUCTION

The cyclocystoids are an enigmatic group of Paleozoic fossils, most of which are known primarily from the well-preserved ring of submarginal plates. They are small disc-shaped objects, apparently without columns and attached by the flat aboral side. Central covering layers, called discs, of both oral and aboral sides, seem to have been weakly calcified. Food-gathering structures are also inadequately represented. Nevertheless, the Cyclocystoidea are extinct echinoderms, which authors generally have classed as Pelmatozoa, considered by some to be related more closely to Cystoidea than to other classes. In the *Treatise,* however, they are transferred to the subphylum Echinozoa.

Although Cyclocystoidea range from Middle Ordovician to Middle Devonian, relatively few specimens are known. This scanty record may be due in part to their small size and poor preservation, but mostly it may be attributed to lack of interest in fossils so imperfectly understood and to consequent perfunctory collecting.

Classification is deficient for two reasons. First, paucity of the record does not reveal much of the diversity that probably existed in the taxon. Second, essential systems of the animal cannot be discerned in sufficient detail to permit clear recognition of differences. Only one genus, *Cyclocystoides,* is accepted by all workers on the group. And this is perhaps a matter of nomenclatorial necessity, rather than taxonomic conviction.

Certain characteristics of cyclocystoids invite comparison with other ancient echinoderms. The discoid shape and the marginal ring of tiny plates forming a "shagreen," imbricating and partly embedded in a flexible integument, are remarkably like those in edrioasteroids. The facets lying within the submarginal ring, small tubercles with surrounding grooves, resemble the brachiole facets of cystoids. The sharp differentiation of oral and aboral surfaces and the ramification of the ambulacral system are more nearly like those in crinoids.

Differences, however, set the cyclocystoids apart as a class discrete from any of these groups. Multiple branching of the ambulacral system is unknown in edrioasteroids; the flattened shape is foreign to cystoids; and the flat aboral (dorsal) surface is unlike that in crinoids. Other points of dissimilarity could be mentioned. Despite our ignorance of parts of their morphology, we can clearly separate cyclocystoids from other echinoderms.

PREVIOUS STUDIES

Contributions to the knowledge of cyclocystoids are few. They are marked by tentative interpretations and confessions of doubt. The first published account of which I know was made by JAMES HALL in FOSTER & WHITNEY's *Report of the Geology of the Lake Superior Land District* (1851). HALL briefly described a specimen from the Escanaba region of Michigan and concluded, "This curious body is evidently Crinoidean. . . . It is possible that it may be the elevated marginal ring of some one of the sessile Crinoids. . . ." In 1858, J. W. SALTER & ELKANAH BILLINGS named *Cyclocystoides* and presented a remarkably good analysis of its organization. They distinguished the "integument of the upper side" from the "integument of the under side," and accurately described the ring of submarginal plates (which they called "marginal plates") as bearing a channel with "marginal excavations" and "connected with the interior by small pores penetrating through the marginal plates." Incorrectly, I believe, they regarded a dissociated tubelike fossil as a part of *Cyclocystoides,* but on the whole their account is more lucid than several which appeared more than half a century later. SALTER & BILLINGS thought that "regarding the affinities of the fossils, the choice seems to be between Star-fishes and Cystideae"; they also compared *Cyclocystoides* with *Amygdalocystites,* now regarded as a paracrinoid.

HALL (8) figured an oral disc with an eccentric oval opening, which he interpreted as the mouth. In a modified version of this figure, F. A. BATHER (1) called this opening the "supposed region of anus." Although incontestable evidence on the location and nature of the anal opening has not yet been offered, I am inclined to agree with BATHER.

S. A. MILLER & C. B. DYER (12) confirmed the presence of ducts or pores through plates of the submarginal ring. In 1882, MILLER created the family Cyclocystoididae, without pertinent distinctions from other families. In 1895, MILLER & W. F. E. GURLEY erected the order Cyclocystoidea, more or less as a convenience in directing attention to the singular features of *Cyclocystoides.* BATHER (1) discussed the genus in his chapter on Edrioasteroidea, stating that it "probably belongs to this class, though not to any of the recognized families." This concept of diverse forms in the edrioasteroids reflects BATHER's conviction that the "Edrioasteroidea are alone among Pelmatozoa in presenting a type of ambulacrum from which the holothurian, stellerid, and echinoid types may readily be derived."

P. E. RAYMOND (14) believed that he could discern small plates covering the facet-bearing channel, or circular canal, in the distal part of the submarginal plates, redescribing and refiguring one of SALTER & BILLINGS' specimens. He paid special attention to the oral disc, and distinguished five main ridges that by successive bifurcations lead to the submarginal ring. Regarding the ambulacral system, he wrote, "These ridges probably cover ducts which lead from the [submarginal] plate to the centre, and the inference might be that through them, food, in water, was carried from the series of collecting basins in the outer circular canal to the mouth, which would be central and beneath the plates of the disk." Concerning the affinities of *Cyclocystoides,* RAYMOND's contribution was less significant. He thought the animal might be a "free Cystidean or Edrioasteroid," "highly specialized root of a free crinoid," or "it is even possible, if one is sufficiently imaginative, to think of this disk as a swimming organ, the method of propulsion being on the same principle as in some of the cephalopods."

In his comprehensive review and revision of the Pelmatozoa, OTTO JAEKEL (9) ignored *Cyclocystoides*.

A. F. FOERSTE (5) assigned the Cyclocystoididae to the Edrioasteroidea, and distinguished two new genera, *Narrawayella* and *Savagella,* and an unnamed genus exemplified by *Cyclocystoides illinoisensis* MILLER & GURLEY, in addition to the type progenus. As discussed later, FOERSTE's new genera have not been awarded general acceptance.

FOERSTE (6) also gave particular attention to *Cyclocystoides huronensis* BILLINGS, in which he described "spout-like appendages" at the border of the submarginal and marginal rings, equal in number to the facets in the channel of the submarginal ring. He described an unnamed species from Tennessee as having marginal plates sloping outward, and went on to say, "This outward sloping of the marginal plates is so frequent in *Cyclocystoides* as to suggest that these plates could be moved at will, either so as to slope inward, thus covering the outer margin of the submarginal ring of plates, or so as to slope outward, exposing the margin." On the basis of this account, FOERSTE has been credited with support of SALTER & BILLINGS' (17) and RAYMOND's (14) contention of plates covering the circular channel in which the facets are located. No author has elaborated on these plates, and no convincing figures have been given to support this very important consideration. Inadvertently, FOERSTE seems to have mistaken the collapsed oral disc for the ventral side of the aboral disc in his figure of *C. huronensis* (6, pl. 6, fig. 3). Ten years later, this figure was copied by BEGG (2, pl. 9, fig. 7), who relied heavily on it for certain aspects of orientation; as a result, parts of BEGG's descriptions are rather confused.

In 1926, E. MALLIEUX reported the occurrence of cyclocystoids in the Devonian of the Ardennes in Belgium, thus greatly extending the known geologic range.

BEGG (2) compared *Cyclocystoides* with the carpoid *Cothurnocystis*, referring to the oral disc as the "lower plate" and the aboral disc as the "upper plate" but describing the submarginal ring as "beyond the spoon-shaped depressions [facets], and sloping downward and outward." At least part of the oral-aboral confusion was undoubtedly occasioned by the preservation of many of his specimens as external molds; yet some of SALTER & BILLINGS' original specimens were similarly preserved, and I am unable to follow some of BEGG's comparisons.

HERTHA SIEVERTS (later SIEVERTS-DORECK) in the same and the following years reviewed BEGG's article. She further proposed that the facets were places for attachment of brachioles.

GERHARD REGNÉLL (15) reviewed assignments of *Cyclocystoides* and described two species from Gotland in considerable detail. Concerning facets, he stated, "The mamillary elevations cannot have been facets of brachioles, as supported by SIEVERTS 1934, nor is there anything to indicate that they are bases of spines. Both theories are made impossible by the fact that the canal, in which they are located, was evidently roofed over by small movable plates." Recently, REGNÉLL (16) seems to have changed his mind, for he wrote, "In the opinion of the present writer, *Cyclocystoides* differs so radically from the edrioasteroids that an attribution to that class is definitely not advisable. . . . The theca of the edrioasteroids is not differentiated into a dorsal and a ventral disk; the ambulacral cover-plates are invariably biserial; the ambulacral system is never tetraradiate; brachioles do not occur in edrioasteroids." The implication is clear that *Cyclocystoides* does possess brachioles.

REGNÉLL (1948) presented a table of occurrences of known species of *Cyclocystoides* both geographically and geologically. He emphasized the North American origin, the early migration to Britain, the Silurian spread to Gotland, and the final appearance in the Devonian of Belgium.

SIEVERTS-DORECK (20) gave an excellent summary of previously described specimens and analysis of morphology, revising ordinal, familial, and generic diagnoses, and describing a new Devonian species in minute detail. Unquestionably, her work is the best founded, most penetrating, and concise of all that have appeared.

To date, many morphological features which one would expect in these fossils have not been substantiated. The peristome and periproct are known only from poorly preserved, inconclusive evidence. The hydropore is not represented by any preserved structures in the circumoral region, where logically it would be expected. Nothing of the gonopore has been found, nor has any suggestion of its nature been made. It would be very helpful to know if the channel in the submarginal plates is roofed by movable plates, as suggested by RAYMOND (14), or bears brachioles, as suggested by SIEVERTS-DORECK (20).

MORPHOLOGY

GENERAL ORGANIZATION

The cyclocystoid bears at least superficial resemblances to the edrioasteroid. Both of these echinoderms, now referred to the Echinozoa, are disc-shaped and both acquire rigidity from peripheral rather than central elements.

The cyclocystoid theca was somewhat like a thin, inflated drum. The **submarginal ring** of stout plates formed relatively rigid sides, the **oral disc** arched across one side (Fig. 140) and the **aboral disc** extended across the opposite side (Fig. 141) like a drumhead. Although the two discs are collapsed in the fossil state and lie in close proximity, presumably space between them accommodated soft organs of the living animal. The oral disc appears to have had greater flexibility or elasticity than the aboral disc, at least in some species. Possibly, it took on the shape of a blister, expanding in response to internal pressure (Fig. 142).

The submarginal ring is made up of numerous thick, complex plates, which have surfaces exposed on both oral and aboral sides of the theca (Fig. 140-142). Aborally, the submarginal plates appear as truncated wedges; orally, in many species they have distal beveled or concave surfaces that together constitute a circular **channel.** In the channel, each plate bears one of three **facets,** presumably for attachment of brachioles. In forms having a well-developed channel, the proximal part of the submarginal plate is elevated, overhanging, and flat-topped. Grooves lead from the facets to ducts or pores penetrating the proximal part of the plate, and these in turn connect wih enclosed passageways just under the surface of the oral disc. These structures—facets, grooves, ducts, and passageways—are considered to be elements of the ambulacral system. The passageways proximally unite as they approach the center of the oral disc, and are properly regarded as **ambulacral grooves.**

Surrounding the submarginal ring and forming the border of the cyclocystoid is the **marginal ring,** a band of small imbricating platelets that were probably embedded in flexible integument (Fig. 140, 141).

Orientation of the cyclocystoid is only inferred. Most paleontologists believe that the oral disc, which contains the hidden ambulacral grooves leading to the mouth, was uppermost, and that the more rigid, flat, aboral disc, which possesses no essential openings, was lowermost in the living animal. Thus, brachioles (or whatever other kind of food-gathering structures were present at the ends of the ambulacral conduits) were on the upper side. The cyclocystoid probably remained attached to the sea bottom in limpet fashion, creating suction under its aboral disc by muscular contraction of the marginal ring. Other aspects of orientation and application of Carpenter letters to the rays have been presented by MOORE & FELL (p. *U*119).

ORAL DISC

The oral disc has two significant characteristics—it contains a complicated ambulacral system and it undoubtedly possessed great elasticity. This is an unusual combination in pelmatozoans but not unexpected in echinozoans. The edrioasteroids, in which some degree of flexibility is indicated by the imbrication of interambulacral plates in many species, seem also to have had limited expansion of the flooring and covering plates of the ambulacra.

On the oral surface, which is the one commonly exposed in a well-preserved

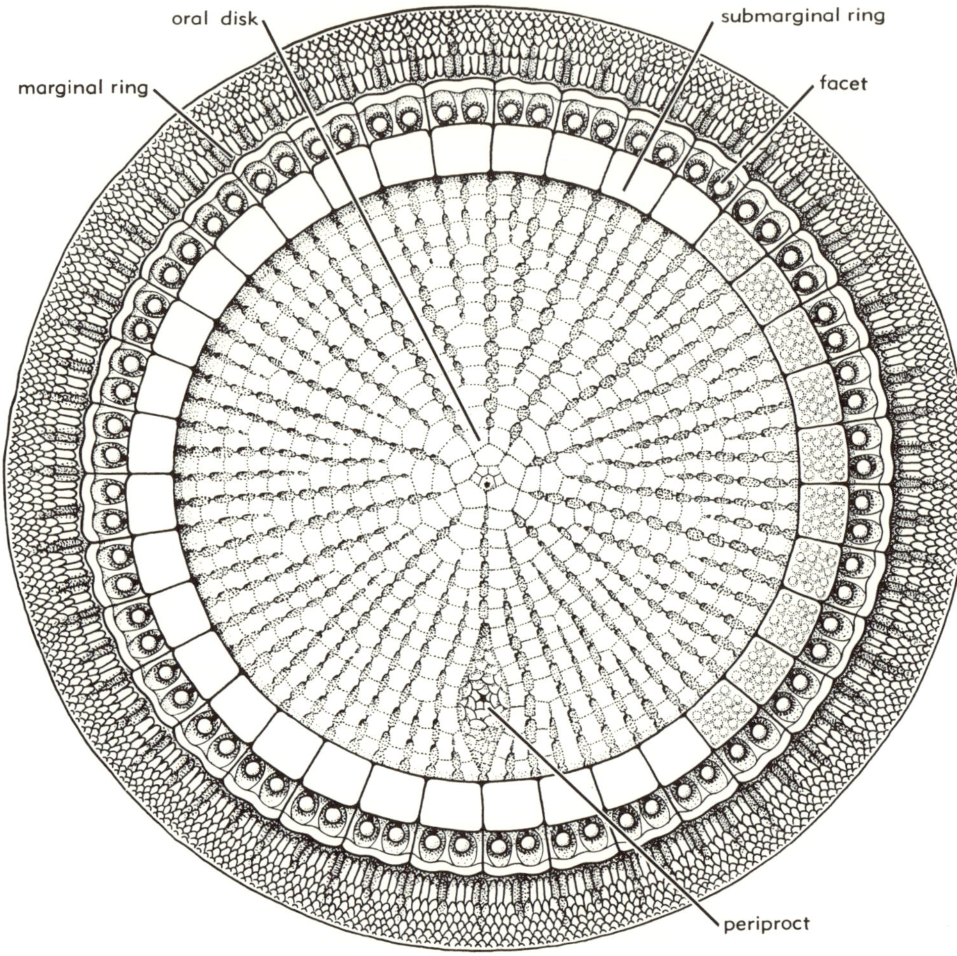

FIG. 140. Hypothetical reconstruction of oral side of *Cyclocystoides halli* BILLINGS, ×4.5 (Kesling, n).

specimen, the dominant feature of the disc is a set of low ridges which bifurcate successively from the center and radiate to the surrounding submarginal ring. This system of branching was well described by RAYMOND (14), "Upon analysis, it is seen that there are one or two ridges in front of each of the submarginal plates, and that each two adjacent ridges quickly unite to form a single ridge. Two of the ridges thus formed unite a little nearer the centre and are joined quite close to the centre by another long ridge formed from four shorter ones." SIEVERTS-DORECK (20) was particularly concerned as to whether the branching was isotomous or heterotomous; unfortunately, so few specimens show the complete pattern that the branching cannot be determined for all species or its taxonomic value tested.

Some details of the plates are not clear, for they are thin and their edges are not sharply delineated, as emphasized by SIEVERTS-DORECK (20). This condition of the plates is much like that in Recent holothurians, and supports the hypothesis that the plates of the oral disc may have been embedded within an integument. At any rate, a sharp difference distinguishes plates of the oral disc and those of the submarginal ring. In his description of *Cyclocystoides halli,* BILLINGS (in SALTER & BILLINGS,

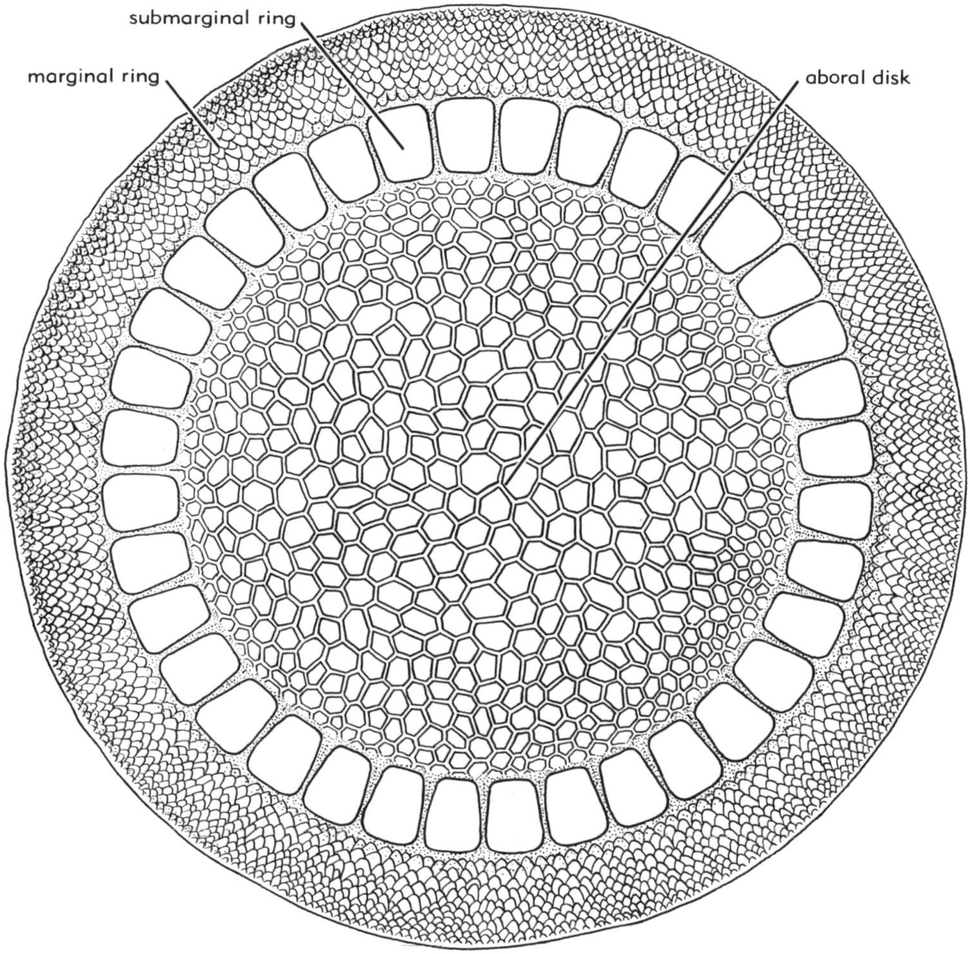

FIG. 141. Hypothetical reconstruction of aboral side of *Cyclocystoides halli* BILLINGS, ×4.5 (Kesling, n).

17) early established that "the integument of the upper side, supposed to be the side on which the marginal plates are excavated, is connected to the inner edge of those plates [submarginal ring] and does not extend over them." This relationship of oral disc and submarginal ring has been confirmed in other species. REGNÉLL (15) reported that in *C. lindstroemi* "the inner margin of the ossicles [submarginal plates] overlap slightly the adjacent rays of the central disk."

The number of ridges at the periphery of the oral disc was at first thought to equal the number of submarginal plates. With respect to *Cyclocystoides davisii,* SALTER (in SALTER & BILLINGS, 17) said, "This species shows the complete surface, on which about as many radiations mark the margin as there are ossicles." In *C. decussatus,* BEGG (2) stated, "The submarginal area is composed of thirty-two plates or ossicles, corresponding to an equal number of rays on the central disc." It seems plausible, however, to assume that all species were constructed similar to *C. devonicus.* In that species, as described by SIEVERTS-DORECK (20) each facet is aligned with a duct through the proximal, elevated part of the submarginal plate and each duct leads to an ambulacral groove. Thus, the peripheral number of ridges equals the number of facets, rather than the number of submarginal plates.

As SIEVERTS-DORECK (20) has pointed out, the plates composing the ridges are

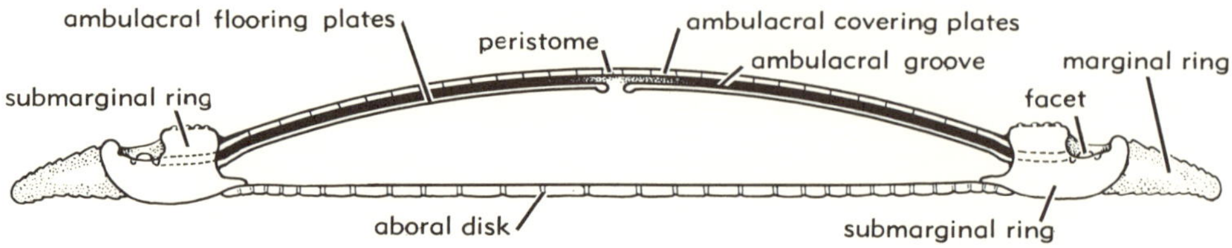

FIG. 142. Hypothetical cross section of *Cyclocystoides halli* BILLINGS, ×4.5 (Kesling, n).

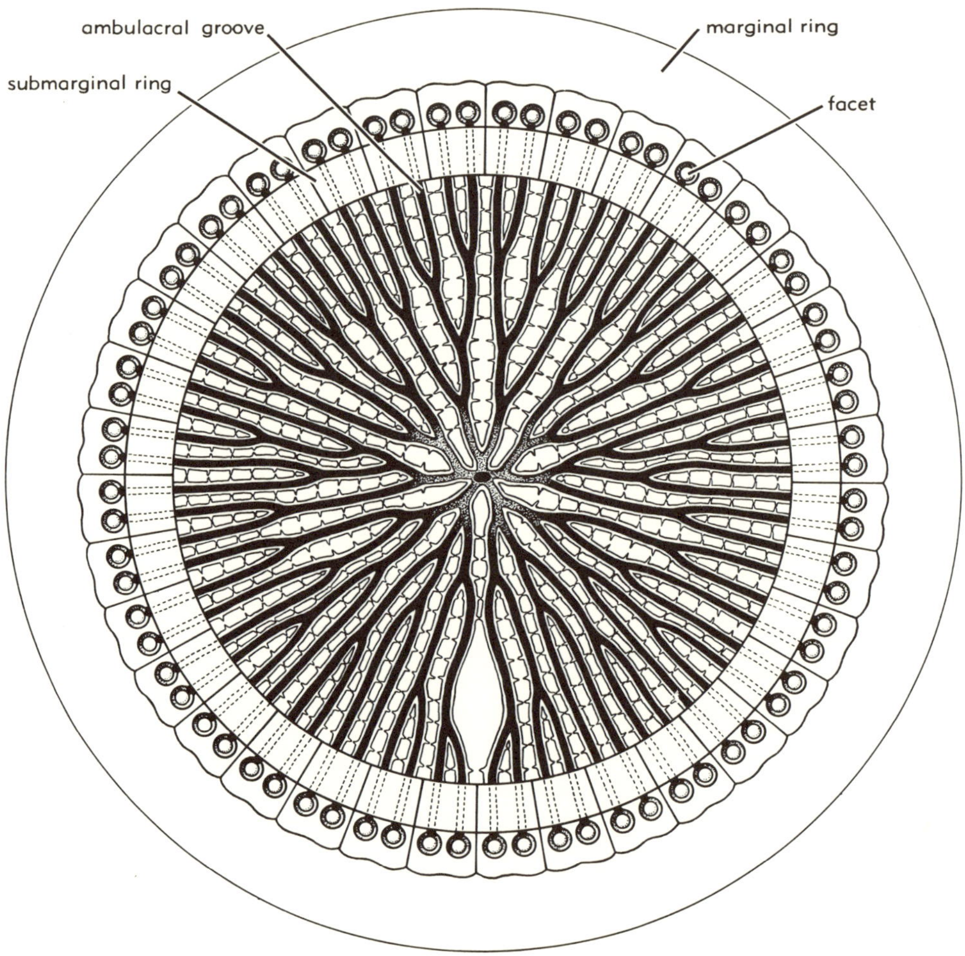

FIG. 143. Hypothetical reconstruction of *Cyclocystoides halli* BILLINGS as viewed orally, ×4.5; ambulacral covering plates removed to show ambulacral grooves (black) (Kesling, n).

uniserial and serve to cover the ambulacral grooves or ducts. Thus, they are **ambulacral covering plates,** fulfilling the same protective function as the biserial covering plates in cystoids. The outer surface of the ambulacral covering plates may vary from one species to another. Those in *Cyclocystoides devonicus* were said by SIEVERTS-DORECK (20) to be somewhat papillate or tuberculate. REGNÉLL (15) stated that "each ray is divided medially by a faint groove . . ." but added, "The rays seem to have been solid." BEGG (2) described *C. decussatus* thus: "Each ray is probably divided medially, for the greater part of its length, by a thin narrow longitudinal ridge." From the sides of each ambulacral covering plate, one or possibly more lateral processes extend to meet similar processes on plates of the adjacent ridge or row, at least in *C. decussatus, C. lindstroemi,* and *C. devonicus.* These processes taper distally, so that they resemble spines. Those of most plates seem to be set directly opposite, but some variations have been reported. Together, the ridges and lateral procsses give the outer surface of the oral disc a reticulate appearance, the ridges being radial and the processes being more or less concentric (Fig. 140).

Little is known about the junctions of the ambulacral plates in each row. FOERSTE (6) stated that in *C. illinoisensis* MILLER & GURLEY, "the ventral or upper disk consists of numerous scutellate plates imbricating toward the centre of the disk." No other author has suggested imbrication in these plates, and one may question whether FOERSTE's specimen was normally preserved.

Between distal parts of adjacent ridges and extending proximally until the ridges join (Fig. 140) is a narrow band of small plates, the **interambulacrals.** Although edges of these plates, like those of the adjoining ambulacral covering plates, are poorly defined, it appears that these interambulacrals are uniserial. They are set somewhat below the general level of ambulacral plates and their sutures may be hidden by spinose processes of the latter.

Neither ambulacral covering plates nor intervening interambulacrals have any pores through them, insofar as known. They seem to have provided a plated cover, ribbed like an umbrella, over soft parts of the animal.

As REGNÉLL (15) and SIEVERTS-DORECK (20) have stressed, the oral disc is composed of more than one layer of plates. To enclose the ambulacral grooves, troughlike **ambulacral flooring plates** are attached under the **ambulacral covering plates** (Fig. 143). In *Cyclocystoides devonicus* no divisions of these flooring plates have been discerned, perhaps because they were composed of continuous, weakly calcified sections of the integument. In this species SIEVERTS-DORECK (20) reported that the flooring plates were supplied with lateral processes like those of the overlying covering plates.

The actual number of ambulacra in the circumoral region may not have been the same in all cyclocystoids. In *Cyclocystoides davisii,* SALTER (in SALTER & BILLINGS, 17) said, "The center of the flat disk is occupied by a star of about eight narrow rays." In *C. decussatus,* BEGG (2) described the central area as having "four rays in the form of a St. Andrews' Cross." RAYMOND (14) stated that in *C. huronensis* BILLINGS "only 5 branches reach the centre." SIEVERTS-DORECK (20) also discerned 5 ambulacra in *C. devonicus,* originating at the center of the oral disc.

The oral region is poorly known, primarily because the plates there are small, thin, and fragile. In my reconstruction (Fig. 140) I have supposed that tiny plates in interambulacral positions covered the peristome, more or less corresponding to those of certain cystoids.

The eccentric space between two ambulacral ridges figured by HALL (8) and supposed by him to be the mouth can scarcely be interpreted thus, inasmuch as it is removed from the place where the ambulacra converge. Instead, as BATHER (1) suggested, it is probably the periproct. No anal pyramid has been discovered in cyclocystoids, and the anus may have been surrounded by tiny plates no larger or thicker than the ambulacral covering plates and interambulacrals.

ABORAL DISC

Underlying the body of the animal, on the aboral or dorsal side of the theca, the aboral disc is a circular layer of plates filling the space wtihin the submarginal ring (Fig. 141). In *Cyclocystoides lindstroemi,* REGNÉLL (15) determined from cross sections that the aboral inner edge of each submarginal plate extends as two lappets on the oral side of the aboral disc, so that on the aboral (or basal) surface of the theca, the aboral disc overlaps the border of the submarginal ring.

Most writers who have commented upon the matter agree that the aboral disc had greater strength and rigidity than the oral disc. In some cyclocystoids, the plates of this disc are definitely thicker and have better-defined edges than those of the oral disc above.

The plates in the aboral disc are of two kinds. Some cyclocystoids (e.g., *Cyclocystoides wrighti* BEGG, *C. halli* BILLINGS) have plates in a mosaic; others (e.g., *C. devonicus* SIEVERTS-DORECK) have them imbricating in a radial arrangement. As SIEVERTS-DORECK (20) believed, mosaic plates probably made a stronger, less flexible layer than did imbricating plates. In both kinds the size of plates decreases away from the center.

The imbricating plates are interpreted to have an arrangement corresponding to the branching of the ambulacra in the oral disc. They seem to have been at least partly embedded in integument, so that the structure is somewhat indistinct. Nevertheless, SIEVERTS-DORECK (20) discerned certain features in *C. devonicus*; she reported that the aboral disc is marked by strong radial ridges, which in places are not in contact but are separated by radial grooves. Around a small central area that cannot be deciphered, five plates appear in the shape of a pentagon. One row of plates that could be traced without interruption to the margin of the aboral disc contains 11 to 13 plates. She distinguished two kinds of plates in the aboral disc; (1) those in elevated rows, each subquadrate or elongate oval to rounded pentagonal or hexagonal in outline, with nodose or tuberculate surface, many of the corners depressed, numerous plates broader distally than proximally, lateral processes extending to processes of adjacent plate rows like those of oral disc, at many places the plates imbricating toward the margin, and (2) elongate oval, rather flat plates corresponding to the interambulacrals of the oral disc. A small structure near the center in an interambulacral plate was questionably called a hydropore by SIEVERTS-DORECK, but this location would indeed seem impossible for intake of water.

Some kind of central structure was previously indicated by RAYMOND (14) who said that in *C. halli* the center of the aboral disc contained "a minute opening, surrounded by an elevated ring of 5 plates. . . ." He added, "The remainder of the disk is covered by small plates which seem to be arranged in a somewhat radial fashion, with larger plates towards the centre and very small ones at the outer margin. Adjoining the sub-marginal ring, there seem to be two very small plates in front of each sub-marginal plate. These small plates do not make a solid covering, but have large pores between them. Around the small mound which resembles an anal pyramid, there are five small, deep depressions, which may indicate the main trunks of the sinuses which extend beneath the integument." It is possible, of course, that the minute central opening and the pores near the submarginal ring were features of preservation in the specimen studied by RAYMOND.

The aboral disc in an unnamed species from Tennessee was described by FOERSTE (6) in somewhat different terms. He stated, "The dorsal [aboral] disk within the submarginal circle consists of numerous erect plates, like fence-palings in form, which incline inward sufficiently to be said to imbricate in that direction. The height of these more central plates is such as to have produced a strong support though still permitting a certain amount of flexibility." It would clarify several points of structure if authors prepared high-quality photographs or accurate drawings to show how plates fit together.

The mosaic type of aboral plates is best exemplified by *Cyclocystoides wrighti,* as illustrated by BEGG (3), in which rather close-fitting plates constitute a pavement. Such an arrangement is indicated in the reconstruction shown in Figure 141.

It might be well to point out that of approximately 21 valid scientific names for cyclocystoids, only six have been applied to fossils for which both oral and aboral sides are known. This helps to explain some of the difficulties in attempting to compare the opposite sides of the same specimen. No evidence has been presented as to whether the imbricating type of aboral plates duplicate the exact pattern of ambulacra in the oral disc or whether they follow a general plan of branching from five central stems.

SUBMARGINAL RING

Because the submarginal ring of plates is the best-preserved part of the cyclocystoid, it has received exceptional attention in definition of species—undue attention in my opinion. According to SIEVERT-DORECK's (20) summary, three species are each known from one specimen showing only the oral side of the submarginal ring, and five species are each known from one specimen showing only the aboral side of the submarginal ring.

Because plates of the submarginal ring are exceptionally thick, they have been referred to as "ossicles" by several authors (Table 1). They extend from one side to the other of the theca, being exposed on both oral and aboral sides (Fig. 140-142). Although not absolutely rigid, they were evidently attached firmly to one another and provided the frame for maintaining the shape of the theca.

The shape of each submarginal plate in radial cross section is not the same in all species. FOERSTE (5) attached considerable importance to this feature in erecting *Narrawayella* and *Savagella*. According to his classification, *Cyclocystoides s.s.* (exemplified by *C. anteceptus* HALL, *C. halli* BILLINGS, *C. bellulus* MILLER & DYER, and others) has submarginal plates with the proximal half strongly elevated above the distal half, *Narrawayella* (exemplified by *C. cincinnatiensis* MILLER & FABER, *C. nitidus*

TABLE 1. *Morphological Terms Used by Various Authors for Cyclocystoidea.*

	Oral Disc	Aboral Disc	Submarginal Ring	Marginal Ring	Facets	Channel	Ambulacral Grooves
SALTER & BILLINGS, 1858	Integument of the upper side	"Underside (at least the flatter and less ornamented side)"	Marginal ossicles	Marginal plating; plated integument	"Circular pit with a tubercle in it"	Tubular channel	"Small pores penetrating through marginal plates," connecting channel with the interior
BATHER, 1900	Ventral membrane	Dorsal membrane	Ring of stout ossicles	Border of smaller plates			
RAYMOND, 1913	Upper side	Lower side	Submarginal ring	"Narrow band . . . of imbricating plates . . . 'shagreen' border"	Spoon-shaped depressions	Circular canal	"Sinuses which extend beneath integument"; "probably pores through plates"
FOERSTE, 1924	Ventral disc; upper disc	Dorsal disc; lower disc	Submarginal ring	Marginal zone of very small plates; marginal plates	"Oval depressions occupied by corresponding oval elevations"	Outer, lower part of plates	
BEGG, 1934	Central disc: lower plate	Upper plate	Submarginal area; ring of ossicles	Marginal zone; "sinuous threads or possibly . . . imbricating plates"	Spoon-shaped depressions	Canal	"Ducts beneath the surface of the disc"
REGNÉLL, 1945	Central disc	Lower plate	Submarginal area; submarginal ossicles	Marginal zone	Mamillary elevations	Canal	"Ducts penetrate the vaults radially to open on the—likewise excavated—inner side of the ossicles"
SIEVERTS-DORECK, 1951	Ventral-Scheibe	Dorsal-Scheibe	Submarginal Ring	Randsaum	Warzenförmige Höcker	Platform; ?Ring-Kanal *or* Höhlung	Ambulakral-System; "Strahlen"

FABER, and *C. mundulus* MILLER & DYER) lacks depressions in the distal halves of the plates, *Savagella* (with type-species *C. ornatus* SAVAGE) shows subtriangular cross sections with a steep inner face on the submarginal ring, and an unnamed genus represented by *C. illinoisensis* has submarginal plates with a "flattened elliptical form" in cross section. SIEVERTS-DORECK (20) described *C. devonicus* as lacking deep depressions in the distal parts of the plates.

The oral side of the submarginal plates is more complex than the aboral side. The sides of each plate converge slightly toward the center of the oral disc, so that adjacent plates have nearly parallel sides. On external molds, the spaces between plates are filled with matrix, which takes the form of radial partitions; this is the preservation in many specimens, the whole of the plates having been dissolved away. The oral part of the submarginal plates is divided into two parts: (1) proximal elevated part, forming a prominent ring, and (2) distal part bearing facets (Fig. 140). The division between the two parts is sharp, tending to be emphasized by the overhanging edge of the proximal part, so that HALL (8) concluded that two circles of plates were present. However, as convincingly shown by REGNÉLL (15) in cross sections, only one circle or ring of plates exists.

In many species, the proximal part of each submarginal plate is elevated as a subquadrate block. The oral surface is variously ornamented. Most species bear numerous low tubercles or papillae (e.g., *Cyclocystoides salteri* HALL, *C. halli* BILLINGS, *C. magnus* MILLER & DYER, *C. decussatus* BEGG, *C. lindstroemi* REGNÉLL, and *C. devonicus* SIEVERTS-DORECK); one has radial grooves dividing the surface into four or five low ribs (*C. ornatus* SAVAGE); another was described by BEGG (3) as having little round punctae *(C. wrighti)*; and still another was said by FOERSTE (5) to be "coarsely pitted" (*C. cincinnatiensis* MILLER & FABER). The ornamented part overhangs both the outer plates of the oral disc on the proximal side and the edge of the rest of the submarginal plates on the distal side in *C. lindstroemi* and probably in some other species, but not in *C. devonicus* and some of the species referred to by FOERSTE (5).

The distal part of the oral side of submarginal plates bear **facets,** small circular to elliptical elevations (Fig. 140). No plate is known which lacks a facet, some plates having only one facet, and some as many as four. In some specimens, each of the plates has two facets. Whether this constancy of facet-to-plate relationship is a character of species or of maturity has not been proved. In *C. halli* and *C. lindstroemi,* facets lie in a circular trough called the **channel.** In *C. devonicus,* however, they are on the shallowly excavated, beveled edge of the plate, called by SIEVERTS-DORECK (20) the **platform.** Those facets within a channel are surrounded by some kind of depression. *C. halli* was described by BILLINGS (in SALTER & BILLINGS, 17) as having the outer half of each submarginal plate "deeply excavate, smooth, divided by a radiating ridge into two shovel-shaped portions, which at their inner base are each deepened into a circular pit, with a tubercle in it." Other authors (Table 1) have referred to similar areas surrounding facets as "spoon-shaped depressions." Even in *C. devonicus* (see Fig. 147,*1-2*), external molds show that the facets are in shallow excavations.

Lack of a well-incised channel in *Cyclocystoides devonicus* led SIEVERTS-DORECK (20) to postulate that parts of submarginal plates under the facets were hollow and that a channel was a secondary feature caused by collapse of the facet-bearing section. She concluded that the channel was not connected with the ambulacral system. The regularity of the channel in specimens of *C. halli* illustrated by SALTER & BILLINGS (17) and by RAYMOND (14), and the lack of any cavities shown in the cross section of *C. lindstroemi* illustrated by REGNÉLL (15) raises considerable doubt about her interpretation. SIEVERTS-DORECK stated, however (my translation), "Should my interpretation not prove correct, then there exists a conspicuous contrast between those species of *Cyclocystoides* which possess a peripheral ring-canal and those which lack it."

Insofar as known, facets are invariably aligned with radial ducts or holes through the proximal part of the submarginal plate. These ducts appear to connect with the ambulacral grooves. Authors agree that food entered the ambulacral system through the channel or platform of the submarginal plates. The presence of "small pores" leading from the channel to the interior of the theca and perforating the submarginal plates was made part of the definition of *Cyclocystoides* by SALTER & BILLINGS (17). REGNÉLL (15) found a groove in *C. lindstroemi* leading from the facet to the duct, but in *C. insularis* he failed to find such a groove.

One of the unsolved problems of cyclocystoids concerns the function of covering plates over the channel, if indeed they exist. Supposition that the channel was covered by movable plates originated with SALTER & BILLINGS (17), who based their interpretation on small isolated plates in the channel of one or two submarginal plates. From his study of the same specimen, RAYMOND (14) also concluded that the channel was roofed over by small plates. FOERSTE (6) described some curious structures just outside the submarginal ring as "spout-like appendages," and proposed that these could be moved at will to enclose or expose the depressions containing the facets. The illustrations of these structures are not sufficiently distinct to show the details of their configuration; possibly, they could be fractured outer edges of spoon-shaped depressions, which in *C. lindstroemi* form an elevated rim.

The general resemblance of the ambulacral system in cyclocystoids to that in cystoids supports their assignment to the Crinozoa, rather than Echinozoa. In cystoids and blastoids, food was gathered by brachioles; in crinoids, the ambulacral system was further branched outside the calyx and food was gathered by pinnules on the arms. Only in edrioasteroids was food gathered directly into the ambulacral grooves. Enclosure of ambulacral grooves in cyclocystoids seems irrefutable evidence that food entered from the channel, specifically at the facets. The form of the facets is strikingly like that in cystoids. It is difficult to conceive of brachioles sufficiently developed to gather adequate food and still so small as to be retractable beneath covering plates of the channel.

Perhaps an alternative interpretation will explain both the small plates and the facets. In some species the facets are not in contact with the ducts, but set a short distance away from the openings. Possibly, small plates served as ambulacral covering plates over this part of the ambulacral system. There is no reason to deny that small plates may have also covered parts of the channel between facets. As pointed out later, the complexity of the submarginal plates may be regarded as evidence of their having evolved by fusion of several plates.

On the sides of the submarginal plates—the surfaces facing the adjacent plates—the oral margin is marked by striae normal to the edge. These have been interpreted as scars of ligaments which bound the plates of the submarginal ring together and yet allowed appreciable flexibility. These markings are prominent on some cyclocystoids, but have not been found in others.

On the aboral side of the theca, the exposed part of the submarginal ring has less radial extent than that on the oral side. Each submarginal plate is slightly convex and trapezoidal in outline. The plates are in contact or nearly in contact only at their distal corners; sides of adjacent plates diverge toward the center of the theca (Fig. 141). As illustrated by REGNÉLL (15), each submarginal plate in *C. lindstroemi* has two lappets concealed by the aboral disc along its proximal edge; probably, some such arrangement was present in other species.

MARGINAL RING

Around the periphery of the theca, the cyclocystoid has a ring of imbricating plates remarkably similar to that present in several edrioasteroids. On both oral and aboral sides, plates decrease in size toward the outside edge. Apparently, they were embedded in a thick integument which formed a seal against the bottom and enabled the cyclocystoid to hold fast by suction. No evidence has been presented to indicate whether the marginal ring was composed of two layers or one. For it to have functioned in attachment, much of it must have consisted of muscles.

AMBULACRA

The ambulacral system began, I believe, with brachioles set in a circle in the distal part of the submarginal ring. Fossils show only the facets (Fig. 143), on which I presume brachioles articulated. From the facets in some species, a groove leads to the duct perforating the proximal part of the submarginal plate. Undoubtedly, this groove was provided with some kind of covering plates. In species which show no such groove, some sort of enclosed passageway probably led from the facet to the duct.

Ducts are said to have a funnel-like adit from the channel. SIEVERTS-DORECK (20) found that in submarginal plates having three ducts, the central one is truly radial and the lateral ducts converge somewhat proximally. Each duct connects with one of the ambulacral grooves under the surface of the oral disc. By successive convergences, the number of ambulacral grooves is reduced to only four or five at the mouth.

UNSUBSTANTIATED AND UNKNOWN STRUCTURES

In this category must be listed the hydropore, gonopore, periproct, and brachioles, structures which might be expected in a pelmatozoan, but not brachioles in an echinozoan. The space between ambulacra noted by HALL (8) and interpreted by BATHER (1) as the anus, very likely was occupied by small, rather undifferentiated plates of the periproct. In comparison with pelmatozoans, one would expect the hydropore and gonopore to be present on the oral side of the theca, probably in the circumoral region.

GLOSSARY OF MORPHOLOGICAL TERMS APPLIED TO CYCLOCYSTOIDEA

Relative importance of terms is indicated by the type in which they appear: first rank by boldface capital letters, second rank by boldface small letters, and third rank (including obsolete terms, terms having cross references, and synonyms) by italic small letters.

aboral. Located away from mouth; used to refer to flat side presumed to have been base of animal; from inferred position of mouth in center of one side, aboral is same as dorsal on cyclocystoids.

ABORAL DISC. Subcircular section of integument and associated plates stretched across dorsal side of submarginal ring, more or less like a drumhead, also known as dorsal disc. Plates within aboral disc may be rather firmly set as a mosaic or flexibly arranged in imbrication, in latter arrangement showing radial symmetry similar to that in oral disc.

ambulacral covering plates. Small plates, probably uniserial, roofing over ambulacral grooves and embedded in integument of oral disc; in at least some species, each plate had lateral processes extending to similar processes on the adjacent ambulacral branch, forming a reticulation to strengthen the disc.

ambulacral flooring plates. Small plates or fused plates serving as internal walls of enclosed ambulacral grooves, trough-shaped and joined by their edges to ambulacral covering plates.

ambulacral groove. Any of branched passageways through which food was conveyed from channel to mouth, aborally bounded by ambulacral flooring plates and orally bounded by ambulacral covering plates; presumably, ambulacral grooves led adorally from brachioles; technically, ducts or perforations through submarginal plates are parts of ambulacral grooves.

AMBULACRUM. Structure in which food was conveyed to mouth, normally applied to one of major structures and all its branches, so that most cyclocystoids may be said to have five ambulacra.

border. See marginal ring.

brachiole. Inferred erect structure by which food was gathered and transmitted to ambulacrum, although none have been found, presence of brachioles is strongly indicated by form of facets in channel of submarginal ring; presumably, these structures were very similar to those in cystoids.

canal. See channel.

central disc. See oral disc.

CHANNEL. Distal part of submarginal ring, bearing facets, in most cyclocystoids troughlike. Varying degrees of indentation exist; if the facet-bearing surface is very shallow, or if it is more or less expressed as a beveled border of the submarginal ring, it is called a platform.

channel covering plate. One of supposed series of plates roofing over channel in some species; existence of covering plates suggested by disarranged small plates in channel of one of type specimens of *Cyclocystoides halli* described by SALTER & BILLINGS (17), who said in their definition of the genus, "Those [oval excavations], in perfect specimens, are covered over by minute polygonal plates, thus forming a tubular channel around the whole animal." From restudy of this specimen, RAYMOND (14) also thought the channel was covered by small plates. Evidence for a continuous roof, which would enclose the facets and make brachioles ineffective, is far from convincing.

covering plate. See ambulacral covering plate, channel covering plate.

dorsal. Located away from mouth; in cyclocystoids synonymous with aboral.

dorsal disc. See aboral disc.

dorsal membrane. See aboral disc.

duct. Perforation through proximal part of submarginal plate, radially aligned with facet and forming part of ambulacral groove.

FACET. Small protuberance, circular or oval, set within channel and connected with ambulacral system. Probably, facets served for attachment of brachioles, like those in cystoids; each facet surrounded by moatlike groove in some species, whole lying within spoon-shaped depression in channel.

hydropore. Small opening very near center of oral disc whereby water was admitted to water-vascular system. Preservation very poor, but hydropore strongly suggested in specimen described and figured by SIEVERTS-DORECK (20), in which it is in interambulacral position.

imbricating. Overlapping, like shingles; applicable to plates in marginal ring and, in some species, to plates in aboral disc.

integument. Supposed exterior layer of tissue in cyclocystoids which secreted plates and probably formed leathery "skin" encasing plates of certain parts; poorly-preserved edges of ambulacral covering plates suggest that much of thickness of oral disc was made up of integument.

interambulacral. One of small plates between ambulacral covering plates, uniserial insofar as known.

lower plate. Used by BEGG (2) for oral disc.

mammillary elevation. *See* facet.

MARGINAL RING. Distal part of cyclocystoid, bordering submarginal ring, composed of small imbricating plates that distally decrease in size. Marginal ring forming part of oral and aboral surfaces; whether made of one or two layers is not clear.

marginal zone. *See* marginal ring.

mosaic. Arrangement of plates more or less in plane, not imbricating and presumably rather rigid; plates of aboral disc in some species disposed as mosaic, in other species imbricating.

mouth. Presumed aperture through which food entered body from ambulacra, situated at center of oral disc, probably covered by small peristomial plates as in cystoids.

ORAL DISC. Subcircular section of integument and associated plates extending across center of oral side and attached to inner surface of submarginal ring, also known as ventral disc; in ambulacral positions, oral disc consists of two layers: ambulacral covering plates and ambulacral flooring plates; in interambulacral positions, disc consists only of interambulacrals. Presumably, mouth was in center of oral disc and anus was offset to one side. In known species, ambulacral covering plates give disc a reticulate appearance, with radial elements made of multiple branching of ridgelike ambulacra and concentric elements made of lateral spinose processes.

ossicle. *See* submarginal plate.

periproct. Presumed circumanal area in oral disc.

plate. Any calcareous secretion, normally flat, in cyclocystoids.

platform. *See* channel.

SUBMARGINAL PLATE. One of thick plates in submarginal ring, exposed on both oral and aboral sides of theca, in many specimens, only submarginal plates preserved; on oral side, distal part of plate bears facets and proximal part is elevated as part of thick rim around oral disc, each submarginal plate perforated by radial ducts leading from facets to ambulacral grooves in oral disc; on aboral side, each plate expressed as truncated wedge diverging proximally from adjacent plate. Because of thick form, submarginal plates called ossicles by many workers.

SUBMARGINAL RING. Prominent ring of submarginal plates, most conspicuous and best-preserved feature of cyclocystoids.

THECA. Enclosure of plates and integument in which body of cyclocystoid was housed; generally applied to all fossilized parts.

upper plate. Used by BEGG (2) for aboral disc.

ventral. Located toward mouth, in cyclocystoids synonymous with oral.

ventral disk. *See* oral disc.

ventral membrane. *See* oral disc.

ONTOGENY

Practically nothing is known about ontogeny of cyclocystoids, but REGNÉLL (15) and SIEVERTS-DORECK (20) have suspected that the number of submarginal plates increases with size and age in each species, and that several of the "species" based solely on number of submarginal plates may represent growth stages of a single species. The latter author also directed attention to a specimen with one submarginal plate having the shape of a thin wedge, as though it were being intercalated into the ring. If plates were added during ontogeny, the process was intimately correlated with branching of the ambulacra, so that the ambulacral grooves joined to the ducts through the submarginal plates.

COMPARISON WITH PELMATOZOANS

Several students of cyclocystoids have been puzzled by the circular, rather than radial, terminus of the ambulacral system and by the great thickness of the submarginal plates, extending from oral to aboral sides of the theca. These have been regarded as anomalous characters in a pelmatozoan.

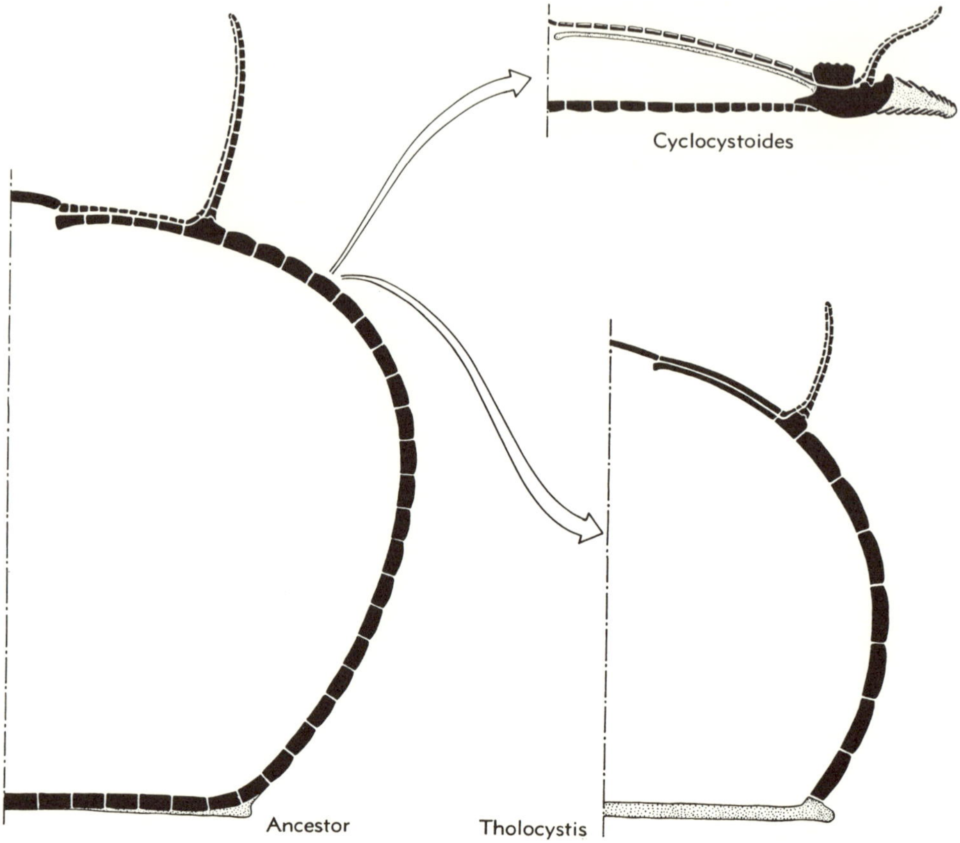

FIG. 144. Hypothetical vertical cross sections of halves of thecae, showing inferred derivation of *Cyclocystoides* and *Tholocystis* from a diplopore-bearing cystoid ancestor. Plates shown in solid black, integument stippled.

Although certain diplopore-bearing cystoids do not duplicate those unusual features they show exceptional similarities. The genus most closely reflecting cyclocystoid organization is *Tholocystis,* of the order Diploporita, superfamily Sphaeronitida, and family Sphaeronitidae (Fig. 144). This cystoid was described by CHAUVEL (4) as shaped like a kettle, with large, orderly disposed radials, orals, and peristomial covering plates corresponding to the lid of the kettle. The broad base of *Tholocystis,* known only from its impression, was probably composed mostly of integument and was called by CHAUVEL the "sole." It is of special interest that *Tholocystis* has (1) branched ambulacra leading to brachioles arranged in a broad-angled pentagram, more or less a circle except for the reentrants necessary for the radials, (2) brachiole facets borne on quadrangular plates, or "adambulacrals," which are thicker than other plates, and (3) a large circular base that was flexible. Presumably, the ambulacral grooves across the oral region were covered by sutures between adjacent oral plates or by small covering plates which left no record.

Cyclocystoides may have descended from a pelmatozoan not very different from *Tholocystis.* The aboral disc and marginal ring of cyclocystoids seem analogous with the base or "sole" of *Tholocystis* and the facet-bearing part of the submarginal ring with the thick quadrangular plates. The ducts through the submarginal plates are novel features for pelmatozoans, but they may have originated by fusion of other plates; their function, protection and enclosure of the ambulacral groove, suggests that the part of the submarginal plate oral to the ducts may have been derived from ambulacral covering plates and that the part aboral to the ducts may have been modified from ambulacral flooring plates. Obviously, *Cyclocystoides* did not attain the exceptional symmetry of *Tholocystis,* and the outer part of its base became more differentiated.

Both *Tholocystis* and *Cyclocystoides* appeared in Middle Ordovician time. The oldest known relative of *Tholocystis* is *Sphaeronites,* from Lower Ordovician strata. Even this genus, however, shows specialization of structures in the circumoral region; the theca is more or less round and attached by the basal surface, which in many specimens retains the imprint of objects to which it was fastened. The Cambrian ancestor of both *Sphaeronites* and *Tholocystis* probably had a theca composed of numerous plates, none of which was highly specialized; such a primitive diploporitan cystoid is conjectured as the possible ancestor of both *Tholocystis* and *Cyclocystoides* (Fig. 144). Whereas no diplopores have been observed in plates of *Cyclocystoides,* integument in which plates of the oral disc are embedded may have fulfilled the same function as diplopores. At any rate, strongest similarities to cyclocystoids occur in this group of sphaeronitid cystoids.

According to this hypothesis, the cyclocystoid is a derivative of the Diploporita in which the theca has undergone extreme oral-aboral compression, the oral region has greatly expanded without plate specialization, submarginal plates have resulted from fusion of "adambulacrals" with proximally adjacent ambulacral flooring and covering plates, the aboral side of the theca has been greatly extended, diplopores have degenerated, and brachioles have migrated outward, assuming circular distribution.

PALEOECOLOGY

Very little can be deduced from occurrences of cyclocystoids. None have been found attached to other organisms. Many have been reported in association with other marine animals, such as corals, brachiopods, and trilobites. No reason exists to suspect that cyclocystoids were not free-living, normal marine creatures.

Because cyclocystoids have been discovered in several kinds of sedimentary strata, it seems likely that they inhabited a variety of depths and thrived under a range of current conditions.

	N. AMERICA	ENGLAND	SCOTLAND	GOTLAND	BELGIUM
Middle Devonian					sp. SIEVERTS-DORECK
Lower Devonian					sp. MAILLIEUX *devonicus* SIEVERTS-DORECK
Ludlovian					
Wenlockian				*insularis* REGNÉLL *lindstroemi* REGNÉLL	
Llandoverian	*ornatus* SAVAGE *huronensis* BILLINGS *illinoisensis* M.&G.	*davisii* SALTER (MS)			
Ashgillian	*cincinnatiensis* M.&F. *minor* MILLER & DYER *mundulus* M.&D. *nitidus* FABER *parvus* M.&D. *bellulus* M.&D. *magnus* M.&D.		*decussatus* BEGG *wrighti* BEGG		
Caradocian	*salteri* HALL	*caractaci* SALTER *marstoni* SALTER (MS)			
Trenton	*raymondi* FOERSTE *anteceptus* HALL *salteri* HALL *halli* BILLINGS *billingsi* WILSON				

FIG. 145. Stratigraphic and geographic distribution of *Cyclocystoides* (modified from Regnéll, 1948).

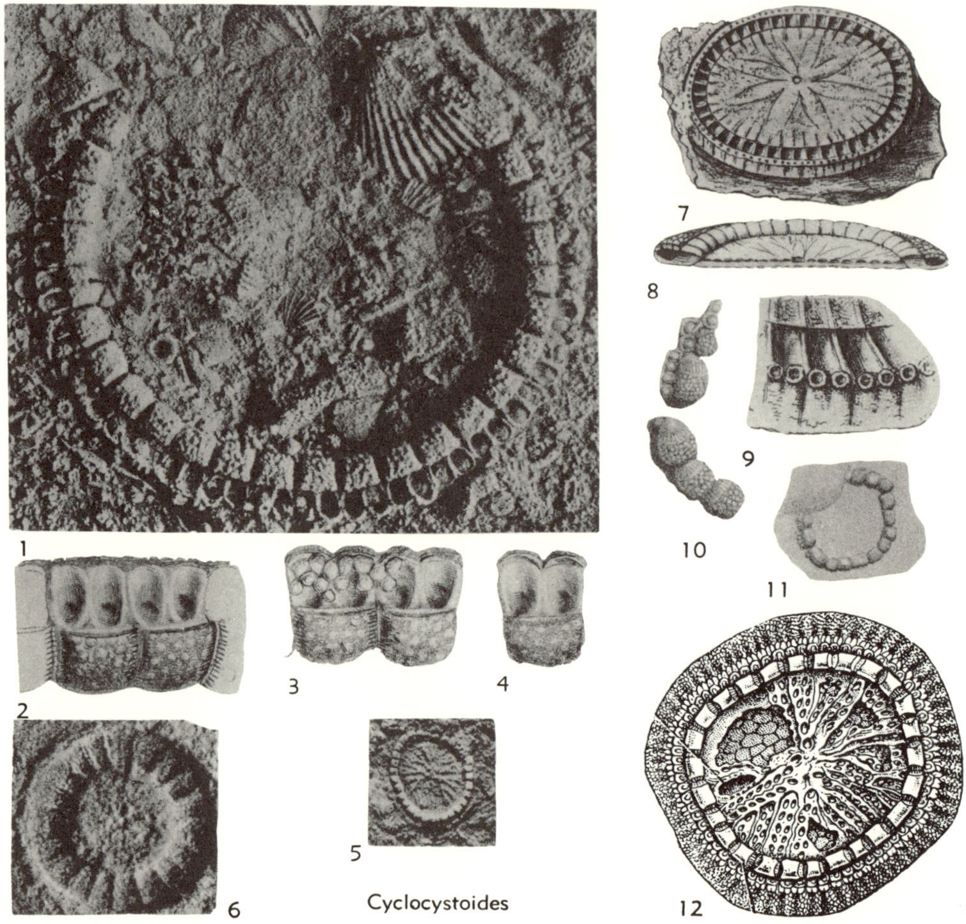

FIG. 146. Cyclocystoididae (p. U206-U209).

STRATIGRAPHIC AND GEOGRAPHIC DISTRIBUTION

The paleontological record of cyclocystoids is fragmentary and interrupted. Nevertheless, the distribution compiled by REGNÉLL (1948) is essentially unaltered today, and probably is nearly correct.

As shown here (Fig. 145), Cyclocystoidea seem to have originated in North America. The oldest known species is *Cyclocystoides billingsi* WILSON (1946), from late Blackriveran or early Trentonian deposits in the St. Lawrence Lowland of Canada. The genus persisted in eastern North America until Early Silurian time; the last survivor on this continent seems to have been *C. illinoisensis* MILLER & GURLEY (1895), in the Girardeau Limestone of Illinois.

The first migration of *Cyclocystoides* occurred in the Caradocian, when it reached England. There it persisted into the Llandoverian. The genus continued to move northward and eastward, reaching Scotland in Ashgillian time, Gotland in Wenlockian time, and Belgium in the Early Devonian. The final stand of *Cyclocystoides* was in Belgium and the Rhineland, where it is present in Lower and Middle Devonian strata.

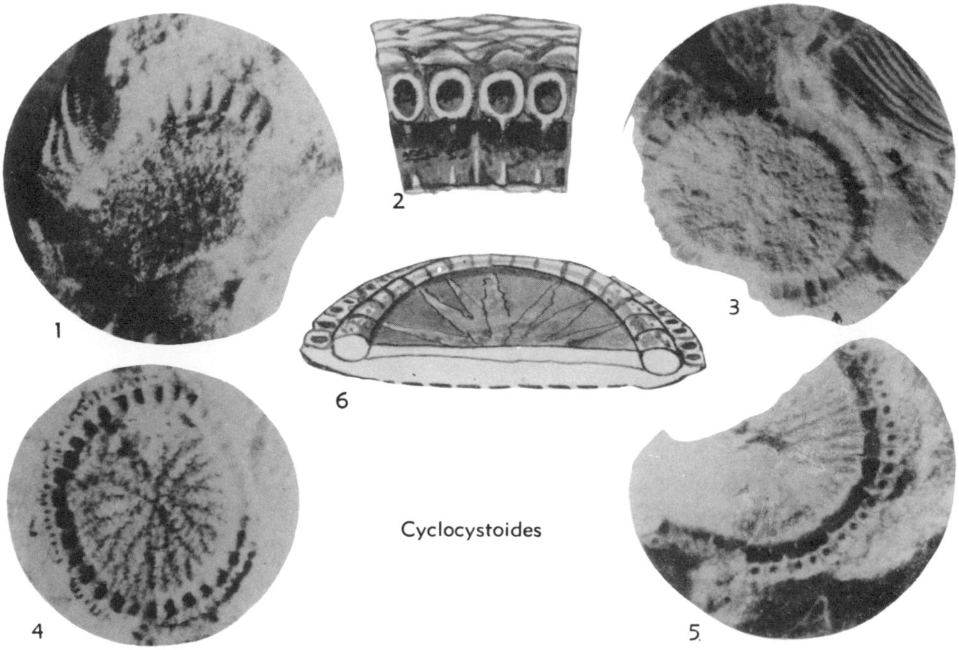

FIG. 147. Cyclocystoididae (p. *U*206-*U*209).

CLASSIFICATION

Admittedly, classification of cyclocystoids is unsatisfactory. Fragmentary nature of most fossils rules out an assessment of characteristics. Whereas workers are aware of variations in plate arrangement in the aboral disc, number and shape of plates in the submarginal ring, depth and form of the channel, number and shape of facets, and ornamentation of submarginal plates, it has been impossible to establish the association of these variations. Many species are known only from one side.

Genera were set up by FOERSTE (5) primarily on shape of submarginal plates in cross section. He distinguished *Narrawayella,* which lacks depressions in the distal halves of the plates, and *Savagella,* which is subtriangular in section, from *Cyclocystoides s.s.,* which has the proximal part strongly elevated and block-shaped. REGNÉLL (15) thought that *Narrawayella* might prove to be a valid genus, but expressed doubt about *Savagella.* SIEVERTS-DORECK (20) placed both of FOERSTE's genera in synonymy with *Cyclocystoides.*

Probably, the cyclocystoids with mosaic plates in the aboral disc are generically different from those with imbricating plates. And forms with deeply excavated channels may prove distinct from those with shallow "platforms." Until the association of characteristics in each species is discovered, it seems best to put all forms into one genus.

SYSTEMATIC DESCRIPTIONS

Class CYCLOCYSTOIDEA Miller & Gurley, 1895

[*nom. transl.* KESLING, herein (*ex* order Cyclocystoidea MILLER & GURLEY, 1895, p. 61)]

Small, disc-shaped echinoderms, lacking columns of any kind, probably attached by suction on nearly flat aboral side. Theca composed of numerous calcareous plates, distinctly separated into central oral and aboral discs, submarginal ring, and marginal ring. Submarginal ring of thick plates, exposed on both oral and aboral surfaces of theca, forming stout frame for attachment of thin oral and aboral discs. Oral disc cir-

FIG. 148. Cyclocystoididae (p. *U*206-*U*209).

cular, flexible, containing very numerous small plates, weakly calcified, serving to cover soft parts of animal and incorporating radiating, multiply branched, enclosed ambulacra. Aboral disc less flexible than oral disc, serving as floor under body. Submarginal plates complex; proximal part of each plate radially perforated; distal part bearing facets on oral surface. Perforation connected with ambulacral passageways and aligned with facets. Marginal ring very flexible, covered with tiny imbricating plates. *M. Ord.-M.Dev.*

Family CYCLOCYSTOIDIDAE S. A. Miller, 1882

[Cyclocystoididae S. A. MILLER, 1882, p. 223]

Characters of class. *M.Ord.-M.Dev.*

Cyclocystoides SALTER & BILLINGS, 1858, p. 86 [**C. halli* BILLINGS, 1858; OD] [=*Narrawayella*

FIG. 149. Cyclocystoididae (p. *U*206-*U*209).

FOERSTE, 1920, p. 59 (type, *Cyclocystoides cincinnatiensis* MILLER & FABER, 1892, p. 84; OD); *Savagella* FOERSTE, 1920, p. 61 (type, *Cyclocystoides ornatus* SAVAGE, 1917, p. 265; OD)]. Oral disc with 4 or 5 ambulacra, each branched several times from circumoral region to periphery, covering most of disc except small area presumed to be periproct. Ambulacra enclosed between flooring plates and covering plates, those of adjacent branches separated by small interambulacral plates. Aboral disc containing plates disposed in either mosaic or imbricating radial pattern. Submarginal plates with shape of truncated wedges, joined together by ligaments, in cross section subtriangular, elliptical, or subquadrate; distal part of oral surface deeply excavate to form circular troughlike channel or shallowly depressed to form platform, each plate bearing 1 to 4 facets, which are radially aligned with perforations through proximal parts of submarginal plates, leading to ambulacral passageways in oral disc. Hydropore, gonopore, and brachioles not known. *M.Ord.-M.Dev.*, N.Am.(Ill.-

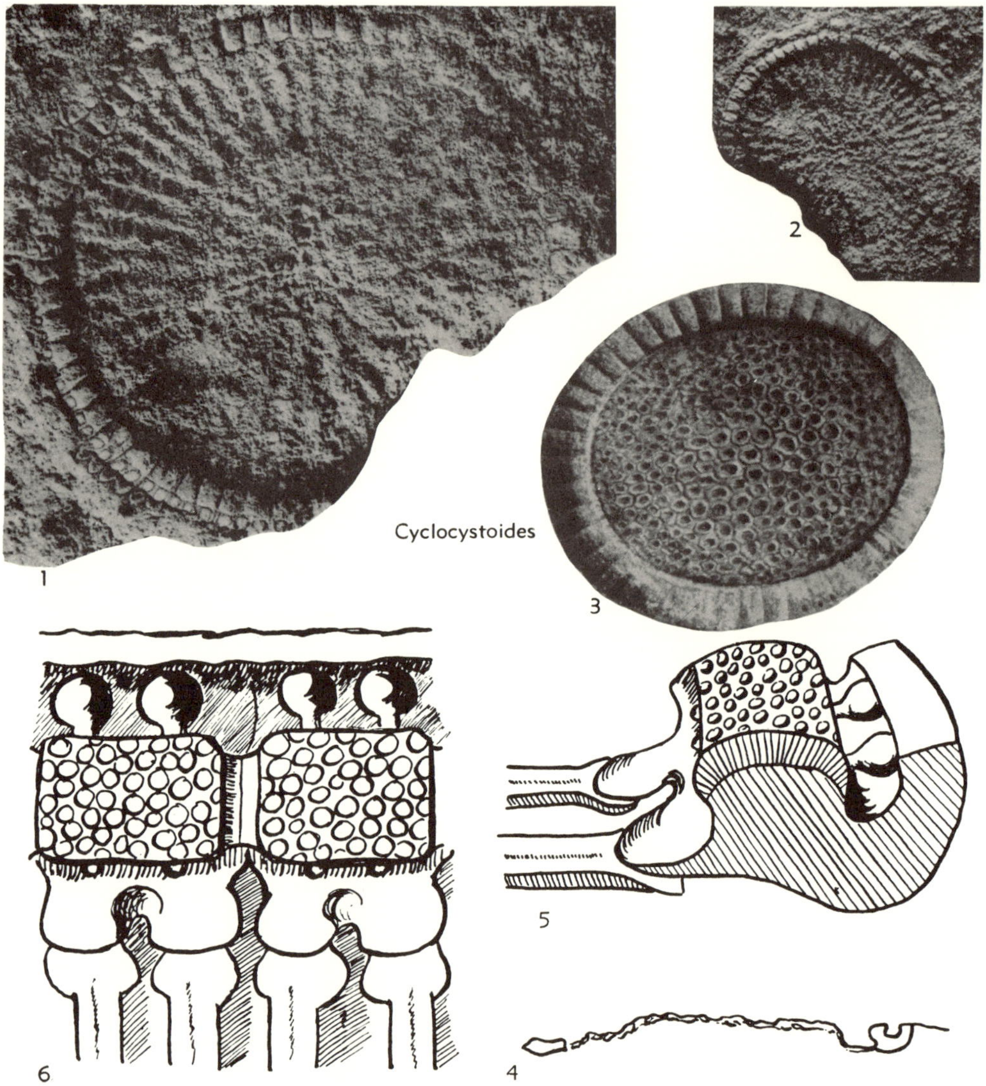

FIG. 150. Cyclocystoididae (p. U206-U209).

Mich.-Ohio-N. Y.-Ont.-Que.)-Eu.(Eng.-Scot.-Gotl.-Belg.-Ger.).——FIG. 146,*1-5*; 147,*6*. **C. halli,* M. Ord.(Trenton.), Can.(Ottawa); 146,*1*, oral side of ring of submarginal plates, 2 spoon-shaped depressions on each plate, vertical marginal plates, ×3 (14); 146,*2-4*, oral side of marginal plates, enlarged (17); 146, *5*, oral view, ×0.2 (14); 147,*6*, oblique view of half of theca (reconstr.), "drawing copied from Salter & Billings, 1858. Here the authors err in showing, by dotted line, the upper plate passing under the lower plate" (Begg has made his correction in this drawing), mag. unknown (2).——FIG. 146,*6*. *C. raymondi* FOERSTE, M.Ord.(Trenton.), Can.; aboral side, ×3.5 (14, wherein this fig. called *C. halli*).——FIG. 146,*10-11*. *C. bellulus* MILLER & DYER, U. Ord., Ohio; *10*, oral side of submarginal plates, enlarged; *11*, oral side, submarginal plates, ×1 (12).——FIG. 146,*7-9*. *C. davisii* SALTER, L.Sil. (Llandover.), S.Wales; *7*, mold of specimen with perfect margin, ×1; *8*, ideal section of *Cyclocystoides*, mag. unknown; *9*, mold of part of margin, enlarged (17).——FIG. 146,*12*. *C. salteri* HALL, M.Ord.(Trenton.), USA(N.Y.); oral side, ×2 (21).——FIG. 147,*1-5*. *C. decussatus* BEGG, U. Ord.(Ashgill.), Scot.; *1*, aboral side, ×2; *2*, drawing of mold of part of marginal and submarginal rings with spoon-shaped depressions, casts of pores,

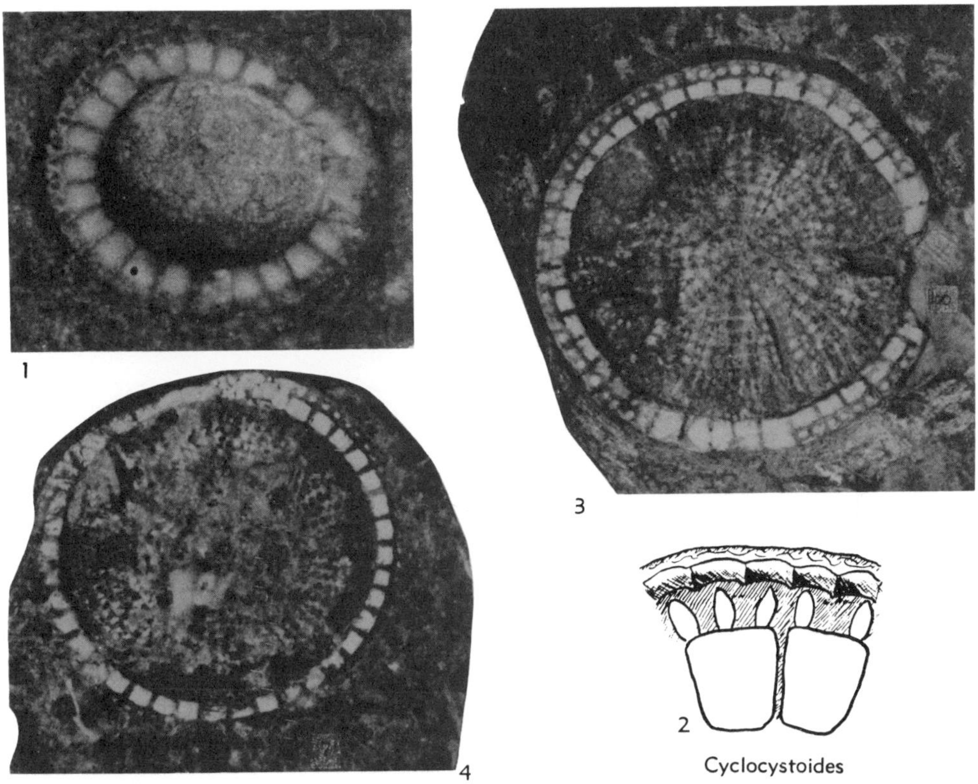

FIG. 151. Cyclocystoididae (p. *U*206-*U*209).

and filling of spaces between submarginal plates, enlarged; *3*, oral side, ×2; *4,5*, molds of oral sides, marginal rings not preserved ×2 (2).——FIG. 148,*1-5*; 149, *1-5*. *C. devonicus* SIEVERTS-DORECK; 148,*1-4*, M.Dev., Ger.(Westphalia); 148, *1*, aboral side, marginal ring, ×6; 148,*2*, mold seen aborally, ×6; 148,*3*, cast of aboral side, ×2; 148,*4*, mold of aboral side, ×2; 148,*5*, upper L. Dev., Ger.(Rhine Prov.); mold of oral side, ×2; 149,*1-3*, M.Dev.(Westphalia); molds of oral sides, ×6, ×5, ×2; 149,*4*, L.Dev., Belg. (Ardennes); mold of oral side, ×2.5; 149,*5*, L.Dev., Ger.(Rhine Prov.); cast of aboral side, ×2 (20).——FIG. 149,*8*. *C. billingsi* WILSON, M. Ord.(Blackriv.-Trenton.), Can.(Ottawa); oral side, ×1 (22).——FIG. 149,*6-7*. *C. magnus* MILLER & DYER, U.Ord., Ohio; *6*, oral side, submarginal plates, enlarged; *7*, oral side, submarginal ring, ×1 (12).——FIG. 150,*3*. *C. wrighti* BEGG, U.Ord. Ashgill.), Girvan, Scot.; aboral side, magn. unknown (20).——FIG. 150,*1-2*. *C. huronensis* BILLINGS, U.Ord.(Richmond.), Rabbit Is., Lake Huron; oral sides, ×2.5 (6), ×1 (20).——FIG. 151,*1-2*. *C. insularis* REGNÉLL, U.Sil., Gotl.; *1*, oral side, ×1; *2*, drawing of oral side of marginal and submarginal plates, ×8 (15).——FIG. 150, *4-6*; 151,*3-4*. *C. lindstroemi* REGNÉLL, U.Sil., Gotl.; 150,*4*, cross section, ×1; 150,*5*, lat., isolated submarginal plate and part of aboral disc, ×8; 150,*6*, oral side of 2 submarginal plates, ×8; 151,*3-4*, oral sides, ×1 (15).

REFERENCES

Bather, F. A.

(1) 1900, *The Edrioasteroida,* in E. R. LANKESTER (ed.), A Treatise on Zoology, Pt. III, The Echinoderma (by F. A. BATHER, J. W. GREGORY, and E. S. GOODRICH), p. 205-16, 8 text-fig., Adam & Charles Black (London).

Begg, J. L.

(2) 1934, *Notes on Cyclocystoides:* Geol. Mag., v. 71, no. 839, p. 220-24, pl. 11.

(3) 1940(1939), *On the genus Cyclocystoides, with a description of a new species from the Ashgillian of Girvan:* Geol. Soc. Glasgow, Trans., v. 20, pt. 1, p. 21-28, pl. 1.

Chauvel, Jean

(4) 1941, *Recherches sur les Cystoïdes et les Carpoïdes Armoricains:* Soc. Géol. et Mineral. Bretagne, Mém. v. 5, p. 1-286, pl. 1-7.

Foerste, A. F.

(5) 1920, *Racine and Cedarville cystids and blastoids with notes on other echinoderms:* Ohio Jour. Sci., v. 21, no. 2, p. 33-78, 4 pl., 4 text-fig.

(6) 1924, *Upper Ordovician faunas of Ontario and Quebec:* Canada, Geol. Survey, Dept. Mines, Mem. 138 (Geol. Ser. No. 121), iv+255 p., 46 pl., 14 text-fig.

Hall, James

(7) 1851, in J. W. Foster & J. D. Whitney, *Report of the geology of the Lake Superior Land District, Pt. II, The iron region:* U.S. 32nd Congress Spec. Session, Sen. Ex. Doc. 4, p. 203-31.

(8) 1872, *Description of new species of Crinoidea and other fossils from strata of the age of the Hudson-River Group and Trenton Limestone:* N.Y. State Museum Nat. History, 24th Ann. Rept. (also N.Y. State Cab. Nat. History, advance sheets 20th Rept., 1866), p. 205-24, pl. 5-8, 3 text-fig.

Jaekel, Otto

(9) 1918, *Phylogenie und System der Pelmatozoen:* Paläont. Zeitschr., v. 3, no. 1, p. 1-128, 114 text-fig.

Kesling, R. V.

(10) 1963, *Morphology and relationships of Cyclocystoides:* Univ. Michigan, Museum Paleontology, Contrib., v. 18, no. 9, p. 157-176, 5 text-fig.

Miller, S. A.

(11) 1882, *Descriptions of three new orders and four new families, in the Class Echinodermata, and eight new species from the Silurian and Devonian formations:* Cincinnati Soc. Nat. History, Jour., v. 5, p. 221-31, pl. 9.

———, & Dyer, C. B.

(12) 1878, *Contributions to Palaeontology:* Cincinnati Soc. Nat. History, Jour., v. 1, no. 1, p. 24-39.

———, & Gurley, W. F. E.

(13) 1895, *Description of new species of Palaeozoic Echinodermata:* Illinois State Museum Nat. History, Bull. 6, 62 p., pl.1-5.

Raymond, P. E.

(14) 1913, *Notes on Cyclocystoides:* Canada, Geol. Survey, Dept. Mines, Victoria Mem. Museum, Bull. 1, p. 23-32, pl. 3 (fig. 1-4), text-fig. 2-3.

Regnéll, Gerhard

(15) 1945, *Non-crinoid Pelmatozoa from the Paleozoic of Sweden:* Lunds Geol.-Mineral Inst., Medd., no. 108, viii+255 p., 15 pl., 30 text-fig.

(16) 1960, *"Intermediate" forms in early Paleozoic echinoderms:* Internatl. Geol. Congress, Rept. 21st Sess., pt. 22, p. 71-80.

Salter, J. W., & Billings, Elkanah

(17) 1858, *On Cyclocystoides, a new genus of Echinodermata from the Lower and Middle Silurian rocks:* Canada, Geol. Survey, Figs. and Descr. Canadian Organic Remains, Decade 3, p. 86-90, pl. 10 bis.

Sieverts [Sieverts-Doreck], Hertha

(18) 1935, *Begg, James L. "On the genus Cyclocystoides (Review):* Neues Jahrb. Mineral., Geol., & Paläont., 1934, p. 975.

(19) 1935, *Begg, J. L. "On the genus Cyclocystoides" (Review):* Zentralblatt, Mineral., Geol., & Paläont., Abt. B: Palaeont. v. 6, no. 1, p. 46 (entry 134).

(20) 1951, *Über Cyclocystoides Salter & Billings und eine neue Art aus dem belgischen und rheinischen Devon:* Senckenbergiana, v. 32, no. 1/4, p. 9-30, 2 pl., 3 text-fig.

ADDITIONAL SOURCES OF ILLUSTRATIONS

(21) **Piveteau, Jean,** 1953

(22) **Wilson, Alice,** 1946

ECHINOIDS

By J. W. Durham, H. B. Fell, A. G. Fischer, P. M. Kier, R. V. Melville, D. L. Pawson, and Carol D. Wagner

CONTENTS

INTRODUCTION

By R. V. Melville and J. W. Durham

[British Embassy, Paris; University of California (Berkeley)]

Echinoids are free-living echinoderms with a test which is subspherical or modified subspherical in shape, built of interlocking calcareous plates. The test bears movable appendages (spines, pedicellariae, spheridia) externally and commonly a masticatory apparatus internally. The mouth is directed toward the substrate. The two principal groups of plates composing the test are known as the apical system and the coronal system, the former invariably including five radially situated ocular plates and five or fewer interradially situated genital plates, and the latter composed of five radial ambulacral and five interradial interambulacral areas built of contiguous meridional columns of plates. In addition, two less conspicuous plate systems, termed peristomial and periproctal, are present. The plates of the ambulacral areas are perforated for the passage of tube feet. The ambulacral plates bordering the peristome are arranged in constant manner according to a plan which has come to be known as LOVÉN's law. An anus is situated either within the apical system or in the posterior interambulacrum. The mouth is nearly always on the lower surface, but, rarely, it is anterior. The mouth and anus are each surrounded by a membrane which usually bears imbricating or dissociated plates of the peristomial and periproctal systems. The gonads are five or fewer and are interradial in position. Radial canals of the water-vascular system are internal to the test in all except a few early genera. All known types are exclusively marine. The geological range of echinoids is Ordovician to Recent.

The basic features of echinoids are so universal and distinctive that the affinity between the most varied types (*Echinus* and *Spatangus*) was recognized by ARISTOTLE, but modifications of these features are very profound. In general, the Echinoidea fall readily into two main morphological groups which formerly were assigned subclass rank (Regularia and Irregularia), each characterized by position of the anus with respect to the apical system. In recent years the so-called Irregularia have been shown to be polyphyletic and the two subclasses have been abandoned. Nevertheless, these familiar terms serve a useful descriptive purpose and are retained here as informal divisions. The term **regular** (or **endocyclic**) is applied to echinoids in which the anus is enclosed in the apical system. In this group the jaw system (Aristotle's lantern) is well developed, pentamerous radial symmetry predominates, and the equatorial (ambital) outline is more or less circular, slightly elliptical, or regularly pentagonal. **Irregular** (or **exocyclic**) echinoids are forms having the anus outside of the apical system. In this group bilateral symmetry predominates, the ambital outline is usually far from circular, and the Aristotle's lantern is absent in many.

The skeleton is composed of thousands of separate skeletal elements, more than 3,000 in *Goniocidaris,* of many different sizes and shapes. Each piece is crystallographically a single crystal. Some have elaborate microscopic detail. The plates of the test, elements of the lantern, and the external appendages account for most of the skeletal parts, but some occur in internal organs as spicules and other isolated elements. All skeletal elements except teeth and spicules have the typical echinodermal meshwork, or **stereom,** through which the living tissue or **stroma** ramifies. The principal structure (**test**) in the form of a shell is composed of ten meridional areas, each including one or more meridional columns of plates. The five alternating areas with plates perforated by pores for passage of the primary tube feet of the water-vascular system are the **ambulacra**; the other five are the **interambulacra.** In post-Paleozoic genera two columns of plates normally are present in each area, but many Paleozoic genera (e.g., *Archaeocidaris, Melonechinus*) have a larger number and are termed **pluriserial.**

The mouth is usually on the lower surface of the test but in some genera is in an anterior position. The lower surface is usually termed the **oral** surface and the opposite side, with the apical system, the **aboral** surface. The greatest circumference of the test in a horizontal sense is termed the **ambitus.** Plates on the ambitus, which are usually the largest ones of the test in regular echinoids, are termed **ambital plates.** New plates are always formed on the aboral surface, at the head of each area. In contrast to all other living echinoderms except the Holothuroidea, the radial water vessel is internal to the test in all except a few early Paleozoic genera.

The sexes are normally separate and reproduction is always sexual. The regenerative powers are strong, new spines replacing those lost by injury or predation and damaged tests being repaired or lost sections healed over.

Symmetry is normally pentameral but rare aberrant individuals may be either completely or partially abnormal (affecting only some systems), with trimerous, tetramerous, and hexamerous types known. R. T. JACKSON (1927) and others have made extensive studies of these variations.

No echinoid is known to live in fresh water or water of very low salinity. Most living species occur in the littoral and sublittoral zones, but a species of *Pourtalesia* has been dredged from a depth of 7,200 m. Various other abyssal and numerous bathyal species have been found.

ANATOMY

By J. Wyatt Durham

[University of California, Berkeley]

The anatomy of the Echinoidea, as enterocoelous coelomate organisms with a mesodermal skeleton and several distinct organ systems, is complex. Because echinoids are readily available, of moderate size, and relatively easy to manipulate, they have been the subjects of much experimental work. Nevertheless, details and functions of some systems, such as the hemal, are still poorly known.

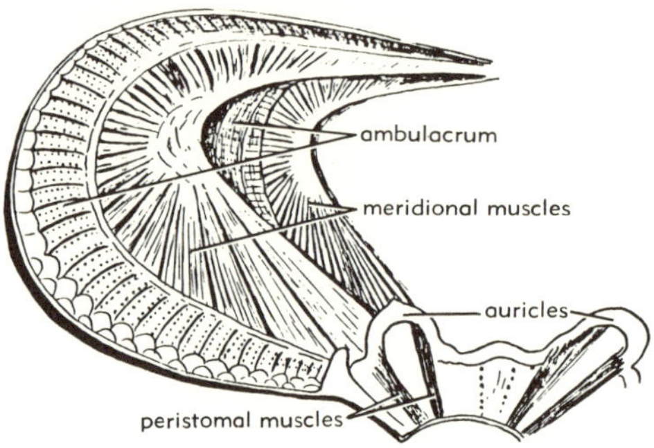

Fig. 152. Meridional muscles of *Asthenosoma* (from 4, after the Sarasins).

The anatomy of echinoids has been considered at length in the modern works of Hyman (1955) and Cuénot (1948) and in a more condensed manner by Nichols (1962) and Ailsa Clark (1962). The material presented here has been condensed largely from these publications and is intended to serve as a background for understanding and interpreting the skeletal remains found in the fossil record. Those interested in more detailed information should consult the above-cited references.

BODY WALL

The body wall of the Echinoidea consists of an external epidermis (**ectoderm**),

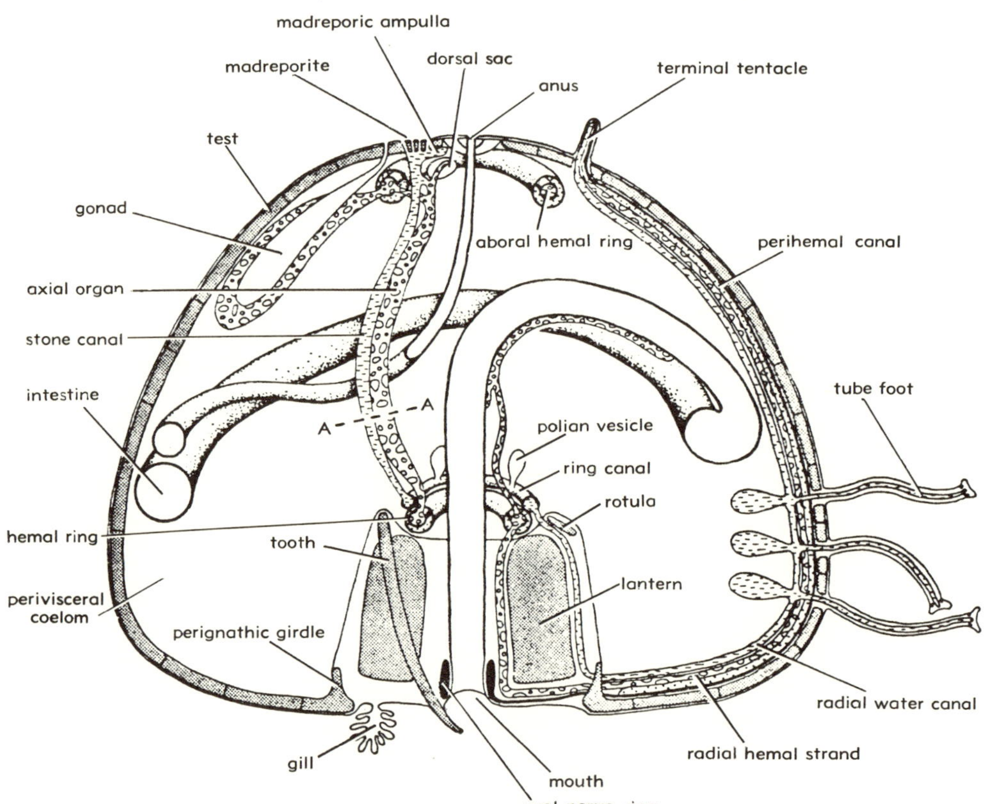

Fig. 153. Diagrammatic vertical section of regular echinoid, based on *Echinus* (5).

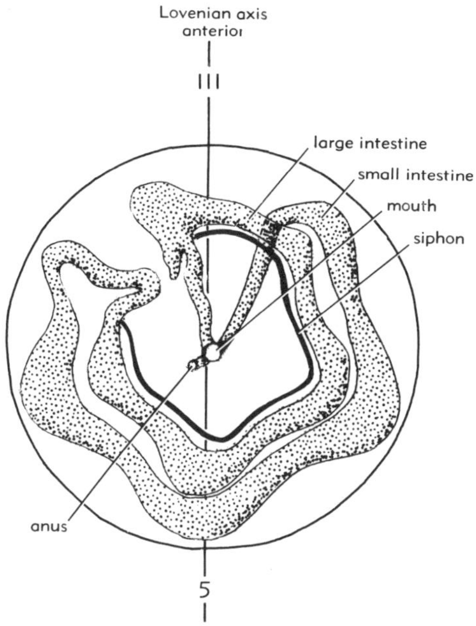

FIG. 154. Intestinal tract of *Echinus* (semischematic), oral view (modified after 3).

a middle dermis (**mesoderm**), and a coelomic lining (**endoderm**). The major portion of the various organ systems is internal to the body wall, but extensions of these systems, such as the tube feet or podia of the water-vascular system, may pass through the wall and be in contact with the external medium. Various appendages, all covered with epidermis, are attached to the body wall. These include the spines (radioles), pedicellariae, spheridia, and gills.

The epidermis is composed of a single layer of cells, and appears to be ciliated throughout, except near the outer ends of mature spines. It is usually said to be provided with a **cuticle.** Although the mesoderm is largely occupied by the calcareous plates of the test, a complex and extensive nerve plexus is present external to the plates and beneath the epidermis. Around the base of the spines and pedicellariae are two sets of muscles and a nerve ring (see Fig. 190). The outer muscle serves to manipulate movements of the spine, whereas the inner (cog muscle) holds the spine rigid if it is touched. When the tubercle supporting a spine is perforate, an elastic ligament extends from the base of the spine into the pit. The attachment of the pedicellariae is similar to that of the spines, but the muscles for the spheridia seem to be simpler in their organization. Branches from the nerve ring extend up the spines and pedicellariae.

The interstices of the plates are filled with connective tissue and stellate cells. In living echinoids no muscles are present in the body wall (the vertical muscles of the echinothurioids not forming part of the body wall proper) except for those around the base of the spines and pedicellariae. Internally, the body wall is covered by a flattened, flagellated epithelium which forms the coelomic lining. In the Echinothurioidea the imbricating plates of the test are controlled by a complex set of meridional muscles (Fig. 152) that protrude into the coelom. Peristomial muscles are also present in this group, extending along the inner surface of the peristomial plates and attaching to the auricles.

BODY CAVITY

The body cavity enclosed by the test of echinoids is divisible into several differently named spaces or hollows (coelomata). The main or **perivisceral coelom** includes most of the interior of the test (Fig. 153). The lantern and immediately adjacent organs of echinoids with external gills are enclosed within the **peripharyngeal coelom** (Fig. 155). In addition, in regular echinoids a small **perianal coelom** surrounds the anus, and is in turn enclosed in a slightly larger **periproctal coelom.** Another ring-shaped sinus, the aboral or **genital coelom,** occurs on the inner surface of the plates of the apical system and encloses the aboral hemal ring and the hemal network supplying the gonads. These aboral coelomic cavities may be fused into a single aboral sinus in irregular echinoids.

The coelomic spaces are filled with a fluid similar to sea water, the coelomic membranes being more or less permeable. The coelomic fluid contains various types of wandering coelomocytes which can move freely throughout the body tissues and organs. Experiments show that the coelomocytes originate from the mesoderm of the body wall. No special excretory organs are present in echinoderms, and it is apparent that the coelomocytes must play a major role in getting rid of waste materials.

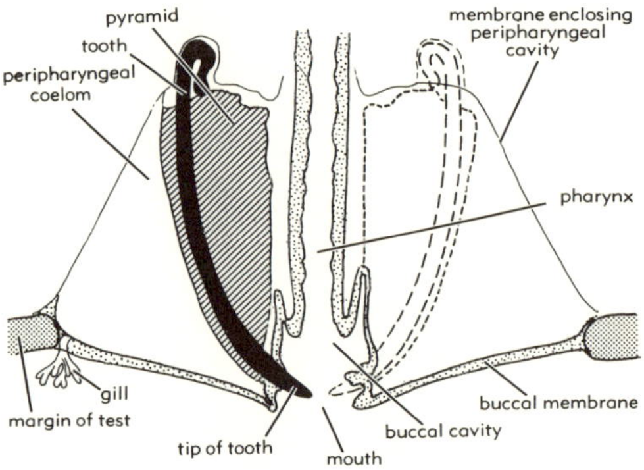

FIG. 155. Vertical section through oral area of *Paracentrotus*, showing peripharyngeal coelom and relationship of teeth to buccal cavity (modified after 3).

The **digestive tract** is prominent (Fig. 153) within the body cavity, extending from the mouth in the middle of the buccal membrane to the anus in the center of the apical system (in regular echinoids). The intestine is twisted in a loop (Fig. 154), being folded back alongside itself for part of the distance and is usually suspended by mesenteries from the interior of the test. The loop is twisted in a counterclockwise direction when viewed from the aboral side. In echinoids provided with a lantern, a small buccal cavity is present just inside the mouth and free ends of the teeth project into it (Fig. 155). A pharynx extends from the buccal cavity through the axis of Aristotle's lantern and then merges immediately into the esophagus. The esophagus then descends adorally and leads into the large intestine. This intestine continues around counterclockwise to the bend in the tract where the small intestine starts and then loops back, eventually leading to the anus. If a lantern is absent (Fig. 156), there is no buccal cavity and the esophagus connects directly to the mouth. A siphon for the passage of water to the posterior end of the large intestine branches off the main tract near the distal end of the esophagus.

The peripharyngeal cavity encloses the complex masticatory apparatus known as **Aristotle's lantern** (see "Morphology" section for description, p. U243) suspended within the test by a series of muscles (Fig. 157), both lantern and muscles being surrounded by a coelomic membrane. The lumen of the external gills (when present) is connected with the peripharyngeal cavity. The compasses and their muscles do not function in mastication but serve in respiration, expanding and contracting the peripharyngeal cavity and thus forcing fluid in and out of the gills. The absence of the compass in clypeasteroids is correlated with the absence of external gills. Stewart's organs, present in cidaroids and echinothurioids, are connected to the peripharyngeal cavity.

WATER-VASCULAR SYSTEM

The ring canal of the **water-vascular system** encircles (Fig. 153) the digestive tract just above the lantern and the hemal ring. The stone canal connects it to the hydropores in the madreporite and is closely associated with the axial organ. The polian vesicles arise from the ring canal in an interradial position. The radial water canals branch from the ring canal in perradial positions, run beneath the rotules and descend the sides of the lantern outside the interpyramidal muscles. The radial canal then runs between the auricles and follows up the interior of the test in perradial position to the ocular pore, passing through the latter and ending as the terminal or ocular tentacle (the so-called "eye"). The primary

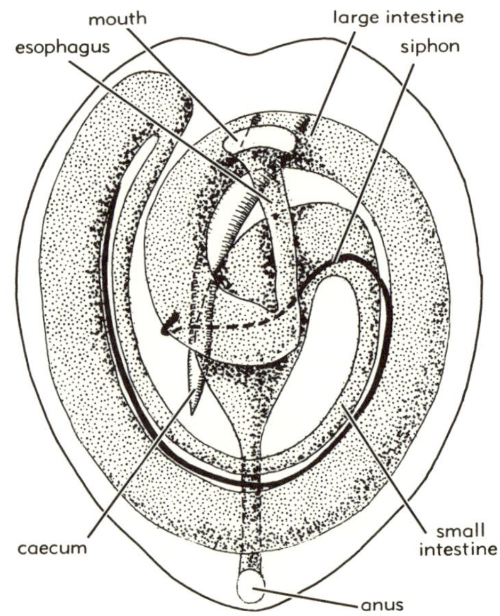

FIG. 156. Intestinal tract of *Meoma* (semischematic), oral view (modified after 4).

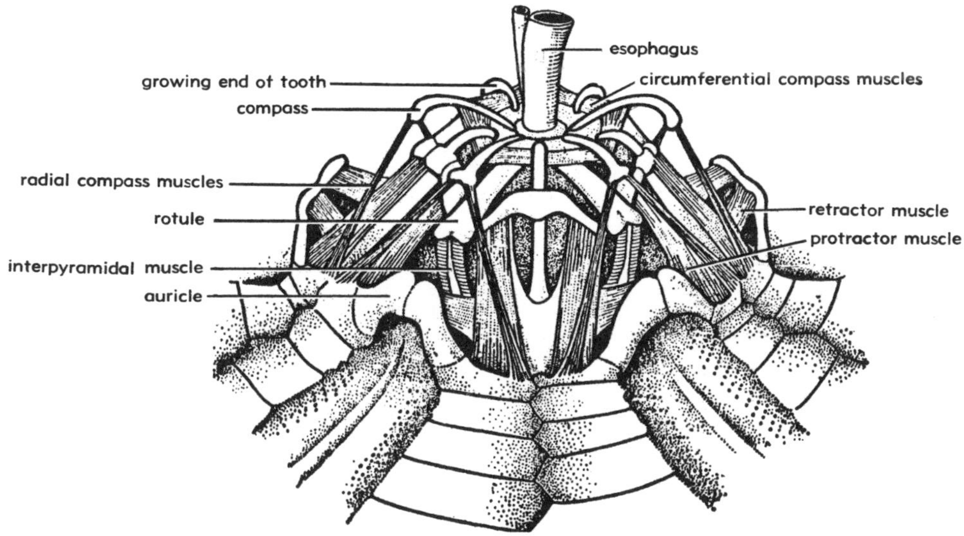

FIG. 157. Oblique view of muscles and lantern of *Echinus esculentus* (3).

podia of the late pluteus and early larval metamorphosis become the terminal tentacle of the adult. Proximally from the auricles the canal gives off an unpaired branch to the buccal tube feet and distally there are alternating branches to each tube foot and its ampulla (Fig. 158). There is a valve just in front of the ampulla. Two canals pass through the test from the ampulla and unite to form the single tube foot or podium. The relationship of the canals leading to the accessory tube feet in the clypeasteroids is uncertain, but it seems probable that they are secondary branches from the primary branches.

In regular echinoids the ambulacral tube feet are undifferentiated, being of the "suckered type," although the suckers may not be present on the younger tube feet near the apical disc. The suckered tube feet are complex (Fig. 159), for the sucker disc bears calcareous strengthening structures, as well as muscles, a sensory ring, and mucus glands. These include a **frame** and (distal to it) the **rosette.** Calcareous spicules and other supporting structures may also be present (Fig. 160-162).

In irregular echinoids the tube feet become specialized for various functions. The primary tube feet within the petals may be bladelike or pinnate and serve for respiration (Fig. 160,*2a*; 161,*4-6*). Adoral tube feet are used for feeding and chemoreceptive purposes (Fig. 160,*2c*; 161,*1,2*), whereas tube feet within the subanal and internal fascioles (Fig. 160,*2b*; 161,*3*) are modified to serve in constructing and maintaining sanitary and respiratory tubes. These latter types, like those of some regular echinoids, may be very extensile, sometimes extending for a distance of as great as five

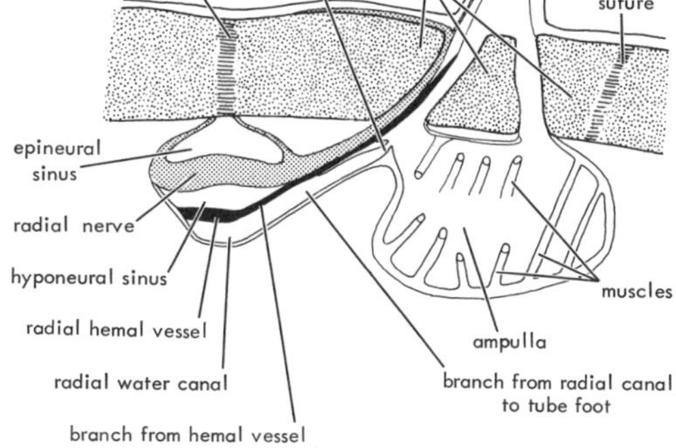

FIG. 158. Transverse section of ambulacrum and tube foot of regular echinoid (modified after 3 and 4).

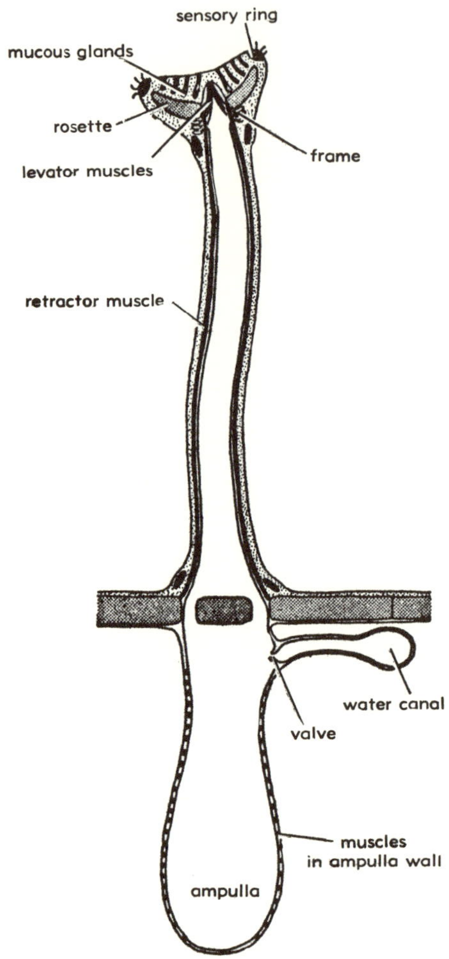

FIG. 159. Diagrammatic longitudinal section of a suckered tube foot (5).

times the diameter of the test. Still other types of tube feet of undetermined function (e.g., umbrella tube feet of *Micropyga*, Fig. 162) are known.

HEMAL SYSTEM

The hemal system of echinoids is closely associated with the water-vascular system. A hemal ring rests on top of the lantern, with the ring canal just above it (Fig. 153). A radial vessel passes down inside the lantern between it and the esophagus, then radially along the inner surface of the buccal membrane to the radial water canal, and finally accompanies it (Fig. 158) in an immediately external position with branches to the tube feet. In the interradial position a branch from the hemal ring leads to the polian vesicles. An aboral hemal ring is also present, located just under the apical system. A canal from the aboral hemal ring leads to the axial organ, where it branches into a highly anastomosing network of small vessels enclosing the organ. These rejoin at the oral end and then connect to the oral hemal ring. Branches from the aboral hemal ring connect with the gonads. A highly complex system of hemal vessels with many fine dendritic ramifications branches off the oral hemal ring and accompanies the intestinal tract. Some uncertainty exists with respect to the status of the radial hyponeural sinuses (Fig. 158). NICHOLS (5) and others consider it to form a separate circulatory system, the **perihemal** system. HYMAN (4) interprets it as a coelomic canal that cushions the radial nerve and perhaps supplies it with nutrition.

The function of the axial organ (or axial gland) has been in dispute, but recently BOOLOTIAN & CAMPBELL (1) have shown by time lapse photography that it pulses several times per minute and is a "primitive heart." Coelomic fluids are moved from

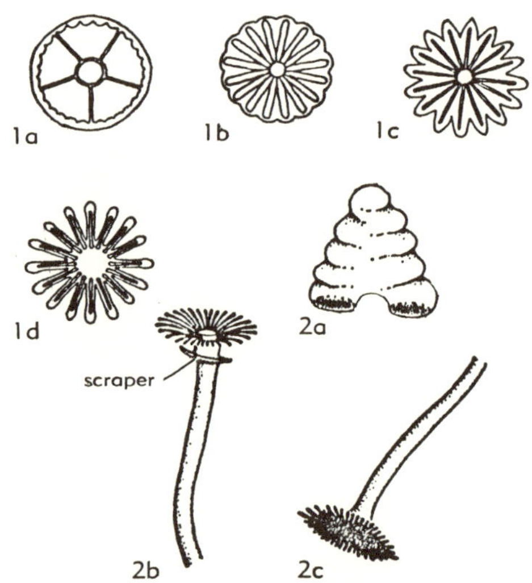

FIG. 160. Top and lateral views of tube foot.——*1.* Plan views of top of suckered tube feet, showing calcareous rosette; *1a, Echinus; 1b, Brissopsis; 1c, Schizaster*; *1d, Echinocardium*.——*2.* Lateral views of terminal disc of tube feet; *2a,* respiratory tube foot of spatangoid; *2b,* funnel-building tube foot of spatangoid; *2c,* feeding tube foot of spatangoid (5).

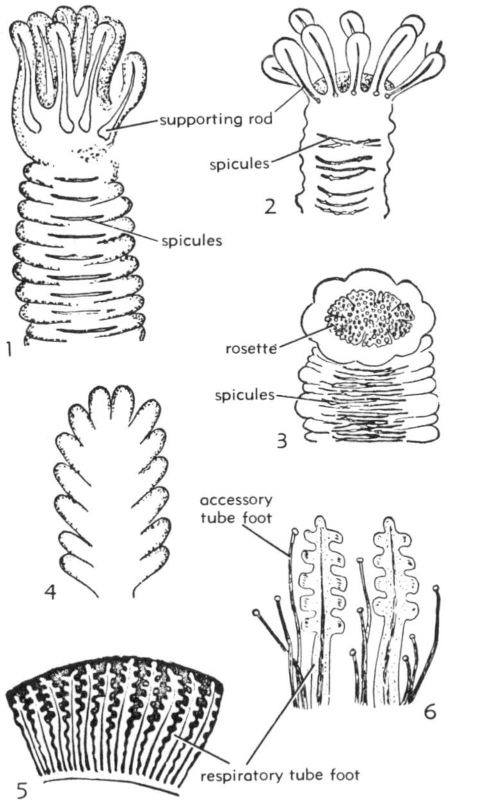

FIG. 161. Tube feet.——*1.* Frontal tube foot of *Echinocardium.*——*2.* Penicillate tube foot of *Palaeotropus.*——*3.* Subanal tube foot of *Palaeotropus.*——*4.* Respiratory tube foot from petal of *Spatangus.*——*5.* Plan view of respiratory tube feet in petal of *Clypeaster.*——*6.* Lateral view of respiratory tube feet and associated accessory tube feet in petal of *Clypeaster* (from 4, after various authors).

the perivisceral cavity into and throughout the hemal system. These authors also observed regular pulsations of the stone canal and verified a direct connection between it and the axial gland. They consider that the axial gland and pulsating stone canal are probably responsible for moving fluids throughout the water-vascular system also.

NERVOUS SYSTEM

The nervous system is external to the hemal system in position (Fig. 158), being separated from it in the radii by the hyponeural sinuses and enclosed externally by the epineural sinus and the body wall. The circumoral nerve ring surrounds the buccal cavity or lower end of the pharynx adjacent to the mouth and inside the lantern. The radial nerves follow the radial water canal and hemal vessels, separated from the latter by the hyponeural sinus. The epineural sinus, lying between the test and the radial nerve, possibly functions as a cushion. The radial nerves give off a branch to each tube foot and other branches that after passing through the pores for the tube feet connect to the extensive subepidermal nerve plexus. Another plexus of nerves from the circumoral ring ascends the digestive tract and forms a layer beneath the epithelial lining. Other branches, in radial positions, lead from the aboral surface of the ring into the muscles of the lantern. An aboral nerve ring with branches leading to the gonads has been observed in a few regular echinoids just below the apical system.

REPRODUCTIVE SYSTEM

The reproductive system consists of gonads suspended by mesenteries from the inner surface of the interambulacra, and connected by a short gonoduct to the pore in the genital plate of the apical system. In regular echinoids there are five gonads, but in irregular echinoids they may be reduced in number to as few as two. *Schizaster*

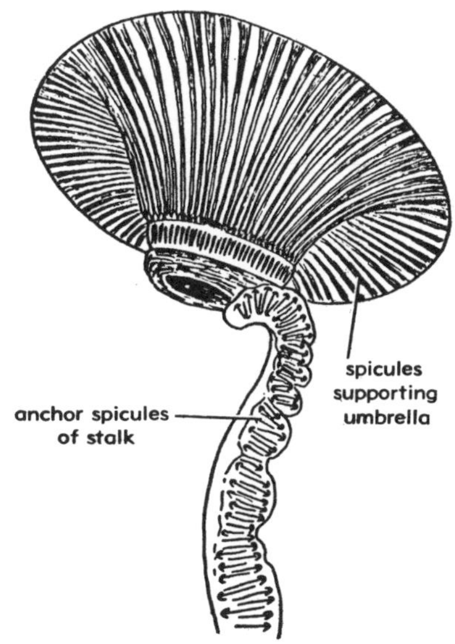

FIG. 162. Umbrella tube foot of *Micropyga* (from 4, after Döderlein).

canaliculatus has gonads in interambulacra 1 and 4 only. The gonoducts are closely associated with the aboral nerve and hemal rings, as well as the aboral coelomic sinus. When ripe, the gonads are very voluminous, but after spawning they are greatly reduced in size. Echinoids are normally dioecious but hermaphroditic individuals have been reported as abnormalities. In most echinoids the sexes cannot be distinguished externally but in some a sexual dimorphism is present. In *Psammechinus miliaris* and a number of other regular echinoids the gonoduct terminates in a short papilla in the males but not in the females. These papillae are present in both sexes of *Echinocyamus pusillus* but those of the males are longer. In some clypeasteroids the genital pore is larger in the females than in males. Eggs and sperm are discharged into the water and fertilization and development follow immediately. In temperate and boreal climates most echinoids seem to have an annual breeding cycle, but some reports indicate the probability of lunar cycles in some species in tropical areas. The spawning period may extend over several months and may take place at different times in different localities. *Lytechinus variegatus* is ripe in March and April in the West Indies, but does not spawn in the Carolinas until June and July. Some echinoids (e.g., *Lytechinus variegatus, Echinus esculentus*) are known to move inshore and aggregate prior to spawning.

Like other echinoderms, echinoids possess considerable regenerative powers. Spines, pedicellariae, and tube feet are readily replaced. Broken spines may have the lost portion replaced. At times, owing to unfavorable conditions, nearly all the spines may be shed and then with the return of a favorable environment, a complete set is regrown. Damage to the test is repaired by filling the wound with coelomocytes, followed by deposition of calcareous plates. Seemingly, however, the new plates are irregularly arranged and do not duplicate the pattern of the lost plates.

REFERENCES

Boolotian, R. A., & Campbell, J. L.

(1) 1964, *A primitive heart in the echinoid Strongylocentrotus purpuratus:* Science, v. 145, p. 173-175.

Clark, Ailsa M.

(2) 1962, *Starfishes and their relations:* 119 p., 16 pl., British Museum (Natural History) (London).

Cuénot, Lucien

(3) 1948, *Echinodermes:* in Traité de Zoologie, P.-P. Grassé (ed.), v. 11, p. 1-363, Masson et Cie. (Paris).

Hyman, Libbie H.

(4) 1955, *The Invertebrates, v. 4, Echinodermata:* 763 p., 280 text-fig., McGraw-Hill Book Co. (New York).

Nichols, David

(5) 1962, *Echinoderms:* 200 p., Hutchinson Univ. Library (London).

SKELETAL MORPHOLOGY

By R. V. Melville and J. W. Durham

[British Embassy, Paris; University of California, Berkeley]

SHAPE

EXTERNAL

Although the external aspect of an echinoid is generally characteristic of the order to which it belongs, the geometrical form alone is not commonly of greater diagnostic value. The primitive shape, as in the order Cidaroida, is spheroidal, with slightly flattened apical and oral surfaces. In some Cidaroida and also in some Paleozoic genera (e.g., *Melonechinus*) the test may be tall, with reduced adapical and adoral surfaces, and in some genera it is vertically fluted. In most other regular echinoids the test is roughly hemispherical, but in the Echinothuriidae it is more or less depressed, as in the majority of Paleozoic forms (Lepidocentridae). In one regular order (Temnopleuroida) the test is deeply sculptured. Among irregular echinoids the test may be depressed-hemispherical or tall and conical (Holectypoida), thin and flattened (Clypeasteroida), heart-shaped or elongate oval (Spatangoida), or bottle-shaped with anterior rostrum (some Holasteroida).

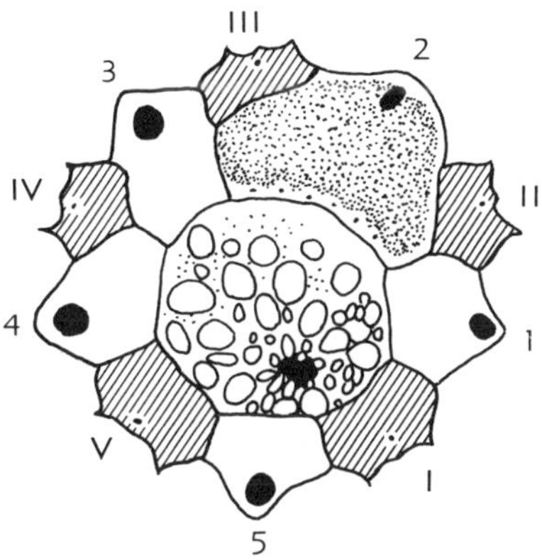

Fig. 163. Apical system of *Strongylocentrotus* showing Lovénian orientation and numerical designations of ocular plates (Roman numerals) and genital plates (Arabic numerals) (9).

INTERNAL

The inner surface of the test in regular echinoids, apart from the perignathic girdle, shows little to indicate the disposition of the soft parts. The mesenteries supporting the intestine are attached to minute projections from the inner surface of the test. In some well-preserved internal molds of fossils these may be seen as pits. In the holectypoid family Discoididae and in one clypeasteroid family (Fibulariidae) the base of the test supports ten simple internal radiating partitions. In other clypeasteroids the test is reinforced by a more or less elaborate internal skeleton of pillars and processes, from the arrangement of which it is possible to deduce the course of the intestine and associated organs and the position of the gonads.

ORIENTATION

In paleontology the Lovénian system (12) of orienting the test has been widely used, particularly because it agrees well with the anteroposterior axis of irregular forms. If an irregular echinoid is placed in the normal position of life, that is, with the apical surface uppermost, the plane of bilateral symmetry passes through the mouth, anus, and apical system. Study of living specimens shows that they move with the mouth in a forward position and with the anus posterior. Within the apical system (Fig. 163) this axis passes through the ocular plate to the left of the madreporite and through the opposite genital plate, and usually corresponds with the longest dimension of bilaterally symmetrical forms. The ocular and genital plates are numbered in relation to this axis, the former with Roman and the latter with Arabic numerals. In aboral view, the right posterior ocular is I, the right anterior II, the anterior III, the left anterior IV, and the left posterior V. The right posterior genital is 1, the right anterior (with the madreporite) 2, the left anterior 3, the left posterior 4, and the posterior 5. The anteroposterior axis thus passes through ocular III and genital 5. The ambulacral and interambulacral areas are similarly designated. Utilizing the madreporite (genital 2) as a point of reference, this same system can be applied to regular echinoids.

In the adoral view of the test, the left posterior ambulacrum is I, and the others are II, III, IV, and V in clockwise succession (Fig. 164). Left and right relationships are reversed with respect to those just noted on the aboral surface. The two columns of each area are designated by the letters *a* and *b* in the same clockwise order. Examination of the first-formed ambulacral plates (on the peristomial edge of the test) shows that five are larger than

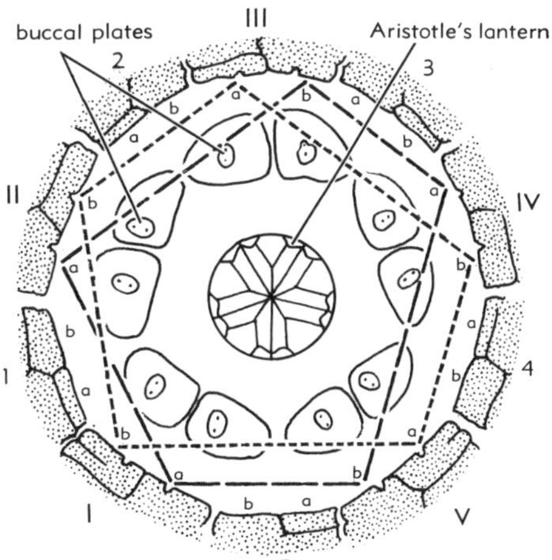

Fig. 164. Diagrammatic oral view of peristomial region of regular echinoid showing Lovénian orientation and numerical designations of ambulacral and interambulacral plates; larger plates of ambulacral tracts at peristome margin joined by long-dashed lines, and smaller plates of these tracts joined by short-dashed lines (12).

others and five smaller. The larger are (with rare inversions) those which begin the columns Ia, IIa, IIIb, IVa, and Vb, joined in Figure 2 by long-dashed lines; the smaller are Ib, IIb, IIIa, IVb, Va, joined in Figure 2 by short-dashed lines. By means of this arrangement, termed *Lovén's law,* which applies to all echinoids, it is usually possible to identify each ambulacrum and interambulacrum and thus to orient the test (Fig. 164). Partial or complete inversion of this sequence may occur in Clypeasteroida.

STRUCTURE

PLATE SYSTEMS

Plates of the echinoid test fall into four categories designated as (1) the apical sys-

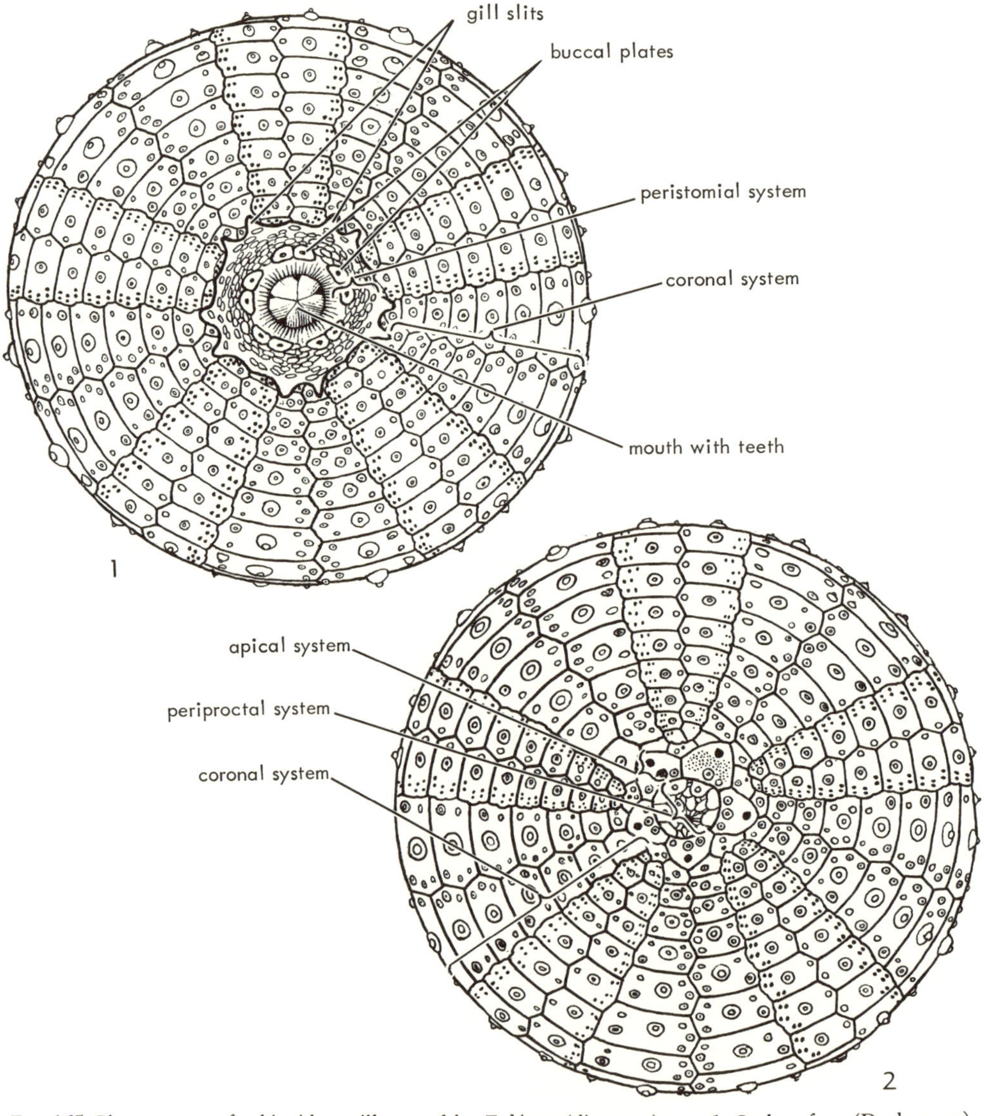

Fig. 165. Plate systems of echinoid test illustrated by *Echinus* (diagram.).——*1.* Oral surface (Durham, n). ——*2.* Aboral surface (33 after MacBride).

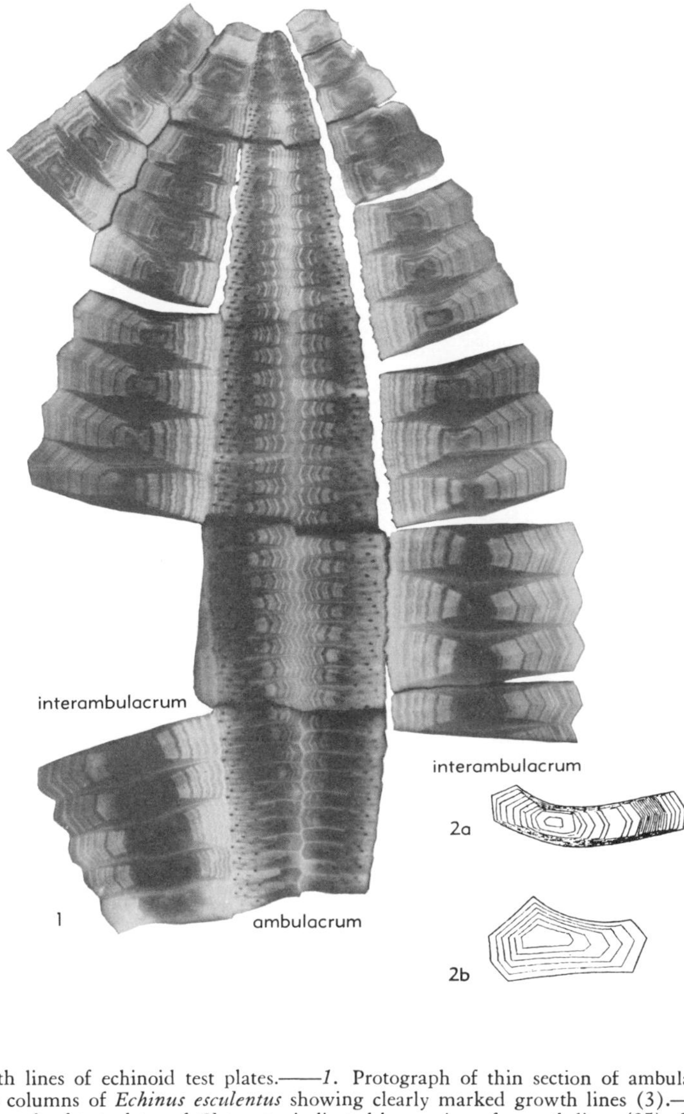

FIG. 166. Growth lines of echinoid test plates.——*1*. Protograph of thin section of ambulacral and interambulacral plate columns of *Echinus esculentus* showing clearly marked growth lines (3).——*2*. Differential peripheral growth of test plates of *Clypeaster* indicated by spacing of growth lines (35); *2a,b,* ambulacral and interambulacral plates of aboral region.

tem, (2) the coronal system, (3) the peristomial system, and (4) the periproctal system (Fig. 165). The coronal plates are formed at or near the borders of the ocular plates of the apical system, and some of the peristomial plates at least are derivative from the coronal plates.

GROWTH AND IMBRICATION

All coronal plates are formed at the edge of the apical system, new plates being inserted between the ocular plates and those previously formed. In the adult stage, therefore, the oldest plates are adjacent to the peristomial margin, and the youngest plates are at the adapical end of each column. The adoral plates of mature individuals were adapical or ambital during earlier stages of growth, and they gradually changed in position from adapical, to ambital, to adoral as new plates were added adjacent to the oculars. As long as the test increases in size, individual plates continue to grow. Within the coronal system a plate never decreases in size during normal growth. In some genera (e.g., *Echinus*) new plates continue to be added throughout life; in others growth is by enlargement of pre-existing plates after a characteristic number of plates is reached. In many echinoids, a great modification in the shape of older plates appears between their first position at the margin of the apical system and their ultimate position on the adoral surface or at the ambitus. The presence of growth lines (Fig. 166,*1*) is easily demonstrated in most echinoids by careful removal of the thin surface layer of the plates. The growth lines demonstrate, contrary to much previous speculation, that no resorption of the plates has occurred, except in a few places around pores for tube feet or in gill slits. Study of the growth lines, particularly in irregular echinoids, reveals the changes in

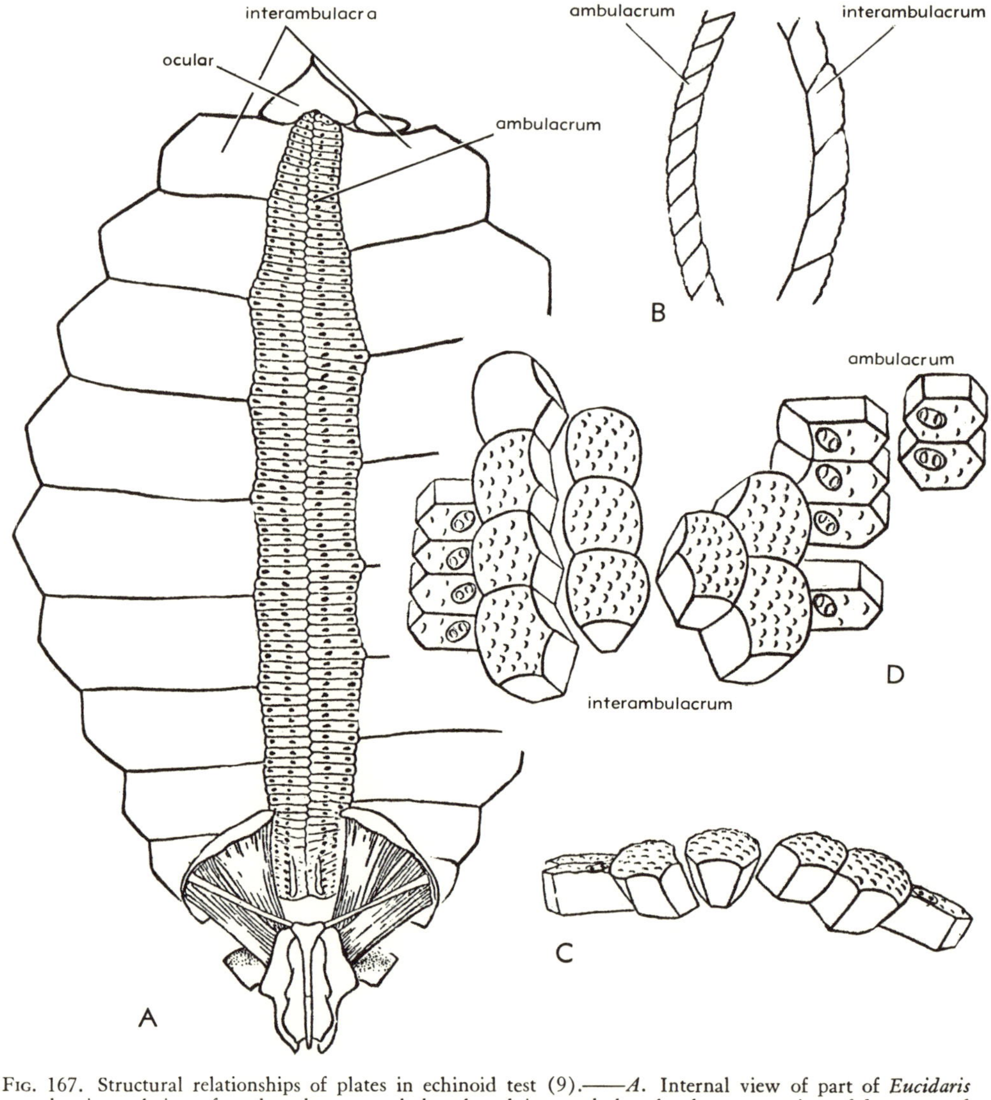

FIG. 167. Structural relationships of plates in echinoid test (9).——*A*. Internal view of part of *Eucidaris* test showing relation of ocular plate to ambulacral and interambulacral columns, portion of lantern and muscles attached to it at bottom of figure, ×2.——*B*. Meridional imbrication of ambulacral and interambulacral plates (diagram.).——*C, D*. Oblique and external views of plate columns in *Lepidesthes* showing imbrication (diagram.).

shape (Fig. 166,*2*), very marked in some species, that plates have undergone as their relative position on the test changed with growth.

Among cidaroids, characterized by narrow simple ambulacra and more or less globular tests, changes in shape with increasing age are minor. Among Paleozoic echinoids it is commonly not individual plates but entire areas that are modified by the introduction of additional columns of small unspecialized plates. In many irregulars the plates, especially of the ambulacra, are differently shaped in different parts of the same area, and the areas differ among themselves also.

Inasmuch as coronal plates originate from the ocular margins, each ocular plate stands at the head of the ambulacral columns of a single ambulacrum and the adjacent halves of the adjoining interambulacra. Thus, in echinoids having only 20 columns of plates, ocular I heads columns 5b, Ia, Ib, and 1a; ocular II heads columns 1b, IIa, IIb, and 2a, and so on (Fig. 167,*A*).

Although the test of most echinoids is rigid, with sutures usually normal to the surface, it may be flexible, with oblique sutures between overlapping plates (as in the Echinothurioida and Recent *Astropyga*). The presence of oblique sutures need not imply flexibility of the test, however. For example, in *Diadema* (Rec.), the sutures are inclined, but the test as a whole is rigid.

Imbrication may occur between plates of a single column, between one column and another, or between one area and another. When ambulacral plates overlap in meridional series, they always do so adorally, that is, each plate overlaps its neighbor toward the peristome. Meridional imbrication in interambulacral columns, on the other hand, is always adapical (Fig. 167,*B*). The adradial columns of ambulacral plates in the Palaechinidae overlap the adradial interambulacral columns. In the Echinocystitoida this is reversed (Fig. 167,*C*), with interambulacra overlapping ambulacra. In the Lepidocentridae the inner interambulacral columns from the interradial line outward bevel over outer columns toward the ambulacra (Fig. 167,*D*).

In most imbrication, the sutural face is merely an inclined plane. In *Astropyga*, however, this surface is S-shaped, which makes parting of the plates from each other possible, though movement of one plate over another is extremely restricted or impossible.

Studies of growth lines in the interambulacral plates demonstrate, despite many previous statements to the contrary, that the ambulacral plates retain a constant position with respect to the adjacent interambulacral plates and that the ambulacral columns do not migrate between the adjacent interambulacral columns.

CRYSTALLOGRAPHIC ORIENTATION

During the last few years many new data, in part unpublished, on crystallography of the echinoid test have been collected, leading to major modifications of many commonly accepted ideas about it. This section on crystallography has been supplied by David M. Raup (Johns Hopkins University) and includes previously unpublished results of his investigations.

Echinoids are typical echinoderms in that nearly all elements of the skeleton appear to be single crystals of calcite. This is true for plates of the corona, spines, spicules of the tube feet, elements of the pedicellariae, and with some exceptions (Prenant, 19), parts of the lantern. The only consistent exception is found in tubercles composed of coarse aggregates of crystals. Nissen (18) has suggested that the typical skeletal part is actually not a continuous crystal lattice but rather a bundle of submicroscopic fibers with parallel orientation. This distinction is not critical for the morphologist at present, and in the summary which follows it is assumed that individual elements are single crystallographic units.

The crystals invariably show a preferred orientation with respect to morphology, but homologous parts in different species commonly show marked differences in orientation. These differences provide a useful taxonomic tool. Laboratory determinations of crystal orientation are not difficult to make. They are most effectively obtained with a petrographic microscope and universal stage, using standard mineralogical techniques (Raup, 20). The crystal orientation is defined by position of the *c*-axis (optic axis) relative to morphology. Orien-

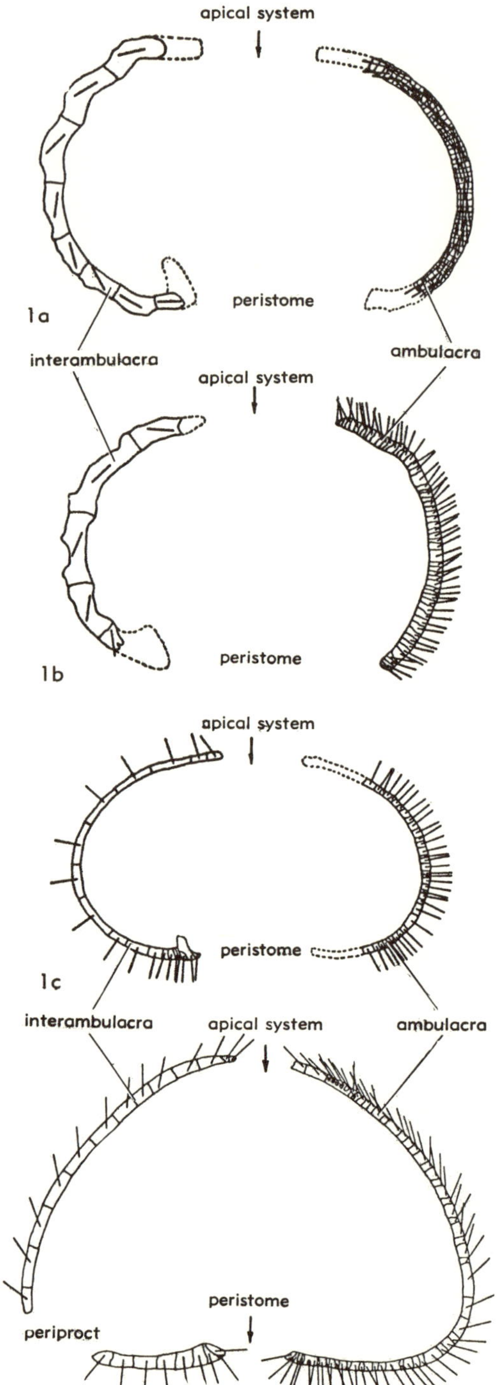

FIG. 168. Orientation of *c*-axis in coronal plates of echinoids (23).——*1.* Common patterns; *1a,* axes all tangential to surface, *Eucidaris thouarsii* (VALENCIENNES); *1b,* ambulacral plates with *c*-axes nearly normal to surface and interambulacral plates with axes tangent, *Plegiocidaris florigemma* (PHILLIPS); *1c,* axes all normal to surface, *Pedina sublaevis* AGASSIZ.——*2.* Ontogenetic variation of *c*-axis orientation, *Conulus albogalerus* LESKE.

tations may be studied in Recent forms, as well as in fossils in which the original calcite is preserved.

Although the literature of echinoid crystallography is relatively large (see RAUP, 20, for partial review) we have systematic knowledge only for the principal plates of the corona, including ocular and genital plates.

Data for ambulacral and interambulacral plates of 120 species have been published (RAUP, 23) and data for an additional 150 species await publication. In general (Fig. 168,*1*), the *c*-axes of these plates are either perpendicular or tangential to the plate surface. If tangential, they parallel the plate columns. Also, the ambulacrals usually have essentially the same crystal orientation as the interambulacrals, although precise determinations often show slight but consistent differences. In a few cidarids and arbaciids, however, the ambulacrals have perpendicular *c*-axes and the interambulacrals have tangential *c*-axes. The only known examples of truly inclined *c*-axes, rather than tangential or perpendicular ones, are found in the Conulidae and in some genera of the Arbaciidae, Toxopneustidae, and Strongylocentrotidae. In these (Fig. 168,*2*) the angle between the *c*-axis and the plate surface is ontogenetically controlled (RAUP, 21). Within limits of precision that have been accepted, calcite orientation is usually stable at the family level and closely related families tend to have the same crystallography. As a taxonomic character, crystallography of ambulacrals and interambulacrals is best applied at relatively high phylogenetic levels.

The crystallographic situation in plates of the apical system is much more complicated. Comparison of work by KIRCHNER (11), LUCAS (13), and JESIONEK-SZYMANSKA (10) reveals many contradictions and uncertainties. A few generalizations are possible, however, based on systematic study of 85 species (RAUP, unpublished) as follows.

(1) The crystal orientations in ocular plates bear no consistent relation to those in genital plates or to those in the rest of the corona, except that species which are aberrant with respect to one tend to be aberrant with respect to the others.

(2) Orientations of plates in the apical system are usually stable at the generic level, but broad spectra of variation are encountered when various genera are compared.

(3) In a given specimen, all ocular plates have similar orientations with respect to morphology. The *c*-axes have azimuths which parallel the plate columns on the corona, but the axes may be perpendicular, tangential, or inclined either toward the center of the apical system or away from it.

(4) The *c*-axes of genital plates 1, 2, and 4 are uniformly almost perpendicular to the plate surface.

(5) The *c*-axes of genital plates 3 and 5 (if present) may be perpendicular, inclined, or tangential. If they are inclined or tangential, their axes run perpendicular rather than parallel to the coronal plate columns. The *c*-axis of G-3 plunges in a counterclockwise direction and that of G-5 plunges in a clockwise direction. This pattern defines a plane of bilateral symmetry which does not coincide with any morphological symmetry axis in the adult but rather appears to reflect VON UBISCH's primordial plane of symmetry and is thus a relic of larval development.

As a tool in taxonomy and phylogeny, crystal orientation patterns in ocular and genital plates seem to have more promise than those of other plates of the corona because of the greater variation between (and to some extent within) genera. This greater variability suggests greater sensitivity to natural selection and thus may ultimately provide answers to basic and as yet unsolved questions of the functional significance of crystal orientations in echinoids. In the meantime, many more measurements on all elements of the echinoid skeleton are needed before the full potential of crystallographic approaches can be realized.

APICAL SYSTEM

The apical system is the focal point of the echinoid test, since plates comprising it are among those first formed at the time of metamorphosis and since it marks the site of origination of coronal plates. The apical system is composed of **ocular** and **genital** plates. The oculars are invariably five in number and they define the five **radii** of the test. The genitals are also originally five (except in some early Paleozoic genera) and they define the five **interradii.** The oculars generally are smaller than the genital plates and each commonly is perforated by a single small pore. The genital plates, except the madreporite, are entire in youth, but with attainment of sexual maturity they are perforated by one or more genital pores. Genital 2 is also perforated by one or more hydropores which serve as portal to the water-vascular system, and this plate is termed the **madreporite.** The madreporite commonly is larger than the other genitals. In regular echinoids the ocular and genital plates form a single or double circlet around the periproctal membrane, which is either naked or covered by periproctal plates. When the oculars are in contact with the periproct, they are described as **insert,** and if all are in contact with the periproct, the apical system is described as **monocyclic** (Fig. 169,*2*). When

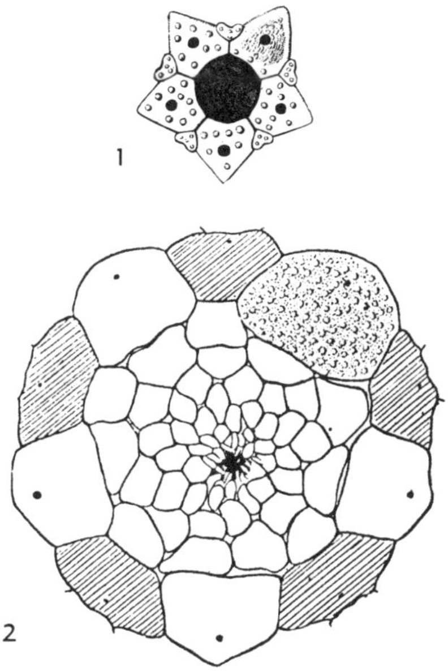

FIG. 169. Types of echinoid apical systems.——*1.* Dicyclic type, *Pseudodiadema* (27b).——*2.* Monocyclic type, *Phyllacanthus* (9).

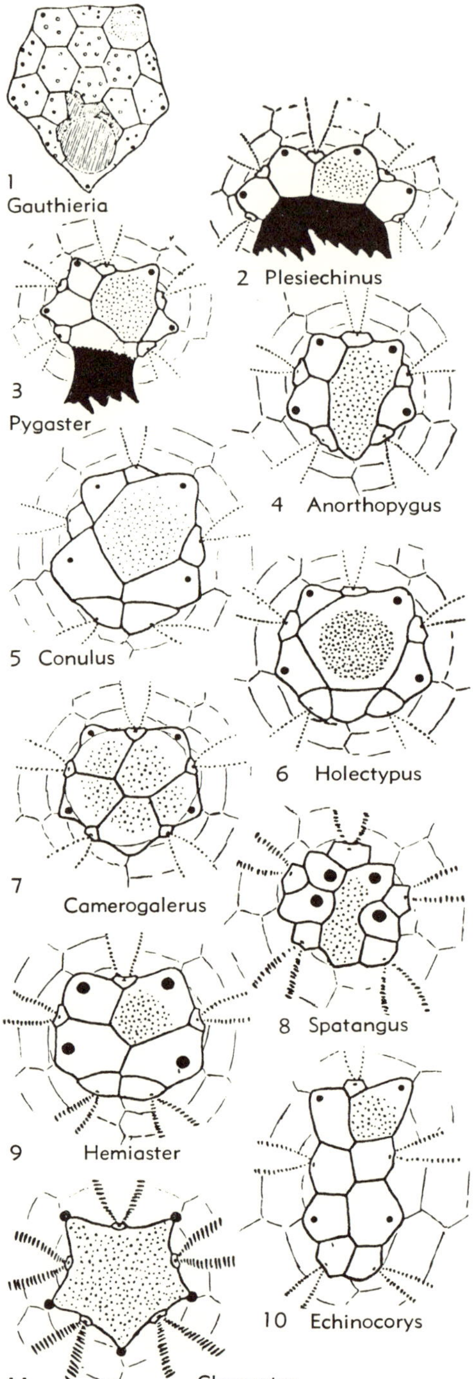

FIG. 170. Periproct and apical systems.——*1.* Posterior migration of periproct in *Gauthieria* (14). ——*2-11.* Apical systems of various genera (29b).

only genitals are in contact with the periproct, the oculars are described as **exsert.** If the oculars are all exsert (Fig. 169,*1*) the apical system is classed as **dicyclic.** If some of the oculars fail to become insert, the last to remain insert are consistently II and III. The number of exsert oculars may be a specific character of the adult, reached at an early stage of growth and, apart from slight variations, does not alter with further growth.

The order of insertion (I, V, IV, II, III), whether it affects all or only some of the oculars, emphasizes bilateral symmetry of the test about the III-5 axis already demonstrated by position of the madreporite. Oculars V and I are symmetrical about the posteror interambulacrum 5, and oculars IV and II are symmetrical about the anterior ambulacrum III.

In irregular echinoids, the posterior genital (5) may be destroyed during movement of the periproct into the posterior interambulacrum, reducing the number of genital plates to four (**tetrabasal**). Alternatively it may be retained or, as in *Echinocardium,* incorporated into the periproctal system of plates. Where genital 5 is retained in the apical system, commonly it has no genital pore and the corresponding gonad has been lost. In other instances, as among clypeasteroids, the genital plates are no longer discrete and have seemingly fused together (**monobasal**) to form a single large central plate, with hydropores present over its entire area in some forms.

Disruption of the apical system characteristic of the irregulars is not brought about abruptly but is foreshadowed in more than one regular genus (e.g., *Heterodiadema, Loriolia, Gauthieria*) (Fig. 170,*1*) which shows the system drawn out posteriorly to a point. In most irregular echinoids, the excentric periproct accompanies a marked bilateral symmetry, but in some genera (e.g., *Plesiechinus, Pygaster*) little apart from the periproct is found to distinguish them from some contemporary regular genera.

In the most primitive type of irregular apical system (*Plesiechinus,* Fig. 170,*2*), the periproct is still in contact with the five ocular and four genital plates arranged in an arc around the anterior edge of the periproct. In *Pygaster* (Fig. 170,*3*) contact with the periproct is maintained, but the oculars and genitals have come together again into a compact group, in which genital 5 is replaced by one or more imperforate plates. In *Anorthopygus* (Fig. 170,*4*) the periproct is completely removed from the apical system and the madreporite has grown backward so as to separate oculars I and V—the **ethmolysian** (ethmolytic) condition. In *Conulus* (Fig. 170,*5*) the apical system is **ethmophract**—that is, the madreporite is separated from oculars I and V, now in contact with each other, by the meeting of genitals 1 and 4, and the fifth genital is not replaced.

In *Coenholectypus* genital 5 is perforate, presumably because it was once associated with a functional gonad. In the closely related *Holectypus* the periproct is far removed, and the apical system has an imperforate plate in the position of the fifth genital (Fig. 170,*6*). In *Camerogalerus* (Fig. 170,*7*) the five genital plates are about equal in size and clearly discrete, and all are perforated by hydropores.

Other irregular echinoids have a marked longitudinal bilateral symmetry accompanied by a separation of the ambulacra into two groups, the anterior three being designated as **trivium** and the posterior two as **bivium.** Nevertheless, the apical system may remain compact and ethmolysian (Fig. 170,*8*), as in some Spatangoida; or compact and ethmophract (Fig. 170,*9*) as in other spatangoids. In other echinoids (e.g., Holasteroida) the apical system becomes elongate (Fig. 170,*10*), derived from an ethmophract condition. Genitals 1 and 4 meet in the mid-line, with oculars II and IV meeting in front of them and I and V behind them. In some (e.g., Collyritidae) the apical system is **disjunct.** The oculars of the trivium, with all four genital plates, remain together anteriorly, whereas oculars of the bivium are situated at some distance posteriorly. The intervening space is occupied by a variable number of narrow supplementary plates which separate the lateral interambulacra along the anteroposterior axis.

Where the apical system is monobasal (Fig. 170,*11*), as in the Clypeasteroida, it is composed of a single, large, central madreporite, pentagonal or stellate in outline, and five minute oculars. The genital pores may open in the angles or midway along the sides of the large central plate, or at some distance outside it in the interambulacra.

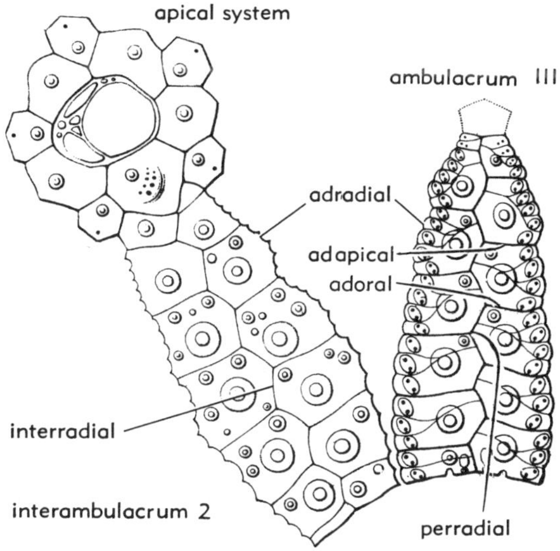

FIG. 171. Nomenclature of sutures in ambulacral and interambulacral areas, young *Strongylocentrotus,* enlarged (diagram.) (12).

CORONAL SYSTEM

General. Plates of the coronal system are arranged in ten meridional areas extending from the edge of the apical system to the edge of the peristome. Of these ten areas, the five ambulacra radiate from the ocular plates and are radial in position, whereas the five interambulacra alternate with them and are interradial in position. Except in some Paleozoic groups and a few aberrant later genera, each area is composed of two columns of alternating plates, giving 20 columns for the entire corona. Each plate is in contact with neighboring plates by means of sutures (Fig. 171). The meridional suture between the two columns of an ambulacrum is termed the **perradial** suture. The meridional suture between the two columns of an interambulacrum is the **interradial** suture. The **adradial** suture is the meridional suture separating an ambulacrum from an interambulacrum. The transverse (horizontal) sutures of the individual plates are distinguished as **adapical** (above,

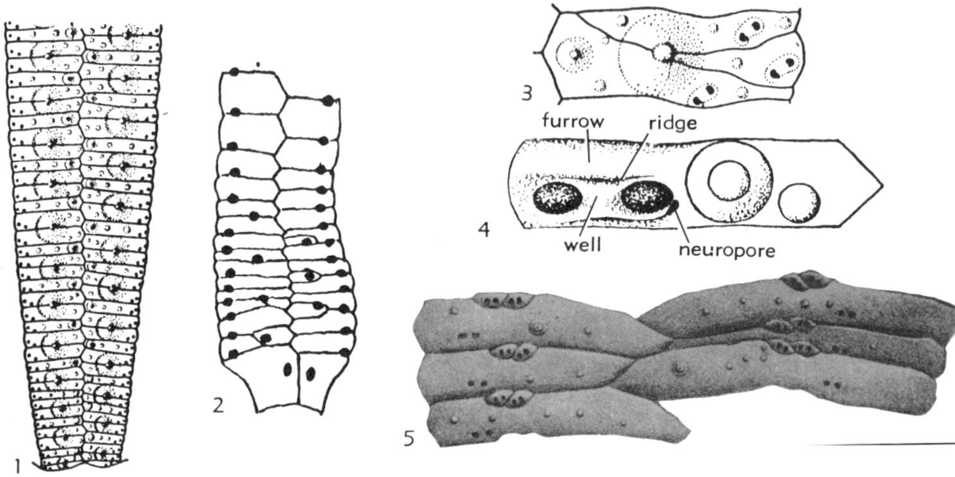

FIG. 172. Morphological features of ambulacral plates.——*1*. Adoral region of ambulacrum of *Orthopsis* showing primary plates (6).——*2*. Occluded plates in phyllode of *Echinanthus* (14).——*3*. Triads of *Echinus* with demiplate (6).——*4*. Cidarid plate with pore pair and neuropore (14).——*5*. Included plates in ambulacrum of *Echinothuria* (14 after Wright).

toward the apical system) and **adoral** (below, toward the peristome). The sutures interlocking with plates of the contiguous column are designated **adapical-transverse** and **adoral-transverse.**

All coronal plates have an external ornament of **tubercles** and **granules.** The larger tubercles support movable spines or **radioles,** and the smaller ones bear minute flexible organs termed **pedicellariae.**

The first-formed **primordial plates** of each area are situated in a row around the peristome in all later echinoids. Each ambulacrum begins with a pair of plates, and each interambulacrum with a single plate immediately followed by a pair. Thus, the initial transverse row includes 15 plates; each succeeding row contains 20 plates. However, in some regular echinoids the first-formed plates are resorbed during growth, so that 20 plates are found in the first (**basicoronal**) surviving row of coronal plates. In rotulinid genera one of the later interradial plates seems to have been inserted into the primordial circle so that it includes 20 plates. In some early Echinocystitoida only ambulacral plates are present on the peristomial margin.

Ambulacra. The ambulacra each consist of two or a larger even number of columns of plates extending from the margin of an ocular plate to the edge of the peristome. In most echinoids each mature plate is perforated by two pores forming a **pore pair.** Each pore pair gives passage to one **tube foot,** which is connected internally with the water-vascular system. The pores of a pair may be similar, round, oval, or pyriform openings, or one may be greatly elongated in comparison with the other. When first deposited, the primordium of a new plate is formed at the edge of the ocular and does not inclose its corresponding tube foot. However, the plate soon grows around the tube foot and shortly thereafter processes grow out from opposite sides of the pore, uniting in the middle and dividing the pore into the two pores characteristic of more mature plates. As the processes grow out, they penetrate the stalk of the foot and divide it into two parts at this point, corresponding to the resulting two pores of the pore pair. A third pore, called **neuropore** (Fig. 172,*4*), may occur near the pore pair. The neuropore serves for passage of the nerve supplying the tube foot. Each plate also bears one or more tubercles and a number of granules.

A **primary** ambulacral plate (Fig. 172,*1*) extends across the entire width of a column, from the perradial to the adradial suture. A **reduced** plate is one which has lost contact with either or both of these sutures. Reduced plates are of three types: **demiplates** (Fig. 172,*3*), which touch the adradial but not the perradial suture; **occluded** plates (Fig. 172,*2*), which touch the perradial but not the adradial suture; and **included** plates (Fig. 172,*5*), which touch neither the adradial nor the perradial suture (plates of the inner columns of the multicolumned Paleozoic genera not considered homologous with reduced plates).

The Cidaroida appear to have the simplest type of ambulacral structure. In them, the ambulacra are usually one-fifth to one-quarter the width of the interambulacra and are built throughout of simple, equal, primary plates. Each plate (Fig. 173) is divided into two parts, an outer **poriferous** area and an inner **interporiferous** area, the latter being about twice the width of the former and nearest to the perradial suture. In the poriferous zones the pore pairs are generally placed in a single meridional series (**uniserial**), but in *Diplocidaris* they are **biserial,** with the pore pair of every alternate plate slightly removed from the adradial suture. The pores of a pair are usually horizontal and equal, but not uncommonly they are oblique in the adapical region and in a few forms throughout the area. They are usually distinctly separated (**nonconjugate**), but in some species they are surrounded by a distinct wall or are associated with an intervening granule, or they may be joined by a groove (**conjugate**).

In interporiferous zones a **marginal tubercle** typically occurs close to the inner pore, the two resulting vertical columns of tubercles usually being conspicuous. A marginal tubercle consists of a low swelling crowned by a rounded boss, or **mamelon,** which is never crenulate and is perforate only in one genus *(Procidaris)*. The remainder of the interporiferous zone is ornamented by a variable number of smaller so-called **inner tubercles** and **granules.**

An ambulacral plate may be an independent **simple** unit with its own tubercle or tubercles (as in Cidaroida); or it may be a **compound** plate, built of a number of primary components or of primary elements combined with reduced plates, with parts bound together by a single tubercle which transgresses the transverse sutures of the component plates. Compound plates are of four basic types as follow.

(1) **Diadematoid** plates are composed of three primary plates, with the middle, larger one reduced in height medially but expanded perradially and adradially (Fig. 174,*1*). Such a group of three is termed a **triad.** In some forms two plates are thus associated and these are then termed **diads.**

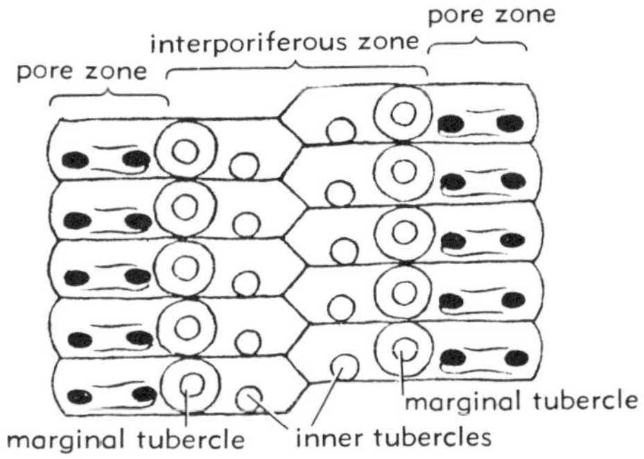

Fig. 173. Cidaroid ambulacrum (diagram.) showing zones and tubercles (14).

(2) **Arbacioid** plates are derivatives of the diadematoid type produced by further perradial expansion of the middle component, so that the adapical and adoral components are reduced to demiplates (Fig. 174,*2*). By the intercalation of additional demiplates the **phymosomatoid** type is evolved (Fig. 174,*3*).

(3) **Echinoid** plates primitively are composed of a single demiplate embraced between two primary plates of which the larger is adoral in position (Fig. 174,*4*). By the intercalation of additional demiplates this gives rise to the most advanced compound plates known (Fig. 174,*5*).

(4) **Echinothurioid** plates, known only in the Echinothuriidae and the rare *Noetlingaster,* are composed of a primary plate with two small included plates on its adoral margin. These are, apparently, derived from the outer demiplates of the arbacioid type (Fig. 174,*2*), and of each apparent pair of included plates, one belongs with the primary plate above and one with the primary plate below.

Each type of compound plate in its simplest form consists of three components, but the arbacioid and echinoid types are capable of much elaboration by incorporation of additional plates.

In many Paleozoic genera, although compound plates are not formed, the same re-

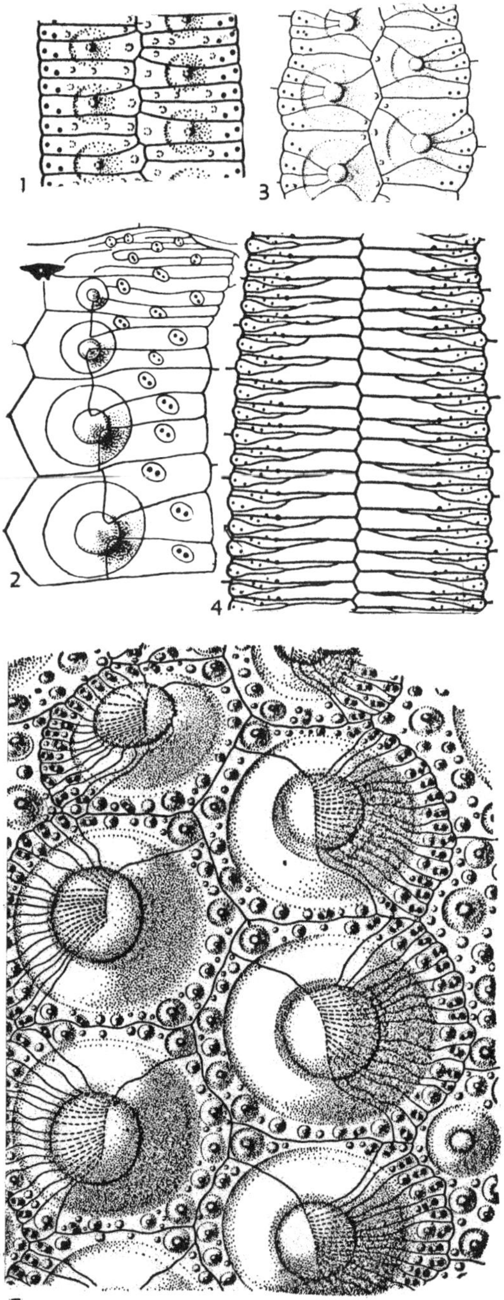

Fig. 174. Morphological features of ambulacral plates.——*1.* Diadematoid triads in *Acrosalenia* (6). ——*2.* Arbacioid triads in *Arbacia* (2).——*3.* Phymosomatoid compound plates of *Phymosoma* (6). ——*4.* Echinoid compound plates of *Echinus* (6). ——*5.* Complex echinoid compound plates of *Heterocentrotus* (6).

sult (i.e., multiplication of tube feet, functionally advantageous to the echinoid) is attained by increasing the number of columns of plates. All plates are simple, without elaborate ornament, and each bears a single pore pair. In the Palaechinidae progressive specialization is recognized from *Palaechinus,* in which the ambulacra are built of two columns of cidaroid primary plates with uniserial pore pairs, to *Melonechinus,* which displays at the ambitus as many as 12 columns of plates with multiserial pore pairs (Fig. 175). In the Echinocystitidae the number of plate columns ranges from two to 20. In the two-columned *Lepidechinus* and *Lepidocidaris* every third plate tends to be larger than others, thus foreshadowing the development of triad groups, but the plates are all simple primaries and the pore pairs are uniserial. In all Paleozoic echinoids, the ornament of ambulacral plates consists of small, imperforate tubercles and granules.

In the many-columned types, it is only the adradial columns that are complete from the apical system to the peristomial margin. The median columns become progressively shorter as the perradius is approached. Correspondingly, it is only at the ambitus that the full complement of columns is found.

In Cidaroida the ambulacra are much narrower than the interambulacra, and as many as 20 ambulacral plates may correspond to each interambulacral plate. However, with increasing complexity and the development of compound plates the two groups tend to reach an equality in size. The mechanical structure of the corona is therewith completely transformed from the cidaroid condition, in which straight or slightly sinuous areas of small, narrow ambulacrals form long lines of weakness, to the phymosomatid or heterocentrotid condition, in which all the meridional sutures run in broad zigzags between large structural units having hexagonal outlines. The ornament of such highly specialized ambulacral compound plates is scarcely distinguishable from that of the interambulacral plates.

The irregular echinoids display ambulacral specializations no less striking than those just described, but their nature is basically different. Although reduced plates

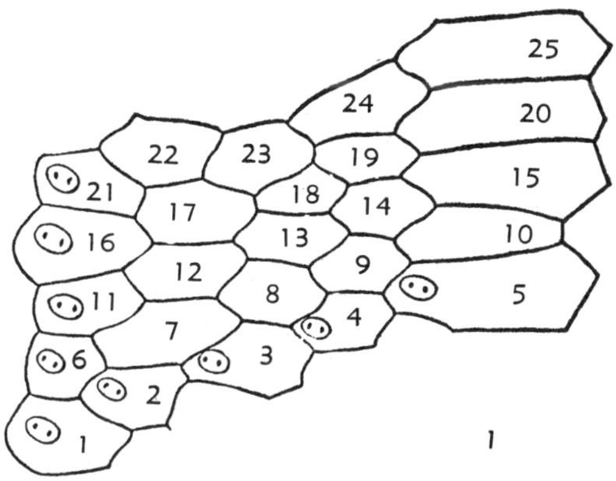

FIG. 175. Part of single ambulacral column of *Melonechinus* (pores omitted in part) [numbers on plates indicate sequence of their addition to test] (9).

occur in all groups, true compound plates are completely unknown. Within the tooth-bearing gnathostomatous irregular echinoids, **pseudocompound** groupings (Fig. 176, *1,2*) occur, but the large primary tubercle that binds the associated elements together in compound plates is lacking. In other irregular echinoids, specialization takes the form of differentiation of various parts in each area and between the areas themselves. Firstly, since the apical system, and still more the peristome, may no longer mark the polar extremities of the test, the ambulacra may not all be of the same length. The peristome is usually anterior in position, and accordingly the ambulacra of the trivium are commonly shorter than those of the bivium. The adapical, ambital, and adoral regions of each area are usually developed differently and in some spatangoids the anterior ambulacrum (III) differs from the paired ambulacra.

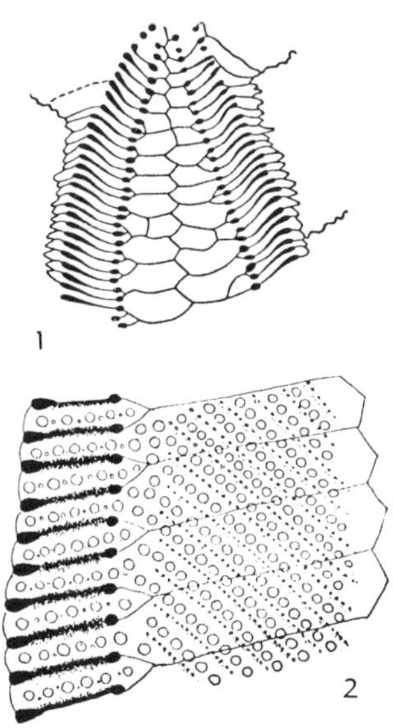

FIG. 176. Petaloid ambulacral plates.——*1*. Pseudocompound plates in petal of *Weisbordella* (26). ——*2*. Combed area in petal of *Arachnoides* (14).

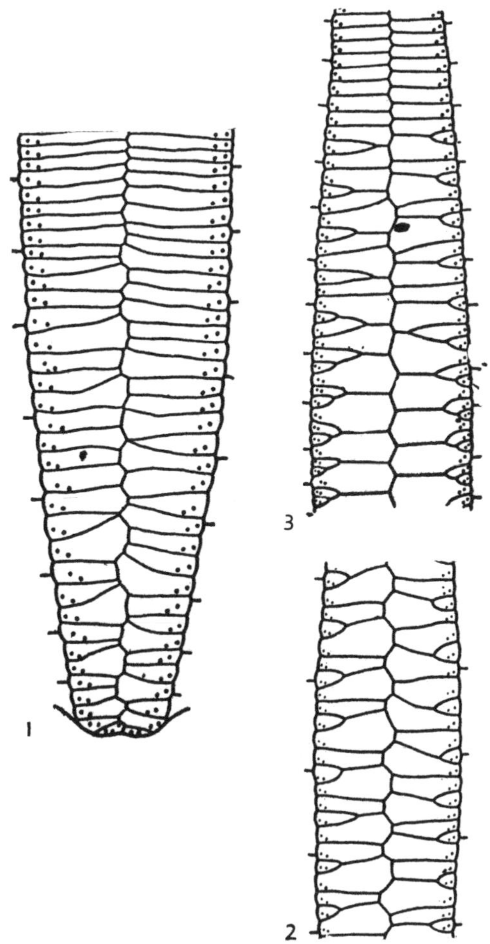

FIG. 177. Types of ambulacral structures in echinoids (6).——*1*. Plesiechinoid type in adoral half of ambulacrum of *Plesiechinus*.——*2*. Pyrinid type in medial part of ambulacrum of *Conulus*.——*3*. Discodeid type in medial part of ambulacrum of *Camerogalerus*.

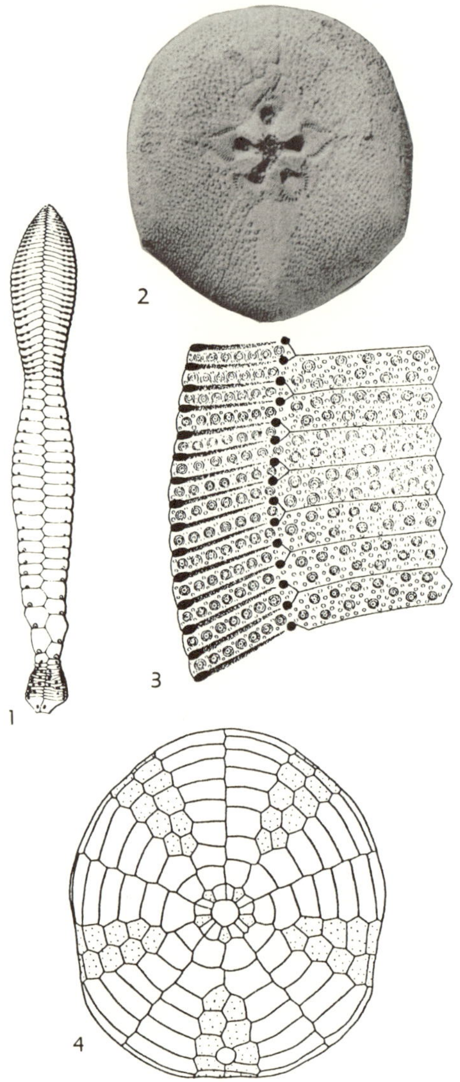

FIG. 178. Plate features of ambulacra and interambulacra in three echinoid genera.——*1. Cassidulus* ambulacrum IV (12).——*2.* Oral surface of *Hardouina* showing floscelle (30).——*3.* Demiplates in petal of *Clypeaster* (14).——*4.* Oral surface of *Clypeaster speciosus* VERRILL (interambulacra stippled), ×0.5 (26).

In the gnathostomous irregulars and Cassiduloida, reduced plates are mostly concentrated in the adoral region, but a few clypeasteroids and spatangoids also have reduced plates in the adapical region.

In the Pygasteroida two types of ambulacral structures are found. In the **plesiechinoid** type (Fig. 177,*1*) diademoid triad-grouping occurs adorally and simple, nearly cidaroid plating ambitally and adapically. Reduced plates are rarely developed and invariably these are demiplates. In the **pygasteroid** type the plates are simple and nearly equal throughout, but the tubercles are rhythmically arranged in triads.

Among the Holectypoida, the ambulacral plates are grouped in triads which become more prominent away from the apical system. The pore pairs tend to group in arcs of three adorally.

The **pyrinoid** type recalls the echinoid type (a demiplate inclosed between two primary plates), but the regular castellation of the perradial suture is noteworthy (Fig. 177,*2*). In the **discoidoid** type simple, nearly cidaroid plating occurs adapically, with pyrinoid triads at the ambitus; adorally, primary plates are separated by pairs of demiplates (Fig. 177,*3*), recalling the condition in *Echinus*.

In the Cassiduloida, regional differentiation of each area is developed (Fig. 178,*1*). There is also a well-marked separation of trivium and bivium. Adapically the ambulacra are **petaloid.** The plates of the petals are of cidaroid dimensions and are expanded horizontally compared with the narrow ambital region. The external pore of each pair may be elongated and the internal one minute, or both may be equal and tear-shaped. Where unequal, the pores are usually conjugate. The petals may contract more or less abruptly at their ambital extremities, being then described as **closed**; or they may be **open,** not markedly contracted. Feebly developed petaloid areas are described as **subpetaloid.** In the ambital region the ambulacral plates are simple, rather high primaries with minute pores. Adorally, a special structure known as the **phyllode** is developed. The areas are abruptly widened and they contract again just before the peristome is reached. Within the expanded phyllodal area the plates may be crowded so intensely that the structure is difficult to resolve. In some primitive and some possibly degenerate forms, it seems to be a derivative of the pyrinid triad, with a castellated median suture. In more advanced forms an alternation of primary and occluded plates is seen, and according to some authors the primary plates are themselves reduced to demiplates so that the phyllode appears to be built of four columns. Among the cassiduloids the phyllodes of earlier species have double pores, but in all Cenozoic and most post-Cenomanian species only single pores are present in ambulacra outside of the petaloid area. Adjacent to the peristome the narrow primordial ambulacrals are deeply sunken between the swollen primordial interambulacrals (**bourrelets**), the whole forming a rosette-like structure known as the **floscelle** (Fig. 178, *2*).

In the Clypeasteroida petals are highly developed, with widely separated, usually conjugate pores. In most the petals are built of primary plates, but in the Clypeasterina they are composed of alternating primaries and demiplates (Fig. 178,*3*), and in the Neolaganidae they may be very complex. At the ambital extremities the petals are usually closed, but rarely they remain open or are even divergent. The tube feet within the petals are respiratory, but minute accessory tube feet occur outside of the petals and extend even into the interambulacral areas. Outside the petals the ambulacra may widen considerably, so much so that the interambulacra may be interrupted on the adoral surface by the meeting of one or more rows of ambulacral plates in the interradial line (Fig. 178,*4*). At the ambitus and on the oral surface, the ambulacra are usually as wide or wider than the interambulacra.

In the Holasteroida and Spatangoida the peristome is always more or less excentric anteriorly. Two chief types of ambulacral structures are seen. In the holasteroids there is only rudimentary adapical subpetaloid development. The plates are high and the small pores are at most only slightly unequal. They may be arranged *en chevron*. Ambitally, the pores are minute, whereas around the mouth they are generally enlarged (Fig. 179,*2*).

In spatangoids the paired areas (I and V; II and IV) normally are markedly petaloid adapically (Fig. 179,*1*). The petals may be flush with the surface of the test or more or less deeply sunken, the plates of one column being larger in some than those of

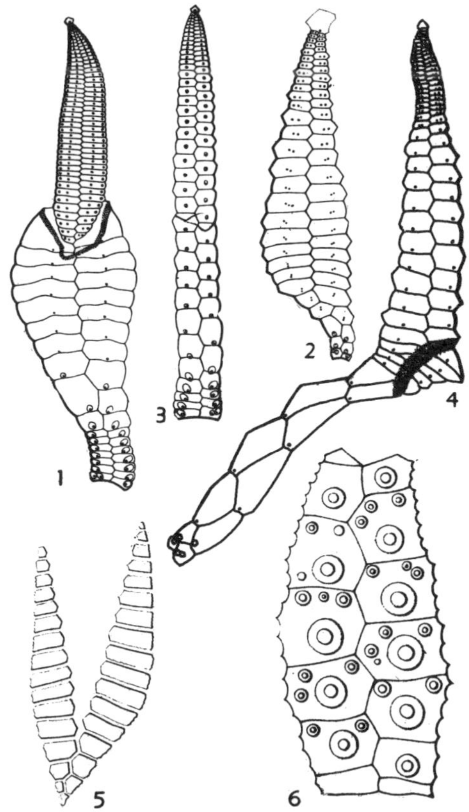

FIG. 179. Structural patterns of ambulacra and interambulacra (12).——*1*. Ambulacrum II of *Meoma* showing petaloid adapical part.——*2*. Ambulacrum II of *Echinocorys* with enlarged pores near peristome.——*3*. Ambulacrum III of *Meoma* showing nonpetaloid adapical part.——*4*. Ambulacrum V of *Micraster* with elongation of plates adorally, ×7.——*5*. Interambulacrum 3 of *Collyrites* with single primordial plate bordering peristome.——*6*. Interambulacrum 3 of *Stronglyocentrotus* with 2 postprimordial plates bordering peristome.

the other. The anterior ambulacrum may be similar to the others, but commonly it is nonpetaloid and less deeply sunken (Fig. 179,*3*). In *Echinocardium* this area is built of a complex series of primary and intercalated reduced plates of all three types, the resulting structure offering a curious similarity to that of the Paleozoic *Lovenechinus*. Demiplates may also be developed in the petals when these are flexuous. If an internal fasciole is present, the structure of parts inclosed by the fasciole may be more complex than that of external portions.

In holasteroids and spatangoids the ambu-

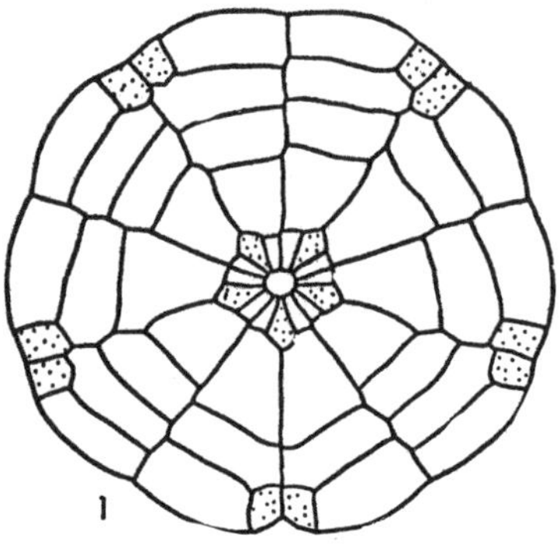

FIG. 180. Oral surface of *Arachnoides* showing discontinuous interambulacra (stippled) (26).

lacral plates of the ambital region are simple and hexagonal in form. On the adoral surface they tend to become narrow and elongate (Fig. 179,*4*). The circumoral tube feet are always large and are grouped into what has been called a phyllodal structure, though not strictly comparable with the phyllodes of Cassiduloida. The ornament of the two posterior areas tends to degenerate into a minute granulation or mammillation, without appendages.

Interambulacra. The interambulacra do not have as intimate and vital a connection with the internal organs as the ambulacra. In most echinoids they bear only the organs of defense (spines, pedicellariae), some of which also have locomotory uses. These appendages are borne on tubercles. Smaller granules without appendages also occur. In clypeasteroid echinoids, accessory tube feet may extend into interambulacral areas. In general, no fundamental difference between ambulacral and interambulacral ornamentation can be pointed out.

Typically, and always in young stages, an interambulacrum begins with a single primordial plate at the peristomial margin (Fig. 179,*5*), though in the adult stage of many regular echinoids this plate has been lost by resorption (Fig. 179,*6*). As with the ambulacra, new plates are formed at margins of the ocular plates. The interambulacra may consist of as many as 14 columns each, but in all post-Paleozoic forms (except a few aberrant genera and some Clypeasteroida) they are built of two columns of alternating plates which usually are continuous from apical system to peristome. In many-columned Paleozoic genera it is only the adradial columns which are continuous. In these, each column approaching the interradial line is progressively shorter. The maximum number of columns is found only at the ambitus.

In all noncidaroid post-Paleozoic regulars and in primitive irregular types each interambulacral plate bordering the peristome is notched at or near each adradial margin by **gill slits** (Fig. 188,*3*), which are smooth incised grooves, each outlined by a raised lip; they serve for accommodation of the external gills or branchiae.

In the Cidaroida and in post-Paleozoic regular echinoids with primitive ambulacral structure, the interambulacra are much broader than the ambulacra, while, in the meridional sense, each interambulacral plate corresponds with a number of ambulacrals. In regular echinoids with compound ambulacral plates, these approach equality in size with the interambulacral plates. In the Holectypoida and Cassiduloida, one interambulacral plate generally corresponds to two or three ambulacral plates adorally and ambitally, and to five or more lathlike plates adapically. In Clypeasteroida, when the interambulacra are broken by adoral expansion of the ambulacra (Figs. 178,*4*; 180) it is always the primordial interambulacral plates which are isolated from the following plates.

In the Holasteroida and Spatangoida, every interambulacral plate is more or less specialized in form. The adoral surface is largely taken up by the posterior interambulacrum and ambulacra I and V (the peristome being always excentric anteriorly). The primordial plate of interambulacrum 5 in primitive forms is narrow and elongate and only just reaches the peristome. In more advanced echinoids it becomes hammer-shaped and expanded into a lip, or **labrum,** which may overhang so as to shield the peristome almost completely. The succeeding plates of this interambulacrum form the **plastron** (**sternum**), bordered on each side by the **periplastronal** areas of ambulacra I and V. In more primitive types, the labrum and sternum are scarcely differentiated. In the Collyritidae, the sternum is built of alternating large pentagonal

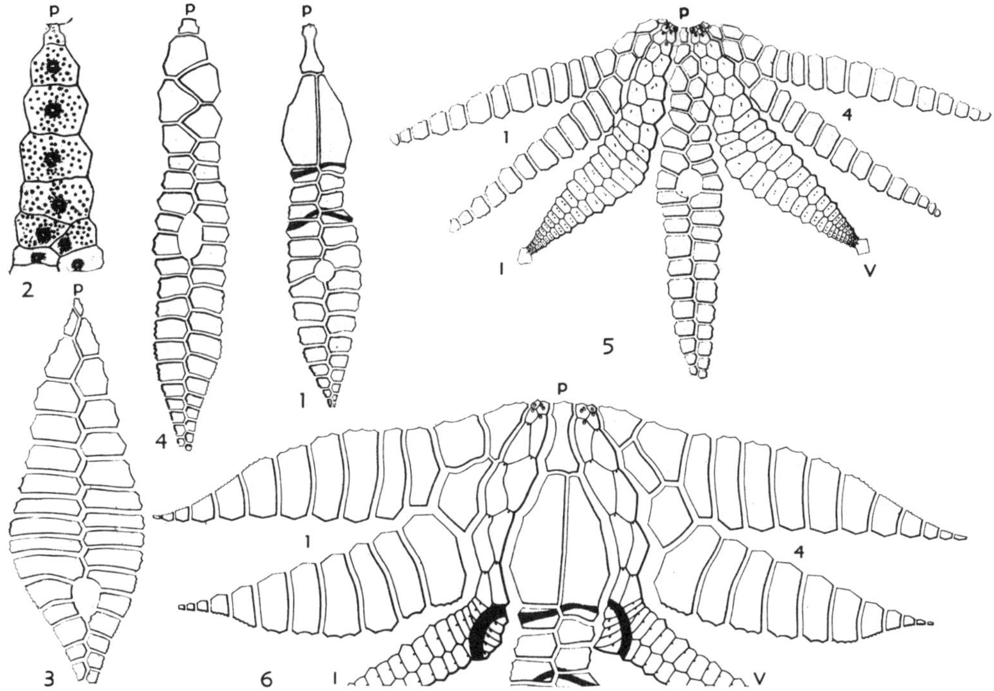

FIG. 181. Interambulacra of sternal region (*p*, peristome).——*1*. Amphisternous plastron of *Brissus* followed by episternous plates (12).——*2*. Meridosternous (and metasternous) plastron of *Cardiotaxis* (31).——*3*. Protosternous interambulacrum 5 of *Collyrites* (12).——*4*. Meridosternous (and orthosternous) interambulacrum 5 of *Holaster* (12).——*5*. Heteronomy between interambulacra 1 and 4 in *Echinocorys* as interpreted by LOVÉN (12).——*6*. Heteronomy between interambulacra 1 and 4 in *Micraster* as interpreted by LOVÉN (12).

plates (**protosternous,** Fig. 181,*3*), which in the Holasteridae (Fig. 181,*4*) come to overlap one another more and more until finally the plastron is built of a single column of large, hexagonal plates (**meridosternous,** Fig. 181,*2*).

In the Spatangoida the labrum is usually prominent. The sternum is built of the first pair of plates behind the labrum (**amphisternous**). These become very large and are separated by a straight, median interradial suture (Fig. 181,*1*). They bear an ornament of closely packed tubercles arranged *en chevron*. The next succeeding plates are termed **episternal** plates.

In the paired inferoambulacra of Spatangoida, the primordial plates are narrow and only just reach the peristome. In advanced forms these become excluded altogether from the peristome, which is thus bounded by the labrum and ten primordial ambulacral plates.

The posterolateral interambulacra of Spatangoida and some Holasteroida are not symmetrical, but area 1 has always one plate less than its vis-à-vis, area 4. This heteronomy differs in pattern between the holasteroids and spatangoids. In the Holasteridae it was interpreted by LOVÉN (1874) as being due to fusion of the first pair of plates in area 1 into a single plate (Fig. 181,*5*), and in the Spatangoida as due to the fusion of the second and third plates of column 1a (Fig. 181,*6*). However, study of the ontogeny of *Echinocardium cordatum* (GORDON, 4) shows that no fusion, but only displacement of plates, takes place in this species.

The character of the interambulacral ornamentation is of great systematic value in groups where the spines are few and differentiated into various kinds. The basic arrangement is best understood in terms of the Cidaroida.

Each interambulacral plate (Fig. 182,*2*)

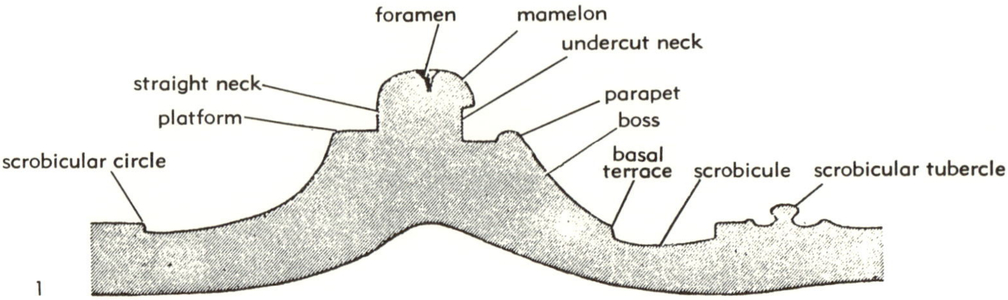

1

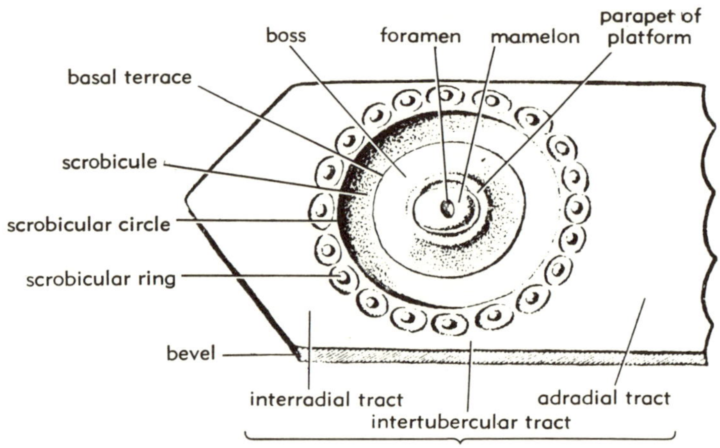

2

FIG. 182. External morphology of interambulacral plates (12).——*1,2*. Vertical section and surface of cidaroid interambulacral plate, enl. (1 is composite showing more than one type of tubercle).

of a cidaroid bears one prominent, more or less central **primary tubercle** with which the large primary spine (**radiole**) is articulated. The tubercle is situated in a smooth area termed **scrobicule** or **areole** which serves for attachment of the muscles that move the spine. The scrobicule is surrounded by a **scrobicular circle** of smaller **secondary tubercles.** The remaining area of the plate is termed the **extrascrobicular** surface. It is covered by very small **miliary tubercles** for the attachment of pedicellariae or very small spines.

The primary tubercle consists of a hemispherical **mamelon** which may be **perforate,** with a circular or elliptical pit, or **imperforate.** In plan the mamelon may be circular or transversely elliptical. The mamelon is set upon a **neck** which may be straight or undercut. The neck rises from the **boss** and stands on a **platform,** which may be either flush or impressed and surrounded by a **parapet.** The parapet (or margin of the flush platform) may be **smooth** (**noncrenulate**) or **crenulate.** The boss may pass down imperceptibly into the scrobicule or it may be limited by a vertical wall, the **basal terrace** (Fig. 182,*1*).

The scrobicule is generally impressed below the level of the extrascrobicular surface. The secondary tubercles of the scrobicular circle are mamelonate and may be perforate or imperforate, smooth or crenulate, and each of them usually has a miniature areole. The scrobicular circle of one plate may be in contact with those of adjoining plates in the same column, so that only a single row of tubercles separates the scrobicules, which are then indicated to be **contiguous**; or the intervening row of secondary tubercles may disappear causing the scrobicules to become **confluent**; or adjacent scrobicular circles may be complete and independent.

The extrascrobicular surface is covered by smaller tubercles and granules. Tubercles, which are mamelonate, may not be easily distinguished from those of the scrobicular circle. They generally diminish in size toward edges and corners of the plate and are interspersed with minute, wartlike granules which bear no appendages.

In primitive euechinoids the interambulacral ornamentation resembles that of Cidaroida, with a single large primary tubercle in a scrobicular circle. Some specialized types (e.g., *Echinus*) possess a larger number of tubercles which are more or less equal in size and arranged in regular horizontal and vertical series. The primary tubercles are usually a little larger than the others, but they can be recognized in any case by the fact that they alone form a complete meridional series, whereas the others (termed also secondary tubercles) are not present on the adapical and adoral plates. With multiplication in number and decrease in size of the tubercles, the finer structures, such as crenulation and scrobiculation, tend to disappear.

In Temnopleuroida the plates of both ambulacra and interambulacra are sculptured and pitted. In the Glyphocyphidae, these pits take the form of rounded depressions along the sutures, while a pattern of raised ridges connects the primary tubercles and may form an intricate network. In the Temnopleuridae the general surface of the plates exhibits sutural grooves and pores, as well as deeper pits and depressions, and the plates are united by dowel-like structures on the sutural faces.

In the majority of Paleozoic echinoids (except genera such as *Miocidaris* and *Archaeocidaris*), the interambulacra bear only small, simple, imperforate tubercles.

In most irregular echinoids the tubercles are numerous, usually about even in size and nearly always perforate. Only in the Holectypoida and Pygasteroida can a regular, serial arrangement be recognized.

In most Holectypoida a continuous meridional series of primary tubercles can be traced, situated a little adoral to the center of each plate. A number of plates (varying

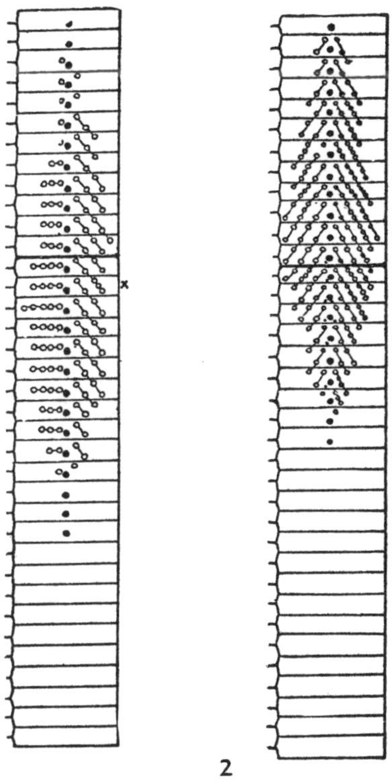

1 2

FIG. 183. Interambulacral tubercle patterns in single plate columns (diagram.) (29a).——*1. Holectypus.* ——*2. Conulus.*

with the genus) at each extremity of the area carry only the primary tubercle, but as the ambitus is approached, the interradial tract of each plate comes to bear a horizontal row of tubercles set adapically just above the primary tubercle; in the adradial tract two such rows appear, one adapical and one adoral in position, and related to each other *en échelon* (Fig. 183,*1*). In some Holectypoida both interradial and adradial tracts bear two rows of tubercles arranged *en échelon* (Fig. 183,*2*). The number of plates at the extremities which bear only the primary tubercle is progressively reduced and number and size of the tubercles tend to increase on the adoral surface while tubercles on the adapical surface are reduced. In the Discoididae and Conulidae a number of minute, sunken tubercles are scattered over the plates of the adoral region. These bear minute, club-shaped glassy radioles of uncertain homology and function.

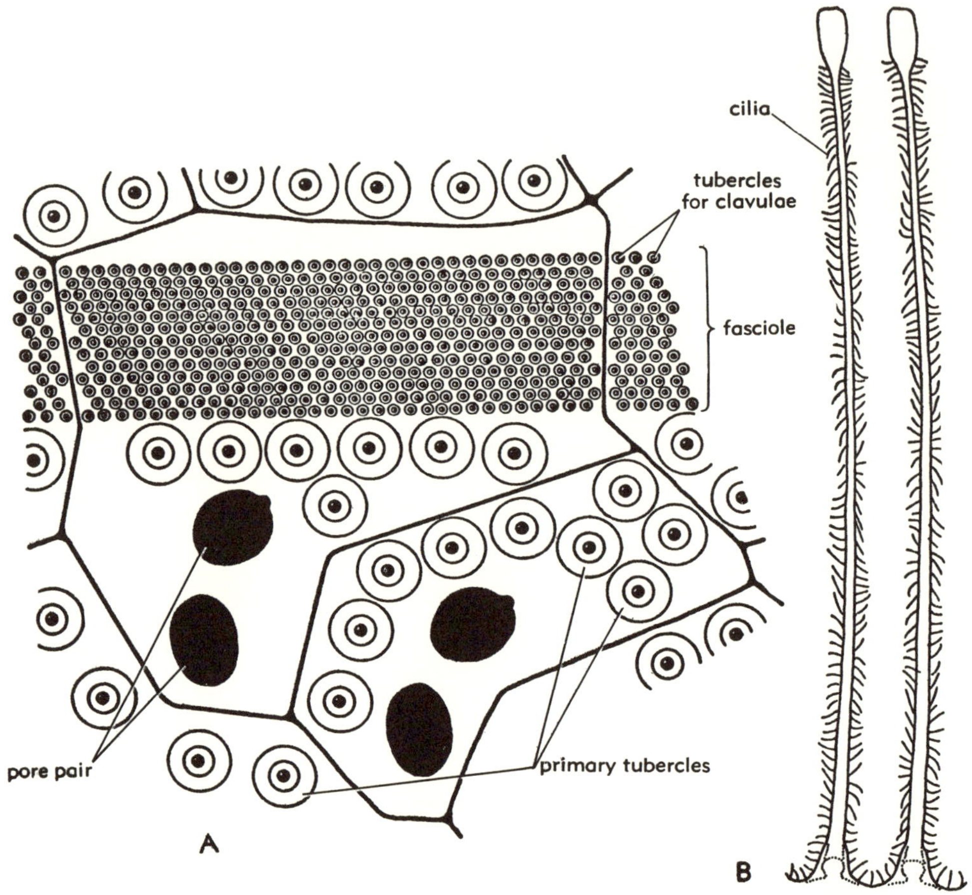

FIG. 184. Fasciole passing across ambulacrum II outside petal of *Echinocardium cordatum* (PENNANT) (16). ——*A*. Detail of fasciole, surface view, enl.——*B*. Side view of 2 clavulae, enl.

In the Cassiduloida, the tubercles are closely crowded and commonly arranged in linear patterns, though no regularity of arrangement has been recognized in some. The primordial interambulacral plates in varying degree are swollen, forming bourrelets covered with equal-sized tubercles. On the adoral surface the interambulacra are commonly somewhat swollen above the sunken ambulacra, giving a cushion-like or **pulvinate** appearance.

In the Clypeasteroida the tubercles are small and closely crowded, usually without apparent pattern. In the Holasteridae the tubercles tend to be small and generally well separated. In the Spatangoida they are small and densely crowded, with notable specializations. In some, adapical parts of the interambulacra bear a few enlarged adradially placed tubercles surrounded by smaller ones. The bilateral symmetry in this group is accompanied by a differentiation of the ornament of the adapical and adoral surfaces. On the adapical surface the tubercles may be arranged in oblique lines streaming forward and downward. On the adoral surface the base of each tubercle may be canted up so as to tilt the boss posteriorly.

Many Spatangoida and some Holasteroida bear one or more bands of small strongly ciliated and modified spines called **clavulae** which serve to create water currents. The bands of fine and dense tuberculation on these areas are termed **fascioles** (Fig. 184). These, when fully developed, are conspicuous narrow bands with fine,

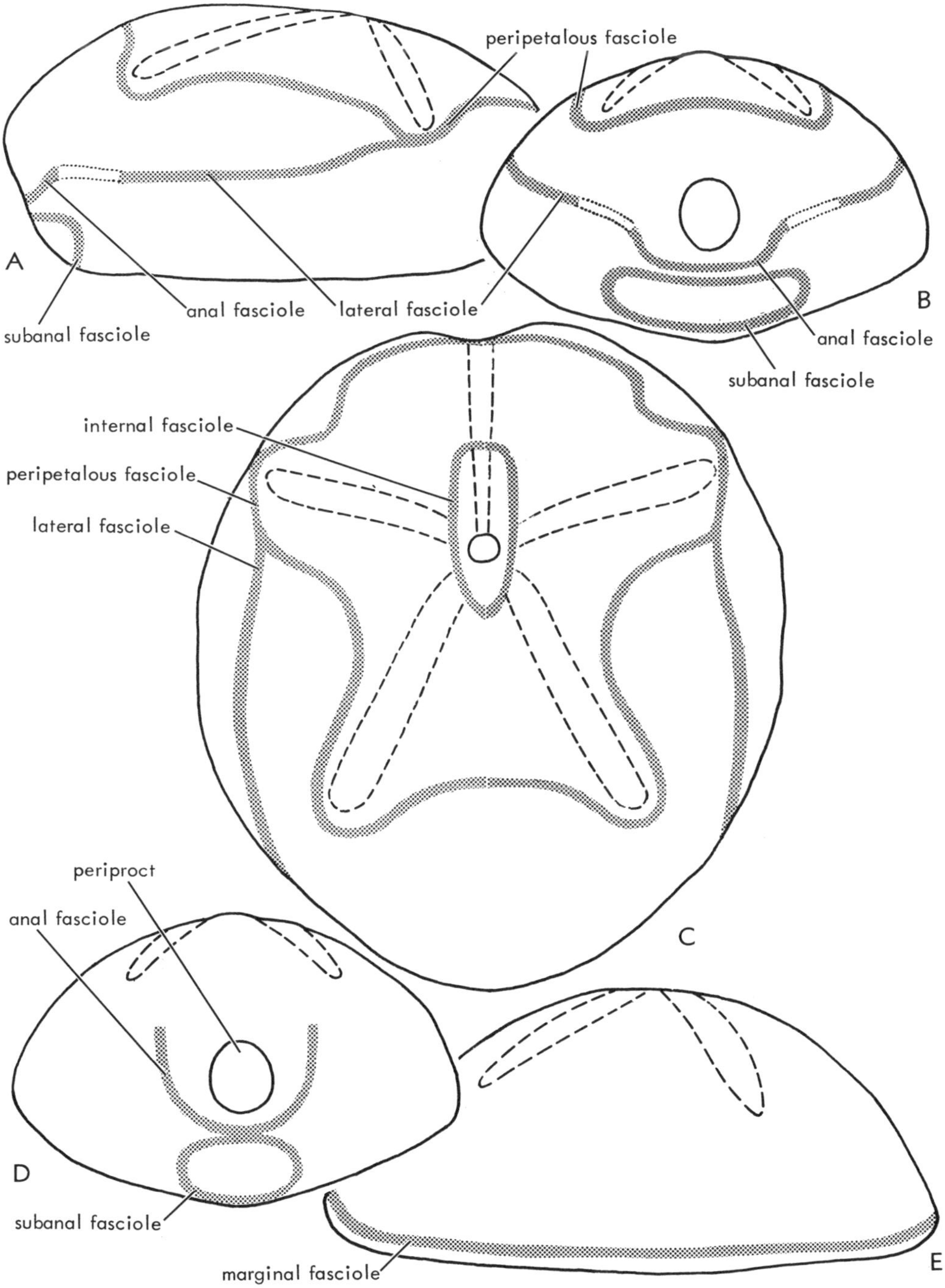

FIG. 185. Fascioles of spatangoid tests (diagram.) (Durham, n).——*A*. Side view of test showing lateral and other fascioles (termed lateroanal fasciole if lateral and anal fascioles are continuous).——*B*. Posterior view of test showing subanal and other fascioles.——*C*. Aboral view of test showing peripetalous and other fascioles.——*D*. Posterior view of test showing anal and subanal fascioles.——*E*. Side view of test showing marginal fasciole.

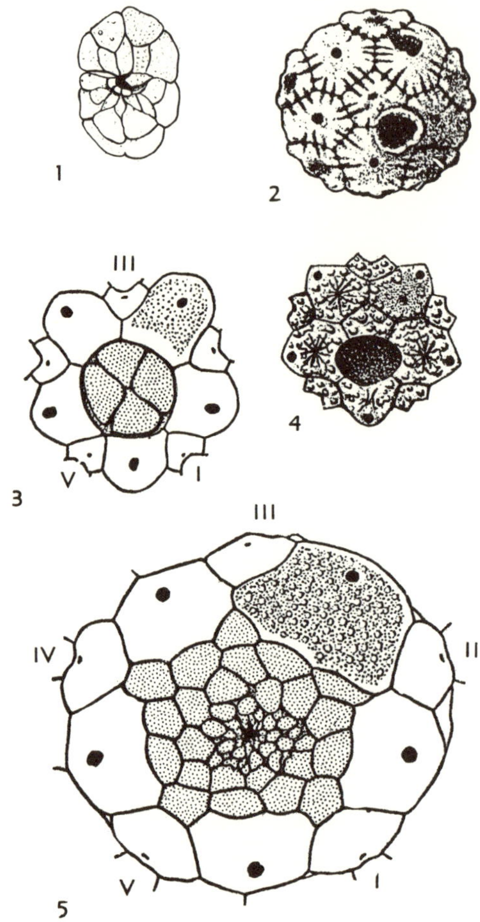

FIG. 186. Plates of periproct and apical system.——*1*. Periproctal plates of *Offaster* (27a).——*2*. Apical system of *Salenia scutigera* GRAY showing large suranal plate and periproct displaced toward ambulacrum I (14).——*3*. Apical system and periproctal plates of *Arbacia punctulata* (LAMARCK) (9).——*4*. Apical system of *Acrosalenia marcoui* COTTEAU showing large suranal plate and posteriorly displaced periproct (14).——*5*. Apical system and periproctal plates of *Eucidaris tribuloides* (LAMARCK) (9).

uniform tubercles passing across both ambulacra and interambulacra. Fascioles vary in position, but, except for a part of the subanal fasciole, are not found on the adoral surface. They inclose various parts of the remaining area of the test and are described according to position as follows.

(1) **Subanal** fasciole, encloses a roughly elliptical area beneath the periproct on the posterior face of the test (e.g., *Micraster*) (Fig. 185,*A,B,D*).

(2) **Marginal** fasciole, encircles the ambitus just above the adoral surface (e.g., *Cardiaster*) (Fig. 185,*E*).

(3) **Peripetalous** fasciole, encloses the petals and apical system (e.g., *Hemiaster*) (Fig. 185,*C*).

(4) **Internal** fasciole, encloses the inner portion of the adapical region, the petals continuing outside of it (e.g., *Echinocardium*) (Fig. 185,*C*).

(5) Paired **lateral** fascioles, run from the peripetalous fasciole to meet beneath the periproct (Fig. 185,*B,C*).

More than one type of fasciole may be found in a single species, but all five types are never found together.

PERIPROCTAL SYSTEM

The flexible membrane through which the anus opens is naked in a few echinoids, but usually it is covered by overlapping or dissociated plates. These tend to be disposed in irregular circlets or a spiral and to diminish in size toward the center (Fig. 186,*5*), but they may exhibit no recognizable pattern. In the Arbaciidae, however, the periproctal plates are nearly always four in number (Fig. 186,*3*). In Acrosaleniidae, Saleniidae, and the young of some other groups a single prominent **suranal** plate pushes the anal opening away from the center of the periproct, in Acrosaleniidae toward interambulacrum 5 and in Salenii-

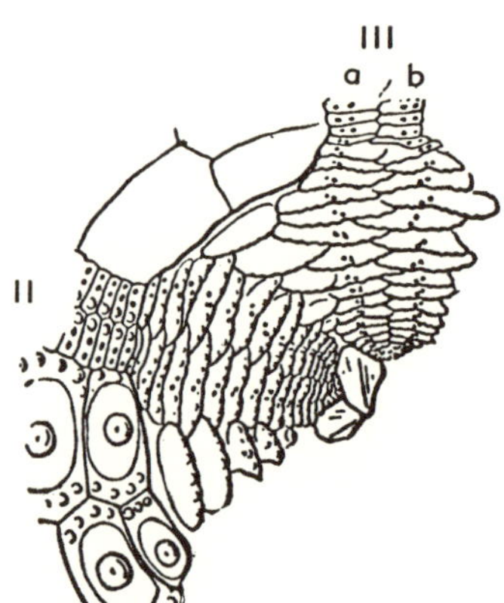

FIG. 187. Peristomial plates of *Eucidaris* (9).

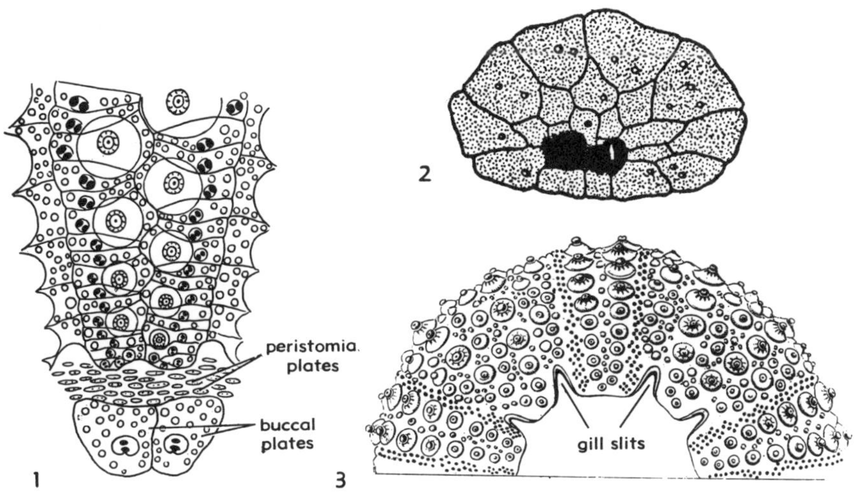

FIG. 188. Morphological features of peristomial region.——*1.* Segment of buccal membrane (toward bottom of figure) of noncidaroid regular echinoid bearing large buccal plates and diminutive peristomial plates, ×4.3 (14).——*2.* Irregular plates of buccal membrane in *Echinocorys* (29b).——*3.* Gill slits around peristome of *Pseudodiadema* (27b).

dae toward ambulacrum I (Figs. 186,*2,4*). In irregular echinoids the periproctal plates are usually small and arranged in indefinite circlets (Fig. 186,*1*). It has been demonstrated in *Echinocardium* that genital 5 becomes part of the periproctal complex of plates. Except in Acrosaleniidae and Saleniidae, however, periproctal plates are rarely preserved in fossils.

PERISTOMIAL SYSTEM

The peristome, or area between the adoral margin of the corona and the mouth opening, is covered by a membrane which usually is more or less completely covered by plates having different origin and arrangement in various groups.

In echinoids such as the Cidaroida, the peristome is covered with regular series of plates corresponding to each area of the corona (Fig. 187). Those corresponding to the ambulacra are ambulacral plates which became incorporated on the membrane as the peristome enlarged; those nearest to the corona each bear a pore pair, the pores of which are situated, not side-by-side as usual in the corona, but one above the other as the mouth is approached. The outer pore comes to lie above the inner and is gradually reduced in size until it disappears altogether, while the inner pore is correspondingly enlarged. The plates corresponding to the interambulacra, usually in single series, have been termed interradial or nonambulacral plates. The latter term should be abandoned, since evidence now available indicates that they must be of interradial origin. They are usually excluded from the mouth opening by the meeting of peristomial ambulacral plates. All peristomial plates are strongly imbricate toward the mouth. Where peristomial ambulacral (but no interradial) plates are present, it seems that the interradials must have been resorbed.

In some echinoids the ambulacral areas are represented on the buccal membrane only by ten large **buccal plates** with radially oriented pores (Fig. 188,*1*). The remaining area is covered by minute plates of uncertain origin. In many irregular types no plates or only small irregular plates occur on the buccal membrane (Fig. 188,*2*).

LANTERN AND GIRDLE

Most Echinoidea possess a complex jaw-apparatus (Fig. 189,*A,E*) consisting of (1) five teeth and their associated structures forming an **Aristotle's lantern,** and (2) outgrowths inside the test of plates at the base of the corona forming the **perignathic girdle** (Fig. 189,*D,F,G*). The lantern is poised within the body of the sea urchin so that

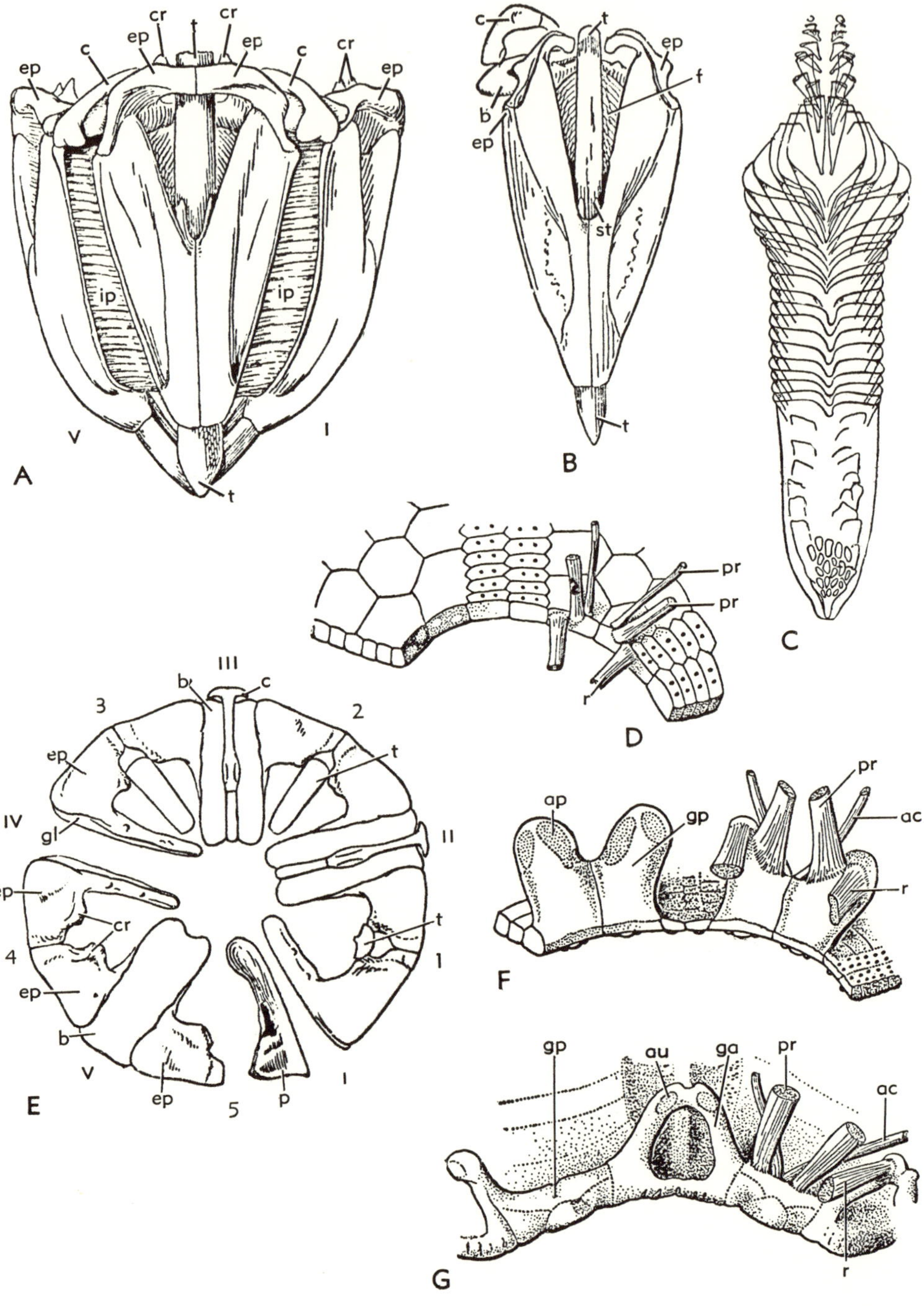

FIG. 189. Morphological features of Aristotle's lantern and adjacent peristome border.——*A*. Lantern of *Strongylocentrotus* viewed from side of interambulacrum 5 (9).——*B*. Pyramid of *Diadema* (9).——*C*. Structure of tooth showing varied form of cones in different parts (8 after Devanese).——*D*. Oblique internal view of peristome border of Paleozoic echinoid showing lantern muscles attached to inner surface of

(*Continued on facing page.*)

points of the teeth protrude through the mouth. The girdle serves for attachment of the muscles which hold the lantern in position and control certain of its movements.

The lantern is made up of 40 calcareous ossicles held together partly by sutures and partly by muscles, as follows: five **teeth** consisting of long, gently curved blades, curled over at the top in a soft, pulpy growing portion and, unlike most of the test, composed of a series of cones, one inside the other (Fig. 189,*C*). Each tooth is enclosed for the greater part of its length in a pyramid which is built of two **demipyramids** joined by a suture (Fig. 189,*B*). The movement of the tooth is guided by a **dental slide** formed by a pair of crests close to the suture. Neighboring pyramids are joined by interpyramidal muscles attached to horizontal corrugations on the external face of each pyramid. The teeth and pyramids are **interradial** in position. The teeth are either grooved or keeled longitudinally.

Each demipyramid is capped by an **epiphysis,** which is united to the demipyramid by suture. The pyramidal suture does not extend to the full height of each pyramid, so that a gape called the **foramen magnum** occurs between the tops of each pair of demipyramids. The epiphyses may or may not meet over the foramen magnum. The top surface of the demipyramids may be smooth or pitted.

The epiphyses of neighboring pyramids are joined by and interlock with a **brace** or **rotula** after the fashion of the shoulder-joint of vertebrates—that is, by the articulation of condyles on the rotula with glenoid cavities on the epiphysis.

The five **compasses** each consist of a slender, curved rod above the rotulae, to which they are attached by ligaments. Each compass consists of two portions, the outer of which is usually bilobed at the end. The rotulae and compasses are radial in position.

These 40 pieces (five teeth, ten demipyramids, ten epiphyses, five rotulae, and five compasses of two parts each) are operated by 60 muscles. These are (1) the ten **protractors,** which push the teeth downward through the mouth; (2) ten **retractors,** which withdraw the teeth and separate their points; (3) ten **radial compass muscles,** which maintain the vertical position of the pyramids and control inclined motion of the whole structure; (4) five powerful **interpyramidal muscles,** which bring the points of the teeth together for biting; (5) 20 **rotula muscles**, which accommodate the rotulae to the movements of the pyramids; and (6) five **circumferential compass muscles** of uncertain function. Some of these muscles work between one lantern piece and another. The others are attached to the perignathic girdle or to the inner surface of the test when no girdle is present. Since the lantern is attached to the test only by these muscles, it is freely suspended and can be moved in various oblique directions.

The perignathic girdle is not developed in a majority of Paleozoic genera (Fig. 189,*D*) nor in the very young stages of Cidaroida. In these forms, the lantern muscles are attached simply to the inner surface of the primordial interambulacral plates. In adult Cidaroida (Fig. 189,*E*) there are prominent, vertical, wing-shaped outgrowths from these plates, the **apophyses,** to which the radial compass muscles, the protractors and retractors are attached. In all other jaw-bearing forms there are also outgrowths from the ambulacral plates at the peristomial margin, termed **auricles** (Diadematoida, Fig. 189,*G*; Pygasteroida, Fig. 190,*2*; Clypeasteroida, Fig. 190,*3*). The auricles are like flattened pillars in shape and are taller than the apophyses. They are joined to the coronal plates and to the apophyses by suture and are usually separated from each other by a space, which may be arched over by supplementary pieces attached to the pillars. In some forms (e.g., Clypeasteroida) they may be joined to each other by suture. When auricles are present,

plates (2).——*E*. Apical view of lantern of *Strongylocentrotus*, areas 2, 3, and III complete, others with various structures removed (9).——*F*. Oblique internal view of peristome edge and girdle of *Cidaris* showing apophyses and attachment of lantern muscles (2).——*G*. Internal view of peristome edge and girdle of *Paracentrotus* (2). [Explanation: *ac*, radial compass muscle; *ap*, protractor muscle attachment scar; *au*, retractor muscle attachment scar; *b*, brace; *c*, compass; *cr*, crest of epiphysis; *ep*, epiphysis; *f*, foramen magnum; *ga*, auricle; *gl*, glenoid cavity; *gp*, apophysis; *ip*, interpyramidal muscle; *p*, top of pyramid; *pr*, protractor muscle; *r*, retractor muscle; *st*, styloid process; *t*, tooth.]

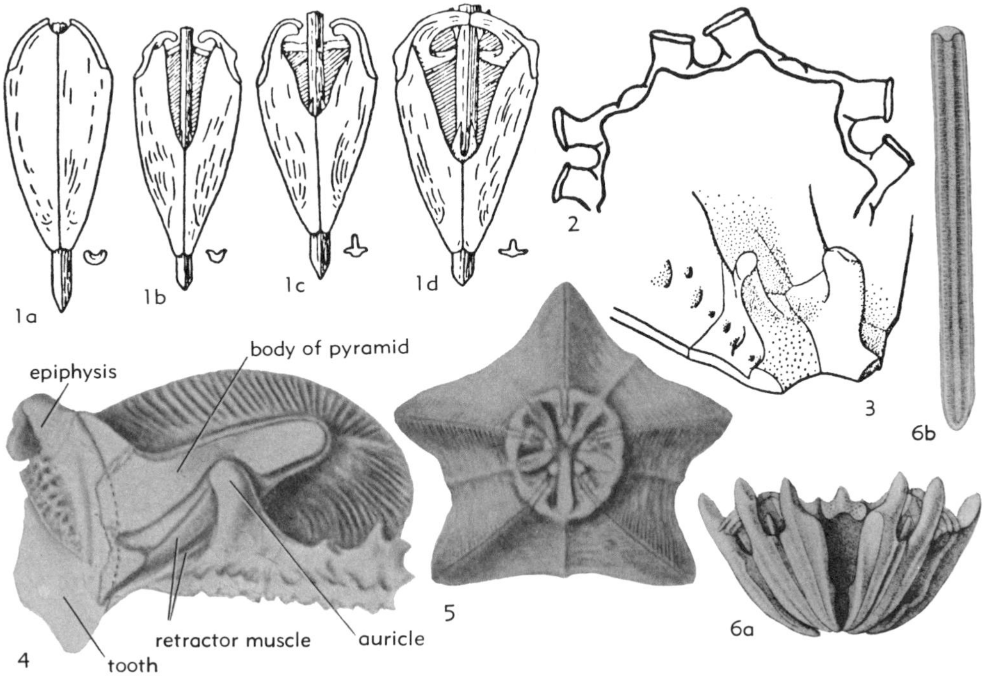

FIG. 190. Morphology of lantern and girdle.——*1.* External view of regular pyramids and cross sections of teeth (7 after Jackson); *1a,* cidaroid; *1b,* aulodont; *1c,* stirodont; *1d,* camarodont.——*2.* Girdle of *Pygaster* (7).——*3.* Auricles of *Clypeaster* (26).——*4.* Side view of pyramid of *Clypeaster* in position adjacent to auricles and tooth (14).——*5.* Lantern of *Clypeaster* viewed from above (14).——*6.* Lantern and tooth of *Echinoneus* (34); *6a,* lantern, ×29; *6b,* tooth, ×87.

the protractor muscles are attached to them and the retractor and radial compass muscles to the apophyses.

Among echinoids the structure of the pyramid (Fig. 190,*1*) and teeth provides useful criteria for separating various higher categories. In Cidaroida the teeth are grooved longitudinally, the foramen magnum is shallow, the epiphyses are narrow and not fused together. The tops of the pyramids are not pitted, and the girdle consists of apophyses only. In the condition termed **aulodont,** the teeth are longitudinally grooved, the foramen magnum is shallow and the epiphyses separate, as in Cidaroida, but the tops of the pyramids are pitted and auricles are developed in the girdle. **Stirodont** is the term applied when the teeth are keeled longitudinally and the foramen magnum is deep. The epiphyses are separate, the tops of the pyramids are pitted, and auricles are present in the girdle. In the **camarodont** condition the teeth are keeled, and the epiphyses meet in a suture over the deep foramen magnum. The tops of the pyramids are pitted and auricles are present in the girdle.

In the Holectypoida the apophyses are feebly developed or absent, and the auricles commonly are supported from behind by buttresses which develop into the massive partition-walls of *Discoides*. In *Echinoneus,* one of the two living holectypoids (Fig. 190,*6*), a complete lantern is formed, but it is never functional, being resorbed before the mouth is opened. It is clearly of the stirodont type, however, as in the fossil *Holectypus* and *Discoides,* with keeled teeth and separate epiphyses. The girdle of *Conoclypus* is supported by a peristomial invagination, so that the auricles are raised well above the floor of the test. The auricles are interradial in position, although their ambulacral origin can be traced. In the Oligopygidae the auricles are likewise interradial but are recumbent on the floor of the test.

A well-developed lantern is present in Clypeasteroida (Fig. 190,*4,5*). It is low, flaring, and somewhat asymmetrical, and compasses are never present. The anterior pair of pyramids and the posterior tooth may be larger than the rest, as in Clypeasteridae, or the reverse may be the case, as in Fibulariidae. The rotulae lie, not above, but impressed between the epiphyses. The auricles are generally fused into a single interradial structure in the more specialized clypeasteroids.

In the Cassiduloida, no trace of lantern or girdle is known in any adult form, though both have been observed in the young of *Echinolampas* and *Conolampas*. Neither lantern nor girdle has been observed in any of the Holasteroida or Spatangoida.

APPENDAGES

GENERAL

The tubercles of the coronal plates bear movable appendages (Fig. 191), each of which articulates with its tubercle by a ball-and-socket joint and is moved by muscles. The three principal types of appendage are **spines** or **radioles, pedicellariae,** and **spheridia.** The spines, like the tubercles on which they rest, may be ranked as **primary, secondary,** and **miliary.**

SPINES

Although the spines are extremely varied in form, from the massive thorny primary spines of Cidaroida to the tessellate spines of *Colobocentrotus* and the short hairlike spines of Spatangoida, they all have a common, basic gross structure which is most easily explained in terms of the cidaroid type.

In the cidaroids each primary spine (Fig. 192,*1*) consists of a distal **shaft,** which may be straight or curved, cylindrical or flattened, stout or slender, club-shaped or bladed, smooth or variously ornamented with secondary spines or thorns which may be arranged in longitudinal rows or in well-spaced whorls (**verticillate**). At the base of the shaft is a smooth **neck** that varies in length relative to the length of the shaft. Below the neck is the **collar,** marked with fine longitudinal striations and flaring out into the **milled ring,** to which are attached

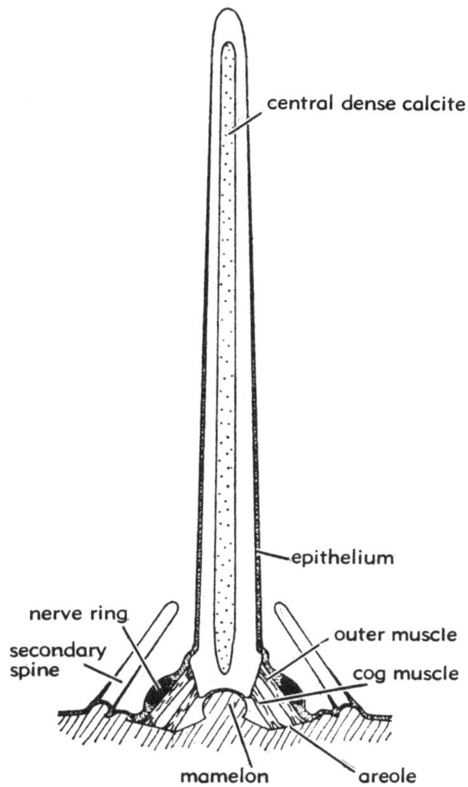

FIG. 191. Diagrammatic section of spine and tubercle of living echinoid (17).

the ligaments and muscles which anchor the spine and control its movements. Below the milled ring is the smooth **base,** hollowed proximally by the **socket** or **acetabulum** which articulates with the mamelon of the tubercle. When the tubercle is perforate, a strand of connective ligament runs to a similar feature in the socket. Spines mounted on crenulate tubercles are correspondingly crenulate around the margin of the socket.

The microscopic structure of the primary interambulacral spines is of great systematic importance in regular echinoids. As seen in transverse section (Fig. 192,*2*) the **central core** or **medulla** consists of an irregular calcareous mesh. Outside the core is a **radiating layer,** built of septa connected by trabeculae. The outer or **cortex** layer is compact but perforated by longitudinal channels (appearing as pores in transverse section), one

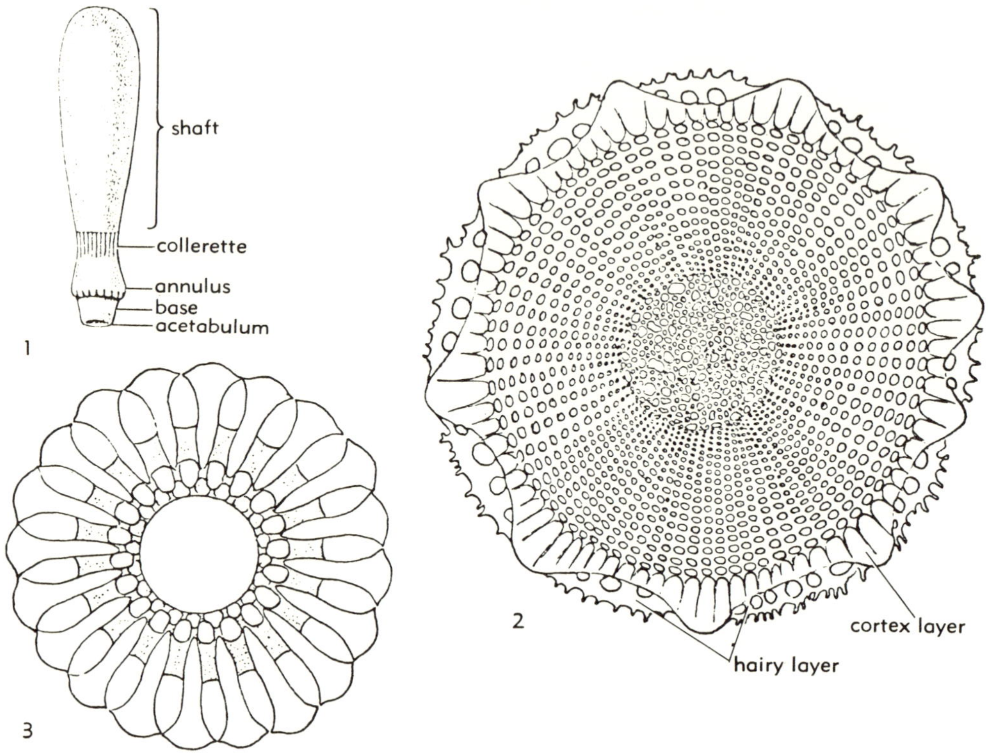

FIG. 192. Morphology of primary echinoid spines.——*1.* Side view of cidaroid spine showing named parts (25).——*2,3.* Cross sections of primary spines of *Stereocidaris,* ×40 (14), and *Centrostephanus,* ×55 (14).

opposite the end of each septum. The cortex does not extend below the neck. From the outer surface of the cortex a great number of fine **hairs** extend; these may be short and thin or long and fluffy, or branching and anastomosing so as to conceal the cortex layer. The tips of the hairs may be plain, singly or doubly hooked, or minutely tuberculate. The septa are bladelike and run the full length of the spine, and it is their exposed edges that give the collar its striated appearance. The hair coat is very rarely preserved in fossils, and even in living forms it is frequently worn off by abrasion.

The secondary (scrobicular) and ambulacral spines differ from the primaries in lacking both the central core and the cortex layer.

In the Echinothuriidae the spines are hollow tubes, the wall of which consists of a single, compact layer perforated by regularly spaced longitudinal canals. Between the canals the wall is raised into longitudinal ridges. The central axial core of larger radioles may be partly filled with an irregular calcareous mesh.

In most other echinoids the spines lack the cortex layer and consist only of the wedge-shaped septa set around an axial cavity (Fig. 192,*3*) which may be partly or wholly filled with a loose calcareous mesh or solid.

The spines of many irregular echinoids are short (except in specialized areas of some spatangoids), thin and longitudinally striate. Many of these forms have a burrowing mode of life, and the spines of the oral surface, especially those of the plastron of Spatangoida, are paddle-shaped and serve for burrowing and locomotion. The spines of some spatangoids are longitudinally fluted and of minaret-like form.

In systematic work, the size, form, and ornamentation of the spines, their general microscopic structure and detailed form and

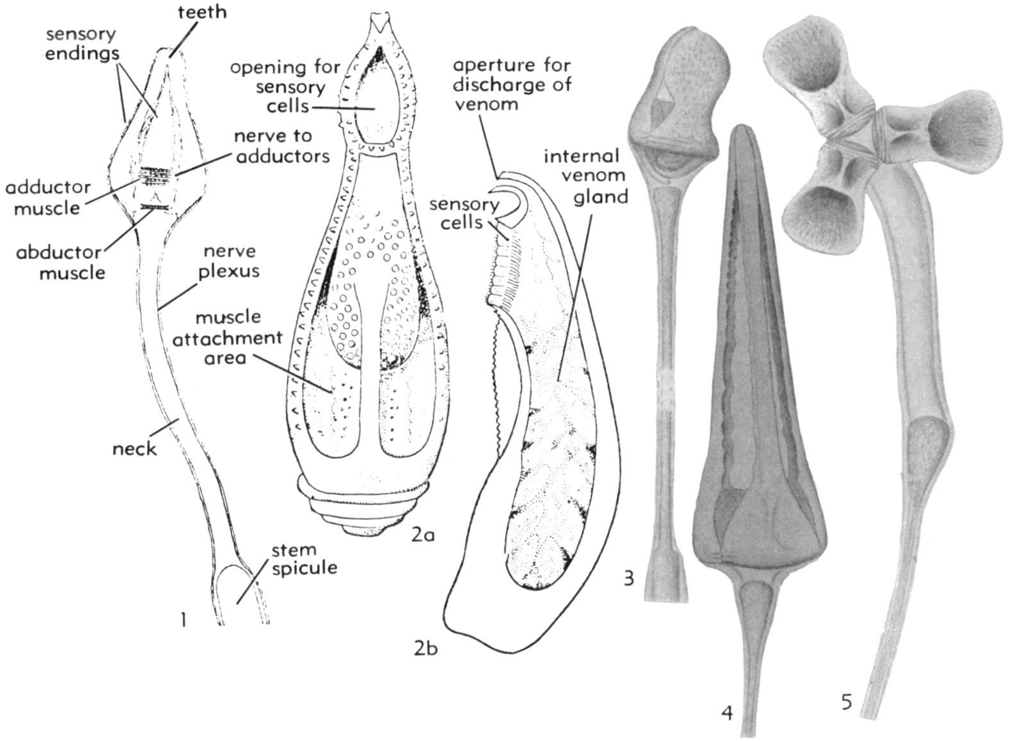

FIG. 193. Morphology of pedicellariae.——*1.* Diagrammatic section of typical pedicellaria of living echinoid (17).——*2.* Valve of globiferous pedicellaria of cidaroid; *2a,b,* internal view and longitudinal section (2 after Mortensen & Prouho).——*3.* Ophicephalous pedicellaria with stalk, ×50 (14).——*4.* Blades of tridentate pedicellaria, ×28 (14).——*5.* Triphyllous pedicellaria with open jaws, ×55 (14).

number of their septa (as well as details of the hair coat in Cidaroida) are all of importance. Unfortunately, fossil spines can rarely be related to particular tests, and in some the internal structure has been obliterated by recrystallization.

PEDICELLARIAE

The minute, grasping, defensive, and scavenging organs known as pedicellariae are attached to small tubercles on the test of echinoids. Each (Fig. 193,*1*) consists of a muscular **stem** stiffened by a calcareous rod and a movable head, usually built of three jawlike **valves.** The whole is enveloped in a muscular sheath and in life is constantly in motion. Four principal types of pedicellariae are recognized.

(1) **Globiferous** pedicellariae (Fig. 193, *2*) possess elongated valves, commonly terminating in a sharp tooth. In Cidaroida and some Spatangoida the valves are hollow and contain poison-secreting glands. In other groups the glands, which may be double, are external to the valves.

(2) **Tridentate** pedicellariae (Fig. 193,*4*) are usually the largest and possess

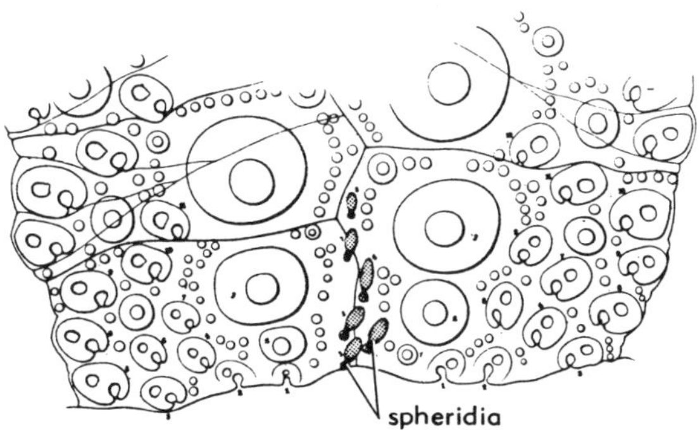

FIG. 194. Adoral region of ambulacrum of *Strongylocentrotus* showing spheridia (12).

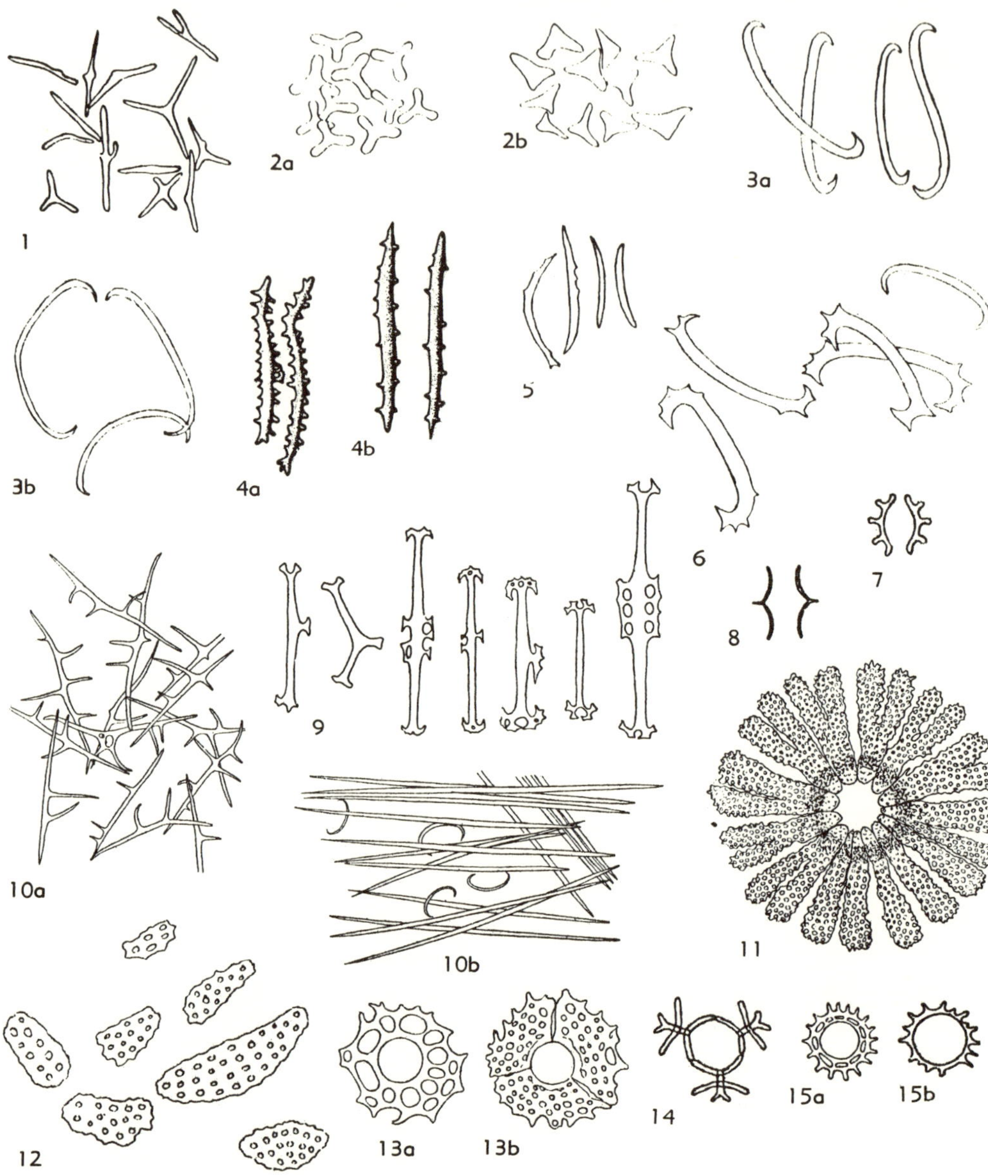

FIG. 195. Calcareous spicules and plates from echinoid tissues.——*1. Aporocidaris milleri* (AGASSIZ), from intestinal wall, ×80 (14).——*2. Goniocidaris* (14); *2a, G. tubaria* (LAMARCK), from intestinal wall, ×80; *2b, G. umbraculum* HUTTON, from intestinal wall, ×80.——*3. Polyechinus agulhensis* (DÖDERLEIN) (14); *3a,b,* from tube foot and buccal membrane, ×200.——*4. Brissopsis* (32a); *4a, B. elongata* MORTENSEN, from tube foot, ×110; *4b, B. lyrifera* FORBES, from tube foot, ×110.——*5. Schizocidaris assimilis* MORTENSEN, from tube foot, ×75 (14).——*6. Selenechinus armatus* (DE MEIJERE), from tube foot, ×180 (14).——*7. Echinodiscus auritus* LESKE, from sucking disc of tube foot, ×250 (14).——*8. Mellitella stokesii* (AGASSIZ), from sucking disc of tube foot, ×280 (14).——*9. Micropyga tuberculata* AGASSIZ, from tube foot, ×120 (14).——*10. Echinometra mathaei* (DE BLAINVILLE) (14); *10a,b,* from water vessel and gonads, ×80, ×100.——*11. Pericosmus melanostomus* MORTENSEN, from sucking disc of tube foot, ×8 (32d).——*12. Arachnoides placenta* (LINNÉ), from buccal membrane, ×80 (14).——*13. Diadema setosum* (LESKE) (32c); *13a,b,* from sucking discs of primordial tube foot and second tube foot, ×200.——*14. Heliophora orbiculus* (LINNÉ), from sucking disc of tube foot, ×300 (14).——*15. Clypeaster* (14); *15a, C. rarispinus* DE MEIJERE, from sucking disc of tube foot, ×175; *15b, C. latissimus* (LAMARCK), from sucking disc of tube foot, ×210.

three long, pointed valves without poison glands or terminal teeth.

(3) **Ophicephalous** pedicellariae (Fig. 193,*3*) are most abundant in the circumoral region. The valves are blunt-ended, with serrate margins, and are hinged together by arc-shaped calcareous rods. They do not occur in Cidaroida.

(4) **Triphyllous** pedicellariae (Fig. 193, *5*) are very small, with flattened, finely-toothed valves not hinged to each other.

An additional type, termed **dactylous,** is found only in the Echinothuriidae. The valves, which are only slightly movable if at all, are three, four, or five in number and are spoon-shaped.

Variation in number of valves is considerable, particularly in the tridentate type, which may have either a lesser or greater number than the usual three. Pedicellariae easily become detached from the test after death with decay of their muscular attachments. Thus they are rarely reported in the fossil state, especially in association with the tests.

SPHERIDIA

The **spheridia** (Fig. 194) are minute, club-shaped calcareous bodies of uncertain function movably attached to minute tubercles. They are confined to the ambulacral areas in the adoral region. They occur in all living groups of Echinoidea except the Cidaroida. They vary greatly in number, from one per ambulacrum (close to the peristomial margin) in *Arbacia* and various clypeasteroids to several on each of the adoral ambulacral plates. They may hang freely down from the surface of the test or may be lodged in grooves or pits or in closed chambers, when they cannot be seen from the exterior.

SPICULES

In addition to skeletal features already described, many echinoids possess calcareous spicules and plates (Fig. 195) imbedded in the soft tissues, for example, in the tube feet, peristomial membrane, and wall of the gut. They are much used in the systematics of Recent forms but have not been identified in the fossil state. Their association with particular fossil tests would in any case be extremely unlikely and there is some risk of confusing them with the spicules of other groups of animals.

LARVAL SKELETON

During larval life the developing young echinoid is very different in appearance from its parents, so much so that the larval stages when first observed were thought to be different animals and were given separate names. In echinoids with a planktonic stage (the majority) after the egg is fertilized, the young echinoid passes through the so-called dipleurula stage before developing into the **pluteus** (or **echinopluteus**) larva. The pluteus (Fig. 196,*A,D*) is strikingly dissimilar to an adult echinoid and has a calcareous skeleton, which, at the time of metamorphosis into the **imago** (Fig. 196,*F*) or immature adult, in part forms the nucleus for some of the adult plates and in part is resorbed.

At first the body of the pluteus has two pairs of upward-projecting **arms** and then an additional four pairs develop. In the irregular echinoids a downward-projecting arm, the **spike,** usually appears also. The body is supported by a more or less complex calcareous basketwork (Fig. 196,*C*), and each of the arms has an axial calcareous rod. In some, the skeleton becomes very complex and ornate (Fig. 196,*E*); in others it is quite simple (Fig. 196,*B*).

No pluteus larvae have yet been reported from the fossil record, but it seems highly probable that they should be present in microfossil assemblages.

GLOSSARY OF MORPHOLOGICAL TERMS APPLIED TO ECHINOIDS

By J. Wyatt Durham and C. D. Wagner

This glossary has been compiled from the text of the section on echinoids. The definitions have benefited from the criticism and suggestions of H. B. Fell, A. G. Fischer, P. M. Kier, R. V. Melville, D. L. Pawson, authors of other parts of this section, and of R. C. Moore, Editor. Suggestions from Leigh W. Mintz have led to clarification of numerous definitions.

abactinal. Aboral or apical aspect; side opposite mouth.

aboral. Side opposite mouth; direction away from mouth.

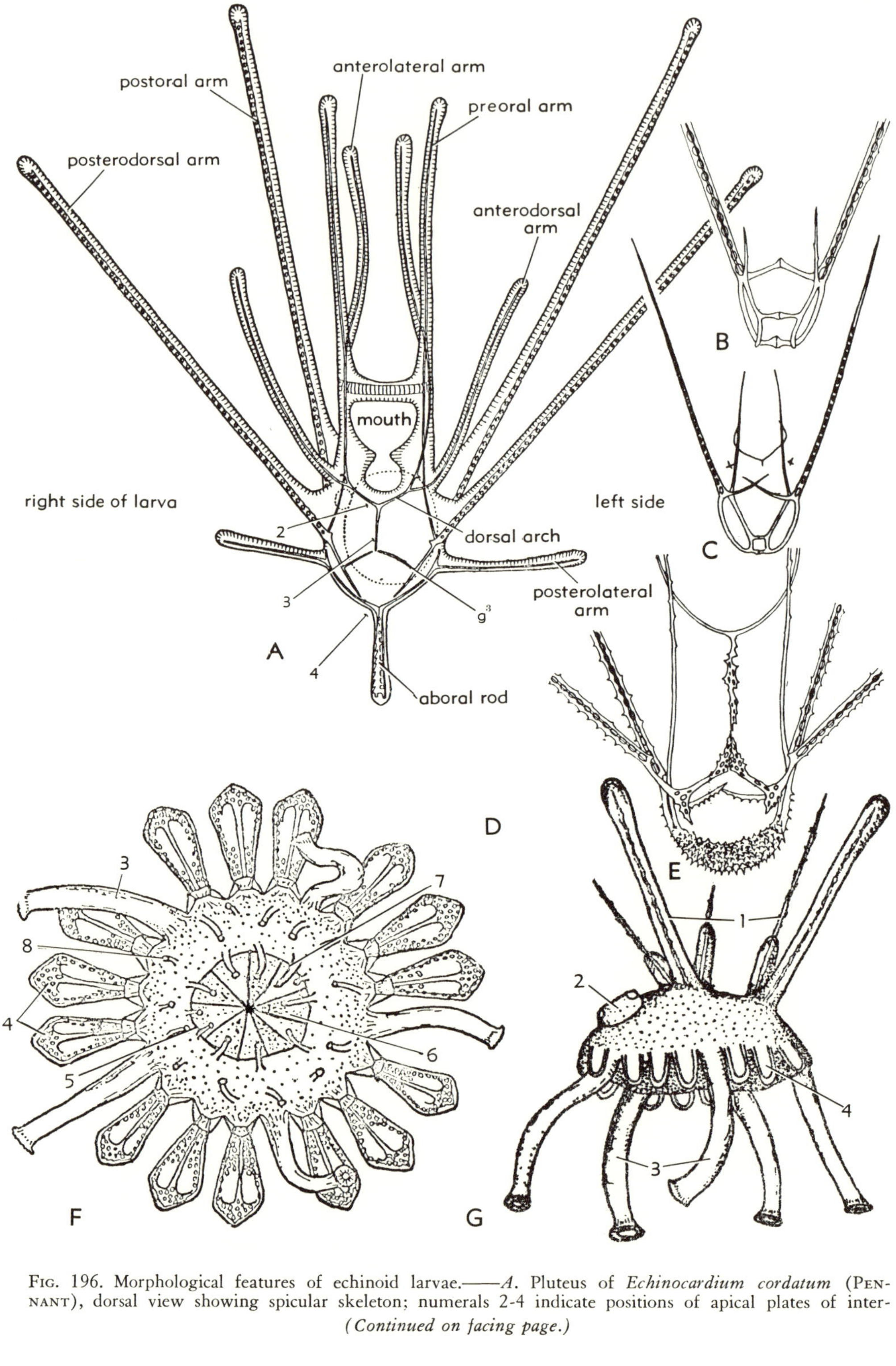

FIG. 196. Morphological features of echinoid larvae.——*A*. Pluteus of *Echinocardium cordatum* (PENNANT), dorsal view showing spicular skeleton; numerals 2-4 indicate positions of apical plates of inter-

(Continued on facing page.)

acetabulum. Concave proximal or articulating end of spine.
actinal. Oral aspect; side on which peristome is situated.
adapical. Toward apical system.
adapical suture. Suture along apical side of coronal plate.
adoral. Toward mouth.
adoral suture. Suture along side of coronal plate nearest mouth.
adradial. Position corresponding to boundary between ambulacral and interambulacral areas.
amb. Abbreviated designation of ambulacrum.
ambitus. Greatest horizontal circumference of test.
ambulacra (sing., **ambulacrum**). Five segments of test extending from apical system to peristome, each underlain by radial water vessel, and alternating with 5 interambulacra; designated by Roman numerals in Lovénian system.
ambulacral. Corresponding in position or pertaining to ambulacra.
ambulacral furrow. Food groove.
ambulacral pore. Opening through ambulacral plate for passage of tube foot.
amphiplacous. With basicoronal interambulacral plate abutting against 2 plates adapically.
amphisternous. With labrum followed by 2 large, more or less equal sternal plates opposite one another.
ampulla (pl., **ampullae**). Contractile chamber of water-vascular system internal to test, forming part of each tube foot complex; of skeleton, *see* camella.
anal fasciole. Fasciole adoral and lateral to periproct; if connecting with lateral fascioles, termed latero-anal fasciole.
anal plates. Paired interambulacral plates in contact with periproct of irregular echinoids; plates of periproctal system.
apex. Highest part of test.
apical system. Plates at aboral terminus of ambulacra and interambulacra including ocular and genital plates (when present); may include one or more complemental plates.
apophysis (pl., **apophyses**). Internal projection from interambulacral basicoronal plates for attachment of muscles supporting lantern.
arbacioid compound plate. Plate with 3 elements, of which adapical and adoral ones are demiplates.
areole. Scrobicule or depression around boss for attachment of muscles controlling movement of spines.
Aristotle's lantern. Structure of 40 or fewer skeletal elements serving for mastication.
aulodont. Descriptive term for lantern with open foramen magnum and with cross section of teeth broadly U-shaped.
auricle. Internal process arising from basicoronal ambulacral plates for attachment of muscles supporting lantern.
base. Portion of spine below milled ring.
basicoronal. Referring to corona at edge of peristome.
bidentate. Type of pedicellaria with head consisting of 2 long pointed valves.
bigeminate. Having 2 pore pairs.
bivium. Two posterior ambulacra.
boss. Part of tubercle below mamelon shaped like truncated cone.
bourrelet. Externally inflated adoral part of interambulacral areas.
brace. Rotula.
branchial slit. Gill slit.
buccal membrane. Tissue between peristomial margin and mouth.
buccal plates. Ten large primordial plates on buccal membrane with pores for buccal tube feet.
buccal system. Peristomial system of plates.
buttress. Ridge of skeletal material extending adapically from auricle on inner surface of test.
camarodont. Descriptive term for lantern with keeled teeth and closed foramen magnum.
camella (pl., **camellae**). Pouchlike ampulla forming bulge on inner wall of the test.
Carpenter system. Ambulacral areas of test designated by capital letters (*A* to *E*) and interambulacral areas by letters of adjacent ambulacra.
catenal plate. Supplementary plate (or plates) along III-5 axis between anterior and posterior portions of apical system where these are disjunct.
clavulae (sing., **clavula**). Small ciliated spines in fascioles.
collar. Smooth tapering portion of spine located above milled ring.
compact. Type of apical system with no separation between anterior and posterior elements.
compass. Slender arched radial rod in ambulacral position at top of lantern.
complemental plate. Supplementary plate (or plates) in apical system.
compound plate. Ambulacral plate unit composed of 2 or more individual plates, each with pore for tube foot, bound together by single large primary tubercle.

ambulacra 2, 3, and 4, respectively, in late larva; g^3 marks posterior extension of right posterodorsal rod, part of which later gives rise to genital 3; outlines of stomach and small "echinus rudiment" indicated by broken lines; enl. (4).——*B.* Spicular skeleton of early pluteus of *Echinothrix diadema* (Linné), full length of arms not shown, ×175 (32b).——*C.* Spicular skeleton of pluteus of *Cyrtechinus verruculatus* (Lütken), ×100 (32b).——*D.* Pluteus of *Diadema setosum* (Leske), ×60 (32b).——*E.* Spicular skeleton of pluteus of *Echinodiscus auritus* (Leske), full length of arms not shown, ×225 (32b).——*F,G.* Lateral and oral views of imago of *Arbacia punctulata* (Lamarck), enl. (1, remnant of pluteal arms; 2, oral lobe of pluteus; 3, primary tube feet; 4, juvenile spines; 5, peristome; 6, teeth; 7, buccal tube feet; 8, postbuccal adult tube feet) (8).

conjugate pores. Pores of pair connected by groove in external surface of test.

corona. Principal skeletal structure excluding apical, periproctal, and peristomial systems, lantern, and appendages; all ambulacra plus interambulacra.

cortex. Differentiated dense outer layer of spine usually bearing ornamentation; nonliving on mature spine.

crenate. Crenulate.

crenulate. Descriptive term for tubercle or acetabulum of spine with ribbed periphery.

dactylous. Type of pedicellaria with spoon-shaped jaws mounted on individual stalks.

demiplate. Ambulacral plate which touches adradial suture but not perradial suture.

demipyramid. One of 10 elements which support teeth in Aristotle's lantern.

diadematoid compound plate. Plate with 3 primary elements, of which the middle one is largest.

dicyclic. Type of apical system with ocular and genital plates in 2 concentric circles, genitals alone in contact with periproctal margin.

disjunct. Type of apical system with anterior part (usually genital plates 1, 2, 3, and 4 and ocular plates II, III, and IV) separated from posterior part (ocular plates I and V, forms with disjunct apical system lacking genital 5).

echinoid compound plate. Plate with 3 or more elements of which adoral and adapical ones are primaries and adoral one is largest.

echinothurioid compound plate. Plate of 3 elements (primary with 2 small included plates on its adoral margin).

endocyclic. With periproct located within oculogenital ring.

endopetalous fasciole. *See* internal fasciole.

epiphysis. Element at top of Aristotle's lantern in interambulacral position.

episternal plates. Second pair of postlabral plates in amphisternous spatangoids.

epistroma. Adventitious skeletal material on outer surface of test plates.

ethmolysian. See ethmolytic.

ethmolytic. Type of apical system in which genital plate 2 extends posteriorly between oculars I and II and genital 1 on one side and oculars III, IV, and V and genitals 3 and 4 on other; genital plate 5 may or may not be present.

ethmophract. Type of apical system in which genital plates 1, 2, 3, and 4 mutually adjoin; genital plate 5 may or may not be present.

exocyclic. With periproct located outside of oculogenital ring.

exsert. With ocular plates not in contact with periproctal margin.

fasciole. Narrow band of small densely ciliated spines (clavulae) in which cilia beat to create currents; on denuded test narrow band of small tubercles which bears such spines.

floscelle. Star-shaped area around peristome formed by phyllodes and bourrelets.

food groove. Narrow grooves leading to peristome in adoral ambulacral areas supplied with specialized tube feet for food gathering and transport; may extend into interambulacral areas and onto aboral surface.

foramen magnum. Space between upper ends of paired demipyramids of lantern.

genital plate. Primordial interradial apical plate usually with one or more pores for discharge of genital products.

genital pore. Opening in genital plate for discharge of reproductive products.

gill slit. Indentation of peristomial margin of interambulacra for passage of stem of external branchia.

globiferous. Type of pedicellaria with 3 valves containing poison glands.

gonopore. See genital pore.

holamphisternous. Type of plastron having symmetrical pair of sternal plates followed by symmetrically paired but otherwise undifferentiated plates.

hypophyllode. Primitive or feebly developed phyllode.

included plate. Ambulacral plate which touches neither perradial nor adradial suture.

insert. With ocular plates in contact with periproctal margin.

interamb. Abbreviated designation of interambulacrum.

interambulacra (sing., **interambulacrum**). Five segments of test extending from apical system to peristome, alternating with ambulacra; designated by Arabic numerals in Lovénian system.

intercalary. Type of apical system in which ocular plates II and IV meet at mid-line so as to separate anterior and posterior portions.

internal fasciole. Fasciole surrounding apical system and crossing all petals.

internal support. Rod- or pillar-like structure between inner oral and inner aboral surfaces of test.

interporiferous zone. Area between 2 inner pore rows of ambulacrum.

interradial suture. Suture between 2 columns of plates in interambulacrum; in forms with more than 2 columns in interambulacral area, suture between 2 middle columns.

interradius. Interambulacrum.

irregular. With periproct located outside oculogenital ring.

keel. Raised or ridged section of corona; also longitudinal ridge on internal side of tooth.

labrum. More or less enlarged and modified liplike primordial plate bordering peristome in interambulacrum 5.

lantern. Structure of 40 or fewer skeletal elements which serves for mastication (same as Aristotle's lantern).

lateral fasciole. Fasciole extending posteriorly from

peripetalous fasciole; termed lateroanal fasciole if connected with anal fasciole.

lateroanal fasciole. Fasciole formed by union of lateral and anal fascioles.

Lovénian symmetry. Bilateral symmetry with respect to plane passing through apical system, peristome, and periproct in irregular echinoids; recognized in regular echinoids chiefly by position of madreporite.

Lovénian system. Numbering system in which ambulacral and interambulacral areas of test are designated by Roman (I-V) and Arabic (1-5) numerals, respectively.

lunule. Opening from aboral surface through oral surface of test at perradial or interradial suture.

madreporite. Plate (or plates) of apical system perforated to provide access to water-vascular system from exterior.

mamelon. Raised, rounded top of tubercle on which spine articulates.

margin. Ambitus of corona in flattened echinoids.

marginal fasciole. Fasciole extending around ambitus.

masticatory apparatus. See Aristotle's lantern.

medulla. Meshlike central core of spines.

meridoplacous. With first adoral interambulacral plate abutting adapically against single plate.

meridosternous. Type of plastron with labrum followed by single large plate.

mesamphisternous. Type of plastron having symmetrical pair of sternal plates followed by alternating plates.

metasternous. Meridosternous plastron with sternal followed by several large plates in single column.

mid-zone. Region of corona midway between the apical system and peristome.

microcanal system. System of canals within plates of corona for passage of secondary branches of water-vascular system leading to accessory tube feet, apparently found only in clypeasteroid echinoids.

miliary. Very small, nonprimary spine.

miliary tubercle. Tubercle supporting miliary spine.

milled ring. Flange near base of spine for attachment of muscles that move spine.

monobasal. Type of apical system with genital plates apparently fused, suture lines being obliterated.

monocyclic. Type of apical system with genital and ocular plates arranged in single ring around periproct.

multiserial. Type of ambulacrum with pore pairs arranged in more than 2 longitudinal series.

neck. Smooth cylindrical portion of primary spine between collar and shaft.

neuropore. Single pore in ambulacral plate for passage of branch of radial nerve to exterior of test.

notch. Indentation of margin of test, usually at perradial suture.

occluded plate. Ambulacral plate which touches perradial but not adradial suture.

ocular plate. Primordial plate of apical system at terminus of ambulacrum, perforated by ocular pore.

ocular pore. Perforation in ocular plate for passage of terminal tentacle.

oculogenital ring. System of ocular and genital plates at apical end of ambulacral and interambulacral areas, surrounding periproct in regular echinoids.

oligoporous. Referring to compound or pseudocompound ambulacral plate with few pore pairs, usually 3.

ophicephalous. Type of pedicellaria with jaws which lock together.

oral. Side of test on which the peristome is located, usually directed toward substrate.

orthosternous. Meridosternous plastron with sternal followed by more or less equal-sized plates in 2 columns.

pedicel. See tube foot.

pedicellariae (sing., **pedicellaria**). Minute stalked specialized grasping or defensive organs articulated on granules.

perforate tubercle. Tubercle with small depression in top for ligament connecting spine with tubercle.

perignathic girdle. Continuous or discontinuous ring of internal processes around peristomial opening for attachment of muscles supporting and controlling lantern.

peripetalous fasciole. Fasciole passing around petals of ambulacra I, II, IV, and V and around or across ambulacral petal III.

peripodium. Raised rim around pore pair on external surface of test.

periproct. Opening in test for anus, covered in life by periproctal membrane, commonly plated.

periproctal system. Collective term for plates on periproctal membrane.

peristome. Opening in test for mouth, covered in life by peristomial membrane, commonly plated.

peristomial system. Collective term for plates on buccal membrane.

perradial. Having meridional position at mid-line of ambulacrum.

perradial suture. Suture between 2 columns of ambulacrum; in forms with more than 2 columns in ambulacrum, suture between 2 middle columns.

petal. Differentiated adapical segment of ambulacrum with tube feet more or less specialized for respiration; pores of pair in petal typically unequal or enlarged.

phyllode. Area of enlarged pores in adoral portion of ambulacra.

pillar. Internal supporting columns between aboral and oral sides of test.

plastron. More or less inflated and enlarged adoral segment of interambulacrum 5.

plate. Individual more or less flattened skeletal element composed of single calcite crystal.

polyporous. Referring to compound or pseudocompound ambulacral plate with many pore pairs, usually 5 or more.

pore pair. Ambulacral pore divided by wall of stereom through which single tube foot passes.

preanal plates. Paired interambulacral plates between

episternal and anal plates in echinoids with plastron.

primary plate. Ambulacral plate that extends from perradial to adradial suture.

primary spine. First-formed and usually largest spine of coronal plates, situated over growth center of plate except on compound ambulacral plates.

primary tubercle. Tubercle articulating with primary spine.

primordial plates. First plates formed following metamorphosis in each plate system.

protamphisternous. Type of plastron with asymmetrical pair of sternal plates followed by alternating plates.

protosternous. Type of plastron with labrum followed by 2 simply alternating, slightly enlarged plates.

pseudocompound. Referring to group of 2 or more associated primary or reduced plates not bound together by primary tubercle.

pyramid. Large beaklike or winglike element of lantern in interambulacral position.

radial. Ambulacral.

radiole. Spine.

reduced plate. Ambulacral plate excluded from perradial suture or adradial suture, or both.

regular. Type of test having periproct within oculogenital ring of apical system.

rostrum. Raised or attenuated area of interambulacrum 5.

rotula (pl., **rotulae**). Massive radial element at top of lantern in ambulacral position.

scrobicule. Depressed ring around base of tubercle for attachment of muscles of spine.

secondary spine. Intermediate-sized spine, later in appearance than primary spine.

secondary tubercle. Tubercle with which secondary spine articulates.

septa. Radial structures external to medulla in spines.

shaft. Main part of spine.

sieve plate. See madreporite.

spheridia. Minute spherically modified spines on short stalks commonly situated adorally in pits near perradial suture.

spicules. Minute calcareous discs or diversely shaped rods imbedded in various tissues of body.

spine. Movable elongated calcareous shaft mounted on tubercle and articulating with it.

stereom. Calcareous mesh of which skeletal elements are composed.

sternal plates. First pair of postlabral plates in echinoids with plastron.

sternum. See plastron.

stirodont. Descriptive term for lantern with keeled teeth and open foramen magnum.

subanal fasciole. Fasciole enclosing more or less elliptical area on posterior face of test below periproct.

subpetaloid. Type of ambulacrum showing tendency toward petaloid development.

suranal plate. First-formed and largest plate of periproctal system; not recognizable in many echinoids.

suture. Narrow zone marking contact between adjacent plates; usually represented by line on surface of test.

terminal tentacle. Terminal podium of radial vessel of water-vascular system extending through ocular pore.

test. Collective term for plates of coronal, apical, periproctal and peristomial systems.

tetrabasal. Type of apical system with 4 separate genital plates (genital 5 not present).

tooth. Calcareous rod located in pyramid in interambulacral position in lantern (upper end of tooth uncalcified).

tridentate. Type of pedicellaria with 3 long, pointed, jawlike valves.

trigeminate. Having 3 pore pairs.

triphyllous. Type of minute pedicellaria with 3 leaflike jaws not hinged to one another.

trivium. Three anterior ambulacra.

tube foot. End of branch of water-vascular system serving for grasping, adhesion, locomotion, respiration, or combination of these.

tubercle. Knoblike structure on outer surface of test plates with which spine articulates.

ultramphisternous. Type of plastron with sternal and succeeding plates symmetrically paired, latter differentiated into episternal, preanal, and anal plates.

uniserial. Referring to ambulacrum with pore pairs in single longitudinal row.

zygopore. Pore pair.

REFERENCES

Clark, Ailsa M.

(1) 1962, *Starfishes and their relations:* 119 p., 16 pl., British Museum (Natural History) (London).

Cuénot, Lucien

(2) 1948, *Anatomie, éthologie et systématique aes Echinodermes:* in Traité de Zoologie, P.-P. Grassé (ed.), v. 11, p. 1-363, Masson et Cie (Paris).

Deutler, F.

(3) 1926, *Über das Wachstum der Seeigelskeletts:* Zool. Jahrb., Abt. Anat. & Ontog., v. 48, p. 119-200, pl. 3-9.

Gordon, Isabella

(4) 1926, *The development of the calcareous test of Echinocardium cordatum:* Royal Soc. London, Philos. Trans., ser. B, v. 215, p. 255-313.

Harvey, E. B.

(5) 1956, *The American Arbacia and other sea urchins:* 298 p., 16 pl., Princeton Univ. Press (Princeton).

Hawkins, H. L.

(6) 1920, *The morphology and evolution of the ambulacrum in the Echinoidea Holectypoida:*

Royal Soc., London, Philos. Trans. ser. B, v. 209, p. 377-480, pl. 61-69.
(7) 1934, *The lantern and girdle of some Recent and fossil Echinoidea:* Same, ser. B, v. 223, p. 617-649, pl. 68-70.

Hyman, Libbie H.
(8) 1955, *The Invertebrates, Echinodermata:* 763 p., 280 text-fig., McGraw-Hill (New York-Toronto-London).

Jackson, R. T.
(9) 1927, *Studies of Arbacia punctulata and allies, and of nonpentamerous Echini:* Boston Soc. Nat. History, Mem., v. 8, p. 437-565.

Jesionek-Szymanska, W.
(10) 1959, *Remarks on the structure of the apical system of irregular echinoids:* Acta Palaeont. Polonica, v. 4, p. 339-352.

Kirchner, G.
(11) 1929, *Über die Lage der optischen Achse in Skeletteilen von Seeigeln:* Zool. Jahrb., Abt. Anat. & Ontog., v. 51, p. 299-312.

Lovén, Sven
(12) 1874, *Études sur les Echinoidées:* [K.] Svenska Vetenskapsakad., Handl., v. 11, p. 1-91, pl. 1-53.

Lucas, G.
(13) 1953, *Existence, dans l'appareil apical des Oursins, de deux sortes de plaques génitales:* Acad. Sci. Paris, Comptes Rendus, v. 237, p. 405-407.

Mortensen, Theodor
(14) 1928-51, *A monograph of the Echinoidea:* v. 1-5 (for details, see p. *U638*).

Nichols, David
(15) 1959, *Mode of life and taxonomy in irregular sea-urchins:* Syst. Assoc., Publ. no. 3, p. 61-80.
(16) 1959, *Changes in the Chalk heart-urchin Micraster interpreted in relation to living forms:* Royal Soc. London, Philos. Trans., ser. B, v. 242, p. 347-437.
(17) 1962, *Echinoderms:* 200 p., Hutchinson Univ. Library (London).

Nissen, H.
(18) 1963, *Röntgengefügeanalyse am Kalzit von Echinodermenskeletten:* Neues Jahrb., Mineral., Geol., & Paläont., v. 117, p. 230-234.

Prenant, M.
(19) 1926, *Note histologique sur la structure et la croissance des dents d'oursins:* Arch. Zoologie Expér. & Gén., v. 65, Notes et Revue, p. 25-38.

Raup, D. M.
(20) 1959, *Crystallography of echinoid calcite:* Jour. Geology, v. 67, p. 661-674.
(21) 1960, *Ontogenetic variation in the crystallography of echinoid calcite:* Jour. Paleontology., v. 34, p. 1041-1050.
(22) 1962, *Crystallographic data in echinoderm classification:* Systematic Zoology, v. 11, p. 97-108.
(23) 1962, *The phylogeny of calcite crystallography in echinoids:* Jour. Paleontology, v. 36, p. 793-810.

Termier, Henri, & Termier, Geneviève
(24) 1953, *Classe des Echinides:* in Traité de Paléontologie, PIVETEAU, JEAN (ed.), v. 3, p. 857-947, Masson et Cie (Paris).

ADDITIONAL SOURCES OF ILLUSTRATIONS

(25) **Bather F. A.,** 1909
(26) **Durham, J. W.,** 1955
(27) **Forbes, E.,** a, 1852; b, 1856
(28) **Gignoux, Maurice,** 1933
(29) **Hawkins, H. L.,** a, 1911; b, 1912
(30) **Kier, P. M.,** 1962
(31) **Lambert, J.,** 1893
(32) **Mortensen, Theodor,** a, 1907; b, 1931; c, 1937; d, 1951
(33) **Parker, J. T., & Haswell, W. A.,** 1943
(34) **Westergren, A. M.,** 1911
(35) **Zoeke, M. E.,** 1952

ECOLOGY AND PALEOECOLOGY

By J. W. DURHAM
[University of California (Berkeley)]

Few detailed studies of the ecology of echinoids have been made. COOKE (3) has given an annotated bibliography of some pertinent papers subsequent to 1930. BRATTSTRÖM (1) has studied factors limiting the distribution of echinoderms at the entrance to the Baltic. NICHOLS (4) has presented an outstanding study of the burrowing habits of *Echinocardium cordatum* and other British spatangoids. Many specific details are available in MORTENSEN's monograph (5) in his discussions of individual species. Some details of ecology are presented in the individual parts of this *Treatise* volume. Other details and broad generalities are given here.

The Echinoidea are aquatic organisms represented by about 800 living species, primarily inhabitants of waters of normal

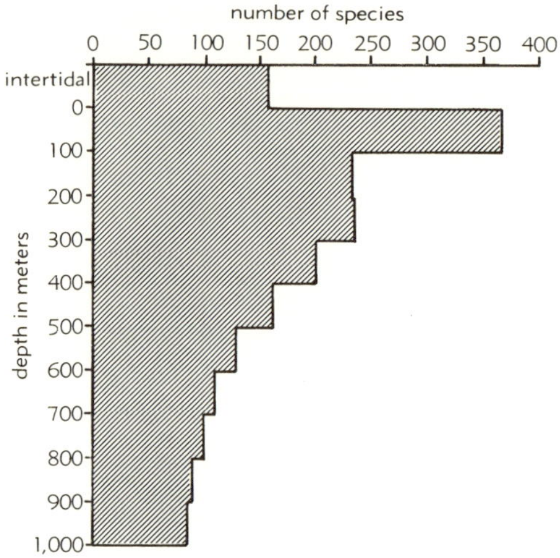

FIG. 197. Distribution of living species of echinoids in upper 1,000 m. of the sea (data from Mortensen, 5).

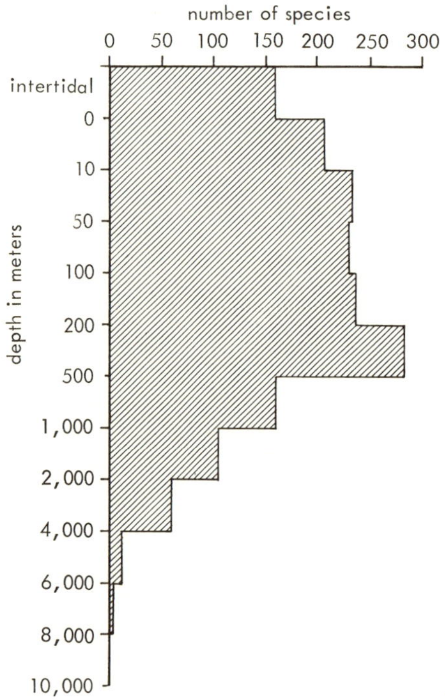

FIG. 198. Distribution of living species of echinoids in the sea (data from Mortensen, 5, and Brun, 1957).

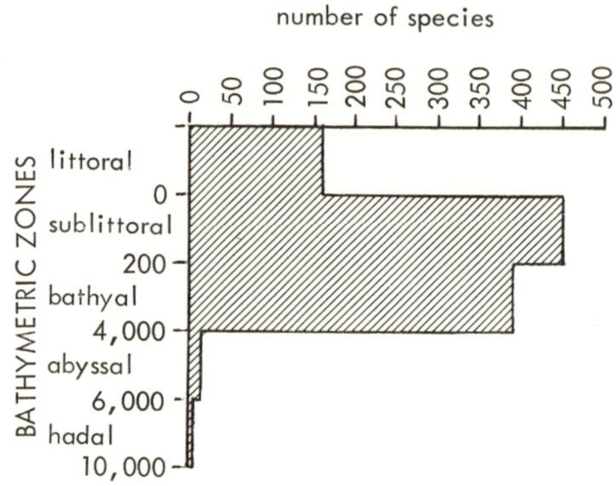

FIG. 199. Distribution of living species of echinoids in the sea according to bathymetric zones (sources as for Fig. 198).

marine salinity. However, a few (e.g., *Mellita quinquiesperforata*) may live in waters with a salinity as low as about 20 per cent (Aransas Bay, Texas). The studies of BRATTSTRÖM (1) show that in the Öresund (at the entrance to the Baltic) *Echinocyamus pusillus* likewise may live in water of about this same low salinity. Others occasionally live where the salinity possibly may range above 40 per cent (e.g., *Lytechinus*, in Laguna Madre, Texas).

Most echinoids live in subtropical and tropical regions where the shallow-water types are highly diversified and abundant. The diversity decreases away from the warmer areas, but *Psammechinus miliaris, Echinus esculentus, E. acutus norvegicus, Echinocyamus pusillus, Echinocardium cordatum, Echinarachnius parma, Schizaster fragilis, Spatangus purpureus*, and *Strongylocentrotus droebachiensis* are known from 70°N. latitude or farther north. An even larger number of echinoids is known from the Antarctic region, where the fauna is characterized by a diversity of cidarids and by the presence of many species that brood their young.

Bathymetrically, living echinoids are distributed (Fig. 197-199) from the littoral[1] to the hadal zones. More than 150 species are known to occur between high and low tides, that is, in the littoral zone. More than 360 species are found in the first 100 m.

[1] Bathymetrical zones referred to are those proposed in the *Treatise on Marine Ecology and Paleocology* (Geol. Soc. America, Mem. 68, 1957).

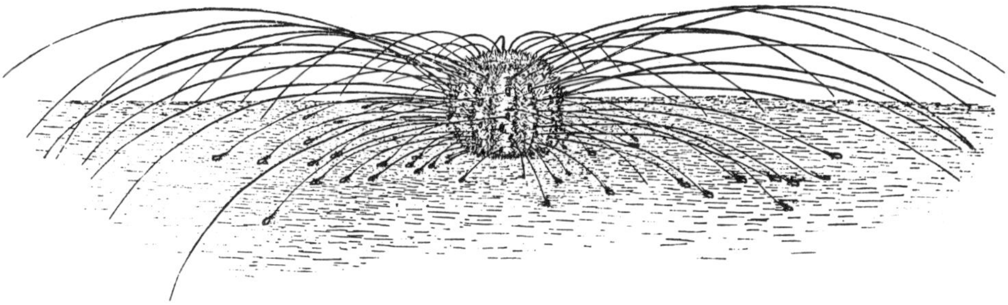

FIG. 200. Diadematoid echinoid *Plesiodiadema indicum* (DÖDERLEIN), showing exceptionally long curved slender spines, ×0.35 (Mortensen, 1923).

below the littoral zone, numbers of species gradually decreasing thereafter with increasing depth. At least two species, members of the genus *Pourtalesia,* are known to occur in the hadal zone, below 6,000 m. One family of regular echinoids (Echinothuriidae) and five families of irregular echinoids (Aeropsidae, Hemiasteridae, Asterostomatidae, Pourtalesiidae, Urechinidae) are represented in the abyssal and hadal zones, that is at depths below 4,000 m. The genera of these families known to occur in these depths are *Aceste, Aeropsis, Ceratophysa, Delopatagus, Helgocystis, Hemiaster, Homolampas, Kamptosoma,* and *Pourtalesia.* Many shallow-water species have a limited bathymetric range.

All known echinoids are benthonic. Many regular types live free on the sea floor, moving about by use of their spines and to some extent their tube feet. Where the substrate is rocky they may utilize cavities and crevices for protection and some (e.g., *Strongylocentrotus purpuratus, Echinostrephus aciculatus*) excavate cavities in rocks. The holes seem to be created by a rotary action of the spines, sometimes aided by gnawing action of the teeth. *S. purpuratus* is even known to bore into steel piling. The burrows of *Echinostrephus* may be as much as 4 inches deep. As the echinoid grows, it enlarges its burrow and in some instances appears to become imprisoned because the aperture is no longer large enough for egress. Some investigations indicate that the formation and utilization of burrows is a protection against wave action and thus is limited to areas of excessive wave action.

Other rocky-shore echinoids (e.g., *Podophora, Colobocentrotus*) protect themselves against wave action by the development of short flat-topped spines on the aboral surface and powerful sucker-like tube feet on the oral surface, enabling them to adhere like limpets to the rocks. The broad, flat shield-shaped spines of the fossil *Anaulocidaris* may indicate a similar habitat.

The aspidodiadematid *Plesiodiadema indicum* (Fig. 200) has very long, downwardly curved primary spines with enlarged tips that seemingly serve to keep it from sinking into the soft substrate on which it dwells. Many echinothuriids have similar hooflike structures on the end of the primary spines of the oral surface. Presumably, these serve the same purpose as those of *Plesiodiadema.*

Many present-day irregular types of echinoids live more or less buried in sand or muddy substrates. Some species of *Clypeaster* and *Encope* may live as much as 5 or 6 inches below the surface in sandy sediments. The minute *Echinocyamus* nestles among particles of sand and gravel. In quiet water *Dendraster excentricus* and rotulinid species assume a steeply inclined position with only the anterior third or fourth of the test buried in sediment, while the rest of the test extends up above the sea floor. Such species may have the food grooves leading to the mouth poorly developed in the anterior portion of the test. However, the same individuals, when disturbed by strong wave action, burrow beneath the surface of the sand. *D. laevis,* which lacks the marked eccentricity of *D. excentricus,* is not known to assume the inclined position.

The deep-water pourtalesiids (e.g., *Echinosigra*), with a funnel-like structure surrounding the peristome and a subanal

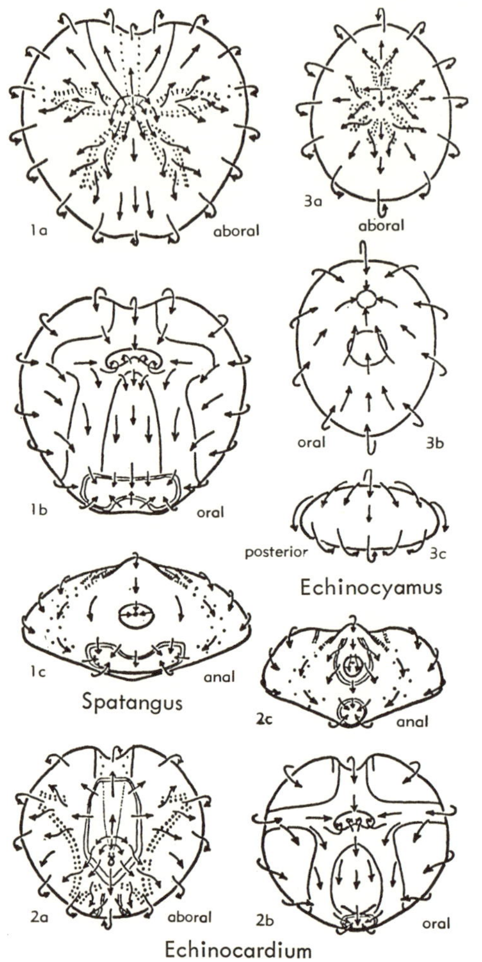

FIG. 201. Courses of ciliary currents (diagram.) on surface of test *(1,2)*, two spatangoids, and *(3)* clypeasteroid (4).——*1a-c. Spatangus purpureus.* ——*2a-c. Echinocardium cordatum.*——*3a-c. Echinocyamus pusillus.*

fasciole, presumably dwell partially submerged in the substrate, with only the upper surface exposed. The extinct holasterid *Hagenowia,* which in shape is suggestive of some pourtalesiids, presumably plowed through the fine calcareous substrate in which it dwelt with only the top of its long "neck," bearing the apical system, exposed.

The development of fascioles in the spatangoids is closely correlated with their burrowing habits and the kind of substrate inhabited. NICHOLS' recent study (4) has greatly clarified the functions of the fascioles, the different types of tube feet, and the various kinds of spines found in this group. The cilia on the clavulae of the fascioles, the bases of the spines, and the intervening epithelium work to create ciliary currents over the surface of the test (Fig. 201). On *Echinocardium, Spatangus,* and *Brissopsis* currents pass outward from the apical system along the ambulacra and over the ambitus (Fig. 202). On the oral surface some currents lead to the mouth, but most run posteriorly to the subanal fasciole, thence away from the test out through the sanitary tube. Presumably, somewhat similar currents are present on most irregular echinoids, varying in pattern according to their habits. Possibly the presence of fascioles indicates active burrowing of one type or another, inasmuch as *Echinocyamus pusillus* (Fig. 201), a nestler, has no fascioles.

Among the burrowing spatangoids studied by NICHOLS (4) the presence of respiratory funnels and sanitary tubes is apparently correlated with kind of substrate and depth

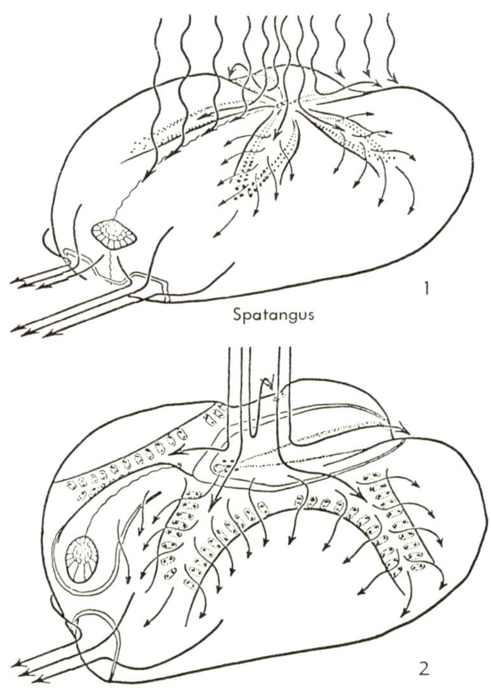

FIG. 202. Principal currents to and from test (diagram.) in two spatangoids (4).——*1. Spatangus purpureus.*——*2. Echinocardium cordatum.*

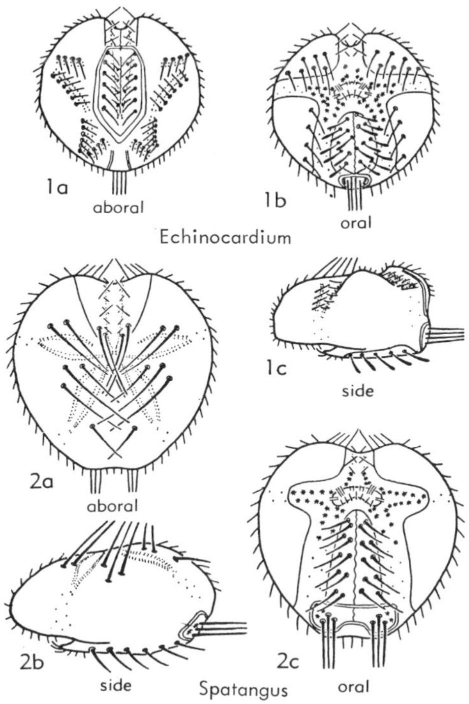

FIG. 203. Arrangement of spines (diagrammatic) in irregular echinoids (only a few of each series of spines shown) (4).——*1. Echinocardium cordatum; 1a-c,* aboral, oral, and lateral views.——*2. Spatangus purpureus; 2a-c,* aboral, lateral, and oral views (4).

of burial. *Echinocardium cordatum,* which burrows to a depth of 18 cm., usually in a sandy substrate, has both a respiratory funnel and a sanitary tube. *Spatangus purpureus,* normally a shell-gravel inhabitant that burrows to a depth of 2 cm. (to top of test), does not maintain a respiratory funnel at all times but regularly has a sanitary tube. *S. raschi,* which plows along the surface of a sandy mud substrate, has the apical surface exposed but utilizes a sanitary tube. NICHOLS considers that inasmuch as *S. purpureus* inhabits only clean shell gravel, enough water is drawn through the interstices of the gravel for respiratory purposes. This species is not equipped with dorsal burrow-building tube feet and could only maintain a respiratory funnel, if at all, by activity of the apical spines. In the species studied, the respiratory funnels and sanitary tubes are built by the combined action of the tube feet and the spines. The spines establish the tubes and funnels by a rotary action, mucus exuded by the tube feet being wiped onto the spines and the spines in turn plastering it onto the wall of the tubes and thereby strengthening them. In *Echinocardium cordatum,* after the animal has burrowed to a depth such that the spines can no longer reach the surface, the upper part of the respiratory funnel is maintained by the tube feet alone.

The clypeasteroid *Echinocyamus pusillus* (Fig. 201,*3*), also studied by NICHOLS, is a minute form which usually is a nestler in shell gravel. It never has been observed to push itself actively into the substrate, but is known to succeed in covering itself by pass-

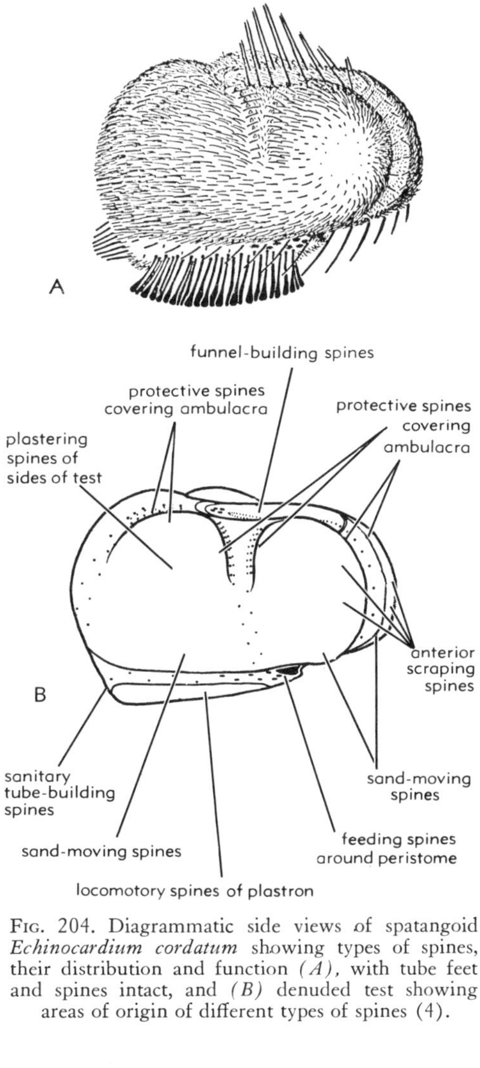

FIG. 204. Diagrammatic side views of spatangoid *Echinocardium cordatum* showing types of spines, their distribution and function *(A),* with tube feet and spines intact, and *(B)* denuded test showing areas of origin of different types of spines (4).

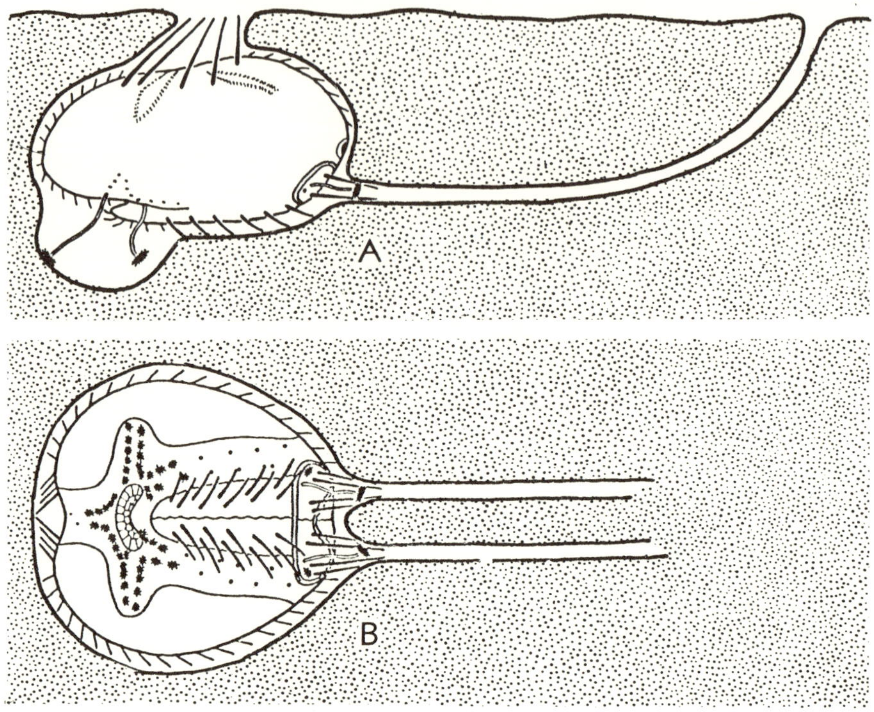

FIG. 205. Spatangoid echinoid *Spatangus purpureus* in newly made burrow, *(A,B)* side and bottom views; respiratory canal maintained by funnel-building spines, feeding excavation with 2 penicillate oral tube feet shown, and double sanitary canal built by sanitary-tube-building spines and penicillate tube feet; respiratory and sanitary canals lose direct communication with surface as animal burrows deeper (4).

ing small particles up onto its upper surface by means of the accessory tube feet. In this animal there are seemingly no feeding or burrow-building tube feet (the function of the accessory tube feet is little known, but they may also serve in food gathering), although there are 10 large sensory tube feet around the peristome.

The function and arrangement of the various kinds of spines have been studied by NICHOLS in *Echinocardium* and *Spatangus* (Fig. 203, 204). (1) The flattened spines on the plastron serve for locomotion, (2) the medium-length lateral spines adjacent to and outside the posterior ambulacra operate to move sand, (3) the short spines around the mouth assist in feeding, (4) the short lateral spines on the sides of the test help in maintaining the walls of the burrow and in passing sand posteriorly, (5) the short spines on the anterior margin of the test scrape material from the front wall of the burrow, (6) the spines adjacent to the sides of the petals form a protective arch over the petals, (7) the long dorsal spines aid in building the respiratory funnel, and (8) the tufts of longer spines within the subanal fasciole build the sanitary tube. The tubercles to which the different kinds of spines are attached are differentiated morphologically, and it seems very possible that the function of the spines and habits of many fossil echinoids can be inferred by comparing them to those living species that have been studied in this manner.

The division of labor among tube feet has also been studied by NICHOLS (Fig. 205, 206). Contrary to the interpretation of previous investigators, he has shown that the very extensile penicillate tube feet of the dorsal region of the anterior ambulacrum of *Echinocardium cordatum,* that often are observed extending out of the respiratory funnel, are not engaged in food gathering but are primarily used in building and maintaining the respiratory funnel. The penicillate tube feet of the phyllodes around the mouth serve for food gathering, while the penicillate tube feet within the subanal fasciole help to build and maintain the sanitary tube that carries waste products away. The tube feet of the petaloid parts of the ambulacra serve for respiration, while the nonspecialized tube feet of the remaining areas apparently serve for sensory purposes only. NICHOLS also has indicated that *Echinocardium cordatum,* which burrows much deeper than the other species studied, has a much greater number of penicillate respiratory funnel and sanitary tube-building tube feet than other species (Table 1). The numbers of tube-building tube feet seemingly also vary according to the kind of substrate in which the echinoid commonly dwells. The pores for the tube feet also vary (Fig. 207) in character according to function of the tube feet. Comparative study of the pores of fossil forms should enable similar inferences regarding the kinds, functions, and numbers of tube feet.

Among extinct echinoids it seems probable that those with elongated flexible tests and short spines (e.g., *Aulechinus*) may not have maintained a regular orientation with respect to the substrate and even may have lain on their sides like holothurians. Types with rigid tests and short spines (e.g., *Bothriocidaris, Melonechinus*) probably were oriented normally with the oral surface adjacent to the substratum but not buried within it. The more flexible lepidocentroids with flattened oral surface (e.g., *Proterocidaris*) probably lived with the oral surface submerged in the substrate and the apical surface projecting above the sea floor.

Like other echinoderms, echinoids tend to be gregarious, or at least to occur in very large numbers in local areas where conditions are favorable. More than 400 individuals of the sand dollar *Dendraster excentricus* were counted in an area of nine square feet in Puget Sound. Similarly SWAN (6) has recorded more than 1,300 specimens of the urchin *Stronglyocentrotus droebachiensis* from an area of three square meters along a rocky shore in Maine. Fossil sand dollars closely related to the living *D. excentricus* in some places form a major constituent of individual beds in the upper

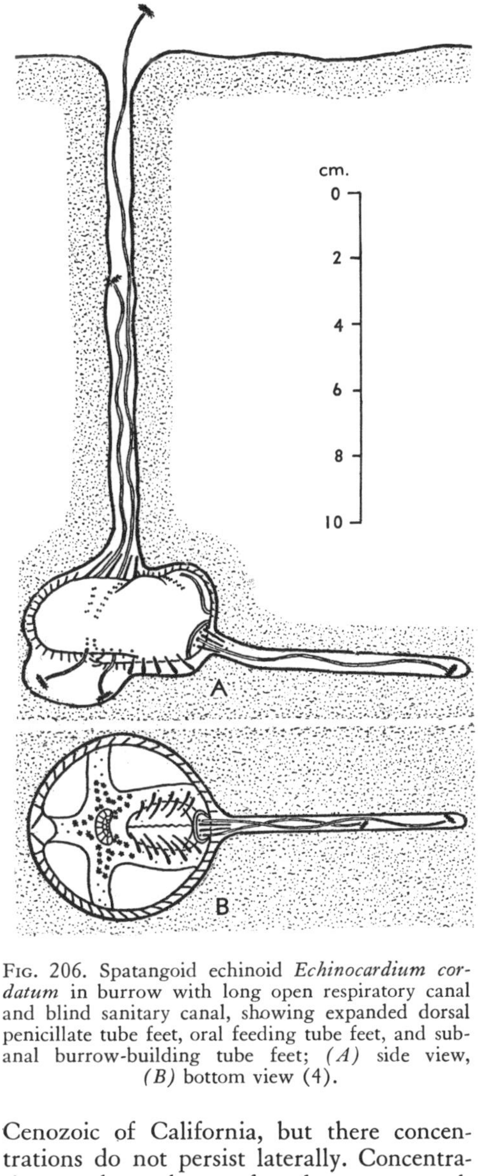

FIG. 206. Spatangoid echinoid *Echinocardium cordatum* in burrow with long open respiratory canal and blind sanitary canal, showing expanded dorsal penicillate tube feet, oral feeding tube feet, and subanal burrow-building tube feet; *(A)* side view, *(B)* bottom view (4).

Cenozoic of California, but there concentrations do not persist laterally. Concentrations such as those referred to are not always present, however, and in some areas no echinoids, or only scattered individuals, occur.

Regular echinoids often seem to show no preferred direction of locomotion but at times a weak preference for movement with ambulacrum III anterior is apparent. Irregular echinoids have a strong preference for movement with ambulacrum III anterior. Some spatangoids seemingly can move in

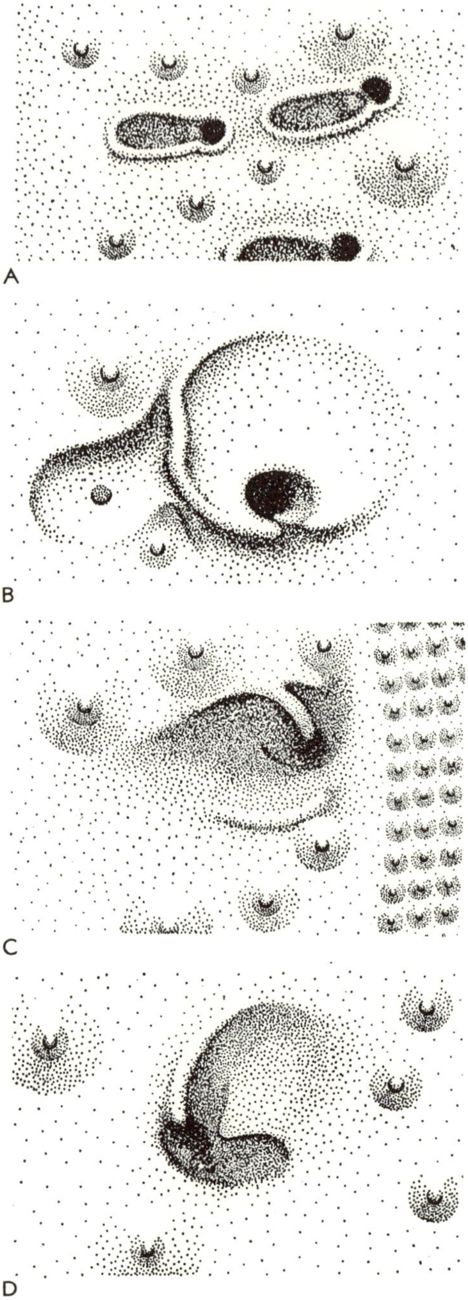

FIG. 207. Pores in test of *Echinocardium cordatum* for various types of tube feet (4).——*A*. Pores for funnel-building tube feet in ambulacrum III, within inner fasciole.——*B*. Pores for oral food-gathering tube foot in adoral plate of ambulacrum II (column b), two tubercles and pit (with stalk) for spheridia to left of wall.——*C*. Pores for subanal burrow-building tube foot in ambulacrum I (column b) within subanal fasciole.——*D*. Pore for sensory tube foot in ambital area of ambulacrum III.

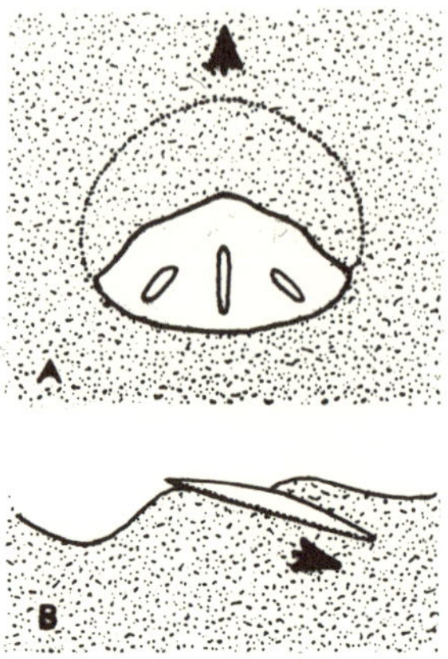

FIG. 208. Sand dollar *Mellita lata* burrowing, *(A)* seen from above, *(B)* side view (2, courtesy Trustees, British Museum, Natural History).

no other direction, but a few clypeasteroids have a limited capacity for movement in other directions. Movement is primarily accomplished by use of the spines but regular echinoids use their tube feet to ascend steeply inclined or vertical surfaces. In sand dollars the tube feet aid in burial by moving sand grains onto the aboral surface. *Lytechinus variegatus,* using its spines, can move along a horizontal sandy surface at a rate of 3 to 5.5 inches per minute. *Mellita quinquiesperforata* may assume a slightly inclined position (Fig. 208) near the surface of a sandy substrate, with the anterior margin buried and the posterior slightly exposed. In this position it may move continuously through the sand at a rate of 0.5 to 1 inch per minute.

Many regular echinoids react negatively to strong light, retreating into shaded areas or cavities during daylight. Others cover the aboral surface with fragments of plants, shells, and pebbles, holding them in place with their tube feet. In some echinoids (e.g., *Diadema*) the spines will quickly cluster and point toward the source of a shadow suddenly cast upon them. Mechanical irritation causes a similar reaction.

TABLE 1. *Number and Kind of Tube Feet in Some British Echinoids (Nichols, 1959)*

	Tube feet, approximate number per animal							
	Respiratory	Feeding	Burrow-building		Sensory only		Depth to top of burrow	Kind of substrate
Spatangus purpureus	200	50	subanal	4	anterior ambulacral lateral periplastronal	32 50 12	2 cm.	shell gravel
S. raschi	200	40	subanal	4	anterior ambulacral lateral periplastronal	54 73 14		sandy mud
Echinocardium cordatum	70	40	subanal anterior ambulacral	 70	anterior ambulacral lateral periplastronal	8 40 16	18 cm.	sand
E. pannatifidum	95	45	subanal anterior ambulacral	4 6	anterior ambulacral lateral periplastronal	24 22 16	3 cm.?	shell gravel
E. flavescens	55	32	subanal anterior ambulacral	4 5	anterior ambulacral lateral periplastronal	14 45 12	3 cm.?	shell gravel
Brissopsis lyrifera	120	35	subanal anterior ambulacral	6 30	anterior ambulacral lateral periplastronal	7 45 22	6-8 cm.	mud
Echinocyamus pusillus	45	none	none		buccal	10	nestler	shell gravel

A few urchins (e.g., *Psammechinus miliaris*) are known to react negatively to gravity, constantly ascending even steeply inclined surfaces if given the opportunity. The purpose of this reaction in nature is uncertain.

Most regular urchins will eat almost any organic material if given the opportunity and the need. However, some (e.g., *Echinus esculentus*) tend to be carnivorous and others (e.g., *Strongylocentrotus*) herbivorous. *Arbacia* has even been known to capture live specimens of the fish *Fundulus* at night. It is believed, however, that the fish were in a weakened condition. Clypeasteroids with well-developed food grooves and small mouths seem to live largely on small organic particles and organisms trapped by mucous strands and carried by ciliary currents or accessory tube feet or both along food grooves to the mouth. The spatangoids seemingly may either ingest large quantities of the substrate and digest the organic material from it, or where the oral food-gathering tube feet are well developed, organic material may be more or less sorted from the nonorganic components before ingestion.

REFERENCES

Brattström, Hans

(1) 1941, *Studien uber die Echinodermen zwischen Skagerrack und Ostsee, besonders des Öresundes, mit einer Übersicht über die physische Geographie:* Undersökningar över Öresund, no. 27, p. 1-329.

Clark, Ailsa M.

(2) 1962, *Starfishes and their relations:* 119 p., 16 pl., British Museum (Natural History) (London).

Cooke, C. W.

(3) 1957, *Echinoids* [annotated bibliography], in Treatise on marine ecology and paleoecology, HEDGPETH, J. W. (ed.): Geol. Soc. America, Mem. 67, v. 1, p. 1191-1192.

Nichols, David

(4) 1959, *Changes in the Chalk heart urchin Micraster interpreted in relation to living forms:* Royal Soc. London, Philos. Trans., ser. B, v. 242, p. 347-437, pl. 9.

Mortensen, Theodor

(5) 1928-51, *A monograph of the Echinoidea:* v. 1-5 (for details, see p. U638).

Swan, Emery

(6) 1958, *Growth and variation in sea urchins of York, Maine:* Jour. Marine Research, v. 17, p. 505-522.

PHYLOGENY AND EVOLUTION

By J. W. Durham
[University of California (Berkeley)]

The presence of a lantern, articulated spines, an internal radial water vessel, and presence of the Lovénian formula in the plates around the peristome place the Ordovician echinoderm *Bothriocidaris* among the echinoids.[1] *Eothuria,* considered by MacBride and Spencer to be a plated holothurian, has the same characters and likewise appears to be an echinoid. Accordingly, the oldest known echinoids (all Ordovician) include *Bothriocidaris* and *Eothuria,* as well as two other genera of undisputed echinoidean character, *Aulechinus* and *Ectinechinus.* The important common characters that stand out from a consideration of these genera are:

1. Both rigid *(Bothriocidaris)* and flexible tests are present, with differentiated ambulacra and interambulacra.
2. The apical system includes a full complement of oculars (except in *Eothuria*), but there is only one genital (none in *Bothriocidaris*).
3. The anus is within the apical system, a periproctal system of plates is present, and the mouth is at the opposite pole.
4. The ambulacra extend from the apical system to the mouth, with ambulacral plates always reaching the peristomial margin. Interambulacral plates may not extend to the margin.
5. The radial water vessel is internal in *Bothriocidaris* and *Eothuria*[2] but rests in a groove open to the exterior (although with internal ampullae) in *Aulechinus* and *Ectinechinus.*
6. The pores for the tube feet are close to the perradial suture in *Aulechinus, Ectinechinus* and *Eothuria,* but nearly medial in *Bothriocidaris.*
7. The ambulacral pores are variable in number, single in *Aulechinus,* double in *Ectinechinus* and *Bothriocidaris,* and multiple in *Eothuria.*
8. *Bothriocidaris* and *Eothuria* each have distinctive types of lanterns differing from that present in *Aulechinus* and *Ectechinus.*
9. All of these genera except *Bothriocidaris* have small apparently undifferentiated spines articulating in pits rather than on tubercles. In *Bothriocidaris* they rest on tubercles.

Review of the above-cited characters makes it seem evident, as it did to MacBride & Spencer, that the echinoid ancestry must extend considerably farther back in geologic time than the Ordovician. Further, the ancestral echinoid must have had the potential to give rise to these distinctive types as well as to all subsequent diverse morphologies present among the Echinoidea. It would seem that this ancestor must have had differentiated ambulacral and interambulacral areas; periproct and peristome at opposite poles; a full complement of ocular plates, but only a single genital plate and thus a single genital gland; articulating spines; internal ampullae for the tube feet; and masticatory apparatus either very simple or lacking. The clearly internal radial water vessel of *Bothriocidaris* casts doubt on the common inference that it was external in the ancestor.

The Lower to Middle Cambrian free-living edrioasteroid *Stromatocystites* has often been suggested as representing the stock from which the echinoids arose. This seems quite improbable inasmuch as the anus is already on the oral surface (i.e., irregular), and the ambulacra are restricted to this surface. In view of the fact that the highly differentiated Edrioasteroidea, Eocrinoidea, and Helicoplacoidea are already present in the lower part of the *Olenellus* Zone of the Lower Cambrian, it may well be that the ancestral echinoid had appeared before Cambrian time. Certainly, the diverse types present in the Ordovician indicate a pre-Ordovician ancestry.

The Early Cambrian Helicoplacoidea had well-differentiated ambulacra and interambulacra, seemingly had the mouth and anus at opposite poles, and thus are marked-

---

[1] Myannil (7) has recently published a description of new and important specimens of *Bothriocidaris* with all new data confirming its place among echinoids.

[2] Erroneously stated to be in an open external furrow by Durham & Melville (1, p. 262) (see MacBride & Spencer, 1938, p. 130, and fig. 8C).

ly different from the contemporaneous Eocrinoidea. This suggests that the echinozoan lineage (2), to which the Helicoplacoidea would seem to belong, had already arisen and that the common ancestor for the Echinodermata is not to be found in the Eocrinoidea, but in some unknown pre-Cambrian type.

Post-Ordovician evolution among the Echinoidea has moved along various paths, often independent of one another. The result has been the diversity of types presented in this volume, as well as others as yet unrecorded. The principal pathways followed seem to have included the following:

1. Adoption of a rigid test by most.
2. Development of anteroposterior orientation and preferred direction of locomotion.
3. Improvement of the water-vascular system by modification and specialization.
4. Modification of appendages for special functions.
5. Increased efficiency of reproduction and protection of young.
6. Specializations for feeding purposes.

Although all known Ordovician echinoids except *Bothriocidaris* had flexible tests, by the end of the Devonian, types with rigid tests (Palaechinoida) were well established, competing with their contemporaries with flexible tests (Echinocystitoida, Archaeocidaridae). The Miocidaridae, with imbricating plates, are the only echinoids reported to have persisted from the Paleo zoic into later geologic time. By the end of the Triassic the derivative Cidaridae with rigidly sutured plates were in full sway. The Cidaroida gave rise to all the subsequent diversity, with most forms characterized by a rigid test. The most notable exception to the generally rigid tests of post-Triassic echinoids occurs among the Echinothurioida, where seemingly, accompanying their exploration of deeper water habitats, calcification was reduced, resulting in a flexible test. In this group, however, imbrication of plates, even though of the same general type as in the Echinocystitoida, is rarely as complete. Commonly gaps of considerable size are present between adjacent plates along the sutural area and they imbricate only near the ends. A very few other living echinoids (e.g., *Astropyga*) have more or less flexible tests, but the rigid test seems to be the most successful, judging by its prevalence among modern echinoids.

Most regular echinoids do not seem to have a preferred direction of locomotion, and movement may begin in any direction, although observations suggest that there is a preferred direction for some. FELL has shown that there is a persistent tendency for the anus to move out of the apical system, ultimately resulting in its localization in interambulacrum 5 and the resultant development of an anteroposterior orientation. Soon after the development of this axis, the test elongated in this same direction. At the same time preferred locomotion along the direction of this axis appeared and with this major step some echinoids began to burrow in the substrate, and a whole new habitat was opened for exploration.

Improvement and modification of the water-vascular system has occurred in many different ways. Seemingly, the first step was migration of the tube feet from perradial to adradial areas of the ambulacra during the lower and middle Paleozoic. At nearly the same time, the Palaechinoida initiated multiplication of the functionally advantageous tube feet by increasing the number of ambulacral columns, along with retention of small plates, while the main stock retained the single column of small ambulacral plates. Subsequently, in the Mesozoic, the formation of compound ambulacral plates permitted strengthening of the test and at the same time ultimately led to multiplication of tube feet to a degree comparable with that of the palaechinoids—compare the ambulacra of *Heterocentrotus* with those of a genus such as *Proterocidaris*.

Accompanying the invasion of substrates permitted by oriented locomotion came a division of responsibilities among the tube feet. On the apical surface some of the tube feet were modified to serve as respiratory organs, and petals developed for accommodation of the elongated bladelike respiratory tube feet. At the same time the external gills present in the regular Euechinoidea and more primitive irregular types were lost. Adorally the tube feet were modified to serve primarily as food-gathering or food-sensing organs, ultimately resulting, in more specialized types, in the formation of phyllodes or pseudophyllodes. Adapically in burrowing types, the tube feet of the anterior interambulacrum adjacent to the apical system became elongated and modified to aid in building a respiratory funnel to the surface, while posteriorly a few tube feet of ambulacra I and V adjacent to the periproct became similarly modified to help in building and maintaining a sanitary tube (8).

Among the clypeasteroids still another modification of the water-vascular system, primarily to aid in food gathering, has occurred. In addition to the regular respiratory tube feet in the petals, minute accessory tube feet are present over much of the surface of the test, even extending into large areas of the interambulacra in some species. An accessory canal system (microcanal system) developed within the calcareous plates of the test in order to accommodate the vessels connecting the accessory tube feet to the radial canals. In some forms these accessory tube feet are exceedingly numerous, several of them occurring around the base of each spine. Possibly interruption of the interambulacral areas on the oral surface of many clypeasteroids is a corollary to the extensive development of accessory tube feet, as is the general narrowing of the interambulacral areas on the oral surface. Although poorly known and little studied as yet, the accessory tube feet seem to function primarily to gather minute organic particles on which these echinoids feed. They become more abundant and usually larger in diameter as the food grooves leading to the mouth are approached.

The earliest echinoids seemingly had small unspecialized spines. No pedicellariae have yet been recognized in association with them. However, the presence of granules, in addition to tubercles, on some of the later Paleozoic genera suggests that pedicellariae, as well as differentiated spines, were present. Pedicellariae have been reported from Mississippian and Pennsylvanian strata by Geis (3) and have been observed on a number of Jurassic echinoids, and it seems that these appendages were well developed and highly specialized by that time. By the Silurian some echinoids already had highly specialized spines (e.g., *Silurocidaris*), and diversification of spines in size and shape for locomotory and protective purposes proceeded rapidly among the Cidaroida. Among the regular echinoids, specialization of the spines on the apical surface for protection took many directions, from the large flat, shield-shaped spines of *Anaulocidaris* to the flat-topped pavement-forming spines of *Colobocentrotus*. In some, as among the Echinothurioida, poison glands are present on the tips of the spines. The primary spines may be greatly elongated (as in the Cidaroida) to aid in locomotion. With the development of the anteroposterior axis in the irregular echinoids came a specialization of spines, especially in spatangoids, according to position on the test. Some became short and paddle-shaped for locomotion through the substrate, others elongated to aid in tube building for respiratory and sanitary purposes, while still others were modified into the highly ciliated clavulae of the fascioles, serving to create water currents over the surface of the test.

Reproduction in most echinoderms takes place by discharge of eggs and sperm into the sea, fertilization and subsequent development taking place by chance outside of the test. Inasmuch as planktonic larval stages occur among all groups of living echinoderms, it seems probable that this was an early development in the common ancestral lineage. Enormous wastage, consequent on utilization of very large numbers of eggs and sperm, occurs in this method of reproduction and a number of echinoids have improved upon it. Most planktonic larvae are planktotrophic, that is, they feed on other smaller organisms. Others, however, depend on yolk stored in the egg (lecithotrophic). In this latter group, fewer and larger eggs are utilized. Direct development, with omission of the planktonic larval stage, occurs in a number of genera. Consequent on this, various means of protecting the young have appeared. In some cidaroids (e.g., *Ctenocidaris*) the young remain on the surface of the test, either around the apical system or around the mouth. The corresponding portion of the test is commonly depressed and the adjacent primary spines arch over the brood.

Among spatangoids such as *Abatus*, the

paired petals of the females may be greatly depressed to serve as brood pouches. In clypeasteroids such as *Echinocyamus,* a brood pouch may develop on the aboral surface of females of some species, while in the fossil *Fossulaster* it is on the oral surface. How the eggs or young get into these protective devices is unknown. As a consequence of these structures, a marked sexual dimorphism is developed in these genera.

All known Ordovician echinoids have relatively small peristomial areas, and it would appear that they must have fed on relatively small organic particles, perhaps obtained by rasping of the teeth of the lantern. Soon, however, a relatively larger peristome appeared in the Cidaroida and it is present in many of their regular descendants. This enabled the ingestion of larger particles and presumably adoption of a more omnivorous diet, as well as protrusion and utilization of the tips of the teeth and lantern for locomotion.

With the development of the anteroposterior axis, oriented locomotion, and invasion of the substrate, a whole new food supply became available. Burrowers like the spatangoids lost the lantern, the mouth remained moderate in size, moved to an anterior position, and food was obtained by ingestion of sediment along with masses of small organic particles gathered together by the specialized tube feet around the mouth. The Clypeasteroida developed accessory tube feet for gathering small organic particles and the food-groove system for conveying it to the central mouth. With increasing efficiency of this system, the mouth decreased in size, and the lantern seemingly was used only for mastication and not for food gathering. Sand dollars such as *Dendraster excentricus* have retained the centrally located mouth but have concentrated the food-gathering apparatus in the posterior portion of the test and assumed a semi-upright position on the sea floor, gathering food from the organic material suspended in the water above the sea floor.

At present the echinoids seem to be quite successful inhabitants of the marine realm, best suited to and as a result most diversified in tropical and warm temperate regions. In view of the present restriction of the tropics as compared to intervals in the late Mesozoic and early Cenozoic, it appears that they are probably more successful now (nearly 800 living species) than at any time in the past. At present they are known from all depths except the hadal trenches and include such forms as the armored *Colobocentrotus,* an inhabitant of the breaker zone, and the bizarre pourtalesiids, that largely live in bathyal depths. Included too are rock-borers (e.g., *Echinostrephus*) and burrowers beneath the sea floor (e.g., *Echinocardium*). The small *Echinocyamus* merely nestles down into the substrate, where its components are large enough to permit such action. However, the majority of echinoids merely move about on the sea floor or nestle in crooks and crannies.

REFERENCES

Durham, J. W., & Melville, R. V.

(1) 1957, *A classification of echinoids:* Jour. Paleontology, v. 31, p. 242-272.

Fell, H. B.

(2) 1963, *The evolution of the echinoderms:* Smithson. Rept. for 1962, p. 457-490, pl. 1-3.

Geis, H. L.

(3) 1936, *Recent and fossil pedicellariae:* Jour. Paleontology, v. 10, p. 427-448, pl. 58-61.

Hawkins, H. L.

(4) 1931, *The first echinoid:* Biol. Reviews, v. 6, p. 443-458.

MacBride, E. W., & Spencer, W. K.

(5) 1938, *Two new Echinoidea, Aulechinus and Ectinechinus, and an adult plated holothurian, Eothuria, from the Upper Ordovician of Girvan, Scotland:* Royal Soc. London, Philos. Trans., ser. B, v. 229, p. 91-136, pl. 10-17.

Mortensen, Theodor

(6) 1930 [1931], *Bothriocidaris and the ancestry of echinoids, a reply:* Vidensk. Medd. fra Dansk Naturh. Foren., v. 90, p. 313-352.

Myannil [Männil], R. M.

(7) 1962, *Taksonomiya i morfologiya roda Bothriocidaris (Echinoidea):* Eesti NSV Teaduste Akad. Geol. Inst., Uurimused, v. 9, p. 143-190, pl. 1-5 [Russian, with Estonian and English summaries.] [*The taxonomy and morphology of Bothriocidaris (Echinoidea).*]

Nichols, David

(8) 1959, *Changes in the Chalk heart-urchin Micraster interpreted in relation to living forms:* Royal Soc. London, Philos. Trans., ser. B, v. 242, p. 347-437.

CLASSIFICATION

By J. WYATT DURHAM
[University of California (Berkeley)]

INTRODUCTION

The history of the classification of echinoids has been reviewed by numerous previous investigators. Most useful are the works of AGASSIZ & DESOR (1846-47), DESOR (1855-58), DUJARDIN & HUPÉ (1862), A. AGASSIZ (1872-74), POMEL (1869; 1883), MORTENSEN (1904), LAMBERT & THIÉRY (1909-25, including a very extensive bibliography, albeit occasionally the citations are inaccurate or too incomplete for easy utilization), COULON (1933), and GIGNOUX (1933). The historical review presented here covers material published through 1965. It deals only with names and concepts of classification that ultimately were used or elevated in rank for use at the ordinal and higher levels. Our legal nomenclature dates from the tenth edition (1758) of *Systema Naturae* in which LINNÉ assigned all recognized echinoids to the genus *Echinus*. Early post-Linnean development of classificatory schemes was characterized by proposal of concepts and names at low-ranking levels, and their subsequent elevation to higher ranks. For example, the subclasses Regularia and Irregularia of many modern works were formally proposed at the family level (LATREILLE, 1825).

The section on the classification used in the *Treatise* is likewise concerned only with classification at the higher levels. Nomenclature and classification used by the individual authors within their various areas of responsibilities is mostly considered in the introductory material to their sections. At the 1953 International Congress of Zoology at Copenhagen proposals were made (ICZN, 1953) to extend the rules of nomenclature to include all higher category names. It proved difficult to obtain agreement on the application of the rules at these levels and in consequence the *Code* does not yet contain these provisions. However, in response to the requests in the *"Copenhagen Decisions"* (ICZN, 1953) DURHAM & MELVILLE (1957) included a list of all higher category names that had come to their attention at that time and designated type genera for them. In order to establish a reference standard to which the higher categories and their names can be unequivocally related it appears highly desirable to have designated types and this practice is continued in this section on classification for all higher categories that are used in this volume.

HISTORICAL REVIEW

In 1758 LINNÉ assigned all living echinoids to the genus *Echinus* (recognizing 17 species), although 24 years earlier (1734), JACOBEUS THEODORUS KLEIN had published the *Naturalis Dispositio Echinodermatum,* a much more advanced classification in which he recognized 22 genera and 12 suprageneric taxa of two ranks. KLEIN's work was so far superior to that of LINNÉ that many Continental authors, led by the French echinologists, have continued to use his names and attribute them to him, even though they antedated the tenth edition of the *Systema Naturae.* This practice was followed by MORTENSEN in parts of his monograph (1928-51), even though at times he appealed to the International Commission on Zoological Nomenclature to establish or maintain other nomenclature that he favored.

It was not until LESKE's (1778) *Additamenta ad Kleinii* that many of KLEIN's names were validated for nomenclatorial purposes. Those of KLEIN's names which were accepted by LESKE in this work legally must be attributed to LESKE as the first post-Linnean author to use them. LESKE did not accept all of KLEIN's nomenclature and systematics, and as a result some of KLEIN's names do not validly appear until later, in the works of others.

KLEIN recognized three major categories among the echinoids, based upon the position of the anus: Anocysti, with anus on the upper surface; Catocysti, with anus on the lower surface; and Pleurocysti, with anus on the side. Also KLEIN proposed the term Echinodermata for the echinoids alone, in recognition of their "spiny skin." It was not until later (beginning with BRUGUIÈRE in 1791) that it was extended to include the other members of the phylum.

After LESKE's work, the first significant additions were those of LAMARCK (1801, 1816), who proposed a number of new genera. PARKINSON (1811) repeated KLEIN's classification and validated more of his pre-Linnaean names. DE BLAINVILLE in 1822 used "Ceratodermaires" as a class for the Echinodermata in the modern sense, and "Cycloides" for the Echinoidea, and divided them into "reguliers" and "irreguliers."

GRAY, in 1825, recognized the difference between regular (families Cidaridae and Echinidae) and irregular types. He placed the latter in his "annectant group," which included the families Scutellidae, Galeritidae, and Spatangidae.

LATREILLE (1825) formally proposed, at the family level, the terms Regularia and Irregularia, although the concepts of these two groups date from CUVIER (1817) who did not formally recognize them in his classification. DE BLAINVILLE (1834) recognized two major groups of echinoids, Echinides excentrostomes and Echinides centrostomes, basing his divisions on the position of the peristome, whether eccentric or central.

The appearance (1835) of L. AGASSIZ's *Prodrome d'une monographie des radiares ou èchinodermes* marked the beginning of a new epoch in the study of echinoids. This vigorous and influential naturalist continued his studies on the group until the publication of the *Catalogue raisonné des Échinides* (with DESOR, in 1847). He left a marked impress on "echinology" and his influence, through his son ALEXANDER, continued into the present century. In the *Prodrome* AGASSIZ recognized three families in his order Echinides, the Spatangues, Clypeástres, and Cidarites. In the *Catalogus Systematicus Ectyporum Echinodermatum* (1840) these names were formally latinized to Spatangoidea, Clypeastroidea and Cidaroidea, but still assigned family rank. This classification was less refined than that of GRAY but did not suffer from the obvious errors of presentation present in the latter's work and was thus more widely accepted.

The series of papers by AGASSIZ and his disciples opened the way to marked and rapid advances in the knowledge and classification of echinoids. AGASSIZ's classification of 1840 was further elaborated in the *Catalogue raisonné des Échinides* by AGASSIZ & DESOR (1846-47). It was modified in this work by the addition of the family "des Cassidulides," the recognition of four "groups" ("Cidarides proprement dit"; "Salenies"; "Echinides"; and "Echinometres") within the cidarids, and the recognition of two additional "groups" ("Echinoneides" and "Nucleolides") within the cassidulids. AGASSIZ & DESOR recognized 101 genera and subgenera and slightly more than 1,000 species, a far cry from the single genus and 17 species of LINNÉ 90 years previously. These authors, like many subsequent French echinologists, failed to latinize their suprageneric taxa, and as a result one can only attribute higher category names to them by invoking Article 11, e, iii of the present Code of Nomenclature (1961). D'ORBIGNY (1851) fully latinized his nomenclature and recognized seven families (Ananchytidae, Spatangidae, Nucleolitidae, Galeritidae, Clypeasteridae, Echinidae, and Cidaritidae) within his order Echinoidea.

F. M'COY was apparently far ahead of most of his contemporaries in his evaluation of the rank to be assigned to higher categories, for in 1849 he proposed the order Perischoechinida for the reception of *Archaeocidaris, Palaechinus,* and *Melonites* [*Melonechinus*], noting that these Paleozoic echinoderms differed from members of the order Echinida by having a great number of rows [columns] of plates in the test.

In 1855 GRAY recognized the Irregularia as a "subdivision" (corresponding to his "annectant group" of 1825) of the Echinida, dividing them into two "sections," Mesostoma (with mouth subcentral) and Apomesostomi (with mouth excentric) on the basis of characters used by KLEIN. He formally recognized the families Scutellidae, Galeritidae, and Echinolampasidae in the first section and Spatangidae and Leskiadae in the second section. In addition, in his "systematic index" (*op. cit.,* p. 65-66) he recognized uncategorized (latinized) subdivisions under each of his families. These (Clypeasterina, Rotulina, Fibularina, Echinoneina, Cassidulina, Echinobrissina, Spatangina, and Brissina) have subsequently formed the basis for additional families and higher-ranking taxa.

DESOR, in his *Synopsis des Echinides fossiles* (1855-58), recognized the "Echinides reguliers ou endocyclique" with two families (Tesselles and Cidarides) and "Echinides irréguliers ou exocycliques" with five families (Galéridées, Dysastéridées, Clypéastroïdes, Cassidulides, and Spatangoides). He based his classification on a totality of characters including the endocyclic or exocyclic character of the periproct, the character of the ambulacra, the position and character of the mouth and the masticatory apparatus, the structure of the apical system, the position of the periproct, the shape and structure of the tubercles and spines, and the kinds of fascioles. DESOR also recognized two or more "tribes" within most of his families.

Although LATREILLE (1825) had formally recognized the distinction between regular and irregular echinoids in 1825, it was not until the time of ALBIN GRAS (1848) that this concept began to be consistently recognized in the classificatory schemes. Authors such as WRIGHT (1855-60), DESOR (1855-58), and DUJARDIN & HUPÉ (1862) followed the lead of GRAS in accepting the significance of this character.

In 1853 ALCIDE D'ORBIGNY prepared the first part of the Cretaceous echinoids for the many-volumed *Paléontologie Française*. On his death (1857) the description of the remaining Cretaceous, Jurassic, and Eocene echinoids was continued by G. COTTEAU and terminated in 1894. This work was largely descriptive and accompanied by numerous illustrations. Because of the seeming excellence of the descriptions and illustrations, the figures have been copied widely and disseminated in systematic literature and have greatly influenced echinoid systematics. Unfortunately, comparison of illustrations with the specimens upon which they were supposed to have been based has shown that there may be but slight resemblance between the two. Characters shown on these illustrations may or may not be present on the specimens. No illustration from this work or other contemporary studies illustrated by the same group of artists should be accepted as valid unless it has been favorably compared with the original (or a photograph of it) or undoubted specimens of the same species.

In England, T. WRIGHT (1855-60), almost as early as D'ORBIGNY, began his extensive studies on British echinoids and other echinoderms with his *Oolitic Echinodermata.* The volume on the Cretaceous was completed in 1871. WRIGHT's illustrations were not as impressive as those in the *Paléontologie Française* and consequently they have not been as extensively copied, but they are more reliable. He recognized 5 families in the suborder Endocyclica and 8 families in the suborder Exocyclica.

ALEXANDER AGASSIZ followed his father's footsteps in the study of echinoids. In 1872-74 he published the important work *Revision of the Echini,* which gathered together an immense amount of information about the morphology and distribution of living echinoids. Included also are exceedingly valuable chronologic lists, a bibliography and a synonymic index. The classification used was basically an updated version of AGASSIZ & DESOR, with the addition of subfamilies, a modernization of the nomenclature, and the utilization of the suborders (of the order Echini) Desmosticha (ambulacra equal and bandlike) and Petalosticha (petaloid ambulacra) of HAECKEL (1866). He also studied the Paleozoic groups for which M'COY had proposed the order Perischoechinida, and which had variously been considered as a group coordinate with the echinoids (M'COY) or referred to the crinoids (L. AGASSIZ), and concluded that they should form a third suborder of the Echini.

The first volume of ZITTEL's influential and impressive *Handbuch der Paläontologie* appeared during the years 1876-80. In this work he recognized the subclasses Palechinidea, and Euechinoidea of the class Echinoidea. He separated the Palechinidea on the basis of their age and the fact that the test was usually composed of more than 20 columns of plates, and recognized the orders Cystocidaridae, Bothriocidaridae, and Perischoechinidae. The subclass Euechinoidea was divided into the orders Regulares and Irregulares (referring to the position of the anus), with 4 families under the first and 2 suborders (Gnathostomata, with jaw apparatus; and Atelostomata, without jaw apparatus) with 6 families under the latter.

In 1869 POMEL recognized two suborders, "Tesselés" and "Échinides non Tesselés," and proposed a twofold classification of the "Échinides Tesselés" or suborder of "true echinoids." On one hand he divided them into three superfamilies: Globiformes, with anus opposite the mouth and orientation indicated by the madreporite; Lampadiformes, with mouth central and anus more or less independent of apical system and indicating orientation but otherwise regular in shape; Spatiformes, with mouth anterior and anus posterior, marking a distinct bilateral symmetry. On the other hand POMEL also noted that the "true echinoids" could be divided into two groups (Atelostomes and Gnathostomes) according to the absence or presence of a dental apparatus and finally concluded that this classification was "more natural" than the other. He noted that the suborder of the Tesselés or Périschéchinides (of M'Coy) was characterized by the numerous columns of plates, with the central ones of each area having a hexagonal shape, and recognized two families (Paléchinides and Mélonéchinides) within it.

Later POMEL (1883), still considering the echinoids as an order, combined the two methods and made the Spatiformes and Lampadiformes families under the Atelostomes; and the Globiformes along with the new group Clypeiformes (removed from the Lampadiformes because of the presence of a masticatory apparatus), families of the Gnathostomes.

A few years later, DUNCAN (1889) reviewed the classification of the echinoids, ranking them as a class, and presented a classificatory scheme that actually regressed from that of ZITTEL in that he did not differentiate the regular and irregular echinoids, recognizing 5 coordinate orders (Cidaroida, Diadematoida, Holectypoida, Clypeastroida, and Spatangoida) within the subclass Euechinoidea, and the new order Plesiocidaroida (for *Tiarechinus*) within the subclass Perischoechinoida, in addition to the three present in ZITTEL's scheme.

In 1895, MUNIER-CHALMAS in BERNARD (1895) presented a classification markedly different from those of his contemporaries. He based it upon the position and character of the dental apparatus, the character of the peristome, the number of columns of interambulacral plates and the character of the apical system. The higher categories of his classification are as follows:

Classification of Echinoids by Munier-Chalmas, 1895

Class Échinides
- Subclass Homognathes, pyramids similar, erect
  - Order Holostomes, peristome entire
    - Suborder Monoplacidés, a single column of interambulacral plates
    - Suborder Polyplacidés, numerous columns of interambulacral plates
    - Suborder Tétraplacidés, 4 columns of interambulacral plates
    - Suborder Diplacidés, 2 columns of interambulacral plates
  - Order Glyphostomes, peristome with gill slits
    - Suborder Glyphostomes Endocycles, regular echinoids
    - Suborder Glyphostomes Exocycles, periproct outside apical system
- Subclass Hétérognathes, pyramids not equal, depressed, apical system fused
- Subclass Atélostomes, no jaw apparatus, periproct outside apical system
  - Order Dysastéridés, apical system disjunct
  - Order Synastérides, apical system not disjunct
    - Group 1, apical system monobasal
    - Group 2, apical system tetrabasal
  - Appended to Échinides:
    - I. Cystoéchinides (for *Cystocidaris*), a transition group between cystoids and echinoids
    - II. Blastoéchinides (for *Tiarechinus*), a group converging toward the blastoids.

HAECKEL (1896, p. 481-489) in his *Systematische Phylogenie der Wirbellosen Thiere (Invertebrata)* presented a classification notable for the introduction of new, as well as a dual, nomenclature. It is as follows:

Classification of Echinoids by Haeckel, 1896

Class Echinoidea
- Subclass Cystechinida
  - Order Promelonaria
  - Order Eumelonaria
- Subclass Palechinida (or Palaeoechinoidea)
  - Order Stenopalmaria
  - Order Eurypalmaria
- Subclass Autechinida (or Euechinoidea)
  - Order Desmosticha (=Cidaronia)
    - Suborder Cidaridaria
    - Suborder Diademaria
  - Order Anthosticha (=Clypeastronia)
    - Suborder Conoclyparia (=Holectypida)
    - Suborder Scutellaria

Order Petalosticha (=Spatangonia)
 Suborder Cassidularia
 Suborder Spatangaria

J. W. GREGORY (1900) presented a classification that recognized three subclasses of echinoids, with the first two based primarily on the internal versus external position of the gills and the third with the periproct outside of the apical system. The major elements of his classification are as follows:

Classification of Echinoids by Gregory, 1900

Class Echinoidea
 Subclass 1, Regularia Endobranchiata
  Order 1, Bothriocidaroida
  Order 2, Cystocidaroida
  Order 3, Cidaroida
  Order 4, Melonitida
  Order 5, Plesiocidaroida
 Subclass 2, Regularia Ectobranchiata
  Order 1, Diademoida
   Suborder 1, Calycina
   Suborder 2, Arbacina
   Suborder 3, Diademina
   Suborder 4, Echinina
 Subclass 3, Irregularia
  Order 1, Gnathostomata
   Suborder 1, Holectypina
   Suborder 2, Clypeastrina
  Order 2, Atelostomata
   Suborder 1, Asternata
   Suborder 2, Sternata

In 1903 DELAGE & HÉROUARD, in their detailed treatment of morphology and anatomy, used a classification in which the primary separation was into the subclasses Reguliers and Irreguliers, based on the position of the periproct with respect to the apical system.

DÖDERLEIN (1906), studying the material collected on the "Deutsches Tiefsee-Expedition," was greatly impressed by the differentiation of the pedicellariae and the light they cast upon relationships. As a result of his studies, he proposed the following classification of Recent echinoids, with characters of the pedicellariae playing an important role in the assemblage of characters used.

Classification of Echinoids by Döderlein, 1906

Subclass Cidariformia
Subclass Diadematiformia
 Order Regularia
  Suborder Diadematina
   Tribe Streptosomata
   Tribe Stereosomata
  Suborder Saleniina
  Suborder Arbaciina
  Suborder Echinina
 Order Irregularia
  Suborder Clypeastroidea
  Suborder Spatangoidea

LAMBERT & THIÉRY published the first part of their compendium *Essai de nomenclature raisonée des Échinides* in 1909, the last part being issued in February 1925. They attempted to list, evaluate, and assign an age to every known species and give a reference to an adequate description of each. The work suffers from contradictions, sometimes from one page to next, and like many similarly long-extended works, the refinement of the systematics varies from part to part. Likewise, the work was not carefully edited and little attention was given to the Code of Nomenclature. However, despite all its faults, it is an invaluable reference and sourcebook, and very influential because of its completeness. The classification used embodies many new names for all categories above species rank and it contains a number of new concepts. The higher categories and their arrangement are as follows:

Classification of Echinoids by Lambert & Thiéry, 1909-1925

Class Echinoidea
 Subclass Gnathostomata
  Order Plagiocysta
    Suborder Cystocidaroida
  Order Endocysta
   Section Homalostomata
    Suborder Bothriocidaroida
    Suborder Perischoechinoidea
    Suborder Cidaroida
   Section Glyphostomata
    Suborder Streptosomata
    Suborder Stereosomata
  Order Exocysta
    Suborder Pileatoida
    Suborder Clypeastroida
 Subclass Atelostomata
  Order Brachygnata
    Suborder Globatoroida
  Order Nodostomata
    Suborder Procassiduloida
    Suborder Spatangoida

In 1896 JACKSON published the first of his detailed morphological studies on echinoids. This work dealt with the Palaechinoida and included a partial classification showing the relationships of Paleozoic echinoids to the cidarids. This classification was essentially that of ZITTEL. Later JACKSON (1912) elaborated and slightly modified his earlier scheme, including all echinoids within it. In this later scheme he eliminated the subclasses Palaechinoida and Euechinoida and recognized seven coordinate orders under the Class Echinoida. The Exocycloida were ranked as an order with 3 suborders (Holectypina, Clypeastrina, and Spatangina). The most notable advance was the recognition within the order Centrechinoida [Diademoida auctores] of three suborders (Aulodonta, Stirodonta, and Camarodonta) based upon the characters of the lantern and teeth.

JACKSON (40, p. 208) considered that *Bothriocidaris* represented the most "primitive type" of echinoid, an interpretation that was followed by many students of the group. Later, MORTENSEN (1928) raised doubts about the interpretation of *Bothriocidaris* and concluded that it was a cystoid. This precipitated a series of papers (espec. BATHER, 1931; CLARK, 1932; HAWKINS, 1931; JACKSON, 1931) defending its position among the Echinoidea, and a reply by MORTENSEN (1931; see also his Monograph, v. 5, p. 565-567, 1951) maintaining his position. DURHAM & MELVILLE (1957) considered that *Bothriocidaris* was a "true echinoid," though not on the direct line to later groups. This view was also held by CUÉNOT (1948), while TERMIER & TERMIER (1953) considered the genus to be of uncertain affinities. Fortunately MYANNIL (1962) has collected additional and well-preserved material demonstrating unequivocally the presence of a system of peristomial plates and a lantern, although the latter is of a somewhat different and more primitive type than that in the main line of echinoid development. Spines that are typically echinoid-like in appearance (59, pl. 4, fig. 2) adhere to at least one specimen. As BATHER noted, an internal radial water vessel must have also been present. The lantern and spines, as well as the characters of the plating of the test, indicate that *Bothriocidaris* is unequivocally an echinoid in terms of current concepts. MYANNIL considered that the Bothriocidaroida, Echinocystitoida, Palaechinoida, and Cidaroida each represent a separate line of descent from the ancestral echinoid.

The first volume of MORTENSEN's monumental *Monograph of the Echinoidea* (5 4to vols., 16 pts.) appeared in 1928, the last being completed in 1951. MORTENSEN thoroughly described every Recent species known to him and reviewed every recorded genus, fossil or living. The systematic position of every genus was analyzed, relationships and suggested derivations were indicated, and the limits and affinities of each family were considered. His major classification of the Echinoidea (restricted) (as summarized in v. 5, pt. 2, p. 565-574, 1951) is as follows:

Classification of Echinoids by Mortensen, 1951

- Class Echinoidea
  - Subclass Regularia
    - Order Melonechinoida
    - Order Megalopoda (referred to echinoids with some uncertainty)
    - Order Lepidocentroida
    - Order Cidaroida
    - Order Aulodonta
    - Order Stirodonta
    - Order Camarodonta
  - Subclass Irregularia
    - Order Holectypoida
    - Order Clypeastroida
    - Order Cassiduloida
    - Order Spatangoida

Suborders were recognized in many of these orders and as indicated in the summary section, as well as in numerous other places in the work, many of the orders were considered to be polyphyletic and merely families grouped together for convenience. Most significant of these conclusions was that the Irregularia had originated from more than one source among the Regularia. Notably he considered: *Pygaster* and its close allies to be derived from the pedinids; *Holectypus* and its close allies to be descendants of the diadematids, probably *Eodiadema*; the conulids and discoidids as derived from "some primitive stirodonts"; and the galeropygids as of probably diadematid origin, perhaps from the genus *Mesodiadema*. MORTENSEN, like BEURLEN (1934), derived his spatangoid suborders Meridosternata and Amphisternata from the collyritid-disasterid stem.

Another noteworthy conclusion of MORTENSEN was derivation of the Echinothuriidae from Paleozoic Lepidocentridae, rather

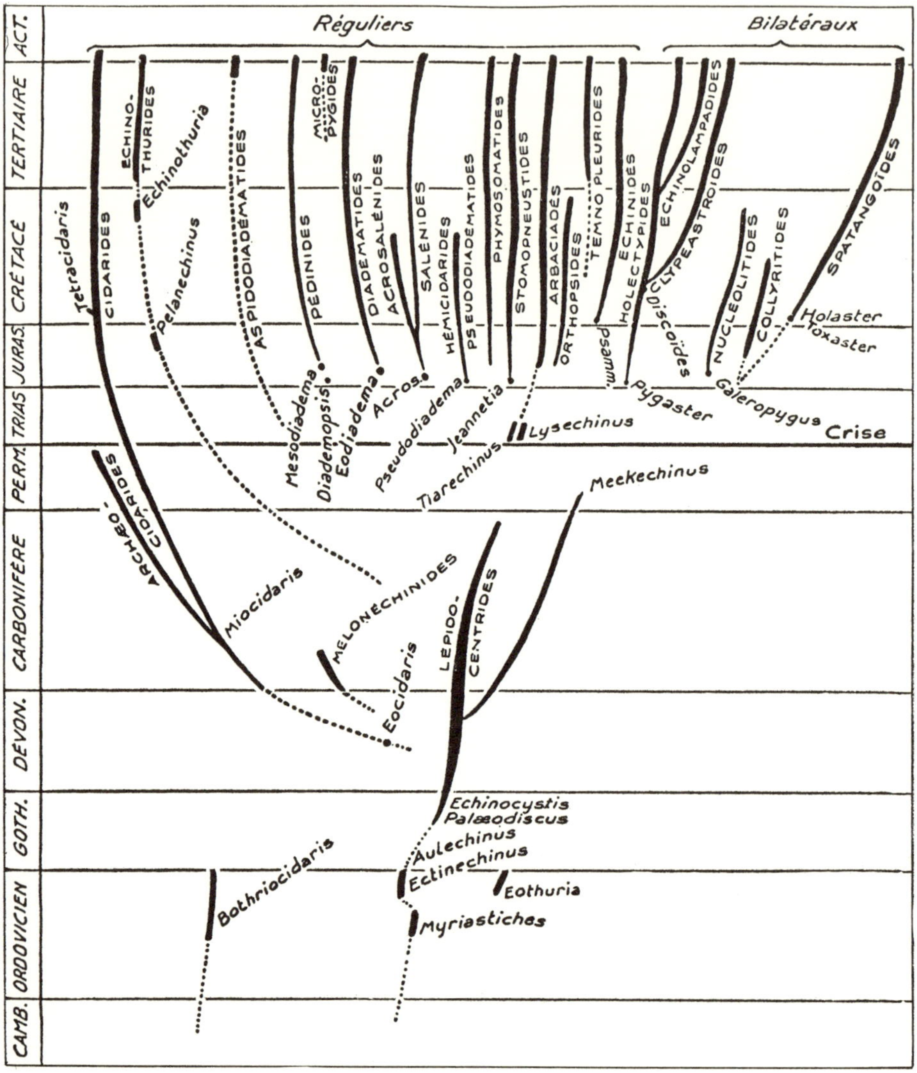

Fig. 209. Cuénot's phylogenetic tree of the Echinoidea (20).

than from a diadematid ancestry. He based this conclusion primarily on the occurrence of imbricating coronal plates in each, "many plated" ambulacra in some echinothurids, the numerous peristomial plates in both families, and the seeming absence of external gills in some echinothurids. He overlooked (or discounted) the fact that if such a derivation is accepted, the complex perignathic girdle of the echinothurids, as well as their external gills, would then have to be of independent origin from the similar structures in the other noncidaroid regular echinoids.

Mortensen did not present any graphic scheme showing his interpretation of the genetic relationships among the families and higher categories of the echinoids. It is difficult to construct a phylogeny on the basis of his work, because at times he made no choice between possible alternatives and in some cases he presented conflicting conclusions in different parts of the work. It is unfortunate that he was not able to summarize his views in a more complete form before his death.

In 1933 GIGNOUX published his *Les oursins réguliers fossiles, evolution et classification.* This excellent descriptive study was primarily concerned with descriptions of morphologic features and their evolutionary development and not with a synthesis of the resultant data into a classificatory hierarchy. He noted that our knowledge of echinoid evolution as documented by the fossil record is very incomplete, and for that reason did not care to propose a "rigid classification." He recognized *Bothriocidaris* as the ancestral echinoid and grouped the regular echinoids in three major categories: Paléchinidés, Cidaridés, and Glyphostomés.

The suborder Urechinina was proposed by H. L. CLARK (1946) for irregular echinoids with a sternum in which the "labrum [is] followed by a single plate." As defined, this suborder includes some but not all of the Holasteroida as proposed subsequently by DURHAM & MELVILLE (1957).

The section on echinoids in the *Traité de Zoologie* (ed. P.-P. GRASSÉ) was prepared by L. CUÉNOT (1948). It is a very clear and careful presentation, emphasizing anatomy and morphology, but includes only representative genera in each group. His phylogenetic tree (Fig. 209) expresses his ideas on relationships clearly, although he did not formally recognize any categories except subclasses (Reguliérs, Bilatéraux) above the family level. Notable is his representation of both the echinothurids and cidarids as representing distinct lineages from the other regular echinoids, deriving each from a Paleozoic ancestry. Although not as positive in his conclusions as MORTENSEN, he noted that two very distinct irregular types (*Pygaster, Galeropygus*) appear early in the Jurassic and suggests that they were derived from different sources. He likewise noted that in several groups of regular echinoids at different times the periproct started to move out of the apical system into interambulacrum 5 (citing the genera *Palaeopedina, Acrosalenia, Pseudosalenia, Hyposalenia, Goniophorus, Gauthieria,* and *Heterodiadema*), thus suggesting that there have been repeated attempts at the development of the irregular condition.

In the *Traité de Paléontologie* (ed. J. PIVETEAU), the section on the Echinoidea was prepared by TERMIER & TERMIER (1953). They were impressed with the plasticity of the echinoid branch of the echinoderms and noted that in consequence there are many obstacles to the establishment of a "rigorous" classification and a clear delineation of affinities. As a result many of their suprafamilial taxa are polyphyletic in origin (Fig. 210). Their order-subclass classification is a but slightly modified version of that used by MORTENSEN. *Bothriocidaris* they considered to be derived from a protocrinid ancestry, while the remaining echinoids were derived from the stromatocystitids. Their phylogeny shows the diadematid, echinothurid and cidarid stocks as directly derived from Paleozoic ancestors among the Lepidocentroida. The Aulodonta include descendants from both cidaroid and lepidocentrid ancestries. The cidarids gave rise to stirodonts twice and to the diadematids once. The Camarodonta are derived from 3 separate ancestries among the Stirodonta. The Irregularia are derived from pseudodiadematid, diadematid, and pedinid ancestors. The Protosternata were derived from cassiduloid and galeropygid stocks, as were the Meridosternata. Only the Cidaroida, Cassiduloidea, and Amphisternata, among post-Paleozoic echinoids, are shown as monophyletic in origin.

In 1960, ANDRÉ DEVRIÈS, as a result of his detailed morphologic studies on the phymosomatids and the genera *Heteraster* and *Toxaster,* presented an evolutionary scheme (Fig. 211) for these groups and their relatives. He considered the cidarids and diadematids to be derived from a Paleozoic lepidocentroid source. The irregulars are likewise polyphyletic: the pygasterids and *Loriolella* arising from a pedinid ancestry; the holectypids and the part of the galeropygids which gave rise to the Cassiduloida being derived from the diadematids; the remaining galeropygids possibly arose directly from a lepidocentroid source, and, in turn, through the collyritids gave rise to the other spatangoids.

DURHAM & MELVILLE (1957) presented a classification (Fig. 212) that conformed to the phylogeny they accepted and that

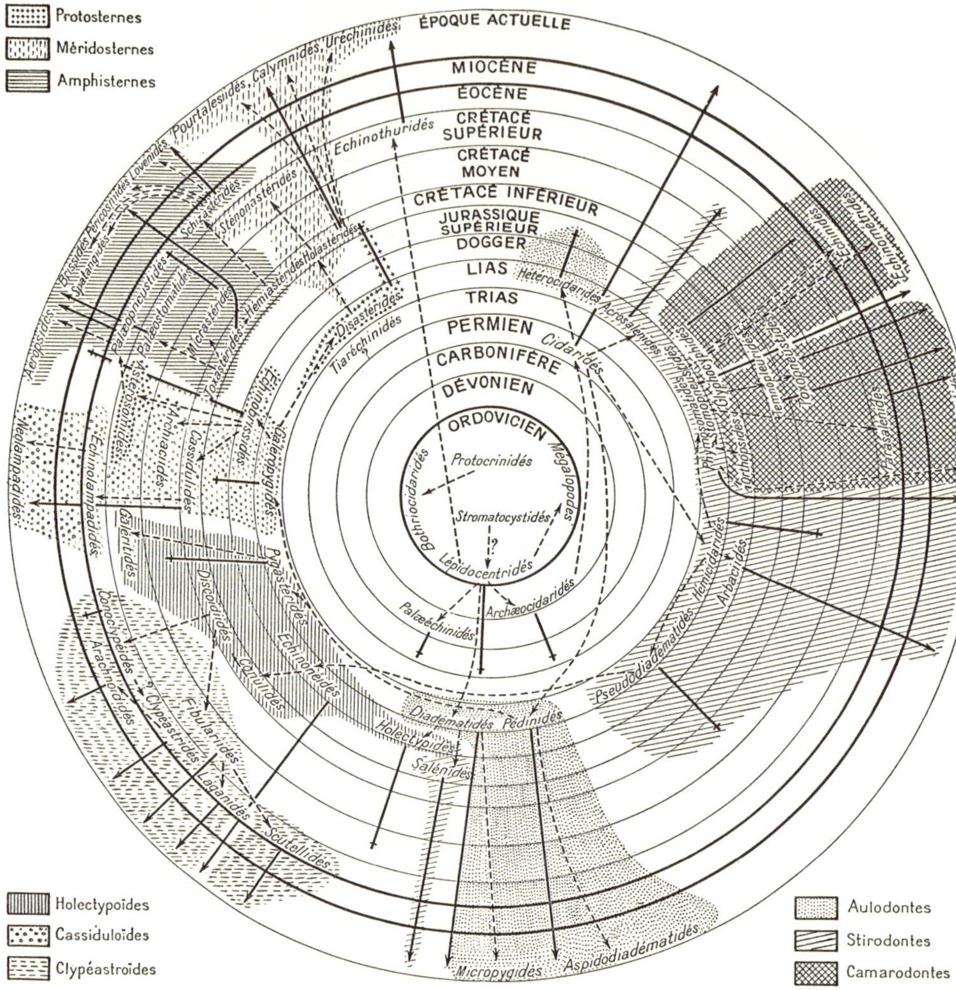

FIG. 210. The TERMIER & TERMIER phylogenetic chart of the Echinoidea (68).

was intended for use in the *Treatise.* In large part, but not completely, their phylogenetic tree agreed with relationships suggested by MORTENSEN in his monograph. As a result, they recognized 18 orders, in contrast to the 11 present in MORTENSEN's classification. They considered that the irregular condition had arisen in at least three different lineages and thus they abandoned this character (recommending usage of "regular" and "irregular" as morphologic terms only) as a basis for differentiating subclasses. Instead, they recognized the Perischoechinoidea and Euechinoidea as subclasses, differentiating them primarily on the lack of a perignathic girdle or of auricles in it, and the absence of gill slits in the Perischoechinoidea; and the presence of a complete girdle and gill slits (or descent from such forms) in the Euechinoidea. The presence or absence of branchial slits was considered (although not so stated) to correspond to the presence or absence of external gills.

DURHAM & MELVILLE considered that the echinothurids were derived from some pedinid stock and that their resemblance to the lepidocentrids was purely a product of adaptive convergence. Thus, contrary to the earlier opinions of MORTENSEN, TERMIER &

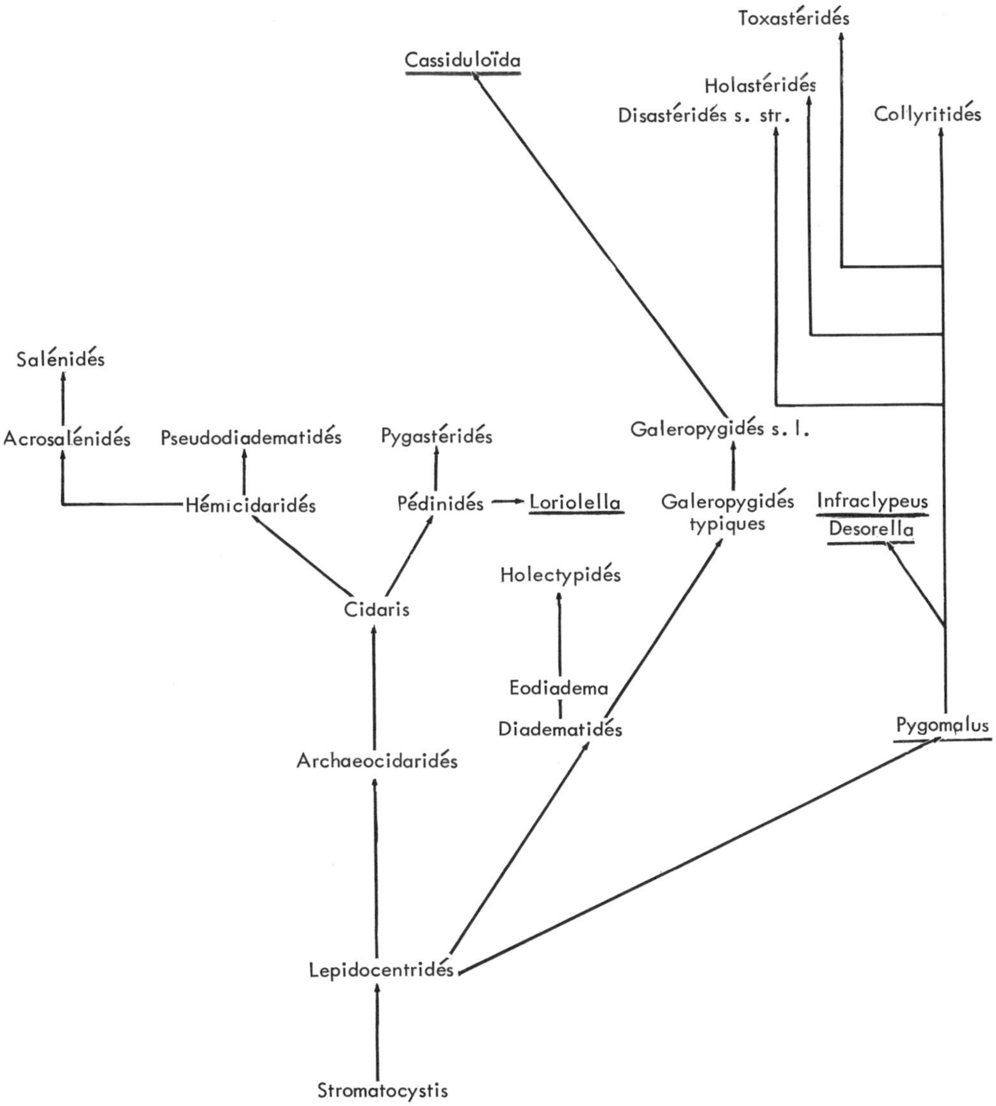

FIG. 211. Evolutionary relationships of certain groups of echinoids as interpreted by DEVRIÈS (24).

TERMIER, CUÉNOT, and the subsequent works of DEVRIÈS and PHILIP, they concluded that the cidarids were the only stock that continued from the Paleozoic into the Mesozoic. They also considered both *Bothriocidaris* and *Eothuria* to be echinoids. Among the euechinoids they recognized four superorders (Diadematacea, Echinacea, Gnathostomata, and Atelostomata), each representing a major branch of the subclass and composed of phylogenetically closely related orders. Subsequent developments have caused some modification of their classification, but for the most part it forms the basis of the scheme used in the present work.

While this volume was in press, PHILIP (1965) published a criticism of the classification proposed by DURHAM & MELVILLE and presented a classification which in his view (p. 44) "approaches that presented by Mortensen in his monograph." PHILIP also

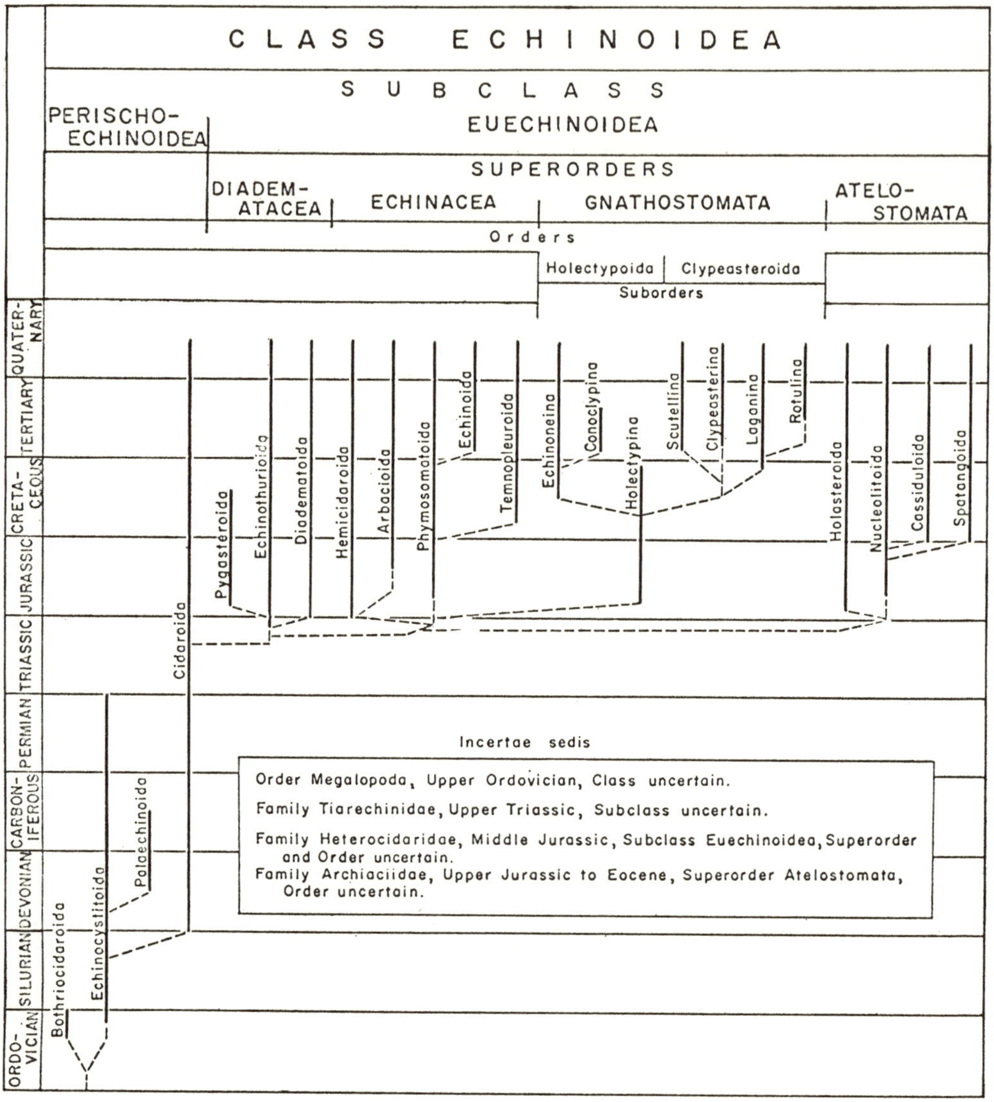

FIG. 212. Phylogeny and classification of the Echinoidea as interpreted by DURHAM & MELVILLE (29).

noted (p. 56) that "these collocations [his] cannot be claimed as more than grades (in the sense of Huxley, 1958)" and (p. 45, *abstract*) that "—the superorders Gnathostomata and Atelostomata are recognized as broad morphological grades probably uniting stocks of different ancestry." PHILIP's classification is thus one of grades and is based on a different philosophy than that used by DURHAM & MELVILLE. The classification of Echinoidea adopted in this volume, like that of DURHAM & MELVILLE, is based on the philosophy that a classification should reflect phylogeny, and thus is largely at variance with that of PHILIP. One of his proposals, the elevation to higher than ordinal rank of a taxon (PHILIP adopted Pseudoechinacea MORTENSEN at superordinal rank) for *Bothriocidaris,* deserves serious consideration, but because of the time of appearance of his paper, this could not be incorporated into the classification used herein.

As an appendix to their paper, DURHAM & MELVILLE (29, p. 262-272), in response to the recommendations of the *Copenhagen Decisions on Zoological Nomenclature* (ICZN, 1953), included a section on "Echinoid Order-Class Group Nomenclature." In compiling their lists of names, they decided to accept authorship and priority as of the first appearance of a name in latinized form, rather than from its first appearance in the vernacular. However, the new zoological Code (1961) contains a proviso (Article 11, e, iii) for names in the family-group category, which grants authorship and priority to the first appearance in the vernacular prior to 1900, providing that it has been so recognized by subsequent authors. If rules are eventually adopted for the higher-ranking groups it appears probable that the same rule would be adopted for them. Under this proviso, many of the authorships and dates credited to names in the DURHAM & MELVILLE list would be superseded.

At the same time DURHAM & MELVILLE designated type genera for all named groups that had come to their attention. Strict application of Article 11,e,iii would invalidate those of their type designations that applied to names which first appeared in the vernacular, so, in continuation of the spirit in which the list was prepared, all such type designations in their list are here designated as likewise applying to the first appearances of those names in the vernacular. DURHAM & MELVILLE, as a result of their compilation also prepared a list of names that they recommended for use in order-class group nomenclature, as follows:

Order-class Group Names Recommended for Use in Echinoid Classification by Durham & Melville, 1957

CLASS NAME
- Echinoidea LESKE, 1778

SUBCLASS NAME
- Endocyclica BRONN, 1860
- Euechinoidea BRONN, 1860
- Exocyclica BRONN, 1860
- Perischoechinoidea M'COY, 1849
- Pseudoechinoidea MORTENSEN, 1935

SUPERORDER NAMES
- Atelostomata ZITTEL, 1879
- Diadematacea DUNCAN, 1889
- Echinacea CLAUS, 1876
- Gnathostomata ZITTEL, 1879

ORDER NAMES
- Arbacioida GREGORY, 1900
- Bothriocidaroida ZITTEL, 1879
- Brachygnata LAMBERT, 1915
- Cassiduloida CLAUS, 1880
- Cidaroida CLAUS, 1880
- Clypeasteroida A. AGASSIZ, 1873
- Diadematoida DUNCAN, 1889
- Echinoida CLAUS, 1876
- Echinocystitoida JACKSON, 1912
- Echinothurioida CLAUS, 1880
- Hemicidaroida BUERLEN, 1937
- Holasteroida DURHAM & MELVILLE, 1957
- Holectypoida DUNCAN, 1889
- Megalopoda MACBRIDE & SPENCER, 1938
- Nucleolitoida HAWKINS, 1920
- Palaechinoida HAECKEL, 1866
- Phymosomatoida MORTENSEN, 1904
- Plesiocidaroida DUNCAN, 1889
- Pygasteroida DURHAM & MELVILLE, 1957
- Spatangoida CLAUS, 1876
- Stereosomata DUNCAN, 1889
- Temnopleuroida MORTENSEN, 1942

SUBORDER NAMES
- Amphisternata MORTENSEN, 1907
- Aspidodiademina MORTENSEN, 1939
- Calycina GREGORY, 1900
- Cassidulina CLAUS, 1880
- Clypeasterina A. AGASSIZ, 1873
- Conoclypina HAECKEL, 1896
- Diademina DUNCAN, 1889
- Echinina CLAUS, 1876
- Echinoneina H. L. CLARK, 1925
- Holectypina DUNCAN, 1889
- Laganina MORTENSEN, 1948
- Meridosternata MORTENSEN, 1907
- Orthopsina MORTENSEN, 1942
- Pedinina MORTENSEN, 1939
- Protosternata MORTENSEN, 1907
- Rotulina DURHAM, 1955
- Scutellina HAECKEL, 1896

TREATISE CLASSIFICATION

The classification used herein (Fig. 213), although based upon that proposed by DURHAM & MELVILLE (29), is a composite of the views and opinions of the authors (J. W. DURHAM, H. B. FELL, A. G. FISCHER, P. M. KIER, R. V. MELVILLE, D. L. PAWSON, and C. D. WAGNER) of the individual parts of the systematic section on the Echinoidea. As a result, some particulars of the expressed classification and relationships are not unanimously agreed upon.

Some of the suggested timings of events expressed on the chart (Fig. 213) and in this section have been modified by DURHAM from those indicated in the individual systematic parts on the basis of the overall perspective of the classification.

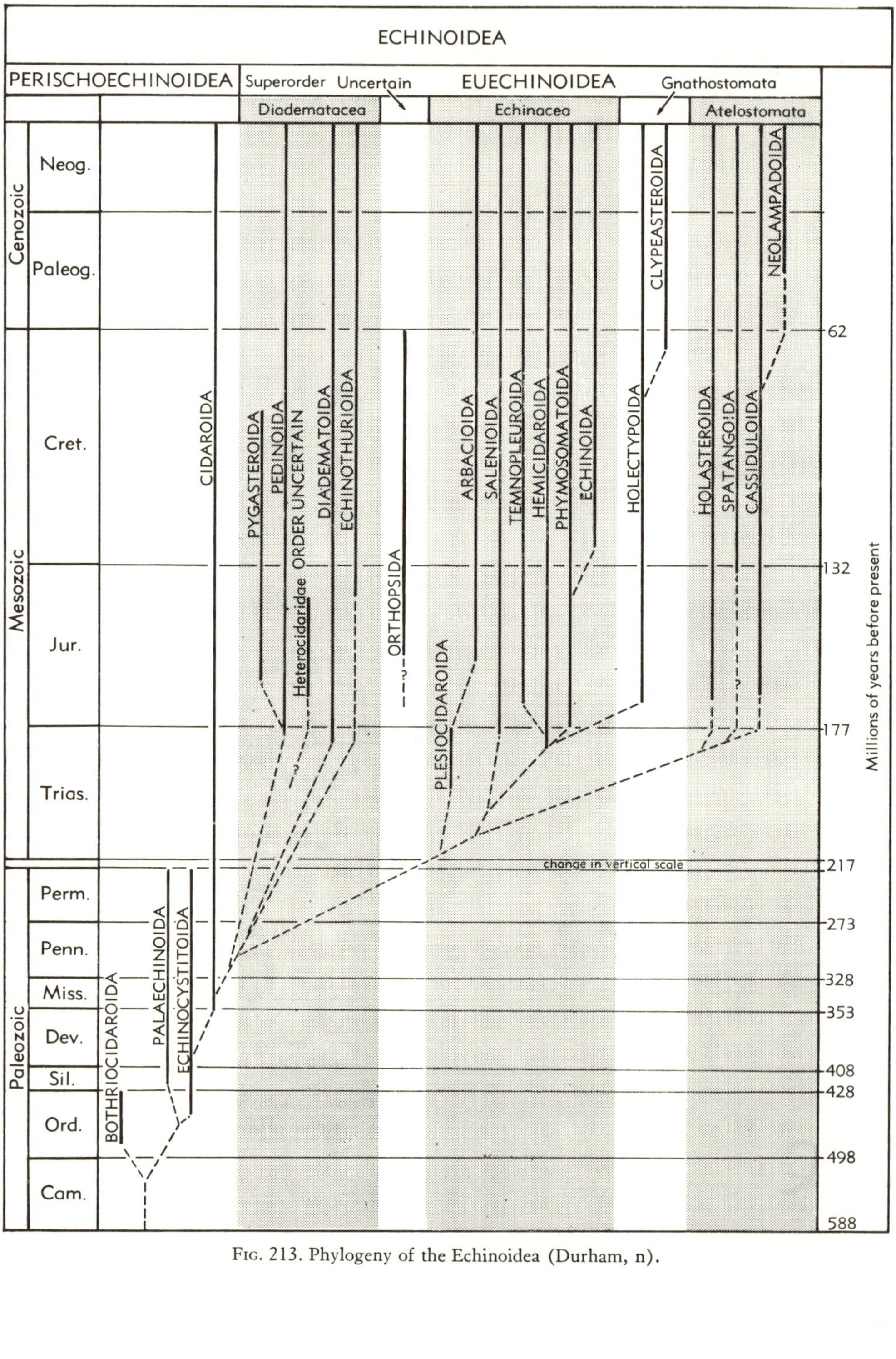

FIG. 213. Phylogeny of the Echinoidea (Durham, n).

Evolution is a branching and diversifying continuum and our classifications are artifacts superimposed upon this continuum as a shorthand for the purposes of communication. The principle that classification must reflect phylogeny is here accepted as an overriding precept. Therefore, every taxon must be monophyletic, and the classification used herein is an attempt to express the genetic relationships within the class as they are now understood or inferred.

It has been stated that a phylogeny must always show derivation of a higher-ranking taxon from one of lower rank. Acceptance of this precept leads ultimately to a single species (or subspecies) as the lowest conceptual taxon as the source of a new stock. Because of incomplete knowledge of the fossil record, it is often difficult or impossible to pinpoint this source, but this does not invalidate the conclusion. Furthermore, although the stated precept is correct in theory, the practical exigencies of graphically presenting the phylogeny of a large and complex group in limited space usually precludes its expression.

One of the major events in the evolution of the echinoids was the acquisition of external gills by some. This event occurred near the end of the Paleozoic, about midway in the course of their history, and apparently it was accomplished at about the same time as auricles appeared in the perignathic girdle. The development of these structures seemingly permitted the deployment of this stock into all the subsequent diversity of Mesozoic and Cenozoic noncidaroid echinoids. Meanwhile the parent stock continued on as the modern Cidaroida. It therefore seems appropriate to recognize two subclasses, the Perischoechinoidea and Euechinoidea, on the basis of these characters.

The **Perischoechinoidea,** as thus conceived, are characterized by absence of external gills (or lack of descent from such forms); corona with ambulacra composed of two to many columns, and interambulacra with one to many columns of plates; apical system endocyclic; ambulacral plates not compound; perignathic girdle absent or composed of apophyses only; lantern present, with grooved teeth. Presumably there were no spheridia or ophicephalous pedicellariae. The name was originally proposed by M'Coy (1849), as the order Perischoechinida, for Paleozoic echinoids, which, as then known, were characterized by having an odd number (3 or more) of columns of plates in the interambulacra. M'Coy regarded them as a group of equal rank with the Echinoidea, which were characterized by an even number of columns in the interambulacra. The concept was ultimately expanded to include the Bothriocidaroida, as well as the Cidaroida, by Durham & Melville (29). As here accepted, it includes the orders Bothriocidaroida, Palaechinoida, Echinocystitoida, and Cidaroida. Durham & Melville (29) designated *Palaechinus* M'Coy as the type-genus.

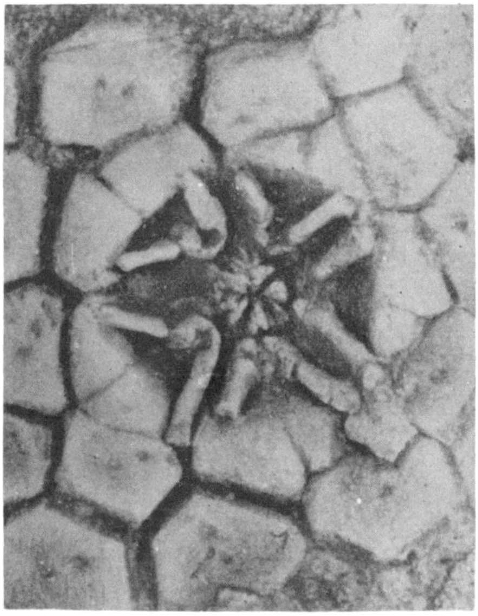

Bothriocidaris

Fig. 214. Internal view of lantern of *Bothriocidaris,* ×6.8 (59, pl. 3, fig. 2).

The **Bothriocidaroida** are characterized by a rigid test, sutures normal to the surface; interambulacra composed of a single column of plates, not reaching peristome; lantern with primitive pyramids and teeth; peristome plated; madreporite radial in position; with imperforate tubercles. The lantern (Fig. 214, 215) and spines clearly indicate that *Bothriocidaris* is an echinoid, while the character of the plates around and on the peristome (Fig. 216, 217), as well as the apparent lack of a differentiated set of genital plates in the apical system, raises doubts as to whether ranking the taxon

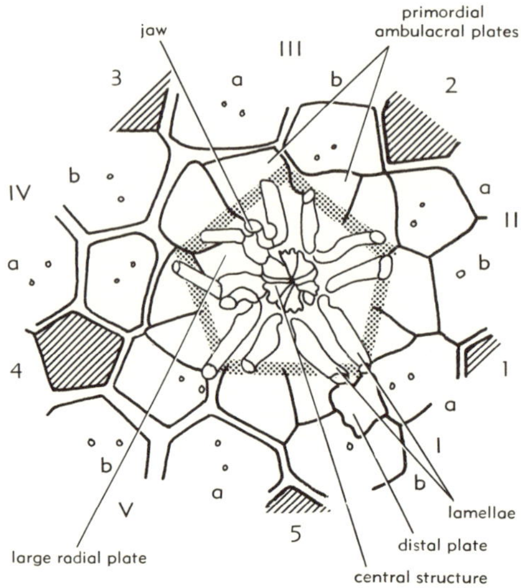

FIG. 215. Diagram of lantern of *Bothriocidaris*, ×6.8 (after 59, fig. 16).

based on it as an order adequately emphasizes its distinctiveness. The group was raised to ordinal rank by ZITTEL in 1879. The order is monotypic, including only the genus *Bothriocidaris* EICHWALD.

The rigid test and characters of the peristome and apical system of the Mid-to-Late Ordovician Bothriocidaroida, when contrasted with the flexible test of the Late Ordovician Echinocystitoida, indicate that the two orders had already greatly diverged from their common ancestor. It appears probable that this separation may have occurred in the Cambrian and that we should look for the ancestral echinoid stock in that interval.

The **Palaechinoida** have a rigid test, ambulacral plates beveling over adjacent interambulacral plates; no perignathic girdle; perforate or imperforate tubercles; and an apical system with both ocular and genital plates. The peristome is covered with ambulacral and interambulacral plates. Members of this group are easily differentiated from their contemporaries by the rigid test. The name was first proposed by HAECKEL (1866); DURHAM & MELVILLE (1957) designated *Palaechinus* M'COY as the type-genus.

The **Echinocystitoida** are characterized by strongly imbricating and flexible tests; ambulacral plates beveling under adjacent interambulacra; ambulacral plates imbricating adorally, interambulacral plates imbricating adapically; apical system with genital plates; absence of perignathic girdle. Most genera have perforate primary tubercles, but in others their character is uncertain. The flexibility of the test and the method of imbrication readily distinguish this Paleozoic group from contemporary palaechinoids. No resorption has been recognized around the peristome and the primordial plates are retained in the test. ZITTEL (1879) proposed the order Cystocidaridae for this group, basing it on his genus *Cystocidaris*, a synonym of *Echinocystites*. JACKSON (1912) proposed Echinocystoida as a substitute name for Cystocidaridae, basing it on the genus *Echinocystites* WYVILLE-THOMSON. DURHAM & MELVILLE (1957) formally designated WYVILLE-THOMSON's genus as the type. They preferred to retain the ordinal name based on the earlier genus, even though it was originally founded on the

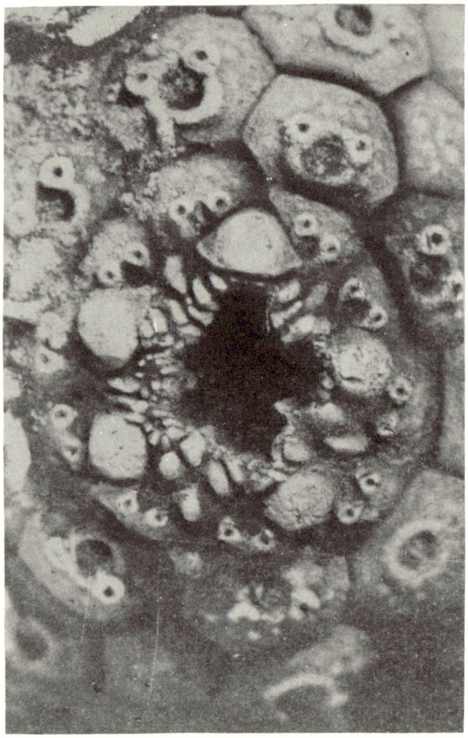

Bothriocidaris

FIG. 216. Peristome and peristomial plates of *Bothriocidaris*, ×10 (59, pl. 5, fig. 1).

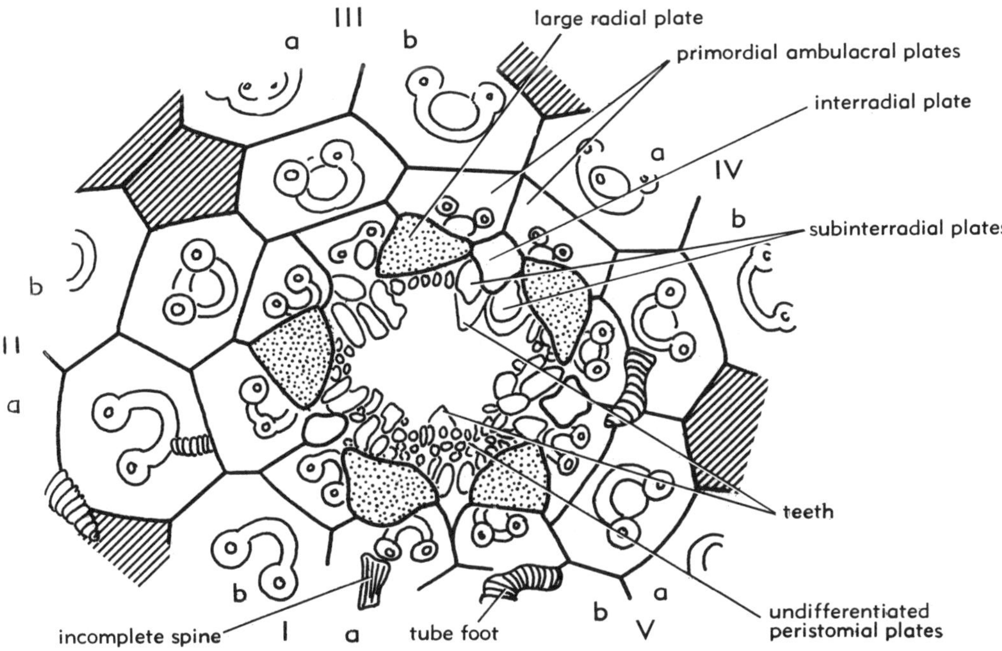

FIG. 217. Diagram of peristome and peristomial plates of *Bothriocidaris*, ×10 (after 59, fig. 12).

erroneous interpretation that the periproct was exocyclic.

The order Megalopoda was established by MacBride & Spencer (1938) for their genus *Eothuria* on the premise that it was a plated holothurian. Mortensen, as well as Durham & Melville, considered that *Eothuria* was probably an echinoid. Accepting this conclusion, Kier (this volume) has referred *Eothuria* to the family Lepidocentridae. Thus, Megalopoda becomes a junior synonym of Echinocystitoida.

The order **Cidaroida** is the one group that can unequivocally be shown to cross the boundary from the Paleozoic into the Mesozoic. This conservative group is characterized by rigid or flexible test; endocyclic periproct; apical system with genital plates; ambulacra of two columns, conspicuously narrower than interambulacra; primary interambulacral tubercles conspicuous, with well-defined areole; primordial interambulacral plates resorbed round peristome; peristomial membrane covered with imbricating plates; perignathic girdle absent or composed of apophyses only; spheridia absent; globiferous and tridentate pedicellariae only; primary spines with a cortex layer. The primary interambulacral tubercles of all Paleozoic members of the order, as well as all post-Paleozoic members, except those of the small family Psychocidaridae, are perforate. All post-Paleozoic cidarids have interambulacra with two (except for the genus *Tetracidaris*) columns of plates and all post-Triassic members have rigid tests. Claus (1880) first used the name (as Cidarideae) at the ordinal level. Durham & Melville formally designated *Cidaris* Leske as the type-genus.

The subclass **Euechinoidea,** as here interpreted, is characterized by presence of external gills or descent from such forms; corona composed of bicolumnar ambulacra and interambulacra, usually rigid; presence of auricles in perignathic girdle, or descent from such forms; lantern with grooved or keeled teeth, or absent; apical system of genital and ocular plates; presence of spheridia and ophicephalous pedicellariae. All unequivocal known members of this subclass are of post-Paleozoic age and most are post-Triassic. The majority, by far, of post-Paleozoic echinoids are referable to this subclass. A few inadequately known Mississippian and Pennsylvanian fossils (Geis,

1936) suggest the possibility that the taxon may have originated as early as mid-Paleozoic, and inferences as to the amount of time necessary for the four superorders to differentiate from one another by the Late Triassic and Liassic lend some support to this suggestion. The subclass Euechinoidea was formalized by BRONN (1860). DURHAM & MELVILLE (1957) designated *Echinus* LINNÉ as the type.

The genera *Tiarechinus* and *Lysechinus* (family Tiarechinidae), from the Late Triassic, appear to lack gill slits (and presumably, therefore, external gills) and thus are not easily referable to the Euechinoidea. However, in many features they are closely similar to some of the modern deep-sea arbaciids. The arbaciids resemble the tiarechinids more than any other group and thus may well be derived from them. Inasmuch as the arbaciids are unequivocal euechinoids, the tiarechinids are referred to the same subclass and superorder, utilizing DUNCAN's order Plesiocidaroida for them.

Within the Euechinoidea 18 orders and one ordinally unassigned family are recognized. On the basis of evolutionary relationships and common possession of morphological features deemed of major significance, all these orders except one (Orthopsida) may be readily grouped in four superorders: Diadematacea, Echinacea, Gnathostomata, and Atelostomata. The Orthopsida are not clearly referable to either the Diadematacea or Echinacea because of inadequate knowledge of their morphology and evolutionary relationships.

The superorders Diadematacea and Echinacea include predominantly radially symmetrical echinoids, many being characterized by compound ambulacral plates. The **Diadematacea** are characterized by perforate tubercles; presence of lantern with open foramen magnum and most with simple grooved teeth; a complete perignathic girdle; presence of gill slits; and ambulacra tending to have compound plates of diadematoid type. The lantern and teeth (except Pygasteroida) are of cidaroid type, suggesting that they originated from the latter group, perhaps in the mid-Paleozoic if the pedicellariae and plates described by GEIS (30) are not referable to some poorly known perischoechinoid. All orders, except the small group of Pygasteroida, are characterized by simple grooved teeth. The pygasteroids have an unusual type of keeled tooth (53) but are otherwise so closely similar to the pedinoids that they are included here.

Of the four orders referred to the Diadematacea, the **Echinothurioida** have been the object of most controversy. They are characterized by a flexible test; spines with hollow axis; noncrenulate tubercles; simple ambulacral plates on peristomial membrane; ambulacral plates compounded on a modified diadematoid plan; grooved teeth; endocyclic periproct. The external gills commonly are inconspicuous and may have been secondarily lost in some, and correspondingly, the gill slits may be poorly defined and difficult to recognize. However, a well-developed perignathic girdle is present. The presence of external gills, a perignathic girdle, and hollow spines indicates strongly that echinothurioids, contrary to the opinion of MORTENSEN and others, are not direct descendants of the Paleozoic echinocystitoids characterized by a flexible test but without perignathic girdle and external gills. The ordinal name was first proposed by CLAUS (1880) as Echinothurideae. DURHAM & MELVILLE (1957) designated *Echinothuria* WOODWARD as the type-genus. Not unexpectedly, considering the fragile test and usual deep-water habitat, the fossil record of the order is poor and the earliest known genus is *Pelanechinus,* from the late Jurassic of England.

The order **Diadematoida** is characterized by a rigid or flexible test; spines with hollow axis; tubercles crenulate or noncrenulate; peristomial membrane without ambulacral or interambulacral plates but with ten buccal plates; ambulacral plates simple or with diadematoid compounding; endocyclic periproct; grooved teeth; external gills and well-developed gill slits. No member of the order is certainly known from the pre-Jurassic, although MORTENSEN has suggested that a spine from the Lower Carboniferous of Germany may belong to a diadematoid. The hollow spines, flexible test of some, and type of ambulacral plating, suggest that the Diadematoida and Echinothurioida are closely related. DUNCAN (1889) first accorded the group ordinal rank. The substitute name Centrechinoida

JACKSON (1912) is unnecessary, inasmuch as the generic name *Diadema* GRAY, 1825, was validated under the Plenary Powers of the International Commission on Zoological Nomenclature (Opinion 206, 1954). DURHAM & MELVILLE (1957) designated *Diadema* GRAY as the type-genus.

The **Pedinoida** have a rigid test; solid spines; noncrenulate tubercles; endocyclic periproct; ambulacra with simple to compound plates; peristomial membrane with 10 buccal plates; shallow gill slits; and grooved teeth. The Late Jurassic echinothurioid genus *Pelanechinus* has many characters in common with some pedinoids, suggesting close relationships between the two orders. The diadematoid type of compound plating and the grooved teeth also indicate its close relationship to the Diadematoida. MORTENSEN (1939) proposed Pedinina as a suborder, whereas FELL (this volume) has elevated it to the rank of an order. DURHAM & MELVILLE (1957) designated *Pedina* L. AGASSIZ as the type-genus. The genus *Diademopsis* occurs in the Late Triassic of western Europe; the order is most diversified in the Jurassic and Cretaceous and at present is represented only by the living *Caenopedina.*

The small extinct order **Pygasteroida** differs from the Pedinoida in having an exocyclic periproct; teeth triangular in cross section, with an unusual buttressed keel; and consistently simple ambulacra. The order ranges through the Jurassic into the Cretaceous (Cenomanian). Immature specimens of *Plesiechinus ornatus* (J. BUCKMAN) are so similar to members of the pedinoid genus *Palaeopedina* that they were assigned to it by LAMBERT & THIÉRY (1910, p. 196), but a series of growth stages shows that they are undeniable pygasteroids. This short-lived group experimented with the irregular condition but for some reason were unsuccessful. Despite their general similarity, the hollow spines, crenulate tubercles, and simple keeled stirodont teeth of *Holectypus* demonstrate that the pygasterids and holectypids are not closely related. The group was given ordinal rank by DURHAM & MELVILLE (1957). *Pygaster* L. AGASSIZ is here designated as the type-genus.

The common possession of external gills, a diadematoid type of ambulacra, and auricles in the perignathic girdle indicate that the Diadematoida, Echinothurioida, and Pedinoida are derived from a common ancestor, which is linked to the Cidaroida by the possession of apophyses in the perignathic girdle and general similarity of the spines and tubercles, as well as the lantern. The Pedinoida (and their derivative, the Pygasteroida) have solid spines and a rigid test, like the Cidaroida, and are thus judged to have departed from the ancestral euechinoid stem earlier than the remaining Diadematacea with hollow spines.

The Jurassic family Heterocidaridae, based on the genus *Heterocidaris* COTTEAU, is characterized by diadematoid ambulacral plating, interambulacra overlapping ambulacra, perforate and crenulate tubercles, indistinct gill slits, and solid spines with a granulated cortex. The lantern is unknown. The general aspect suggests the Diadematoida but the solid spines preclude reference to this order. At present the order is left uncertain, but its assignment to the Diadematacea appears reasonably secure.

Simple keeled teeth differentiate members of the superorder Echinacea from those of the Diadematacea. The **Echinacea** are characterized by a rigid test; endocyclic periproct; solid spines; well-developed external gills and gill slits; peristomial membrane with 10 buccal plates; ambulacra simple or with various types of compound plating; lantern present, with simple keeled teeth; and girdle with well-developed auricles. The simple keeled teeth characteristic of the order differ markedly from those of pygasteroids with the keel bolstered by lateral buttresses.

The genus *Tiarechinus,* here assigned to the order Plesiocidaroida, appears in the Late Triassic (Carnian) and thus, if its relationship to the Arbacioida is correctly inferred, is the oldest undoubted euechinoid, as well as the oldest member of the superorder Echinacea. The genus *Acrosalenia* appears to be represented in the latest Triassic (Rhaetic) and is thus the earliest undoubted member of the Echinacea that can be assigned to an order (Salenioida). By the end of the Jurassic four more echinacean orders (Arbacioida, Hemicidaroida, Temnopleuroida, Phymosomatoida) had appeared and the last, the Echinoida, appeared within the Cretaceous. CLAUS (1876) used Echinideae as a taxon at the subordinal level. DURHAM & MELVILLE (1957) emended it to Echinacea and designated *Echinus* LINNÉ as the type-genus.

The enigmatic small group of the Tiarechinidae are characterized by a rigid small test; imperforate tubercles, noncrenulate; primordial interambulacral plate persistent, followed by three plates; apical system very large; peristome large; gill slits absent (or indistinct?). Although not used by MORTENSEN or DURHAM & MELVILLE, DUNCAN'S (1889) order **Plesiocidaroida** is here employed for this family. The peculiar arrangement of plates in the interambulacra is like that in the newly metamorphosed imago of *Arbacia* and the general morphology is suggestive of the arbacioids. They are known only from the Triassic. On this basis it is considered that the arbacioids may have been derived from the Triassic tiarechinids.

The **Arbacioida** are characterized by a rigid test, ambulacral plating simple to compound, when compound of the arbacioid type; primary tubercles imperforate, noncrenulate; spines smooth, with some development of cortex; periproct with 4 or 5 plates; lantern stirodont; primordial plates persistent; epistroma usually present. The group first appeared in the mid-Jurassic and has continued on to the Recent. GREGORY (1900) first ranked the group as a suborder (Arbacina). *Arbacia* GRAY was designated as the type by DURHAM & MELVILLE (1957).

The suborder Calycina GREGORY (1900) has priority over Salenina DELAGE & HÉROUARD (1903) and was recommended by DURHAM & MELVILLE (1957), but Salenioida is preferred by FELL & PAWSON for use in the *Treatise* because GREGORY'S name was not based on any included taxon. The **Salenioida** are characterized by a rigid test of cidaroid aspect; stirodont lantern; ambulacra simple or with diadematoid compounding of plates; apical system with one or more large suranal plates; primary tubercles usually crenulate; and spines with collar and cortex. The apical system with suranal plates and posteriorly positioned periproct are easily recognized and distinctive features. The genus *Acrosalenia* is uncertainly identified in the Late Triassic, the group continuing on to the Recent. The group is elevated to ordinal rank by FELL & PAWSON (this volume). *Salenia* GRAY was designated by DURHAM & MELVILLE (1957) as the type-genus.

The **Hemicidaroida** are characterized by a stirodont lantern; apical system without large suranal plates; ambulacral plates diadematoid, simple adapically in some; primary tubercles perforate, mostly crenulate; and peristome large, with conspicuous gill slits. Two genera (*Hessotiara* and *Pseudodiadema*) representing two families are present in the basal Jurassic (Hettangian), while the genus *Diplopodia* is reported from the Rhaetic, indicating that the order must have had a prior origin. No members are known in the post-Cretaceous. Hemicidarina was proposed by BEURLEN (1937) as a suborder. DURHAM & MELVILLE (1957) designated *Hemicidaris* L. AGASSIZ as the type-genus and ranked the assemblage as an order. The family Pseudodiadematidae is assigned to the order by FELL & PAWSON (this volume), although DURHAM & MELVILLE had included it in the Phymosomatoida.

The lack of large suranal plates and imperforate primary tubercles distinguish the **Phymosomatoida** from the Hemicidaroida. The order is characterized by a stirodont lantern; apical system without large suranal plates; primary tubercles imperforate; ambulacral plates simple or diadematoid. The genus *Jeannetia* appears in the Hettangian of France. The genera *Glyptocidaris* (Phymosomatidae) and *Stomopneustes* (Stomechinidae) are the only two representatives in the Recent; the order was most diversified in the Cretaceous. The order is presumably derived from the Hemicidaroida by loss of the perforate condition of the primary tubercles. The Phymosomina were proposed as a "tribe" by MORTENSEN (1904), later considered as a suborder in his monograph, and elevated to an order by DURHAM & MELVILLE (1957), who designated *Phymosoma* HAIME as the type-genus.

The Temnopleuroida and Echinoida, in contrast to the other Echinacea, have a closed foramen magnum in the lantern. The **Temnopleuroida** are characterized by a camarodont lantern; test usually sculptured, if not, gill slits sharp and deep; ambulacral plates compound, diadematoid or echinoid in character. *Glyptodiadema* appears in the

Pliensbachian, but the order is most diversified in the Recent. It seems possible that the order may have had its origin in the Pseudodiadematidae (order Hemicidaroida). The group was given subordinal rank (as Temnopleurina) by MORTENSEN in his monograph (1942) and was elevated to ordinal rank by DURHAM & MELVILLE (1957) who designated *Temnopleurus* DUNCAN as the type.

The order **Echinoida** is characterized by a camarodont lantern; nonsculptured test; shallow gill slits; imperforate and noncrenulate tubercles; and compound ambulacral plates of echinoid type. The genus *Spaniocyphus,* referred to the order by FELL & PAWSON (this volume) occurs in the Valanginian of Europe, while *Trochoechinus* is found in the Upper Cretaceous, and *Psammechinus* may occur in the Upper Cretaceous. The genus *Echinometra* is first recorded in the Paleocene. The maximum diversity is in the Recent. MORTENSEN (1951) suggested that the order may have been derived from the Stomechinidae. CLAUS (1876) ranked the group as a suborder, MORTENSEN in his monograph considered it as a suborder. DURHAM & MELVILLE elevated it to ordinal rank and designated *Echinus* LINNÉ as the type-genus.

The small order **Orthopsida** is of uncertain superordinal assignment. It is characterized by a camarodont lantern; rigid test; ambulacral plates simple or with a few triads (29, p. 254, text-fig. 1-c); tubercles perforate, noncrenulate. The genus *Dubarechinus* occurs in the Domerian of Morocco, while the order becomes extinct by the end of the Cretaceous. FELL & PAWSON (this volume) suggest that the group may be independently derived from some aulodont ancestry, although DURHAM & MELVILLE had suggested a possible hemicidarid derivation. Orthopsina was proposed as a suborder by MORTENSEN (1942) and is elevated to ordinal rank by FELL & PAWSON (this volume). DURHAM & MELVILLE designated *Orthopsis* COTTEAU as the type-genus.

The two remaining superorders, Gnathostomata and Atelostomata, in contrast to the preceding euechinoids (except the order Pygasteroida), are characterized by an exocyclic periproct and a lack of truly compound plates in the ambulacra. A posteriorly eccentric but still endocyclic position of the periproct is found in several distinct groups of the Euechinoida, such as *Palaeopedina* of the Pedinoida, *Heterodiadema* and *Pseudodiadema* of the Hemicidaroida, and *Phymosoma* of the Phymosomatoida. As far as can be recognized, this exocyclic tendency was not very successful in the Diadematacea, resulting only in the short-lived Pygasteroida, but in the Echinacea it was very advantageous, resulting in the gnathostomous order Holectypoida and ultimately the very successful, mainly Cenozoic, Clypeasteroida or sand dollars.

The origins of the **Atelostomata** are less clearly evident but the fact that young *Apatopygus recens, Echinolampas depressa,* and *Conolampas sigsbei* possess lanterns (MORTENSEN, 1948, p. 266, 305) and auricles is evidence that they must be descended from a euechinoid ancestry. The teeth in young *Apatopygus* are keeled, indicating that they are probably derived from an echinacean ancestry or the Echinacea themselves. However, the common occurrence of hollow spines, as well as perforate and crenulate tubercles in both the Atelostomata and Gnathostomata, likewise suggests a close affinity to the Diadematoida. An alternative possibility to derivation from the echinacean stem (as shown in Fig. 213) is to postulate that the stirodont type of lantern and teeth has arisen more than once and that the Holectypoida were derived from a diadematoid ancestor. However, the general morphology of the test is more suggestive of a pseudodiadematid origin and this source is here accepted.

The **Gnathostomata** are characterized by a rigid corona; exocyclic apical system; adults mostly with lantern and keeled teeth; apical system and peristome approximately opposite; spines hollow; primary tubercles usually perforate and crenulate. The genus *Holectypus* appears in the Domerian and in the position of the periproct is already well removed from a regular echinoid ancestor, suggesting that intermediates may well be searched for in the Triassic. Gnathostomata was proposed by ZITTEL (1879) as a suborder. DURHAM & MELVILLE (1957) employed it as a superorder, designating *Clypeaster* LAMARCK as the type.

The Gnathostomata are easily separated

into two groups, the Holectypoida and Clypeasteroida, here accepted as of ordinal rank. The **Holectypoida** are characterized by interambulacra wider than ambulacra; apical system monobasal or not; absence of a microcanal system in walls of test; teeth keeled and with lateral flanges when present. The morphologic diversity is very great and wide morphologic gaps exist between known taxa, suggesting that knowledge of their fossil record is very uneven. The obvious gaps in the record of morphologic types make interpretation of evolutionary relationships hazardous, but WAGNER & DURHAM (this volume) have recognized the same three suborders as DURHAM & MELVILLE (1957), albeit with a differing content of families. The suborder Holectypina has orderly ornament, distinct gill slits, and radially positioned auricles. The suborder Echinoneina (H. L. CLARK, 1925; type-genus *Echinoneus* LESKE, 1778, SD DURHAM & MELVILLE, 1957) lacks orderly ornament, has lantern and girdle rudimentary or lacking in adults, and no gill slits. The suborder Conoclypina (ZITTEL, 1879; type-genus *Conoclypus* L. AGASSIZ, here designated) is characterized by petaloid or subpetaloid ambulacra, interradially positioned auricles, monobasal apical system, and lack of orderly ornament.

The suborder Conoclypina (families Oligopygidae and Conoclypidae) and the members here referred to it was assigned to the Cassiduloida (and thus by inference to the superorder Atelostomata of the present classification) by MORTENSEN and PHILIP (1963). The monobasal apical system and petaloid ambulacra support such an assignment but the well-developed lantern and interradial auricles (in genera where the internal structures have been studied, *Conoclypus, Oligopygus, Bonaireaster*) are here considered to be more important indicators of affinities. Petals and the monobasal apical system are known to have developed independently in the Clypeasteroida and Cassiduloida and thus cannot be considered as indicative of affinities. All known unequivocal cassiduloids, from the earliest *(Galeropygus)* on, lack a lantern in the adult and thus unless paedogenesis or some similar process is invoked, the presence of a lantern in the Conoclypina could not be explained if they were assigned to the Cassiduloida.

As here interpreted, the Holectypoida appeared in the Liassic, had their heyday in the Cretaceous, decreased rapidly by the Eocene, and are represented in the Recent only by two closely related genera. DUNCAN (1889) proposed the Holectypoida as an order and DURHAM & MELVILLE (1957) designated *Holectypus* DESOR as its type.

The order **Clypeasteroida** is characterized by petaloid ambulacra; ambulacra never narrower than interambulacra on oral surface; monobasal apical system; lantern without compass; peristome small, without gill slits; teeth keeled, without lateral flanges; presence of small accessory tube feet occurring outside of petals. Four suborders can be recognized. The suborder Clypeasterina has demiplates in petals, discontinuous interambulacra terminated adapically by a pair of plates, and separate auricles. The suborder Laganina MORTENSEN (1948) with *Laganum* LINK designated by DURHAM & MELVILLE (1957) as type, has narrow, continuous interambulacra on the oral surface, terminated adapically by a single plate, and fused auricles in an interradial position. The suborder Scutellina HAECKEL (1896) (as suborder Scutellaria), with *Scutella* LAMARCK designated by DURHAM & MELVILLE (1957) as type, is characterized by fused interradial auricles, interambulacra terminating adapically in a pair of plates, and absence of demiplates in petals. The suborder Rotulina DURHAM (1955), with *Rotula* SCHUMACHER designated by DURHAM & MELVILLE (1957) as type, has interambulacra terminating adapically in a series of single plates, fused auricles, no demiplates in petals, and posterior of test dentate or digitate.

The genera *Fibularia* and *Echinocyamus* have been reported in the late Senonian. The order is abundantly represented in the Cenozoic, with a maximum in the Miocene and a slight decrease to the Recent. The order appears to have been derived from some member of the Holectypina in the Upper Cretaceous. A. AGASSIZ (1873) proposed the Clypeastridae as a suborder, whereas CLAUS (1876) considered the taxon as an order (named Clypeastridea). DUR-

HAM & MELVILLE (1957) designated *Clypeaster* LAMARCK as the type-genus.

The superorder **Atelostomata** is characterized by rigid (albeit in some species very fragile) test; exocyclic periproct; absence of lantern, girdle, and branchial slits in adult; interambulacra wider than amulacra on oral surface; peristome small; primary spines hollow; and tubercles usually perforate and crenulate. Although DURHAM & MELVILLE (1957) recognized four orders (Cassiduloida, Nucleolitoida, Holasteroida, Spatangoida) subsequent studies by KIER have resulted in the union of two (Nucleolitoida, Cassiduloida). PHILIP (1963) has proposed the suborder Neolampadina (here elevated to ordinal rank) for the family Neolampadidae; thus four orders of Atelostomata are recognized in the *Treatise.*

The genus *Pygomalus* is known from the Sinemurian and *Galeropygus* from the Domerian. These two genera were already widely separated in their detailed morphology and respectively they represent the orders Holasteroida and Cassiduloida. They are so divergent from contemporary Euechinoida and Gnathostomata that it is difficult to suggest a possible ancestor. The distinctiveness of their morphology (small peristome, no lantern, and lack of large primary tubercles on interambulacral plates) suggests that their ancestry has long previously branched off the euechinoid line. The divergent apical systems and character of the ambulacra likewise suggest that the separation of the two orders had occurred at some time prior to the Early Jurassic.

The source of the third order, the Spatangoida, is uncertain but it seems to lie in one of the other two orders. As a corollary to these postulated prior ancestries the occurrence of atelostomate echinoids is to be expected in the Triassic and perhaps even in the latest Paleozoic.

The genus *Loriolella* FUCINI, which has been considered by some as a type intermediate between *Galeropygus* and its endocyclic ancestor is excluded from the Cassiduloida (which includes *Galeropygus*) by KIER because it lacks phyllodes, bourrelets, and petaloid ambulacra. *Loriolella* occurs in the Domerian, contemporaneous with the earliest known *Galeropygus*. It is here retained in the superorder but left unassigned to family or order. There seem to be traces of faint gill slits around the small peristome on some specimens. If correctly interpreted, this indicates that the genus retained external gills and would strongly support the interpretation that it was a persistent representative of the intermediates between the euechinoids and the atelostomates.

The **Cassiduloida,** as interpreted by KIER, are characterized by a compact apical system, adapically petaloid ambulacra, presence of phyllodes and bourrelets, and absence of fascioles. They are first known in the Liassic, are very abundant in the Early Tertiary and have since declined to a few living species. The name Cassidulideae was proposed by CLAUS (1880). DURHAM & MELVILLE designated *Cassidulus* LAMARCK as the type-genus. KIER (1962) showed that the change from a tetrabasal to monobasal apical system occurred in more than one stock and negated the principal basis used by DURHAM & MELVILLE to separate their orders Cassiduloida and Nucleolitoida.

The family Neolampadidae was considered by MORTENSEN to be closely allied to the cassiduloids. However, the restricted concept of the latter order employed by KIER excludes them from that taxon. PHILIP (1963) established the suborder Neolampadina for the family and at the same time showed that it was represented in the Paleogene, describing two new genera *(Pisolampas, Notolampas)* that occur as fossils. PHILIP's suborder is here elevated to ordinal rank as **Neolampadoida.** It is characterized by simple ambulacral pores, nonpetaloid ambulacra, and floscelle absent or weakly developed.

The order **Holasteroida** is characterized by absence of floscelle; no fifth genital plate; apical system typically elongate or disjunct; plastron lacking to meridosternous; paired petals not impressed. The elongate to disjunct apical system is the most notable feature of the order, readily separating it from the Cassiduloida and Spatangoida. As conceived by DURHAM & MELVILLE (1957), this assemblage includes the suborders Protosternata and Meridosternata of MORTENSEN's usage. The oldest known atelostomate echinoid, the Liassic *Pygomaulus,* is referable to this order. Members are common in the Jurassic and Cretaceous but are less abundant in the known Tertiary faunas. Living representatives are deep-water inhabitants with thin and fragile tests, types that are not ordinarily found or recognized in the fossil record. It is here considered, as noted above, that the Holasteroida and Cassiduloida are derived from a common pre-Liassic ancestor. The order was established by DURHAM & MELVILLE (1957) and *Holaster* L. AGASSIZ is here designated as the type.

The order **Spatangoida** is characterized by amphisternous plastron; compact apical system; posterior periproct; presence of phyllodes; and absence of bourrelets. The Spatangoida, as restricted by DURHAM & MELVILLE (1957) and here accepted, include only the Amphisternata of MORTENSEN and are thus a much more limited assemblage than that of other classifications. They appeared at the beginning of the Cretaceous, attained their greatest diversity during the Eocene, and have steadily declined since then. The Spatangoida were first elevated (as Spatangoidea) to a suprafamilial rank by CLAUS (1876). DURHAM & MELVILLE designated *Spatangus* GRAY as the type.

The source of the Spatangoida as here understood is obscure. MORTENSEN (1951, v. 5, pt. 2, p. 571) sought their origins, in part, among his Disasteridae (Collyritidae +Disasteridae of this volume) by "a kind of retrograde development" and, in part, among the Conoclypidae (which he derived from the Echinobrissidae). TERMIER & TERMIER (1953, p. 881) seemingly thought that the source was among their "Echinobrissides" (Nucleolitidae of this volume). DEVRIÈS (1960), on the basis of his studies of the toxasterids, considered that they arose from the collyritids. In particular, he (*op. cit.*, p. 177) would derive the genus *Holaster* and the toxasterids from a common source in the collyritids, seemingly from some form such as *Acrolusia*. These interpretations are primarily based on the ease of progression from the simple protosternous plastron of the collyritids to the more specialized amphisternous plastron, and they overlook the difficulties in deriving the compact apical system from the disjunct apical system of the collyritids. It is difficult to postulate an appropriate genetic mechanism to account for the necessary reversal of the collyritine trend to disassociation of the apical system, and in the absence of undoubted intermediates between the two types, it seems preferable to appeal to the inadequacy of our knowledge of the fossil record and suggest that the source lay somewhere in the more primitive members of the atelostomate stock. DEVRIÈS suggested that his *Toxaster lafittei,* from the Berriasian, represents the "connecting link" between the toxasterids, *Holaster*, and the collyritids. However, the height of the plates in the ambulacra and the position of the pores in the plates, in combination with the elongate apical system, suggest that his species is incorrectly assigned to *Toxaster* and is not the long sought-for intermediate.

REFERENCES

Agassiz, Alexander

(1) 1872 [1872-74], *Revision of the Echini:* Harvard Univ., Museum Comp. Zoology, Mem., v. 3, p. 1-762, pl. 1-38 (total 94).

Agassiz, Louis

(2) 1835, *Prodrome d'une monographie des radiares ou échinodermes:* Soc. Neuchâtel Sci. Nat., Mém., v. 1, p. 168-199.

(3) 1840, *Catalogus systematicus ectyporum echinodermatum fossilium musei Neocomensis:* 20 p., Oliv. Petitpierre (Neocomi Helvetorum).

———, & Desor, P. J. E.

(4) 1846-47, *Catalogue raisonné des familles, des genres, et des espèces de la classe des échinodermes:* Ann. Sci. Nat., ser. 3, Zool., v. 6, p. 305-374, pl. 15-16 (1846); Same [title changed to "Catalogue raisonné des espèces, des genres, et des familles d'échinides"], v. 7, p. 129-168 (1847); v. 8, p. 5-35, 355-380 (1847).

Bather, F. A.

(5) 1931, *What is Bothriocidaris?:* Paläont. Zeitschr., v. 13, p. 55-60.

Bernard, Félix

(6) 1895, *Eléments de paléontologie:* v. 1, pt. 1, 512 p., J. B. Baillière et fils (Paris).

Beurlen, Karl

(7) 1934, *Monographie der Echinoiden-Familie Collyritidae D'Orbigny:* Palaeontographica, v. 80, Abt. A, p. 41-194.

(8) 1936, *Die Saleniiden und ihre Bedeutung für die Stammesgeschichte der Echinoiden:* Paläont. Zeitschr., v. 18, p. 17-38.

(9) 1937, *Revision der Seeigel aus dem norddeutschen Jura. Teil 2, Die regulären*

Seeigel: Preuss. Geol. Landesanst., Abhandl., new ser., v. 174, p. 6-149, pl. 1.

Blainville, H. D. de

(10) 1822, *De l'organization des animaux:* 574 p., F. G. Levrault (Paris).

(11) 1834, *Manuel d'actinologie ou de zoophytologie:* 694 p., atlas, 100 pl., F. G. Levrault (Paris).

Bronn, H. G.

(12) 1860, *Klassen und Ordnungen des Thier-Reichs. v. 1, Amorphozoen:* 434 p., 48 pl. (Leipzig and Heidelberg).

Carus, J. V.

(13) 1863, in J. V. Carus & C. E. A. Gerstaecker, *Handbuch der Zoologie:* v. 2, vii+642 p., Wilhelm Engelmann (Leipzig).

Clark, H. L.

(14) 1925, *A catalog of the Recent sea-urchins in the British Museum:* 278 p., 12 pl., British Museum, Nat. History (London).

(15) 1946, *The echinoderm fauna of Australia; its composition and its origin:* Carnegie Inst. Washington, Publ. 566, p. 567.

(16) 1932, *The ancestry of Echini:* Science, new ser., v. 76, no. 1982, p. 591-593.

Claus, C. F. W.

(17) 1876-80, *Grundzüge der Zoologie:* (a) 3rd ed. (1876), v. 1, xii+1254 p.; (b) 4th ed. (1880), v. 1, vii+821 p., v. 2, iv+522 p. (Marburg and Leipzig).

Cotteau, G. H.

(18) 1862-94, *Paléontologie française. Description des animaux invertébrés:* (a) Terrains crétacés, v. 7 (1861-67), Échinides, 894 p., pl. 1007-1204, 1204 bis, 1087 bis; (b) Terrains jurassiques, v. 9 (1867-74), Échinides Irréguliers, 552 p., atlas, p. 1-142; (c) Terrains jurassiques, v. 10, pt. 1 (1875-80), Echinides Reguliers, 468 p., pl. 143-262; (d) Terrains jurassiques, v. 10, pt. 2 (1880-85), 960 p., pl. 263-520; (e) Terrains tertiaires, v. 1 (1885-89), Échinides Eocènes, 672 p., pl. 1-200; (f) Terrains tertiaires, v. 2 (1889-94), 789 p., pl. 201-384, G. Masson (Paris).

Coulon, L. M.

(19) 1933, *Les classifications diverses des échinides:* Soc. Sci. Nat. Elbeuf, Bull., 20 p.

Cuénot, Lucien

(20) 1948, *Anatomie, éthologie et systématique des échinodermes:* in Traité de Zoologie, P.-P. Grassé (ed.), v. 11, p. 1-275, text-fig. 1-312, Masson et Cie (Paris).

Cuvier, G. L. C. F. D.

(21) 1817, *Le regne animal distribué d'après son organisation:* v. 4, viii+255 p., Deterville (Paris).

Delage, Yves & Hérouard, Edgard

(22) 1903, *Les échinodermes:* Traité de zoologie concrète, v. 3, x+495 p., 53 pl., Ch. Reinwald (Paris).

Desor, Edouard

(23) 1855-58, *Synopsis des échinides fossiles:* 490 p., 44 pl., Ch. Reinwald (Paris), Kriedel & Niedner (Wiesbaden).

Devriès, André

(24) 1960, *Contribution à l'étude de quelques groupes d'échinides fossiles d'Algérie:* Publ. Serv. Carte Géol. Algérie, new ser., Paléont., Mém. 3, p. 1-278, pl. 1-39.

Döderlein, Ludwig

(25) 1906, *Die Echinoiden der deutschen Tiefsee-Expedition:* Deutsche Tiefsee-Exped. 1898-99, v. 5, p. 63-290, pl. 9-50.

Dujardin, M. F., & Hupé, M. H.

(26) 1862, *Histoire naturelle des zoophytes échinodermes:* 627 p., atlas, 10 pl., Librairie Encyclopédique de Roret (Paris).

Duncan, P. M.

(27) 1889, *A revision of the genera and great groups of the Echinoidea:* Linnean Soc. London (Zool.), Jour., v. 23, p. 1-311.

Durham, J. W.

(28) 1955, *Classification of clypeasteroid echinoids:* California Univ., Publ. Geol. Sci., v. 31, p. 73-198, pl. 3-4, 38 text-fig.

———, & Melville, R. V.

(29) 1957, *A classification of echinoids:* Jour. Paleontology, v. 31, p. 242-272, text-fig. 1-9.

Geis, H. L.

(30) 1936, *Recent and fossil pedicellariae:* Jour. Paleontology, v. 10, p. 427-448, pl. 58-61.

Gignoux, Maurice

(31) 1933, *Les oursins réguliers fossiles, évolution et classification:* Grenoble Univ., Lab. Géol., Travaux, v. 17, pt. 2, p. 1-125.

Gras, C. J. A.

(32) 1848, *Description des oursins fossiles du département de l'Isère:* Soc. Sci. Isère (Grenoble), Bull., v. 4, 54 p., 4 pl.

Gray, J. E.

(33) 1825, *An attempt to divide the Echinida, or sea eggs, into natural families:* Ann. Philos., new ser., v. 10, p. 423-431.

Gregory, J. W.

(34) 1900, *Echinoidea: in* A treatise on zoology, E. R. Lankester (ed.), Part III, the Echinoderma, p. 282-332, text-fig. 1-47, A. & C. Black (London).

Haeckel, Ernst

(35) 1896, *Systematische Phylogenie der wirbellosen Thiere (Invertebrata):* v. 2, xviii+720 p., Georg Reimer (Berlin).

Hawkins, H. L.

(36) 1931, *The first echinoid:* Biol. Reviews, v. 6, p. 443-458.

International Commission on Zoological Nomenclature

(37) 1953, *Copenhagen decisions on zoological nomenclature:* 135 p. (London).

(38) 1964, *International code of zoological nomenclature (revised edition):* 175 p. (London).

Jackson, R. T.

(39) 1896, *Studies of Palaeechinoida:* Geol. Soc. America, Bull., v. 7, p. 171-254, pl. 2-9, 5 text-fig.

(40) 1912, *Phylogeny of the Echini, with a revision of Paleozoic species:* Boston Soc. Nat. History, Mem., v. 7, p. 1-490, pl. 1-76, 256 text-fig.

(41) 1929, *The status of Bothriocidaris:* Harvard Univ., Museum Comp. Zoology, Bull., v. 69, p. 481-512.

Kier, P. M.

(42) 1962, *Revision of the cassiduloid echinoids:* Smithsonian Misc. Coll., v. 144, no. 3, 262 p., 44 pl., 184 text-fig.

Klein, J. T.

(43) 1734, *Naturalis dispositio Echinodermatum:* 78 p., 36 pl. (Gedani).

Lamarck, J. B. P. M. de

(44) 1801, *Système des animaux sans vertèbres:* 432 p., Deterville (Paris).

(45) 1816, *Histoire naturelle des animaux sans vertèbres:* v. 3, 770 p., Verdiere (Paris).

Lambert, Jules

(46) 1918, *Considerations sur la classification des échinides atelostomes:* Soc. Acad. l'Aube, Mém., v. 82, p. 25-70.

———, & Thiéry, Paul

(47) 1909-1925, *Essai de nomenclature raisonnée des échinides:* 607 p., 15 pl. (Chaumont).

Latreille, P. A.

(48) 1825, *Familles naturelles du régne animal:* 570 (Paris).

Leske, N. G.

(49) 1778, *Jacobi Theodori Klein naturalis dispositio echinodermatum . . . , edita et descriptionibus novisque inventis et synonomis auctorem aucta:* 278 p., 54 pl. (Leipzig) [Additamenta ad Kleinii dispositionem echinodermatum].

Linné, Carl von [Linnaeus, Carolus]

(50) 1758, *Systema Naturae . . .:* ed. 10, v. 1, 824 p., Laurent Salvii (Holmiae).

MacBride, E. W., & Spencer, W. K.

(51) 1938, *Two new Echinoidea, Aulechinus and Ectinechinus, and an adult plated holothurian, Eothuria, from the Upper Ordovician of Girvan, Scotland:* Royal Soc. London, Philos. Trans., ser. B, v. 229, p. 91-136, pl. 10-17.

M'Coy, Frederick

(52) 1849, *On some new Palaeozoic Echinodermata:* Ann. & Mag. Nat. History, ser. 2, v. 3, p. 244-254.

Melville, R. V.

(53) 1961, *Dentition and relationships of the echinoid genus Pygaster J. L. R. Agassiz, 1836:* Palaeontology, v. 4, p. 243-246, pl. 28-29.

Mercier, Jean

(54) 1932, *Etudes sur les échinides du Bathonian:* Soc. Linnéenne Normandie, Mém., new ser., Géol., v. 2, p. 1-273, pl. 1-11.

Mortensen, Theodor

(55) 1904, *The Danish expedition to Siam, 1899-1900. II. Echinoidea* (1): [K.] Danske Vidensk. Selsk. Skrift., ser. 7, v. 1, p. 1-124, pl. 1-7, map.

(56) 1928, *Bothriocidaris and the origin of echinoids:* Vidensk. Medd. Dansk Naturh. Foren., v. 86, p. 93-124.

(57) 1928-1951, *A monograph of the Echinoidea:* v. 1-5 (for details, see p. *U*638).

(58) 1930 [1931], *Bothriocidaris and the ancestry of echinoids, a reply:* Vidensk. Medd. Dansk Naturh. Foren., v. 90, p. 313-352.

Myannil [Männil], R. M.

(59) 1962, *Taksonomiya i morfologiya roda Bothriocidaris (Echinoidea):* Eesti NSV Teaduste Akad. Geol. Inst., Uurimused, v. 9, p. 143-190, pl. 1-5, 22 text-fig. (Russian, with Estonian and English summaries). [*The taxonomy and morphology of Bothriocidaris (Echinoidea).*]

Nichols, David

(60) 1959, *Mode of life and taxonomy in irregular sea-urchins* in *Function and taxonomic importance:* Syst. Assoc., Publ. no. 3, p. 61-80.

Orbigny, Alcide D. d'

(61) 1851, *Cours élémentaire de paléontologie et de géologie stratigraphique:* v. 2, p. 1-382, Victor Masson (Paris).

(62) 1853-55 [1854-60], *Echinodermes:* in Paléontologie française, Terrain crétacé: ser. 1, v. 6, p. 1-432, pl. 801-908 (d'Orbigny, 1854-59); p. 433-596, pl. 909-1006 (Cotteau, 1860).

Parkinson, James

(63) 1811, *Organic remains of a former world:* v. 3, xv+479 p., 22 pl., Sherwood, Neely, & Jones (London).

(64) 1822, *An introduction ot the study of fossil organic remains:* vii+346 p., 10 pl. (London).

Philip, G. M.

(65) 1965, *Classification of echinoids:* Jour. Paleontology, v. 39, p. 45-62, text-fig. 1-4.

Pomel, Auguste

(66) 1869, *Revue des échinodermes et de leur classification pour servir d'introduction à l'étude des fossiles:* 67 p., Deyrolle (Paris).

(67) 1883, *Classification méthodique et genera des échinides vivants et fossiles:* 131 p., 1 pl., Typ. Adolphe Jourdan (Alger).

Termier, Henri, & Termier, Geneviève

(68) 1953, *Classe des échinides:* in Traité de paléontologie, JEAN PIVETEAU (ed.), v. 3, p. 857-947, Masson et Cie. (Paris).

Wright, Thomas

(69) 1855-60, *A monograph on the British fossil Echinodermata of the Oolitic Formations:* Palaeontograph. Soc. London, Mon., v. 1, 469 p., 43 pl.

Zittel, K. A. von

(70) 1876-80, *Handbuch der Paläontologie:* v. 1, 765 p., 558 text-fig., R. Oldenbourg (München & Leipzig).

OUTLINE OF CLASSIFICATION

By LAVON MCCORMICK and R. C. MOORE

[University of Kansas]

The following outline of the classification of the Echinoidea summarizes taxonomic relationships, geologic occurrence, and numbers of recognized genera and subgenera in each family group and higher-rank taxon. Where a single number is given, it refers to genera; where two numbers are given, the second indicates subgenera. Authorship of the systematic descriptions is indicated by recording with each division the initial letters of the author's name as shown by the tabulation below:

Authorship of Systematic Descriptions

Durham, J. W.	D
Fell, H. B.	FE
Fischer, A. G.	FI
Kier, P. M.	K
Melville, R. V.	M
Pawson, D. L.	P
Wagner, C. D.	W

Main Divisions of Echinoidea

Echinoidea *(class)* (768;78; excludes doubtful genera). *Ord.-Rec.*
- Perischoechinoidea *(subclass)* (99;5). *Ord.-Rec.* (D-FE-K)
  - Bothriocidaroida *(order)* (1). *Ord.* (K)
    - Bothriocidaridae (1). *Ord.* (K)
  - Echinocystitoida *(order)* (24). *Ord.-Perm.* (K)
    - Echinocystitidae (5). *Sil.-Perm.* (K)
    - Lepidesthidae (2). *Dev.-Perm.* (K)
    - Lepidocentridae (15). *Ord.-Miss.* (K)
    - Family Uncertain (2). *Dev.* (K)
  - Palaechinoida *(order)* (10). *Sil.-Perm.* (K)
    - Palaechinidae (5). *Miss.* (K)
    - Cravenechinidae (3). *Sil.-Perm.* (K)
    - Family Uncertain (2). *?Sil., L.Carb.* (K)
  - Cidaroida *(order)* (63;5). *U.Sil.-Rec.* (FE)
    - Archaeocidaridae (6). *?U.Sil., Dev.-Perm.* (FE)
    - Miocidaridae (7). *L.Carb.-L.Jur.* (FE)
    - Cidaridae (42;5). *U.Trias.-Rec.* (FE)
      - Histocidarinae (4). *Jur.-Rec.* (FE)
      - Ctenocidarinae (8). *?Eoc., Rec.* (FE)
      - Goniocidarinae (4;5). *Eoc.-Rec.* (FE)
      - Stereocidarinae (4). *U.Jur.-Rec.* (FE)
      - Rhabdocidarinae (9). *L.Jur.-Rec.* (FE)
      - Cidarinae (13). *U.Trias.(Rhaet.)-Rec.* (FE)
    - Psychocidaridae (6). *U.Jur.-Rec.* (FE)
    - Diplocidaridae (2). *Jur.-Cret.* (FE)
  - Order and Family Uncertain (1). *Sil.* (K)
- Euechinoidea *(subclass)* (669;73). *?Carb., U. Trias.-Rec.* (D-FE-FI-K-M-P-W)
  - Diadematacea *(superorder)* (56;2). *?L.Carb., U. Trias.-Rec.* (FE-M)
    - Echinothurioida *(order)* (12). *U.Jur.-Rec.* (FE)
      - Echinothuriidae (12). *U.Jur.(Oxford.)-Rec.* (FE)
        - Echinothuriinae (8). *U.Cret.(Senon.)-Rec.* (FE)
        - Phormosomatinae (3). *Rec.* (FE)
        - Pelanechininae (1). *U.Jur.* (FE)
    - Diadematoida *(order)* (23). *?L.Carb., U.Trias.-Rec.* (FE)
      - Diadematidae (11). *L.Jur.-Rec.* (FE)
      - Lissodiadematidae (1). *Rec.* (FE)
      - Micropygidae (1). *Rec.* (FE)
      - Aspidodiadematidae (4). *?U.Jur., Rec.* (FE)
      - Family Uncertain (6). *?L.Carb., Jur.-Cret.* (FE)
    - Pedinoida *(order)* (16;2). *U.Trias.-Rec.* (FE)
      - Pedinidae (15;2). *U.Trias.(Rhaet.)-Rec.* (FE)
      - Family Uncertain (1). *U.Jur.(Oxford.).* (FE)
    - Pygasteroida *(order)* (3). *L.Jur.-U.Cret.* (M)
      - Pygasteridae (3). *L.Jur.-U.Cret.* (M)
    - Order Uncertain (1). *Jur.(Domer.-Oxford.).* (FE)
      - Heterocidaridae (1). *Jur.(Domer.-Oxford.).* (FE)

Order and Family Uncertain (1). *Trias.* (FE)
Echinacea *(superorder)* (195;14). *U.Trias., L. Jur.-Rec.* (D-FE-P)
Salenioida *(order)* (16). *?U.Trias., L.Jur.-Rec.* (FE)
Acrosaleniidae (6). *?U.Trias., L.Jur.-U.Cret.* (FE)
Saleniidae (10). *U.Jur.-Rec.* (FE)
Saleniinae (5). *L.Cret.-Rec.* (FE)
Hyposaleniinae (5). *U.Jur.-U.Cret.* (FE)
Hemicidaroida *(order)* (25). *U.Trias.-U.Cret.* (FE)
Hemicidaridae (10). *L.Jur.-U.Cret. (Cenoman.).* (FE)
Pseudodiadematidae (12). *U.Trias.-U.Cret.* (FE)
Family Uncertain (3). *U.Jur.-U.Cret.* (FE)
Phymosomatoida *(order)* (38;6). *L.Jur.-Rec.* (FE)
Phymosomatidae (18;4). *L.Jur.-Rec.* (FE)
Stomechinidae (19;2). *L.Jur.-Rec.* (FE)
Family Uncertain (1). *U.Cret.(Maastricht.).* (FE)
Arbacioida *(order)* (20;4). *M.Jur.(Bathon.)-Rec.* (FE)
Arbaciidae (20;4). *M.Jur.-Rec.* (FE)
Temnopleuroida *(order)* (62;4). *L.Jur.-Rec.* (FE)
Glyphocyphidae (9;2). *L.Jur.-Eoc.* (FE)
Temnopleuridae (39;2). *U.Cret.(Cenoman.)-Rec.* (FE)
Toxopneustidae (13). *?Cret.-?Oligo., Mio.-Rec.* (FE)
Family Uncertain (1). *?Paleoc., Eoc.-U. Oligo.* (FE)
Echinoida *(order)* (32). *?U.Cret.(Cenoman.), Paleoc.-Rec.* (FE-P)
Echinidae (13). *?U.Cret.(Cenoman.), Mio.-Rec.* (FE-P)
Echinometridae (12). *Paleoc.-Rec.* (FE-P)
Strongylocentrotidae (3). *Mio.-Rec.* (FE-P)
Parasaleniidae (2). *Eoc.-Rec.* (FE-P)
Family Uncertain (2). *L.Cret.-Mio.* (FE-P)
Plesiocidaroida *(order)* (2). *U.Trias.(Carn.).* (D-FE)
Tiarechinidae (2). *U.Trias.* (FE)
Superorder Uncertain (Echinacea or Diademata-cea) (6). *L.Jur.-U.Cret.* (FE)
Orthopsida *(order)* (6). *L.Jur.-U.Cret.* (FE)
Orthopsidae (6). *L.Jur.-U.Cret.* (FE)
Gnathostomata *(superorder)* (103). *Jur.-Rec.* (D-W)
Holectypoida *(order)* (30). *L.Jur.(Pliensbach.)-Rec.* (W-D)
Holectypina *(suborder)* (10). *L.Jur. (Pliensbach.)-U.Cret.(Senon.).* (D-W)
Holectypidae (5). *L.Jur.(Pleinsbach.)-U. Cret.(Senon.).* (D-W)
Anorthopygidae (1). *Cret.(Alb.-Cenoman.).* (D-W)
Discoididae (4). *Cret.* (D-W)
Echinoneina *(suborder)* (10). *M.Jur.(Callov.)-Rec.* (D-W)
Echinoneidae (3). *U.Cret.-Rec.* (D-W)
Conulidae (4). *M.Jur.(Callov.)-Eoc.* (D-W)
Galeritidae (1). *U.Cret.(Senon.).* (D-W)
Family Uncertain (2). *L.Cret.(Hauteriv.)-U. Cret.(Senon.).* (D-W)
Conoclypina *(suborder)* (8). *U.Cret.(Senon.)-Mio.* (D-W)
Conoclypidae (2). *Eoc.-Mio.* (D-W)
Oligopygidae (6). *U.Cret.(Senon.)-Oligo.* (D-W)
Suborder and Family Uncertain (2). *U.Cret.-Oligo.* (D-W)
Clypeasteroida *(order)* (73). *U.Cret. (Maastricht.)-Rec.* (D)
Clypeasterina *(suborder)* (7). *U.Eoc.-Rec.* (D)
Clypeasteridae (1). *U.Eoc.(Auvers.)-Rec.* (D)
Arachnoididae (6). *Oligo.-Rec.* (D)
Arachnoidinae (2). *Oligo.-Rec.* (D)
Ammotrophinae (3). *Mio.-Rec.* (D)
Subfamily Uncertain (1). *?L.Mio.* (D)
Laganina *(suborder)* (28). *U.Cret.(Senon.)-Rec.* (D)
Fibulariidae (14). *U.Cret.(Senon.)-Rec.* (D)
Laganidae (8). *Eoc.-Rec.* (D)
Neolaganidae (6). *Eoc.-Oligo.* (D)
Scutellina *(suborder)* (32). *Eoc.-Rec.* (D)
Scutellidae (3). *Oligo-Mio.* (D)
Protoscutellidae (3). *Eoc.* (D)
Eoscutellidae (1). *Eoc.* (D)
Dendrasteridae (4). *Plio-Rec.* (D)
Echinarachniidae (7). *Oligo-Rec.* (D)
Monophorasteridae (3). *Mio.* (D)
Mellitidae (4). *L.Mio.-Rec.* (D)
Astriclypeidae (3). *Oligo-Rec.* (D)
Abertellidae (1). *Mio.* (D)
Scutasteridae (1). *L.Mio.* (D)
Family Uncertain (2). *U.Eoc.-Oligo.* (D)
Rotulina *(suborder)* (3). *Mio.-Rec.* (D)
Rotulidae (3). *Mio.-Rec.* (D)
Suborder and Family Uncertain (3). *M.Eoc.-Mio.* (D)
Atelostomata *(superorder)* (304;57). *Jur.-Rec.* (D-FI-K-W)
Cassiduloida *(order)* (69;3). *Jur.-Rec.* (K)
Galeropygidae (2). *Jur.* (K)
Clypeidae (5;3). *Jur.-U.Cret.* (K)
Nucleolitidae (12). *M.Jur.-U.Cret.* (K)
Echinolampadidae (6). *Cret.(Cenoman.)-Rec.* (K)
Faujasiidae (12). *U.Cret.-Eoc.* (K)
Archiaciidae (2). *Cret.* (K)
Cassidulidae (6). *L.Cret.-Rec.* (K)
Clypeolampadidae (2). *U.Cret.(Cenoman.-Maastricht.).* (K)
Pliolampadidae (12). *U.Cret.(Senon.)-Rec.* (K)
Apatopygidae (1). *Neog.* (K)
Family Uncertain (9). *Jur.-Rec.* (K)

Holasteroida *(order)* (81). *L.Jur.-Rec.* (D-W)
Collyritidae (8). *L.Jur.-L.Cret.* (D-W)
Disasteridae (9). *M.Jur.-L.Cret.* (D-W)
Holasteridae (38). *L.Cret.-Rec.* (D-W)
Urechinidae (6). *?U.Eoc., Mio.-Rec.* (D-W)
Calymnidae (1). *Rec.* (D-W)
Pourtalesiidae (7). *Rec.* (D-W)
Stenonasteridae (1). *U.Cret.* (D-W)
Somaliasteridae (4). *U.Cret.(Senon.)-Paleoc.* (D-W)
Family Uncertain (7). *Cret.-Eoc.* (D-W)
Spatangoida *(order)* (147;54). *L.Cret. (Berrias.)-Rec.* (FI)
Toxasterina *(suborder)* (13). *L.Cret. (Berrias.)-Rec.* (FI)
Toxasteridae (13). *L.Cret.(Berrias.)-Rec.* (FI)
Hemiasterina *(suborder)* (42;26). *L.Cret. (Apt.)-Rec.* (FI)
Hemiasteridae (15;9). *L.Cret.(Apt.)-Rec.* (FI)
Palaeostomatidae (5). *U.Cret.-Rec.* (FI)
Pericosmidae (1;3). *Eoc.Rec.* (FI)
Schizasteridae (19;14). *U.Cret.(Cenoman.)-Rec.* (FI)
Aeropsidae (2). *Rec.* (FI)
Micrasterina *(suborder)* (59;26). *Cret. (Cenoman.)-Rec.* (FI)
Micrasteridae (4;2). *U.Cret.(Cenoman.)-Eoc.* (FI)
Brissidae (40;16). *U.Cret.(Santon.)-Rec.* (FI)
Spatangidae (8;6). *Eoc.-Rec.* (FI)
Loveniidae (7;2). *Eoc-Rec.* (FI)
Asterostomatina *(suborder)* (23;2). *Eoc.-Rec.* (FI)
Asterostomatidae (23;2). *Eoc.-Rec.* (FI)
Suborder and Family Uncertain (10). *Cret.-Mio.* (FI)
Neolampadoida *(order)* (7). *U.Eoc.-Rec.* (D-W)
Neolampadidae (7). *U.Eoc.-Rec.* (D-W)
Gnathostomata or Atelostomata, Order Uncertain (5). *Jur.* (D-W)
Doubtful genera of regular echinoids (14). *Perm.-Jur.* (FE)
[Numerical tabulations do not include 4 genera added late.]

SYSTEMATIC DESCRIPTIONS

By J. W. Durham, H. B. Fell, A. G. Fischer, P. M. Kier, R. V. Melville, D. L. Pawson, and C. D. Wagner

Class ECHINOIDEA Leske, 1778

[*nom. transl.* Bronn, 1860, p. 295 (*ex* order Echinoidea d'Orbigny, 1852, p. 114, *nom. correct. pro* ordo Echinus Leske, 1778, p. xvi)] [=Echinides Lamarck, 1801; order Echini Goldfuss, 1820; order Cycloïdes de Blainville, 1822; Echinida Fleming, 1822; order Echinata Fischer de Waldheim, 1823; order Echinoida Latreille, 1825; order Echinidea de Blainville, 1834; order Pedicellata Griffith & Pidgeon, 1834; order Cirrhi-Spinigrada Forbes, 1841; order Adostella Austin & Austin, 1842; Echinodea Düben & Koren, 1846] [Diagnosis prepared by J. Wyatt Durham. Research on authorship and synonymy by J. W. Durham, R. V. Melville, & H. B. Fell]

Free-living Echinodermata with subspherical or modified subspherical test, built of interlocking calcareous plates and bearing movable appendages (spines, pedicellariae, spheridia) externally; equipped with masticatory apparatus or descended from forms having one; mouth directed toward substrate; two principal groups of plates comprising apical and coronal systems, apical system invariably including five radially situated ocular plates and five or fewer interradially situated genital plates, and coronal system being composed of five radial ambulacral and five interradial interambulacral areas built of contiguous meridional columns of plates; in addition, two less conspicuous plate systems, termed peristomial and periproctal, are present. Plates of ambulacral areas perforated for passage of tube feet. Primordial ambulacral plates bordering peristome conforming to Lovén's law. Mouth nearly always on lower surface, rarely anterior; anus present in all individuals, situated entirely or partly within apical system, or outside of it in posterior interambulacrum; mouth and anus each surrounded by membrane which usually bears imbricating or dissociated plates of peristomial and periproctal systems. Gonads five or fewer, interradial in position. Radial canals of water-vascular system internal to test. [All known types are exclusively marine in habitat.] *Ord.-Rec.*

Subclass PERISCHOECHINOIDEA M'Coy, 1849

[*nom. transl.* Durham & Melville, 1957 (*ex nom. correct.* Bronn, 1860, *pro* order Perischoechinida M'Coy, 1949)] [Diagnosis prepared by J. W. Durham]

Regular (endocyclic) echinoids with interambulacra of one to many columns; ambulacra of two to 20 columns, without compound plates; perignathic girdle of apophyses only or none; teeth grooved; no gill slits, spheridia, or ophiocephalous pedicellariae. *Ord.-Rec.*

NONCIDAROID PALEOZOIC ECHINOIDS

By Porter M. Kier

[Smithsonian Institution, U.S. National Museum]

INTRODUCTION

This section includes all noncidaroid regular Paleozoic echinoids which are grouped in the three orders designated as Bothriocidaroida, Echinocystitoida, and Palaechinoida. The Bothriocidaroida is an aberrant order known only from the Ordovician. It consists of a single genus characterized by nonimbricating plates, lack of genital plates, and single-columned interambulacra. The Echinocystitoida, which range from Ordovician to Permian, have strongly imbricate plates, ambulacral plates that bevel under the interambulacra, and genital plates. The Palaechinoida, known from the Silurian to the Pennsylvanian, have slightly imbricating plates with ambulacra beveling over the interambulacra. The latter two orders reached their maximum in numbers during the Mississippian and became extinct before the end of the Permian. Most echinoid workers believe that they were not the ancestors of any post-Paleozoic echinoids.

EVOLUTIONARY TRENDS

Many evolutionary trends are apparent in the Palaechinoida and Echinocystitoida. Because only one genus of one age is known in the Bothriocidaroida, no trends are known in this order.

ADORAL EXPANSION OF AMBULACRA

The adoral ambulacra are enlarged in later genera of the Echinocystitoida. In Ordovician, Silurian, and Devonian genera (e.g., *Aulechinus,* Fig. 218,*1*), the adoral ambulacral plates are similar in size and arrangement to the adapical, whereas in most Mississippian and later genera the adoral ambulacra are greatly expanded, with far larger plates, larger and more widely spaced pores, and in many species more columns (e.g., *Proterocidaris,* Fig. 218,*2*; *Pronechinus,* Fig. 218,*3*). The expansion of the adoral ambulacra and increase in size of the pores would increase the locomotory and food-gathering capabilities of the echinoid.

SHIFT OF AMBULACRAL PORES

In earlier echinoids (e.g., *Ectinechinus,* Fig. 219,*1*) the pores are adjacent to the perradial suture, but by the Late Silurian, the pores shifted somewhat away from the perradial suture (e.g., *Palaeodiscus,* Fig. 219,*2*). By the Early Devonian the pores were midway between the perradial and adradial sutures (e.g., *Porechinus,* Fig. 219, *3*) and by the Middle Devonian (e.g., *Lepidocentrus,* Fig. 219,*4*) the pores were nearer to the adradial suture. In all Mississippian genera with two columns in each ambula-

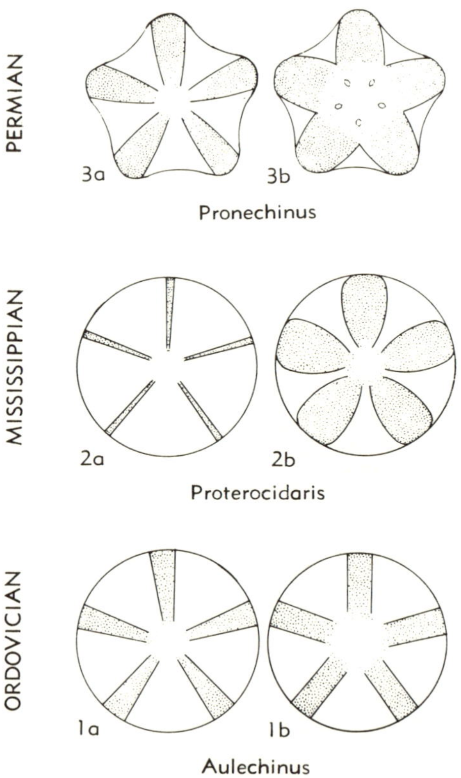

Fig. 218. Aboral *(1a,2a,3a)* and oral *(1b,2b,3b)* views showing increase in area of adapical ambulacra (shaded) in the Echinocystitoida (Kier, n).

crum, the pore pairs are adjacent to the adradial suture (e.g., *Pholidoechinus,* Fig. 219,5). This shift of the pores away from the perradial suture would decrease the amount of strain on a particular area of the test when the tube feet were contracted while attached to an object, for instead of all strain being concentrated along the perradial suture, it shifted to two areas as far distant as possible. When the tube feet were used for respiration, this separation would increase the area from which oxygen could be extracted.

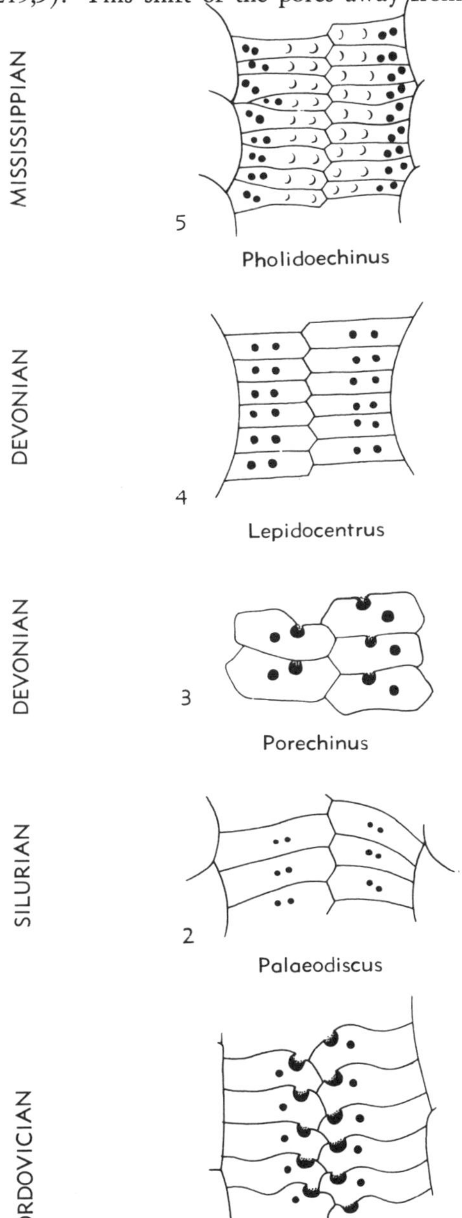

FIG. 219. View of portion of ambulacral areas showing shift of pores toward adradial suture (Kier, n).

WATER-VASCULAR SYSTEM

In Ordovician echinoids the radial water vessel is usually completely enclosed but in Silurian echinoids only the lower parts of the enclosure are still present. These remnants of the enclosures are absent in post-Mississippian species. Accordingly, loss of internal enclosure for the water-vascular system may be noted as an evolutionary trend in the noncidaroid Paleozoic echinoids.

PLATE COLUMNS IN AMBULACRA

The evolution of noncidaroid echinoids is marked by an increase in the number of ambulacral plate columns in all families having more than two columns in each ambulacrum. All Ordovician species have only two columns, but the Late Silurian *Echinocystites* and Early Devonian *Rhenechinus* have four. Among the Mississippian echinocystitids no species has fewer than six columns. In the Palaechinidae, the number of columns increased from two in *Palaechinus* to more than ten in *Melonechinus.* The increase in number of ambulacral columns enlarged the number of tube feet, and presumably it was advantageous to have more tube feet to aid in locomotion, food gathering, and respiration.

INTERAMBULACRAL PLATES

Earliest echinoids (e.g., *Aulechinus*) had irregularly arranged interambulacral plates, with no perceptible disposition in columns. Also, each plate was irregular in outline. By Late Silurian time the plates had become regularly arranged in some genera (e.g., *Myriastiches*), though they remained irregular in others (e.g., *Echinocystites*). All Devonian genera, except *Albertechinus,* have regular columns, and after the Devonian all genera have regular columns.

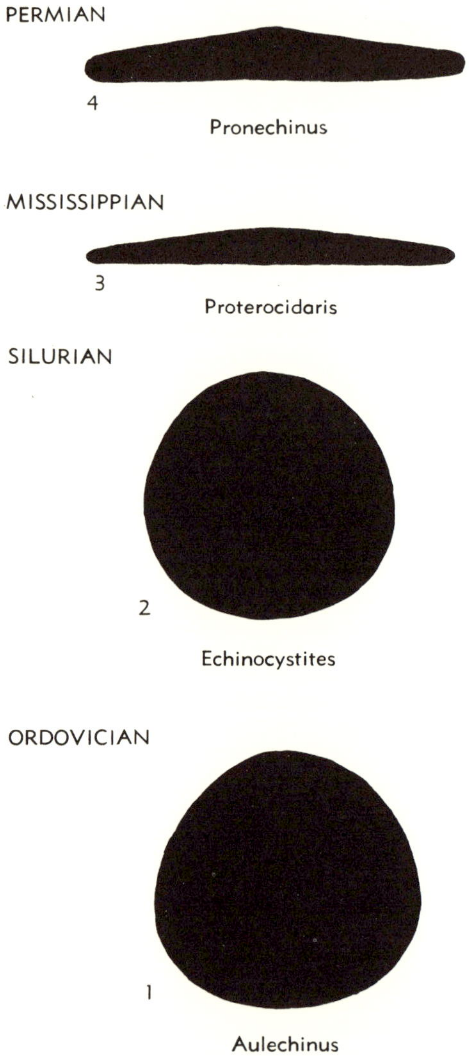

FIG. 220. Lateral profiles of four representative genera of the Echinocystitoida showing trend toward flattening of test (Kier, n).

FLATTENING OF TEST IN ECHINOCYSTITOIDA

In earliest genera (e.g., *Aulechinus,* Fig. 220,*1*; *Echinocystites,* Fig. 220,*2*) the test was high, but by the Mississippian it was low in some genera (e.g., *Proterocidaris,* Fig. 220,*3*). In both genera introduced in the Permian (*Pronechinus,* Fig. 220,*4,* and *Meekechinus*), the test was very flattened.

NUMBER OF THECAL PLATES

The number of plates in echinoid tests increased during geologic time. Ordovician genera have relatively few plates, not more than 1,000 in a single specimen, but by Silurian time 1,500 plates were present in *Echinocystites,* and by the Mississippian some specimens had as many as 3,650 plates.

SIZE OF THECA

The noncidaroid echinoids increased in size throughout Paleozoic time. Ordovician genera are all small, Silurian and Devonian genera larger, and finally in the Mississippian, echinoids reached their maximum size.

COMPLEXITY OF LANTERN

Braces, compasses, and epiphyses are absent in Ordovician echinoid genera, but by

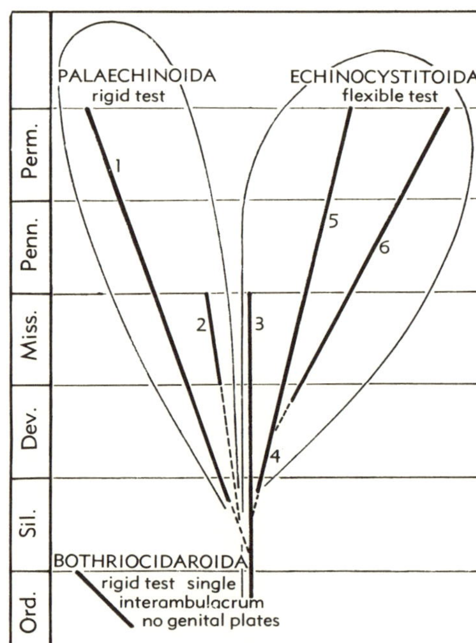

1. CRAVENECHINIDAE adradial plates enlarged
2. PALAECHINIDAE adradial plates not enlarged
3. LEPIDOCENTRIDAE only two ambulacral columns
4. more than two ambulacral columns
5. ECHINOCYSTITIDAE ambulacrum enlarged adorally
6. LEPIDESTHIDAE ambulacrum not enlarged adorally

FIG. 221. Phylogeny of noncidaroid Paleozoic echinoids (Kier, n).

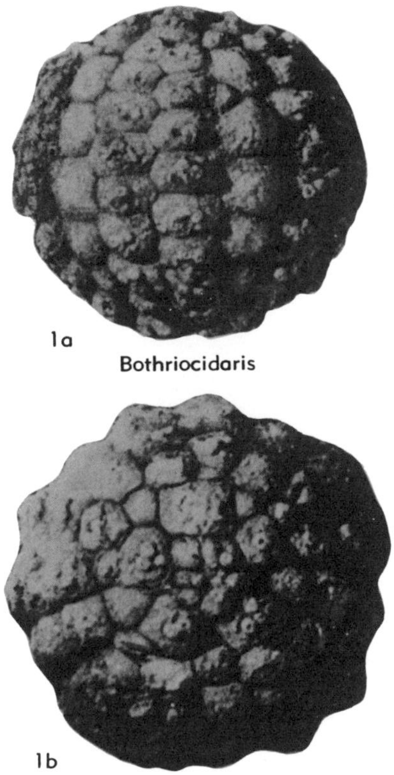

FIG. 222. Bothriocidaridae (p. *U*301).

Silurian time they were present in some species. Thus, complexity in structure of the lantern increased in the course of evolution.

COMPLEXITY OF SPINES

Among Ordovician echinoids the spines were small, undifferentiated, and situated in pits with no tubercles. By the Late Silurian, the spines were differentiated in types, and tubercles usually were present.

PHYLOGENY

The phylogeny of the noncidaroid Paleozoic echinoids (Fig. 221) has been described at the generic level by KIER (99).

Order BOTHRIOCIDAROIDA Zittel, 1879

Plates of test thick, not imbricating; ambulacrum with 2 columns, terminating in single nonporiferous plate; interambulacrum with single column not reaching margin of peristome; no genital plates, 5 oculars, one being madreporite. *Ord.*

Family BOTHRIOCIDARIDAE Klem, 1904

Characters of order. *Ord.*

Bothriocidaris EICHWALD, 1859, p. 654 [**B. globulus*; SD LAMBERT & THIÉRY, 1910]. Characters of order. *Ord.*, Eng.——FIG. 222,*1*. **B. globulus*; *1a,b*, lat. and oral views, ×3, ×3.5 (138).

Order ECHINOCYSTITOIDA Jackson, 1912

Plates strongly imbricate, ambulacral plates bevel under interambulacra, imbricate adorally, interambulacral plates imbricate adapically; interambulacra reaching peristome; genital plates; no perignathic girdle. *Ord.-Perm.*

Family ECHINOCYSTITIDAE Gregory, 1897

[Echinocystitidae GREGORY, 1897, p. 133]

More than two columns in each ambulacrum, ambulacral plates enlarged adorally in all Mississippian and later genera. *Sil.-Perm.*

Echinocystites THOMSON, 1861, p. 106 [**E. pomum*; SD GREGORY, 1897, p. 133] [=*Cystocidaris* ZITTEL, 1879, p. 480 (obj.)]. Plates very imbricate; ambulacrum adorally with 4 columns, adapically occluded plates alternate with primaries; interambulacrum with many irregular columns, small perforate primary tubercle on each plate. *Sil.*, Eng. ——FIG. 223,*1*. **E. pomum*; *1a*, apical part of test, ×2; *1b,c*, ambs, ×4 (83). [=*Echinocystis* GREGORY, 1897, p. 124 *(nom. van.)*.]

Pronechinus KIER, 1965, p. 00 [**P. anatoliensis*; OD]. Outer columns of adoral ambulacra with large plates with peripodia alternating with small plates without peripodia; median columns composed of small plates. *Perm.*, Turkey.——FIG. 223, *2*. **P. anatoliensis*; *2a,b*, part of oral surface showing lantern, aboral surface, ×1.3 (Kier, n).

Perischocidaris NEUMAYR, 1881, p. 174 [**Archaeocidaris harteiana* BAILEY, 1874, p. 42; OD, M] [=*Prosechinus* POMEL, 1883, p. 113 (obj.); *Homotoechus* SOLLAS, 1892, p. 152 (obj.)]. Ambulacrum with 6 columns (in only known species), median columns elevated; interambulacra wider than ambulacra, with 5 columns in each area; large, perforate, primary tubercles on some interambulacral plates; amount of imbrication not known. *L.Carb.*, Ire.——FIG. 223,*3*. **P. harteiana* (BAILEY); aboral surface, ×0.7 (87).

Proterocidaris de Koninck, 1882, p. 514 [**P. giganteus*; OD, M] [=*Fournierechinus* Jackson, 1929, p. 67 (type, *F. deneensis*); ?*Jacksonechinus* Lambert, 1936, p. 39 (type, *J. andrewi* Lambert, 1936, p. 39); *Eupholidocidaris* Kier, 1956, p. 15 (type, *E. brightoni*)]. Low, large ambulacra and interambulacra with more than 2 columns; adorally ambulacra much more developed; plates of equal size; small perforate primary tubercles on some interambulacral and ambulacral plates. *Miss.*-

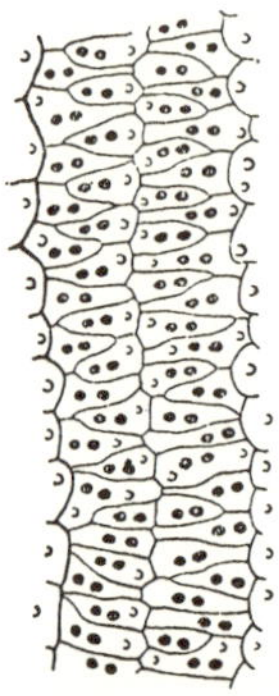

1b

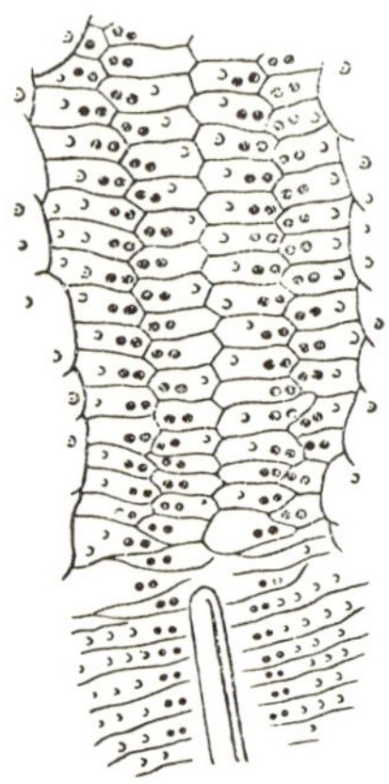

1c

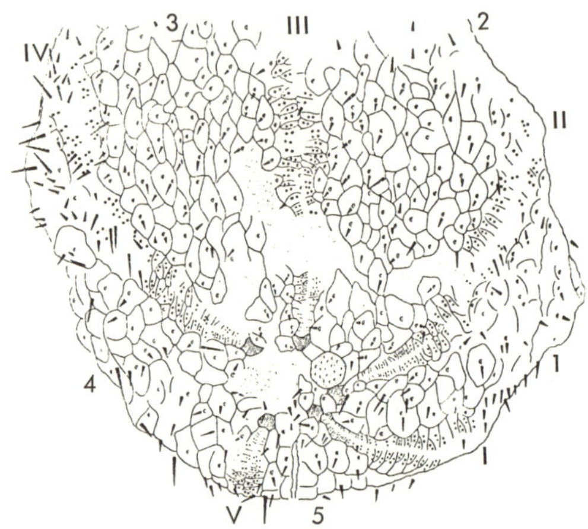

1a Echinocystites

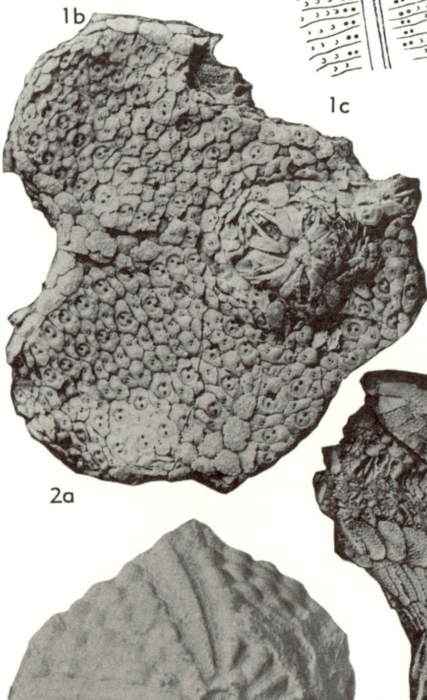

2a

2b Pronechinus

Perischocidaris 3

Penn., N.Am.-Eu.-?N.Afr.(Egypt).——Fig. 224,*1*. **P. giganteus*, L.Carb., Belg.; *1a,b*, aboral and oral faces, ×0.25, ×0.29 (97).

Rhenechinus Dehm, 1953 [**R. hopstatteri*; OD]. Ambulacrum with 4 columns, occluded plates alternating with primaries; interambulacrum with many regular columns. *Dev.*, Ger.——Fig. 224,*2*. **R. hopstatteri*; *2a*, part of amb, ×2.5; *2b*, part of oral surface, ×1 (42).

Family LEPIDESTHIDAE Jackson, 1896

[Lepidesthidae Jackson, 1896, p. 206]

More than 2 columns in each ambulacrum, ambulacral plates not enlarged adorally. *Dev.-Perm.*

Lepidesthes Meek & Worthen, 1868, p. 522 [**L. coreyi*; OD, M] [=*Hybochinus* Worthen & Miller, 1883, p. 331 (type, *H. spectabilis*)]. Test

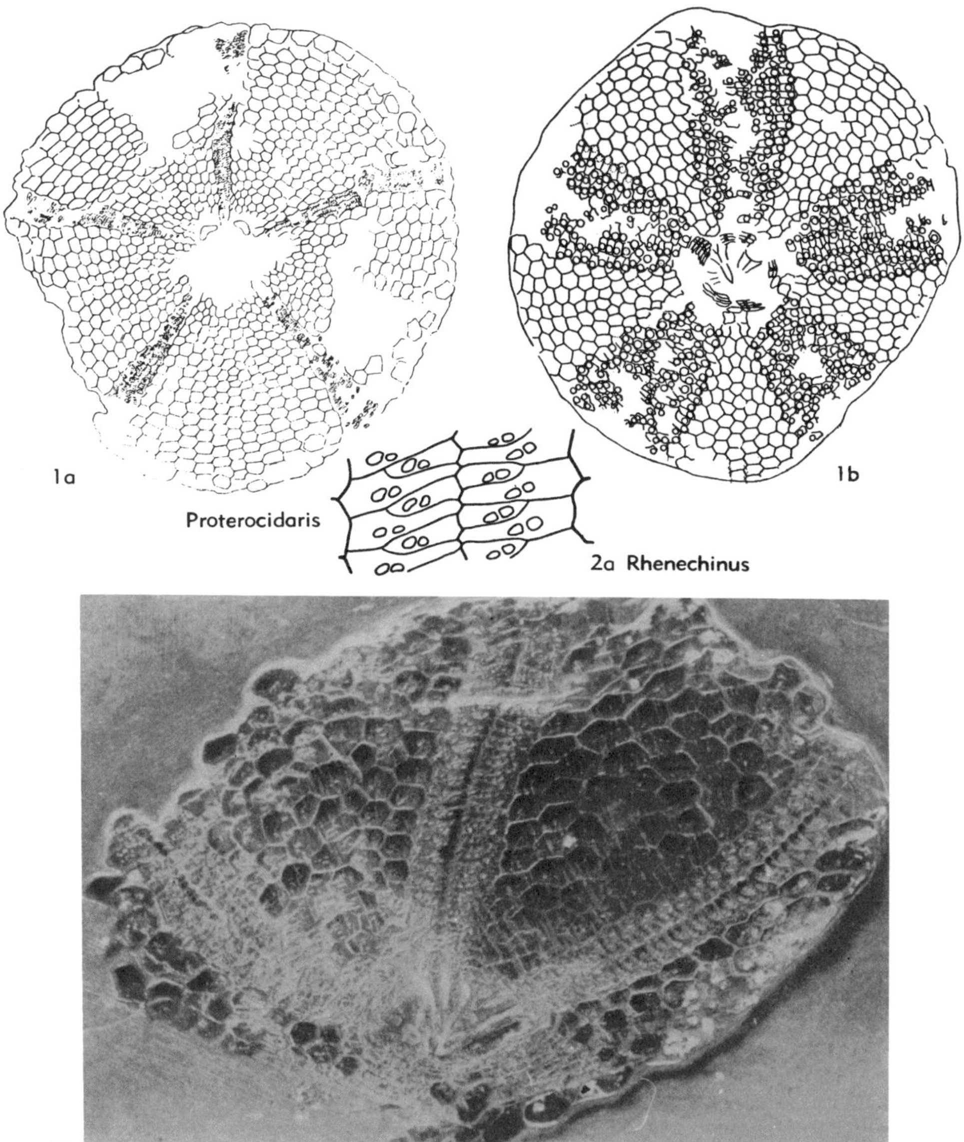

Fig. 224. Echinocystitidae (p. *U*302-*U*303).

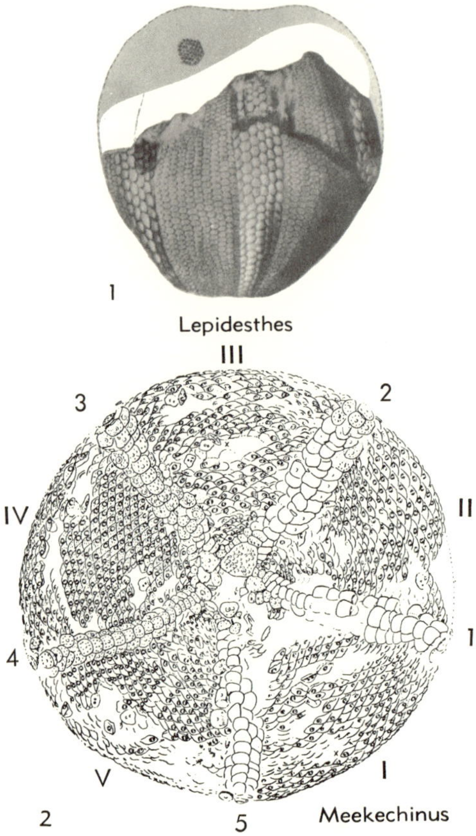

FIG. 225. Lepidesthidae (p. U303-U304).

high, plates strongly imbricate, ambulacra with many columns; interambulacra with few; ambulacra wider at mid-zone than interambulacra; no primary tubercles. *Dev.-Penn.*, Eu.-USSR-N. Am.-N.Afr.——FIG. 225,*1*. **L. coreyi*, Miss., USA (Ind.); lat. view, interamb. column central, ×0.8 (87).

Meekechinus JACKSON, 1912, p. 442 [**M. elegans*; OD]. Plates strongly imbricate; ambulacra with many columns; interambulacra with few; ambulacra much wider at mid-zone; small perforate primary tubercle on ambulacral and interambulacral plates; teeth serrated. *Perm.*, USA(Kans.).——FIG. 225,*2*. **M. elegans*; aboral view, ×1.5 (87).

Family LEPIDOCENTRIDAE Lovén, 1874

[Lepidocentridae LOVÉN, 1874, p. 39]

Only 2 columns of plates in each ambulacrum. *Ord.-Miss.*

Lepidocentrus MÜLLER, 1857, p. 258 [**L. eifelianus*; OD, M]. Pore pairs uniserial, all in contact with interambulacrum, adoral plates similar to adapical; interambulacrum wide, with many regular columns; small perforate primary tubercles on most of interambulacral plates. *Dev.-Miss.*, Eu.-N.Am.——FIG. 226,*3*. *L. mülleri* SCHULTZE, Dev., Ger.; lat. view, part of test showing amb and 2 interambs, ×1 (87).

Albertechinus STEARN, 1956, p. 741 [**A. montanus*; OD]. Ambulacra narrow, plates low, pore pairs uniserial, radial vessel almost isolated from interior by internal processes; interambulacra with many irregularly arranged plates; large primary tubercle on some interambulacral plates. *Dev.*, Can.——FIG. 226,*4*. **A. montanus*; part of crushed surface showing narrow amb and adjoining interambs, ×1 (156).

Aulechinus BATHER & SPENCER, 1934, p. 558 [**A. grayae*; OD]. Plates of test strongly imbricate, ambulacrum with deep median groove, radial vessel internally enclosed; ambulacral plates overlapping perradially, not fused; podial pores large, single or incompletely divided, notched; interambulacrum with numerous irregularly arranged plates; no primary tubercles. *Ord.*, Eng.——FIG. 226,*2*. **A. grayae*; lat. view, summit above (*M*, madreporite), ×2 (130).

Deneechinus JACKSON, 1929, p. 22 [**D. tenuisponus*; OD]. Test large, low; interambulacrum with many columns; each plate with several small perforate primary tubercles. *L.Carb.*, Belg.

Ectinechinus MACBRIDE & SPENCER, 1938, p. 95 [**E. lamonti*; OD]. Test elongated, plates strongly imbricate; ambulacrum with no median groove, radial vessel internally enclosed; podial pores double or incompletely divided; ambulacral plates overlapping perradially, not fused; interambulacrum with numerous irregularly arranged plates; no primary tubercles. *Ord.*, Eng.——FIG. 226,*1*. **E. lamonti*; lat. view, summit above, ×2 (130).

Eothuria MACBRIDE & SPENCER, 1938, p. 95 [**E. beggi*; OD]. Elongate test; plates strongly imbricated, ambulacrum curved, perradial groove slight or absent, radial vessel not internally enclosed, ambulacral plates overlapping perradially, not fused, each pore with many openings; interambulacrum with numerous more or less irregularly arranged plates; no primary tubercles; jaws reduced or absent. *Ord.*, Scot.(Girvan).——FIG. 226,*5*. **E. beggi*; *5a*, oblique lat. view showing apical disc; *5b*, similar view showing peristome, ×2 (130).

[MACBRIDE & SPENCER considered *Eothuria* to be a holothurian and not an echinoid. However, it has so many characters similar to *Aulechinus* and *Ectinechinus*, such as similar ambulacral and interambulacral plate arrangement with the same imbrication, and presence of a single genital plate, that it is here considered classifiable as an echinoid. It differs only from other echinoids in having many pores in each ambulacral plate, in having no oculars, and in having valvelike plates at the mouth in place of 5 teeth. It appears to be an offshoot from the *Aulechinus* and *Ectinechinus* line.]

Hyattechinus JACKSON, 1912, p. 291 [**H. beecheri*; OD]. Test low; ambulacra adorally wide, with

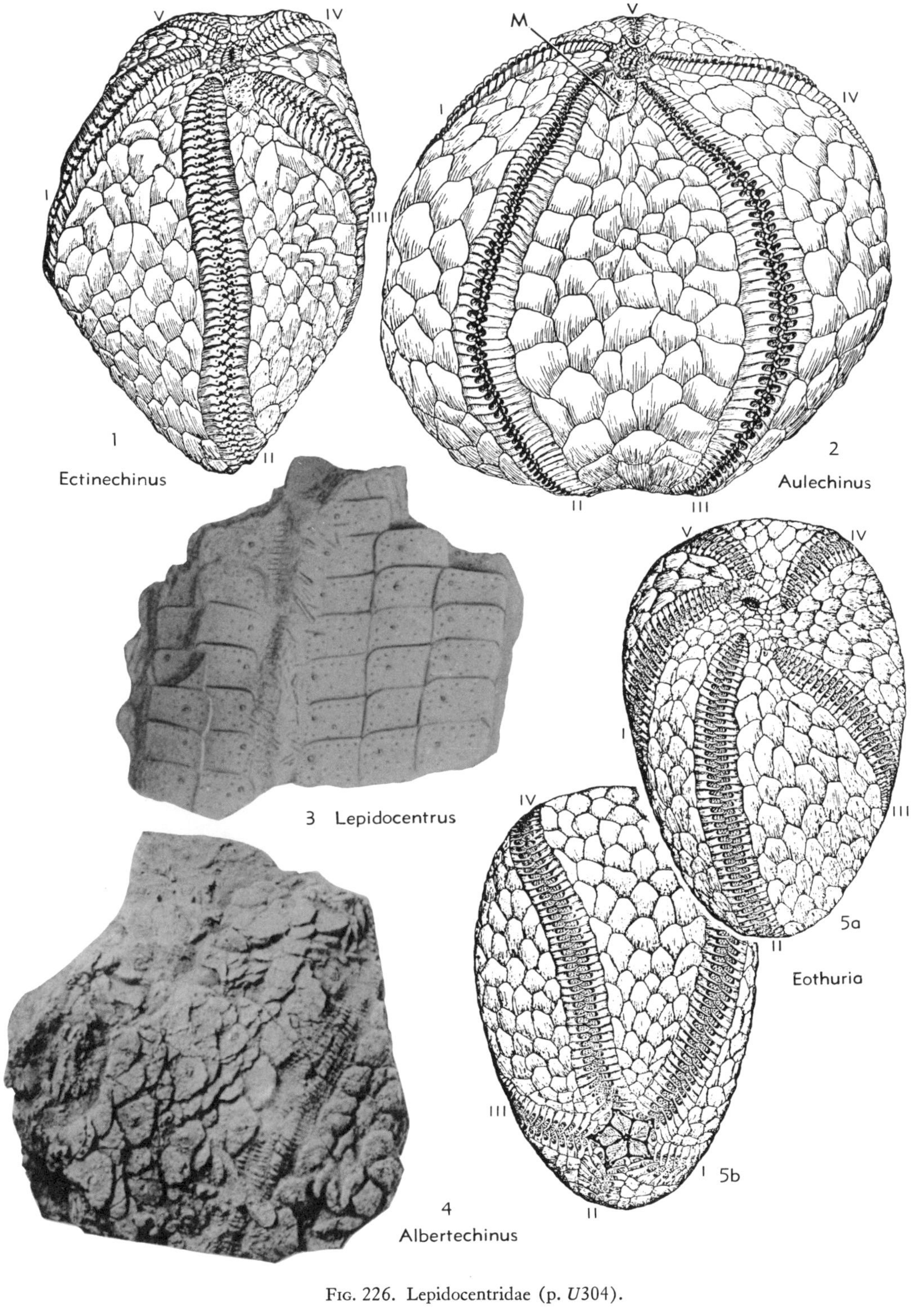

Fig. 226. Lepidocentridae (p. *U*304).

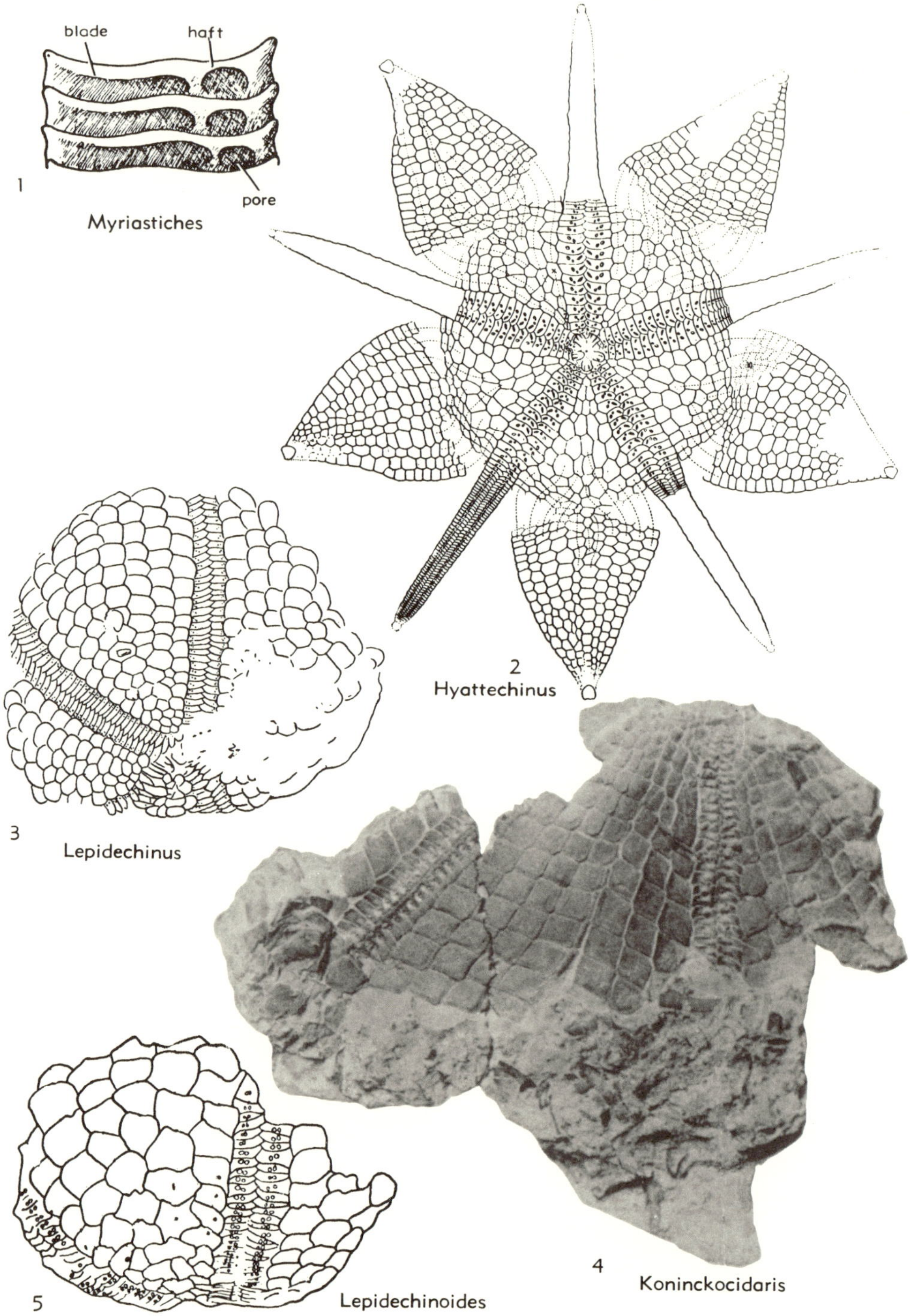

FIG. 227. Lepidocentridae (p. *U*304, *U*307).

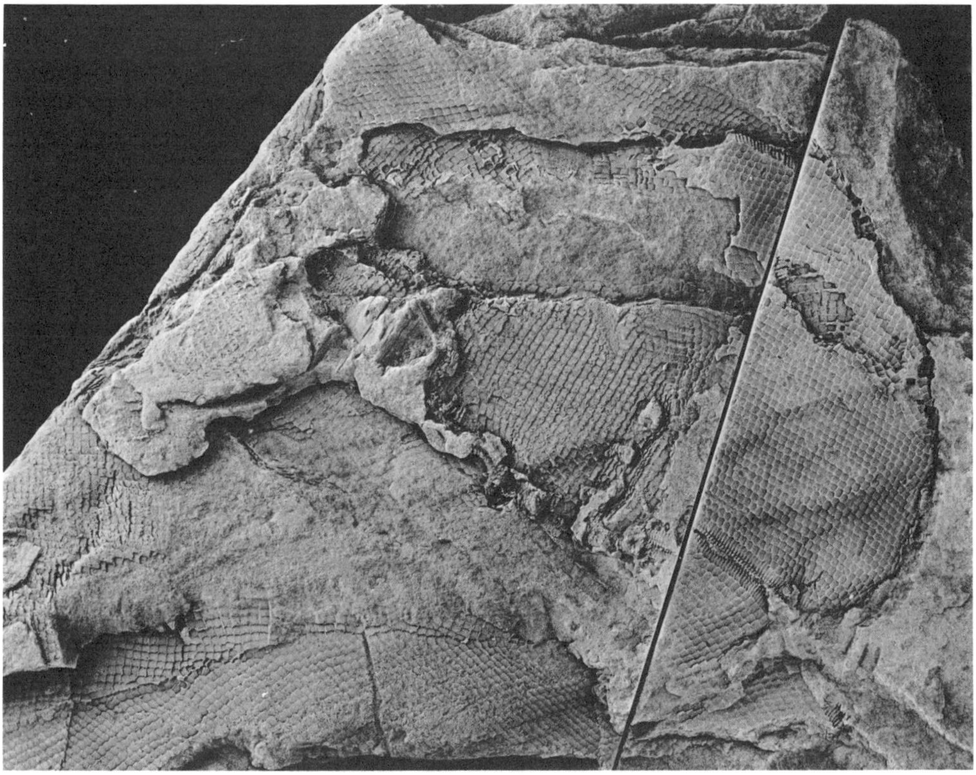

1 Myriastiches

FIG. 228. Lepidocentridae (p. *U*307).

large peripodia, adapically narrow without peripodia; internal spinose processes on adoral ambulacral plates; interambulacrum with many columns, small perforate tubercles on some interambulacral plates. *L.Carb.(Miss.),* Eu.-N.Am.——FIG. 227,*2*. **H. beecheri,* USA(Pa.); plate diagram extending from centrally placed peristome, ×5.5 (87).

Koninckocidaris DOLLO & BUISSERET, 1888, p. 959 [**K. cotteaui*; SD, M]. Ambulacrum with high plates, pore pairs uniserial, internal ridge along perradial suture; interambulacrum with many regular columns; no primary tubercles. *Sil., ?L. Carb.,* N.Am.-?Eu.——FIG. 227,*4*. *K. silurica* JACKSON, Sil., USA(N.Y.); aboral part of test showing broad interambs and narrow ambs, ×1 (87).

Lepidechinoides OLSSON, 1912, p. 442 [**L. ithacensis*; SD, M]. Adoral plates similar to adapical; internally ambulacral plates opposite horizontal ambulacral sutures expanded laterally, fan-shaped; internal spinose processes on adoral portion of ambulacrum; interambulacrum with many columns; no primary tubercles. *Dev.,* N.Am.——FIG. 227,*5*. **L. ithacensis,* USA(N.Y.); adoral part of test showing wide interambs and narrow ambs, ×2 (25).

Lepidechinus HALL, 1861, p. 18 [**L. imbricatus*; OD, M] [=*Rhoechinus* KEEPING, 1876, p. 37 (type, *R. irregularis*)]. Adoral ambulacral plates similar to adapical; interambulacrum wide with many columns; only secondary tubercles. *Miss.,* N.Am.-Eu.——FIG. 227,*3*. **L. imbricatus,* Miss., USA(Iowa); adoral part of test, oblique view, ×1.9 (87).

Myriastiches SOLLAS, 1899, p. 700 [**M. gigas*; OD]. Pore pairs uniserial, near perradial suture, through plates; ambulacral, interambulacral plates small; interambulacrum with many small plates in regular columns (more than 32 in type-species); small spines, no tubercles. *Sil.,* Eng.——FIG. 227,*1*; 228, *1*. * *M. gigas*; 227,*1*, part of amb, ×18 (130); 228,*1*, part of oral surface, ×0.65 (Kier, n).

Palaeodiscus SALTER, 1857, p. 332 [**P. ferox*; OD, M]. Test very flexible; ambulacrum with external median groove, internal processes present but not covering radial canal, pore pairs uniserial, piercing ambulacral plates; interambulacrum with many regular columns; no primary tubercles. *Sil.,*

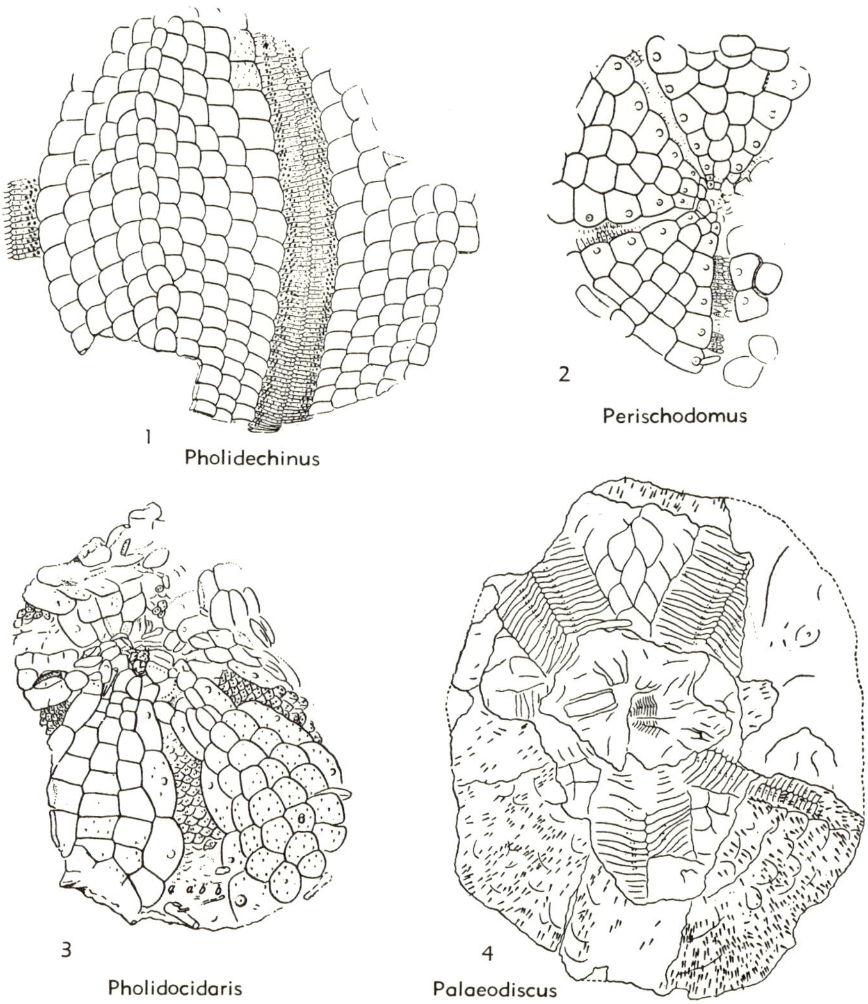

FIG. 229. Lepidocentridae (p. *U*307-*U*309).

Eng.——FIG. 229,*4*. **P. ferox;* peristomial region, ×2 (83).

Perischodomus M'COY, 1849, p. 253 [**P. biserialis*; OD, M] [=*Tretechinus* TORNQUIST, 1897, p. 784 (type, *Perischodomus illinoisensis* WORTHEN & MILLER, 1883, p. 333)]. Pore pairs biserial at ambitus, where some of ambulacral plates not in contact with interambulacra; adoral plates larger than adapical; interambulacrum wide with many columns; primary perforate tubercle on some interambulacral plates. *L.Carb.(Miss.)*, Eu.-N.Am. ——FIG. 229,*2*. **P. biserialis*, Ire.; aboral surface, ×0.95 (87).

Pholidechinus JACKSON, 1912, p. 299 [**P. brauni*; OD]. Adoral ambulacral plates similar in size to adapical, low; pore pairs uniserial to slightly biserial; interambulacrum with many regular columns; no primary tubercles. *Miss.*, USA(Ind.). ——FIG. 229,*1*. **P. brauni*; lat. view showing wide interambs and narrow ambs, ×1.05 (87).

Pholidocidaris MEEK & WORTHEN, 1869, p. 78 (*nom. conserv.* ICZN, 1955) [**Lepidocentrus irregularis* MEEK & WORTHEN, 1869, p. 78] [=*Protoechinus* AUSTIN, 1860, p. 446 (type, *P. anceps*); ?*Protocidaris* WHIDBORNE, 1898, p. 202 (type, *Eocidaris? acuaria* WHIDBORNE, 1896, p. 376)]. Ambulacra adorally much more developed; adapically adambulacral plates much larger than other interambulacral plates; adorally all interambulacral plates of same size; large perforate primary tubercle on adapical adambulacral and all adoral interambulacral plates; secondary tubercle on other plates. *?Dev., L.Carb.(Miss.)*, Eu.-N.Am.——FIG. 229,*3*. **P. irregularis* (MEEK

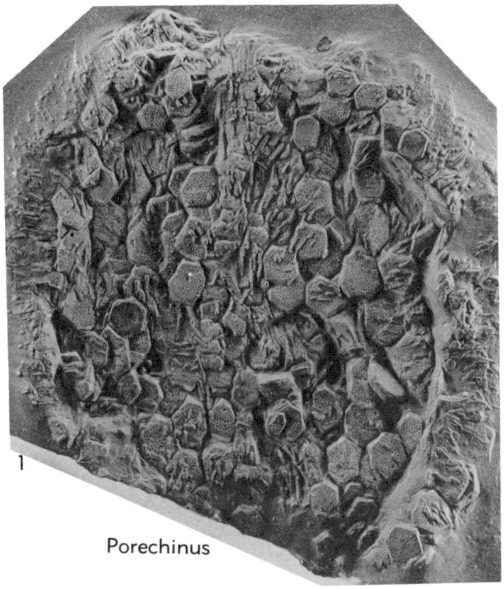

FIG. 230. Lepidocentridae (p. U309).

& WORTHEN), Miss., USA(Ind.); oblique view of apical region, ×0.9 (87).

Porechinus DEHM, 1961, p. 4 [**P. porosus*; OD]. Pore pairs uniserial, oblique, situated near middle of plate; inner pore of pair on edge of plate, not closed; interambulacrum with many columns; no primary tubercles. *Dev.*, Ger.——FIG. 230,*1*. **P. porosus*, interamb near summit, ambs at left and right, ×1 (Kier, n).

FAMILY UNCERTAIN

Devonocidaris THOMAS, 1920, p. 212 [**D. jacksoni* THOMAS, 1924, p. 500; SM THOMAS, 1924]. Plates thin; perforate primary tubercle on interambulacral and larger ambulacral plates; teeth with median furrow on outer side. *Dev.*, Eu.-N.Am.

Tornquistellus BERG, 1899, p. 77 [**Leptechinus gracilis* TORNQUIST, 1897, p. 785] [=*Leptechinus* TORNQUIST, 1897, p. 785 (obj.) (*non Leptechinus* GAUTHIER, 1869)]. Known only from isolated interambulacral plates, thin, flat, wider than high, secondary tubercles. *Dev.*, Ger.

Order PALAECHINOIDA Haeckel, 1866

Plates not strongly imbricate, ambulacral plates bevel over interambulacra; interambulacrum of one or more than 2 columns; no perignathic girdle. *Sil.-Perm.*

Family PALAECHINIDAE M'Coy, 1849

[Palaechinidae M'COY, 1849, p. 253] [=Palaeechinidae JACKSON, 1912, p. 302]

Enlarged adradial ambulacral plates; interambulacrum with more than 2 columns; imperforate tubercles. *Miss.*

Palaechinus M'COY, 1844, p. 172 [**P. ellipticus*; SD LAMBERT & THIÉRY, 1910, p. 119] [=*Palechinus* SCOULER in GRIFFITH, 1840 (*nom. nud.*); *Palechtinus* FISCHER DE WALDHEIM, 1848, p. 247 (*nom. null.*); *Palaeechinus* LOVÉN, 1874, p. 40 (*nom. van.*)]. Ambulacrum with 2 columns of plates, pore pairs uniserial to slightly biserial. *L.Carb.(Miss.)*, Eu.-N.Am.——FIG. 231,*1*. *P. canadensis* KIER, Miss., Can.; *1a*, lat. view showing interamb, ×1; *1b*, apical region, ×1; *1c*, apical disc, ×5; *1d*, amb, ×4 (92).

Lovenechinus JACKSON, 1912, p. 324 [*nom. conserv.* ICZN, 1955] [**Oligoporus missouriensis* JACKSON, 1896, p. 184; OD] [=*Eriechinus* POMEL, 1883, p. 114 (type, *Palaechinus sphaericus* M'COY, 1844, p. 172); *Typhlechinus* NEUMAYR, 1889, p. 363 (type, *Palaechinus sphaericus* M'COY, 1844, p. 172)]. Ambulacrum with 4 columns of plates, consisting of 2 columns of narrow demiplates, 2 columns of wider occluded plates; pore pairs biserial. *L. Carb. (Miss.)*, Eu.-N. Am.-China.——FIG. 232,*1*. **L. missouriensis* (JACKSON), USA (Mo.); *1a*, aboral view of int. mold, ×0.5; *1b*, amb and interamb plates, ×1.3 (87).

Maccoya POMEL, 1869, p. 46 [**Palaechinus gigas* M'COY, 1844; OD]. Ambulacrum with 2 columns of plates, all in contact at median suture, at adradial suture every other plate nearly excluded from contact with interambulacra; pore pairs biserial. *L.Carb.(Miss.)*, Eu.-N.Am.——FIG. 231, *2*. *M. sphaerica* (M'COY), Ire.; *2a*, oblique aboral view of test, ×1.2; *2b*, amb and interamb plates, ×2.6 (87).

Melonechinus MEEK & WORTHEN, 1861, p. 396 [**Melonites multipora* OWEN & NORWOOD, 1846, p. 225; OD] [=*Melonites* NORWOOD & OWEN, 1846, p. 225 (*non* LAMARCK, 1822, p. 615); *Melechinus* QUENSTEDT, 1875, p. 381 (obj.); *?Donbassechinus* FAAS, 1941, p. 73 (type, *D. kumpani*)]. Ambulacrum with more than 4 columns of plates, consisting of 2 columns of narrow demiplates, 2 columns of wider occluded plates, and in addition one or more irregular columns of isolated plates between demi- and occluded plates; pore pairs multiserial. *L.Carb.(Miss.)*, Eu.-N.Am.-USSR-China.——FIG. 231,*3*; 232,*2*. **M. multiporus* (OWEN & NORWOOD), USA(Mo.); 231,*3*, amb with bordering rows of interamb plates, ×2.3; 232,*2*, aboral view of test, ×1 (87).

Oligoporus MEEK & WORTHEN, 1862, p. 472 [**Melonites danae* MEEK & WORTHEN, 1861, p. 397; OD, M] [=*Melonopsis* MEEK & WORTHEN, 1866, p. 249 (obj.)]. Ambulacrum with 4 columns of plates, consisting of 2 columns of wider occluded plates, and in addition scattered isolated plates in middle line of each half area; pore pairs multiserial. *Miss.*, N.Am.——FIG. 233,*1*. **O. danae* (MEEK & WORTHEN), USA(Iowa); amb with bordering interamb plates at right, ×2.4 (87).

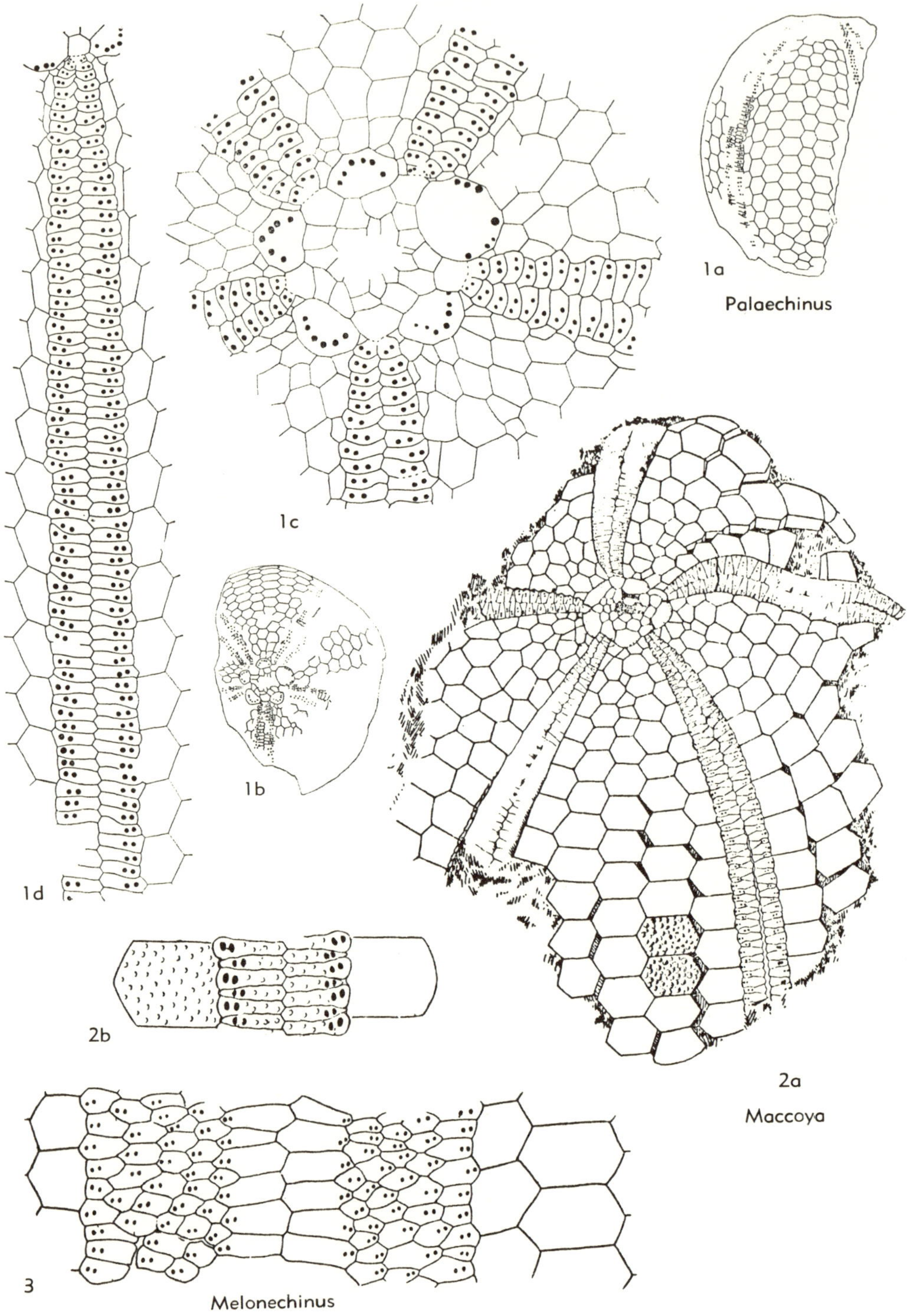

FIG. 231. Palaechinidae (p. U309).

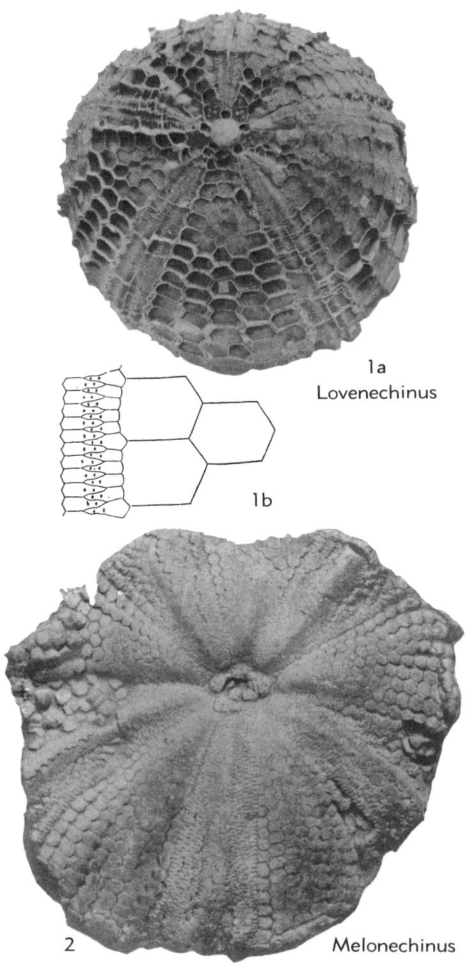

FIG. 232. Palaechinidae (p. *U*309).

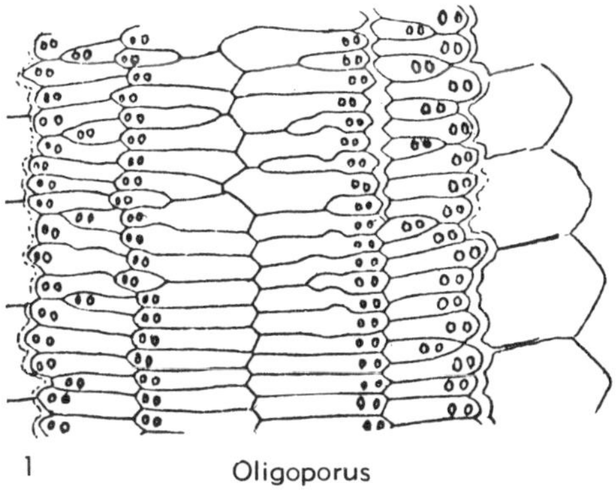

FIG. 233. Palaechinidae (p. *U*309).

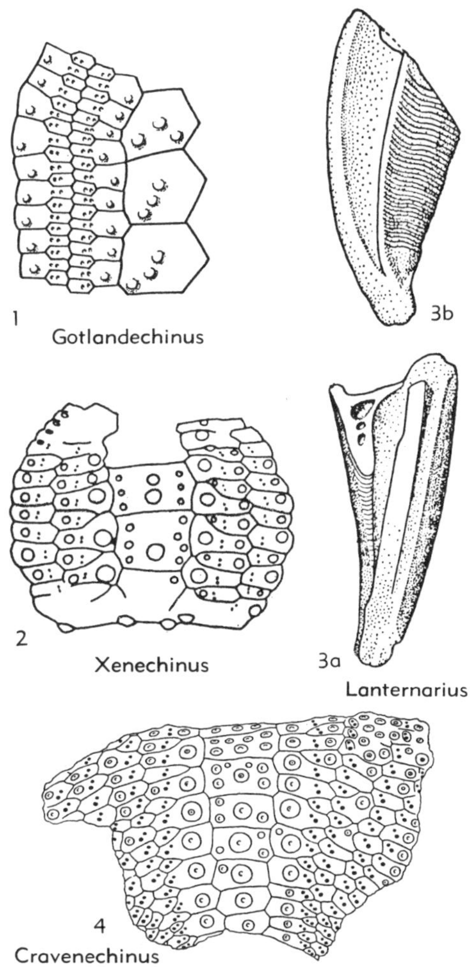

FIG. 234. Cravenechinidae *(1-2,4)*; Order and Family Uncertain *(3)* (p. *U*311-*U*312, *U*329).

Family CRAVENECHINIDAE Hawkins, 1946

[Cravenechinidae HAWKINS, 1946, p. 195]

Interambulacrum with one or 4 columns; adradial ambulacral plates enlarged; small perforate or imperforate primary tubercles on ambulacral and interambulacral plates. *Sil.-Perm.*

Cravenechinus HAWKINS, 1946, p. 195 [**C. uniserialis*; OD]. Ambulacrum with 8 series of pore pairs; perforate primary tubercles. *L.Carb.*, Eng.——FIG. 234,*4*. **C. uniserialis*; part of amb column, ×2 (82).

Gotlandechinus REGNÉLL, 1956, p. 158 [**G. balticus*; OD]. Ambulacrum with 4 columns of plates, pore pairs uniserial; 4 columns in each interambulacrum. *Sil.*, Sweden.——FIG. 234,*1*. **G. balticus*, Gotl.; amb column with interamb plates at right, ×2.5 (148).

Xenechinus KIER, 1958, p. 889 [**X. parvus*; OD]. Ambulacrum with 4 series of pore pairs; covered passageway on interior for radial vessel; imperforate primary tubercles. *Perm.*, USA(Tex.).——FIG. 234,*2*. **X. parvus*; interamb bordered by amb columns, ×8 (95).

FAMILY UNCERTAIN

Wrighthia *(sic)* POMEL, 1869, p. 46 [**Palechinus phillipsiae* FORBES, 1848; OD, M] [=*Wrightella* POMEL, 1883, p. 115 (obj.)]. Known only from one poorly described specimen showing portion of ambulacrum and interambulacrum. *?Sil.*, Eng.

Xysteria POMEL, 1883, p. 114 [**Palaechinus konigii* M'COY, 1844; OD, M]. Known only from few isolated interambulacral plates covered with secondary tubercles. *L.Carb.*, Ire.

CIDAROIDS

By H. BARRACLOUGH FELL

[Victoria University of Wellington, New Zealand; transferred to Harvard University]

INTRODUCTION

The cidaroids are the only surviving echinoids with authenticated Paleozoic representatives, and they are believed to be ancestral to all other surviving echinoids. They are here regarded as an order of the subclass Perischoechinoidea, the three extinct Paleozoic orders of which have already been treated in this volume. Cidaroids are of exceptional interest as archaic living echinoids with generalized morphological features.

To judge by their fossil abundance, cidaroids reached a peak of development in Mesozoic seas, and thereafter declined in Europe, the Mediterranean, and North America, playing only a minor role in the faunas of those regions after the Eocene. However, the group still flourishes in the Indo-West-Pacific, and another quite distinctive assemblage is conspicuous in the seas of Antarctica, where it apparently evolved in prolonged isolation.

Specimens of Indo-Pacific cidaroids were brought to Europe by travelers in the 17th and 18th centuries. These echinoids, by their exquisite symmetry and evident similarity to fossils, evoked the admiration of amateur naturalists, such as Queen LOUISA ULRIKA, of Sweden, whose cabinets were afterwards to be studied by LINNÉ and others.

The widespread extinction of northern cidaroid faunas, coupled with their survival in the Indo-West-Pacific regions, has long presented something of a mystery. However, it may well have been that the impoverishment of the Tertiary cidaroid fauna of Europe was the result of the southeastward retreat of Tethys, the cidaroid fauna of that Mesozoic sea being now part of the inheritance of the present-day Indian Ocean, and neighboring seas. On this interpretation, the existing cidaroid fauna of the Indo-West-Pacific is merely part of a continuing succession of Tethyan stocks and Tethys itself is simply a former northwestern extension of the Indian Ocean. The existing North Atlantic cidaroid fauna seems to be a late derivative of a small tropical Caribbean nucleus of genera derived from the Indo-Pacific. Whether or not the supposed recency of the Atlantic Ocean can be justified, such supposition offers a more reasonable interpretation of the changes in cidaroid faunas of North Atlantic regions than the earlier hypothesis of northern extinctions with simultaneous (and inexplicable) Indo-Pacific survivals.

CLASSIFICATION AND EVOLUTIONARY TRENDS

The oldest undoubted cidaroids are the Early Carboniferous Archaeocidaridae. These forms have a pluriserial structure of the interambulacra, a feature seen also in other Perischoechinoidea which were their precursors, and this structure doubtless was inherited from them. From these early cidaroids probably arose the family Miocidaridae, of which the oldest undoubted members are of Permian age (though a Carboniferous miocidarid seems to be represented by the incompletely known *"Miocidaris" cannoni* of North America). The Miocidaridae resemble the Archaeocidaridae in having a semiflexible test, in which the interambulacral plates overlap upon the ambulacra, but differ from them (and by the same token resemble modern cidaroids) in having biserial interambulacra. By late Triassic times the first representatives of the modern Cidaridae had appeared; these had a rigid test, but retained the simple uniserial arrangement of the ambulacral pores seen in the Miocidaridae and the perforate tubercles of the latter. By the Jurassic the other two families of cidaroids had appeared, the Psychocidaridae, with imperforate tubercles (still surviving in the North Pacific), and the extinct Diplocidaridae, with biserial arrangement of the ambulacral pores. The Archaeocidaridae vanished from the fossil record after the Permian, the Miocidaridae after the Jurassic, and Diplocidaridae apparently did not survive the Early Cretaceous. Thus, from mid-Cretaceous time onward the fossil cidaroids are represented only by families which still flourish today. Most cidaroids are now restricted to the tropics, especially of the Indo-West-Pacific region.

ECOLOGY

The structure of Paleozoic and early Mesozoic cidaroids parallels that of living forms sufficiently to imply that their ecology would not have differed materially from that of extant representatives of the order. The living forms are inactive echinoids, moving only slowly, mainly by using the large spines as levers or stilts. Shallow-water cidaroids hide during daylight hours in crevices or under stones. Large muscles occur at the base of the spines in one Antarctic genus *(Homalocidaris),* and it has been supposed that this implies a more active mode of progression (136a); recently, however, living specimens freshly dredged from the floor of the Ross Sea show that the animal, at least when lying on the ship's deck, scarcely moved the spines at all, and it is accordingly doubtful whether any cidaroid can move rapidly.

Cidaroids live at almost all depths, at least down to 4,000 m. Most species prefer hard bottom, such as reefs, and it is probable that short-spined forms, even in deeper offshore waters, contrive to make use of shell beds as temporary hard bottom. Forms with long slender spines seem to tolerate soft mud, and deep-sea cidaroids develop such spines. The shallow-water reef-dwelling genera do not conceal themselves by holding other objects over the test with the tube feet, as do other regular echinoids; this is because the tube feet of the aboral side have only vestigial or modified suctorial discs, serving as respiratory organs.

Cidaroids feed upon available bottom animals, including mollusks, tubicolous annelids, polyzoans, foraminifers, and sponges. Their teeth are strong enough to crush the hard parts of such organisms.

Development may be direct, the young stages being carried on the aboral side (e.g., *Austrocidaris,* see Fig. 241,*3*), or on the oral side (e.g., *Eurocidaris, Goniocidaris*); or indirect development, involving a pluteus larva, may occur, as described in the chapter on "Ontogeny" (*Treatise* Part S).

Various commensals, especially sponges, polyzoans, and cirripeds, may occur on the primary spines. Foraminifers and annelids occur among the secondary spines, and a holothuroid *(Taeniogyrus cidaris)* coils itself around the primary spines of *Stylocidaris.* Parasitic gastropods, and other parasites, occur either externally or as endoparasites. Remains of such organisms may be found with fossil cidaroids, or the evidence of their work may be apparent as borings in fossil cidaroid skeletal parts.

Certain genera of extant cidaroids (e.g., *Eucidaris, Phyllacanthus*) exhibit marked preference for seas in which the surface temperature does not go beyond definite limits throughout the year; among genera named their distribution at present falls within the winter isotherms for approximately 15° C. On the basis of such distribution patterns, estimates of Tertiary sea temperatures have been made, yielding results not inconsistent with other data (56).

MORPHOLOGY OF HARD PARTS

Cidaroids, in common with other echinoids, have a complex skeleton, the individual parts of which may exhibit specific or higher taxonomic characters, or both. In *Goniocidaris,* for example, there are at least 3,000 separate skeletal elements of some 60 different shapes and sizes, some with elab-

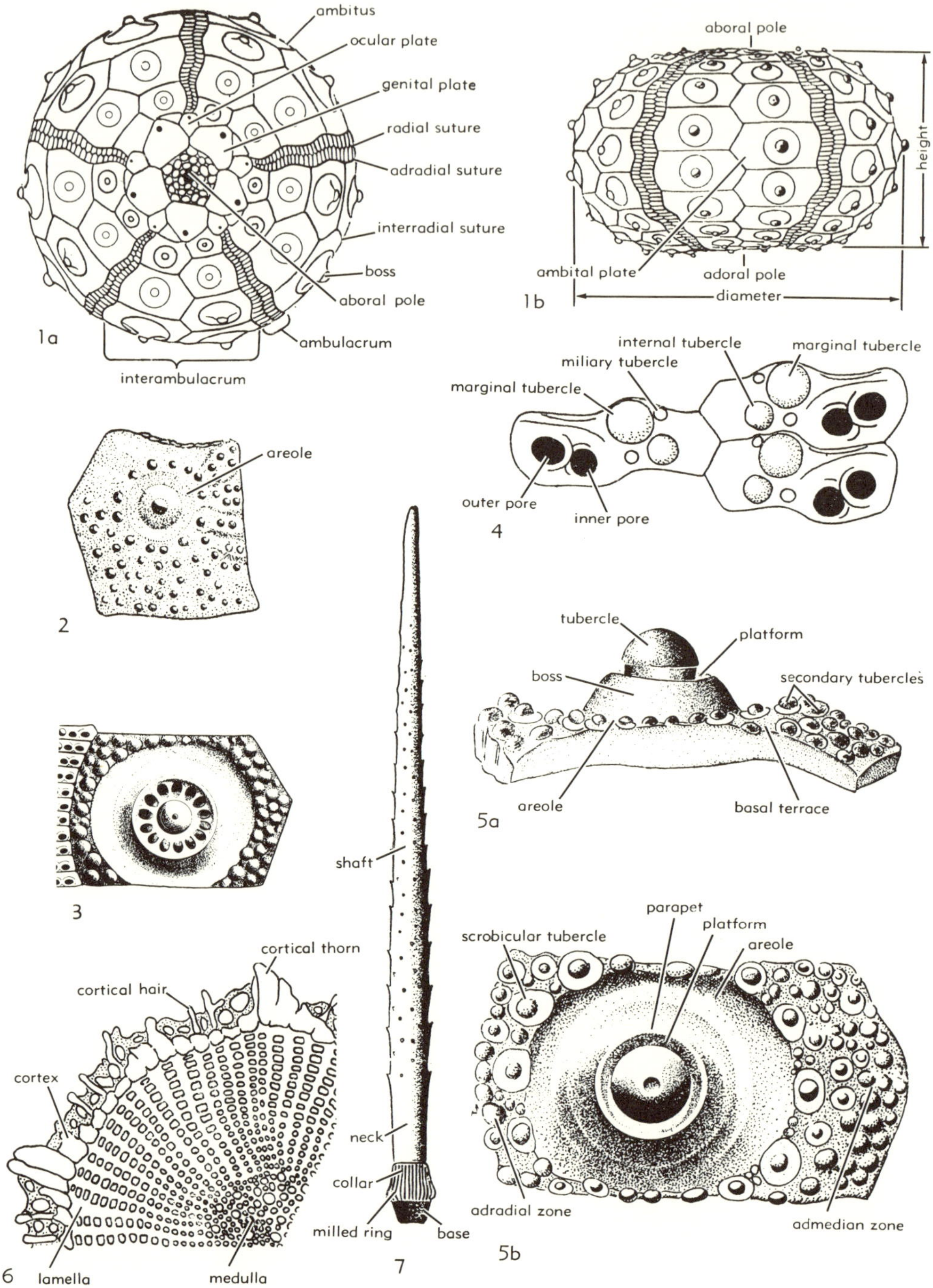

FIG. 235. Morphological features of cidaroid echinoids (Fell, n).——*1a,b.* Aboral and lateral views of test.——*2,3.* Interambulacral plates.——*4.* Part of ambulacrum.——*5a,b.* Side and external views of interambulacral plate.——*6.* Part of cross section of spine, enl.——*7.* Side view of spine.

orate microscopic detail. Of these, only 600, at most, would normally be of any paleontological significance—that is, several associated plates or rarely a single plate or spine sufficient to yield a generic or specific determination. In general, recognition is difficult from isolated skeletal pieces. Again, some forms cannot be placed in correct systematic position on the basis of the skeleton alone, even when complete, because pedicellariae are required to yield a final determination; fortunately, not many cidaroid genera are so difficult to identify, though it is probable that many fossil genera would be subdivided if adequate information on the nature of their pedicellariae or larvae was available. Although pedicellariae are occasionally found on fossils, as a rule, the only parts of the cidaroid skeleton of value in paleontology are whole or partial tests, ambulacral and interambulacral plates, and primary spines. These structures are briefly explained in the following paragraphs, illustrated by Figure 235.

The **test** (Fig. 235,*1a,b*) is the outer (though mesodermal) shell formed by vertical columns of plates. All Cenozoic genera have 20 columns of plates, but a larger number occur in some Paleozoic cidaroids (e.g., Archaeocidaridae, see Fig. 237), and in the Cretaceous genus *Tetracidaris* (see 254,*3*), such genera being termed **pluriserial.** A pluriserial condition occurs temporarily in the young stage of some living cidaroids, and fossils suggest that the pluriserial pattern is the original one for the order. The upper surface of the test is termed **aboral,** or adapical, and the lower surface **oral.** The major circumference of the test, always horizontal, is termed the **ambitus,** and the adjacent skeletal plates are called **ambital plates.** Thus the whole test may be thought of as a globe, in which the plates are arranged in vertical meridians, the ambitus forming the equator. Of the vertical series of plates, five double (or multiple) columns carry pores for tube feet, and are termed **ambulacra,** or in abbreviated form **ambs.** Five intervening double (or multiple) columns of plates, carrying large tubercles, are termed the **interambulacra,** or in short **interambs.**

The ambulacra (Fig. 235,*4*) are divisible into an outer **poriferous area,** and an inner, broader **interporiferous area.** The poriferous area is marked on each component amb plate by the presence of two similar pores, termed a **pore pair.** These correspond to the points of ingress and egress of ambulacral fluid circulating in the tube foot of the living animal; thus each pore pair corresponds to one tube foot. It is of systematic importance whether the pore pair is horizontally or obliquely placed, and whether the two pores are each provided with a distinct wall (**nonconjugate pores**), or united by a common depressed groove (**conjugate pores**). Nonconjugate pores are illustrated in Figure 235,*4*, conjugate pores are shown in Figure 247, *1e*. The interporiferous area carries various rounded **tubercles,** of which usually one on each plate is distinguished as a larger **marginal tubercle,** adjoining the pore pair. The vertical series of such marginal tubercles is termed the **marginal series;** it may be straight, or sinuous, as also the vertical series of pore pairs. The other tubercles are termed the inner tubercles. All tubercles, in life, carry small spines, distinguished as **ambulacral spines.** Very small **miliary tubercles** may also occur for the attachment of **pedicellariae,** which are small grasping or toilet organs of complex structure, but their paleontological significance is slight. A third pore may occur near the pore pair; this is for the emergence of the nerve supplying the tube foot and is termed the **neuropore.** A neuropore, if enlarged, may form a significant systematic character in some Antarctic genera, but it is not at present known in fossils.

Each interambulacral plate bears a single large prominence, the **boss** (Fig. 235,*2,3,5*). Surmounting the boss is a **platform** on which is placed a central rounded **tubercle** (or **mamelon**). The tubercle articulates with a hemispherical cup on the base of a large spine, called the **primary spine** (Fig. 235,*7*). In most cidaroids a central **perforation** on the tubercle provides in life for a strand of connective ligament which runs to a corresponding perforation in the cup of the spine. In the Psychocidaridae no perforation is seen on the tubercles (at least of the adoral hemisphere), such tubercles being termed **imperforate** (e.g., *Tylocidaris,* see Fig. 252,*3*). The platform is commonly surrounded by a low **parapet,** and its surface may be **crenulate** (e.g., *Histocidaris,* Fig. 235,*4*) or **noncrenulate** (e.g., *Goniocidaris,* Fig. 235,*5*). The boss is surrounded by a broad, saucer-shaped shallow depression, the **areole,** which is devoid of sculpture. The outer margin of the areole, usually somewhat more deeply depressed, is termed the **scrobicule.** It serves as the region for the origin of muscles which move the spine. Surrounding the scrobicule and situated at a higher level than the rest of the plate is a more or less continuous ring of smaller tubercles, the **scrobicular tubercles.** Each of these usually has a miniature areole, and in life it carries a small spine, the **scrobicular spine.** The rest of the plate is usually covered by many small tubercles termed **secondary tubercles** and **miliary tubercles.** In some genera the scrobicular ring of tubercles is incomplete above and below, so that the areoles of adjacent plates are **confluent** (e.g., *Rhabdocidaris,* see Fig. 245,*2*), whereas in other genera (e.g., *Stereocidaris,* see Fig. 242,*3*), the areoles are widely separated. Characters such as these aid identification of isolated plates of fossils. The genus *Dicyclocidaris* is peculiar in having an inner ring of tubercles within the areole (see Fig. 239,*2b*) and *Porocidaris* has radially placed slots in the same position (see Fig. 244,*2f*); these seem to be specialized features of the muscle-attachment surface.

Plates are separated by **sutures,** rigid in post-Triassic genera, but imbricated and evidently movable in earlier cidaroids, where the test must have been more or less flexible. The latter feature accounts for the dislocation of plates in *Archaeocidaris* (see Fig. 237,*3a*), for example, and to lesser extent in *Dicyclocidaris* (see Fig. 239,*2c*). Imbricating test plates occur in other perischoechinoid orders of the Paleozoic and are also seen in the extant orders Diadematoida and Echinothurioida, where the condition may well be an inheritance from Paleozoic ancestors, as inferred in the case of the Cidaroida.

The **spines,** or radioles, comprise the large **primary spines** carried by the **primary tubercles,** and smaller **secondary spines** carried on **secondary** (and **scrobicular**) **tubercles.** Usually only the former are of paleontological significance. The base of a primary spine articulates with the tubercle of its plate by means of the cup-shaped depression in its lower surface, termed the **acetabulum.** Spines carried on crenulate tubercles are correspondingly crenulate around the margin of the acetabulum (e.g., *Histocidaris,* see Fig. 240,*1f*). Above the base is a more or less conspicuous **milled ring** and **collar,** both of which are striated, to serve as attachments for muscles arising from the areole (Fig. 235,*7*). In some another distinct region occurs just beyond the collar, termed the **neck.** In all, the greater part of the spine is distinguished as the **shaft;** this may be cylindrical, flattened, fluted, smooth, or thorny; it usually tapers but it may expand into a blade, or into a hollow trumpet-like structure, or the whole organ may be modified into a cup or umbrella. These variations are often of value in narrowing the field of possible affinity of fossil fragments, as can be observed by studying the illustrations here given in the systematic section. The microscopic structure of spines is complex, and usually preserved in fossils, even when secondarily impregnated by mineral calcite. It is studied by transverse sections. At the center is the **medulla** (Fig. 235,*6*). Radiating from it are many fine vertical **lamellae** or septa, united to each other by intervening **trabeculae.** The whole structure forms a 3-dimensional mesh, the **stereom,** through which the living tissue (**stroma**) ramifies. Nearly always an outer, denser, zone is observed, the **cortex,** also of calcite, on which **flutings** or **cortical hairs** (of calcite) or other structures may develop; this is nonliving in the adult, and consequently epizoic animals can adhere to the spines of cidaroids. The microscopic detail of the transverse section often aids in identification of spines.

The usual zoological keys to the identification of cidaroids are difficult to apply to fossils, for the pedicellariae and other finer details are almost always lost. If, however, attention is paid to the sum total of available characters exhibited by spines, and amb and interamb plates, as given in the diagnoses which follow, many fossils can be classified in taxa which are unlikely to differ substantially from those used for living forms.

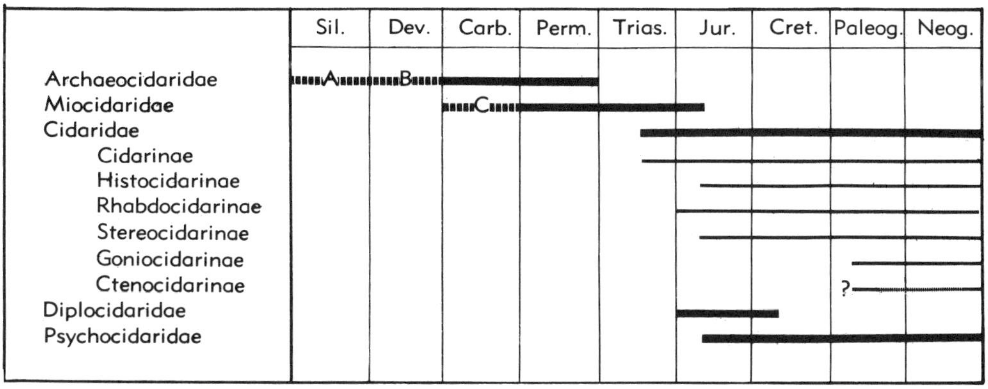

FIG. 236. Stratigraphic distribution of cidaroid family and subfamily assemblages. [*A* represents *Siluro-cidaris* only, archaeocidarid affinity uncertain; *B* represents *Nortonechinus* and *Xenocidaris*, archaeocidarid affinity uncertain; *C* represents *"Miocidaris" cannoni* JACKSON, miocidarid affinity very probable but genus uncertain.] (Fell, n.)

STRATIGRAPHICAL DISTRIBUTION

The known stratigraphical range of the families and subfamilies of Cidaroida is shown in Figure 236. Owing to the present inability to classify fossil genera on the basis of pedicellariae, the fossil genera are doubtless more inclusive than Recent ones, and would be more subdivided if their pedicellariae were available. It has therefore seemed undesirable to indicate the number of genera known for each family and subfamily by variation in the thickness of lines showing their time range, for such treatment of the data would probably be misleading. Accordingly, Figure 236 shows no more than the recorded time ranges.

Order CIDAROIDA Claus, 1880

[*nom. correct.* MORTENSEN, 1935 (*pro* Cidaroidea MORTENSEN, 1928)] [=Cidaridae CLAUS, 1880]

Test subspherical, radially symmetrical, rigid or with imbricating plates; base resorbed during growth. Ambs of 2 columns; each plate with a single pore pair, not uniting in compound plates, though grouped in diads or triads in some. Interambs conspicuously wider than ambs, of 2 or more columns; each interamb plate with one enlarged primary tubercle, carrying an enlarged corticate (rarely decorticate) primary spine; areole conspicuous, usually defined by scrobicular ring of secondary tubercles. Lantern present; teeth not keeled. Peristome covered (in life) by imbricating plates; no gills or gill slits. Apical system enclosing periproct. No spheridia. Pedicellariae of 2 types; globiferous, with a median venom cell; and tridentate (usually lacking). *U.Sil.-Rec.*

Family ARCHAEOCIDARIDAE M'Coy, 1844

Test flexible; interamb plates pluriserial (in 4 or more columns), imbricating over amb plates at adradial sutures; primary tubercles perforate, noncrenulate; amb pores uniserial. *U.Dev.-Perm.*

Archaeocidaris M'COY, 1844, p. 173 [**Cidaris urii* FLEMING, 1828; OD, M] [*nom. conserv.* ICZN, Op. 370, 1955] [=*Echinocrinus* L. AGASSIZ, 1841 (obj.) (suppressed ICZN Op. 370, 1955); *Palaeocidaris* DESOR, 1846 (type, *Cidarites nerei* MÜNSTER, 18??); *Eocidaris* DESOR, 1856 (type, *Cidaris laevispina* SANDBERGER, 18??); *Cidarotropus* POMEL, 1883 (type, *Archaeocidaris wortheni* HALL, 18??; *Permocidaris* LAMBERT, 1899 (type, *Cidaris forbesiana* DE KONINCK, 18??)]. Test subspherical, probably depressed adorally and adapically. Amb plates tending to form triads, with irregular enlargement of each successive third plate. Interamb plates in 4 columns (at least ambitally), interradial series imbricating more or less upon adradial and adradial more or less upon amb series. Primary spines smooth, striate, or spinulose, or with lateral expansions, but without terminal clavate or discoid shaft; cortex reduced (or ?absent), medulla (in some or all) hollow. *L. Carb.*, Eu., N.Am.; *Perm.*, N.India.——FIG. 237,*1*. *A. immanis* KIER, Penn., Okla.; *1a*, test, lat., ×0.7; *1b*, amb, ×2.1 (largest known cidaroid).

——Fig. 237,2. *A. blairi* (Miller), Miss., USA (Mo.); oral view, lantern, ×1.4.——Fig. 237,3. *A. aliquantula* Kier, Miss., USA (Iowa); *3a,b,* interamb plates, spines, ×2.9 (96).——Fig. 237,4. *A. rossica* (von Buch), Carb., USSR; *4a,b,* test aboral, oral, ×1.4 (87).

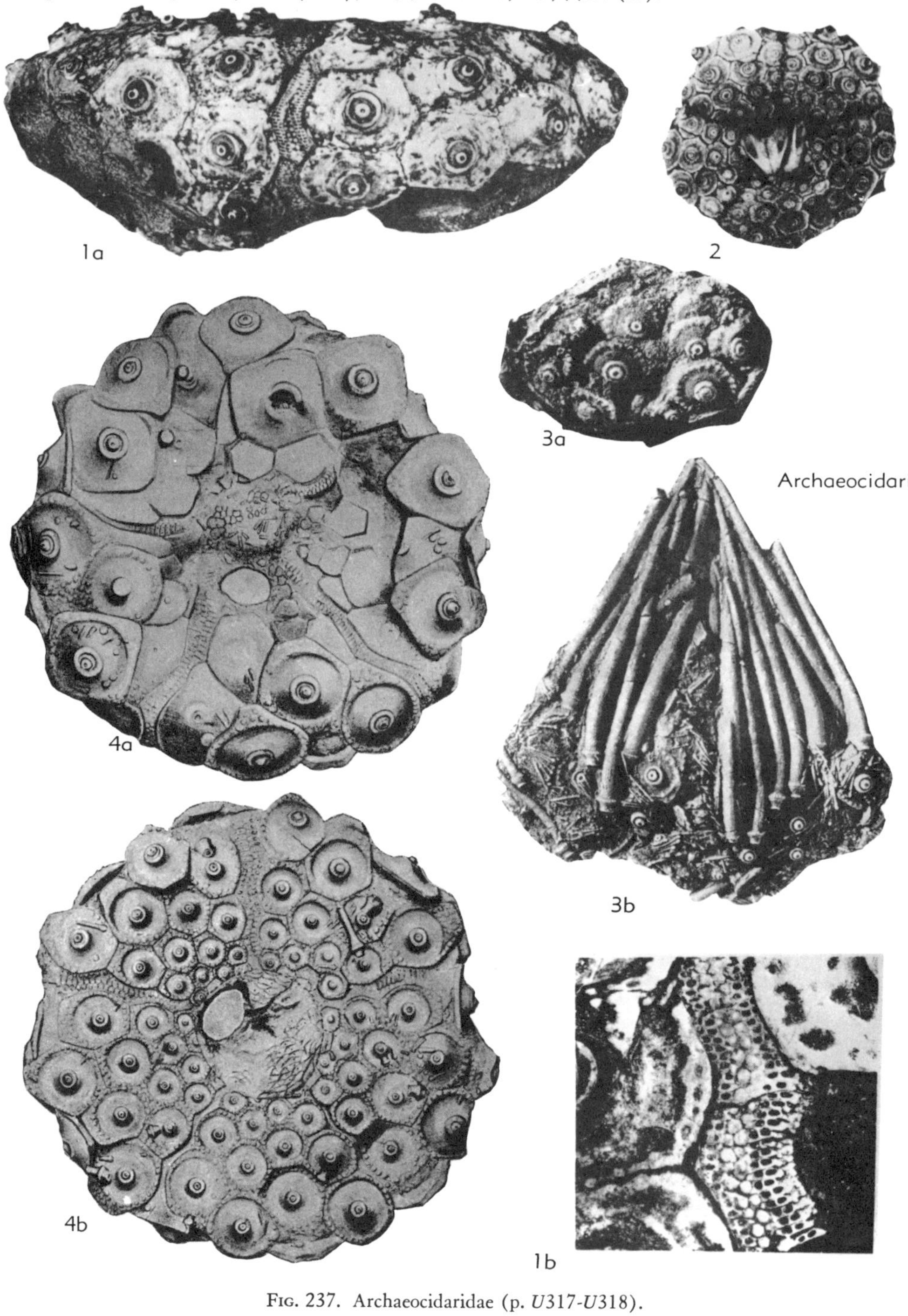

Fig. 237. Archaeocidaridae (p. U317-U318).

Lepidocidaris Meek & Worthen, 1873, p. 478 [**L. squamosa*; OD, M]. Like *Archaeocidaris,* but amb triads more conspicuous and more regular, and interamb plates in 6 to 8 columns, with scrobicular ring surrounded by raised tumid area. *Miss.,* N. Am.——Fig. 238,5. **L. squamosa,* USA(Ill.); *5a,b,* parts of test, ×2.8, ×0.7 (87).

Nortonechinus Thomas, 1920, p. 481 [**N. welleri;*

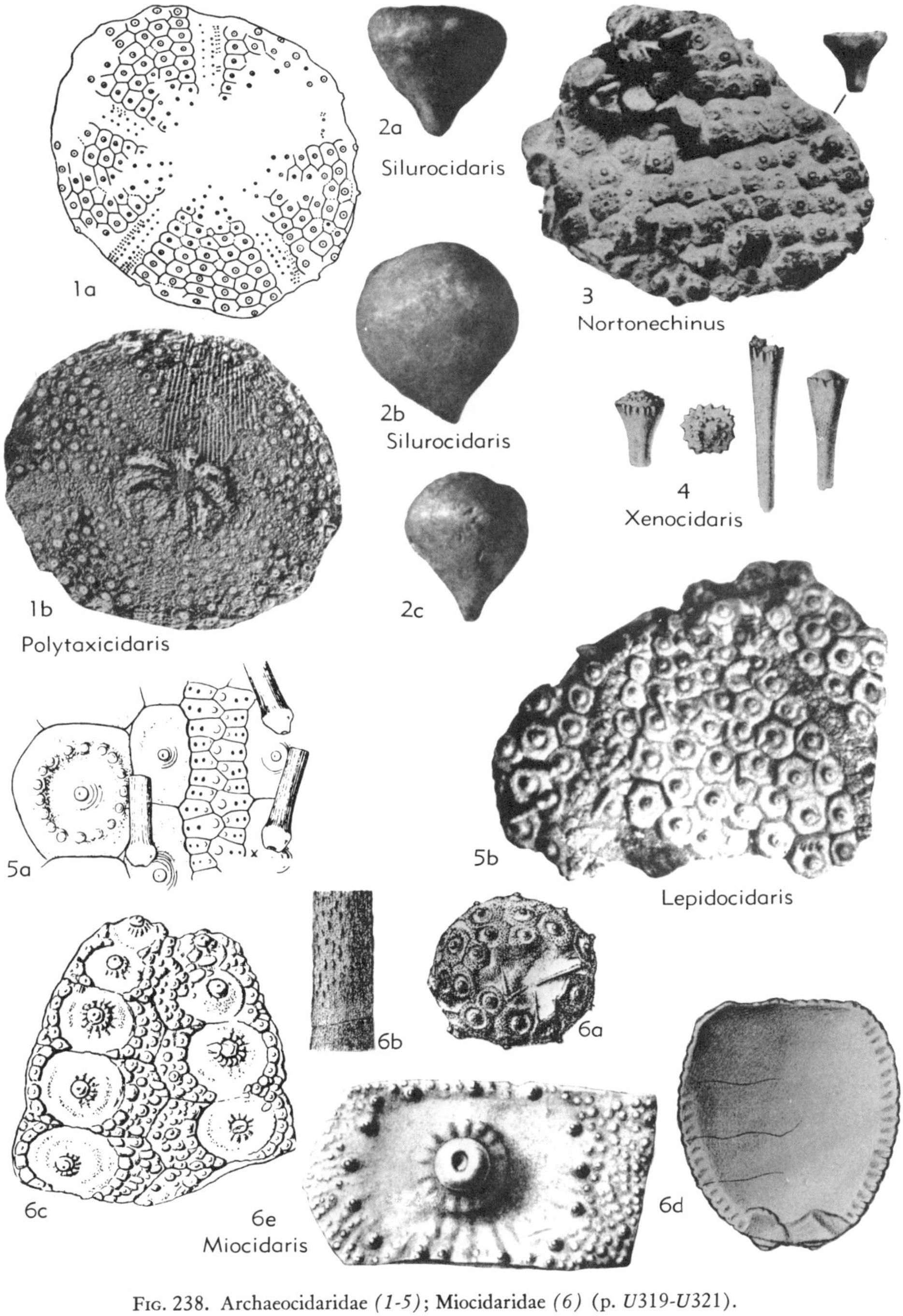

Fig. 238. Archaeocidaridae *(1-5)*; Miocidaridae *(6)* (p. *U*319-*U*321).

OD]. Primary spines short, shaft expanded into terminal disc; interambs as in *Lepidocidaris*; ambs unknown. *U.Dev.*, N.Am.(Iowa).——Fig. 238,*3*. **N. welleri*; interamb plates, ×2 (162).

Polytaxicidaris Kier, 1958, p. 10 [**P. dyeri*; OD]. Like *Lepidocidaris* but amb primary tubercles mostly perforate, no triads; and scrobicular ring marginal, with no surrounding tumid area. *Miss.*, N.Am.(Ind.).——Fig. 238,*1*. **P. dyeri*; *1a,b*, holotype, oral, ×1.4 (96).

?**Silurocidaris** Regnéll, 1956, p. 165 [**S. clavata*; OD]. Primary spines clavate, shaft spheroidal; test unknown. *U.Sil.*, Sweden.——Fig. 238,*2*. **S. clavata*; *2a-c*, spine shafts, ×2.8 (94a).

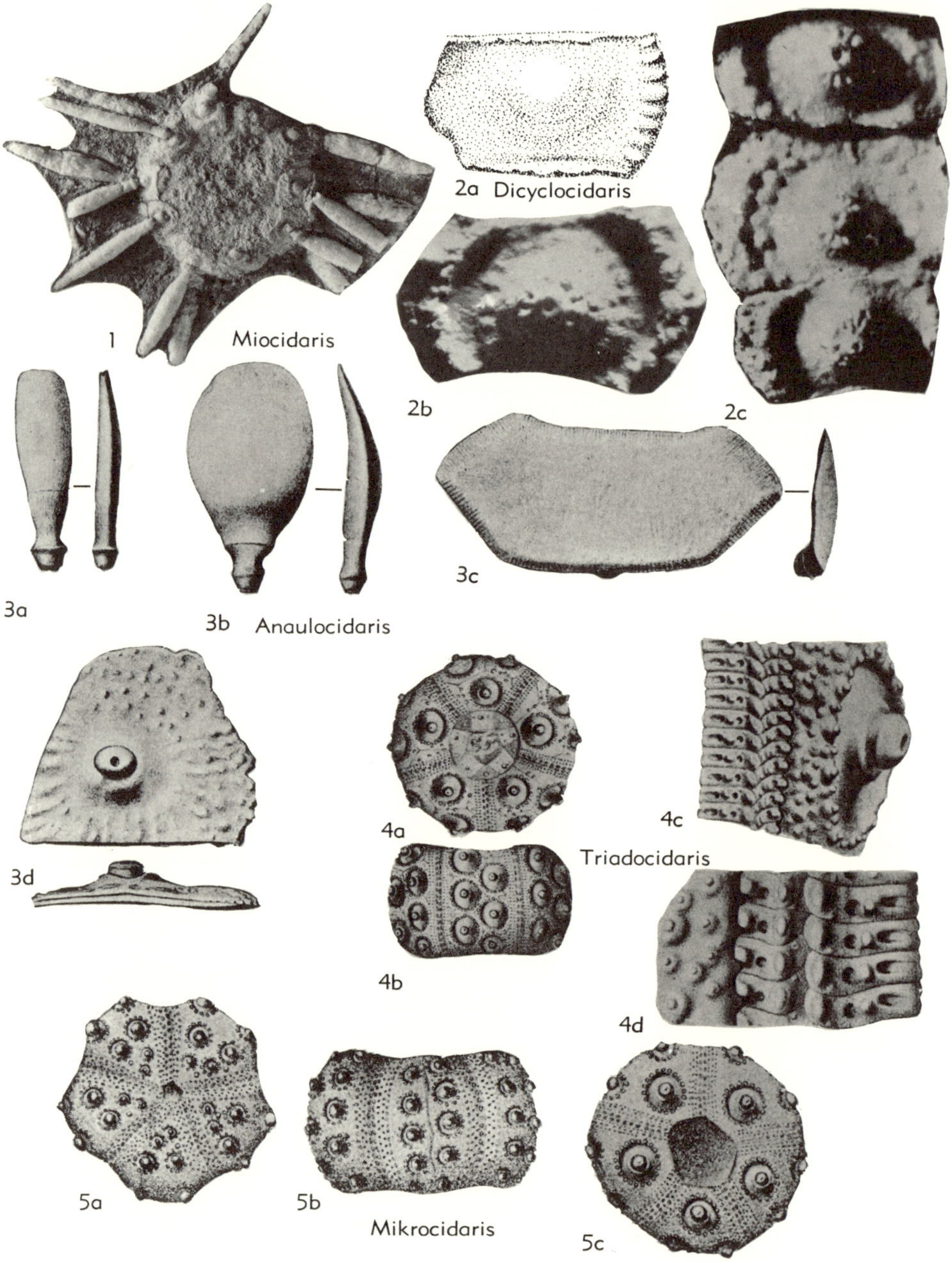

Fig. 239. Miocidaridae (p. U321).

?**Xenocidaris** SCHULTZE, 1866, p. 126 [*X. clavigera; OD, M]. Primary spines as in *Nortonechinus,* but more slender and with terminal spinulose ring. Test unknown. *M.Dev.,* Eu.(Ger.); *U.Dev.,* N.Am. (Iowa).——FIG. 238,*4*. *X. clavigera,* M.Dev., Ger.; spines, ×1.4 (136a).

Family MIOCIDARIDAE Durham & Melville, 1957

[*nom. nov. pro* Streptocidaridae LAMBERT, 1900=Streptocidarinae MORTENSEN, 1928 (*nom. correct.,* not based upon any generic name)]

Test partly flexible, interamb plates in 2 columns, plates imbricating adorally and adradially, beveled over adjoining amb plates. Primary tubercles perforate. Amb plates uniserial, with nonconjugate pores, not forming diads or triads. *L.Carb.-L.Jur.*

Miocidaris DÖDERLEIN, 1887, p. 40 [*M. cassiani* BATHER, 1909 (*non. subst. pro Cidaris klipsteini* DESOR); SD BATHER, 1909, p. 84] [=*Eotiaris* LAMBERT, 1900]. Test small to medium-sized. Areoles well developed, more or less confluent. Primary tubercles perforate, crenulate, bearing cylindrical spines with granular ornament on shaft. Ambs narrow, plates probably supported in membrane, overlapped by denticulate, beveled adradial margin of interamb plates. Tridentate pedicellariae present, with globular head and elongate jaws (known in *M. lorioli* only). *?L.Carb.,* N.Am.; *Perm.-L.Jur.,* Eu.-Asia.——FIG. 238,*6a,b*. *M. lorioli* LAMBERT & THIÉRY, L.Jur.(Hettang.), Fr.; *6a,* test, aboral, ×0.7; *6b,* spine, ×11 (115).——FIG. 238,*6c,d*. *M. keyserlingi* (GEINITZ), Perm., Hung.; *6c,d,* interamb ext. and int. aspects, ×4.2 (10).——FIG. 238,*6e*. *M. planus* BATHER, U.Trias. (Carn.), Hung.; interamb plate, ×3.5 (10).——FIG. 239,*1*. *M. pakistanensis* LINCK, L.Trias., Pak.; test with spines, aboral, ×0.7 (118).

Anaulocidaris ZITTEL, 1879, p. 486 [*Cidaris buchi* MÜNSTER, 1843; SD BATHER, 1909, p. 168]. Interambulacral plates thin, imbricating adradial edge not denticulate, areoles indistinct, without scrobicular tubercles. Primary tubercles noncrenulate, without parapet or basal terrace. Primary spines strongly depressed, shield-shaped, imbricating so that upper edges are covered by lower edges of spines immediately adapical, completely obscuring aboral hemisphere of test. *U.Trias.(Carn.),* Eu.——FIG. 239,*3*. *A. testudo* BATHER, Hung.; *3a-c,* spines in ext. and lat. aspects, ×2.3; *3d,* interamb plate, ext. and lat. aspects, ×3.3 (11).

?**Aplocidaris** LAMBERT & THIÉRY, 1909, p. 31 [*Cidaris helenae* DE REGNY, 1903; OD]. Test very small, interamb plates with noncrenulate primary tubercle, and single ring of scrobicular tubercles, areole ill-defined; no other secondary tubercles. Ambs very narrow, without tubercles, sutures oblique. Primary spines compressed, with denticulate margins. [Possibly juvenile stage of some miocidarid. Family assignment doubtful owing to lack of imbricating test plates.] *U.Trias.(Nor.),* Eu.

Dicyclocidaris FELL, 1950, p. 83 [*D. denticulata;* OD]. Like *Miocidaris,* but with 2 rings of scrobicular tubercles on ambital plates, inner ring developed on floor of areole; spines cylindrical, thorny. *U.Trias.(Carn.),* N.Z.——FIG. 239,*2*. *D. denticulata; 2a-c,* int. and ext. aspects of interamb plates, ×5.3 (54).

?**Mikrocidaris** DÖDERLEIN, 1887, p. 39 [*Cidaris pentagona* MÜNSTER, 1843; SD LAMBERT & THIÉRY, 1909, p. 140] [=*Microcidaris* LAMBERT & THIÉRY, 1909, p. 140 (*nom. van.*)]. As *Triadocidaris,* but test very small (3-5 mm.), and test plates not beveled. *U.Trias.(Carn.),* Eu.——FIG. 239,*5a,b*. *M. pentagona* (MÜNSTER), Aus.; *5a,b,* test, aboral, lat., ×5.4.——FIG. 239,*5c*. *M. venusta* (MÜNSTER), Aus.; test, aboral, ×7 (116). [Family assignment doubtful owing to lack of imbricating test plates.]

Pachycidaris THIÉRY, 1928, p. 179 [*P. thieryi* COLLIGNON & LAMBERT, 1928 (=*Cidaris spinosa* COTTEAU, *partim, non C. spinosa* AGASSIZ); OD]. Primary spines very thorny. Ambulacral marginal tubercles developed irregularly. *U.Jur. (Oxford.),* Eu.

Triadocidaris DÖDERLEIN, 1887, p. 39 [*Cidaris subsimilis* MÜNSTER, 1843; SD BATHER, 1909, p. 79]. Test as *Miocidaris,* but tubercles noncrenulate, and uppermost interamb plate hypertrophied in some. *U.Trias.(Carn.),* Eu.——FIG. 239,*4a,b*. *T. subsimilis,* Hung.; *4a,b,* test aboral, lat., ×0.7 (116).——FIG. 239,*4c,d*. *T. persimilis* BATHER, Hung.; *4c,* amb, interamb, ×6.5; *4d,* amb plates, ×13 (11).

Family CIDARIDAE Gray, 1825

Test rigid. Interambulacral plates in 2 columns. Primary tubercles perforate. Ambulacral pore pairs uniserial aborally, but in some exhibiting pluriserial tendencies on adoral region, though never forming compound plates. *U.Trias.-Rec.*

Subfamily HISTOCIDARINAE Mortensen, 1928

[*nom. transl.* FELL, herein (*ex* Histocidarina MORTENSEN, 1928)]

Primary tubercles strongly crenulate. Primary spines cylindrical, not clavate; either smooth or thorny. Secondary spines elongate, erect, more or less flattened, but not squamiform. Pores nonconjugate, conspicuous wall separating inner from outer pore. Tridentate pedicellariae present, globiferous pedicellariae wanting. *Jur.-Rec.*

Histocidaris MORTENSEN, 1903, p. 22 [*Porocidaris elegans* A. AGASSIZ, 1879; OD]. Test usually well arched, flattened at apex and peristome. Interambulacral plates broader than high, their areoles large, more or less confluent below ambitus. Apical plates partly naked, especially genital plates; genital pore usually entirely enclosed by genital plate. Primary spines cylindrical throughout most or all of their length, tip tapering or widening in some; collar short (not more than *ca.* 3 mm.), not swollen midway. *Oligo.*, N.Z.; *Mio.*, N.Afr.; *Rec.*, IndoPac.-Carib.——FIG. 240,*1*. *H. mackayi* FELL, M.Oligo., N.Z.; *1a*, test (holotype) aboral, ×1.35; *1b*, test (paratype), lat., ×1.35; *1c*, in-

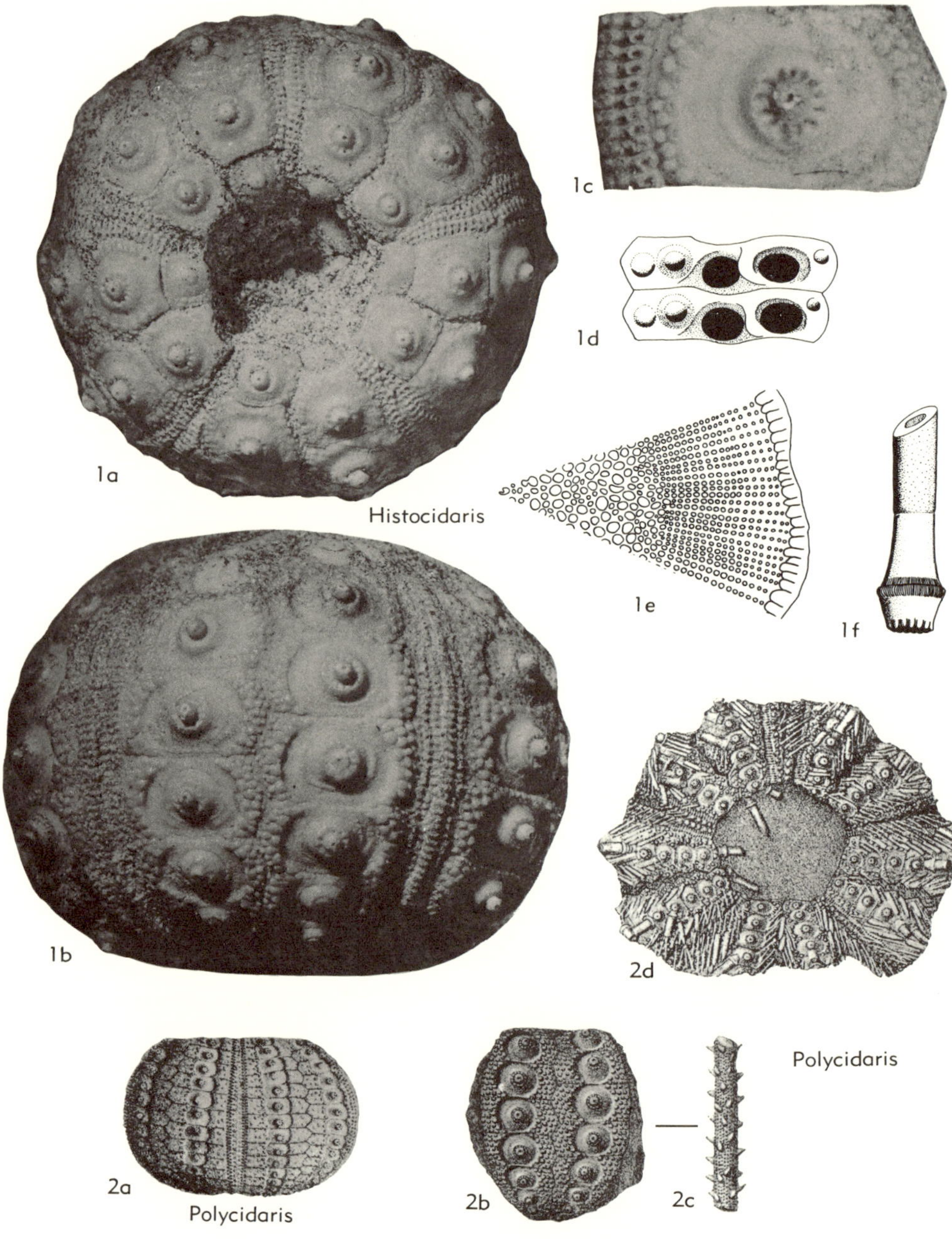

FIG. 240. Cidaridae (Histocidarinae) (p. *U322-U323*).

teramb plate with attached amb plates, ×3.3; *1d,* amb plate, ×13.5; *1e,* transv. sec. of spine, ×27; *1f,* base of spine, ×2.7 (56).

Polycidaris QUENSTEDT, 1858, p. 644 [**Cidarites multiceps* QUENSTEDT, 1858] [*non Polycidaris* BETTONI, 1900 (=*Loriolella* FUCINI, 1904)]. Interambulacral plates relatively numerous (up to 15), and broader than high; areoles all confluent, scrobicular rings reduced to undulating line of tuberculation on either side of areole series. Ambulacra straight. Primary spines probably slender, thorny. *Jur.(Bajoc.-Oxford.),* Eu.——FIG. 240, *2a.* **P. multiceps* (QUENSTEDT), Oxford., Ger.; lat., ×0.8 (145).——FIG. 240,*2b,c. P. legayi* SAUVAGE, Portland., Fr.; *2b,* interamb, ×0.8; *2c,* spine, ×2.7 (27c).——FIG. 240,*2d. P. spinulosa* ROEMER, Bajoc., Fr.; peristome and adoral part of test, with adhering spines, ×1.35 (27c).

Poriocidaris MORTENSEN, 1909, p. 53 [**Porocidaris purpurata* WYVILLE THOMSON, 1874; OD] [=*Porocidaris* A. AGASSIZ, 1872 (*non* DESOR, 1855)]. Like *Histocidaris* but primary spines tapering and with long collar (10-20 mm.), especially apical spines, where collar may occupy 0.3 length of whole spine. Tridentate pedicellariae with only 2 valves. *Rec.,* Atl.-?Ind.O.

Procidaris POMEL, 1883, p. 109 [**Cidaris edwardsi* WRIGHT, 1855; SD MORTENSEN, 1928]. Like *Polycidaris* but marginal ambulacral tubercles perforate. Primary spines slender, cylindrical, smooth. *L.Jur.-M.Jur.,* Eu.——FIG. 241,*1,2.* **P. edwardsi,* L.Jur.(Charmouth.), Eng.; *1,* test, lat., ×1; *2,* test plates, ambs at right, interamb at left, ×2.7 (172).

Subfamily CTENOCIDARINAE Mortensen, 1928

[*nom. transl.* FELL, herein (*ex* Ctenocidarina MORTENSEN, 1928)]

Primary tubercles noncrenulate on oral hemisphere, subcrenulate or noncrenulate on aboral hemisphere. No horizontal sutural grooves. Ambulacral pore pairs placed obliquely on plate; 2 pores confluent or separated by very narrow wall, thickness of which does not exceed 0.5 of pore diameter. No tridentate pedicellariae. Large and small globiferous pedicellariae present. *?Eoc.,* Patagonia; *Rec.,* Antarctic-N.Z.-S.Pac.

Ctenocidaris MORTENSEN, 1910, p. 3 [**C. speciosa*; OD]. Test low. Interambulacra with or without ill-defined, naked, slightly sunken median area; areoles rather deep, proximal 4 or 5 usually confluent; all primary tubercles noncrenulate; scrobicular tubercles not conspicuously differentiated from other secondaries. Ambulacral pores commonly confluent, neuropore inconspicuous. Apical system and peristome approximately half of horizontal diameter; ocular pore usually surrounded by circular wall; female genital pore not entirely enclosed by genital plate. Primary spines cylindrical, with numerous thorns irregularly scattered on shaft or (less commonly) arranged in longitudinal rows; oral primaries coarsely serrate or spatulate, not spearhead-shaped. Secondary spines clavate. Globiferous pedicellariae without end tooth. *Rec.,* Antarctic.

Aporocidaris AGASSIZ & CLARK, 1907, p. 36 [**Porocidaris milleri* A. AGASSIZ, 1898; OD]. Like *Ctenocidaris* but test high, arched or conical, very thin and fragile. Areoles shallow. Neuropore conspicuous. Median ambulacral area rather naked, slightly sunken, weak grooves in median area at terminations of horizontal ambulacral sutures. Primary spines long, slender, with sparse, irregular thorns; oral primaries spearhead-shaped, serrate, distally curved; secondary spines erect, cylindrical or weakly clavate. *Rec.,* Antarctic (littoral)-Pac. (abyssal).

Austrocidaris H. L. CLARK, 1907, p. 212 [**Temnocidaris canaliculata* A. AGASSIZ, 1863; OD] [*non Temnocidaris* COTTEAU, 1863]. Like *Ctenocidaris* but conspicuous, sunken, median furrow present in both interambulacra and ambulacra. Areoles well separated, deep; primary tubercles subcrenulate aborally, noncrenulate adorally. Pores not confluent. Primary spines with rather smooth, slender, cylindrical shaft; secondary spines flattened, somewhat adpressed. *?Eoc.,* Patagonia; *Rec.,* subantarctic.——FIG. 241,*3.* **A. canaliculata* (A. AGASSIZ), Rec., subantarctic; lat., with juveniles adhering to adapical region, ×1.35 (163).

Eurocidaris MORTENSEN, 1909, p. 30 [**Cidaris nutrix* WYVILLE THOMSON, 1876; OD]. Like *Ctenocidaris* but having flattened scrobicular spines. *Rec.,* Antarctic.

Homalocidaris MORTENSEN, 1928, p. 67 [**Austrocidaris gigantea* H. L. CLARK, 1925; OD]. Like *Ctenocidaris* but interambulacra without slightly sunken, naked median area; areoles large, deep, separated by narrow ridges. Ambulacral pores confluent. Primary spines cylindrical, shaft smooth, milled ring very conspicuous. *Rec.,* Antarctic.

Notocidaris MORTENSEN, 1909, p. 17 [**Goniocidaris mortenseni* KOEHLER, 1900; OD]. Like *Ctenocidaris,* but differing in primary spines, which are more or less compressed, with longitudinal rows of spinules on shaft or longitudinal ridges; oral primaries conspicuously spearhead-shaped. [*Ctenocidaris* has, *Notocidaris* lacks, a coat of calcareous hairs on the shaft, a difference unlikely to have paleontological significance, however.] *?Plio.,* N.Z.; *Rec.,* Antarctic.

Ogmocidaris MORTENSEN, 1921, p. 151 [**O. benhami*; OD]. Like *Austrocidaris* but small globiferous pedicellariae with end tooth and developing in adult stage small adapical umbrella-shaped spines, other spines elongate, cylindrical, and slender. [The occurrence (here first recorded) of umbrella-shaped adapical spines in this genus, and the character of the small globiferous pedicellariae, suggest that it is transitional to the Goniocidarinae, to which subfamily the genus might with equal right be referred.] *Rec.*, N.Z.——FIG. 242,*1*. **O. benhami*; *1a*, interamb plate with ambs at left, ×6.7; *1b*, amb plates, ×13.5 (56).

Rhynchocidaris MORTENSEN, 1909, p. 5 [**R. triplopora;* OD]. Like *Ctenocidaris* but with neuropore

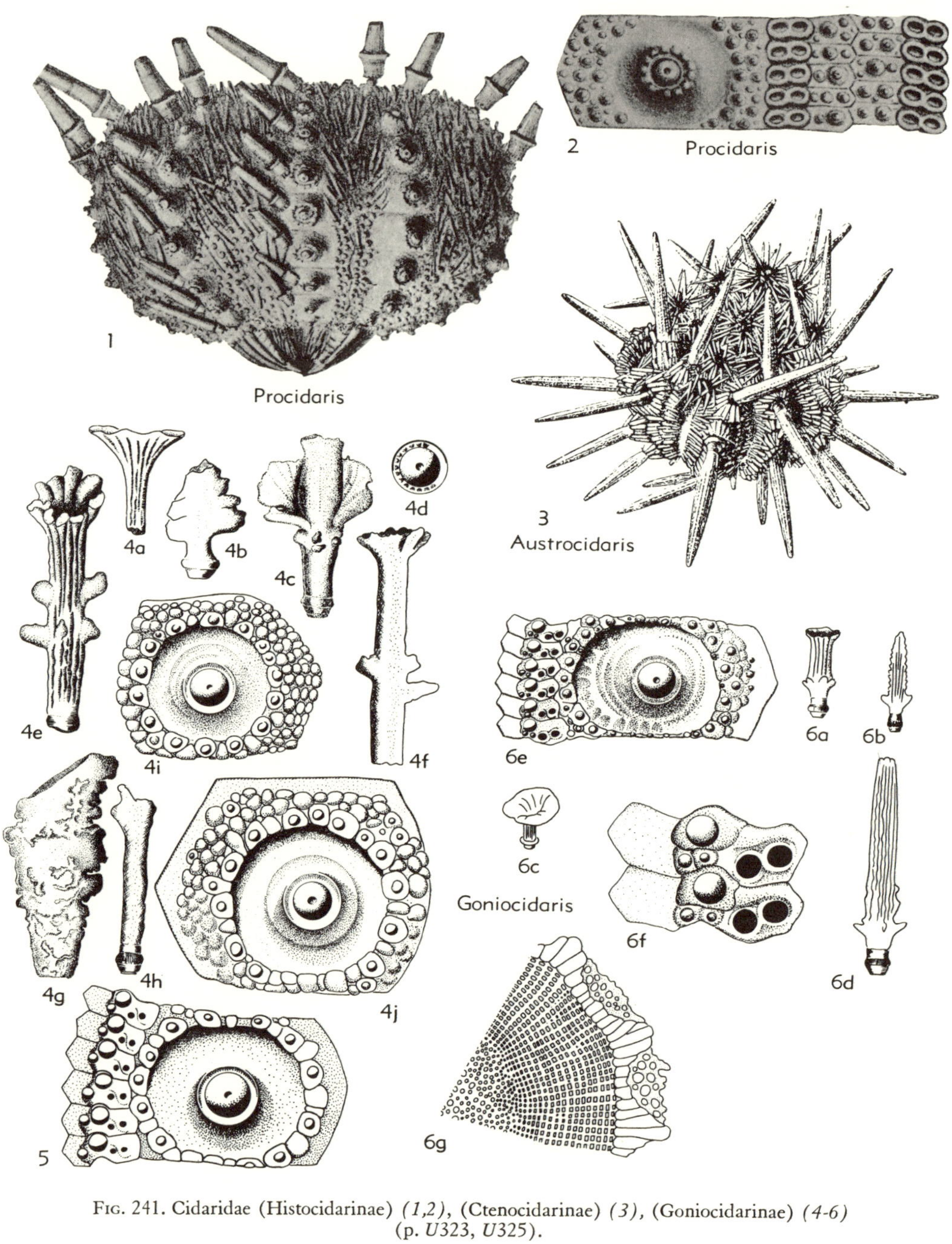

FIG. 241. Cidaridae (Histocidarinae) *(1,2)*, (Ctenocidarinae) *(3)*, (Goniocidarinae) *(4-6)* (p. *U*323, *U*325).

so strongly developed that ambulacral plates (on adoral side at least) appear to carry pore-triplets in place of pore pairs. *Rec.*, Antarctic.

Subfamily GONIOCIDARINAE Mortensen, 1928

[*nom. transl.* FELL, herein (*ex* Goniocidarina MORTENSEN, 1928)]

Test usually flattened above and below. Grooves or pits, or both, on horizontal sutures of interambulacra and ambulacra, commonly confluent with median vertical grooves. Primary tubercles adorally noncrenulate, aborally noncrenulate or weakly subcrenulate. Pores nonconjugate, close together. Tridentate pedicellariae unknown. Globiferous pedicellariae present, small forms with end tooth, large forms without it. *Eoc.-Rec.*

Goniocidaris DESOR in AGASSIZ & DESOR, 1846, p. 337 [**Cidarites tubaria* LAMARCK, 1816; SD MORTENSEN, 1928, p. 150] [=*Stephanocidaris* A. AGASSIZ, 1863 (*non* AGASSIZ, 1872); *Adelcidaris* COTTON & GODFREY, 1942 (*nom. van.*)]. Median parts of horizontal sutures sunken or naked, forming conspicuous grooves in interambulacra and ambulacra; grooves commonly confluent with vertical furrow or median suture. Pores horizontal or, less commonly, oblique. Adapical primary spines with terminal disc or cup (developed only at maturity). *Eoc.*, Australia; *Rec.*, Indo-W.Pac.

G. (Goniocidaris) DESOR, 1846; *emend.* MORTENSEN, 1928 [**Cidarites tubaria* LAMARCK, 1816; SD MORTENSEN, 1928, p. 150]. Primary spines without basal disc, having instead basal spurs; shaft with coarse ridges or thorns. *Eoc.*, S.Australia; *Oligo.*, N.Z.; *Mio.*, India-Australia; *Plio.*, Iran-Australia; *Rec.*, Indo-W.Pac., incl. SE.Afr.——FIG. 241,*4*. *G. hebe* FELL, U.Oligo., N.Z.; *4a-h*, spines, ×1.35; *4i,j*, interamb plates, ×4.——FIG. 241,*5*. *G. pusilla* FELL, L.Mio., N.Z.; amb plates (left) and interamb, ×10.——FIG. 241,*6*. *G. umbraculum* HUTTON, Rec., N.Z.; *6a-d*, spines, ×2; *6e*, amb plates (left) and interamb, ×4; *6f*, amb plate, ×13.5; *6g*, transv. sec. of spine, ×27 (56).

G. (Aspidocidaris) MORTENSEN, 1928, p. 67 [**Goniocidaris alba*; OD]. Basal disc, and usually also very large terminal disc, present on primary spines; secondary spines short, flattened, with transversely straight-cut termination. *Rec.*, Japan-Indon.-Australia-N.Z.——FIG. 242,*2*. *A. parasol* FELL, Rec., N.Z.; test with spines (holotype), ×1.35 (Fell, n).

G. (Cyrtocidaris) MORTENSEN, 1927, p. 264 [**Goniocidaris tenuispina*; OD]. Basal disc, and usually also large terminal disc with indented edge, present on primary spines; shaft long, slender, thorny, secondary spines long, slender, pointed. *Rec.*, Philip.

G. (Discocidaris) DÖDERLEIN, 1885, p. 80 [**D. mikado*; OD]. Discs repeated serially at intervals along proximal part of shaft of ambital primary spines. *Rec.*, Japan-Indon.

G. (Petalocidaris) MORTENSEN, 1903, p. 18 [**Goniocidaris florigera* A. AGASSIZ, 1879; OD]. Primary spines with basal disc, shaft coarsely thorny; secondary spines flattened, thorny at base. *Rec.*, Japan-Indon.

Psilocidaris MORTENSEN, 1927, p. 282 [**P. echinulata*; OD]. Like *Schizocidaris* but primary spines long (3 to 5 times test horizontal diameter) and slender and lacking calcareous investment of shaft. Apical system conspicuously larger than peristome. *Rec.*, Indon. (abyssal).

Rhopalocidaris MORTENSEN, 1927, p. 272 [**Cidaris hirsutispinus* DE MEIJERE, 1904; OD]. Like *Psilocidaris* but secondary spines clavate, and apical system not conspicuously larger than peristome. Small abyssal forms less than 20 mm. diameter. *Rec.*, Japan-Indon.

Schizocidaris MORTENSEN, 1903, p. 25 [**S. assimilis*; OD]. Neither basal nor apical discs, shaft covered by smooth, glabrous calcareous investment which is supported between thorns by erect, columnar, calcareous cortical hairs. Secondary spines smooth, flattened, not clavate. *Rec.*, Indon.

Subfamily STEREOCIDARINAE Lambert, 1900

[*nom. transl.* FELL, herein (*ex* Stereocidaridae LAMBERT, 1900)]

Test robust, usually well arched. Interambulacra with more or less distinct horizontal sutural grooves on aboral side, upper areoles and tubercles commonly more or less rudimentary; interambulacral plates tending to be higher than broad on aboral side so that areoles there tend to be widely separated. Pores nonconjugate. Upper primary spines more or less reduced; secondary spines flattened, more or less adpressed. Globiferous pedicellariae usually without end tooth. *U.Jur.-Rec.*

Stereocidaris POMEL, 1883, p. 110 [**Cidaris cretosa* MANTELL, 1835; SD LAMBERT & THIÉRY, 1909 (Feb., p. 31; *non* Mar., 1909, where *C. merceyi* was designated, p. 152)] [=*Typocidaris* POMEL, 1883 (type, *Cidaris malum* A. GRAS); *Phalacrocidaris* LAMBERT, 1902 (type, *Dorocidaris japonica* DÖDERLEIN, 1885); *Anomocidaris* AGASSIZ & CLARK, 1907 (type, *Cidaris tenuispinus* YOSHIWARA, 1898)]. Test robust. Interambulacral plates high, especially aboral ones; rarely more than 7 plates in column, upper 1 to 3 having areole, tubercle, and spine more or less rudimentary; areoles deep, well separated, even on adoral side; intervening tumid surfaces densely covered by secondary and miliary tubercles. Primary tubercles noncrenulate or aborally subcrenulate. Ambulacra usually sinuate; pores nonconjugate. Primary spines with neck approximately twice length of collar, shaft commonly flaring toward tip; scrobicular spines flattened, adpressed around areole; other secondary spines squamiform or spiniform, densely packed. Tridentate pedicellariae present or absent; large globiferous pedicellariae without end-tooth, small globiferous pedicellariae with or without them. *Cret.*, Eu.-N.Am.; *Eoc.*, Eu.-Australia; *Oligo.*, N.Z.; *Mio.*, Australia-?Indon.; *Plio.*, Australia-N.Z.; *Rec.*, Indo-W.Pac. incl. SE. Afr., but not yet known from Australasia, where it may be expected). [Note on synonymy: *Typocidaris* POMEL, 1883, p. 111 (type, *Cidaris malum* A. GRASSÉ) is

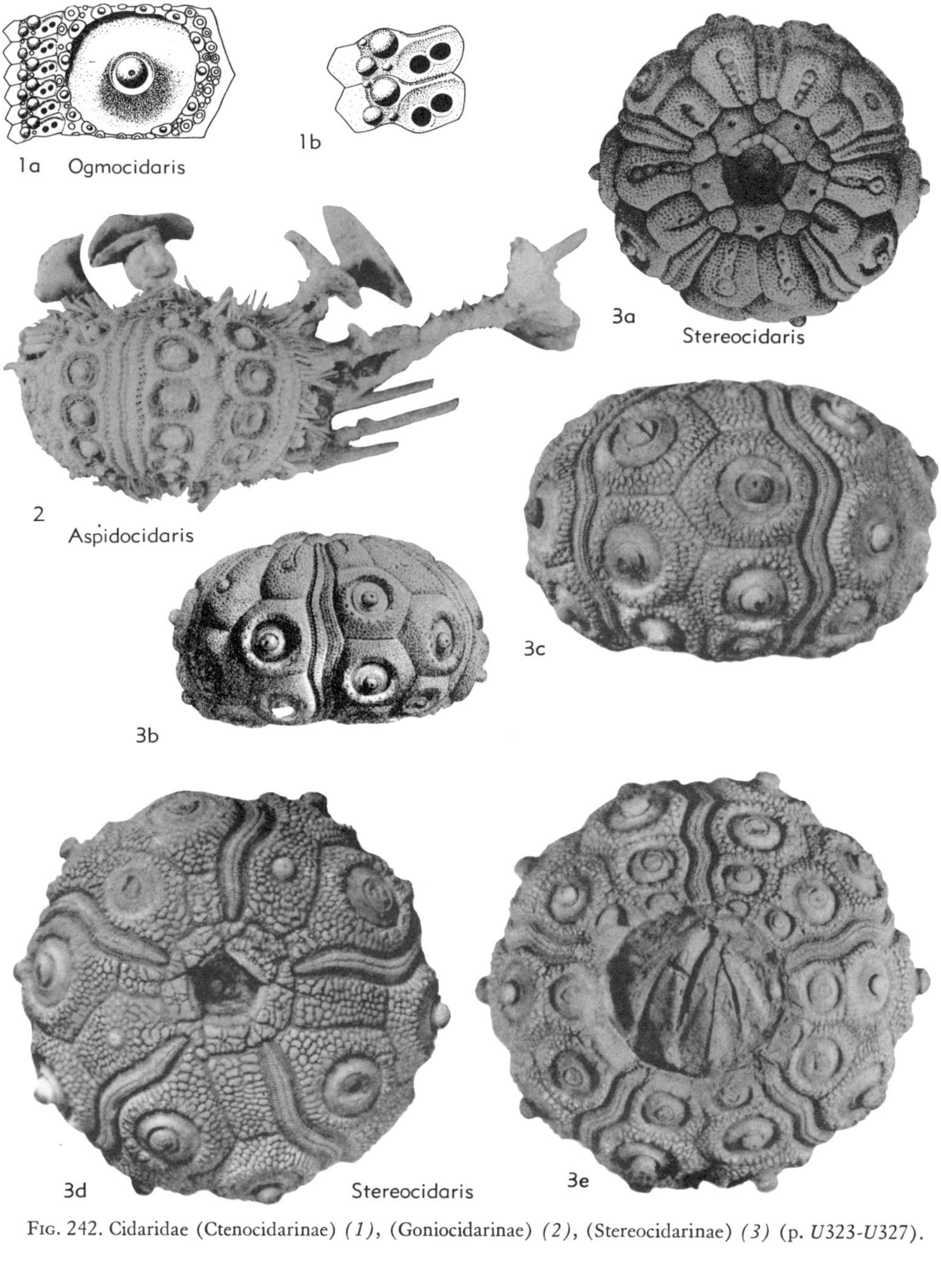

FIG. 242. Cidaridae (Ctenocidarinae) *(1)*, (Goniocidarinae) *(2)*, (Stereocidarinae) *(3)* (p. U323-U327).

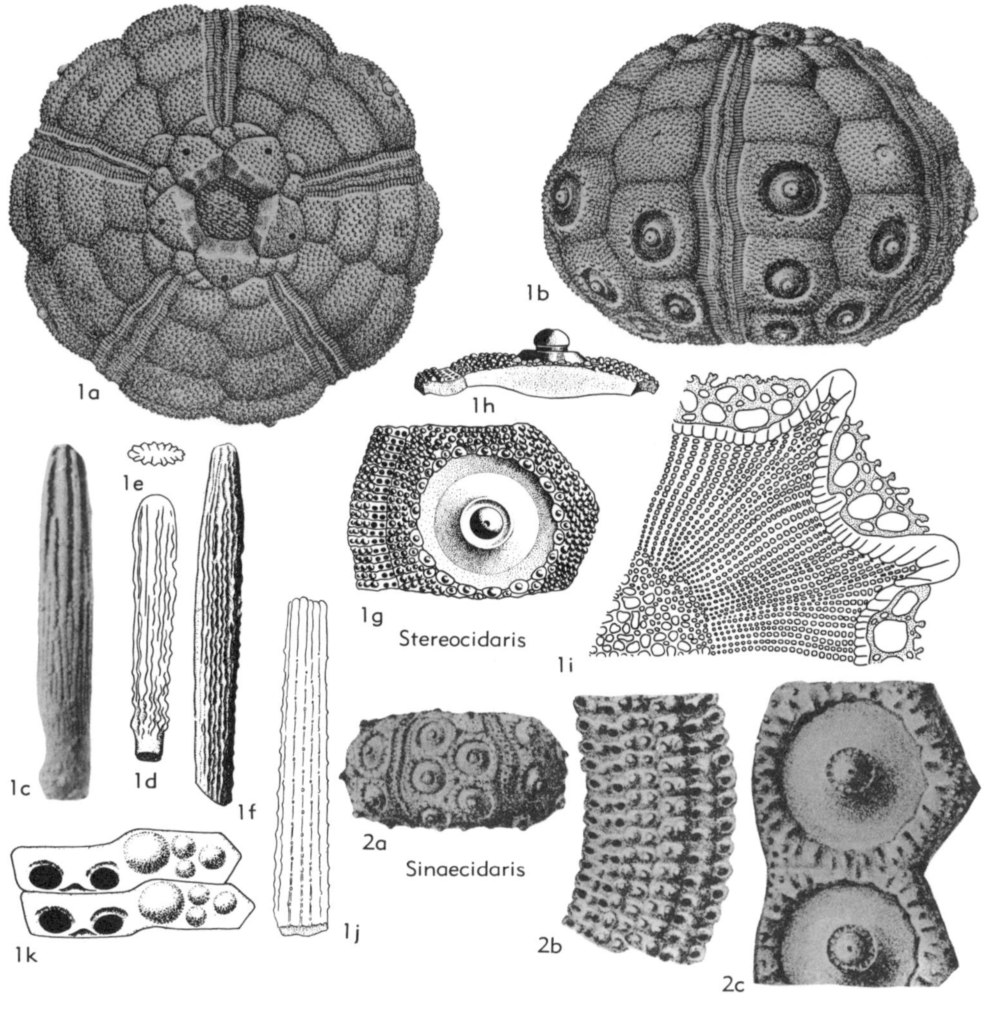

FIG. 243. Cidaridae (Stereocidarinae) (p. *U325-U327*).

founded on forms with grooved sutures and fully developed upper areoles; *Phalacrocidaris* LAMBERT, 1902 (type, *Dorocidaris japonica* DÖDERLEIN) for forms having atrophied upper areoles but no sutural grooves as in *S. merceyi* (COTTEAU). *Anomocidaris* AGASSIZ & CLARK, 1907, is a junior objective synonym of *Phalacrocidaris*. The characters supposed to distinguish *Typocidaris* and *Phalacrocidaris*, however, intergrade, and these genera must be united with *Stereocidaris*.]——FIG. 242,*3a,b*. *S. sceptifera* (MANTELL), U.Cret., Eng.; *3a,b*, test, aboral, lat., ×0.7 (173).——FIG. 242,*3c-e*; 243,*1c*. *S. hudspethensis* COOKE, M.Cret., N.Am.(Tex.); 242,*3c-e*, test, lat., aboral, oral, ×0.7 (23); 243,*1c*, spine, ×1.35 (23).——FIG. 243,*1a,b*. *S. merceyi* (COTTEAU), M.Cret., Fr.; aboral, lat., ×1 (27a).——FIG. 243,*1d-i*. *S. hutchinsoni* FELL, L.Pleist., N.Z.; *1d-f*, spines, ×1.35; *1g,h*, interamb and amb plates, ×2; *1i*, transv. sec. of spine, ×27 (56).——FIG. 243,*1j,k*. *S. striata* (HUTTON), L.Oligo., N.Z.; *1j*, spine, ×2.7; *1k*, amb plates, ×13.5 (56).

Compsocidaris IKEDA, 1939, p. 160 [**C. pyrsacantha*; OD]. Like *Stereocidaris* but test with only sparsely scattered secondary tubercles instead of dense tuberculation. Primary spines cylindrical, ridged, not thorny. *Rec.*, Bonin Is.

Sinaecidaris FOURTAU, 1921, p. 9 [**S. gauthieri*; OD]. Like *Stereocidaris* but primary tubercles distinctly crenulate throughout interambulacra. *Cret.*, Egypt.——FIG. 243,*2*. **S. gauthieri*; *2a*, test lat., ×1; *2b*, amb plates, ×6.7; *2c*, interamb plates, ×2.7 (65).

Temnocidaris Cotteau, 1863, p. 355 [**T. magnifica*; OD] [*non Temnocidaris* A. Agassiz, 1863 (=*Austrocidaris* H. L. Clark, 1907)]. Numerous conspicuous grooves and porelike impressions scattered over all coronal plates (probably accommodating globiferous pedicellariae); horizontal sutural grooves, and vertical zigzag groove on interambulacral mid-line. Primary tubercles noncrenulate; pores subconjugate; primary spines cylindrical, slender, finely thorny. *U.Cret.*, Eu.——Fig. 244,*1*. **T. magnifica,* Fr.; *1a,* test (holotype), lat., ×0.85; *1b,* amb plates, ×3.5; *1c,* amb plates, ×8.5; *1d,e,* interamb plates, ×0.85, ×3.5 (27).

Subfamily RHABDOCIDARINAE Lambert, 1900

[*emend.* Fell, herein]

Test robust, without sutural grooves. Pores conjugate or subconjugate. Primary spines large, robust. *L.Jur.-Rec.*

Rhabdocidaris Desor, 1855, p. 39 [**Cidaris orbigny-*

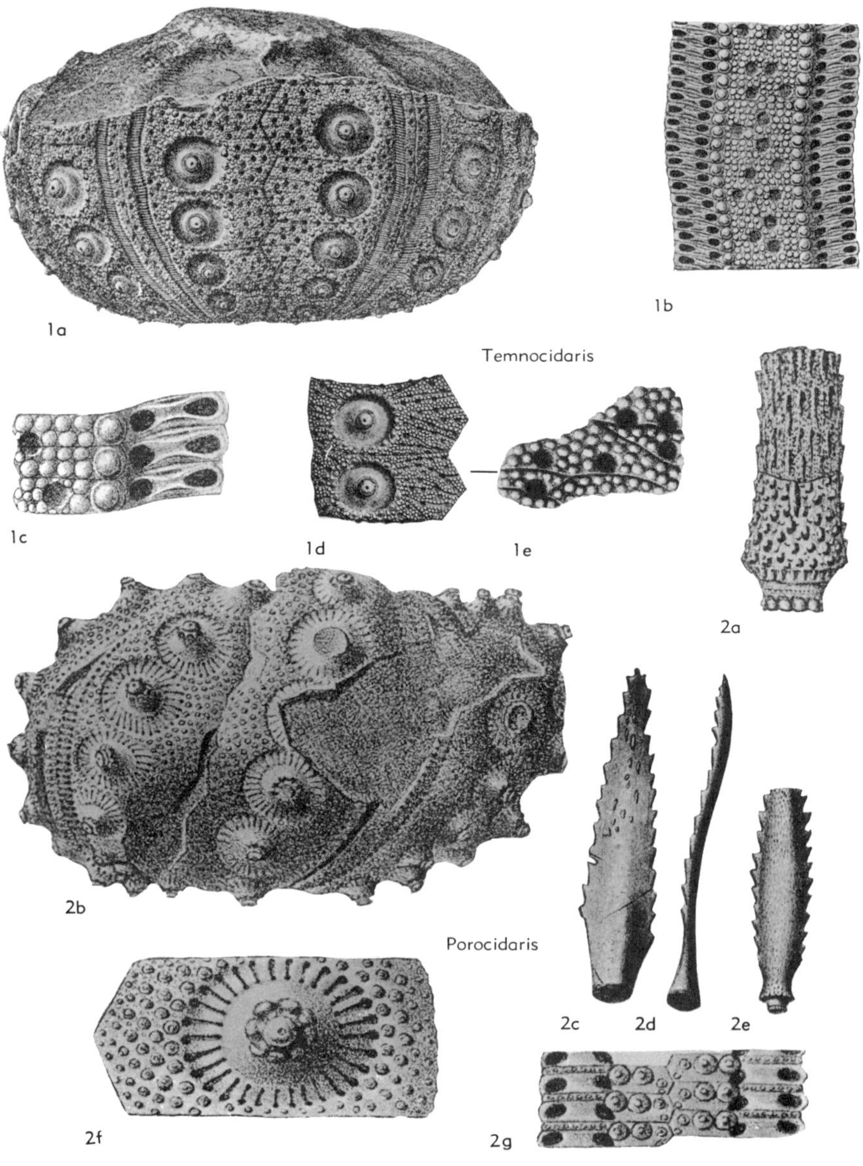

Fig. 244. Cidaridae (Stereocidarinae) *(1)*, (Rhabdocidarinae) *(2)* (p. *U*328, *U*330).

ana L. AGASSIZ; SD LAMBERT & THIÉRY, 1910, p. 136]. Test spherical, slightly flattened at apex and peristome, usually large (diam. ± 100 mm.). Areoles circular, shallow, not confluent; primary tubercles strongly crenulate. Ambulacra sinuate, pores conjugate. Primary spines long, typically depressed and expanded to form broad, obcordate or fan-shaped plates, shaft bearing longitudinal, radiating series of thorns. *L.Jur.-Eoc.*, Eu.——FIG. 245,*1*. **R. orbignyana* (AGASSIZ), Jur., Fr.; *1a*, test, lat., ×0.75; *1b,c*, spines, ×0.75.——FIG. 245,*2*. *R. rhodani* COTTEAU, Jur., Fr.; interamb plates, ×0.9.——FIG. 245,*3*. *R. copeoides* DESOR, Jur., Fr.; *3a,b*, spines, ×0.75 (27d).

Actinocidaris MORTENSEN, 1928, p. 73 [**Phyllacanthus thomasi* A. AGASSIZ & H. L. CLARK, 1907; OD]. Like *Prionocidaris* but primary spines thick, cylindrical or fusiform, finely tuberculated, surface of shaft covered by coarse reticulate layer formed by thick, anastomosing cortical, calcareous hairs. Pores subconjugate. *Rec.*, Hawaii.

Chondrocidaris A. AGASSIZ, 1863, p. 18 [**C. gigantea*; OD]. Test low, height half of diameter, flattened above, sides arched. Areoles shallow, not

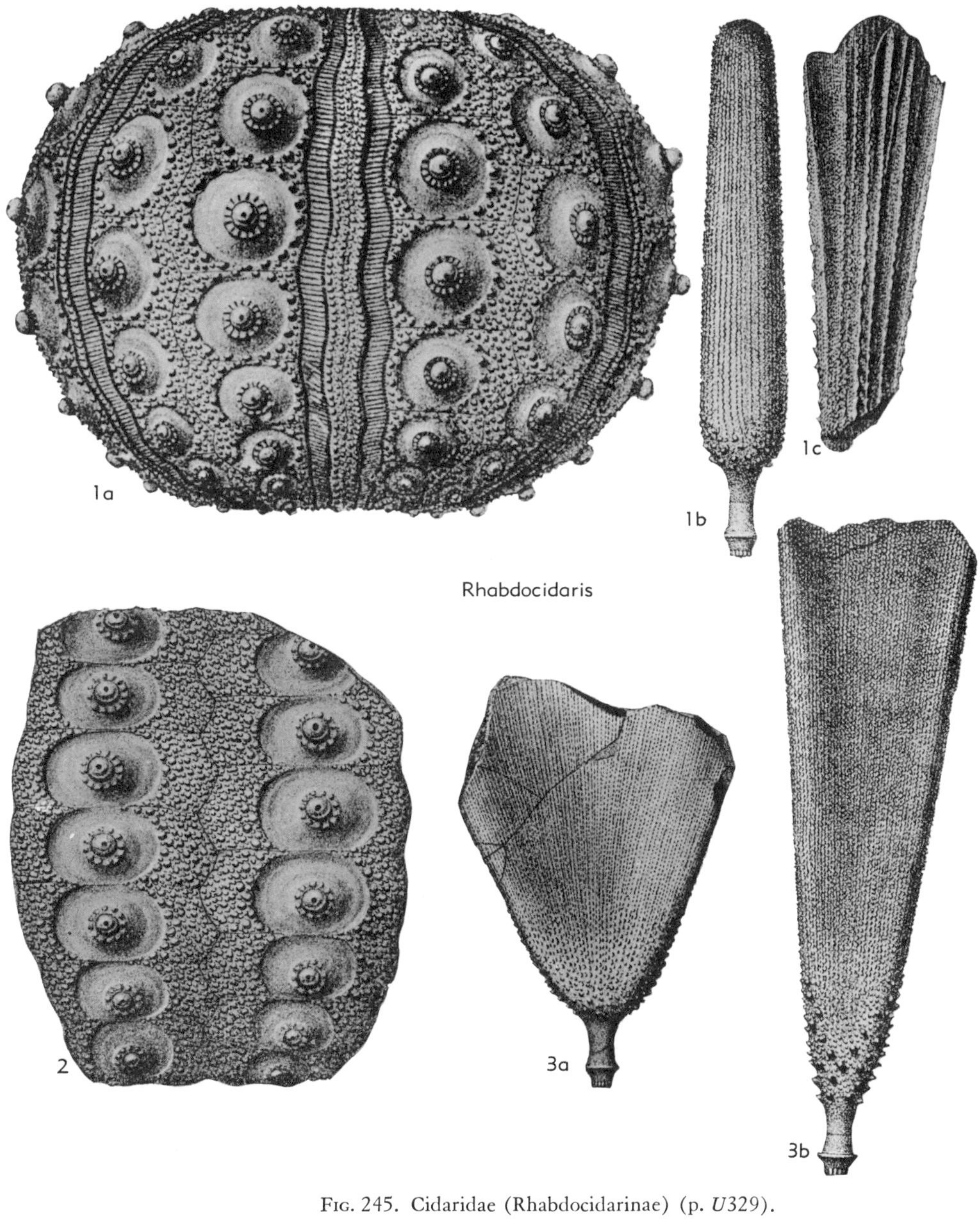

FIG. 245. Cidaridae (Rhabdocidarinae) (p. *U*329).

confluent, primary tubercles noncrenulate. Ambulacra straight or sinuate, pores conjugate. Madreporite not conspicuously enlarged. Whole surface of corona, save for areoles and poriferous area, densely coated in uniform, small, hyaline, more or less spinuliform tubercles; areoles carry small, conical, adpressed secondary spines, attached to sides (not tops) of tubercles. Primary spines thick, commonly with dense hair at base of shaft, with coarse thorns which tend to unite into longitudinal lamellae, latter spreading fanlike at tips; cortex thin. Large globiferous pedicellariae without end-tooth, small ones with them; tridentate pedicellariae present. *Oligo.*, Australia; *Mio.*, Australia-Fiji-Indon.-Madag.; *Rec.*, IndoPac. (incl. Hawaii but not New Zealand).

Megacidaris Thiéry, 1928, p. 180 [**Cidarites horrida* Merian, 1880; OD]. Sutures oblique. Pores conjugate. *M.Jur.(Bajoc.)*, Eu.

Parhabdocidaris Thiéry, 1928, p. 181 [**Rhabdocidaris varusensis* Cotteau, 1880; OD] [=*Parrhabdocidaris* Mortensen, 1928 *(nom. van.)*]. Like *Rhabdocidaris* but ambulacral plates irregularly arranged, some as diads, some in simple series, with corresponding irregularity of marginal tubercles. *U.Jur.*, Eu.

Phyllacanthus Brandt, 1835, p. 67 [**Cidarites (Phyllacanthus) dubia*; OD, M] [*nom. conserv.* ICZN, Op. 208, 1954] [=*Leiocidaris* Desor, 1855 (type, *Cidaris imperialis* Lamarck, 1816); *Aulacocidaris* Lambert, 1903 (type, *A. lamberti* (Savin, 1903)]. Test spherical or low, usually flattened above, sides arched. Areoles well separated, central part elevated, carrying prominent, noncrenulate primary tubercle. Madreporite conspicuously larger than other genital plates, encroaching on small periproct. Scrobicular tubercles conspicuously larger than other secondaries, usually with distinct elevation on side toward areole. Pores conjugate, but with wall elevated aborally. Primary spines cylindrical, thick, robust, with fine granules arranged in regular longitudinal series on shaft; cortex thick; primary radial lamellae (as seen in transverse section) arising in fanlike clusters from projecting portions of medulla. Secondary spines broad, flat, squamiform, closely adpressed. Globiferous pedicellariae without end-tooth. *Oligo.*, Australia-N.Z.; *Mio.*, Australia-N.Z.-Fiji Is.-India; *Plio.*, Australia; *Rec.*, Australia (5 of 6 species), IndoPac.——Fig. 246, *1. P. wellmanae* Fell, U.Mio., N.Z.; *1a,b,* part of test, lat., and aboral, ×1.3; *1c,* amb plates, ×13.——Fig. 246,*2. P. titan* Fell, U.Oligo.-L. Mio., N.Z. (*2a,b,* U.Oligo.; *2c-f,* L.Mio.); *2a-f,* spines (*2a,* ×4; *2b,* ×1.3; *2c-f,* ×0.7); *2g,* spine transv. sec., ×26; *2h,i,* amb and interamb plates, ×1.7; *2j,* amb plates, ×10 (56).

Plococidaris Mortensen, 1909, p. 51 [**Cidarites verticillata* Lamarck, 1816; OD, M]. Like *Prionocidaris* but primary spines verticillate. Pores subconjugate. *Mio.*, India-Madag.; *Rec.*, IndoPac. (but not N.Z.).

Porocidaris Desor, 1855, p. 46 [**Cidarites schmidelii* Münster, 1843; SD Lambert & Thiéry, 1910, p. 108]. Like *Rhabdocidaris* but ambulacra straight. Areoles circular and shallow (as in *Rhabdocidaris*) but confluent, each with circle of radiating, porelike or slitlike marginal depressions. Primary spines partly known only, orals flattened, coarsely serrate, collar more or less tuberculate. *Eoc.*, Eu.-N.Afr.——Fig. 244,*2. *P. schmidelii* (Münster), Egypt; *2a,* base of spine, ×2.8 (27e); *2b,* test, lat., ×1.1; *2c-e,* spines, ×1.1 (44); *2f,* interamb plate, ×2.1; *2g,* amb plates, ×8.5 (121). [=*Procidaris* Pávay, 1875, p. 230 *(nom. van.)*.]

Prionocidaris A. Agassiz, 1863, p. 18 [**Cidarites pistillaris* Lamarck, 1816; OD] [=*Stephanocidaris* A. Agassiz, 1872 (*non* 1863); *Schleinitzia* Studer, 1880 (type, *S. crenularis*); *?Pleurocidaris* Pomel, 1883 (type not designated, genus of doubtful validity)]. Test arched or low, more or less flattened at apex, thin and somewhat fragile. Primary tubercles noncrenulate adorally, weakly subcrenulate or noncrenulate aborally; areoles shallow, well separated save for lowermost 2 or 3, which may be confluent. Pores distinctly conjugate or subconjugate. Primary spines usually long, tapering, with coarse thorns in longitudinal series; less commonly cylindrical, smooth or widened distally, or with thorns arranged in whorls; cortex thin; oral primaries with relatively long collar, tipped by rudimentary shaft. Secondary spines not adpressed, larger ones flattened, smaller ones spiniform. Tridentate pedicellariae slender; large globiferous pedicellariae without end tooth, or wanting; small globiferous with end tooth. *U. Cret.*, Eu.; *Eoc.*, Eu.-India-Australia-N.Z.; *Oligo.*, N.Z.; *Mio.*, Medit.(Sardinia-Malta)-Australia; *Rec.*, IndoPac. (but not N.Z.).——Fig. 247,*1a-d. P. marshalli* Fell, M.Eoc., N.Z.; *1a-c,* spines, ×1.5; *1d,* part of transv. sec. of spine, ×30 (56).——Fig. 247,*1e-g. P. canaliculata* (Duncan & Sladen), Eoc., N.India; *1e,* amb plates, ×3.75; *1f,* test lat., ×0.75; *1g,* interamb plates, ×1.5 (47).——Fig. 247,*1h. P. sismondai* (K. Mayer), Mio., Eu. (Sardinia); test lat., with spines, ×0.7 (107).——Fig. 247,*1i. P. mitchelli* (Emmons), M.Eoc., USA (N. Car.); test, oral, ×0.75 (22).

Subfamily CIDARINAE Gray, 1825

[*nom. transl.* Fell, herein (*ex* Cidaridae Gray, 1825)] [=Cidarina Mortensen, 1928, *plus* Stylocidarina Mortensen, 1928 *(partim)*]

Corona without sutural pits or grooves. Primary tubercles crenulate or noncrenulate; but if crenulate, primary spines short and thick. Pores horizontal, nonconjugate. Globiferous and tridentate pedicellariae. *U. Trias.(Rhaet.)-Rec.*

Cidaris Leske, 1778, p. 74 [**Echinus cidaris* Linné, 1758, approx. limited in 1761] [=*Cidarites* auctt. (*non* Lamarck, 1816) (=*Phyllacanthus* Brandt, 1835); *Orthocidaris* A. Agassiz, 1863 (type, *Cidaris papillata* Leske, 1878); *Dorocidaris* A. Agassiz, 1869 (type, *Cidaris papillata* Leske, 1878); *Papula* Bayle, 1878]. Areoles generally deep, well separated; primary tubercles noncrenulate adorally, aborally noncrenulate or (exceptionally) subcrenulate. Primary spines with more or less distinct longitudinal regular rows of spinules, sometimes forming ridges. Oral primaries flattened, smooth, slightly serrate. Large and small globiferous pedicellariae with end tooth; tridentate pedicellariae present. *Rec.,* Atl.-Medit.-Ind.O.——Fig. 247,*2*. **C. cidaris* (Linné); test, lat., ×0.75 (136a).

[It is unfortunate that the primary type of the order and family, *Cidaris cidaris* (Linné), should belong to a small group of Atlantic forms characterized by the presence of end teeth on large and small globiferous pedicellariae,

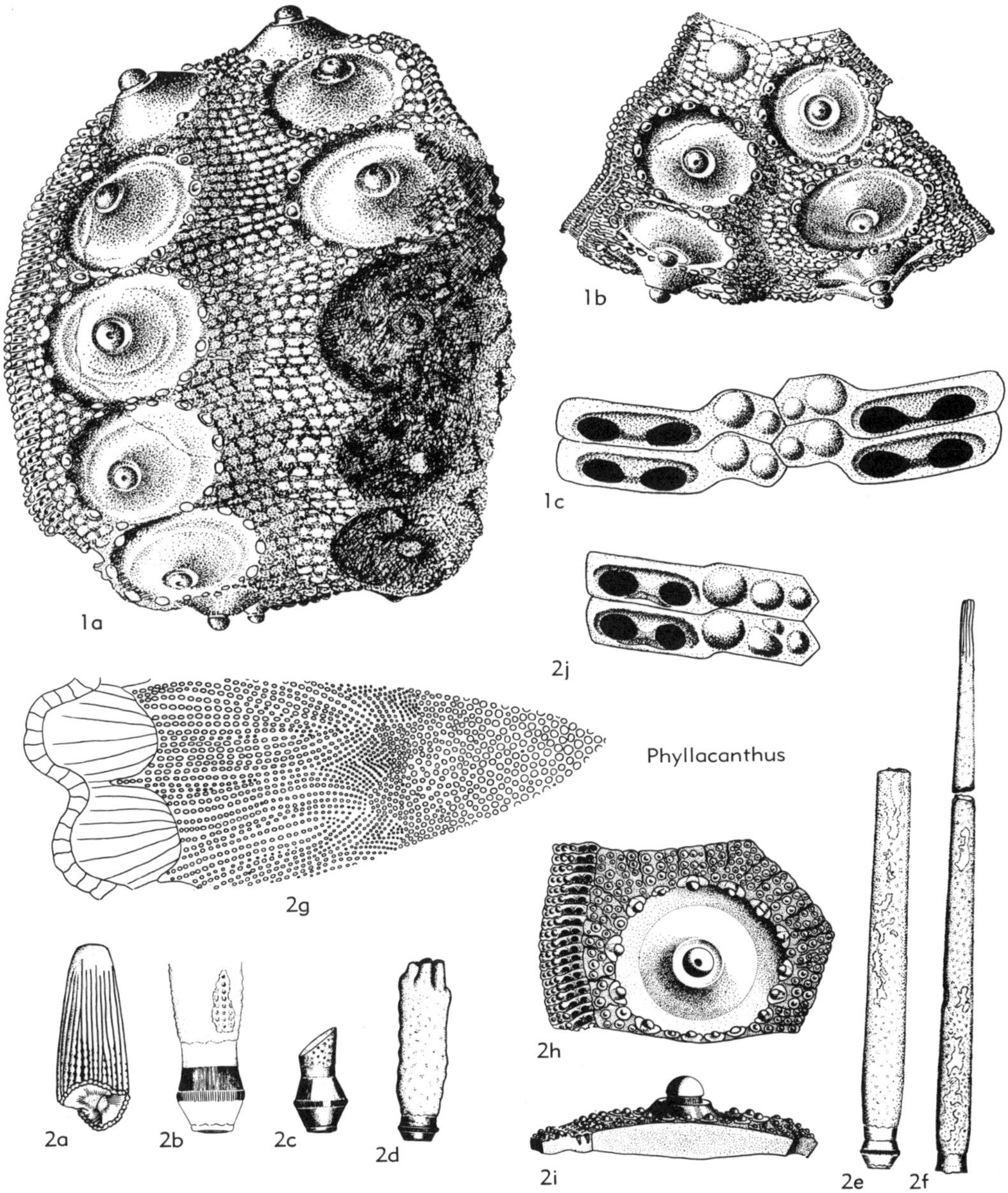

Fig. 246. Cidaridae (Rhabdocidarinae) (p. *U*330).

features of very slight paleontological significance; in fact, no valid fossil species has yet been described for the genus *Cidaris*. On the other hand, since it is the primary type-genus, it has long been the practice to assign to it all fossil species whose precise generic position has not been established. In this sense the name is without taxonomic validity, and should therefore be distinguished in some way—such as "*Cidaris*," or "*Cidaris*" s.l. (=*Cidarites* auctt., *non Cidaris* LESKE).] [=*Cidarites* LAMARCK, 1816, p. 52 (=jr. hom., ICZN Art. 20, 56b)—MOORE.]

Alpicidaris LAMBERT, 1910, p. 4 [**A. cureti*; OD]. Ambulacral plates tending to form diads or triads, especially adorally, where one marginal tubercle occurs on every 2nd or 3rd plate. Primary tubercles noncrenulate; areoles transversely oval, not well separated, more or less confluent below ambitus. Spines unknown. *L.Cret.(Hauteriv.)*, Eu.——FIG. 248,*1*. **A. cureti*, Fr.; *1a*, test, lat., ×1.1; *1b*, amb plates, ×2.2 (115).

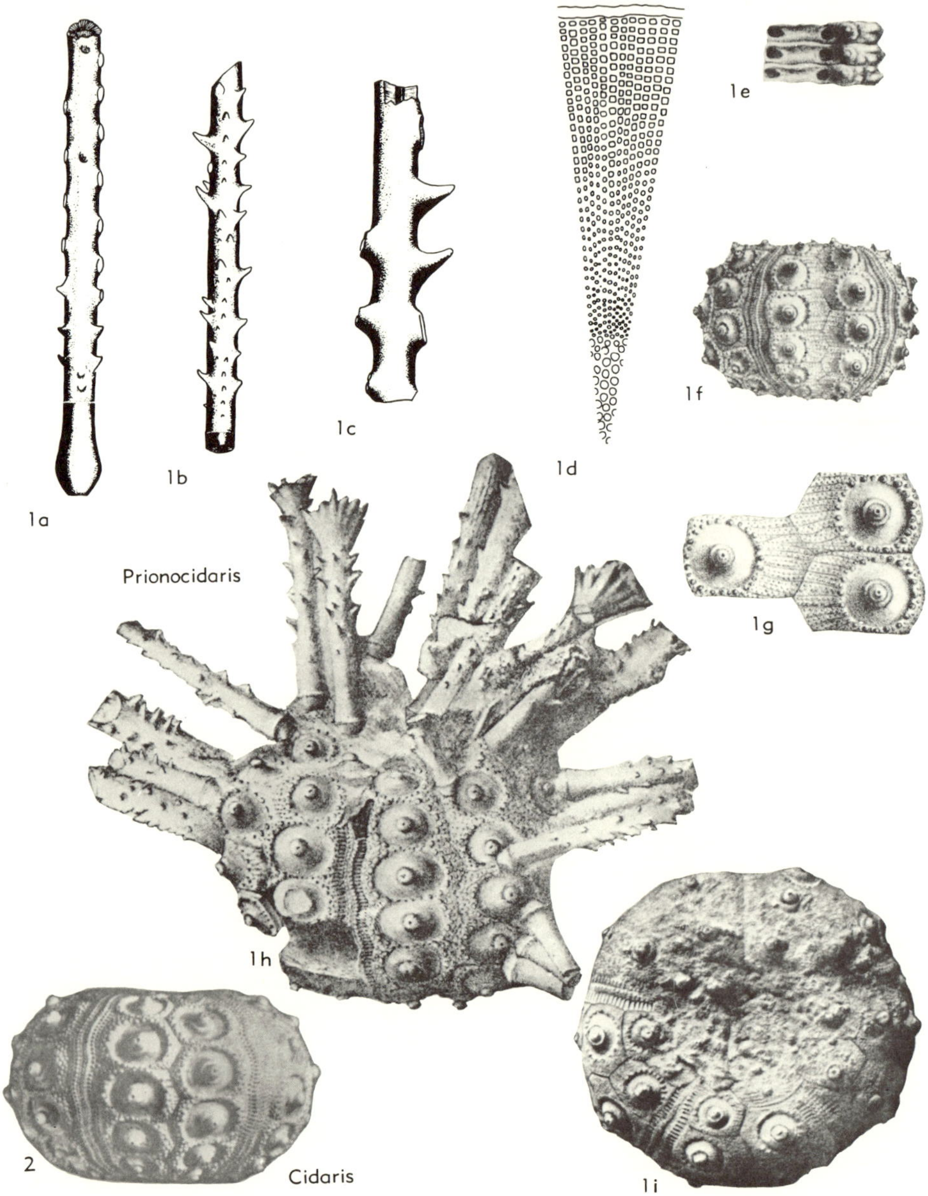

FIG. 247. Cidaridae (Rhabdocidarinae) *(1)*, (Cidarinae) *(2)* (p. *U330-U331*).

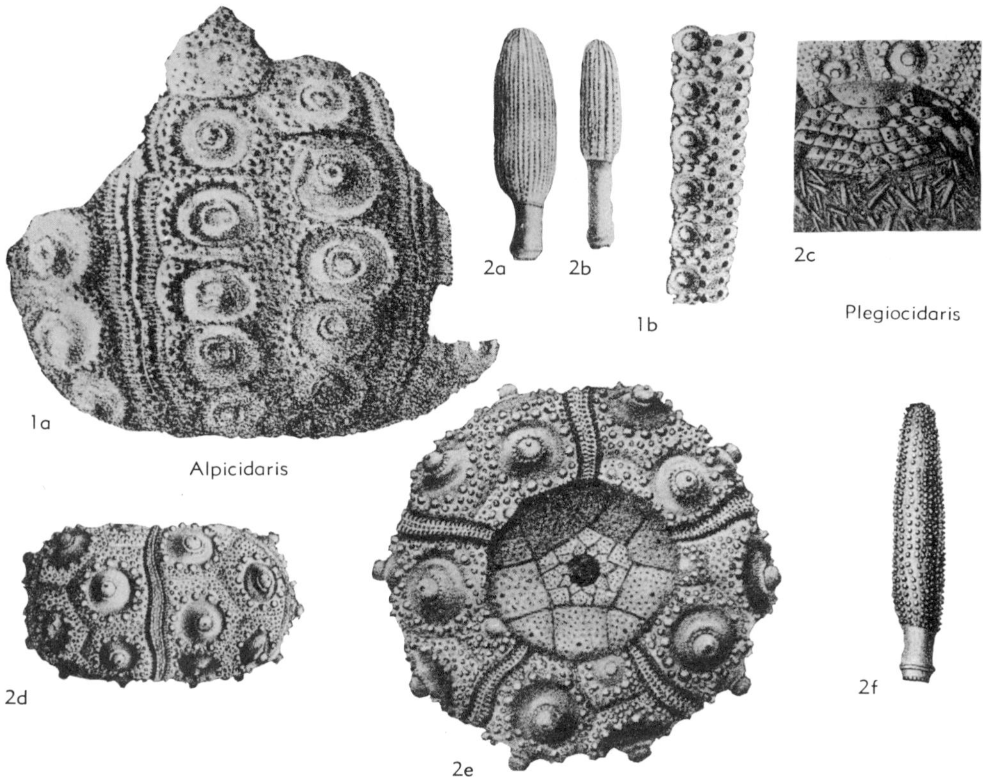

FIG. 248. Cidaridae (Cidarinae) (p. *U332*, *U337*).

Balanocidaris LAMBERT, 1910, p. 4 [**Cidaris glandifera* MÜNSTER, 1843; OD]. Primary tubercles noncrenulate. Primary spines glandiform. Ambulacra sinuous, pore-zone narrow, interporiferous area broad, densely tuberculate, tubercles arranged in uniform longitudinal and horizontal rows. *U.Trias. (Carn.)*, Eu.; *Jur.*, Eu.-?Calif.; *U.Cret.*, Eu.-?Calif. ——FIG. 249,*2a,b*. *?B. pleracantha* (AGASSIZ), U.Cret., Fr.; *2a,b*, spines, ×0.9 (44). ——FIG. 249,*2c,d*. *?B. californica* (CLARK), M. Jur., USA(Calif.); *2c,d*, spines, ×1.8 (22).——FIG. 249,*2e,f*. **B. glandifera* (MÜNSTER), Jur., Ger.; *2e*, spine, ×0.9 (44); *2f*, amb plates, ×1.8 (27c).——FIG. 249,*2g*. *B. roysii* (DESOR), Jur., Fr.; spine, ×1.8 (27c).

Calocidaris H. L. CLARK, 1907, p. 211 [**Dorocidaris micans* MORTENSEN, 1903; OD]. Like *Cidaris* but shaft of primary spines smooth, glabrous; all primary tubercles noncrenulate. *Rec.*, W.Indies.

Centrocidaris A. AGASSIZ, 1904, p. 32 [**Goniocidaris doederleini* A. AGASSIZ, 1898; OD]. Like *Stylocidaris* but ambulacra broad (*ca.* half as wide as interamb); median ambulacral and interambulacral areas bare, somewhat depressed, but not forming well-defined grooves; primary spines with smooth, glabrous surface; large globiferous pedicellariae with or without end tooth. *Rec.*, W.C.Am. (Cocos I.-Galapagos Is.).

Cyathocidaris LAMBERT, 1910, p. 12 [**Cidaris cyathifera* AGASSIZ; OD]. Corona high, subspherical. Interambulacral plates high, primary tubercles noncrenulate. Ambulacra sinuous, simple throughout. Primary spines of aboral side cup-shaped or trumpet-shaped, some with central prominence. *U.Cret.*, Eu.-Antarctic; *Eoc.*, *?Mio.*, Eu.——FIG. 249,*1a-c*. **C. cyathifera* (AGASSIZ), U.Cret. (Senon.), N.Fr.; *1a-c*, spines, ×0.9 (27a).——FIG. 249,*1d*. *C. erebus* LAMBERT, U.Cret., Antarctic, spine, ×0.9 (108).——FIG. 249,*1e*. *C. crateriformis* (GÜMBEL), Eoc., Fr.; spine, ×0.9 (27e).——FIG. 249,*1f-h*. *C. nordenskioldi* LAMBERT, Cret., Antarctic; *1f-h*, spines, ×0.9 (108).

Eucidaris POMEL, 1883, p. 109 [**Cidarites metularia* LAMARCK, 1816; SD CLARK & BATHER, 1909, p. 88] [=*Cidaris* A. AGASSIZ, 1872 (*non* LESKE, 1778); *Gymnocidaris* A. AGASSIZ, 1863 (type, *Cidarites metularia* LAMARCK, 1816) (*non* L. AGASSIZ, 1838)]. Like *Stylocidaris* but madreporite slightly larger than other genital plates; primary spines typically cylindrical, truncate, otherwise fusiform or clavate; shaft abruptly trun-

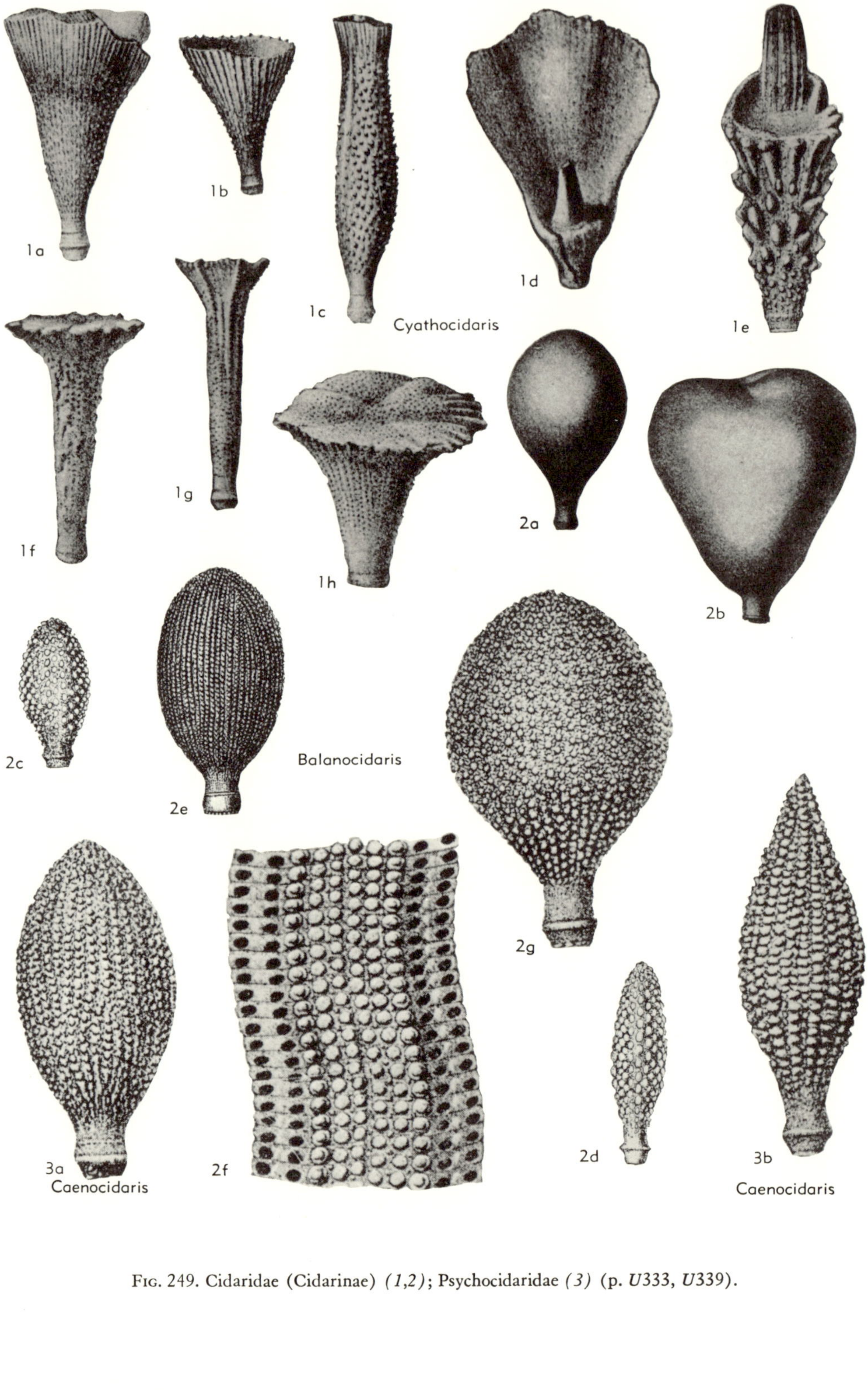

FIG. 249. Cidaridae (Cidarinae) *(1,2)*; Psychocidaridae *(3)* (p. *U*333, *U*339).

cate, terminating in crown with central prominence, and with low, rounded warts disposed in regular, longitudinal series; secondary spines adpressed; tridentate pedicellariae of 2 types, valves either straight or curved. *U.Eoc.,* N.Z.; *Oligo.,* N.Z.; *Mio.,* Fiji-?Australia-Calif.-W.Indies; *Plio.,* Calif.; *Rec.,* trop. and subtrop.——FIG. 250,*1a-f*. *E. strobilata* FELL, U.Eoc.-L.Oligo., N.Z. (*1-3,* U. Eoc.; *4-6,* L.Oligo.); *1a-f,* spines, ×2.8.——FIG. 250,*1g,h*. *E. coralloides* FELL, L.Oligo., N.Z.; *1g,h,* spine, lat. and top, ×2.1 (56).

Hesperocidaris MORTENSEN, 1928, p. 73 [**Dorocidaris panamensis* A. AGASSIZ, 1898; OD]. Like *Stylocidaris* but primary spines cylindrical, not tapering, some conspicuously expanded at tip, with low granules arranged in longitudinal series. *?Eoc.,* Calif.; *Rec.,* Calif.-W.Panama-Ecuador.

Kionocidaris MORTENSEN, 1932, p. 165 [**K. striata*; OD]. Like *Stylocidaris* but primary spines columnar, slightly tapering, with about 25 shallow flutes on shaft, low elevations between flutes and regular longitudinal series of pores in surface of distal region. *Rec.,* Natal.

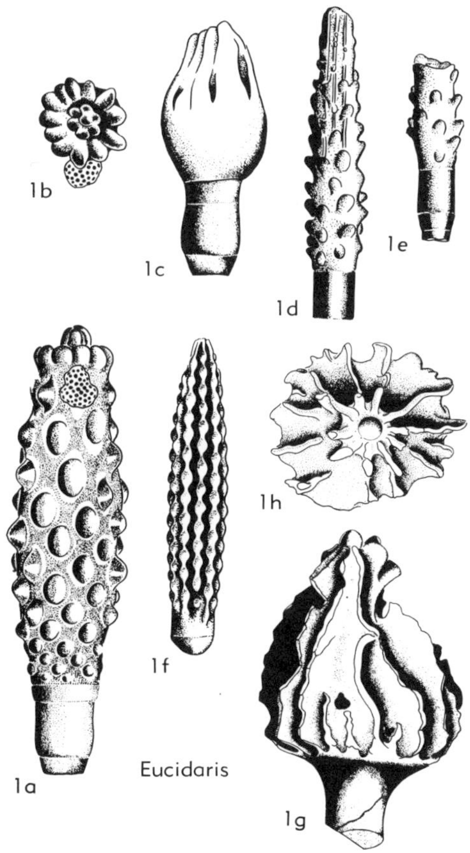

FIG. 250. Cidaridae (Cidarinae) (p. *U333, U335*).

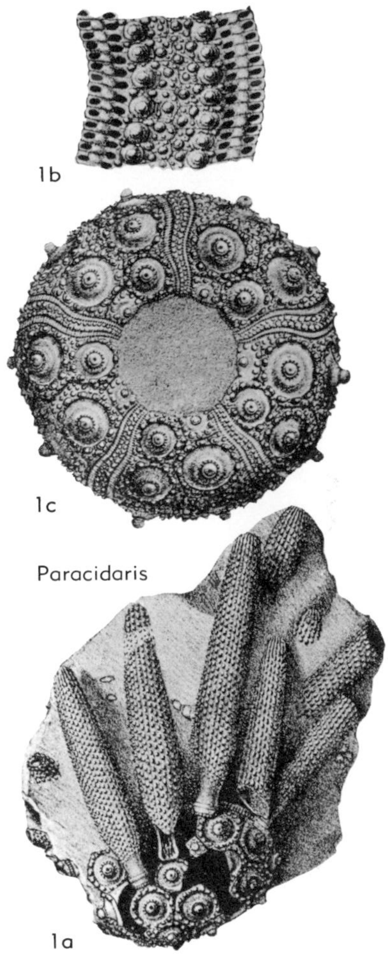

FIG. 251. Cidaridae (Cidarinae) (p. *U335, U337*).

Lissocidaris MORTENSEN, 1939, p. 11 [**L. fusca*; OD]. Like *Calocidaris,* but cortex layer of spines covered by thick anastomosing layer of hairs which coalesce to produce smooth, glabrous investment about shaft. *Rec.,* Ind.O. (Maldive Is.).

Paracidaris POMEL, 1883, p. 109 [**Cidarites florigemma* PHILLIPS, 1829; SD LAMBERT & THIÉRY, 1910, p. 135]. Ambulacral plates simple but tending to form diads, especially adorally, where one marginal tubercle occurs on only every alternate plate. Primary tubercles crenulate; areoles rounded, well separated; scrobicular tubercles larger than other secondaries; primary spines thick, clavate or fusiform, collar short, shaft with coarse, uniform spinules in regular longitudinal series. *U. Trias.(Rhaet.)-U.Jur.(Portland.),* Eu.——FIG. 251, *1.* **P. florigemma,* U.Jur.(Oxford.), Fr.; *1a,* test

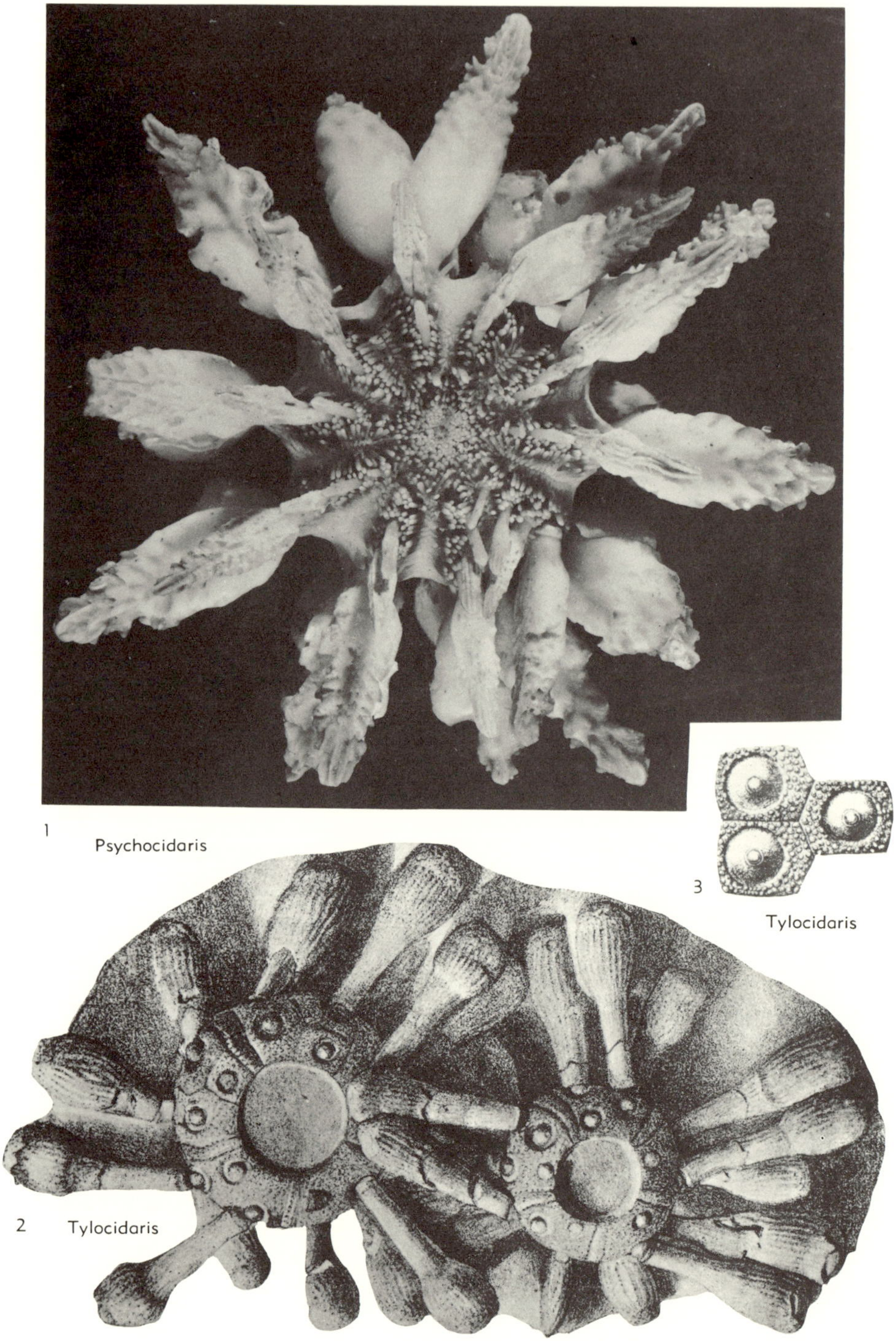

1 Psychocidaris

2 Tylocidaris

3 Tylocidaris

Fig. 252. Psychocidaridae (p. U338-U339).

with spines, ×0.7; *1b,* amb plates, ×3.5; *1c,* test, aboral, ×1.4 (27c).

Plegiocidaris POMEL, 1883, p. 109 [**Echinus coronatus* VON SCHLOTHEIM, 1820; SD LAMBERT & THIÉRY, 1910, p. 135]. Like *Paracidaris* but ambulacra simple throughout corona, becoming pluriserial only on peristome. Primary radioles with long cylindrical collar, and cylindrical shaft as long as collar or 2 or 3 times longer, diameter of shaft twice that of collar; transition from collar to shaft oblique, with abrupt change in diameter of spine. *U.Trias.(Nor.)-U.Jur.*, Eu.——FIG. 248, *2a-e.* **P. coronata* (VON SCHLOTHEIM), U.Jur. (Oxford.), Ger.; *2a,b,* spines, ×1; *2c,* peristome, ×2.25; *2d,e,* test, lat., aboral, ×1, ×1.3 (145). ——FIG. 248,*2f. P. cervicalis* AGASSIZ, U.Jur.(Oxford.), Fr.; spine, ×0.75 (44).

Stylocidaris MORTENSEN, 1909, p. 52 [**Cidaris*

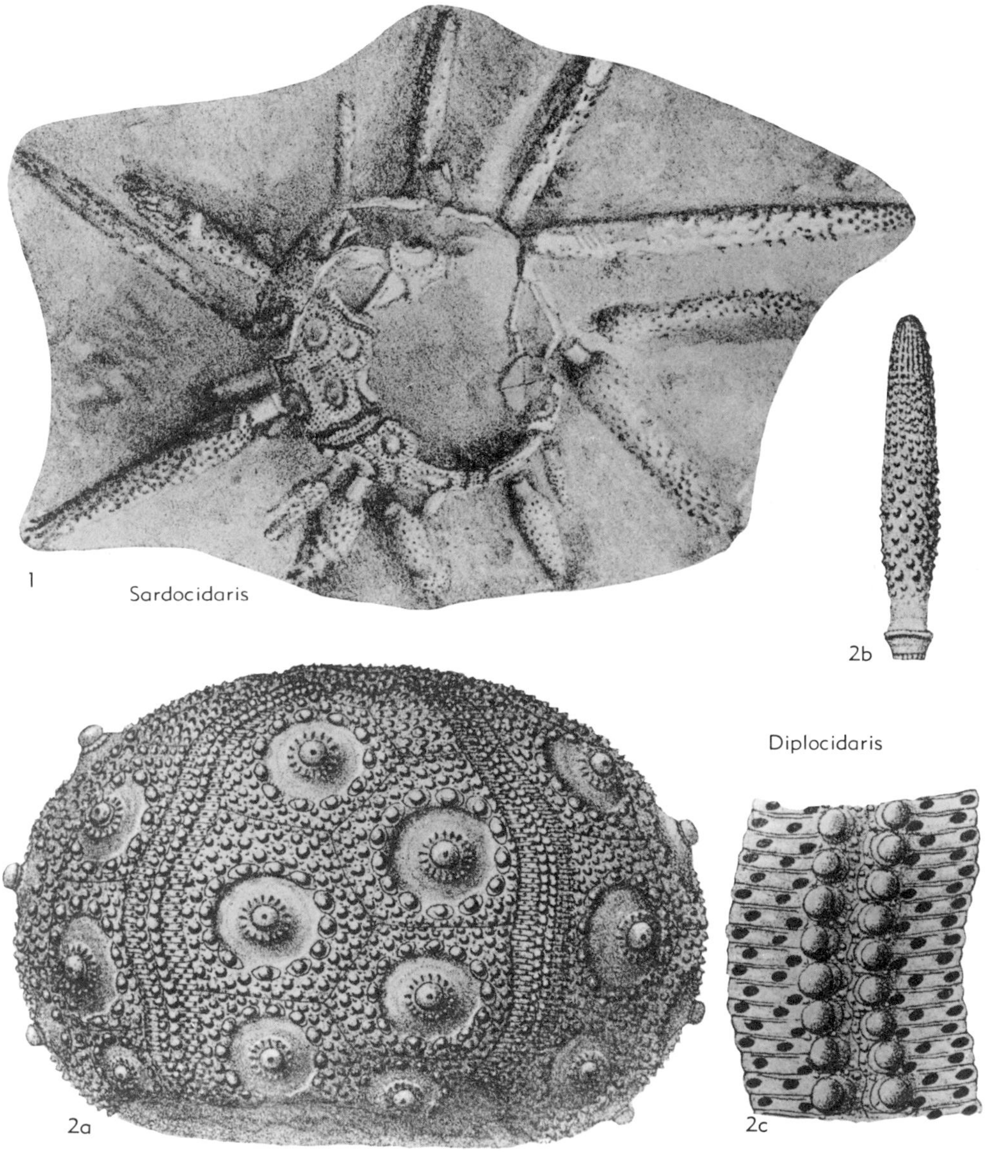

FIG. 253. Psychocidaridae *(1)*; Diplocidaridae *(2)* (p. *U*339).

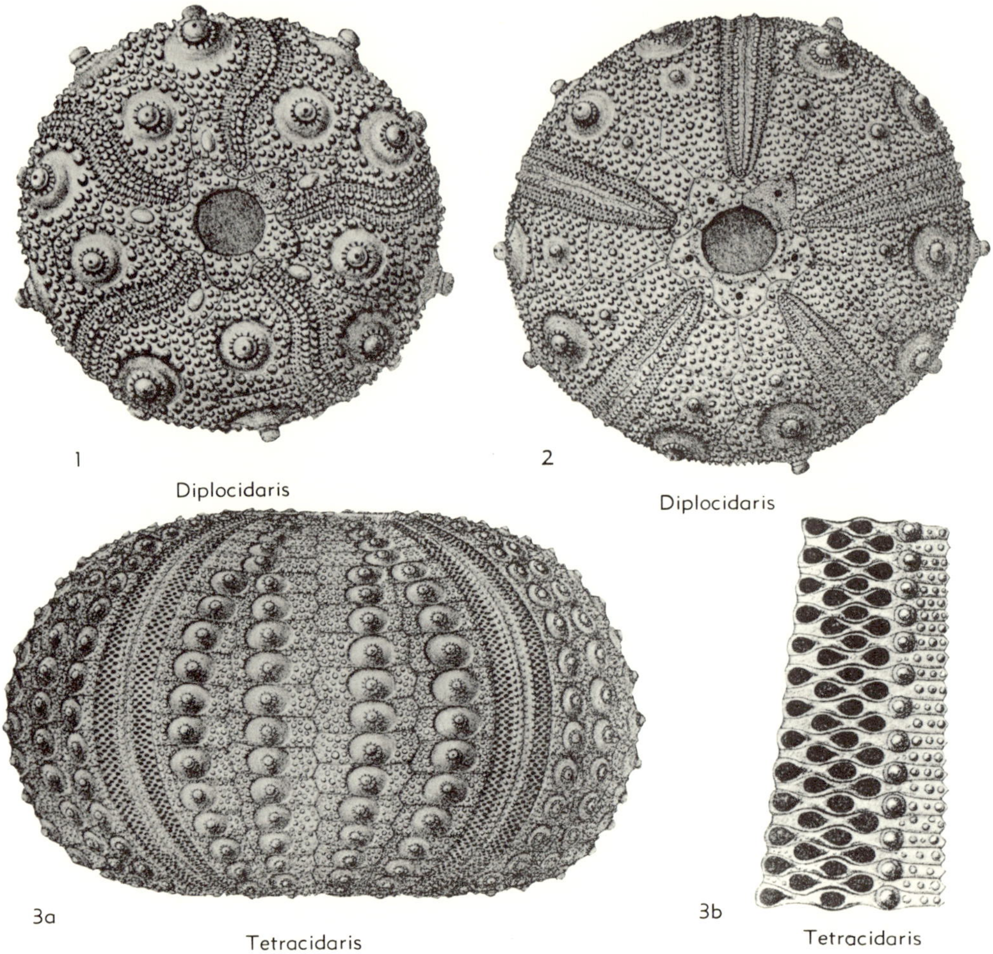

FIG. 254. Diplocidaridae (p. U339).

affinis PHILIPPI, 1845; OD]. Test usually flattened above and below. Areoles deep or shallow, well separated, only proximal 2-3 confluent; primary tubercles noncrenulate adorally, some weakly subcrenulate aborally, especially in young stages, otherwise noncrenulate aborally. Primary spines usually tapering to point; secondary spines more or less flattened, not adpressed. Tridentate pedicellariae slender; large globiferous without small end-tooth, globiferous with them. *Mio.,* Medit.; *Rec.,* trop. and subtrop. seas.

Tretocidaris MORTENSEN, 1903, p. 28 [**Dorocidaris bartletti* A. AGASSIZ, 1880; OD]. Like *Cidaris* but areoles very shallow, well separated; primary tubercles all noncrenulate. Primary spines cylindrical, coarsely thorny, spinules in distinct longitudinal ridges, end in some forms conspicuously widened; shorter spines commonly more or less cup-shaped. *Rec.,* trop. Atl.

Family PSYCHOCIDARIDAE Ikeda, 1936

Primary tubercles imperforate. Pores nonconjugate, commonly separated by prominent wall. No peristomial interradial plates; peristomial ambulacral plates in double series. *U.Jur.-Rec.*

Psychocidaris IKEDA, 1935, p. 386 [**P. ohshimai*; OD]. Test low, of moderate size, flattened above and below, plates robust with rigid sutures. Apical system nearly half of horizontal diameter. Madreporite scarcely larger than other genital plates. Areoles large, sunken; primary tubercles large, hyaline, noncrenulate; few upper ones (immature) weakly perforate. Upper primary spines glandiform, ambital primaries longer, somewhat flattened, oral primaries cylindrical; all primaries, save orals, with coarse irregular thorns; cortex bulky, spongy, containing cavities (dry spines having density of *ca.* 0.6). Large globiferous pedicellariae without poison chamber. *Rec.*, Bonin Is. (180 m.).——FIG. 252,*1*. **P. ohshimai*; oral view of test with spines, ×0.9 (Fell, n).

Anisocidaris THIÉRY, 1928, p. 181 [**Cidaris bajocensis* COTTEAU, 1880; OD]. Like *Caenocidaris*, but pore zones widened, pores nonconjugate. *M. Jur.(Bajoc.)*, Eu.

Caenocidaris THIÉRY, 1928, p. 180 [**Cidaris cucumifera* AGASSIZ, 18??; OD]. Like *Merocidaris* but adoral ambulacral plates biserial. *M.Jur.(Bajoc.)*, Eu.——FIG. 249,*3*. **C. cucumifera* (AGASSIZ), Fr.; *3a,b*, spines, ×1.8 (27c).

Merocidaris THIÉRY, 1928, p. 180 [**Cidaris honorinae* COTTEAU, 1880; OD]. Adoral primary tubercles imperforate, noncrenulate; aboral primary tubercles perforate, crenulate. *Jur.(Bajoc.-Kimmeridg.)*, Eu.

Sardocidaris LAMBERT, 1907, p. 22 [**S. piae*; OD]. Like *Psychocidaris* but primary spines long, cylindrical, tapering, not glandiform. *M.Cret.*, N. Afr.; *Mio.*, Eu.——FIG. 253,*1*. **S. piae*, L.Mio., Eu.(Sardinia); test with spines, ×1 (107).

Tylocidaris POMEL, 1838, p. 109 [**Cidaris clavigera* KÖNIG, 1820; SD LAMBERT & THIÉRY, 1910, p. 156]. Like *Psychocidaris* but primary spines with fine spinules, not coarse thorns, on glandiform and other spines, spinules scattered irregularly or in regular longitudinal series. *U.Cret.*, Eu.(Eng.-Fr.-Denm.); *L.Eoc.*, N.Am.——FIG. 252,*2*. **T. clavigera* (KÖNIG), Eng.; tests with spines, oral, ×0.9 (173).——FIG. 252,*3*. *T. walcotti* (CLARK), L.Eoc., USA(N.J.); interamb, ×0.9 (22).

Family DIPLOCIDARIDAE Gregory, 1900

Plates of ambulacra arranged in diads throughout, marginal tubercles developed on alternate plates only, pore zone biserial. Primary tubercles perforate. *Jur.-Cret.*

Diplocidaris DESOR, 1855, p. 44 [**Cidaris gigantea* AGASSIZ, 18??; SD LAMBERT & THIÉRY, 1910, p. 138] [=*Alternocidaris* QUENSTEDT, 1873 (obj.)]. Interambulacral plates in 2 columns throughout; primary tubercles crenulate. Pores conjugate or subconjugate, uppermost more or less rudimentary. Primary spines cylindrical, short, granulated. *L.Jur.-L.Cret.*, Eu.——FIG. 253,*2*. **D. gigantea*, U.Jur. (Kimmeridg.), Fr.; *2a,b*, test lat., spine, ×1; *2c*, amb plates, ×4.——FIG. 254,*1*. *D. miranda* (AGASSIZ), U.Jur.(Kimmeridg.), Fr.; test, aboral, ×0.8.——FIG. 254,*2*. *D. etalloni* DE LORIOL, U. Jur.(Raurac.), Fr.; test, aboral, ×0.7 (27c). [=?*Diplotiaris* QUENSTEDT, 1872, p. 36.]

Tetracidaris COTTEAU, 1872, p. 445 [**T. reynesi*; OD]. Interambulacral plates in 4 columns at ambitus, in 2 columns adapically and adorally. Pores distant but nonconjugate. Primary tubercles crenulate. Primary spines cylindrical, slender. *L.Cret. (Neocom.)*, Eu.——FIG. 254,*3*. **T. reynesi*, Fr.; *3a*, test, lat., ×0.8; *3b*, amb plates, ×4 (28).

Order and Family UNCERTAIN

Lanternarius REGNÉLL, 1956, p. 171 [**L. latens*; OD]. Known only from lantern; pyramids large, adperradial walls adapically with internal excavation; epiphyses narrow, but uniting over moderately deep foramen magnum. *Sil.*, Sweden (Gotl.).——FIG. 234,*3*. **L. latens*; *3a,b*, parts of lantern, ×3 (148).

EUECHINOIDS

By H. BARRACLOUGH FELL (with sections by R. V. MELVILLE and by H. B. FELL and D. L. PAWSON)

[Victoria University of Wellington, New Zealand; transferred to Harvard University]

Subclass EUECHINOIDEA Bronn, 1860

[Diagnosis and discussion by H. B. FELL]

Test composed of 5 bicolumnar ambs and 5 alternating bicolumnar interambs. Plates imbricating, or joined by flexible integument, or (more usually) united by rigid sutures. Anus and surrounding periproct either placed within apical system (endocyclic condition) or secondarily translocated to interamb 5, which is termed posterior interamb (exocyclic condition). Lantern present or absent, or present only in juvenile stage. Gills and gill slits present or absent. Spheridia present. Pedicellariae present, including ophicephalous types. *?Carb., U. Trias.-Rec.*

The included groups are here classified as 18 orders which, following proposals of DURHAM & MELVILLE (52), may be arranged in four superorders, namely Diadematacea, Echinacea, Gnathostomata, and Atelostomata. The evolutionary trends exhibited are so diverse as to require separate treatment under the subsidiary taxa. In general, however, the first two superorders are characterized by remaining predominantly radially symmetrical subspherical forms, similar to the Cidaroida, but differing conspicuously in exhibiting a persistent trend toward compounding of the ambulacral plates, especially in groups of three.

The Diadematacea achieve this compounding by a different method from that observed in the Echinacea; the details are given under the superorders. In addition, many of the Echinacea achieve a greater complexity of jaw structure than is found in the Diadematacea.

The other two superorders, Gnathostomata and Atelostomata, retain simple amb plates, without compounding of the elements, but differ in structure of the dental apparatus, which tends to disappear in the postembryonic stages of development, and in the degree to which a secondary bilateral symmetry comes to be superimposed upon the earlier radial symmetry. All members of the Gnathostomata and Atelostomata are exocyclic, that is, the anus and periproct have entered the posterior interamb. One group of Diadematacea, however, the Pygasteroida, also shows an early exocyclic condition, and the structure of the jaws and other features suggest that the Pygasteroida do not share a common derivation with that of the other exocyclic groups.

This implies that the older classification, in which all exocyclic echinoids were grouped in one taxon, the Irregularia, is unlikely to reflect natural relationships, and accordingly should be discarded (52). In practice, however, the older term Irregularia does virtually correspond to the two taxa Gnathostomata and Atelostomata, here recognized, differing only by including the Pygasteroida, a very small group. Similarly, the older taxon Regularia (including all endocyclic forms, i.e., forms in which the anus remained within the apical system) corresponds to the groups Perischoechinoidea, Diadematacea, and Echinacea, excluding only the Pygasteroida.

Although the classification here employed seems superficially to differ widely from that in general use hitherto, including that used in MORTENSEN's monograph (136), the differences are more apparent than real, for, as DURHAM & MELVILLE (65) have stressed, MORTENSEN himself demonstrated that the probable affinities of the various taxa are those implicit in the arrangement here presented.

DIADEMATACEA

By H. BARRACLOUGH FELL

INTRODUCTION

The Diadematacea comprise an assemblage of Euechinoidea all characterized by retaining a lantern of the same type as found in the Cidaroida (aulodont dentition). Most members of the group also have well-developed gills and gill slits, and the tubercles are always perforate. The amb plates in primitive, older genera retain the simple structure seen in cidaroids, but in most Diadematacea they tend to become compounded; in the latter case, the compounding follows the so-called diadematoid or arbacioid patterns, as defined below.

The assemblage is here regarded as comprising four orders, grouped to form a taxon of higher status to which DURHAM & MELVILLE (65) assigned the rank of superorder. The origins of the Diadematacea are not entirely clear, but recent opinion derives all from a cidaroid ancestry (65). This topic is referred to below (p. U344), where a polyphyletic derivation is considered as possible, though improbable. The earliest Diadematacea, so far known, are of Late Triassic age, though it is possible that some fragmentary Lower Carboniferous fossils may be referable to the group. The three surviving orders are mainly deep-water forms of cosmopolitan distribution, though one family, the Diadematidae, comprises littoral and sublittoral representatives found in tropical and subtropical waters, especially of the Indo-Pacific. One order (Pygasteroida) is extinct. The extant members are all more or less venomous, the spines either carrying terminal venom glands or producing a toxic secretion over the shaft itself, which is sharp and fragile and capable of penetrating the skin of predators.

MORPHOLOGY

The test and other skeletal features have fundamentally the structure seen in Cidaroida; hence the only aspects requiring particular mention are those which differ from that order.

GENERAL NATURE OF TEST

The body shape is subspherical, ranging from high subconical forms among the Pedinoida (e.g., *Leiopedina,* see Fig. 267,*3*) to depressed hemispherical or rotular forms, as in the other orders. In the Pygasteroida (see Fig. 270, 271) the anus and periproct lie partly or entirely in one interamb, giving an axis of bilateral symmetry, but despite this, the overall radial symmetry remains conspicuous, and the body does not lose its hemispherical or rotular form, though it may be truncated posteriorly in *Pygaster* (see Fig. 271).

The size of the body ranges from small forms of *ca.* 10 mm. horizontal diameter, to large ones up to *ca.* 150 mm. diameter (e.g., *Micropyga, Asthenosoma*).

As in some of the Paleozoic orders, a persistent tendency to develop imbricating plates is seen, imparting more or less flexibility to the test. This is especially marked in the order Echinothurioida and in some families of Diadematoida, especially the Diadematidae and Micropygidae. In the orders Pedinoida and Pygasteroida such flexibility is lacking.

APICAL SYSTEM

The apical system shows very great variation, and is evidently indicative of some major trends of evolution within the Diadematacea. A completely monocyclic apex characterizes two families of Diadematoida, namely the Aspidodiadematidae and Micropygidae. In these forms the ocular and genital plates are all broadly in contact with the centrally placed periproct; an example is *Tiaridia,* an early aspidodiadematid (see Fig. 261).

In the family Diadematidae the apex is typically monocyclic also, but in the type-genus *Diadema* the anterior oculars (especially oculars II and III) tend to become exsert, probably indicating an incipient rearward movement of the periproct. The same tendency, more strongly expressed, is seen in the order Pygasteroida, where all the anterior oculars (II, III, and IV) become exsert, the posterior oculars I and V are not only in direct contact with the periproct, but are actually thrust aside by it as it moves into the posterior interamb (interamb 5) (see Fig. 271).

An intermediate condition is seen in the Lissodiadematidae, among the diadematoid families, where the posterior oculars I and V become insert, oculars II, III, and IV being exsert; or only ocular II may be exsert. All examples cited so far may be regarded as forming a sequence indicative of a persistent tendency for the periproct and anus to move backward along the axis amb III-interamb 5. There remain two orders where no such trend is conspicuous. One of these is the Echinothurioida, where the apical system is initially dicyclic, but with increasing age all oculars become insert, thus transforming the apex to a monocyclic type. A further development occurs in some echinothurioid genera, by which the oculars and genitals become separated, by resorption of their adjacent borders, or insertion of other tissue between them.

Present opinion derives the Echinothurioida from Cidaroida (52). Since the development of Cidaroida shows that the dicyclic apex is secondary to the monocyclic pattern, it appears unlikely that the initial dicyclic condition of Echinothurioida could be primitive. In such case, the adult monocyclic apex in Echinothurioida must be a tertiary condition, produced by resorption of the adapical margins of the genital plates, the periproct apparently remaining central. An alternative hypothesis, however, derives the Echinothurioida, not from Cidaroida, but from some other Paleozoic order—for example, the lepidocentrid Echinocystitoida (as postulated by MORTENSEN, 136b); so far as can be gathered from the fragmentary remains of Paleozoic echinocystitoids, the oculars seem to have been broadly insert (136b), though exceptionally they were exsert, as in *Lepidechinus* (87). The latter case, then, might imply that the exsert oculars of young Echinothurioida could be primitive. On the whole, however, the balance of evidence from other characters is not in favor of MORTENSEN's hypothesis, and hence it seems more likely that the dicyclic juvenile apex of Echinothurioida is a secondary condition, perhaps derived from a cidaroid ancestry.

Similar considerations apply to the remaining diadematacean order, the Pedinoida; here the apex is typically dicyclic throughout life, all oculars being broadly exsert (see Fig. 263,*1c*). It is possible that in some pedinoid genera (e.g., *Echinopedina*) the posterior oculars became insert, but the evidence at present is insufficient. Such insertion of the posterior oculars would not be surprising, however, for other evidence suggests that the Pygasteroida arose from a pedinoid ancestry, which would imply a backward movement of the anus and periproct in some pedinoids, with consequent insertion of oculars I and V.

PERISTOME

The peristome, like the apical system, shows wide variation in structure, and here again we may detect major evolutionary trends. In all Diadematacea there is evidence of the development of gills, but in some of the Echinothurioida the gills are secondarily lost during growth, and in no member of that order are the gills strongly developed. Thus, the peristome in Echinothurioida never shows very conspicuous gill slits, and in adults of some genera there may be no evident gill slits at all. Clearly such cases are secondary and not comparable with the condition in Cidaroida, where the absence of gill slits is due to the complete lack of gills from that order.

In other orders of Diadematacea gill slits are generally conspicuous, though exceptionally they may be inconspicuous or lacking, as in the pygasteroid *Pileus* (see Fig. 270,*3a*). The major divergence in peristome structure concerns the plating, however, and relationship of the ambs. In the Echinothurioida, as also in Cidaroida, but no other group, the ambs and interambs extend across the peristome, the peristomial amb plates carrying pore pairs, arranged like spokes in continuation of the ambs on the test itself (see Fig. 256,*2a*).

In all other Diadematacea the peristome lacks such plates, and instead has five pairs of oral plates. Such oral plates occur in the juvenile stages of Echinothurioida (136b, 52), and hence it may be inferred that the multiplated peristome of adult Echinothurioida is a secondary condition, and does not indicate a direct derivation of that order from the Paleozoic Echinocystitoida, where multiplated peristome structure also occurs (52).

AMBULACRA

The ambulacra exhibit a persistent tendency toward compounding of the plates by fusion. The compounding follows a pattern, termed diadematoid, in which successive groups of three plates unite to form one plate, the middle element being always larger than the other two elements. In simplest cases the individual amb plates remain distinguishable, each carrying a single pore pair, though a large primary tubercle usually develops only on every third plate; an example is *Plesiodiadema* (see Fig. 261,*1b*). Each of the plates carrying an enlarged tubercle is called a primary amb plate, and its unmodified adjacent neighbors are termed secondary plates.

A group of three, that is, a primary together with the adjacent secondaries on either side of it, is termed a triad. Occasionally, two instead of three plates are associated, and such pairs are termed diads. A more advanced example of compounding is seen in *Astropyga* (see Fig. 259,*5*), where the two secondaries have fused to the intervening primary, and the compound plate consequently carries three successive pore pairs arranged in a short arc. Such plates are termed trigeminate. Numerous other examples are given in illustrations accompanying this chapter, and in all cases the primary is the middle element.

Sometimes the compounding may proceed further, leading to polyporous amb plates. The exact manner in which this occurs varies. An example of a compound polyporous plate derived from the fusion of three trigeminate plates is seen in *Pelanechinus* (see Fig. 256,*2b*). This also illustrates a further process, by which some of the secondary plates become restricted to the outer part of the compound plate; that is, they fail to make contact with the radial mid-line and suture. Such secondaries are termed demiplates. Demiplates are also seen in trigeminate plates (e.g., *Phormosoma,* Fig. 257,*2*), where a large primary is flanked by a demiplate on either side. This latter pattern is sometimes termed arbacioid (for it also occurs in the order Arbacioida), but it is obviously only a variety of diadematoid pattern.

Complex patterns derived from diadematoid or arbacioid compounding are illustrated in the Echinothurioida (see Fig. 257), where the demiplates may be excluded from both the adradial and perradial margins of the amb, and may be very reduced in size. Such occluded demiplates may give a multicolumnar or pluriserial aspect to the amb in which they occur, but study of the patterns seen in the Echinothurioida makes it clear that the resemblance to pluriserial Paleozoic genera is superficial, and that the fundamental pattern is diadematoid. Hence the view advanced by Mortensen (136b), which related the Echinothurioida to the pluriserial perischoechinoids of the Paleozoic, does not bear close examination, and has been rejected by Durham & Melville (52). There remain, however, more puzzling Echinothurioida (e.g., *Sperosoma*; see Fig. 257,*7*) in which the complexity of amb structure invites closer comparison with that of Paleozoic Echinocystitoida than with other Echinothurioida, and perhaps the notion of affinity between the Echinothurioida and the pluriserial Paleozoic echinoids cannot altogether be ruled out.

INTERAMBULACRA

The interambs present relatively little complexity, the main features which vary being the number of primary tubercles developed on each plate, and the number of vertical series in which the tubercles are arranged. It may be presumed that the more primitive forms would inherit a single primary tubercle on each interamb plate, as seen in cidaroids, and that the development of enlarged secondary tubercles (or additional primaries, since the two categories are not separable) is a secondary feature.

As in the Cidaroida, the primary tubercles can be either crenulate or noncrenulate. In all Diadematacea the tubercles are perforate.

The primary spines of Diadematacea are nearly always hollow, owing to the incomplete development of the medulla; this feature recalls the Archaeocidaridae. A notable exception is the order Pedinoida, where the spines are solid (at least in those cases where spines are known). The family Aspidodiadematidae is unique in having septate medullary structure; hence the central lumen of spines is divided by transverse sheets of stereom into a series of loculi. In some Echinothurioida the medulla may be almost entirely filled in, producing a secondarily solid spine; an example is *Araeosoma thetidis* (see Fig. 256,*1*). The external form of the primary spines may be significant in classification. Among Echinothurioida, for example, the subfamily Echinothuriinae is characterized by primary spines on the oral surface with terminal hooves (see Fig. 256,*1c,e*). Many of the diadematoid genera carry primary and secondary spines which have spinous processes arranged in spirals.

SPHERIDIA AND PEDICELLARIAE

Minute organs of balance termed spheridia, and pedicellariae of various types, occur on the amb plates in Diadematacea. In general their paleontological significance is slight, and they are seldom found on fossils, though the pits in which spheridia lie are sometimes observable; for example, in *Plesiodiadema* (see Fig. 261,*1b*) a spheridian pit occurs on the lower edge of the lowest plate of each triad, and similar pits are seen in the corresponding position in *Aspidodiadema* (see Fig. 261,*2*).

LANTERN

In the lantern, the structure is essentially as in Cidaroida, the teeth carrying no keel, and the foramen magnum being open above; this condition is described as aulodont. The perignathic girdle is also developed, the auricles comparable with those of cidaroids, though secondarily reduced in the post-Mesozoic Echinothurioida. This latter feature points to a cidaroid ancestry rather than to derivation from Echinocystitoida (where auricles are unknown).

CLASSIFICATION AND EVOLUTIONARY TRENDS

In the light of the foregoing review, the main groups of Diadematacea may be distinguished as follows: One stock, the order Echinothurioida, retains a peristome of cidaroid type, on which uncompounded poriferous amb plates form radial series (separated by interradial series), the test amb plates being variously compounded on a diadematoid plan, the test flexible, the spines hollow, and tubercles noncrenulate. The group has apparently given rise to no derivatives.

Another ordinal assemblage, the Diadematoida, shares diadematoid amb-compounding tendencies and hollow spines, but differs in having five pairs of oral plates on the peristome. Two families of Diadematoida have noncrenulate tubercles and a more or less flexible test; these are the Lissodiadematidae and Micropygidae, both deep-water groups. Two other families, Diadematidae and Aspidodiadematidae, have crenulate tubercles, and a more or less imbricate test (flexible only in the Diadematidae); they are distinguished mainly by their spines which are septate in the Aspidodiadematidae, nonseptate in the Diadematidae.

A third ordinal stock, the Pedinoida, differs from the preceding groups in having solid spines and a rigid test; the tubercles are noncrenulate. The diadematoid structure of the ambs, and the supraordinal characters point to a common origin with the other orders.

The remaining order, Pygasteroida, comprises a small group of genera with an exocyclic periproct, resembling the Pedinoida in other features (the nature of the dentition being uncertain, however); these are inferred to have arisen from some pedinoid ancestry (52). Both the Pedinoida and Pygasteroida are uniform assemblages, each comprising a single family. With exception of the Echinothurioida, the structure of the apex in Diadematacea points to an incipient anteroposterior axis along which bilateral symmetry becomes increasingly evident in the later families, Lissodiadematidae, Diadematidae, and in the Pygasteroida.

From the morphological features reviewed above it appears that origins of the Diadematacea probably lie among some Paleozoic cidaroid stock, and that the assemblage is more uniform than Mortensen (136b) believed. Durham & Melville (52) drew attention to the fact that the oldest known echinothurioid *(Pelanechinus)* more closely resembled the Pedinoida than do extant Echinothurioida, and the amb structure of *Pelanechinus* is clearly diadematoid (see Fig. 256,*2*). It is therefore reasonable to infer a common ancestry for the orders Diadematoida, Pedinoida, and Echinothurioida. The order Pygasteroida appears to be a derivative of the Pedinoida.

ECOLOGY

Except for the Diadematidae, and the genus *Asthenosoma* among the Echinothurioida, the Diadematacea are predominantly deep-water forms. This fact, combined with the more or less imbricate and consequently fragile test structure, makes them rather rare as fossils. Most extant forms appear to be scavengers rather than predators, and the deep-water genera feed largely on detrital material, including leaves and wood of terrestrial plants, swept out to sea by rivers, subsequently sinking to the bottom. The Diadematidae are essentially littoral and sublittoral forms in tropical and subtropical seas, some genera (e.g., *Diadema, Echinothrix*) inhabiting coral reefs. These have very long, brittle, hollow spines able to puncture the skin of predators and when broken off, producing dangerous wounds on account of toxins in their substance. Various fishes and crustaceans shelter between the longer spines, secure from attack by larger animals. Parasites include gastropods and a crab *(Eumedon)* which enters and inhabits the rectum of *Echinothrix.* The reef-dwelling Diadematidae feed on tubicolous and encrusting organisms attached to the coral or substrate. Some Diadematidae have light-sensitive areas on the skin of the aboral and oral surfaces of the test, sometimes conspicuous on account of

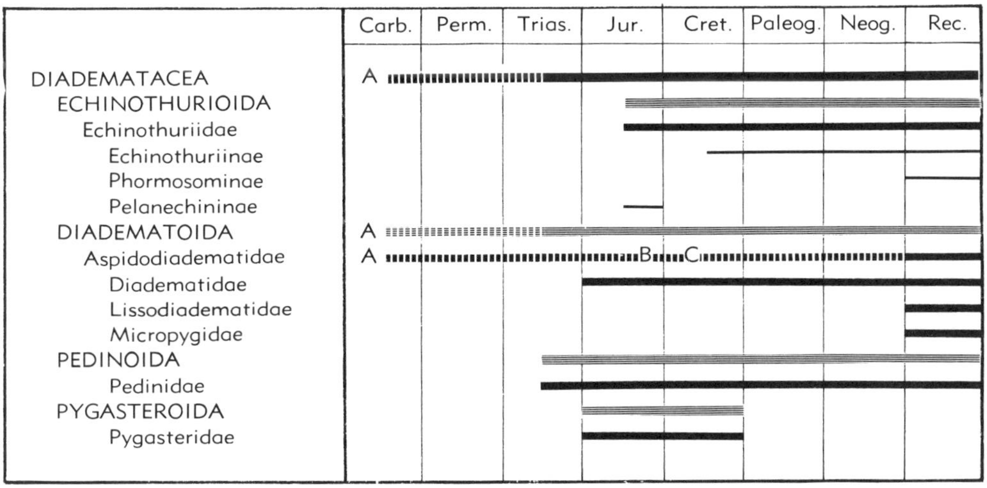

FIG. 255. Stratigraphic distribution of Diadematacea. [*A*, genus indet.; *B*, *Eosalenia* only; *C*, *Tiaridia* only; aspidodiadematid affinity of *A* possible, of *B* and *C* probable.]

the blue pigment which they contain. They are recorded as capable of detecting the shadows of approaching objects, toward which the defensive spines can be directed. The dermis which invests the whole test and spines is usually a blackish or dark purplish color.

The Echinothurioida are also venomous, but for a different reason, namely the development of poison glands on the spines, especially secondary spines. These may terminate in an acuminate tip, which is invested by venom glands (see Fig. 257,*6*). The most dangerous species are those of the shallow-water genus *Asthenosoma,* which is often entangled in fishermen's nets in the tropical and subtropical Indo-Pacific; cases of fatal injury to man are recorded from Japan (GISLÉN, 1933). *Asthenosoma,* unlike the deep-water genera of Echinothurioida, feeds on organic detritus available in littoral situations, and MORTENSEN (136b) observed intertidal specimens in Indonesia feeding selectively upon feces derived from village sewage.

Little is known of the ecology of Pedinoida. The extant species are exclusively archibenthal, and it is inferred that the fossils probably frequented similar environments on the continental slope, down to 2,000 m. The living forms apparently feed on bottom detritus and organisms such as foraminifers. The fossil genus *Pedina* was occasionally gregarious (e.g., *P. sublaevis*); species of the surviving genus *Caenopedina* appear to be rare and solitary.

Of the Pygasteroida nothing is known as to their ecology, but from the general similarity of the test structure to that of the surviving nucleolitid *Apatopygus,* it may be inferred that they probably rested upon soft bottom (shell grit or sand), and fed upon detritus.

STRATIGRAPHICAL DISTRIBUTION

The recorded time ranges of the families and subfamilies are indicated in Figure 255. Owing to the fragility of the test, it is impracticable to represent the peaking of genera by variation in thickness of the lines, for sampling on the basis of fossils is by no means comparable with that on the basis of extant forms, and such comparison would give misleading emphasis to Recent genera.

Superorder DIADEMATACEA Duncan, 1889

[=Aulodonta *sensu* MORTENSEN, 1940, *plus* Echinothuriidae THOMSON, 1872, *plus* Pygasteridae DUNCAN, 1900]

Test rigid or flexible, plates united by sutures or by membranous interstices, or imbricating beveled margins; without conspicuous bilateral symmetry. Primary tubercles perforate. Periproct endocyclic or exocyclic. Perignathic girdle complete in adult. Lantern well developed at all stages, aulodont. Gills and gill slits normally present in adult (exceptionally lacking, as secondary feature). Amb plates simple or (more usually) compounded in diadematoid or arbacioid groups, or in more complex arrangements derived from diadematoid pattern. *?L.Carb., U.Trias.-Rec.*

Order ECHINOTHURIOIDA Claus, 1880

Test low hemispherical to rotular, flexible, with imbricating plates or interstitial membranous junctions. Ambs and interambs extending in bicolumnar series across peristome. Peristomial amb plates simple, amb plates of test compounded in diadematoid (or derived) patterns, in some simulating pluriserial columns. Periproct endocyclic. Apical system dicyclic in young stages, becoming monocyclic in adult; five genital pores. Tubercles noncrenulate. Spines striate, usually hollow, exceptionally with secondary medullary infilling. Gills and gill slits inconspicuous, or lost in adult. Spheridia present aborally and adorally, located in pits beside pore pairs. Pedicellariae present, including tridentate, tridactylous,, triphyllous, and ophicephalous types. *U.Jur.-Rec.*

Family ECHINOTHURIIDAE Thomson, 1872

Characters of order. *U.Jur.(Oxford.)-Rec.*

Subfamily ECHINOTHURIINAE Thomson, 1872

[=Asthenosominae MORTENSEN, 1934]

Ambs trigeminate. Adoral primary spines with terminal hoof. Teeth with rounded tip. *U.Cret.(Senon.)-Rec.*

Echinothuria WOODWARD, 1863, p. 327 [**E. floris*; OD]. Test large (to 100 mm. diameter). Ambs and interambs of equal width, almost devoid of tubercles. Amb plates comprising enlarged primaries, with alternating demiplates located in pairs about midway across horizontal suture; primaries with or without primary tubercle, and few scattered secondary tubercles. No membranous interstices between succeeding test plates. Apical system, oral surface and spines unknown. *U.Cret. (Senon.)*, Eng.——FIG. 256,*3*. **E. floris*; *3a*, test aboral, ×1; *3b*, amb aboral, ×4 (173).

Araeosoma MORTENSEN, 1903, p. 53 [**Calveria fenestratum* THOMSON, 1872, p. 494; OD]. Test large, depressed. Conspicuous membranous interstices between plates, especially of aboral surface. Primary amb plates entire, much larger than demiplates. Pores on oral surface arranged in three series on either side of the interporiferous area. *?U.Cret.(Senon.)*, Eu.(Denm.-Fr.); *Plio.*, N.Z.; *Rec.*, IndoPac.-Atl., shelf-archibenthal. —— FIG. 256,*1a,b*. *A.* sp. aff. *A. thetidis* (H. L. CLARK), Plio., N.Z.; *1a*, transv. sec. of spine, ×85; *1b*, fragments of associated adapical test plates, ×7 (here first recorded).——FIG. 256,*1c*. *A. violaceum* MORTENSEN, Rec., Atl.; adoral spine shaft and hoof, ×8.5 (136b).——FIG. 256,*1d*. *A.? bruennichi* RAVN, U.Cret., Denm.; hoof, ×8.5 (147).——FIG. 256,*1e*, *A.? mortenseni* RAVN, U. Cret., Denm.; adoral spine shaft and hoof, ×8.5 (147).——FIG. 257,*6*. *A. thetidis* (H. L. CLARK), Rec., N.Atl. (360 m.); secondary spine with venom sac, ×29 (136b).

Asthenosoma GRUBE, 1868, p. 42 [**A. varium*; OD] [=*Cyanosoma* SARASIN, 1886 (type, *C. urens*, =*A. varium*)]. Like *Araeosoma* but pores arranged in 3 dense series on both adoral and aboral surfaces of test, and more slender and elongate hoof on adoral spines. Aboral spines completely invested by dermis, secondary spines with venom glands strongly developed, capable of inflicting lethal stings. *U.Cret.*, Eu.; *Rec.*, IndoPac., littoral and sublittoral.——FIG. 257,*8a,b*. **A. varium*, Ind.O.; *8a,b*, amb (aboral), juv. and adult, ×2.8 (136b).——FIG. 257,*8c*. *A. striatissimum* RAVN, *U.Cret.*, Denm., hoof, ×8.5 (147).

Calveriosoma MORTENSEN, 1934, p. 163 [**Calveria hystrix* THOMSON, 1872; OD] [=*Calveria* THOMSON, 1872 (preocc.)]. Like *Araeosoma* but membranous interstices between test plates conspicuous only on oral side. *Rec.*, N.Atl.-N.Pac. (160-1,800 m.).——FIG. 257,*4*. **C. hystrix* (THOMSON), N.Atl.; amb plates, ×4.3 (136b).

Hapalosoma MORTENSEN, 1903, p. 64 [**Asthenosoma pellucidum* A. AGASSIZ, 1879; OD]. Like *Araeosoma* but lacking primary tubercles from admedian regions of amb and interamb plates. *Rec.*, Indon.-Japan (shelf-archibenthal). —— FIG. 257,*3*. **H. pellucidum* (AGASSIZ), Indon.(Kei Is.); amb, adapical, ×5.7 (3).

Hygrosoma MORTENSEN, 1903, p. 59 [**Phormosoma petersii* A. AGASSIZ, 1880; OD]. Pores on oral surface arranged in single series placed near abradial margin of amb. Aboral demiplates small, wholly enclosed within primary plates, which alone reach abradial margin of amb. *Rec.*, IndoPac.-Atl. (200-3,000 m.).

Sperosoma KOEHLER, 1897, p. 304 [**Sperosoma grimaldii*; OD]. Test large. Primary amb plates divided into 2 parts by longitudinal (adradial) fissure, adradial portion carrying pore pair, adradial nonporiferous area. *Rec.*, IndoPac.-Atl. (300-2,300 m.).——FIG. 257,*7*. **S. grimaldii*, E.Atl.; amb adoral, ×4.3 (136b).

Tromikosoma MORTENSEN, 1903, p. 62 [**T. koeh-*

leri; OD] [=*Echinosoma* POMEL, 1883, p. 108 (*non* AUDINET-SERVILLE, 1839; *nec* SEMPER, 1868; *nec* WOLLASTON, 1854)]. Pores on oral surface arranged in single series placed near adradial margin of amb. Demiplates on aboral surface forming transverse pairs, outer member reaching ab-

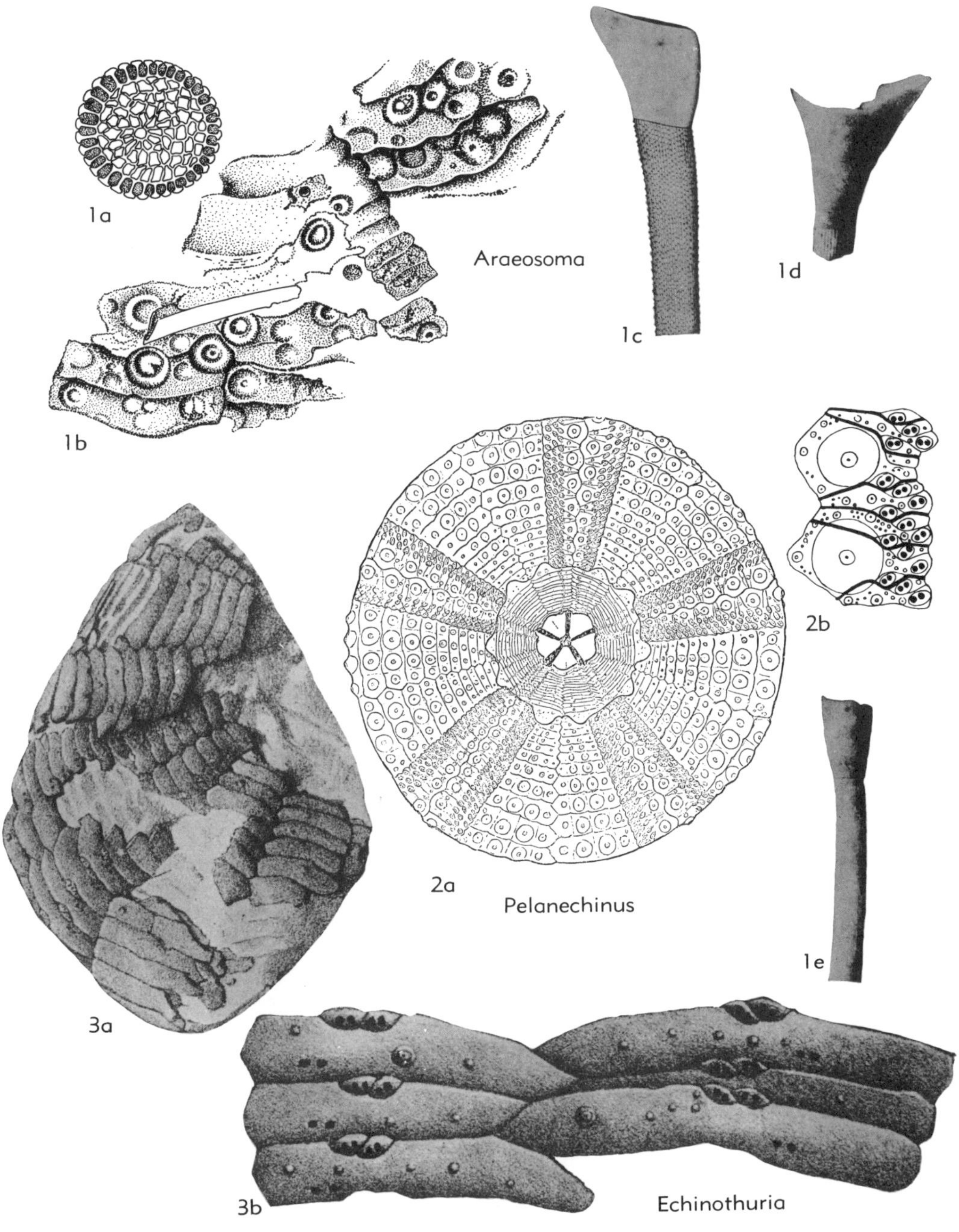

FIG. 256. Echinothuriidae (Echinothuriinae) *(1,3)*, (Pelanechininae) *(2)* (p. *U*346, *U*349-*U*350).

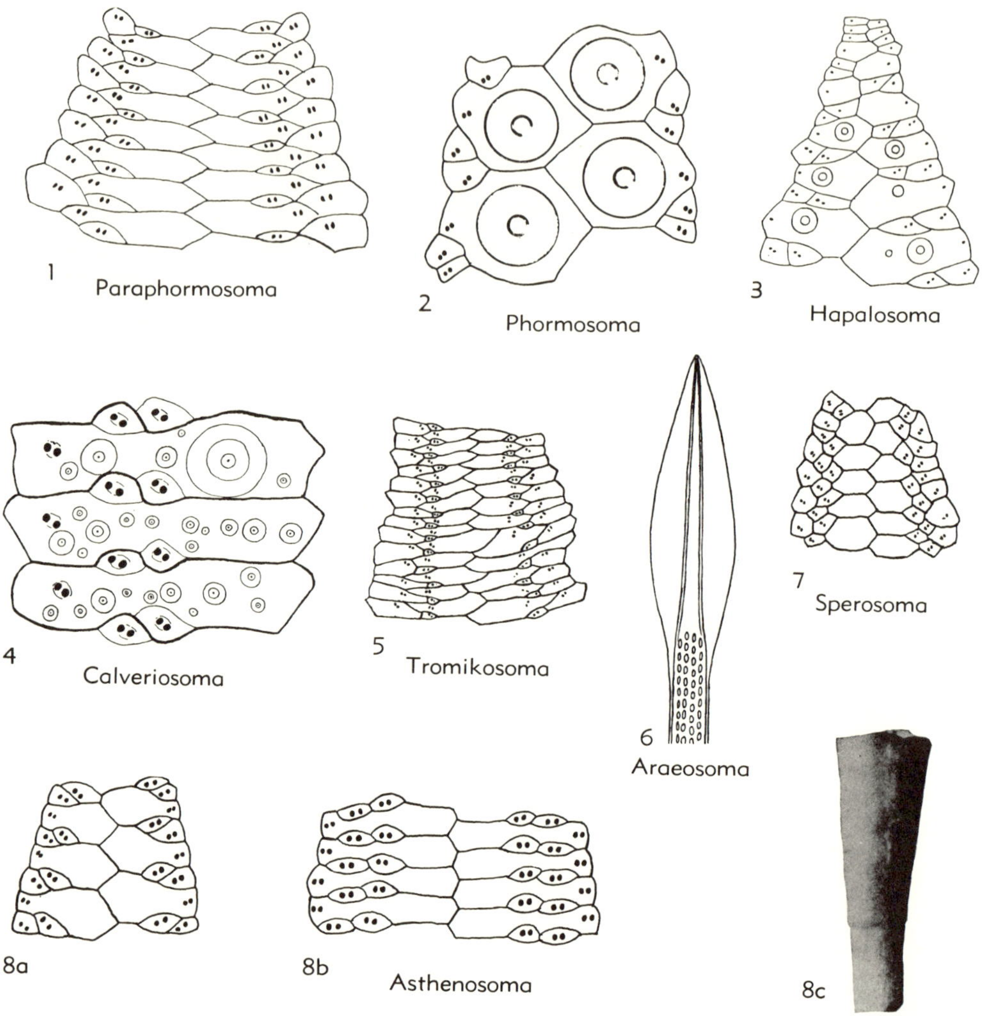

FIG. 257. Echinothuriidae (Echinothuriinae) *(3-8)*, (Phormosomatinae) *(1-2)* (p. *U*346-*U*348).

radial margin of amb. *Rec.,* IndoPac.-Atl. (deep water only, 850-3,500 m.).——FIG. 257,*5*. *T. panamense* A. AGASSIZ, E.Pac. (2,400-3,300 m.); amb, ×2.9 (3).

Subfamily PHORMOSOMATINAE Mortensen, 1934

[*nom. correct.* FELL, herein (*pro* Phormosominae MORTENSEN, 1934, p. 162)]

Ambs polyporous, plates triple-comclavate, without terminal hoof. Teeth acuminate. *Rec.*

Phormosoma THOMSON, 1872, p. 493 [**P. placenta*; OD]. Areoles of oral surface large and deep, with raised margin, giving honeycomb appearance to lower side of test. *Rec.,* IndoPac.-Atl. (shelf-abyssal).——FIG. 257,*2*. **P. bursarium* A. AGASSIZ, IndoPac.; amb adoral, ×2.9 (136).

Hemiphormosoma MORTENSEN, 1934, p. 162 [**H. paucispinum*; OD]. Distal (ambital) plates alone having enlarged areoles. Primary tubercles not forming regular series. *Rec.,* Indon. (4,000 m.).

Paraphormosoma MORTENSEN, 1934, p. 162 [**Phormosoma alternans* DE MEIJERE, 1902; OD]. Areoles not enlarged. Primary tubercles arranged in regular series from aboral to oral surface. *Rec.,* Indon.(archibenth.).——FIG. 257,*1*. **P. alternans*; amb, ×3.6 (136b).

Subfamily PELANECHININAE Groom, 1887

[*nom. transl.* GREGORY, 1900 (*ex* Pelanechinidae GROOM, 1887)]

Ambs polyporous, plates triple-compounded from successive diadematoid triads. *U.Jur.*, Eu.

Pelanechinus KEEPING, 1878, p. 924 [**Hemipedina corallina* WRIGHT, 1856; OD, M]. Test to 100 mm. diameter. Amb plates bearing one enlarged primary tubercle, its areole formed (more or less) by 3 primary amb elements, with which are associated 6 other demiplates; pores of each compound amb plate arranged in 3 arcs of 3. Up to 4 primary tubercles on ambital interamb plates. Conspicuous gill clefts. Plates of peristome in regular series, ambs here with pores in double series. Pedicellariae tridentate (coarse and slender), possibly also ophicephalous. *U.Jur.(Oxford.)*, Eng.——FIG. 256,2. **P. corallinus*; *2a*, test adoral, ×0.57; *2b*, amb plates, ×2.9 (77).

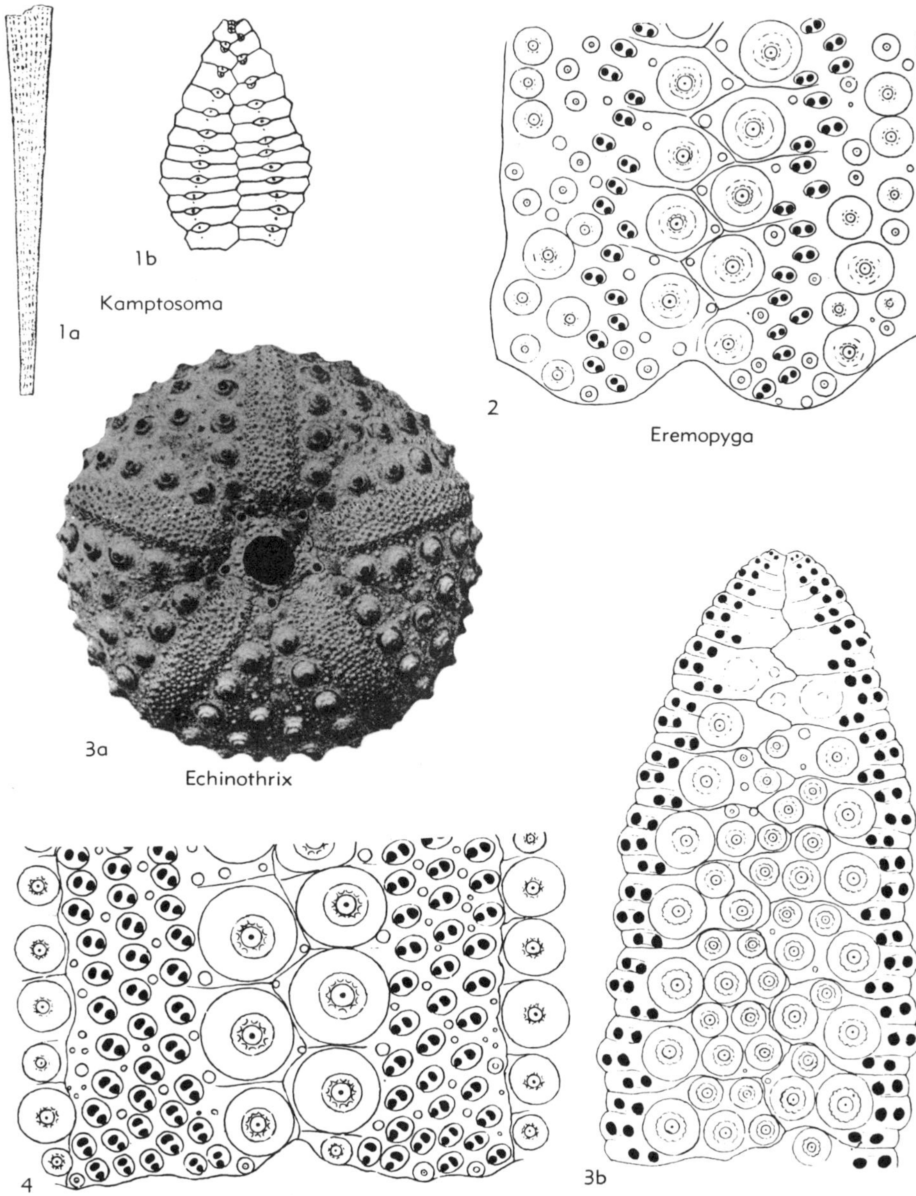

FIG. 258. Diadematidae (*U*350-*U*352).

Order DIADEMATOIDA Duncan, 1889

[=Diademaria HAECKEL, 1896; Diademida DELAGE & HÉROUARD, 1903; Centrechinoida JACKSON, 1912)]

Test subspherical, depressed hemispherical or pentagonal; rigid or flexible, plates usually imbricated internally. Ambs and interambs not extending across peristome, ambs composed of simple or compound diadematoid plates. Five pairs of oral plates on peristomial membrane. Gills present, gill slits conspicuously notching peristomial margin. Periproct endocyclic. Apical system monocyclic, or with anterior oculars (II, III, IV) exsert; 5 genital pores. Tubercles crenulate or noncrenulate. Spines hollow, cylindrical, typically verticillate. Spheridium on each compound amb plate (lacking in some adapically). Pedicellariae present, including tridentate, triphyllous, ophicephalous, and rarely globiferous types. *?L.Carb., U.Trias.-Rec.*

Family DIADEMATIDAE Gray, 1855

[*nom. correct.* ZITTEL, 1879, *pro* Diademadae GRAY, 1855] [=Centrechinidae JACKSON, 1912]

Test of moderate to large size, usually somewhat flattened, commonly rather flexible. Primary tubercles crenulate. Primary and secondary spines usually hollow and verticillate, not divided internally by transverse septa. *L.Jur.-Rec.*

Diadema GRAY, 1825, p. 246 [**Echinometra setosa* LESKE, 1778; ICZN Op. 206, 1954] [=*Centrechinus* JACKSON, 1912 (obj.)]. Test large (to 110 m.), subhemispherical, depressed, not flexible. Primary amb tubercles conspicuous in 2 regular series. Ambulacral spines not conspicuously unlike others; primary spines of oral surface not clavate, not expanded distally, or invested by dermis. No globiferous pedicellariae; no spines on buccal plates. Oculars I, IV, and V insert. *?U.Cret.*, Eu., *Rec.*, IndoPac.-Atl. —— FIG. 258,*4*. **D. setosum* (LESKE), IndoW.Pac., littoral; amb adoral, ×6.2 (136c).——FIG. 259,*4*. *D.? ebroicense* CAFFIN, U.Cret.(Cenon.), Fr.; *4a,b,* spines, ×2.2 (17).

Astropyga GRAY, 1825, p. 426 [**Cidaris radiata* LESKE, 1778; OD]. Like *Eremopyga,* but porezones more or less widened adorally, aboral surface of test not conspicuously naked. Spines with central lumen filled by loose mesh of stereom. *Rec.*, IndoPac.-Carib.——FIG. 259,*5*. *A. pulvinata* (LAMARCK), Gulf Panama; amb aboral, ×5 (136c).

Centrostephanus PETERS, 1855, p. 109 [**Diadema longispina* PHILIPPI, 1845; OD] [=*Thrichodiadema* A. AGASSIZ, 1863 (type, *T. rodgersii,* p. 354); *Echinodiadema* VERRILL, 1867 (*non* COTTEAU, 1869) (type, *E. coronata,* p. 580)]. Like *Diadema* but retaining embryonic adoral amb plate in adult unresorbed, with pore pair intact; globiferous pedicellariae occur, and spines on buccal plates. *?Mio., Plio.,* S.Eu., *Rec.*, IndoPac.-Atl. ——FIG. 259,*2*. *C. rubricingulus* H. L. CLARK, Carib.; amb adoral, ×10 (136c).

Chaetodiadema MORTENSEN, 1903, p. 1 [**C. granulatum*; OD]. Test rounded, depressed; somewhat flexible. Amb plates imbricated adorally, interamb plates imbricated aborally (observable only internally, where directions appear reversed). Pore pairs on adoral side arranged in single series. Tubercles greatly reduced adorally. *Rec.*, IndoW. Pac.——FIG. 259,*1*. **C. granulatum*; *1a,b,* amb, adoral, aboral, ×3.6 (136c).

Echinothrix PETERS, 1853, p. 484 [**Echinus calamaris* PALLAS, 1774; SD MORTENSEN, 1940, p. 283] [=*Garelia* GRAY, 1855 (type, *G. aequalis,* p. 38, =*E. calamaris* PALLAS); *Savignya* DESOR, 1855 (obj.)]. Like *Diadema,* but ambs widened adapically, primary amb tubercles inconspicuous, not forming 2 regular series. Aboral ambulacral spines small, setiform, barbed distally. *Rec.*, Indo-W.Pac.——FIG. 258,*3*. *E. diadema* LINNÉ; *3a,* test (holotype in collection of Queen Louisa Ulrika), aboral, ×0.75 (129); *3b,* amb, aboral, ×5.5 (136c).

Eodiadema DUNCAN, 1889, p. 339 [**E. granulatum* WILSON, 1889, pl. 10, fig. 5; OD]. Test small to medium. Ambs composed of simple plates, pores arranged in single straight series, except near peristome where they are arranged in triads and form incipient compound trigeminate plates of diademoid type. Amb primary tubercles perforate, one to each plate, except adorally, where only middle plate of each triad carries tubercle. Oculars I and V insert. Spines long and slender, verticillate (not known in type-species). *L.Jur.*, Eu.——FIG. 260,*1*. **E. granulatum,* Eng.; *1a,* test, lat., ×0.75; *1b,* amb and interamb, ×6; *1c,* peristome, ×2.25; *1d,* apical system, ×3 (170).

Eremopyga AGASSIZ & CLARK, 1908, p. 110 [**Astropyga denudata* DE MEIJERE, 1902; OD]. Like *Chaetodiadema,* but pore zones adorally arranged in triads, aboral plates almost naked. Spines hollow. *Rec.*, China-Indon.——FIG. 258,*2*. **E. denudata* (DE MEIJERE), Malaya (70-275 m.); amb adoral, ×5.6 (136c).

Goniodiadema MORTENSEN, 1939, p. 549 [**G. mauritiense*; OD]. Test large (horiz. diam. to 110 mm.), depressed though not flattened, somewhat flexible; ambitus pentagonal, rounded angles formed by interambs. Pore zones uniserial adorally. *Rec.*, Maurit.(archibenthal).

Kamptosoma MORTENSEN, 1903, p. 60 [**Phormosoma asterias* A. AGASSIZ, 1881; OD] [=*Cuenotia* LAMBERT & THIÉRY, 1914 (obj.)]. Amb plates of diademoid triads only at extreme adapical region, elsewhere forming diads, with alternate plates occluded as small demiplates. Ambital primary spines flattened and expanded distally, others cylindrical, hollow, more or less thorny. [The affinities of this genus are obscure, and hitherto it has been classified as an aberrant echinothurioid; the crenulate tubercles suggest, however, that it is a

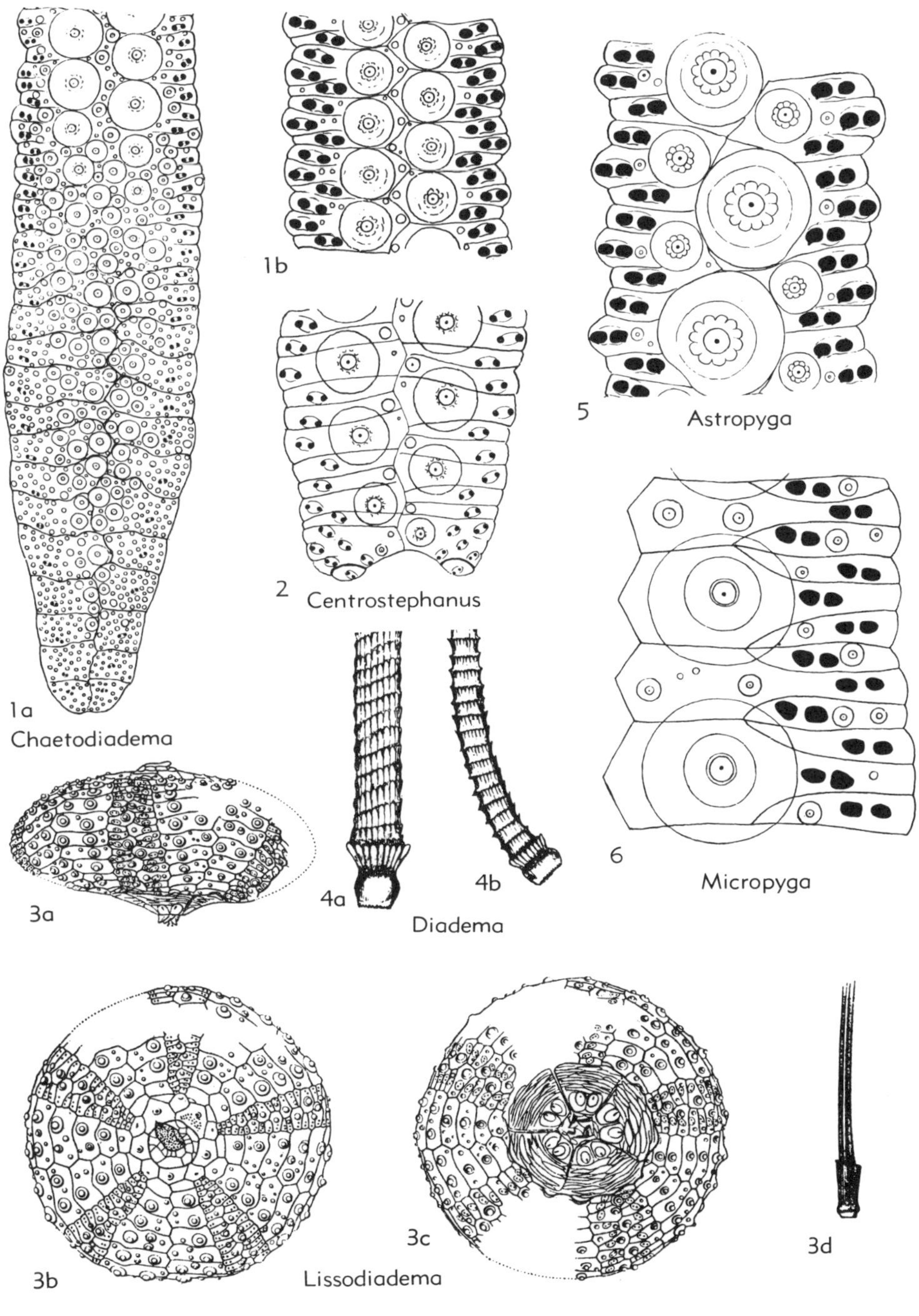

FIG. 259. Diadematidae *(1-2,4-5)*; Lissodiadematidae *(3)*; Micropygidae *(6)* (p. *U*350, *U*352).

diadematid adapted for abyssal life.] *Rec.,* C.Pac.-SE.Pac.(abyssal, 4,000-5,000 m.).——FIG. 258,*1*. **K. asterias* (A. AGASSIZ), SE.Pac.; *1a,* ambital radiole, ×19; *1b,* amb adapical region with ambital and part of subambital region, ×1.5 (3).

Kierechinus PHILIP, 1963, p. 1104 [**Pedinopsis melo* KIER, 1957, p. 845; OD]. Like *Pedinothuria,* but with about 12 primary tubercles on each interamb plate. Radioles imperfectly verticillate. *L.Eoc.,* Somalia (Auradu), N.Afr.

Palaeodiadema POMEL, 1887, p. 318 [**Pseudodiadema fragile* WILTSHIRE; OD] [=*Helikodiadema* GREGORY, 1896 (obj.) *(nom. van.)*]. Test fragile, of moderate size, low, flattened above, slightly rounded below, plates apparently somewhat imbricated. Amb plates simple aborally, trigeminate adorally, where pores are arranged in arcs of 3. Primary tubercles of ambs and interambs in distinct longitudinal series, secondary tubercles of interambs few, and not distinctly arranged in series; plates otherwise rather naked. Apical system large, probably monocyclic. Peristome large, gill slits indistinct. Radioles slender, verticillate. *Cret.(Cenoman.-Senon.),* Eu.-N.Afr.——FIG. 260, *3*. **P. fragile* (WILTSHIRE), Senon., Eng.; *3a,* amb adoral, ×7.5 (74); *3b,* spine, ×4.5 (173); *3c,d,* test, aboral, oral, ×1.9 (173).

Pedinothuria GREGORY, 1897, p. 119 [**P. cidaroides*; OD]. Test small (*ca.* 12 mm. horizontal diameter), depressed, flattened above and below. Amb pores uniserial aborally, becoming biserial at ambitus, and assuming trigeminate arcs of 3 adorally; occluded demiplate in each triad below ambitus. Primary tubercles small on ambs, occurring on alternate, or every 3rd plate. Primary tubercles large on interambs, areole occupying most of plate. Few secondary tubercles. Apical system large. Peristome smaller than apical system, with very deep gill slits. Spines unknown. *M.Jur.(Bathon.), ?U.Jur.(Sequan.),* Eu.(Ger.-Fr.). ——FIG. 260,*2*. **P. cidaroides,* Ger.; *2a,b,* test (holotype), lat., oral, ×3.4; *2c-e,* amb plates, aboral, ambital, adoral, all ×19 (74).

Family LISSODIADEMATIDAE Fell, n. fam.

Primary tubercles noncrenulate. Primary spines hollow, not verticillate. Outer tube feet not specialized. *Rec.*

Lissodiadema MORTENSEN, 1903, p. 393 [**L. lorioli*; OD] [=*Leptodiadema* AGASSIZ & CLARK, 1907 (type, *L. purpureum*]. Test small (25 mm. diam.), flattened above and below, delicate, somewhat flexible by imbrication of alternate interamb plates across interradius. Amb plates trigeminate, pores forming straight series. Spines smooth, slightly curved basally. *Rec.,* Indon.-Hawaii (littoral).——FIG. 259,*3*. *L. purpureum* (AGASSIZ & CLARK), Hawaii; *3a-c,* test, lat., aboral, oral, ×4.3; *3d,* spine, ×4.3 (4).

Family MICROPYGIDAE Mortensen, 1904

Primary tubercles noncrenulate. Primary spines hollow, not verticillate. Pore pairs forming double series; outer series of tube feet converted into umbrella-shaped structures containing anchor-shaped spicules. Amb plate trigeminate, but arcs of pore pairs of successive compound plates reversed, so that successive primaries have pores displaced alternately inward or outward. *Rec.*

Micropyga A. AGASSIZ, 1879, p. 274 [**M. tuberculata*; OD] [=*Rotapedina* LAMBERT & THIÉRY, 1914 (obj.) *(nom. van.)*]. Test large (up to 140 mm. diam.), low hemispherical, flattened adorally, flexible. Primary tubercles in conspicuous longitudinal series. Gill clefts deep. Spines of moderate length, thorny. *Rec.,* IndoW.Pac.——FIG. 259,*6*. **M. tuberculata,* Indon., amb, ×5.8 (136c).

Family ASPIDODIADEMATIDAE Duncan, 1889

[=Aspidodiademinae DELAGE & HÉROUARD, 1903; Aspidodiademina MORTENSEN, 1939]

Test of small to moderate size (20 to 40 mm. diam.), fragile, ovoid or spherical, interamb plates slightly imbricating adorally, as seen from within. Apical system monocyclic. Primary tubercles crenulate. Primary spines hollow, central lumen transversely divided by delicate fenestrated plates which are connected by vertical calcareous strands, verticillate, ambital spines elongated and curved so that apex is directed downward. Apical system monocyclic. *?U.Jur., Rec.*

Aspidodiadema A. AGASSIZ, 1878, p. 188 [**A. tonsum*; OD]. Amb plates compounded in normal diadematoid triads, of which primary median components are much larger than upper and lower elements, and median primary tubercles resemble primary interamb tubercles in size and form; secondary components of triads without primary tubercles. *Rec.,* IndoPac.-Carib.(archibenthal).——FIG. 261,*2*. **A. tonsum,* Indon.; amb at ambitus, ×6.5 (136c).

Eosalenia LAMBERT, 1905, p. 311 [**E. miranda*; OD]. Test as *Aspidodiadema,* but adapical amb plates arranged in diads, with primary tubercle on every second plate (as in Saleniidae). Apical system large, but not otherwise known. Spines unknown. *U.Jur.(Bathon.),* Fr.——FIG. 261,*4*. **E. miranda; 4a-c,* test, lat., oral, aboral, ×1.1; *4d,* amb, ×5.7; *4e,* interamb plates, ×4.3 (115).

Plesiodiadema POMEL, 1883, p. 106 [**Aspidodiadema microtuberculatum* A. AGASSIZ, 1879; OD] [=*Dermatodiadema* A. AGASSIZ, 1898 *(nom.*

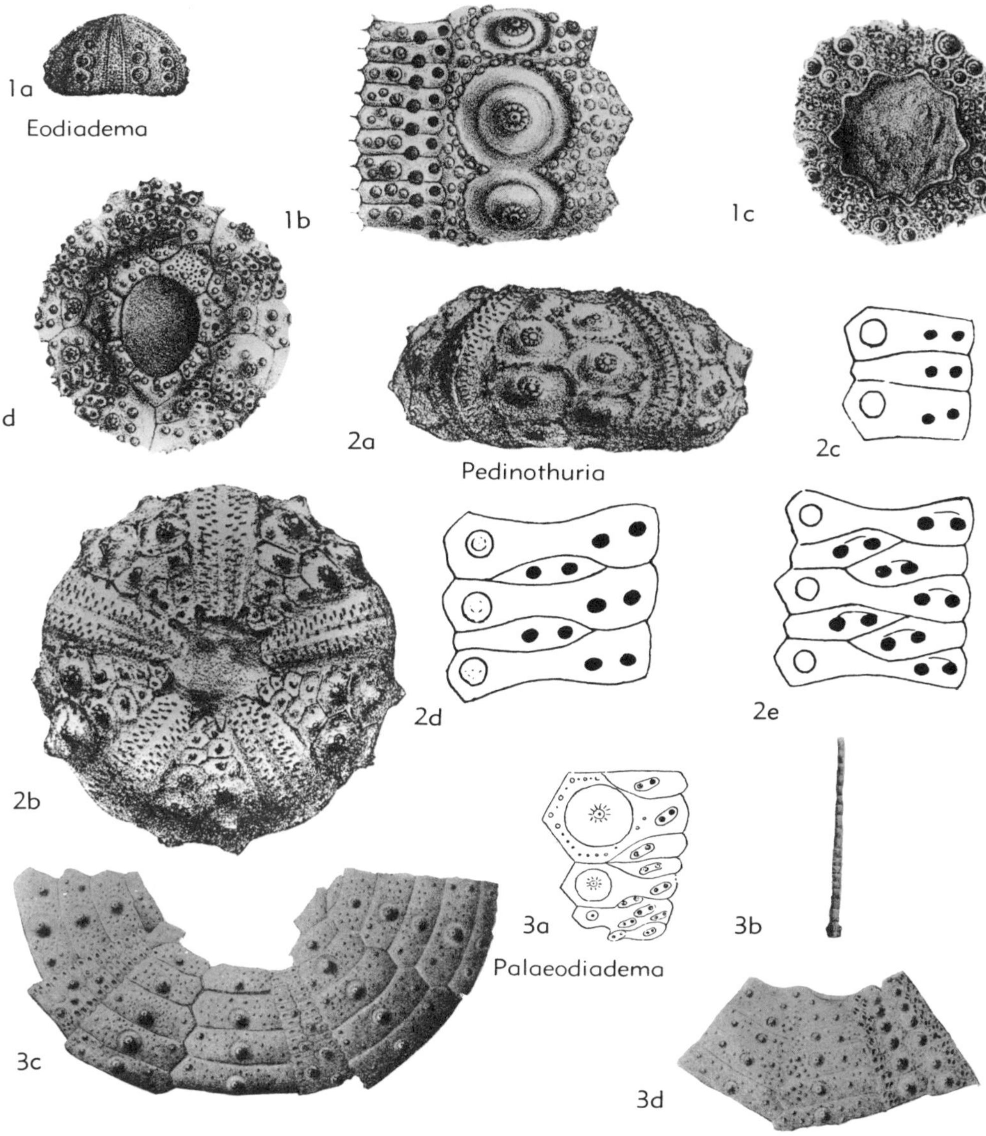

FIG. 260. Diadematidae (p. *U350, U352*).

van.)] [*non Plesiodiadema* DUNCAN, 1885, p. 433 (=*Polydiadema* LAMBERT, 1888)]. Like *Aspidodiadema,* but amb plates all (or nearly all) simple, arranged in triads, primary tubercle of each median plate not enlarged, other 2 plates of triad without primary tubercle. *Rec.,* IndoPac.-Atl. (archibenthal-abyssal, 300-3,900 m.).——FIG. 261, *1a. P. indicum* (DÖDERLEIN), Indon., 300-520 m.; long. sec. of primary spine, ×36 (136c).——FIG. 261,*1b. P. antillarum* (A. AGASSIZ), trop. Atl., 750-3,000 m., adoral amb, ×13 (136c).

Tiaridia POMEL, 1883, p. 97 [**Hemicidaris batnensis* COTTEAU; OD, M]. Test like *Aspidodiadema,* but adapical amb plates like *Plesiodiadema* (with primary tubercle on every third plate). Apical system as *Aspidodiadema,* but genitals extending outward into interambs. *U.Cret.(Cenoman.),* N. Afr.(Alg.).——FIG. 261,*3.* **T. batnensis* (COT-

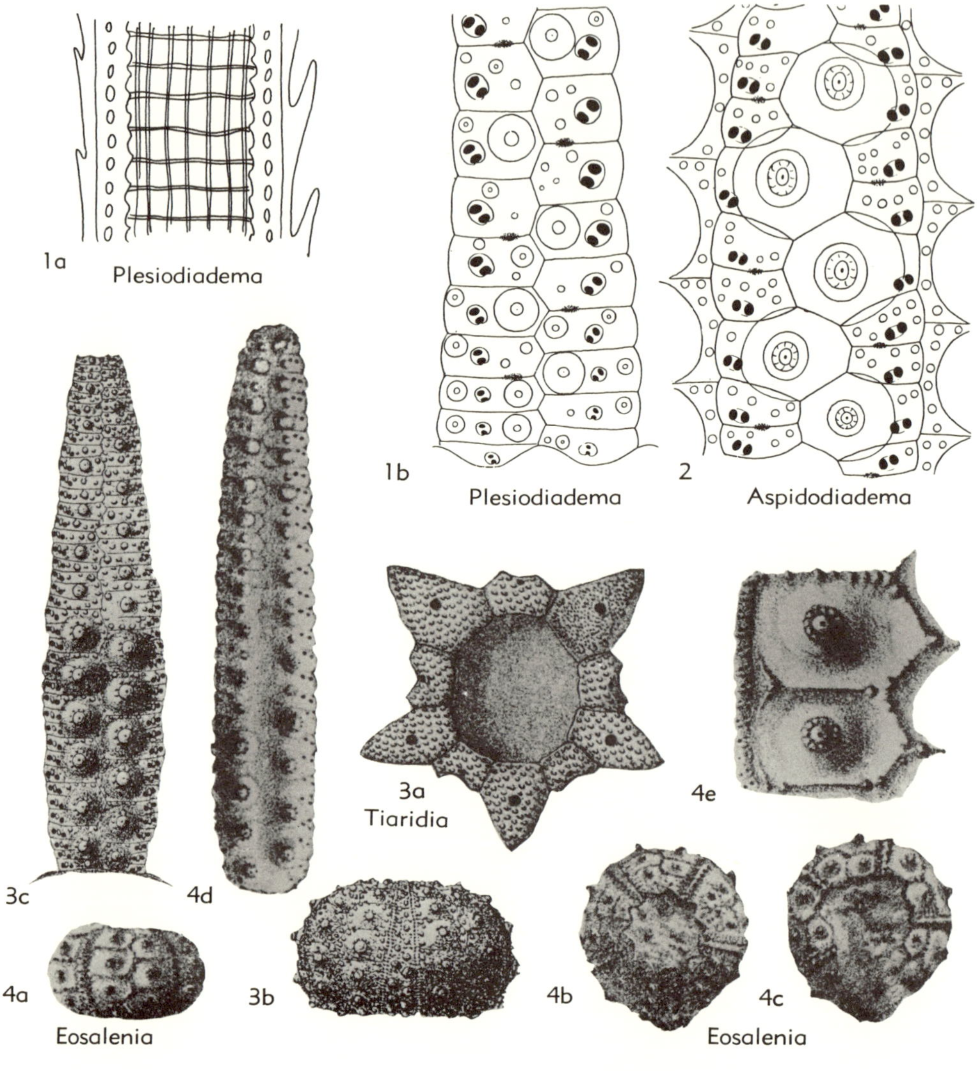

FIG. 261. Aspidodiadematidae (p. *U352-U354*).

TEAU); *3a*, apical system, ×4.3 (136c); *3b*, test, lat., ×1.1; *3c*, amb, ×5 (31).

Family UNCERTAIN

The following genera cannot at present be assigned to families, but are probably diadematoids.

Ancylocidaris MILLER, 1929, p. 334 [**A. spenceri*; OD]. Test low, small (9 mm. horiz. diam.). Amb plates simple, with oblique pore pair close to abradial edge. Interamb plates chevron-shaped, angle directed adorally. Other features unknown. *Jur.*, N.Am.

Endeodiadema DE LORIOL, 1890, p. 90 [**E. lepidum*; OD]. Like *Eodiadema* but amb plates all simple (not forming triads adorally) and pore pairs set obliquely. *M.Jur.(Callov.)*, Eu.(Port.).——FIG. 262,*5*. **E. lepidum*; *5a,b*, interamb, amb, ×7.8 (124).

Engelia TORNQUIST, 1908, p. 408 [**Cidaris laqueata* QUENSTEDT, 1875; OD]. Amb plates with pore zone situated some distance from abradial margin, which carries marginal series of small tubercles. *L.Jur.*, Eu.(Ger.).——FIG. 262,*3*. **E. laqueata* (QUENSTEDT); *3a*, amb (*fide* TORNQUIST), ×2.3 (164); *3b*, amb (*fide* QUENSTEDT), ×3.9

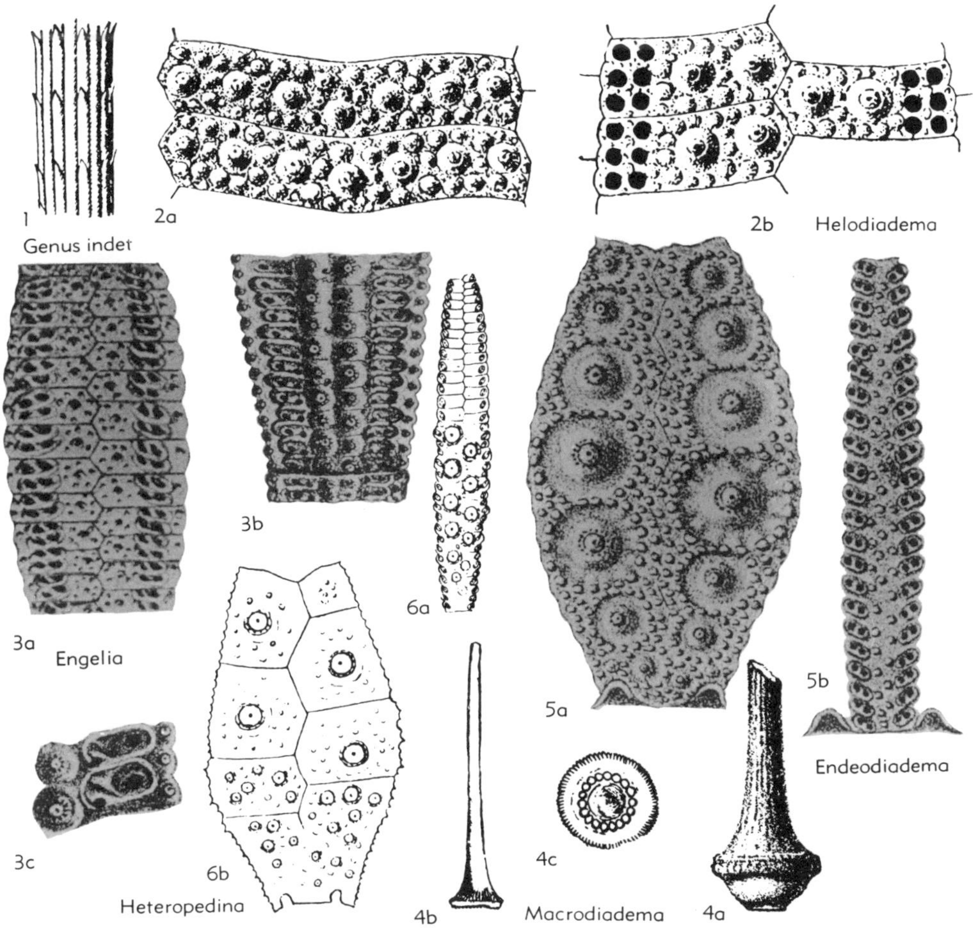

FIG. 262. Family Uncertain (p. *U*354-*U*357).

(145); *3c,* amb plate (*fide* QUENSTEDT), ×6.2 (145).

[Other characters given by QUENSTEDT (145) for the type-species include perforate, crenulate primary tubercles on every second amb plate (as in Fig. 262,*3b*), but TORNQUIST (164) illustrated the tubercles as imperforate noncrenulate (as in Fig. 263,*3a*), and a detail published by QUENSTEDT (145) shows large primary crenulate perforate tubercles to every amb plate. These contradictions prohibit exact classification of *Engelia.* MORTENSEN (136c) suggested that QUENSTEDT's original material contained parts of three genera, and he nominated the illustration here shown in Fig. 262,*3a,* as the type of *Engelia,* leaving unsettled whether or not it is identical with all, or only part, or no part, of QUENSTEDT's original material.]

Helodiadema MORTENSEN, 1939, p. 550 [**Cottaldia rotula* W. B. CLARK; OD]. Test small, globular. Ambs trigeminate, tubercles perforate, crenulate, 2 or 3 in horizontal row in each compound plate. Interambs with up to 11 tubercles on each plate, in horizontal series, secondaries not distinguishable from primary tubercles, all crenulate, perforate. Plates otherwise covered by granulation. Apical system dicyclic. Peristome large, gill slits small. Radioles unknown. *L.Cret.,* N.Am.——FIG. 262,2. **H. rotula* (CLARK), Washita, USA(Tex.); *2a,b,* interamb, amb, ×14 (22).

Heteropedina MICHALET, 1895, p. 71 [**H. moteti*; OD]. Test small (20 mm.). Adoral amb plates trigeminate, carrying well-developed crenulate perforate tubercle; adapical amb plates simple primaries, without primary tubercles, and almost devoid of secondary tubercles. Pore zones almost straight. Adoral interamb plates with irregularly arranged small primary tubercles, perforate and finely crenulate, similar to amb primary tubercles; aboral interamb plates each with one large tubercle which is perforate (possibly crenulate nature doubtful); uppermost plate of each interamb without tubercle. Apical system and peristome large.

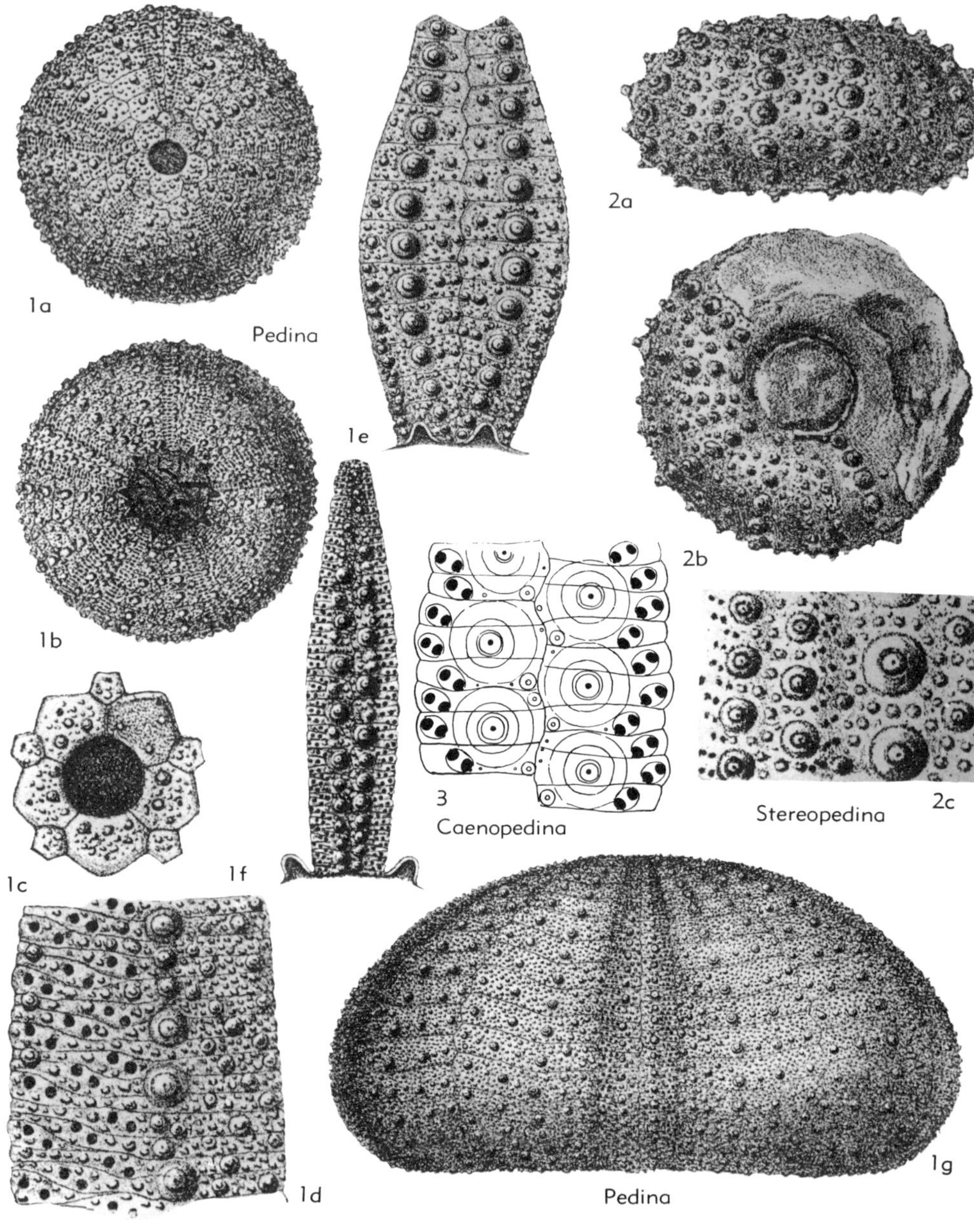

FIG. 263. Pedinidae (p. U357).

Spines unknown. *U.Jur.(Bathon.)*, Fr.——FIG. 262,6. **H. moteti*; *6a,b*, amb, interamb, ×3.1 (136c).

Macrodiadema LAMBERT, 1897, p. 6 [**M. ciplyensis*; OD]. Primary radiole with slender, finely striated shaft, and broad, short basal region, margin of acetabulum crenulate. Test and other structures unknown. *U.Cret.*, Fr.——FIG. 262,*4a*. **M. ciplyensis*; spine, lat., ×1.8 (103).——FIG. 262, *4b,c*. *M. buccinifera* (CAFFIN); *4b*, spine, lat., ×1.8; *4c*, base from below, ×3.5 (17).

Genus indet., ?fam. Aspidodiadematidae. *L.Carb.*,

Ger.——FIG. 262,*1*. Spine, ×5.5 (152). [Suggests an aspidodiadematid, but internal structure of spines unknown. No diadematoid is with certainty known from the Paleozoic. MORTENSEN (136c), however, is inclined to see in this fossil some possible archaic diadematoid, rather than the perischoechinoid *Pholidocidaris*, to which SCHMIDT (152) has attributed it.]

Order PEDINOIDA Mortensen, 1939

[*nom. transl.* FELL, herein (*ex* suborder Pedinina MORTENSEN, 1939, p. 547)]

Test subspherical, high subconical to depressed hemispherical or rotular; rigid though fragile, plates not imbricating. Ambs and interambs not extending across peristome. Ambs composed of simple or compounded diadematoid plates. Five pairs of oral plates on peristomial membrane. Gills present; gill slits shallow, notching peristomial margin. Periproct endocyclic. Apical system dicyclic; 5 genital pores. Tubercles noncrenulate. Spines finely striate, more or less thorny (but not verticillate); primaries solid, secondaries hollow. Spheridia unknown in fossils (but placed near tube foot of adoral secondary element of amb plates in living genus *Caenopedina*, where they continue to the adapical region). Pedicellariae including globiferous, ophicephalous, and tridentate types. *U.Trias.-Rec.*

Family PEDINIDAE Pomel, 1883

[*nom. transl.* GREGORY, 1900 (*ex* Pedininae, *nom. correct.* DUNCAN, 1889, *pro les* Pediniens POMEL, 1883)] [=Propedinidae LAMBERT, 1937 (*nom. van.*, based on jr. obj. syn. of *Pedina*)]

Characters of order. *U.Trias.(Rhaet.)-Rec.*

Pedina L. AGASSIZ, 1838, p. 4 [**P. sublaevis*; SD SAVIN, 1905] [=*Megapedina* LAMBERT & THIÉRY, 1910 (type, *Pedina charmassei* COTTEAU, 1885); *Propedina* LAMBERT & THIÉRY, 1925 (obj.) (*nom. van.*); *Hectopedina* THIÉRY, 1928; *Atlasaster* LAMBERT, 1931]. Test medium-sized to large (100 mm. horiz. diam.), more or less depressed, rather fragile. Pore pairs arranged in arcs of 3, adoral pair outermost. Primary ambulacral tubercles occurring on both oral and aboral hemispheres. Scattered secondary tubercles. *Jur.(Pliensbach.-Oxford.)*, Eu., Madag.; *Mio.*, S.Am.

P. (Pedina). Single series of primary tubercles in each column, but primary ambulacral tubercles lacking from some compound plates and interambulacral ones not contiguous throughout. *Jur.(Pliensbach.-Oxford.)*, Eu.-Madag.——FIG. 263, *1a-f*. **P. (P.) sublaevis*, Callov., Fr.; *1a,b*, test aboral, adoral, ×1.2; *1c*, apical system, ×2.7; *1d*, amb detail, ×6; *1e,f*, interamb, amb, ×2 (27c).——FIG. 263,*1g*. *P. (P.) gigas* A. AGASSIZ, Bathon., Fr.; test, lat., ×0.85 (27c).

P. (Stereopedina) DE LORIOL, 1902, p. 11 [**Stereopedina ameghinoi*; OD, M]. Like *P. (Pedina)*, but primary ambulacral tubercles forming regular vertical series, present on all amb plates. *Mio.*, S.Am.(Patagonia).——FIG. 263,*2*. **P. (S.) ameghinoi* (DE LORIOL); *2a,b*, test, lat., aboral, ×1.7; *2c*, detail of test, ×3.5 (125).

Caenopedina A. AGASSIZ, 1869, p. 256 [**C. cubensis*; OD]. Like *Hemipedina*, but pore zones in arcs of 3 throughout. Test not exceeding 40 mm. horiz. diam. Generally brightly colored, mainly deep-sea forms, usually with bands of pigment on finely thorny spines. *Rec.*, IndoPac.-Carib. (20-2,000 m.).——FIG. 263,*3*. *C. diomedeae* MORTENSEN, Gulf of Panama (840 m.); amb, ×5.5 (136c). [=*Coenopedina* POMEL, 1883, p. 99 (*nom. null.*); *Coenodiadema* BATHER, 1900, p. 86 (*nom. null.*).]

Diademopsis DESOR, 1855, p. 79 [**Echinus serialis* L. AGASSIZ, 1840; SD BATHER, 1909, p. 109] [=*Hecistocyphus* POMEL, 1883 (type, *Diademopsis bonissenti* COTTEAU)]. Test small to medium-sized, low hemispherical or rotular. Amb plates compound trigeminate adorally, simple aborally but every 3rd plate with primary tubercle. Pore zones straight, except near peristome, where forming arcs of 3. All ambulacral components reaching radial mid-line. Interamb plates low and broad, with more than one series of enlarged tubercles, secondaries resembling primary and forming series parallel to primary series. *U.Trias.(Rhaet.)-U.Jur.(Kimmeridg.)*, Eu.(Fr.-Eng.-Ger.-Switz.-Italy)-N.Afr.——FIG. 264,*2*. **D. serialis* (L. AGASSIZ); *2a-c*, test lat., oral, aboral, ×0.85 (44); *2d*, interamb, ×2.1 (27c).——FIG. 265,*2a,b*. *D. michelini* (COTTEAU), L.Jur.(Hettang.), Switz.; *2a,b*, interamb, amb, ×1.8 (27c).——FIG. 265,*2c*. *D. micropora* (L. AGASSIZ), Rhaet., Fr.; interamb, ×1.8 (27c).——FIG. 265,*2d*. *D. heeri* MERIAN, Rhaet., Fr.; test with spines, ×0.9 (44).

Echinopedina COTTEAU, 1866, p. 117 [**Echinus gacheti* DESMOULINS; OD] [=*Hebertia* LAMBERT, 1910, p. 2 (obj.); (*non Hebertia* MICHELIN, 1859, =*Echinopsis* L. AGASSIZ, 1840)]. Test subspherical, of moderate size (to 40 mm. diam.). Pore zones arranged in arcs of 3, of which adapical pair is very slightly nearer adradial margin than others (hence obscurely inverse), but not forming 3 vertical series. Only single vertical series of primary tubercles in amb sand interambs. *Eoc.*, Eu., N.Afr.-Carib.——FIG. 267,*1*. **E. gacheti* (DESMOULINS), Fr.; test, lat., ×1.2 (31).

Hemipedina WRIGHT, 1855, p. 2 [**Pedina etheridgei* WRIGHT, 1854; SD LAMBERT, 1900, p. 6] [=*Leiodiadema* QUENSTEDT, 1873 (obj.); *Miopedina* POMEL, 1883 (type, *Hemipedina tubercu-*

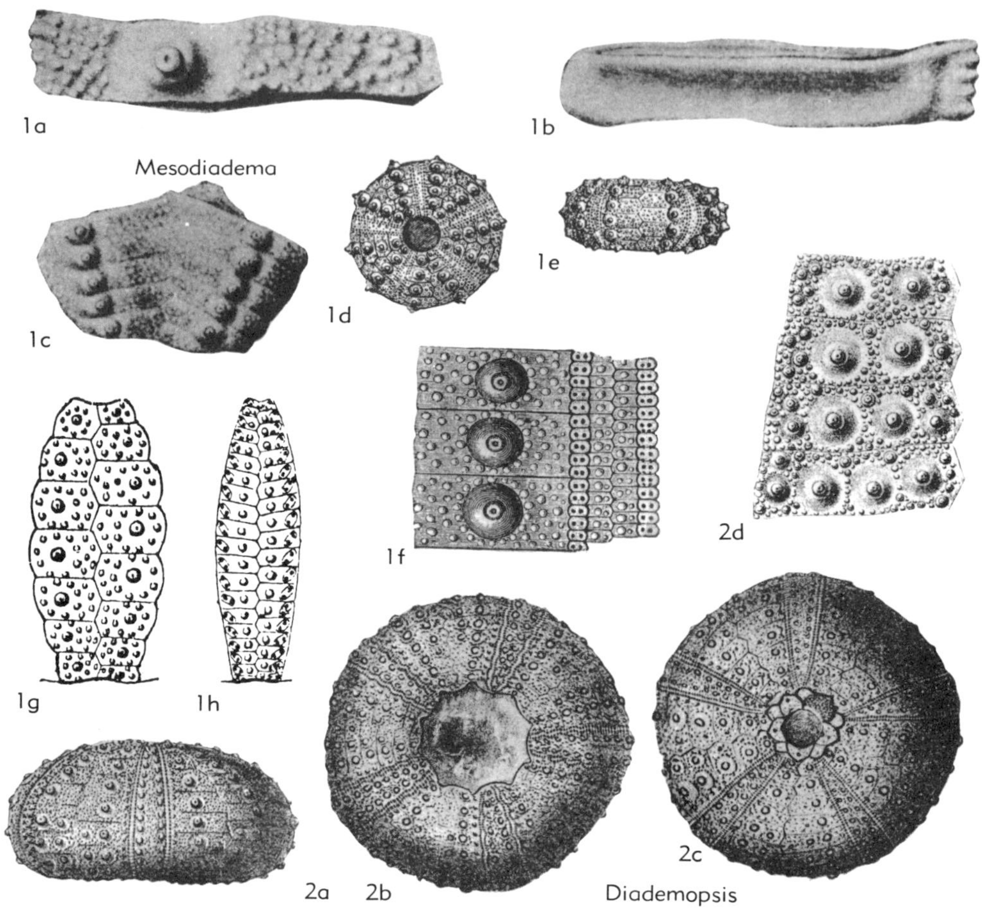

FIG. 264. Pedinidae (p. *U357-U359*).

losa WRIGHT, 1860; *Archaeodiadema* GREGORY, 1896 (type, *A. thompsoni*)]. Like *Diademopsis,* but interamb plates high and narrow, with only single series of primary tubercles, single large primary tubercle and areole of each interamb plate lying near its center. *L.Jur.(Pliensbach.)-U.Cret. (Cenoman.),* Eu. (Eng.-Fr.-Ger.-Switz.)-?Iran.——FIG. 266,*2a-d.* **H. etheridgei,* J.Jur.(Toarc.), Eng.; *2a,* apical system (holotype), ×7.5 (136c); *2b-d,* test, lat., aboral, oral, ×4.2 (172).——FIG. 266,*2e*; 267,*5*. *H. tuberculosa* WRIGHT, Oxford., Eng.; 266,*2e,* test aboral, with spines, ×1.25 (172); 267,*5a,b,* amb, interamb, ×4 (27c). [=*Psilosalenia* QUENSTEDT, 1873, p. 256.]

Leiopedina COTTEAU, 1866, p. 114 [**Codechinus tallevignesi* COTTEAU, 1856; OD, M] [=*Chrysomelon* LAUBE, 1868, p. 13 (obj.)]. Test large (to 60 mm. diam.), as high as broad, or higher, subglobular to subconical. Pore zones broad, pore pairs in oblique arcs of 3, in inverse sequence (adapical pore outermost), forming 3 well-defined vertical series. Outermost pore pairs not differing from others. *Eoc.(Lutet.),* Eu.(Fr.-Ger.).——FIG. 267,*3*. **L. tallevignesi* (COTTEAU), Fr.; *3a,* amb (adoral), ×2 (27e); *3b,* amb plates, ×3.3 (136c); *3e,* test, lat., ×0.9 (31).

Loriolipedina LAMBERT, 1910, p. 133 [**L. alpina*; OD]. Like *Leiopedina,* but outermost pore pairs of each series elongate. *Eoc.,* Eu.——FIG. 267,*6*. **L. alpina,* Switz.; *6a,* amb and interamb plate, ×2; *6b,* test, lat., ×0.7 (120).

Mesodiadema NEUMAYR, 1889, p. 372 [**Hemipedina marconissae* DESOR, 1858; OD]. Test small, depressed. Pore pairs in single straight series. Amb plates simple, uniform, not forming triads, each with single small primary tubercle. Interamb plates each with single small primary tubercle, placed near adradial edge, remainder of amb plate covered by granulation. Peristome small. Apical system and spines unknown. *?U.Trias.,* Hung.; *L.*

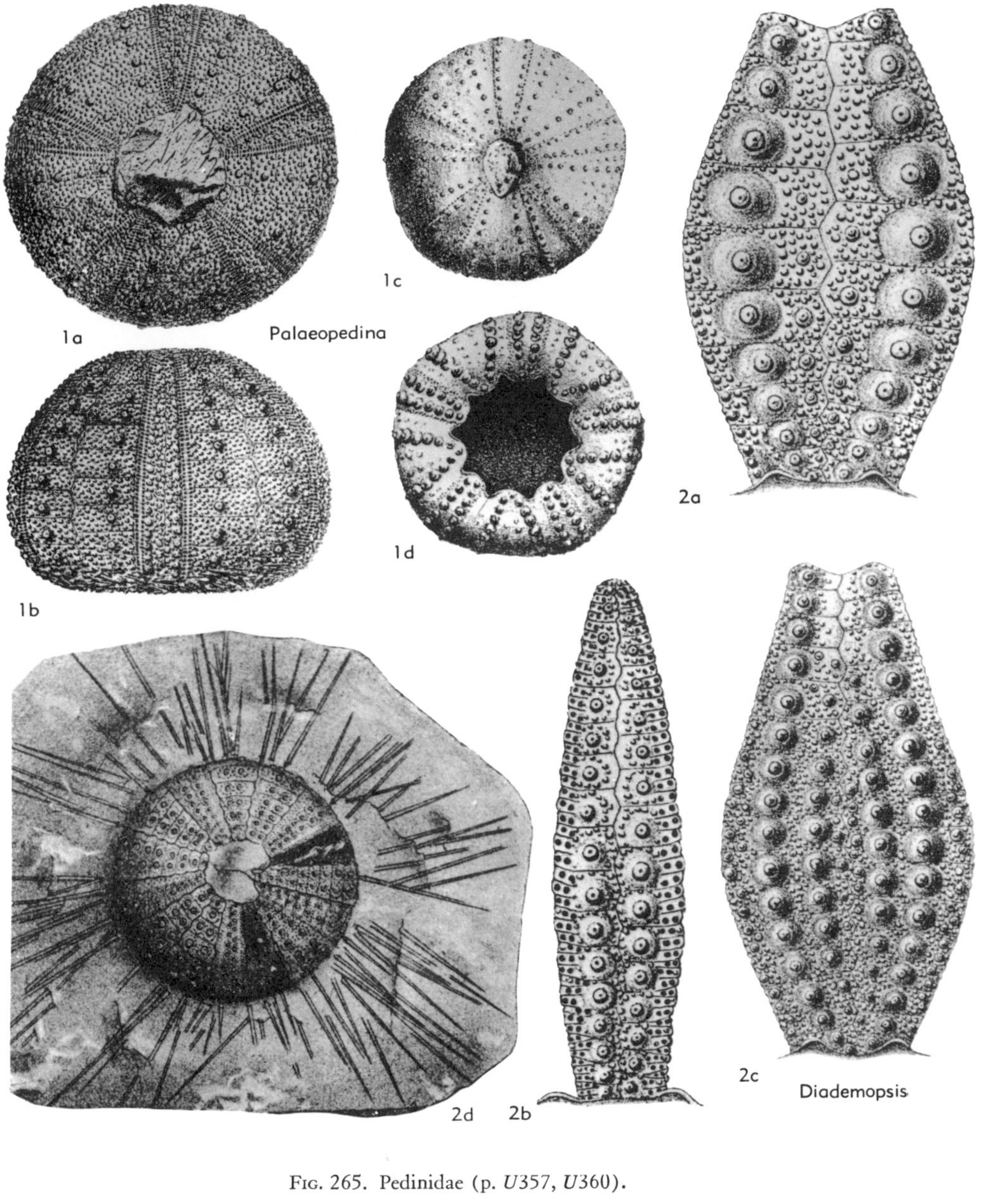

FIG. 265. Pedinidae (p. *U357*, *U360*).

Jur.(Toarc.), Fr.-N.Italy.——FIG. 264,*1a-c. M. latum* BATHER, ?U.Trias. (Carn.), Hung.; *1a,b,* interamb plates, ext., int., ×8.5; *1c,* interamb (adoral), ×3.5 (11).——FIG. 264,*1d-f. *M. marconissae* (DESOR), L.Jur., N.Italy; *1d,e,* test, aboral, lat., ×1.4; *1f,* detail of amb and interamb, ×5.6 (139).——FIG. 264,*1g,h. M. angeliacense* VALETTE, L.Jur.(Charmouth.), Fr.; *1g,h,* interamb, amb (latter with oblique pores), ×? (165).

Micropedina COTTEAU, 1866, p. 822 [**Echinus olisiponensis* FORBES, 1850; OD]. Test small to medium-sized (25-40 mm.), subglobular to subconical. Pore pairs arranged in arcs of 3, of which adapical pair lies outermost (inverse). Numerous enlarged tubercles in both ambs and interambs, forming many series. Peristome small. Spines slender, longitudinally striate. *U.Cret.(Cenoman.-Senon.),* Eu.-N.Afr.-India.——FIG. 267,*2a-d. M. olisiponensis cotteaui* COQUAND, Cenoman., Fr.; *2a,* amb, aboral, ×?; *2b,c,* test, aboral, lat., ×1.2;

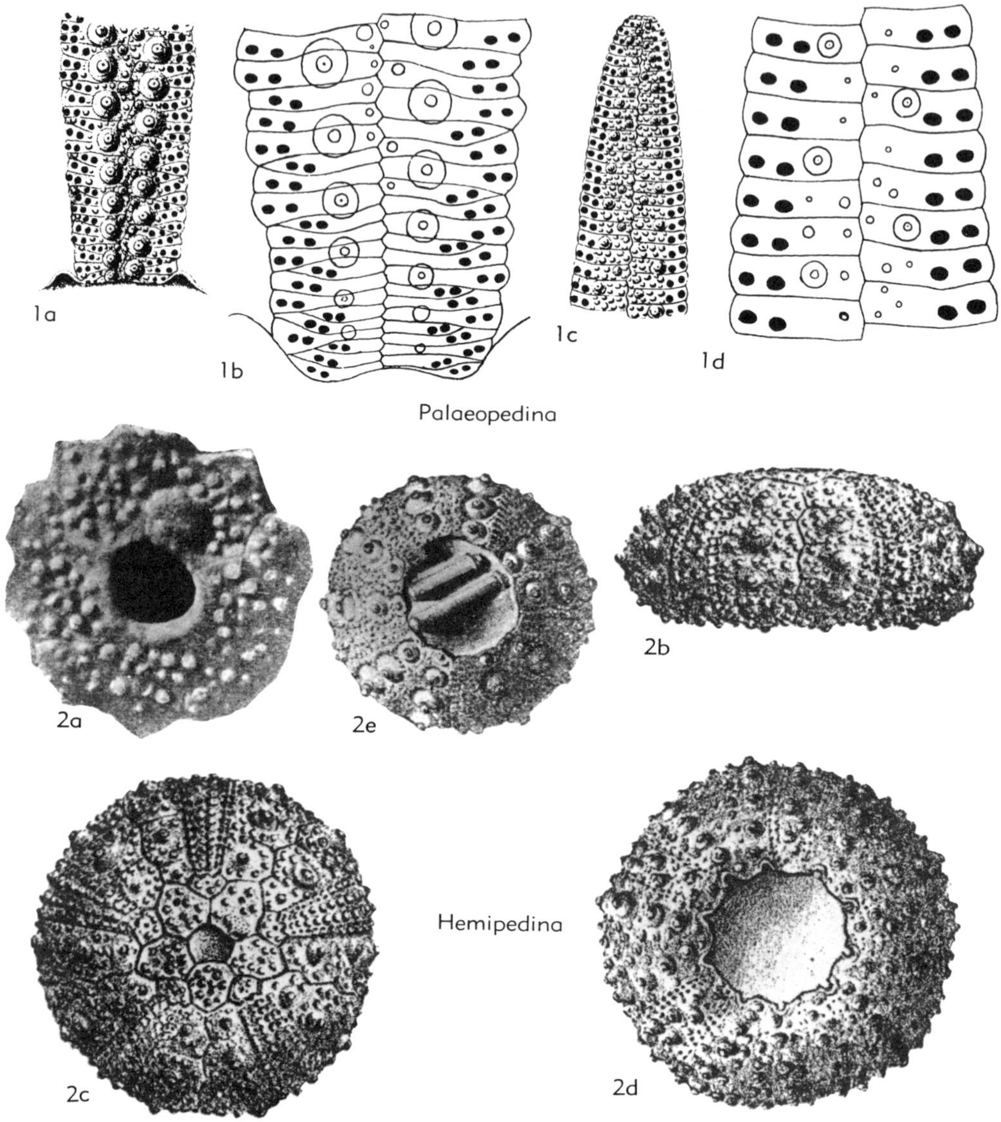

FIG. 266. Pedinidae (p. *U357-U358, U360*).

2d, apical system, ×5.5 (27a).——FIG. 267,*2e,f.* **M. olisiponensis* (FORBES), Cenoman., Port.; *2e,f,* amb, interamb, ×7 (157).

Palaeopedina LAMBERT, 1900, p. 22 [**Diadema globulus* L. AGASSIZ, 18??; OD]. Test small to medium-sized, subglobular or subhemispherical. Apical system elongate in anteroposterior axis, periproct displaced toward interamb 5; suranal plate may be present. Primary tubercles in single series, without conspicuous areoles, but secondary tuberculation present. [Young stages of *Plesiechinus* (Pygasteroida) resemble *Palaeopedina,* and erroneously have been referred to the genus (52).] *L.Jur.(Hettang.),* Eu.——FIG. 265,*1a,b,*; 266,*1.* **P. globulus* (AGASSIZ), S.Fr.; 265,*1a,b,* test, aboral, lat., ×0.9 (27c); 266, *1a,b,* amb adoral, ×2.1, ×5 (27c); 266,*1c,d,* amb aboral, ×2.1, ×5 (136c).——FIG. 265,*1c,d. P. bonei* (WRIGHT), Bajoc., Eng.; *1c,d,* test, aboral, oral, ×1.8 (172). [*P. bonei* is regarded by BATHER (1909) as probably a young *Pygaster*; the similarity suggests probable close relationship of pedinids and pygasterids.].

Phalacropedina LAMBERT, 1900, p. 30 [**Hemi-*

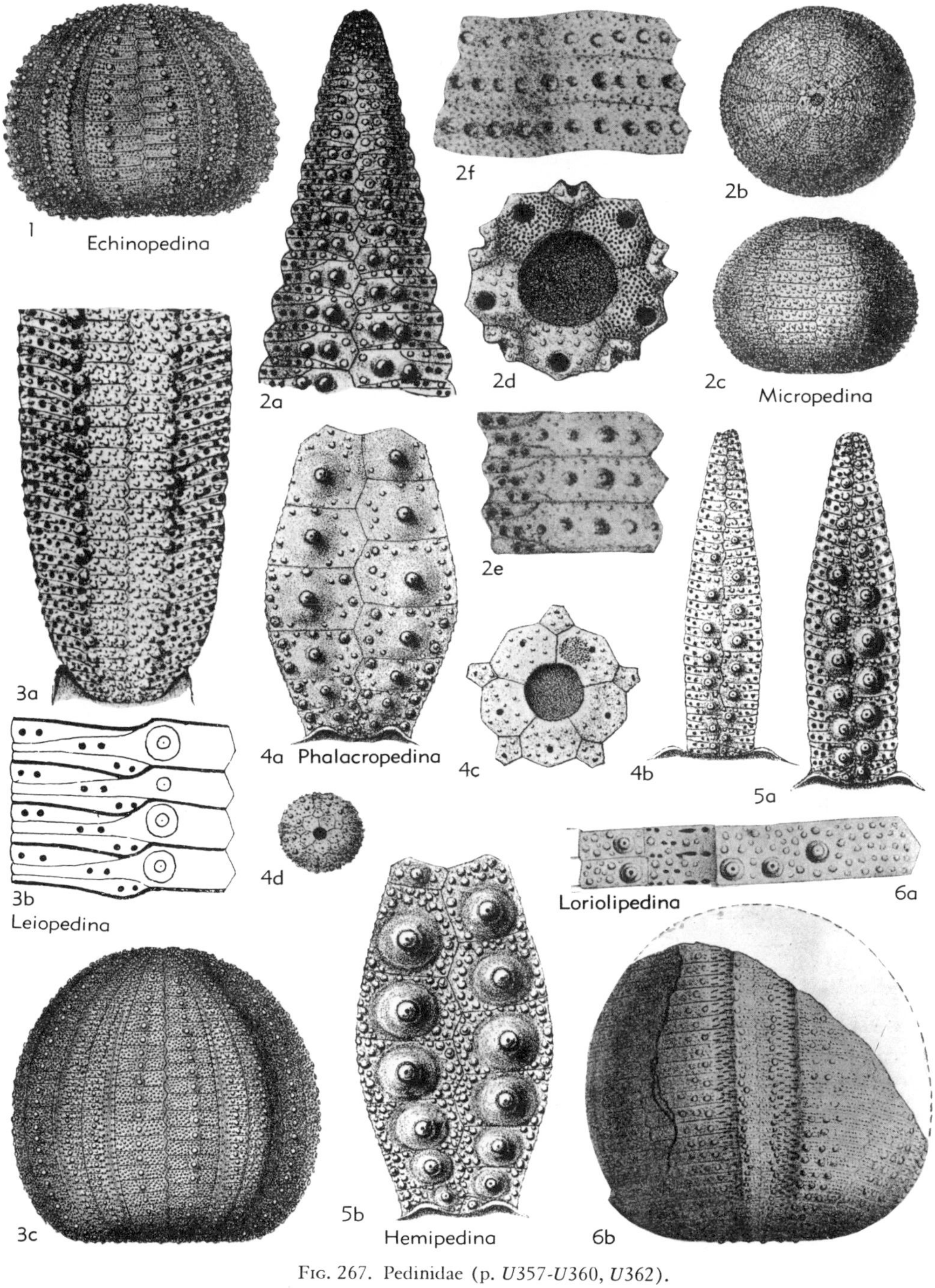

FIG. 267. Pedinidae (p. *U*357-*U*360, *U*362).

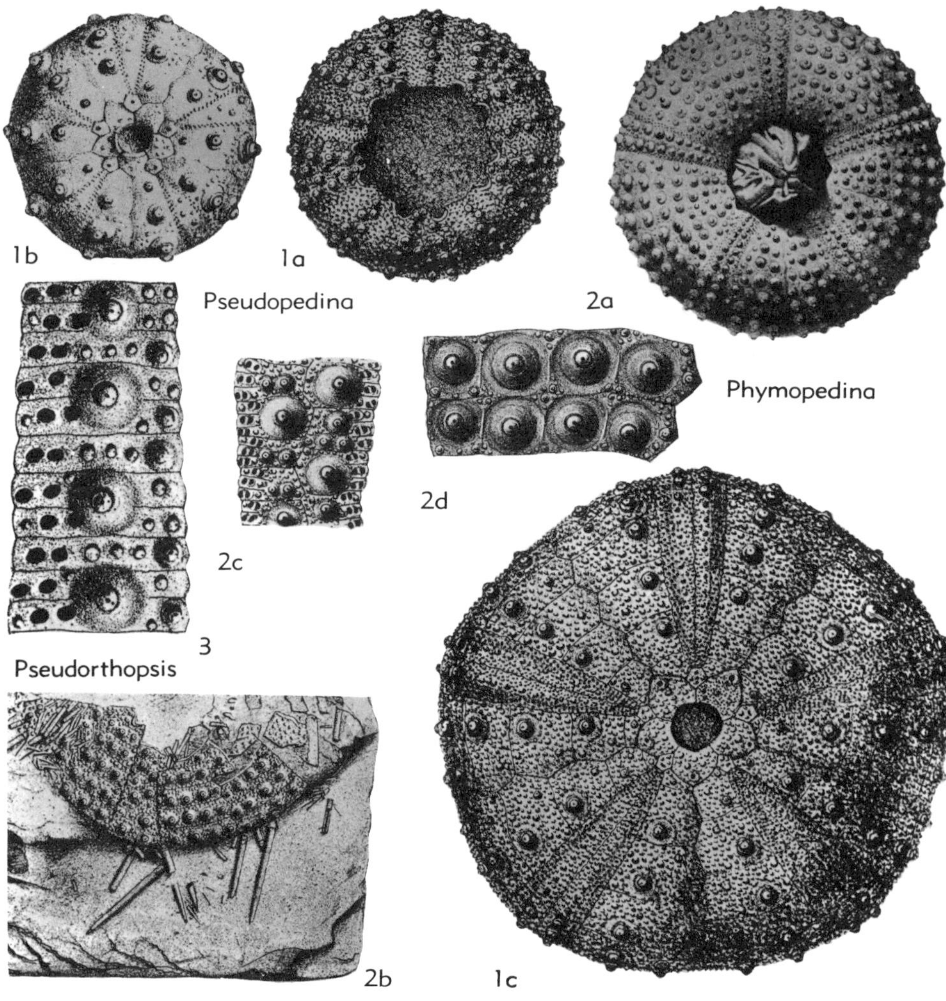

FIG. 268. Pedinidae (p. U362-U365).

pedina guerangeri COTTEAU, 1858; OD]. Like *Hemipedina,* but secondary tubercles very sparse, plates thus being almost naked, and primary areoles indistinct. *U.Jur.(Oxford.-Kimmeridg.),* Eu.——FIG. 267,*4*. **P. guerangeri* (COTTEAU), Oxford., Fr.; *4a-c,* interamb, amb, apical system, ×2.7; *4d,* test, aboral, ×0.9 (27c).

Phymopedina POMEL, 1883, p. 100 [**Hemipedina marchamensis* WRIGHT, 1855; SD LAMBERT, 1900, p. 28]. Test large (horiz. diam. *ca.* 70 mm.), hemispherical, depressed. Amb plates trigeminate, pores in indistinct arcs of 3, arcs more oblique adorally and crowded at peristome; adoral pore pair of each triad placed outermost. Primary amb tubercles as large as interamb primaries; amb primaries in single vertical series; either one enlarged primary on every compound plate or larger and smaller primary tubercles on alternate plates. Interamb primary and secondary tubercles similar, forming transverse series on each plate, and up to 4 vertical series on each column. Spines finely striated. *Jur.(Bathon.-Portland.),* W.Eu.——FIG. 268,*2a*. **P. marchamensis* (WRIGHT), Oxford, Eng.; test, oral, ×0.55 (172).——FIG. 268,*2b,c*. *P. legayi* (COTTEAU), Bathon., Fr.; *2b,* test, aboral, with spines, ×0.55; *2c,* amb detail, ×1.5 (27c). ——FIG. 268,*2d*. *P. bouchardi* (WRIGHT), Portland., Eng.; ambital interamb plates, ×1.5 (27c).

Pseudorthopsis SÁNCHEZ ROIG, 1949, p. 37 [**Echinopedina cubensis* COTTEAU, 1881, pl. 1; OD].

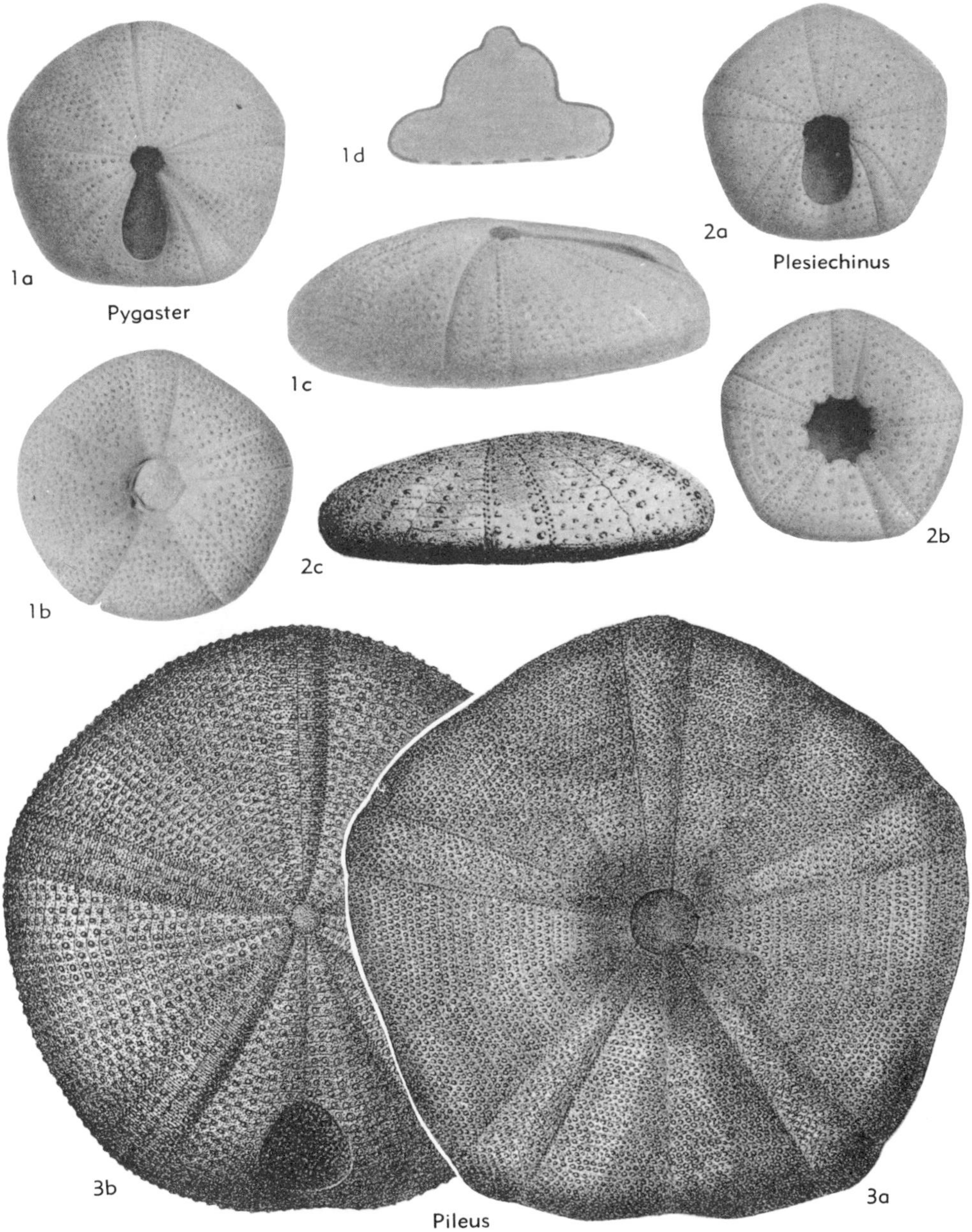

FIG. 269. Pygasteridae (p. U365).

Like *Echinopedina,* but pore pairs not inverse. *Eoc.,* Cuba.——FIG. 268,*3.* **P. cubensis* (COTTEAU); amb, ×4 (30).

Pseudopedina COTTEAU, 1858, p. 9 [**Pedina bakeri* WRIGHT, 1854; OD]. Like *Pedina,* but without aboral primary ambulacral tubercles. *M.Jur. (Bajoc.-Bathon.),* Eu.——FIG. 268,*1a,b.* **P. bakeri* (WRIGHT), Bajoc., Eng.; *1a,* test, oral, ×0.9

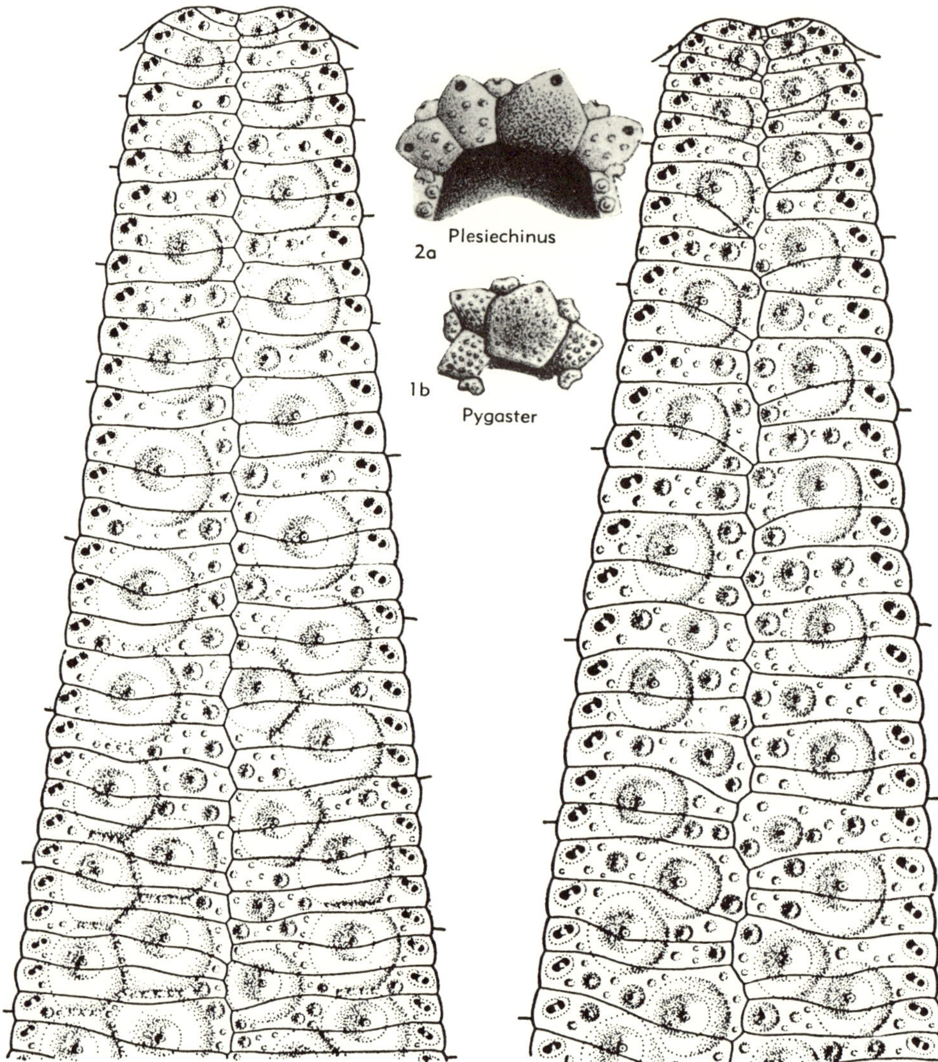

FIG. 270. Pygasteridae (p. *U*365).

(27a); *1b,* test, aboral, immature, ×1.8 (172). ——FIG. 268,*1c*. *P. divionensis* (MICHELIN), Bajoc., Fr.; test aboral, ×0.9 (27a).

Stenechinus ARNOLD & CLARK, 1927, p. 13 [**S. regularis*; OD]. Test subglobular, of moderate size. Pore zones in narrow, straight series, not widening at peristome. Amb plates trigeminate, median component bearing large primary tubercle close to pore pair, and adoral components on ambital amb plates excluded from radial mid-line (hence demiplates). Interamb plates each with large central primary tubercle, and scattered secondary tubercles. ?*Eoc.,* Jamaica.

Family UNCERTAIN

The following genus is insufficiently known, but may provisionally be associated with the Pedinidae.

Leptocidaris QUENSTEDT, 1858, p. 644 [**L. triceps*;

OD]. Test small, depressed subspherical. Ambs trigeminate (?diadematoid), amb plates without primary tubercle except 3 ambital plates, each of which has single conspicuous tubercle. Interambs with single vertical series of primary tubercles on each column. Apical system, peristome and spines unknown. *U.Jur.(Oxford.)*, Ger.

The following genera have earlier been associated with the Pedinidae, but are here excluded from the family as it now seems probable that the tubercles are crenulate.

Dumblea CRAGIN, 1893 [**D. symmetrica*; OD]. Here assigned to Pseudodiadematidae (see p. U386).

Farquharsonia CURRIE, 1927 [**F. somaliensis*; OD]. Here listed as *incertae sedis.*

Order PYGASTEROIDA Durham & Melville, 1957

[=Pileatoida LAMBERT, 1900 *(partim)*] [Materials for this order prepared by R. V. MELVILLE]

Medium-sized to large Diadematacea with rigid corona; apical system with four or five genital plates and four genital pores; periproct outside apical system; ambulacra lacking compound plates; interambulacral tubercles smooth, in regular vertical and horizontal series; radioles with solid axis; girdle composed of strong auricles suported by buttresses, apophyses rudimentary; teeth triangular in section, keeled and buttressed. *L.Jur.-U.Cret.*

Members of this order have been discussed by HAWKINS (1911; 1918; 1920), while the reasons for establishing it as a separate order are given by DURHAM & MELVILLE (1957) and MELVILLE (1962).

Family PYGASTERIDAE Lambert, 1900

Characters of order. *L. Jur.-U.Cret.*

Pygaster J. L. R. AGASSIZ, 1836, p. 18 [**Clypeus semisulcatus* PHILLIPS, 1829, SD SAVIN, 1905, p. 187] [=*Macropygus* DESOR, 1857 (type, *Pygaster truncatus* AGASSIZ); *?Echinoclypus* POMEL, 1869; *Megapygus* HAWKINS, 1912 (type, *Pygaster umbrella*)]. Medium-sized, more or less depressed; apical system in contact with periproct, but genital 5 missing, and genital 3 with oculars II, III, and IV excluded; pore pairs uniserial throughout, or very weakly triserial adorally; elongated pits in transverse sutures of adoral interambulacral plates. *M. Jur. (Bajoc.)-U. Cret. (Cenoman.)*, Eu.——FIG. 269,*1a-c*; 270,*1*. **P. semisulcatus* (PHILLIPS), U. Oxford., Eng.; 269,*1a,b*, aboral, oral, ×0.7 (172); 269,*1c*, lat., ×1.1 (172); 270,*1a*, adoral portion of ambulacrum, enl. (81); 270,*1b*, apical system, ×2.5 (172).——FIG. 269,*1d*. *P. trigeri* COTTEAU, BATHON., Fr.; cross section of tooth, ×45 (131).

Plesiechinus POMEL, 1883 [**Pygaster macrostoma* WRIGHT, 1861; SD HAWKINS, 1917, p. 167]. Differs from *Pygaster* in having plates of apical system arranged in arc around apical edge of periproct, with which all 4 genitals (but not oculars II, III, IV) are in contact; and in more pronounced triserial arrangement of adoral pore pairs. *Jur. (Pliensbach.-Bathon.)*, Eu., W. N. Am. —— FIG. 269,*2*; 270,*2a*. **P. macrostoma* (WRIGHT), Bathon., Eng.; 269,*2a,b*, aboral, oral, ×0.75; 269,*2c*, lat., ×1 (172); 270,*2a*, apical system, enlarged (172).——FIG. 270,*2b*. *P. ornatus* (BUCKMAN), Bajoc., Eng.; adoral part of ambulacrum, enl. (81).

Pileus DESOR, 1856 [**Pygaster pileus* J. L. R. AGASSIZ, 1847; OD]. Large, high-arched or subconical in profile; apical system compact, with imperforate genital 5; periproct separated from apical system, low on adapical surface; peristome smaller than in *Pygaster*; pore pairs biserial or irregularly triserial adapically, minute and nearly uniserial adorally. *U.Jur.(Oxford.)*, Eu.——FIG. 269,*3*. **P. pileus* (AGASSIZ), Oxford., Fr. *(3a)*, Port. *(3b)*; *3a*, oral, ×0.7 (27b); *3b*, aboral, ×0.7 (124).

Order UNCERTAIN

Family HETEROCIDARIDAE Mortensen, 1934

Ambs straight, comprising trigeminate (or ?polyporous) diadematoid plates. Interambs broad, each plate with 2 to 4 large primary tubercles arranged in horizontal row; tubercles also forming vertical series, with 6 to 8 such vertical series at ambitus. Adradial margin of interamb plates apparently imbricating upon amb plates. All primary tubercles perforate, crenulate. Peristome large, pentagonal, gill slits indistinct (or ?absent). Primary spines of cidaroid type, with granulated cortex. Lantern and apical system unknown. [The general aspect of the test recalls the Diadematoida, but the cidaroid character of the spines is opposed to such affinity, pointing rather to Hemicidaroida or even Cidaroida; no precise relationships can be suggested until the lantern structure is known.] *Jur.(Domer.-Oxford.).*

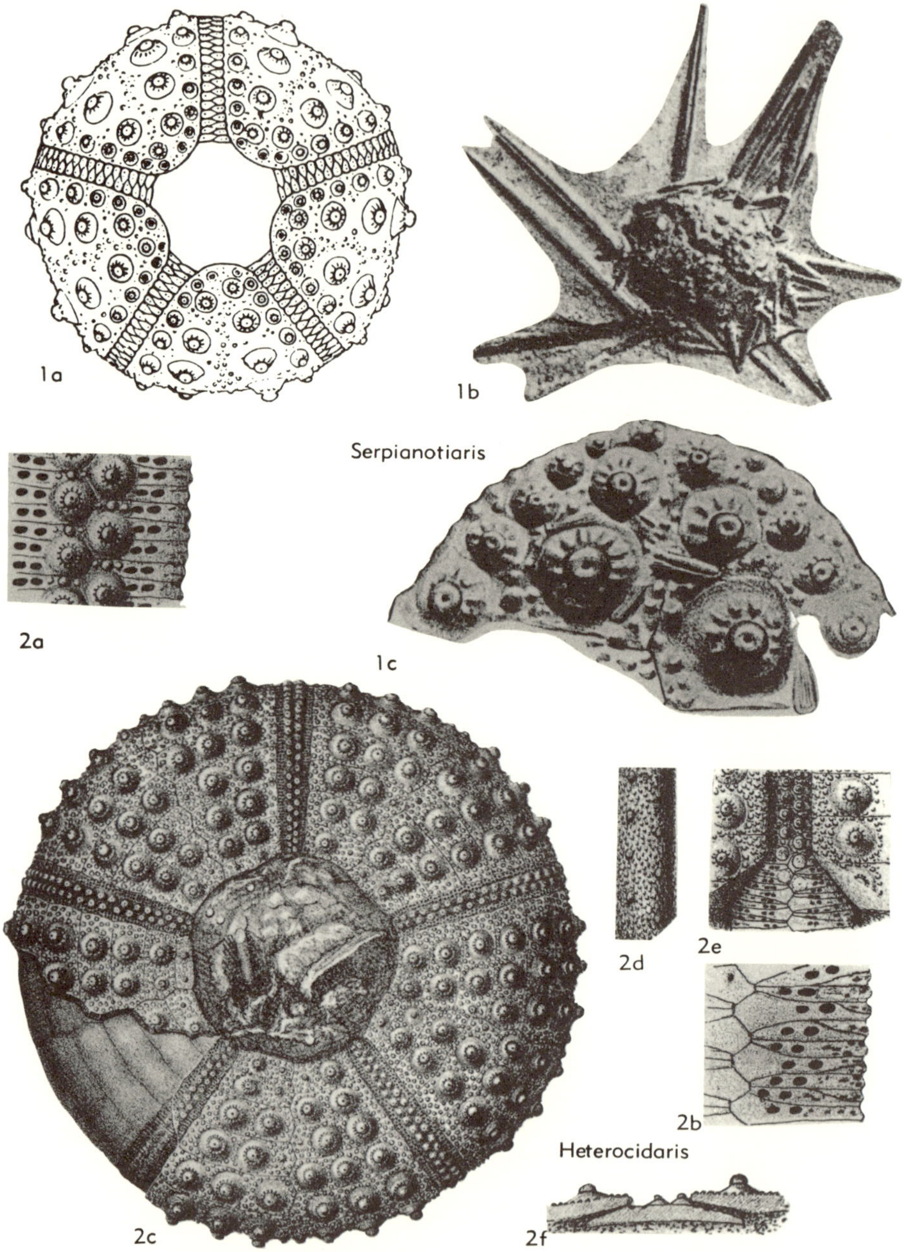

FIG. 271. Heterocidaridae *(2)*; Order and Family Uncertain *(1)* (p. *U366-U367*).

Heterocidaris COTTEAU, 1860, p. 17, 378 [**H. trigeri*; OD] [*non Heterocidaris* HALL, 1861 (=*Pholidocidaris* MEEK & WORTHEN)] [=*Heteroechinus* QUENSTEDT, 1874, p. 370]. Test low hemispherical, flattened below, large (more than 100 mm. diam.); other characters as for family. *Jur.(Domer.-Oxford.)*, Eu.-N.Am.——FIG. 271,*2*. **H. trigeri*, Bajoc., Fr.; *2a,b*, amb ext., int., ×4; *2c*, test aboral, ×0.6 (36); *2d*, spine fragment, ×2 (27c); *2e,f*, amb ext. (with supposedly imbricating interamb plates), and transv. sec., ×2 (26).

Order and Family UNCERTAIN

Serpianotiaris JEANNET in PEYER, 1933, p. 1 [**S. hescheleri*; OD]. Test small (diam. 15 mm.), hemispherical. Ambs simple, without compound plates, narrow. Interamb plates imbricating over ambs at adradial margin, each with large abradially placed primary tubercle; additional adradial series of secondary tubercles, more numerous than primary tubercles, forming vertical series adjacent to amb. Tubercles perforate, crenulate. Primary spines slender, without cortex. *Trias.*, Switz.——FIG. 271,*1*. **S. hescheleri*; *1a*, test, oral (restored), ×4; *1b*, test with spines, ×1.8 (89); *1c*, interamb (adoral), ×12 (89).

Printed in the U.S.A. by Malloy Lithographing, Inc., Ann Arbor, Michigan